ESSENTIAL UNIVERSITY PHYSICS

VOLUME 1 CHAPTERS 1–19

THIRD EDITION

Richard Wolfson
Middlebury College

PEARSON

Executive Editor: Nancy Whilton
Project Manager: Katie Conley
Development Editors: John Murdzek, Matt Walker
Editorial Assistant: Sarah Kaubisch
Development Manager: Cathy Murphy
Project Management Team Lead: Kristen Flathman
Compositor: Lumina Datamatics, Inc.
Design Manager: Marilyn Perry
Illustrators: Rolin Graphics
Rights & Permissions Management: Timothy Nicholls
Photo Researcher: Stephen Merland, Jen Simmons
Manufacturing Buyer: Maura Zaldivar-Garcia
Marketing Manager: Will Moore
Cover Photo Credit: 68/Ocean/Corbis

CIP data is on file with the Library of Congress.

1 2 3 4 5 6 7 8 9 10—CKV—18 17 16 15 14

www.pearsonhighered.com

ISBN 10: 0-321-99372-1; ISBN 13: 978-0-321-99372-4

PhET Simulations

Available in the Pearson eText and in the Study Area of MasteringPhysics MP

Chapter	PhET	Page
1	Estimation	8
2	The Moving Man	16
3	Vector Addition	33
3	Projectile Motion	39
3	Ladybug Motion 2D	43
3	Motion in 2D	43
5	The Ramp	73
5	Forces in 1 Dimension	80
5	Friction	80
6	The Ramp	92, 96
7	Calculus Grapher	120
7	Energy Skate Park	120
8	Gravity Force Lab	131
8	My Solar System	132
8	Gravity and Orbits	132
9	Collisions (Introduction)	158
9	Collisions (Advanced)	159
10	Ladybug Revolution	169
10	Torque (Torque)	173
10	Torque (Moment of Inertia)	173
11	Simplified MRI	196
13	Masses and Springs	224
13	Pendulum Lab	229
14	Wave on a String	247, 255
14	Sound	251
14	Fourier: Making Waves	252
14	Wave Interference	253
15	Balloons and Buoyancy	269
16	Blackbody Spectrum	293
16	The Greenhouse Effect	296
17	Gas Properties	303
17	States of Matter	307
19	Reversible Reactions	344
20	Balloons and Static Electricity	356
20	Charges and Fields	359, 362
20	Electricity Field Hockey	366
22	Calculus Grapher	409
22	Charges and Fields	409
24	Conductivity	435
24	Semiconductors	437
24	Resistance in a Wire	439
24	Ohm's Law	439
25	Battery Resistor Circuit	450
25	Circuit Construction Kit (DC Only)	451, 456
25	Signal Circuit	451
26	Magnet and Compass	470
26	Magnets and Electromagnets	483
27	Farady's Law	498
27	Farady's Electromagentic Lab	502
27	Generator	506
28	Circuit Construction Kit (AC + DC)	526, 530, 533
29	Radio Waves and Electromagnetic Fields	556
29	Radiating Charge	556
29	Optical Tweezers and Applications	559
30	Bending Light (Intro)	569
30	Bending Light (Prism Break)	572
31	Geometric Optics	585
32	Wave Interference: Light	601
34	Blackbody Spectrum	648
34	Photoelectric Effect	651
34	Neon Lights & Other Discharge Lamps	654
34	Quantum Wave Interference	659
35	Quantum Bound State: One Well	671, 675
35	Quantum Tunneling and Wave Packets	677
36	Quantum Bound State: One Well: 3D Coulomb	685
36	Build an Atom	693
37	Quantum Bound States: Two Wells (Molecular Bonding)	703
37	Band Structure	709
37	Semiconductors	712
38	Rutherford Scattering	721
38	Simplified MRI	724
38	Radioactive Dating Game	726
38	Alpha Decay	729
38	Beta Decay	729
38	Nuclear Fission: One Nucleus	734
38	Nuclear Fission: Chain Reaction	735
38	Nuclear Fission: Nuclear Reactor	736

Video Tutor Demonstrations

Video tutor demonstrations can be accessed by scanning the QR codes in the textbook using a smartphone. They are also available in the Study Area and Instructor's Resource Area on MasteringPhysics and in the eText.

Brief Contents

About the Author

Richard Wolfson

Richard Wolfson is the Benjamin F. Wissler Professor of Physics at Middlebury College, where he has taught since 1976. He did undergraduate work at MIT and Swarthmore College, and he holds an M.S. degree from the University of Michigan and Ph.D. from Dartmouth. His ongoing research on the Sun's corona and climate change has taken him to sabbaticals at the National Center for Atmospheric Research in Boulder, Colorado; St. Andrews University in Scotland; and Stanford University.

Rich is a committed and passionate teacher. This is reflected in his many publications for students and the general public, including the video series *Einstein's Relativity and the Quantum Revolution: Modern Physics for Nonscientists* (The Teaching Company, 1999), *Physics in Your Life* (The Teaching Company, 2004), *Physics and Our Universe: How It All Works* (The Teaching Company, 2011), and *Understanding Modern Electronics* (The Teaching Company, 2014); books *Nuclear Choices: A Citizen's Guide to Nuclear Technology* (MIT Press, 1993), *Simply Einstein: Relativity Demystified* (W. W. Norton, 2003), and *Energy, Environment, and Climate* (W. W. Norton, 2012); and articles for *Scientific American* and the *World Book Encyclopedia.*

Outside of his research and teaching, Rich enjoys hiking, canoeing, gardening, cooking, and watercolor painting.

Preface to the Instructor

Introductory physics texts have grown ever larger, more massive, more encyclopedic, more colorful, and more expensive. *Essential University Physics* bucks that trend—without compromising coverage, pedagogy, or quality. The text benefits from the author's three decades of teaching introductory physics, seeing firsthand the difficulties and misconceptions that students face as well as the "Got It!" moments when big ideas become clear. It also builds on the author's honing multiple editions of a previous calculus-based textbook and on feedback from hundreds of instructors and students.

Goals of This Book

Physics is the fundamental science, at once fascinating, challenging, and subtle—and yet simple in a way that reflects the few basic principles that govern the physical universe. My goal is to bring this sense of physics alive for students in a range of academic disciplines who need a solid calculus-based physics course—whether they're engineers, physics majors, premeds, biologists, chemists, geologists, mathematicians, computer scientists, or other majors. My own courses are populated by just such a variety of students, and among my greatest joys as a teacher is having students who took a course only because it was required say afterward that they really enjoyed their exposure to the ideas of physics. More specifically, my goals include:

- Helping students build the analytical and quantitative skills and confidence needed to apply physics in problem solving for science and engineering.
- Addressing key misconceptions and helping students build a stronger conceptual understanding.
- Helping students see the relevance and excitement of the physics they're studying with contemporary applications in science, technology, and everyday life.
- Helping students develop an appreciation of the physical universe at its most fundamental level.
- Engaging students with an informal, conversational writing style that balances precision with approachability.

New to the Third Edition

The overall theme for this third-edition revision is to present a more unified view of physics, emphasizing "big ideas" and the connections among different topics covered throughout the book. We've also updated material and features based on feedback from instructors, students, and reviewers. A modest growth, averaging about one page per chapter, allows for expanded coverage of topics where additional elaboration seemed warranted. Several chapters have had major rewrites of key physics topics. We've also made a number of additions and modifications aimed at improving students' understanding, increasing relevancy, and offering expanded problem-solving opportunities.

- Chapter opening pages have been redesigned to include explicit connections, both textual and graphic, with preceding and subsequent chapters.
- The presentation of **energy and work** in Chapters 6 and 7 has been extensively rewritten with a clearer invocation of **systems concepts**. Internal energy is introduced much earlier in the book, and potential energy is carefully presented as a property not of objects but of systems. Two new sections in Chapter 7 emphasize the universality of energy conservation, including the role of internal energy in systems subject to dissipative forces. Forward references tie this material to the chapters on thermodynamics, electromagnetism, and relativity. The updated treatment of energy also allows the text to make a closer connection between the conservation laws for energy and momentum.

- The presentation of **magnetic flux and Faraday's law** in Chapter 27 has been recast so as to distinguish motional emf from emfs induced by changing magnetic fields—including Einstein's observation about induction, which is presented as a forward-looking connection to Chapter 33.
- There is more emphasis on calculus in earlier chapters, allowing instructors who wish to do so to use calculus approaches to topics that are usually introduced algebraically. We've also added more calculus-based problems. However, we continue to emphasize the standard approach in the main text for those who teach the course with a calculus corequisite or otherwise want to go slowly with more challenging math.
- A host of **new applications** connects the physics concepts that students are learning with contemporary technological and biomedical innovations, as well as recent scientific discoveries. A sample of new applications includes Inertial Guidance Systems, Vehicle Stability Control, Climate Modeling, Electrophoresis, MEMS (Microelectromechanical Systems), The Taser, Uninterruptible Power Supplies, Geomagnetic Storms, PET Scans, Noise-Cancelling Headphones, Femtosecond Chemistry, Windows on the Universe, and many more.
- Additional **worked examples** have been added in areas where students show the need for more practice in problem solving. Many of these are not just artificial textbook problems but are based on contemporary science and technology, such as the Mars *Curiosity* rover landing, the Fukushima accident, and the Chelyabinsk meteor. Following user requests, we've added an example of a collision in the center-of-mass reference frame.
- New GOT IT? boxes, now in nearly every section of every chapter, provide quick checks on students' conceptual understanding. Many of the GOT IT? questions have been formatted as Clicker questions, available on the Instructor's Resource DVD and in the Instructor's Resource Area in Mastering.
- End-of chapter problem sets have been extensively revised:
  - Each EOC problem set has at least 10 percent new or substantially revised problems.
  - More "For Thought and Discussion Questions" have been added.
  - Nearly every chapter has more intermediate-level problems.
  - More calculus-based problems have been added.
  - Every chapter now has at least one data problem, designed to help students develop strong quantitative reasoning skills. These problems present a data table and require students to determine appropriate functions of the data to plot in order to achieve a linear relationship and from that to find values of physical quantities involved in the experiment from which the data were taken.
  - New tags have been added to label appropriate problems. These include CH (challenge), ENV (environmental), and DATA, and they join the previous BIO and COMP (computer) problem tags.
- QR codes in margins allow students to use smartphones or other devices for immediate access to video tutor demonstrations that illustrate selected concepts while challenging students to interact with the video by predicting outcomes of simple experiments.
- References to PhET simulations appear in the margins where appropriate.
- As with earlier revisions, we've incorporated new research results, new applications of physics principles, and findings from physics education research.

Pedagogical Innovations

This book is *concise*, but it's also *progressive* in its embrace of proven techniques from physics education research and *strategic* in its approach to learning physics. Chapter 1 introduces the IDEA framework for problem solving, and every one of the book's subsequent **worked examples** employs this framework. IDEA—an acronym for Identify, Develop, Evaluate, Assess—is not a "cookbook" method for students to apply mindlessly, but rather a tool for organizing students' thinking and discouraging equation hunting. It begins with an interpretation of the problem and an identification of the key physics concepts involved; develops a plan for reaching the solution; carries out the mathematical evaluation; and assesses the solution to see that it makes sense, to compare the example with others, and to mine additional insights into physics. In nearly all of the text's worked examples, the Develop phase includes making a drawing, and most of these use a hand-drawn style to encourage students to make their own drawings—a step that research suggests they often skip. IDEA provides a common approach to all physics problem solving, an approach that emphasizes the conceptual unity of physics and helps break the typical student view of physics as a hodgepodge of equations and unrelated ideas. In addition to IDEA-based worked examples, other pedagogical features include:

- **Problem-Solving Strategy boxes** that follow the IDEA framework to provide detailed guidance for specific classes of physics problems, such as Newton's second law, conservation of energy, thermal-energy balance, Gauss's law, or multiloop circuits.
- **Tactics boxes** that reinforce specific essential skills such as differentiation, setting up integrals, vector products, drawing free-body diagrams, simplifying series and parallel circuits, or ray tracing.
- **QR codes** in the textbook allow students to link to video tutor demonstrations as they read, using their smartphones. These "Pause and predict" videos of key physics concepts ask students to submit a prediction before they see the outcome. The videos are also available in the Study Area of Mastering and in the Pearson eText.
- **GOT IT? boxes** that provide quick checks for students to test their conceptual understanding. Many of these use a multiple-choice or quantitative ranking format to probe student misconceptions and facilitate their use with classroom-response systems. Many new GOT IT? boxes have been added in the third edition, and now nearly every section of every chapter has at least one GOT IT? box.
- **Tips** that provide helpful problem-solving hints or warn against common pitfalls and misconceptions.
- **Chapter openers** that include a graphical indication of where the chapter lies in sequence as well as three columns of points that help make connections with other material throughout the book. These include a backward-looking "What You Know," "What You're Learning" for the present chapter, and a forward-looking "How You'll Use It." Each chapter also includes an opening photo, captioned with a question whose answer should be evident after the student has completed the chapter.
- **Applications**, self-contained presentations typically shorter than half a page, provide interesting and contemporary instances of physics in the real world, such as bicycle stability; flywheel energy storage; laser vision correction; ultracapacitors; noise-cancelling headphones; wind energy; magnetic resonance imaging; smartphone gyroscopes; combined-cycle power generation; circuit models of the cell membrane; CD, DVD, and Blu-ray technologies; radiocarbon dating; and many, many more.
- **For Thought and Discussion** questions at the end of each chapter designed for peer learning or for self-study to enhance students' conceptual understanding of physics.
- **Annotated figures** that adopt the research-based approach of including simple "instructor's voice" commentary to help students read and interpret pictorial and graphical information.
- **End-of-chapter** problems that begin with simpler exercises keyed to individual chapter sections and ramp up to more challenging and often multistep problems that synthesize chapter material. Context-rich problems focusing on real-world situations are interspersed throughout each problem set.
- **Chapter summaries** that combine text, art, and equations to provide a synthesized overview of each chapter. Each summary is hierarchical, beginning with the chapter's "big ideas," then focusing on key concepts and equations, and ending with a list of "applications"—specific instances or applications of the physics presented in the chapter.

Organization

This contemporary book is *concise*, *strategic*, and *progressive*, but it's *traditional* in its organization. Following the introductory Chapter 1, the book is divided into six parts. Part One (Chapters 2–12) develops the basic concepts of mechanics, including Newton's laws and conservation principles as applied to single particles and multiparticle systems. Part Two (Chapters 13–15) extends mechanics to oscillations, waves, and fluids. Part Three (Chapters 16–19) covers thermodynamics. Part Four (Chapters 20–29) deals with electricity and magnetism. Part Five (Chapters 30–32) treats optics, first in the geometrical optics approximation and then including wave phenomena. Part Six (Chapters 33–39) introduces relativity and quantum physics. Each part begins with a brief description of its coverage, and ends with a conceptual summary and a challenge problem that synthesizes ideas from several chapters.

Essential University Physics is available in two paperback volumes, so students can purchase only what they need—making the low-cost aspect of this text even more attractive. Volume 1 includes Parts One, Two, and Three, mechanics through thermodynamics. Volume 2 contains Parts Four, Five, and Six, electricity and magnetism along with optics and modern physics.

Instructor Supplements

NOTE: For convenience, all of the following instructor supplements (except the Instructor's Resource DVD) can be downloaded from the Instructor's Resource Area of MasteringPhysics® (www.masteringphysics.com) as well as from the Instructor's Resource Center on www.pearsonhighered.com/irc.

- The **Instructor's Solutions Manual** (ISBN 0-133-85713-1) contains solutions to all end-of-chapter exercises and problems, written in the Interpret/Develop/Evaluate/Assess (IDEA) problem-solving framework. The solutions are provided in PDF and editable Microsoft® Word formats for Mac and PC, with equations in MathType.
- The **Instructor's Resource DVD** (ISBN 0-133-85714-X) provides all the figures, photos, and tables from the text in JPEG format. All the problem-solving strategies, Tactics Boxes, key equations, and chapter summaries are provided in PDF and editable Microsoft® Word formats with equations in MathType. Each chapter also has a set of PowerPoint® lecture outlines and questions including the new GOT IT! Clickers. A comprehensive library of more than 220 applets from **ActivPhysics OnLine™**, a suite of over 70 PhET simulations, and 40 video tutor demonstrations are also included. Also, the complete Instructor's Solutions Manual is provided in both Word and PDF formats.
- MP **MasteringPhysics®** (www.masteringphysics.com) is the most advanced physics homework and tutorial system available. This online homework and tutoring system guides students through the toughest topics in physics with self-paced tutorials that provide individualized coaching. These assignable, in-depth tutorials are designed to coach students with hints and feedback specific to their individual errors. Instructors can also assign end-of-chapter problems from every chapter, including multiple-choice questions, section-specific exercises, and general problems. Quantitative problems can be assigned with numerical answers and randomized values (with sig fig feedback) or solutions. This third edition includes nearly 400 new problems written by the author explictly for use with MasteringPhysics.
- **Learning Catalytics** is a "bring your own device" student engagement, assessment, and classroom intelligence system that is based on cutting-edge research, innovation, and implementation of interactive teaching and peer instruction. With Learning Catalytics pre-lecture questions, you can see what students do and don't understand and adjust lectures accordingly.
- **Pearson eText** is available either automatically when MasteringPhysics® is packaged with new books or as a purchased upgrade online. Users can search for words or phrases, create notes, highlight text, bookmark sections, click on definitions to key terms, and launch PhET simulations and video tutor demonstrations as they read. Professors also have the ability to annotate the text for their course and hide chapters not covered in their syllabi.
- The **Test Bank** (ISBN 0-133-85715-8) contains more than 2000 multiple-choice, true-false, and conceptual questions in TestGen® and Microsoft Word® formats for Mac and PC users. More than half of the questions can be assigned with randomized numerical values.

Student Supplements

- **MasteringPhysics®** (www.masteringphysics.com) is the most advanced physics homework and tutorial system available. This online homework and tutoring system guides students through the most important topics in physics with self-paced tutorials that provide individualized coaching. These assignable, in-depth tutorials are designed to coach students with hints and feedback specific to their individual errors. Instructors can also assign end-of-chapter problems from every chapter including multiple-choice questions, section-specific exercises, and general problems. Quantitative problems can be assigned with numerical answers and randomized values (with sig fig feedback) or solutions.
- **Pearson eText** is available through MasteringPhysics®, either automatically when MasteringPhysics® is packaged with new books or as a purchased upgrade online. Allowing students access to the text wherever they have access to the Internet, Pearson eText comprises the full text with additional interactive features. Users can search for words or phrases, create notes, highlight text, bookmark sections, click on definitions to key terms, and launch PhET simulations and video tutor demonstrations as they read.

Acknowledgments

A project of this magnitude isn't the work of its author alone. First and foremost among those I thank for their contributions are the now several thousand students I've taught in calculus-based introductory physics courses at Middlebury College. Over the years your questions have taught me how to convey physics ideas in many different ways appropriate to your diverse learning styles. You've helped identify the "sticking points" that challenge introductory physics students, and you've showed me ways to help you avoid and "unlearn" the misconceptions that many students bring to introductory physics.

Thanks also to the numerous instructors and students from around the world who have contributed valuable suggestions for improvement of this text. I've heard you, and you'll find many of your ideas implemented in this third edition of *Essential University Physics*. And special thanks to my Middlebury physics colleagues who have taught from this text and who contribute valuable advice and insights on a regular basis: Jeff Dunham, Anne Goodsell, Noah Graham, Steve Ratcliff, and Susan Watson.

Experienced physics instructors thoroughly reviewed every chapter of this book, and reviewers' comments resulted in substantive changes—and sometimes in major rewrites—to the first drafts of the manuscript. We list all these reviewers below. But first, special thanks are due to several individuals who made exceptional contributions to the quality and in some cases the very existence of this book. First is Professor Jay Pasachoff of Williams College, whose willingness more than three decades ago to take a chance on an inexperienced coauthor has made writing introductory physics a large part of my professional career. Dr. Adam Black, former physics editor at Pearson, had the vision to see promise in a new introductory text that would respond to the rising chorus of complaints about massive, encyclopedic, and expensive physics texts. Brad Patterson, developmental editor for the first edition, brought his graduate-level knowledge of physics to a role that made him a real collaborator. Brad is responsible for many of the book's innovative features, and it was a pleasure to work with him. John Murdzek and Matt Walker continued with Brad's excellent tradition of developmental editing on this third edition. We've gone to great lengths to make this book as error-free as possible, and much of the credit for that happy situation goes to Sen-Ben Liao, who solved every new and revised homework problem and updated the solutions manual.

I also wish to thank Nancy Whilton and Katie Conley at Pearson Education, and Haylee Schwenk at Lumina Datamatics, for their highly professional efforts in shepherding this book through its vigorous production schedule. Finally, as always, I thank my family, my colleagues, and my students for the patience they showed during the intensive process of writing and revising this book.

Reviewers

John R. Albright, *Purdue University–Calumet*
Rama Bansil, *Boston University*
Richard Barber, *Santa Clara University*
Linda S. Barton, *Rochester Institute of Technology*
Rasheed Bashirov, *Albertson College of Idaho*
Chris Berven, *University of Idaho*
David Bixler, *Angelo State University*
Ben Bromley, *University of Utah*
Charles Burkhardt, *St. Louis Community College*
Susan Cable, *Central Florida Community College*
George T. Carlson, Jr., *West Virginia Institute of Technology–West Virginia University*
Catherine Check, *Rock Valley College*
Norbert Chencinski, *College of Staten Island*
Carl Covatto, *Arizona State University*
David Donnelly, *Texas State University–San Marcos*
David G. Ellis, *University of Toledo*
Tim Farris, *Volunteer State Community College*
Paula Fekete, *Hunter College of The City University of New York*

Idan Ginsburg, *Harvard University*
James Goff, *Pima Community College*
Austin Hedeman, *University of California–Berkeley*
Andrew Hirsch, *Purdue University*
Mark Hollabaugh, *Normandale Community College*
Eric Hudson, *Pennsylvania State University*
Rex W. Joyner, *Indiana Institute of Technology*
Nikos Kalogeropoulos, *Borough of Manhattan Community College–The City University of New York*
Viken Kiledjian, *East Los Angeles College*
Kevin T. Kilty, *Laramie County Community College*
Duane Larson, *Bevill State Community College*
Kenneth W. McLaughlin, *Loras College*
Tom Marvin, *Southern Oregon University*
Perry S. Mason, *Lubbock Christian University*
Mark Masters, *Indiana University–Purdue University Fort Wayne*
Jonathan Mitschele, *Saint Joseph's College*
Gregor Novak, *United States Air Force Academy*
Richard Olenick, *University of Dallas*
Robert Philbin, *Trinidad State Junior College*
Russell Poch, *Howard Community College*
Steven Pollock, *Colorado University–Boulder*
Richard Price, *University of Texas at Brownsville*
James Rabchuk, *Western Illinois University*
George Schmiedeshoff, *Occidental College*
Natalia Semushkina, *Shippensburg University of Pennsylvania*
Anwar Shiekh, *Dine College*
David Slimmer, *Lander University*
Chris Sorensen, *Kansas State University*
Ronald G. Tabak, *Youngstown State University*
Gajendra Tulsian, *Daytona Beach Community College*
Brigita Urbanc, *Drexel University*
Henry Weigel, *Arapahoe Community College*
Arthur W. Wiggins, *Oakland Community College*
Fredy Zypman, *Yeshiva University*

Preface to the Student

Welcome to physics! Maybe you're taking introductory physics because you're majoring in a field of science or engineering that requires a semester or two of physics. Maybe you're premed, and you know that medical schools are increasingly interested in seeing calculus-based physics on your transcript. Perhaps you're really gung-ho and plan to major in physics. Or maybe you want to study physics further as a minor associated with related fields like math or chemistry or to complement a discipline like economics, environmental studies, or even music. Perhaps you had a great high-school physics course, and you're eager to continue. Maybe high-school physics was an academic disaster for you, and you're approaching this course with trepidation. Or perhaps this is your first experience with physics. Whatever your reason for taking introductory physics, welcome!

And whatever your reason, my goals for you are similar: I'd like to help you develop an understanding and appreciation of the physical universe at a deep and fundamental level; I'd like you to become aware of the broad range of natural and technological phenomena that physics can explain; and I'd like to help you strengthen your analytic and quantitative problem-solving skills. Even if you're studying physics only because it's a requirement, I want to help you engage the subject and come away with an appreciation for this fundamental science and its wide applicability. One of my greatest joys as a physics teacher is having students tell me after the course that they had taken it only because it was required, but found they really enjoyed their exposure to the ideas of physics.

Physics is fundamental. To understand physics is to understand how the world works, both in everyday life and on scales of time and space so small and so large as to defy intuition. For that reason I hope you'll find physics fascinating. But you'll also find it challenging. Learning physics will challenge you with the need for precise thinking and language; with subtle interpretations of even commonplace phenomena; and with the need for skillful application of mathematics. But there's also a simplicity to physics, a simplicity that results because there are in physics only a very few really basic principles to learn. Those succinct principles encompass a universe of natural phenomena and technological applications.

I've been teaching introductory physics for decades, and this book distills everything my students have taught me about the many different ways to approach physics; about the subtle misconceptions students often bring to physics; about the ideas and types of problems that present the greatest challenges; and about ways to make physics engaging, exciting, and relevant to your life and interests.

I have some specific advice for you that grows out of my long experience teaching introductory physics. Keeping this advice in mind will make physics easier (but not necessarily easy!), more interesting, and, I hope, more fun:

- *Read* each chapter thoroughly and carefully before you attempt to work any problem assignments. I've written this text with an informal, conversational style to make it engaging. It's not a reference work to be left alone until you need some specific piece of information; rather, it's an unfolding "story" of physics—its big ideas and their applications in quantitative problem solving. You may think physics is hard because it's mathematical, but in my long experience I've found that failure to *read* thoroughly is the biggest single reason for difficulties in introductory physics.
- *Look for the big ideas*. Physics isn't a hodgepodge of different phenomena, laws, and equations to memorize. Rather, it's a few big ideas from which flow myriad applications, examples, and special cases. In particular, don't think of physics as a jumble of equations that you choose among when solving a problem. Rather, identify those few big ideas and the equations that represent them, and try to see how seemingly distinct examples and special cases relate to the big ideas.
- *When working problems, re-read* the appropriate sections of the text, paying particular attention to the worked examples. Follow the IDEA strategy described in Chapter 1 and used in every subsequent worked example. Don't skimp on the final Assess step. Always ask: Does this answer make sense? How can I understand my answer in relation to the big principles of physics? How was this problem like others I've worked, or like examples in the text?
- *Don't confuse physics with math*. Mathematics is a tool, not an end in itself. Equations in physics aren't abstract math, but statements about the physical world. Be sure you understand each equation for what it says about physics, not just as an equality between mathematical terms.
- *Work with others*. Getting together informally in a room with a blackboard is a great way to explore physics, to clarify your ideas and help others clarify theirs, and to learn from your peers. I urge you to discuss physics problems together with your classmates, to contemplate together the "For Thought and Discussion" questions at the end of each chapter, and to engage one another in lively dialog as you grow your understanding of physics, the fundamental science.

Detailed Contents

Volume 1 contains Chapters 1–19
Volume 2 contains Chapters 20–39

Doing Physics

What You Know

- You're coming to this course with a solid background in algebra, geometry, and trigonometry.
- You may have had calculus, or you'll be starting it concurrently.
- You don't need to have taken physics to get a full understanding from this book.

What You're Learning

- This chapter gives you an overview of physics and its subfields, which together describe the entire physical universe.
- You'll learn the basis of the SI system of measurement units.
- You'll learn to express and manipulate numbers used in quantitative science.
- You'll learn to deal with precision and uncertainty.
- You'll develop a skill for making quick estimates.
- You'll learn how to extract information from experimental data.
- You'll see a strategy for solving physics problems.

How You'll Use It

- Skills and knowledge that you develop in this chapter will serve you throughout your study of physics.
- You'll be able to express quantitative answers to physics problems in scientific notation, with the correct units and the appropriate uncertainty expressed through significant figures.
- Being able to make quick estimates will help you gauge the sizes of physical effects and will help you recognize whether your quantitative answers make sense.
- The problem-solving strategy you'll learn here will serve you in the many physics problems that you'll work in order to really learn physics.

Which realms of physics are involved in the workings of your DVD player?

You slip a DVD into your player and settle in to watch a movie. The DVD spins, and a precisely focused laser beam "reads" its content. Electronic circuitry processes the information, sending it to your video display and to loudspeakers that turn electrical signals into sound waves. Every step of the way, principles of physics govern the delivery of the movie from DVD to you.

1.1 Realms of Physics

That DVD player is a metaphor for all of **physics**—the science that describes the fundamental workings of physical reality. Physics explains natural phenomena ranging from the behavior of atoms and molecules to thunderstorms and rainbows and on to the evolution of stars, galaxies, and the universe itself. Technological applications of physics are the basis for everything from microelectronics to medical imaging to cars, airplanes, and space flight.

At its most fundamental, physics provides a nearly unified description of all physical phenomena. However, it's convenient to divide physics into distinct realms (Fig. 1.1). Your DVD player encompasses essentially all those realms. **Mechanics**, the branch of physics that deals with motion, describes the spinning disc. Mechanics also explains the motion of a car, the orbits of the planets, and the stability of a skyscraper. Part 1 of this book deals with the basic ideas of mechanics.

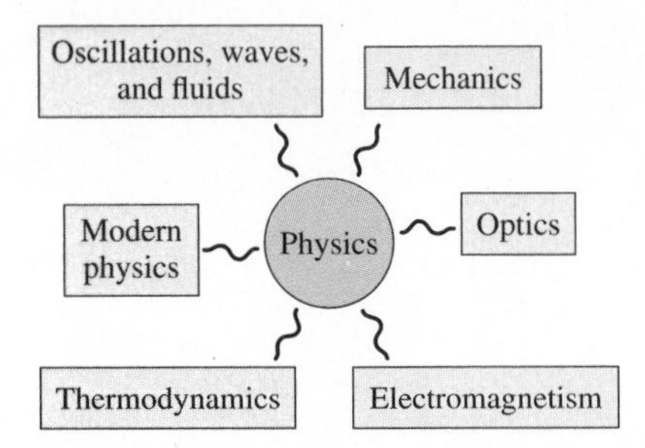

FIGURE 1.1 Realms of physics.

Those sound waves coming from your loudspeakers represent **wave motion**. Other examples include the ocean waves that pound Earth's coastlines, the wave of standing spectators that sweeps through a football stadium, and the undulations of Earth's crust that spread the energy of an earthquake. Part 2 of this book covers wave motion and other phenomena involving the motion of fluids like air and water.

When you burn your own DVD, the high temperature produced by an intensely focused laser beam alters the material properties of a writable DVD, thus storing video or computer information. That's an example of **thermodynamics**—the study of heat and its effects on matter. Thermodynamics also describes the delicate balance of energy-transfer processes that keeps our planet at a habitable temperature and puts serious constraints on our ability to meet the burgeoning energy demands of modern society. Part 3 comprises four chapters on thermodynamics.

An electric motor spins your DVD, converting electrical energy to the energy of motion. Electric motors are ubiquitous in modern society, running everything from subway trains and hybrid cars, to elevators and washing machines, to insulin pumps and artificial hearts. Conversely, electric generators convert the energy of motion to electricity, providing virtually all of our electrical energy. Motors and generators are two applications of **electromagnetism** in modern technology. Others include computers, audiovisual electronics, microwave ovens, digital watches, and even the humble lightbulb; without these electromagnetic technologies our lives would be very different. Equally electromagnetic are all the wireless technologies that enable modern communications, from satellite TV to cell phones to wireless computer networks, mice, and keyboards. And even light itself is an electromagnetic phenomenon. Part 4 presents the principles of electromagnetism and their many applications.

The precise focusing of laser light in your DVD player allows hours of video to fit on a small plastic disc. The details and limitations of that focusing are governed by the principles of **optics**, the study of light and its behavior. Applications of optics range from simple magnifiers to contact lenses to sophisticated instruments such as microscopes, telescopes, and spectrometers. Optical fibers carry your e-mail, web pages, and music downloads over the global Internet. Natural optical systems include your eye and the raindrops that deflect sunlight to form rainbows. Part 5 of the book explores optical principles and their applications.

That laser light in your DVD player is an example of an electromagnetic wave, but an atomic-level look at the light's interaction with matter reveals particle-like "bundles" of electromagnetic energy. This is the realm of **quantum physics**, which deals with the often counterintuitive behavior of matter and energy at the atomic level. Quantum phenomena also explain how that DVD laser works and, more profoundly, the structure of atoms and the periodic arrangement of the elements that is the basis of all chemistry. Quantum physics is one of the two great developments of **modern physics**. The other is Einstein's **theory of relativity**. Relativity and quantum physics arose during the 20th century, and together they've radically altered our commonsense notions of time, space, and causality. Part 6 of the book surveys the ideas of modern physics, ending with what we do—and don't—know about the history, future, and composition of the entire universe.

CONCEPTUAL EXAMPLE 1.1 Car Physics

Name some systems in your car that exemplify the different realms of physics.

EVALUATE *Mechanics* is easy; the car is fundamentally a mechanical system whose purpose is motion. Details include starting, stopping, cornering, as well as a host of other motions within mechanical subsystems. Your car's springs and shock absorbers constitute an *oscillatory* system engineered to give a comfortable ride. The car's engine is a prime example of a *thermodynamic* system, converting the energy of burning gasoline into the car's motion. *Electromagnetic* systems range from the starter motor and spark plugs to sophisticated electronic devices that monitor and optimize engine performance. *Optical* principles govern rear- and side-view mirrors and headlights. Increasingly, optical fibers transmit information to critical safety systems. *Modern physics* is less obvious in your car, but ultimately, everything from the chemical reactions of burning gasoline to the atomic-scale operation of automotive electronics is governed by its principles.

1.2 Measurements and Units

"A long way" means different things to a sedentary person, a marathon runner, a pilot, and an astronaut. We need to quantify our measurements. Science uses the **metric system**, with fundamental quantities length, mass, and time measured in meters, kilograms, and seconds, respectively. The modern version of the metric system is **SI**, for Système International d'Unités (International System of Units), which incorporates scientifically precise definitions of the fundamental quantities.

The three fundamental quantities were originally defined in reference to nature: the meter in terms of Earth's size, the kilogram as an amount of water, and the second by the length of the day. For length and mass, these were later replaced by specific artifacts—a bar whose length was defined as 1 meter and a cylinder whose mass defined the kilogram. But natural standards like the day's length can change, as can the properties of artifacts. So early SI definitions gave way to **operational definitions**, which are measurement standards based on laboratory procedures. Such standards have the advantage that scientists anywhere can reproduce them. By the late 20th century, two of the three fundamental units—the meter and the second—had operational definitions, but the kilogram did not.

A special type of operational definition involves giving an exact value to a particular constant of nature—a quantity formerly subject to experimental determination and with a stated uncertainty in its value. As described below, the meter was the first such unit to be defined in this way. By the early 21st century, it was clear that defining units in terms of fundamental, invariant physical constants was the best way to ensure long-term stability of the SI unit system. Currently, SI is undergoing a sweeping revision, which will result in redefining the kilogram and three of the four remaining so-called base units with definitions that lock in exact values of fundamental constants. These so-called **explicit-constant** definitions will have similar wording, making explicit that the unit in question follows from the defined value of the particular physical constant.

APPLICATION Units Matter: A Bad Day on Mars

In September 1999, the Mars Climate Orbiter was destroyed when the spacecraft passed through Mars's atmosphere and experienced stresses and heating it was not designed to tolerate. Why did this $125-million craft enter the Martian atmosphere when it was supposed to remain in the vacuum of space? NASA identified the root cause as a failure to convert the English units one team used to specify rocket thrust to the SI units another team expected. Units matter!

Length

The **meter** was first defined as one ten-millionth of the distance from Earth's equator to the North Pole. In 1889 a standard meter was fabricated to replace the Earth-based unit, and in 1960 that gave way to a standard based on the wavelength of light. By the 1970s, the speed of light had become one of the most precisely determined quantities. As a result, the meter was redefined in 1983 as the distance light travels in vacuum in 1/299,792,458 of a second. The effect of this definition is to make the speed of light a defined quantity: 299,792,458 m/s. Thus, the meter became the first SI unit to be based on a defined value for a fundamental constant. The new SI definitions won't change the meter but will reword its definition to make it of the explicit-constant type:

> The meter, symbol m, is the unit of length; its magnitude is set by fixing the numerical value of the speed of light in vacuum to be equal to exactly 299,792,458 when it is expressed in the SI unit m/s.

Time

The **second** used to be defined by Earth's rotation, but that's not constant, so it was later redefined as a specific fraction of the year 1900. An operational definition followed in 1967, associating the second with the radiation emitted by a particular atomic process. The new definition will keep the essence of that operational definition but reworded in the explicit-constant style:

> The second, symbol s, is the unit of time; its magnitude is set by fixing the numerical value of the ground-state hyperfine splitting frequency of the cesium-133 atom, at rest and at a temperature of 0 K, to be exactly 9,192,631,770 when it is expressed in the SI unit s^{-1}, which is equal to Hz.

The device that implements this definition—which will seem less obscure once you've studied some atomic physics—is called an *atomic clock*. Here the phrase "equal to Hz" introduces the unit hertz (Hz) for frequency—the number of cycles of a repeating process that occur each second.

Mass

Since 1889, the kilogram has been defined as the mass of a single artifact—the international prototype kilogram, a platinum–iridium cylinder kept in a vault at the International Bureau of Weights and Measures in Sèvres, France. Not only is this artifact-based standard awkward to access, but comparison measurements have revealed tiny yet growing mass discrepancies between the international prototype kilogram and secondary mass standards based on it.

In the current SI revision, the kilogram will become the last of the SI base units to be defined operationally, with a new explicit-constant definition resulting from fixing the value of *Planck's constant*, h, a fundamental constant of nature related to the "graininess" of physical quantities at the atomic and subatomic levels. The units of Planck's constant involve seconds, meters, and kilograms, and giving h an exact value actually sets the value of $1\ \mathrm{s}^{-1} \cdot \mathrm{m}^2 \cdot \mathrm{kg}$. But with the meter and second already defined, fixing the unit $\mathrm{s}^{-1} \cdot \mathrm{m}^2 \cdot \mathrm{kg}$ then determines the kilogram. A device that implements this definition is the *watt balance*, which balances an unknown mass against forces resulting from electrical effects whose magnitude, in turn, can be related to Planck's constant. The new formal definition of the kilogram will be similar to the explicit-constant definitions of the meter and second, but the exact value of Planck's constant is yet to be established.

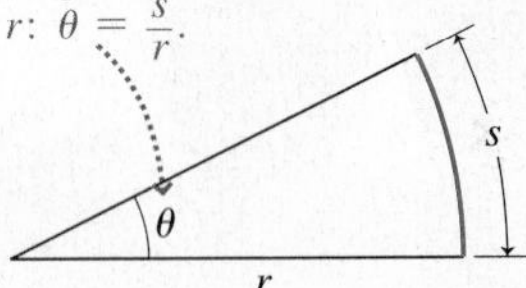

FIGURE 1.2 The radian is the SI unit of angle.

Other SI Units

The SI includes seven independent base units: In addition to the meter, second, and kilogram, there are the ampere (A) for electric current, the kelvin (K) for temperature, the mole (mol) for the amount of a substance, and the candela (cd) for luminosity. We'll introduce these units later, as needed. In the ongoing SI revision these will be given new, explicit-constant definitions; for all but the candela, this involves fixing the values of fundamental physical constants. In addition to the seven physical base units, two supplementary units define geometrical measures of angle: the radian (rad) for ordinary angles (Fig. 1.2) and the steradian (sr) for solid angles. Units for all other physical quantities are derived from the base units.

Table 1.1 SI Prefixes

Prefix	Symbol	Power
yotta	Y	10^{24}
zetta	Z	10^{21}
exa	E	10^{18}
peta	P	10^{15}
tera	T	10^{12}
giga	G	10^{9}
mega	M	10^{6}
kilo	k	10^{3}
hecto	h	10^{2}
deca	da	10^{1}
—	—	10^{0}
deci	d	10^{-1}
centi	c	10^{-2}
milli	m	10^{-3}
micro	µ	10^{-6}
nano	n	10^{-9}
pico	p	10^{-12}
femto	f	10^{-15}
atto	a	10^{-18}
zepto	z	10^{-21}
yocto	y	10^{-24}

SI Prefixes

You could specify the length of a bacterium (e.g., 0.00001 m) or the distance to the next city (e.g., 58,000 m) in meters, but the results are unwieldy—too small in the first case and too large in the latter. So we use prefixes to indicate multiples of the SI base units. For example, the prefix k (for "kilo") means 1000; 1 km is 1000 m, and the distance to the next city is 58 km. Similarly, the prefix μ (the lowercase Greek "mu") means "micro," or 10^{-6}. So our bacterium is 10 µm long. The SI prefixes are listed in Table 1.1, which is repeated inside the front cover. We'll use the prefixes routinely in examples and problems, and we'll often express answers using SI prefixes, without doing an explicit unit conversion.

When two units are used together, a hyphen appears between them—for example, newton-meter. Each unit has a symbol, such as m for meter or N for newton (the SI unit of force). Symbols are ordinarily lowercase, but those named after people are uppercase. Thus "newton" is written with a small "n" but its symbol is a capital N. The exception is the unit of volume, the liter; since the lowercase "l" is easily confused with the number 1, the symbol for liter is a capital L. When two units are multiplied, their symbols are separated by a centered dot: $\mathrm{N} \cdot \mathrm{m}$ for newton-meter. Division of units is expressed by using the slash (/) or writing with the denominator unit raised to the -1 power. Thus the SI unit of speed is the meter per second, written m/s or $\mathrm{m} \cdot \mathrm{s}^{-1}$.

EXAMPLE 1.1 Changing Units: Speed Limits

Express a 65 mi/h speed limit in meters per second.

EVALUATE According to Appendix C, 1 mi = 1609 m, so we can multiply miles by the ratio 1609 m/mi to get meters. Similarly, we use the conversion factor 3600 s/h to convert hours to seconds. Combining these two conversions gives

$$65\ \text{mi/h} = \left(\frac{65\ \cancel{\text{mi}}}{\cancel{\text{h}}}\right)\left(\frac{1609\ \text{m}}{\cancel{\text{mi}}}\right)\left(\frac{1\ \cancel{\text{h}}}{3600\ \text{s}}\right) = 29\ \text{m/s}$$

■

Other Unit Systems

The inches, feet, yards, miles, and pounds of the so-called English system still dominate measurement in the United States. Other non-SI units such as the hour are often mixed with English or SI units, as with speed limits in miles per hour or kilometers per hour. In some areas of physics there are good reasons for using non-SI units. We'll discuss these as the need arises and will occasionally use non-SI units in examples and problems. We'll also often find it convenient to use degrees rather than radians for angles. The vast majority of examples and problems in this book, however, use strictly SI units.

Changing Units

Sometimes we need to change from one unit system to another—for example, from English to SI. Appendix C contains tables for converting among unit systems; you should familiarize yourself with this and the other appendices and refer to them often.

For example, Appendix C shows that 1 ft = 0.3048 m. Since 1 ft and 0.3048 m represent the same physical distance, multiplying any distance by their ratio will change the units but not the actual physical distance. Thus the height of Dubai's Burj Khalifa (Fig. 1.3)—the world's tallest structure—is 2717 ft or

$$(2717\ \text{ft})\left(\frac{0.3048\ \text{m}}{1\ \text{ft}}\right) = 828.1\ \text{m}$$

Often you'll need to change several units in the same expression. Keeping track of the units through a chain of multiplications helps prevent you from carelessly inverting any of the conversion factors. A numerical answer cannot be correct unless it has the right units!

FIGURE 1.3 Dubai's Burj Khalifa is the world's tallest structure.

GOT IT? 1.1 A Canadian speed limit of 50 km/h is closest to which U.S. limit expressed in miles per hour? (a) 60 mph; (b) 45 mph; (c) 30 mph

1.3 Working with Numbers

Scientific Notation

The range of measured quantities in the universe is enormous; lengths alone go from about 1/1,000,000,000,000,000 m for the radius of a proton to 1,000,000,000,000,000,000,000 m for the size of a galaxy; our telescopes see 100,000 times farther still. Therefore, we frequently express numbers in **scientific notation**, where a reasonable-size number is multiplied by a power of 10. For example, 4185 is 4.185×10^3 and 0.00012 is 1.2×10^{-4}. Table 1.2 suggests the vast range of measurements for the fundamental quantities of length, time, and mass. Take a minute (about 10^2 heartbeats, or 3×10^{-8} of a typical human lifespan) to peruse this table along with Fig. 1.4.

This galaxy is 10^{21} m across and has a mass of $\sim 10^{42}$ kg.

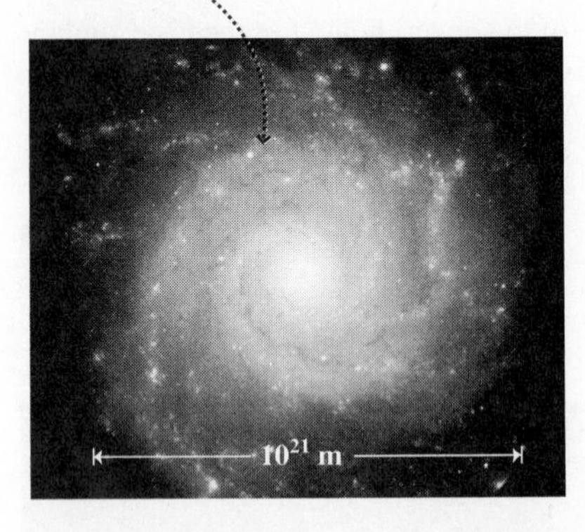

Your movie is stored on a DVD in "pits" only 4×10^{-7} m in size.

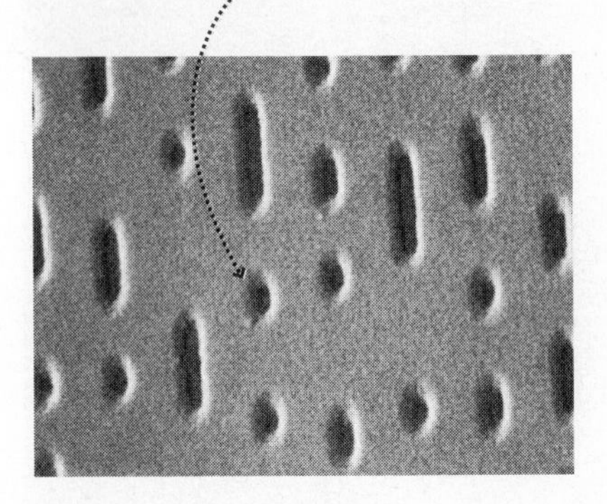

FIGURE 1.4 Large and small.

Table 1.2 Distances, Times, and Masses (rounded to one significant figure)

Radius of observable universe	1×10^{26} m
Earth's radius	6×10^{6} m
Tallest mountain	9×10^{3} m
Height of person	2 m
Diameter of red blood cell	1×10^{-5} m
Size of proton	1×10^{-15} m
Age of universe	4×10^{17} s
Earth's orbital period (1 year)	3×10^{7} s
Human heartbeat	1 s
Wave period, microwave oven	5×10^{-10} s
Time for light to cross a proton	3×10^{-24} s
Mass of Milky Way galaxy	1×10^{42} kg
Mass of mountain	1×10^{18} kg
Mass of human	70 kg
Mass of red blood cell	1×10^{-13} kg
Mass of uranium atom	4×10^{-25} kg
Mass of electron	1×10^{-30} kg

Scientific calculators handle numbers in scientific notation. But straightforward rules allow you to manipulate scientific notation if you don't have such a calculator handy.

TACTICS 1.1 Using Scientific Notation

Addition/Subtraction

To add (or subtract) numbers in scientific notation, first give them the same exponent and then add (or subtract):

$$3.75 \times 10^{6} + 5.2 \times 10^{5} = 3.75 \times 10^{6} + 0.52 \times 10^{6} = 4.27 \times 10^{6}$$

Multiplication/Division

To multiply (or divide) numbers in scientific notation, multiply (or divide) the digits and add (or subtract) the exponents:

$$(3.0 \times 10^{8}\ \text{m/s})(2.1 \times 10^{-10}\ \text{s}) = (3.0)(2.1) \times 10^{8+(-10)}\ \text{m} = 6.3 \times 10^{-2}\ \text{m}$$

Powers/Roots

To raise numbers in scientific notation to any power, raise the digits to the given power and multiply the exponent by the power:

$$\sqrt{(3.61 \times 10^{4})^{3}} = \sqrt{3.61^{3} \times 10^{(4)(3)}} = (47.04 \times 10^{12})^{1/2}$$
$$= \sqrt{47.04} \times 10^{(12)(1/2)} = 6.86 \times 10^{6}$$

EXAMPLE 1.2 Scientific Notation: Tsunami Warnings

Earthquake-generated tsunamis are so devastating because the entire ocean, from surface to bottom, participates in the wave motion. The speed of such waves is given by $v = \sqrt{gh}$, where $g = 9.8\ \text{m/s}^2$ is the gravitational acceleration and h is the depth in meters. Determine a tsunami's speed in 3.0-km-deep water.

EVALUATE That 3.0-km depth is 3.0×10^{3} m, so we have

$$v = \sqrt{gh} = [(9.8\ \text{m/s}^2)(3.0 \times 10^{3}\ \text{m})]^{1/2} = (29.4 \times 10^{3}\ \text{m}^2/\text{s}^2)^{1/2}$$
$$= (2.94 \times 10^{4}\ \text{m}^2/\text{s}^2)^{1/2} = \sqrt{2.94} \times 10^{2}\ \text{m/s} = 1.7 \times 10^{2}\ \text{m/s}$$

where we wrote $29.4 \times 10^{3}\ \text{m}^2/\text{s}^2$ as $2.94 \times 10^{4}\ \text{m}^2/\text{s}^2$ in the second line in order to calculate the square root more easily. Converting the speed to km/h gives

$$1.7 \times 10^{2}\ \text{m/s} = \left(\frac{1.7 \times 10^{2}\ \cancel{\text{m}}}{\cancel{\text{s}}}\right)\left(\frac{1\ \text{km}}{1.0 \times 10^{3}\ \cancel{\text{m}}}\right)\left(\frac{3.6 \times 10^{3}\ \cancel{\text{s}}}{\text{h}}\right)$$
$$= 6.1 \times 10^{2}\ \text{km/h}$$

This speed—about 600 km/h—shows why even distant coastlines have little time to prepare for the arrival of a tsunami. ■

Significant Figures

How precise is that 1.7×10^2 m/s we calculated in Example 1.2? The two **significant figures** in this number imply that the value is closer to 1.7 than to 1.6 or 1.8. The fewer significant figures, the less precisely we can claim to know a given quantity.

In Example 1.2 we were, in fact, given two significant figures for both quantities. The mere act of calculating can't add precision, so we rounded our answer to two significant figures as well. Calculators and computers often give numbers with many figures, but most of those are usually meaningless.

What's Earth's circumference? It's $2\pi R_E$, and π is approximately 3.14159. . . . But if you only know Earth's radius as 6.37×10^6 m, knowing π to more significant figures doesn't mean you can claim to know the circumference any more precisely. This example suggests a rule for handling calculations involving numbers with different precisions:

> In multiplication and division, the answer should have the same number of significant figures as the least precise of the quantities entering the calculation.

You're engineering an access ramp to a bridge whose main span is 1.248 km long. The ramp will be 65.4 m long. What will be the overall length? A simple calculation gives 1.248 km + 0.0654 km = 1.3134 km. How should you round this? You know the bridge length to ± 0.001 km, so an addition this small is significant. Therefore, your answer should have three digits to the right of the decimal point, giving 1.313 km. Thus:

> In addition and subtraction, the answer should have the same number of digits to the right of the decimal point as the term in the sum or difference that has the smallest number of digits to the right of the decimal point.

In subtraction, this rule can quickly lead to loss of precision, as Example 1.3 illustrates.

EXAMPLE 1.3 Significant Figures: Nuclear Fuel

A uranium fuel rod is 3.241 m long before it's inserted in a nuclear reactor. After insertion, heat from the nuclear reaction has increased its length to 3.249 m. What's the increase in its length?

EVALUATE Subtraction gives 3.249 m − 3.241 m = 0.008 m or 8 mm. Should this be 8 mm or 8.000 mm? Just 8 mm. Subtraction affected only the last digit of the four-significant-figure lengths, leaving only one significant figure in the answer. ■

✓TIP Intermediate Results

Although it's important that your final answer reflect the precision of the numbers that went into it, any intermediate results should have at least one extra significant figure. Otherwise, rounding of intermediate results could alter your answer.

GOT IT? 1.2 Rank the numbers according to (1) their size and (2) the number of significant figures. Some may be of equal rank. 0.0008, 3.14×10^7, 2.998×10^{-9}, 55×10^6, 0.041×10^9

What about whole numbers ending in zero, like 60, 300, or 410? How many significant figures do they have? Strictly speaking, 60 and 300 have only one significant figure, while 410 has two. If you want to express the number 60 to two significant figures, you should write 6.0×10^1; similarly, 300 to three significant figures would be 3.00×10^2, and 410 to three significant figures would be 4.10×10^2.

Working with Data

In physics, in other sciences, and even in nonscience fields, you'll find yourself working with data—numbers that come from real-world measurements. One important use of data in the sciences is to confirm hypotheses about relations between physical quantities. Scientific hypotheses can generally be described quantitatively using equations, which often give or can be manipulated to give a linear relationship between quantities. Plotting such data and fitting a line through the data points—using procedures such as regression analysis, least-squares fitting, or even "eyeballing" a best-fit line—can confirm the hypothesis and give useful information about the phenomena under study. You'll probably have opportunities to do such data fitting in your physics lab and in other science courses. Because it's so important in experimental science, we've included at least one data problem with each chapter. Example 1.4 shows a typical example of fitting data to a straight line.

EXAMPLE 1.4 Data Analysis: A Falling Ball

As you'll see in Chapter 2, the distance fallen by an object dropped from rest should increase in proportion to the square of the time since it was dropped; the proportionality should be half the acceleration due to gravity. The table shows actual data from measurements on a falling ball. Determine a quantity such that, when you plot fall distance y against it, you should get a straight line. Make the plot, fit a straight line, and from its slope determine an approximate value for the gravitational acceleration.

Time (s)	0.500	1.00	1.50	2.00	2.50	3.00
Distance (m)	1.12	5.30	12.2	18.5	34.1	43.6

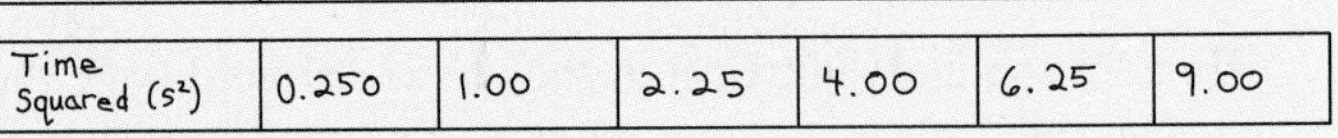

Time Squared (s²)	0.250	1.00	2.25	4.00	6.25	9.00

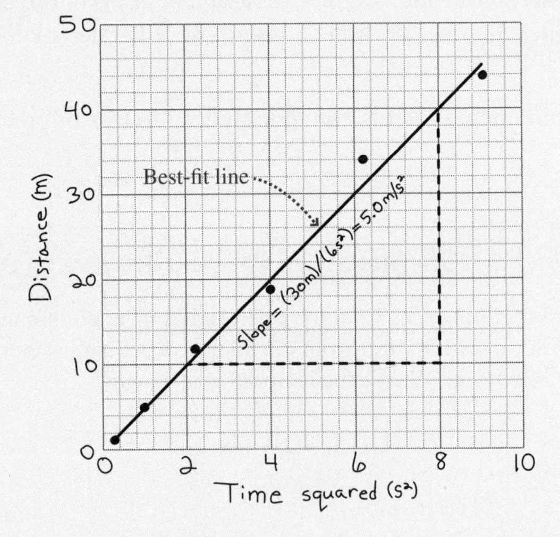

FIGURE 1.5 Our graph for Example 1.4. We "eyeballed" the best-fit line using a ruler; note that it doesn't go through particular points but tries to capture the average trend of all the data points.

EVALUATE We're told that the fall distance y should be proportional to the square of the time; thus we choose to plot y versus t^2. So we've added a row to the table, listing the values of t^2. Figure 1.5 is our plot. Although we did this one by hand, on graph paper, you could use a spreadsheet or other program to make your plot. A spreadsheet program would offer the option to draw a best-fit line and give its slope, but a hand-drawn line, "eyeballed" to catch the general trend of the data points, works surprisingly well. We've indicated such a line, and the figure shows that its slope is very nearly 5.0 m/s².

ASSESS The fact that our data points lie very nearly on a straight line confirms the hypothesis that fall distance should be proportional to time squared. Real data almost never lie exactly on a theoretically predicted line or curve. A more sophisticated analysis would show error bars, indicating the measurement uncertainty in each data point. Because our line's measured slope is supposed to be half the gravitational acceleration, our analysis suggests a gravitational acceleration of about 10 m/s². This is close to the commonly used value of 9.8 m/s². ■

MP

PhET: Estimation

Estimation

Some problems in physics and engineering call for precise numerical answers. We need to know exactly how long to fire a rocket to put a space probe on course toward a distant planet, or exactly what size to cut the tiny quartz crystal whose vibrations set the pulse of a digital watch. But for many other purposes, we need only a rough idea of the size of a physical effect. And rough estimates help check whether the results of more difficult calculations make sense.

EXAMPLE 1.5 Estimation: Counting Brain Cells

Estimate the mass of your brain and the number of cells it contains.

EVALUATE My head is about 6 in. or 15 cm wide, but there's a lot of skull bone in there, so maybe my brain is about 10 cm or 0.1 m across. I don't know its exact shape, but for estimating, I'll take it to be a cube. Then its volume is $(10\text{ cm})^3 = 1000\text{ cm}^3$, or 10^{-3} m^3. I'm mostly water, and water's density is 1 gram per cubic centimeter (1 g/cm^3), so my 1000-cm^3 brain has a mass of about 1 kg.

How big is a brain cell? I don't know, but Table 1.2 lists the diameter of a red blood cell as about 10^{-5} m. If brain cells are roughly the same size, then each cell has a volume of approximately $(10^{-5}\text{ m})^3 = 10^{-15}\text{ m}^3$. Then the number of cells in my 10^{-3}-m^3 brain is roughly

$$N = \frac{10^{-3}\text{ m}^3/\text{brain}}{10^{-15}\text{ m}^3/\text{cell}} = 10^{12}\text{ cells/brain}$$

Crude though they are, these estimates aren't bad. The average adult brain's mass is about 1.3 kg, and it contains at least 10^{11} cells (Fig. 1.6).

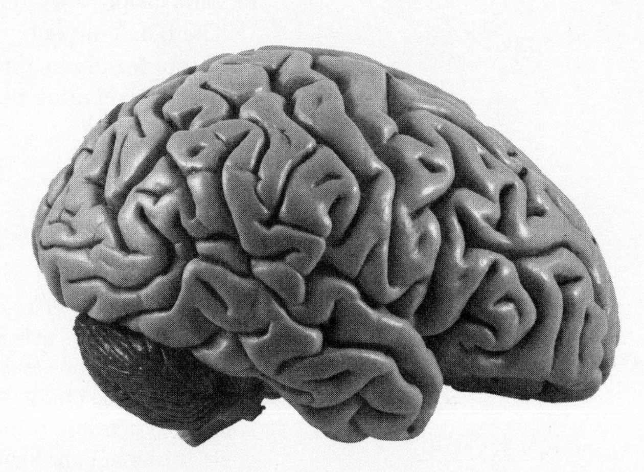

FIGURE 1.6 The average human brain contains more than 10^{11} cells.

1.4 Strategies for Learning Physics

You can learn *about* physics, and you can learn to *do* physics. This book is for science and engineering students, so it emphasizes both. Learning about physics will help you appreciate the role of this fundamental science in explaining both natural and technological phenomena. Learning to do physics will make you adept at solving quantitative problems—finding answers to questions about how the natural world works and about how we forge the technologies at the heart of modern society.

Physics: Challenge and Simplicity

Physics problems can be challenging, calling for clever insight and mathematical agility. That challenge is what gives physics a reputation as a difficult subject. But underlying all of physics is only a handful of basic principles. Because physics is so fundamental, it's also inherently simple. There are only a few basic ideas to learn; if you really understand those, you can apply them in a wide variety of situations. These ideas and their applications are all connected, and we'll emphasize those connections and the underlying simplicity of physics by reminding you how the many examples, applications, and problems are manifestations of the same few basic principles. If you approach physics as a hodgepodge of unrelated laws and equations, you'll miss the point and make things difficult. But if you look for the basic principles, for connections among seemingly unrelated phenomena and problems, then you'll discover the underlying simplicity that reflects the scope and power of physics—the fundamental science.

Problem Solving: The IDEA Strategy

Solving a quantitative physics problem always starts with basic principles or concepts and ends with a precise answer expressed as either a numerical quantity or an algebraic expression. Whatever the principle, whatever the realm of physics, and whatever the specific situation, the path from principle to answer follows four simple steps—steps that make up a comprehensive strategy for approaching all problems in physics. Their acronym, IDEA, will help you remember these steps, and they'll be reinforced as we apply them over and over again in worked examples throughout the book. We'll generally write all four steps separately, although the examples in this chapter cut right to the EVALUATE phase. And in some chapters we'll introduce versions of this strategy tailored to specific material.

The IDEA strategy isn't a "cookbook" formula for working physics problems. Rather, it's a tool for organizing your thoughts, clarifying your conceptual understanding, developing and executing plans for solving problems, and assessing your answers. Here's the big IDEA:

PROBLEM-SOLVING STRATEGY 1.1 Physics Problems

INTERPRET The first step is to *interpret* the problem to be sure you know what it's asking. Then *identify* the applicable concepts and principles—Newton's laws of motion, conservation of energy, the first law of thermodynamics, Gauss's law, and so forth. Also *identify* the players in the situation—the object whose motion you're asked to describe, the forces acting, the thermodynamic system you're to analyze, the charges that produce an electric field, the components in an electric circuit, the light rays that will help you locate an image, and so on.

DEVELOP The second step is to *develop* a plan for solving the problem. It's always helpful and often essential to *draw* a diagram showing the situation. Your drawing should indicate objects, forces, and other physical entities. Labeling masses, positions, forces, velocities, heat flows, electric or magnetic fields, and other quantities will be a big help. Next, *determine* the relevant mathematical formulas—namely, those that contain the quantities you're given in the problem as well as the unknown(s) you're solving for. Don't just grab equations—rather, think about how each reflects the underlying concepts and principles that you've identified as applying to this problem. The plan you develop might include calculating intermediate quantities, finding values in a table or in one of this text's several appendices, or even solving a preliminary problem whose answer you need in order to get your final result.

EVALUATE Physics problems have numerical or symbolic answers, and you need to *evaluate* your answer. In this step you *execute* your plan, going in sequence through the steps you've outlined. Here's where your math skills come in. Use algebra, trig, or calculus, as needed, to solve your equations. It's a good idea to keep all numerical quantities, whether known or not, in symbolic form as you work through the solution of your problem. At the end you can plug in numbers and work the arithmetic to *evaluate* the numerical answer, if the problem calls for one.

ASSESS Don't be satisfied with your answer until you *assess* whether it makes sense! Are the units correct? Do the numbers sound reasonable? Does the algebraic form of your answer work in obvious special cases, like perhaps "turning off" gravity or making an object's mass zero or infinite? Checking special cases not only helps you decide whether your answer makes sense but also can give you insights into the underlying physics. In worked examples, we'll often use this step to enhance your knowledge of physics by relating the example to other applications of physics.

Don't memorize the IDEA problem-solving strategy. Instead, grow to understand it as you see it applied in examples and as you apply it yourself in working end-of-chapter problems. This book has a number of additional features and supplements, discussed in the Preface, to help you develop your problem-solving skills.

CHAPTER 1 SUMMARY

Big Idea

Physics is the fundamental science. It's convenient to consider several realms of physics, which together describe all that's known about physical reality:

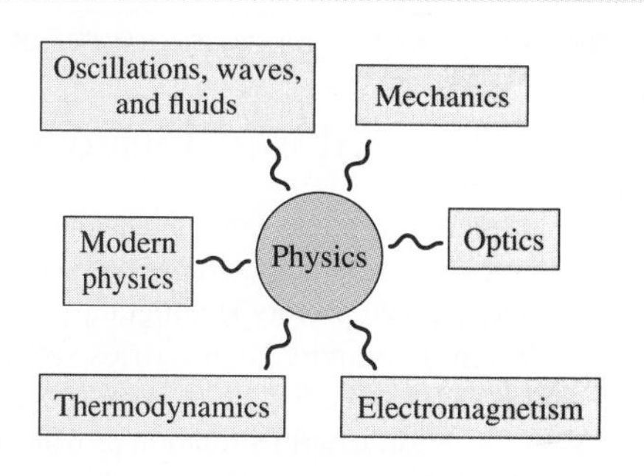

Key Concepts and Equations

Numbers describing physical quantities must have units. The SI unit system comprises seven fundamental units:

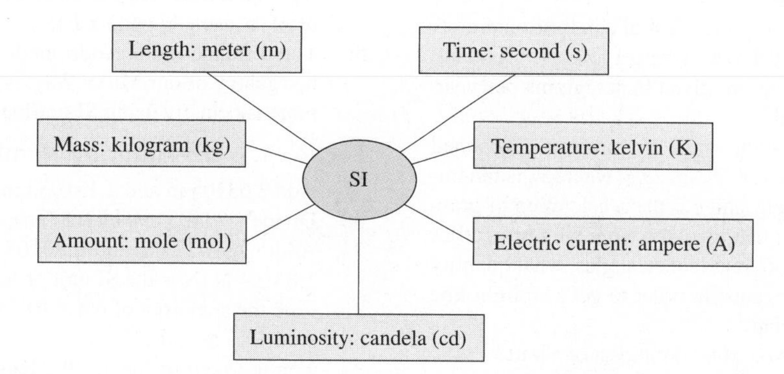

In addition, physics uses geometric measures of angle.

Numbers are often written with prefixes or in scientific notation to express powers of 10. Precision is shown by the number of significant figures:

Power of 10

Earth's radius $6.37 \times 10^6\,\text{m} = 6.37\,\text{Mm}$

Three significant figures   SI prefix for $\times 10^6$

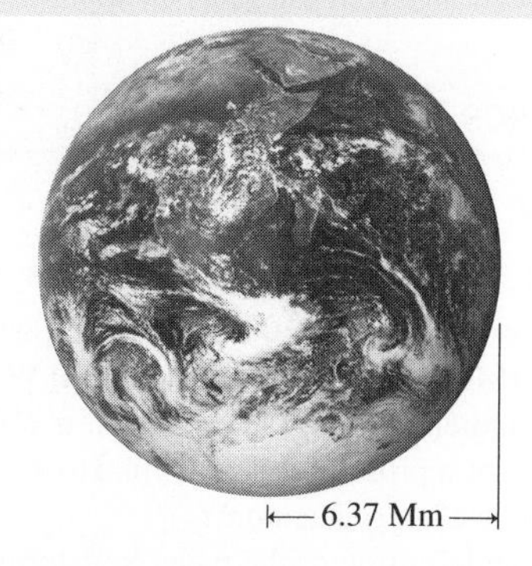

Applications

The IDEA strategy for solving physics problems consists of four steps: Interpret, Develop, Evaluate, and Assess. Estimation and data analysis are additional skills that help with physics.

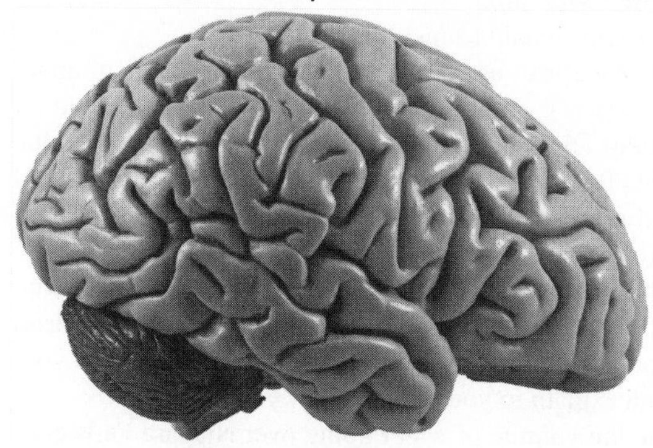

$$N = \frac{10^{-3}\ \text{m}^3/\text{brain}}{10^{-15}\ \text{m}^3/\text{cell}} = 10^{12}\ \text{cells/brain}$$

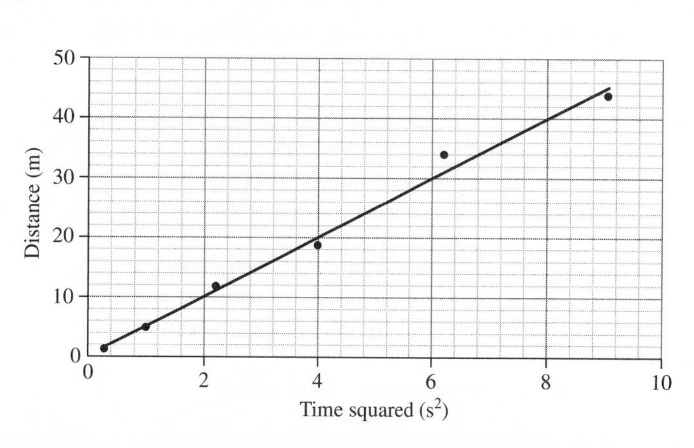

For homework assigned on MasteringPhysics, go to www.masteringphysics.com

BIO *Biology and/or medicine-related problems* **DATA** *Data problems* **ENV** *Environmental problems* **CH** *Challenge problems* **COMP** *Computer problems*

For Thought and Discussion

1. Explain why measurement standards based on laboratory procedures are preferable to those based on specific objects such as the international prototype kilogram.
2. When a computer that carries seven significant figures adds 1.000000 and 2.5×10^{-15}, what's its answer? Why?
3. Why doesn't Earth's rotation provide a suitable time standard?
4. To raise a power of 10 to another power, you multiply the exponent by the power. Explain why this works.
5. What facts might a scientist use in estimating Earth's age?
6. How would you determine the length of a curved line?
7. Write $1/x$ as x to some power.
8. Emissions of carbon dioxide from fossil-fuel combustion are often expressed in gigatonnes per year, where 1 tonne = 1000 kg. But sometimes CO_2 emissions are given in petagrams per year. How are the two units related?
9. In Chapter 3, you'll learn that the range of a projectile launched over level ground is given by $x = v_0^2 \sin 2\theta/g$, where v_0 is the initial speed, θ is the launch angle, and g is the acceleration of gravity. If you did an experiment that involved launching projectiles with the same speed v_0 but different launch angles, what quantity would you plot the range x against in order to get a straight line and thus verify this relationship?
10. What is meant by an *explicit-constant* definition of a unit?
11. You're asked to make a rough estimate of the total mass of all the students in your university. You report your answer as 1.16×10^6 kg. Why isn't this an appropriate answer?

Exercises and Problems

Exercises

Section 1.2 Measurements and Units

12. The power output of a typical large power plant is 1000 megawatts (MW). Express this result in (a) W, (b) kW, and (c) GW.
13. The diameter of a hydrogen atom is about 0.1 nm, and the diameter of a proton is about 1 fm. How many times bigger than a proton is a hydrogen atom?
14. Use the definition of the meter to determine how far light travels in 1 ns.
15. In nanoseconds, how long is the period of the cesium-133 radiation used to define the second?
16. Lake Baikal in Siberia holds the world's largest quantity of fresh water, about 14 Eg. How many kilograms is that?
17. A hydrogen atom is about 0.1 nm in diameter. How many hydrogen atoms lined up side by side would make a line 1 cm long?
18. How long a piece of wire would you need to form a circular arc subtending an angle of 1.4 rad, if the radius of the arc is 8.1 cm?
19. Making a turn, a jetliner flies 2.1 km on a circular path of radius 3.4 km. Through what angle does it turn?
20. A car is moving at 35.0 mi/h. Express its speed in (a) m/s and (b) ft/s.
21. You have postage for a 1-oz letter but only a metric scale. What's the maximum mass your letter can have, in grams?
22. A year is very nearly $\pi \times 10^7$ s. By what percentage is this figure in error?
23. How many cubic centimeters are in a cubic meter?
24. Since the start of the industrial era, humankind has emitted about half an exagram of carbon to the atmosphere. What's that in tonnes (t, where 1 t = 1000 kg)?
25. A gallon of paint covers 350 ft^2. What's its coverage in m^2/L?
26. Highways in Canada have speed limits of 100 km/h. How does this compare with the 65 mi/h speed limit common in the United States?
27. One m/s is how many km/h?
28. A 3.0-lb box of grass seed will seed 2100 ft^2 of lawn. Express this coverage in m^2/kg.
29. A radian is how many degrees?
30. Convert the following to SI units: (a) 55 mi/h; (b) 40.0 km/h; (c) 1 week (take that 1 as an exact number); (d) the period of Mars's orbit (consult Appendix E).
31. The distance to the Andromeda galaxy, the nearest large neighbor galaxy of our Milky Way, is about 2.4×10^{22} m. Express this more succinctly using SI prefixes.

Section 1.3 Working with Numbers

32. Add 3.63105 m and 2.13103 km.
33. Divide 4.23103 m/s by 0.57 ms, and express your answer in m/s^2.
34. Add 5.131022 cm and 6.83103 mm, and multiply the result by 1.83104 N (N is the SI unit of force).
35. Find the cube root of 6.4×10^{19} without a calculator.
36. Add 1.46 m and 2.3 cm.
37. You're asked to specify the length of an updated aircraft model for a sales brochure. The original plane was 41 m long; the new model has a 3.6-cm-long radio antenna added to its nose. What length do you put in the brochure?
38. Repeat the preceding exercise, this time using 41.05 m as the airplane's original length.

Problems

39. To see why it's important to carry more digits in intermediate calculations, determine $(\sqrt{3})^3$ to three significant figures in two ways: (a) Find $\sqrt{3}$ and round to three significant figures, then cube and again round; and (b) find $\sqrt{3}$ to four significant figures, then cube and round to three significant figures.
40. You've been hired as an environmental watchdog for a big-city newspaper. You're asked to estimate the number of trees that go into one day's printing, given that half the newsprint comes from recycling, the rest from new wood pulp. What do you report?
41. The average dairy cow produces about 10^4 kg of milk per year. Estimate the number of dairy cows needed to keep the United States supplied with milk.
42. How many Earths would fit inside the Sun?
43. The average American uses electrical energy at the rate of about 1.5 kilowatts (kW). Solar energy reaches Earth's surface at an average rate of about 300 watts on every square meter (a value that accounts for night and clouds). What fraction of the United States' land area would have to be covered with 20% efficient solar cells to provide all of our electrical energy?
44. You're writing a biography of the physicist Enrico Fermi, who was fond of estimation problems. Here's one problem Fermi posed: What's the number of piano tuners in Chicago? Give your estimate, and explain to your readers how you got it.
45. (a) Estimate the volume of water going over Niagara Falls each second. (b) The falls provides the outlet for Lake Erie; if the

falls were shut off, estimate how long it would take Lake Erie to rise 1 m.

46. Estimate the number of air molecules in your dorm room.
47. A human hair is about 100 μm across. Estimate the number of hairs in a typical braid.
48. You're working in the fraud protection division of a credit-card company, and you're asked to estimate the chances that a 16-digit number chosen at random will be a valid credit-card number. What do you answer?
49. Bubble gum's density is about 1 g/cm^3. You blow an 8-g wad of gum into a bubble 10 cm in diameter. What's the bubble's thickness? (*Hint*: Think about spreading the bubble into a flat sheet. The surface area of a sphere is $4\pi r^2$.)
50. The Moon barely covers the Sun during a solar eclipse. Given that Moon and Sun are, respectively, 4×10^5 km and 1.5×10^8 km from Earth, determine how much bigger the Sun's diameter is than the Moon's. If the Moon's radius is 1800 km, how big is the Sun?
51. The semiconductor chip at the heart of a personal computer is a square 4 mm on a side and contains 10^{10} electronic components. (a) What's the size of each component, assuming they're square? (b) If a calculation requires that electrical impulses traverse 10^4 components on the chip, each a million times, how many such calculations can the computer perform each second? (*Hint*: The maximum speed of an electrical impulse is about two-thirds the speed of light.)
52. Estimate the number of (a) atoms and (b) cells in your body.
53. When we write the number 3.6 as typical of a number with two significant figures, we're saying that the actual value is closer to 3.6 than to 3.5 or 3.7; that is, the actual value lies between 3.55 and 3.65. Show that the percent uncertainty implied by such two-significant-figure precision varies with the value of the number, being the lowest for numbers beginning with 9 and the highest for numbers beginning with 1. In particular, what is the percent uncertainty implied by the numbers (a) 1.1, (b) 5.0, and (c) 9.9?
54. Continental drift occurs at about the rate your fingernails grow. Estimate the age of the Atlantic Ocean, given that the eastern and western hemispheres have been drifting apart.
55. You're driving into Canada and trying to decide whether to fill your gas tank before or after crossing the border. Gas in the United States costs \$3.67/gallon, in Canada it's \$1.32/L, and the Canadian dollar is worth 95¢ in U.S. currency. Where should you fill up?
56. In the 1908 London Olympics, the intended 26-mile marathon was extended 385 yards to put the end in front of the royal reviewing stand. This distance subsequently became standard. What's the marathon distance in kilometers, to the nearest meter?
57. ENV An environmental group is lobbying to shut down a coal-burning power plant that produces electrical energy at the rate of 1 GW (a watt, W, is a unit of power—the rate of energy production or consumption). They suggest replacing the plant with wind turbines that can produce 1.5 MW each but that, due to intermittent wind, average only 30% of that power. Estimate the number of wind turbines needed.
58. If you're working from the print version of this book, estimate the thickness of each page.
59. BIO Estimate the area of skin on your body.
60. Estimate the mass of water in the world's oceans, and express it with SI prefixes.
61. Express the following with appropriate units and significant figures: (a) 1.0 m plus 1 mm, (b) 1.0 m times 1 mm, (c) 1.0 m minus 999 mm, and (d) 1.0 m divided by 999 mm.
62. You're shopping for a new computer, and a salesperson claims the microprocessor chip in the model you're looking at contains 50 billion electronic components. The chip measures 5 mm on a side and uses 14-nm technology, meaning each component is 14 nm across. Is the salesperson right?
63. Café Milagro sells coffee online. A half-kilogram bag of coffee costs \$8.95, excluding shipping. If you order six bags, the shipping costs \$6.90. What's the cost per bag when you include shipping?
64. The world consumes energy at the rate of about 500 EJ per year, where the joule (J) is the SI energy unit. Convert this figure to watts (W), where 1 W = 1 J/s, and then estimate the average per capita energy consumption rate in watts.
65. DATA The volume of a sphere is given by $V=\frac{4}{3}\pi r^3$, where r is the sphere's radius. For solid spheres with the same density—made, for example, from the same material—mass is proportional to volume. The table below lists measures of diameter and mass for different steel balls. (a) Determine a quantity which, when you plot mass against it, should yield a straight line. (b) Make your plot, establish a best-fit line, and determine its slope (which in this case is proportional to the spheres' density).

Diameter (cm)	0.75	1.00	1.54	2.16	2.54
Mass (g)	1.81	3.95	15.8	38.6	68.2

Passage Problems

BIO The human body contains about 10^{14} cells, and the diameter of a typical cell is about 10 μm Like all ordinary matter, cells are made of atoms; a typical atomic diameter is 0.1 nm.

66. How does the number of atoms in a cell compare with the number of cells in the body?
    a. greater
    b. smaller
    c. about the same
67. The volume of a cell is about
    a. 10^{-10} m^3.
    b. 10^{-15} m^3.
    c. 10^{-20} m^3.
    d. 10^{-30} m^3.
68. The mass of a cell is about
    a. 10^{-10} kg.
    b. 10^{-12} kg.
    c. 10^{-14} kg.
    d. 10^{-16} kg.
69. The number of atoms in the body is closest to
    a. 10^{14}.
    b. 10^{20}.
    c. 10^{30}.
    d. 10^{40}.

Answers to Chapter Questions

Answer to Chapter Opening Question

All of them!

Answers to GOT IT? Questions

1.1 (c)

1.2 (1) 2.998×10^{-9}, 0.0008, 3.14×10^7, 0.041×10^9, 55×10^6
(2) 0.0008, 0.041×10^9 and 55×10^6 (with two significant figures each), 3.14×10^7, 2.998×10^{-9}

PART ONE

OVERVIEW

Mechanics

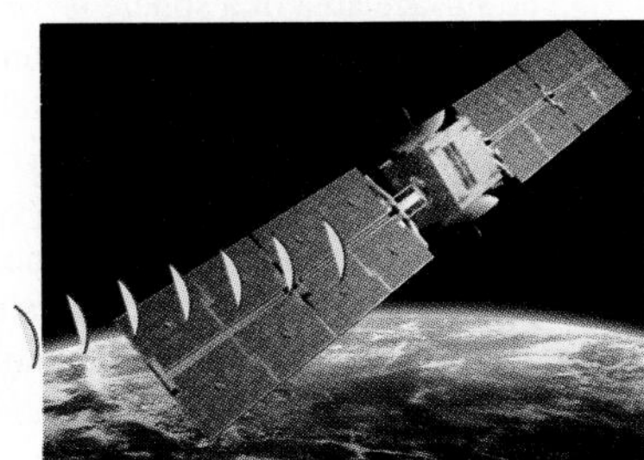

A hiker checks her position using signals from GPS satellites.

A wilderness hiker uses the Global Positioning System to follow her chosen route. A farmer plows a field with centimeter-scale precision, guided by GPS and saving precious fuel as a result. One scientist uses GPS to track endangered elephants, another to study the accelerated flow of glaciers as Earth's climate warms. Our deep understanding of motion is what lets us use a constellation of satellites, 20,000 km up and moving faster than 10,000 km/h, to find positions on Earth so precisely.

Motion occurs at all scales, from the intricate dance of molecules at the heart of life's cellular mechanics, to the everyday motion of cars, baseballs, and our own bodies, to the trajectories of GPS and TV satellites and of spacecraft exploring the distant planets, to the stately motions of the celestial bodies themselves and the overall expansion of the universe. The study of motion is called **mechanics**. The 11 chapters of Part 1 introduce the physics of motion, first for individual bodies and then for complicated systems whose constituent parts move relative to one another.

We explore motion here from the viewpoint of Newtonian mechanics, which applies accurately in all cases except the subatomic realm and when relative speeds approach that of light. The Newtonian mechanics of Part 1 provides the groundwork for much of the material in subsequent parts, until, in the book's final chapters, we extend mechanics into the subatomic and high-speed realms.

Motion in a Straight Line

What You Know

- You've learned the units for basic physical quantities.
- You understand the SI unit system, especially units for length, time, and mass.
- You can express numbers in scientific notation and using SI prefixes.
- You can handle precision and accuracy through significant figures.
- You can make order-of-magnitude estimates.
- You've learned the IDEA problem-solving strategy.

What You're Learning

- You'll learn the fundamental concepts used to describe motion: position, velocity, and acceleration—restricted in this chapter to motion in one dimension.
- You'll learn to distinguish average from instantaneous values.
- You'll see how calculus is used to establish instantaneous values.
- You'll learn to describe motion resulting from constant acceleration, including the important case of objects moving under the influence of gravity near Earth's surface.

How You'll Use It

- One-dimensional motion will be your stepping stone to richer and more complex motion in two and three dimensions, which you'll see in Chapter 3.
- Your understanding of acceleration will help you adopt the Newtonian view of motion, introduced in Chapter 4 and elaborated in Chapter 5.
- You'll encounter analogies to Chapter 2's motion concepts in Chapter 10's treatment of rotational motion.
- You'll apply motion concepts to systems of particles in Chapter 9.
- You'll continue to encounter motion concepts throughout the book, even beyond Part 1.

Electrons swarming around atomic nuclei, cars speeding along a highway, blood coursing through your veins, galaxies rushing apart in the expanding universe—all these are examples of matter in motion. The study of motion without regard to its cause is called **kinematics** (from the Greek "kinema," or motion, as in motion pictures). This chapter deals with the simplest case: a single object moving in a straight line. Later, we generalize to motion in more dimensions and with more complicated objects. But the basic concepts and mathematical techniques we develop here continue to apply.

2.1 Average Motion

You drive 15 minutes to a pizza place 10 km away, grab your pizza, and return home in another 15 minutes. You've traveled a total distance of 20 km, and the trip took half an hour, so your **average speed**—distance divided by time—was 40 kilometers per hour. To describe your motion more precisely, we introduce the quantity x that gives your position at any time t. We then define **displacement**, Δx, as the net change in

The server tosses the tennis ball straight up and hits it on its way down. Right at its peak height, the ball has zero velocity, but what's its acceleration?

Video Tutor Demo | Balls Take High and Low Tracks

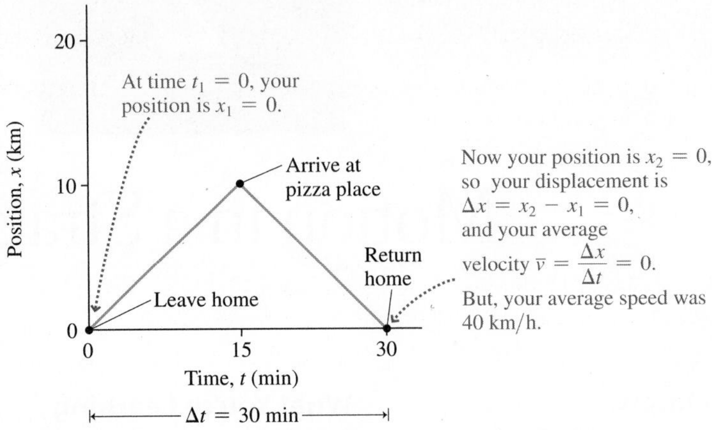

FIGURE 2.1 Position versus time for the pizza trip.

PhET: The Moving Man

position: $\Delta x = x_2 - x_1$, where x_1 and x_2 are your starting and ending positions, respectively. Your **average velocity**, $\bar{v}$, is displacement divided by the time interval:

$$\bar{v} = \frac{\Delta x}{\Delta t} \quad \text{(average velocity)} \tag{2.1}$$

where $\Delta t = t_2 - t_1$ is the interval between your ending and starting times. The bar in $\bar{v}$ indicates an average quantity (and is read "v bar"). The symbol Δ (capital Greek delta) stands for "the change in." For the round trip to the pizza place, your overall displacement was zero and therefore your average velocity was also zero—even though your average *speed* was not (Fig. 2.1).

Directions and Coordinate Systems

It matters whether you go north or south, east or west. Displacement therefore includes not only *how far* but also *in what direction*. For motion in a straight line, we can describe both properties by taking position coordinates x to be positive going in one direction from some origin, and negative in the other. This gives us a one-dimensional **coordinate system**. The choice of coordinate system—both of origin and of which direction is positive—is entirely up to you. The coordinate system isn't physically real; it's just a convenience we create to help in the mathematical description of motion.

Figure 2.2 shows some Midwestern cities that lie on a north–south line. We've established a coordinate system with northward direction positive and origin at Kansas City. Arrows show displacements from Houston to Des Moines and from International Falls to Des Moines; the former is approximately +1300 km, and the latter is approximately −750 km, with the minus sign indicating a southward direction. Suppose the Houston-to-Des Moines trip takes 2.6 hours by plane; then the average velocity is $(1300\text{ km})/(2.6\text{ h}) = 500\text{ km/h}$. If the International Falls-to-Des Moines trip takes 10 h by car, then the average velocity is $(-750\text{ km})/(10\text{ h}) = -75\text{ km/h}$; again, the minus sign indicates southward.

In calculating average velocity, all that matters is the overall displacement. Maybe that trip from Houston to Des Moines was a nonstop flight going 500 km/h. Or maybe it involved a faster plane that stopped for half an hour in Kansas City. Maybe the plane even went first to Minneapolis, then backtracked to Des Moines. No matter: The displacement remains 1300 km and, as long as the total time is 2.6 h, the average velocity remains 500 km/h.

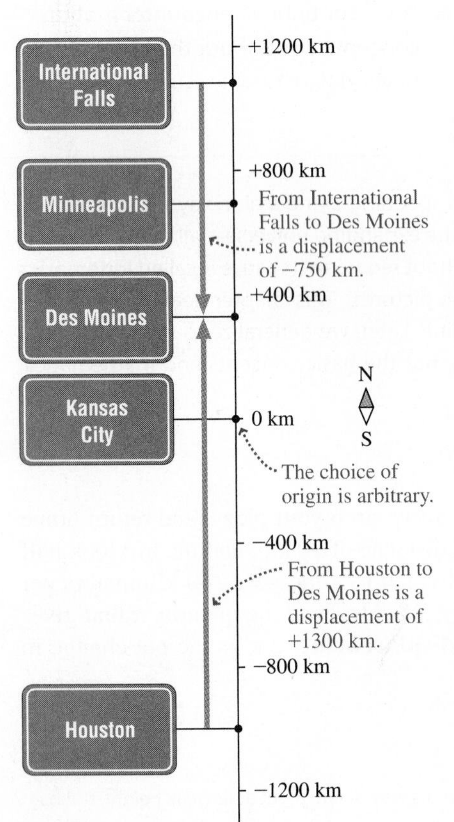

FIGURE 2.2 Describing motion in the central United States.

GOT IT? 2.1 We just described three trips from Houston to Des Moines: (a) direct; (b) with a stop in Kansas City; and (c) via Minneapolis. For which of these trips is the average speed the same as the average velocity? Where the two differ, which is greater?

EXAMPLE 2.1 Speed and Velocity: Flying with a Connection

To get a cheap flight from Houston to Kansas City—a distance of 1000 km—you have to connect in Minneapolis, 700 km north of Kansas City. The flight to Minneapolis takes 2.2 h, then you have a 30-min layover, and then a 1.3-h flight to Kansas City. What are your average velocity and your average speed on this trip?

INTERPRET We interpret this as a one-dimensional kinematics problem involving the distinction between velocity and speed, and we identify three distinct travel segments: the two flights and the layover. We identify the key concepts as speed and velocity; their distinction is clear from our pizza example.

DEVELOP Figure 2.2 is our drawing. We determine that Equation 2.1, $\bar{v} = \Delta x/\Delta t$, will give the average velocity, and that the average speed is the total distance divided by the total time. We develop our plan: Find the displacement and the total time, and use those values to get the average velocity; then find the total distance traveled and use that along with the total time to get the average speed.

EVALUATE You start in Houston and end up in Kansas City, for a displacement of 1000 km—regardless of how far you actually traveled. The total time for the three segments is $\Delta t = 2.2\text{ h} + 0.50\text{ h} + 1.3\text{ h} = 4.0\text{ h}$. Then the average velocity, from Equation 2.1, is

$$\bar{v} = \frac{\Delta x}{\Delta t} = \frac{1000\text{ km}}{4.0\text{ h}} = 250\text{ km/h}$$

However, that Minneapolis connection means you've gone an extra 2×700 km, for a total distance of 2400 km in 4 hours. Thus your average speed is $(2400\text{ km})/(4.0\text{ h}) = 600$ km/h, more than twice your average velocity.

ASSESS Make sense? Average velocity depends only on the net displacement between the starting and ending points. Average speed takes into account the actual distance you travel—which can be a lot longer on a circuitous trip like this one. So it's entirely reasonable that the average speed should be greater. ■

2.2 Instantaneous Velocity

Geologists determine the velocity of a lava flow by dropping a stick into the lava and timing how long it takes the stick to go a known distance (Fig. 2.3*a*). Dividing the distance by the time then gives the average velocity. But did the lava flow faster at the beginning of the interval? Or did it speed up and slow down again? To understand motion fully, including how it changes with time, we need to know the velocity at each instant.

Geologists could explore that detail with a series of observations taken over smaller intervals of time and distance (Fig. 2.3*b*). As the size of the intervals shrinks, a more detailed picture of the motion emerges. In the limit of very small intervals, we're measuring the velocity at a single instant. This is the **instantaneous velocity**, or simply the **velocity**. The magnitude of the instantaneous velocity is the **instantaneous speed**.

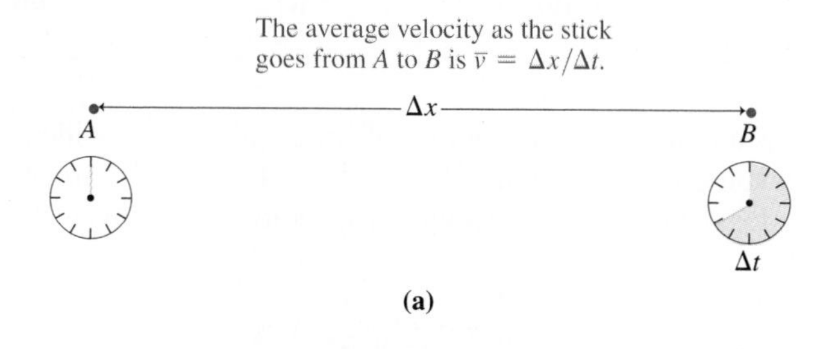

FIGURE 2.3 Determining the velocity of a lava flow.

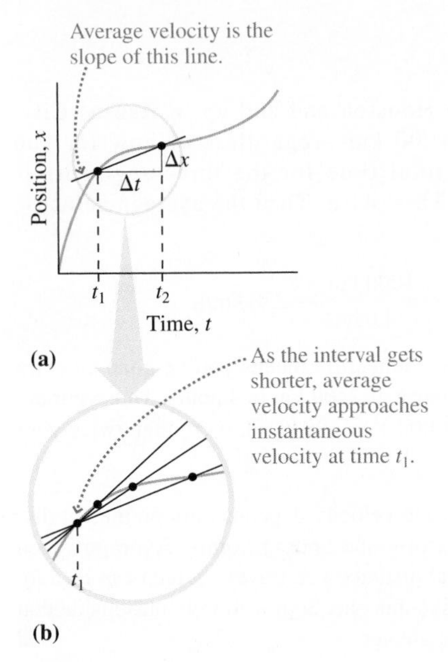

FIGURE 2.4 Position-versus-time graph for the motion in Fig. 2.3.

You might object that it's impossible to achieve that limit of an arbitrarily small time interval. With observational measurements that's true, but calculus lets us go there. Figure 2.4*a* is a plot of position versus time for the stick in the lava flow shown in Fig. 2.3. Where the curve is steep, the position changes rapidly with time—so the velocity is greater. Where the curve is flatter, the velocity is lower. Study the clocks in Fig. 2.3*b* and you'll see that the stick starts out moving rapidly, then slows, and then speeds up a bit at the end. The curve in Fig. 2.4*a* reflects this behavior.

Suppose we want the instantaneous velocity at the time marked t_1 in Fig. 2.4*a*. We can approximate this quantity by measuring the displacement Δx over the interval Δt between t_1 and some later time t_2: the ratio $\Delta x/\Delta t$ is then the average velocity over this interval. Note that this ratio is the slope of a line drawn through points on the curve that mark the ends of the interval.

Figure 2.4*b* shows what happens as we make the time interval Δt arbitrarily small: Eventually, the line between the two points becomes indistinguishable from the tangent line to the curve. That tangent line has the same slope as the curve right at the point we're interested in, and therefore it defines the instantaneous velocity at that point. We write this mathematically by saying that the instantaneous velocity is the limit, as the time interval Δt becomes arbitrarily close to zero, of the ratio of displacement Δx to Δt:

$$v = \lim_{\Delta t \to 0} \frac{\Delta x}{\Delta t} \tag{2.2a}$$

You can imagine making the interval Δt as close to zero as you like, getting ever better approximations to the instantaneous velocity. Given a graph of position versus time, an easy approach is to "eyeball" the tangent line to the graph at a point you're interested in; its slope is the instantaneous velocity (Fig. 2.5).

GOT IT? 2.2 The figures show position-versus-time graphs for four objects. Which object is moving with constant speed? Which reverses direction? Which starts slowly and then speeds up?

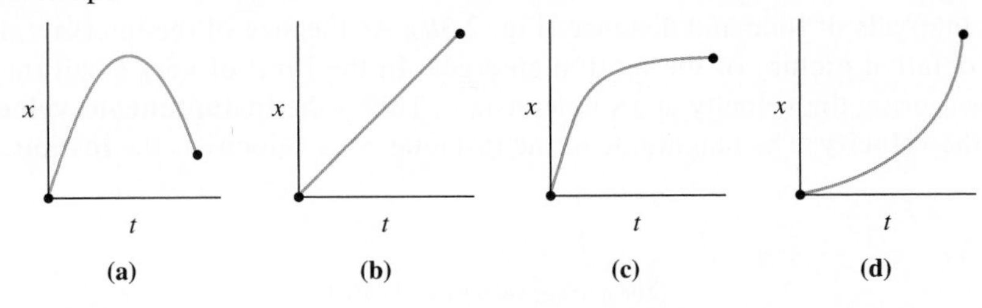

Given position as a mathematical function of time, calculus provides a quick way to find instantaneous velocity. In calculus, the result of the limiting process described in Equation 2.2a is called the **derivative** of x with respect to t and is given the symbol dx/dt:

$$\frac{dx}{dt} = \lim_{\Delta t \to 0} \frac{\Delta x}{\Delta t}$$

The quantities dx and dt are called **infinitesimals**; they represent vanishingly small quantities that result from the limiting process. We can then write Equation 2.2a as

$$v = \frac{dx}{dt} \quad \text{(instantaneous velocity)} \tag{2.2b}$$

Given position x as a function of time t, calculus shows how to find the velocity $v = dx/dt$. Consult Tactics 2.1 if you haven't yet seen derivatives in your calculus class or if you need a refresher.

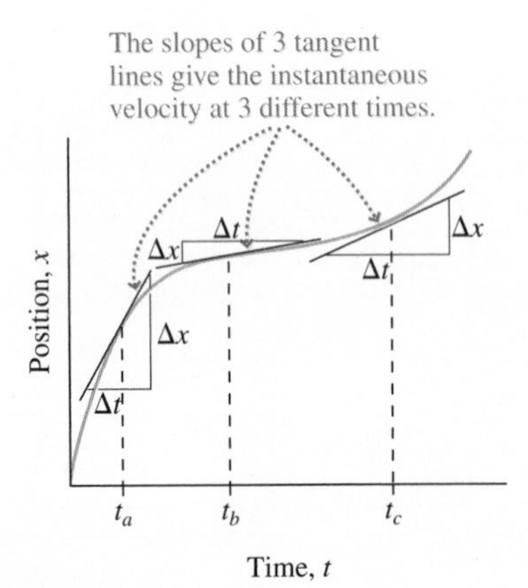

FIGURE 2.5 The instantaneous velocity is the slope of the tangent line.

TACTICS 2.1 Taking Derivatives

You don't have to go through an elaborate limiting process every time you want to find an instantaneous velocity. That's because calculus provides formulas for the derivatives of common functions. For example, any function of the form $x = bt^n$, where b and n are constants, has the derivative

$$\frac{dx}{dt} = nbt^{n-1} \tag{2.3}$$

Appendix A lists derivatives of other common functions.

EXAMPLE 2.2 Instantaneous Velocity: A Rocket Ascends

The altitude of a rocket in the first half-minute of its ascent is given by $x = bt^2$, where the constant b is 2.90 m/s². Find a general expression for the rocket's velocity as a function of time and from it the instantaneous velocity at $t = 20$ s. Also find an expression for the average velocity, and compare your two velocity expressions.

INTERPRET We interpret this as a problem involving the comparison of two distinct but related concepts: instantaneous velocity and average velocity. We identify the rocket as the object whose velocities we're interested in.

DEVELOP Equation 2.2b, $v = dx/dt$, gives the instantaneous velocity and Equation 2.1, $\bar{v} = \Delta x/\Delta t$, gives the average velocity. Our plan is to use Equation 2.3, $dx/dt = nbt^{n-1}$, to evaluate the derivative that gives the instantaneous velocity. Then we can use Equation 2.1 for the average velocity, but first we'll need to determine the displacement from the equation we're given for the rocket's position.

EVALUATE Applying Equation 2.2b with position given by $x = bt^2$ and using Equation 2.3 to evaluate the derivative, we have

$$v = \frac{dx}{dt} = \frac{d(bt^2)}{dt} = 2bt$$

for the instantaneous velocity. Evaluating at $t = 20$ s with $b = 2.90$ m/s² gives $v = 116$ m/s. For the average velocity we need the total displacement at 20 s. Since $x = bt^2$, Equation 2.1 gives

$$\bar{v} = \frac{\Delta x}{\Delta t} = \frac{bt^2}{t} = bt$$

where we've used $x = bt^2$ for Δx and t for Δt because both position and time are taken to be zero at liftoff. Comparison with our earlier result shows that the average velocity from liftoff to any particular time is exactly half the instantaneous velocity at that time.

ASSESS Make sense? Yes: The rocket's speed is always increasing, so its velocity at the end of any time interval is greater than the average velocity over that interval. The fact that the average velocity is exactly half the instantaneous velocity results from the quadratic (t^2) dependence of position on time.

✓TIP Language

Language often holds clues to the meaning of physical concepts. In this example we speak of the *instantaneous* velocity *at* a particular time. That wording should remind you of the limiting process that focuses on a single instant. In contrast, we speak of the *average* velocity *over* a time interval, since averaging explicitly involves a range of times.

2.3 Acceleration

When velocity changes, as in Example 2.2, an object is said to undergo **acceleration**. Quantitatively, we define acceleration as the rate of change of velocity, just as we defined velocity as the rate of change of position. The **average acceleration** over a time interval Δt is

$$\bar{a} = \frac{\Delta v}{\Delta t} \quad \text{(average acceleration)} \tag{2.4}$$

where Δv is the change in velocity and the bar on $\bar{a}$ indicates that this is an average value. Just as we defined instantaneous velocity through a limiting procedure, we define **instantaneous acceleration** as

$$a = \lim_{\Delta t \to 0} \frac{\Delta v}{\Delta t} = \frac{dv}{dt} \quad \text{(instantaneous acceleration)} \tag{2.5}$$

As we did with velocity, we also use the term *acceleration* alone to mean instantaneous acceleration.

In one-dimensional motion, acceleration is either in the direction of the velocity or opposite it. In the former case the accelerating object speeds up, whereas in the latter it slows (Fig. 2.6). Although slowing is sometimes called *deceleration*, it's simpler to use

PhET: Calculus Grapher

When a and v have the same direction, the car speeds up.

(a)

When a is opposite v, the car slows.

(b)

FIGURE 2.6 Acceleration and velocity.

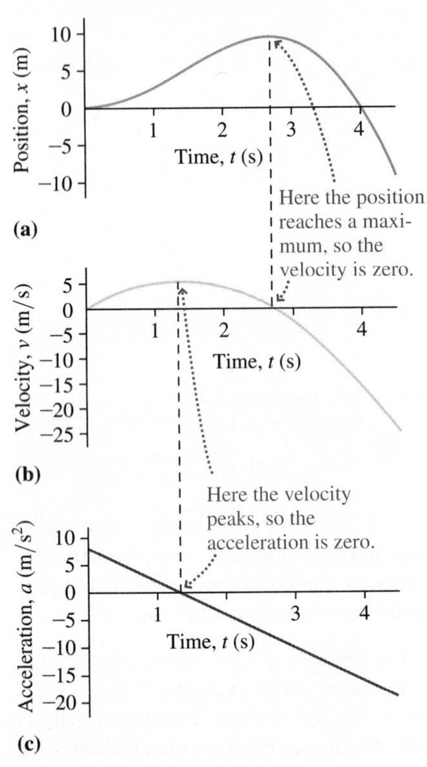

FIGURE 2.7 (a) Position, (b) velocity, and (c) acceleration versus time.

acceleration to describe the time rate of change of velocity no matter what's happening. With two-dimensional motion, we'll find much richer relationships between the directions of velocity and acceleration.

Since acceleration is the rate of change of velocity, its units are (distance per time) per time, or distance/time^2. In SI, that's m/s^2. Sometimes acceleration is given in mixed units; for example, a car going from 0 to 60 mi/h in 10 s has an average acceleration of 6 mi/h/s.

Position, Velocity, and Acceleration

Figure 2.7 shows graphs of position, velocity, and acceleration for an object undergoing one-dimensional motion. In Fig. 2.7*a*, the rise and fall of the position-versus-time curve shows that the object first moves away from the origin, reverses, then reaches the origin again at $t = 4$ s. It then continues moving into the region $x < 0$. Velocity, shown in Fig. 2.7*b*, is the slope of the position-versus-time curve in Fig. 2.7*a*. Note that the magnitude of the velocity (that is, the speed) is large where the curve in Fig. 2.7*a* is steep—that is, where position is changing most rapidly. At the peak of the position curve, the object is momentarily at rest as it reverses, so there the position curve is flat and the velocity is zero. After the object reverses, at about 2.7 s, it's heading in the negative *x*-direction and so its velocity is negative.

Just as velocity is the slope of the position-versus-time curve, acceleration is the slope of the velocity-versus-time curve. Initially that slope is positive—velocity is increasing—but eventually it peaks at the point of maximum velocity and zero acceleration and then it decreases. That velocity decrease corresponds to a negative acceleration, as shown clearly in the region of Fig. 2.7*c* beyond about 1.3 s.

CONCEPTUAL EXAMPLE 2.1 Acceleration Without Velocity?

Can an object be accelerating even though it's not moving?

EVALUATE Figure 2.7 shows that velocity is the *slope* of the position curve—and the slope depends on how the position is *changing*, not on its actual value. Similarly, acceleration depends only on the *rate of change* of velocity, not on velocity itself. So there's no intrinsic reason why there can't be acceleration at an instant when velocity is zero.

ASSESS Figure 2.8, which shows a ball thrown straight up, is a case in point. Right at the peak of its flight, the ball's velocity is instantaneously zero. But just before the peak it's moving upward, and just after it's moving downward. No matter how small a time interval you consider, the velocity is always changing. Therefore, the ball is accelerating, even right at the instant its velocity is zero.

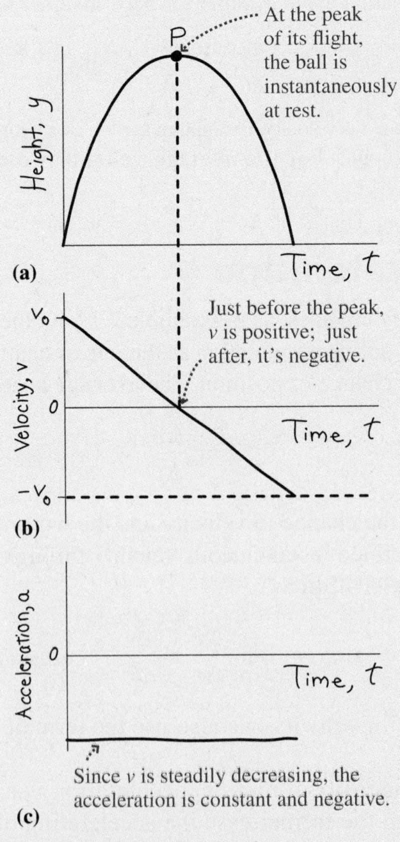

FIGURE 2.8 Our sketch for Conceptual Example 2.1.

MAKING THE CONNECTION Just 0.010 s before it peaks, the ball in Fig. 2.8 is moving upward at 0.098 m/s; 0.010 s after it peaks, it's moving downward with the same speed. What's its average acceleration over this 0.02-s interval?

EVALUATE Equation 2.4 gives the average acceleration: $\bar{a} = \Delta v/\Delta t = (-0.098\text{ m/s} - 0.098\text{ m/s})/(0.020\text{ s}) = -9.8\text{ m/s}^2$. Here we've implicitly chosen a coordinate system with a positive upward direction, so both the final velocity and the acceleration are negative. The time interval is so small that our result must be close to the instantaneous acceleration right at the peak—when the velocity is zero. You might recognize 9.8 m/s^2 as the acceleration due to the Earth's gravity.

Acceleration is the rate of change of velocity, and velocity is the rate of change of position. That makes acceleration the rate of change of the rate of change of position. Mathematically, acceleration is the **second derivative** of position with respect to time. Symbolically, we write the second derivative as d^2x/dt^2; this is just a symbol and doesn't mean that anything is actually squared. Then the relationship among acceleration, velocity, and position can be written

$$a = \frac{dv}{dt} = \frac{d}{dt}\left(\frac{dx}{dt}\right) = \frac{d^2x}{dt^2} \quad (2.6)$$

Equation 2.6 expresses acceleration in terms of position through the calculus operation of taking the second derivative. If you've studied integrals in calculus, you can see that it should be possible to go the opposite way, finding position as a function of time given acceleration as a function of time. In Section 2.4 we'll do this for the special case of constant acceleration, although there we'll take an algebra-based approach; Problem 87 obtains the same results using calculus. We'll take a quick look at nonconstant acceleration in Section 2.6. The Application on this page provides an important technology that finds an object's position from its acceleration.

GOT IT? 2.3 An elevator is going up at constant speed, slows to a stop, then starts down and soon reaches the same constant speed it had going up. Is the elevator's average acceleration between its upward and downward constant-speed motions (a) zero, (b) downward, (c) first upward and then downward, or (d) first downward and then upward?

2.4 Constant Acceleration

The description of motion has an especially simple form when acceleration is constant. Suppose an object starts at time $t = 0$ with some initial velocity v_0 and constant acceleration a. Later, at some time t, it has velocity v. Because the acceleration doesn't change, its average and instantaneous values are identical, so we can write

$$a = \bar{a} = \frac{\Delta v}{\Delta t} = \frac{v - v_0}{t - 0}$$

or, rearranging,

$$v = v_0 + at \qquad \text{(for constant acceleration only)} \quad (2.7)$$

This equation says that the velocity changes from its initial value by an amount that is the product of acceleration and time.

✓TIP Know Your Limits

Many equations we develop are special cases of more general laws, and they're limited to special circumstances. Equation 2.7 is a case in point: It applies *only when acceleration is constant.*

Having determined velocity as a function of time, we now consider position. With constant acceleration, velocity increases steadily—and thus the average velocity over an interval is the average of the velocities at the beginning and the end of that interval. So we can write

$$\bar{v} = \tfrac{1}{2}(v_0 + v) \quad (2.8)$$

for the average velocity over the interval from $t = 0$ to some later time when the velocity is v. We can also write the average velocity as the change in position divided by the time interval. Suppose that at time 0 our object was at position x_0. Then its average velocity over a time interval from 0 to time t is

$$\bar{v} = \frac{\Delta x}{\Delta t} = \frac{x - x_0}{t - 0}$$

APPLICATION Inertial Guidance

Given an object's initial position and velocity, and its subsequent acceleration—which may vary with time—it's possible to invert Equation 2.6 and solve for position (more on the mathematics of this inversion in Section 2.6). *Inertial guidance systems*, also called *inertial navigation systems*, exploit this principle to allow submarines, ships, and airplanes to keep track of their locations based solely on internal measurements of their own acceleration. This frees them from the need for external positioning references such as the Global Positioning System (GPS), radar, or direct observation. Inertial guidance is especially important for submarines, which usually can't access external sources for information about their positions. In the one-dimensional motion of this chapter, an inertial guidance system would consist of a single accelerometer whose reading is tracked continually. In practical systems, three accelerometers at right angles track acceleration in all three dimensions. Information from on-board gyroscopes registers orientation, so the system "knows" the changing directions of the three accelerations.

Early inertial guidance systems were heavy and expensive, but the miniaturization of accelerometers and gyroscopes—so that they're now in every smartphone—has enabled smaller and less expensive inertial guidance systems. The photo shows a complete inertial navigation system developed by the U.S. Defense Advanced Research Projects Agency (DARPA) for use in locations where GPS signals aren't available; it's so small that it fits within the Lincoln Memorial on a penny!

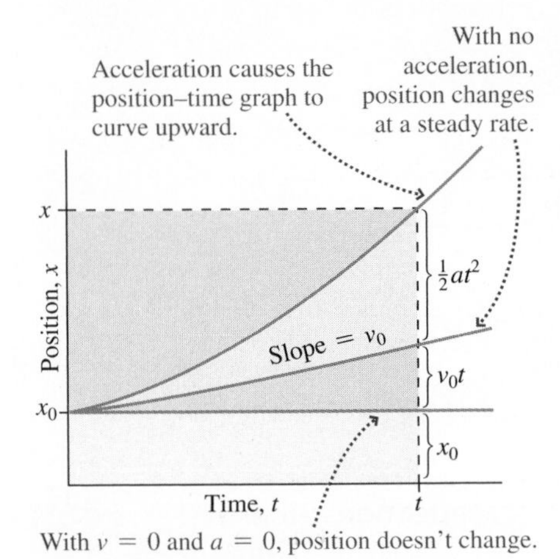

FIGURE 2.9 Meaning of the terms in Equation 2.10.

where x is the object's position at time t. Equating this expression for $\bar{v}$ with the expression in Equation 2.8 gives

$$x = x_0 + \bar{v}t = x_0 + \tfrac{1}{2}(v_0 + v)t \qquad (2.9)$$

But we already found the instantaneous velocity v that appears in this expression; it's given by Equation 2.7. Substituting and simplifying then give the position as a function of time:

$$x = x_0 + v_0t + \tfrac{1}{2}at^2 \quad \text{(for constant acceleration only)} \qquad (2.10)$$

Does Equation 2.10 make sense? With no acceleration ($a = 0$), position would increase linearly with time, at a rate given by the initial velocity v_0. With constant acceleration, the additional term $\frac{1}{2}at^2$ describes the effect of the ever-changing velocity; time is squared because the longer the object travels, the faster it moves, so the more distance it covers in a given time. Figure 2.9 shows the meaning of the terms in Equation 2.10.

How much runway do I need to land a jetliner, given touchdown speed and a constant acceleration? A question like this involves position, velocity, and acceleration without explicit mention of time. So we solve Equation 2.7 for time, $t = (v - v_0)/a$, and substitute this expression for t in Equation 2.9 to write

$$x - x_0 = \frac{1}{2}\frac{(v_0 + v)(v - v_0)}{a}$$

or, since $(a + b)(a - b) = a^2 - b^2$,

$$v^2 = v_0^2 + 2a(x - x_0) \qquad (2.11)$$

Table 2.1 Equations of Motion for Constant Acceleration

Equation	Contains	Number
$v = v_0 + at$	v, a, t; no x	2.7
$x = x_0 + \frac{1}{2}(v_0 + v)t$	x, v, t; no a	2.9
$x = x_0 + v_0t + \frac{1}{2}at^2$	x, a, t; no v	2.10
$v^2 = v_0^2 + 2a(x - x_0)$	x, v, a; no t	2.11

Equations 2.7, 2.9, 2.10, and 2.11 link all possible combinations of position, velocity, and acceleration for motion with constant acceleration. We summarize them in Table 2.1, and remind you that they apply *only* in the case of constant acceleration.

Although we derived these equations algebraically, we could instead have used calculus. Problem 87 takes this approach in getting from Equation 2.7 to Equation 2.10.

Using the Equations of Motion

The equations in Table 2.1 fully describe motion under constant acceleration. Don't regard them as separate laws, but recognize them as complementary descriptions of a single underlying phenomenon—one-dimensional motion with *constant acceleration*. Having several equations provides convenient starting points for approaching problems. Don't memorize these equations, but grow familiar with them as you work problems. We now offer a strategy for solving problems about one-dimensional motion with *constant acceleration* using these equations.

PROBLEM-SOLVING STRATEGY 2.1 Motion with Constant Acceleration

INTERPRET Interpret the problem to be sure it asks about motion with *constant acceleration*. Next, identify the object(s) whose motion you're interested in.

DEVELOP Draw a diagram with appropriate labels, and choose a coordinate system. For instance, sketch the initial and final physical situations, or draw a position-versus-time graph. Then determine which equations of motion from Table 2.1 contain the quantities you're given and will be easiest to solve for the unknown(s).

EVALUATE Solve the equations in symbolic form and then evaluate numerical quantities.

ASSESS Does your answer make sense? Are the units correct? Do the numbers sound reasonable? What happens in special cases—for example, when a distance, velocity, acceleration, or time becomes very large or very small?

The next two examples are typical of problems involving constant acceleration. Example 2.3 is a straightforward application of the equations we've just derived to a single object. Example 2.4 involves two objects, in which case we need to write equations describing the motions of both objects.

EXAMPLE 2.3 Motion with Constant Acceleration: Landing a Jetliner

A jetliner touches down at 270 km/h. The plane then decelerates (i.e., undergoes acceleration directed opposite its velocity) at 4.5 m/s^2. What's the minimum runway length on which this aircraft can land?

INTERPRET We interpret this as a problem involving one-dimensional motion with constant acceleration and identify the airplane as the object of interest.

DEVELOP We determine that Equation 2.11, $v^2 = v_0^2 + 2a(x - x_0)$, relates distance, velocity, and acceleration; so our plan is to solve that equation for the minimum runway length. We want the airplane to come to a stop, so the final velocity v is 0, and v_0 is the initial touchdown velocity. If x_0 is the touchdown point, then the quantity $x - x_0$ is the distance we're interested in; we'll call this Δx.

EVALUATE Setting $v = 0$ and solving Equation 2.11 then give

$$\Delta x = \frac{-v_0^2}{2a} = \frac{-[(270\text{ km/h})(1000\text{ m/km})(1/3600\text{ h/s})]^2}{(2)(-4.5\text{ m/s}^2)} = 625\text{ m}$$

Note that we used a negative value for the acceleration because the plane's acceleration is directed opposite its velocity—which we chose as the positive x-direction. We also converted the speed to m/s for compatibility with the SI units given for acceleration.

ASSESS Make sense? That 625 m is just over one-third of a mile, which seems a bit short. However, this is an absolute minimum with no margin of safety. For full-size jetliners, the standard for minimum landing runway length is about 5000 feet or 1.5 km.

✓TIP Be Careful with Mixed Units

Frequently problems are stated in units other than SI. Although it's possible to work consistently in other units, when in doubt, convert to SI. In this problem, the acceleration is originally in SI units but the velocity isn't—a sure indication of the need for conversion.

■

EXAMPLE 2.4 Motion with Two Objects: Speed Trap!

A speeding motorist zooms through a 50 km/h zone at 75 km/h (that's 21 m/s) without noticing a stationary police car. The police officer immediately heads after the speeder, accelerating at 2.5 m/s^2. When the officer catches up to the speeder, how far down the road are they, and how fast is the police car going?

INTERPRET We interpret this as *two* problems involving one-dimensional motion with constant acceleration. We identify the objects in question as the speeding car and the police car. Their motions are related because we're interested in the point where the two coincide.

DEVELOP It's helpful to draw a sketch showing qualitatively the position-versus-time graphs for the two cars. Since the speeding car moves with constant speed, its graph is a straight line. The police car is accelerating from rest, so its graph starts flat and gets increasingly steeper. Our sketch in Fig. 2.10 shows clearly the point we're interested in, when the two cars coincide for the second time. Equation 2.10, $x = x_0 + v_0t + \frac{1}{2}at^2$, gives position versus time with constant acceleration. Our plan is (1) to write versions of this equation specialized to each car, (2) to equate the resulting position expressions to find the time when the cars coincide, and (3) to find the corresponding position and the police car's velocity. For the latter we'll use Equation 2.7, $v = v_0 + at$.

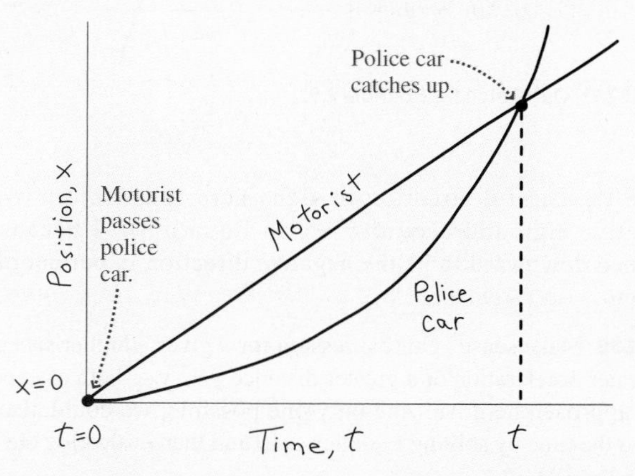

FIGURE 2.10 Our sketch of position versus time for the cars in Example 2.4.

EVALUATE Let's take the origin to be the point where the speeder passes the police car and $t = 0$ to be the corresponding time, as marked in Fig. 2.10. Then $x_0 = 0$ in Equation 2.10 for both cars, while the speeder has no acceleration and the police car has no initial velocity. Thus our two versions of Equation 2.10 are

$$x_s = v_{s0}t \text{ (speeder)} \quad \text{and} \quad x_p = \tfrac{1}{2}a_pt^2 \text{ (police car)}$$

Equating x_s and x_p tells when the speeder and the police car are at the same place, so we write $v_{s0}t = \frac{1}{2}a_pt^2$. This equation is satisfied when $t = 0$ or $t = 2v_{s0}/a_p$. Why two answers? We asked for *any* times when the two cars are in the same place. That includes the initial encounter at $t = 0$ as well as the later time $t = 2v_{s0}/a_p$ when the police car catches the speeder; both points are shown on our sketch. *Where* does this occur? We can evaluate using $t = 2v_{s0}/a_p$ in the speeder's equation:

$$x_s = v_{s0}t = v_{s0}\frac{2v_{s0}}{a_p} = \frac{2v_{s0}^2}{a_p} = \frac{(2)(21\text{ m/s})^2}{2.5\text{ m/s}^2} = 350\text{ m}$$

Equation 2.7 then gives the police car's speed at this time:

$$v_p = a_pt = a_p\frac{2v_{s0}}{a_p} = 2v_{s0} = 150\text{ km/h}$$

ASSESS Make sense? As Fig. 2.10 shows, the police car starts from rest and undergoes constant acceleration, so it has to be going faster at the point where the two cars meet. In fact, it's going twice as fast—again, as in Example 2.2, that's because the police car's position depends quadratically on time. That quadratic dependence also tells us that the police car's position-versus-time graph in Fig. 2.10 is a parabola.

■

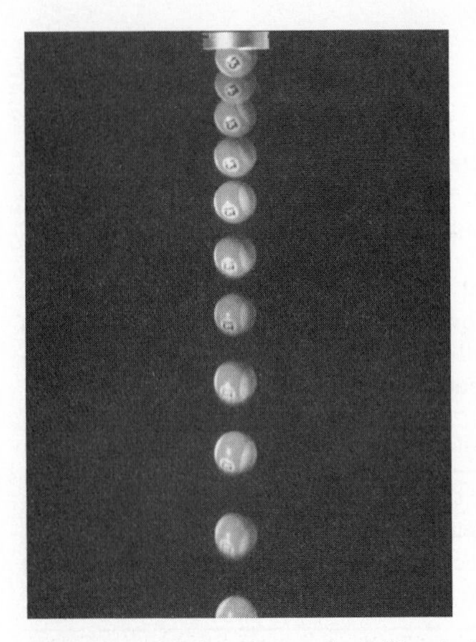

FIGURE 2.11 Strobe photo of a falling ball. Successive images are farther apart, showing that the ball is accelerating.

GOT IT? 2.4 The police car in Example 2.4 starts with zero velocity and is going at twice the car's velocity when it catches up to the car. So at some intermediate instant it must be going at the same velocity as the car. Is that instant (a) halfway between the times when the two cars coincide, (b) closer to the time when the speeder passes the stationary police car, or (c) closer to the time when the police car catches the speeder?

2.5 The Acceleration of Gravity

Drop an object, and it falls at an increasing rate, accelerating because of gravity (Fig. 2.11). The acceleration is constant for objects falling near Earth's surface, and furthermore it has the same value for all objects. This value, the **acceleration of gravity**, is designated g and is approximately 9.8 m/s^2 near Earth's surface.

The acceleration of gravity applies strictly only in **free fall**—motion under the influence of gravity alone. Air resistance, in particular, may dramatically alter the motion, giving the false impression that gravity acts differently on lighter and heavier objects. As early as the year 1600, Galileo is reputed to have shown that all objects have the same acceleration by dropping objects off the Leaning Tower of Pisa. Astronauts have verified that a feather and a hammer fall with the same acceleration on the airless Moon—although that acceleration is less than on Earth.

Although g is approximately constant near Earth's surface, it varies slightly with latitude and even local geology. The variation with altitude becomes substantial over distances of tens to hundreds of kilometers. But nearer Earth's surface it's a good approximation to take g as strictly constant. Then an object in free fall undergoes constant acceleration, and the equations of Table 2.1 apply. In working gravitational problems, we usually replace x with y to designate the vertical direction. If we make the arbitrary but common choice that the upward direction is positive, then acceleration a becomes $-g$ because the acceleration is downward.

EXAMPLE 2.5 Constant Acceleration due to Gravity: Cliff Diving

A diver drops from a 10-m-high cliff. At what speed does he enter the water, and how long is he in the air?

INTERPRET This is a case of constant acceleration due to gravity, and the diver is the object of interest. The diver drops a known distance starting from rest, and we want to know the speed and time when he hits the water.

DEVELOP Figure 2.12 is a sketch showing what the diver's position versus time should look like. We've incorporated what we know: the initial position 10 m above the water, the start from rest, and the downward acceleration that results in a parabolic position-versus-time curve. Given the dive height, Equation 2.11 determines the speed v. Following our newly adopted convention that y designates the vertical direction, we write Equation 2.11 as $v^2 = v_0^2 + 2a(y - y_0)$. Since the diver starts from rest, $v_0 = 0$ and the equation becomes $v^2 = -2g(y - y_0)$. So our plan is first to solve for the speed at the water; then use Equation 2.7, $v = v_0 + at$, to get the time.

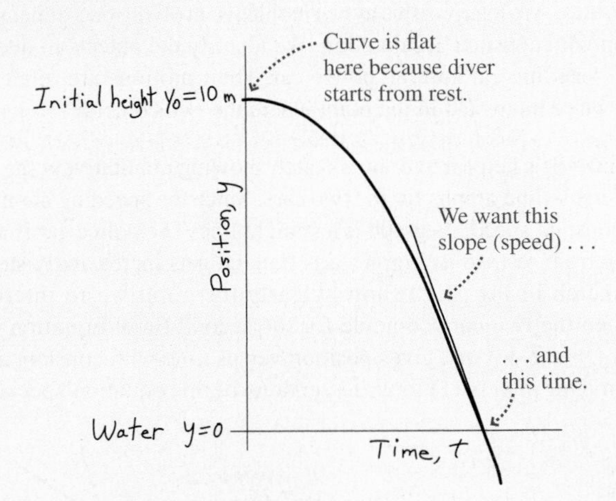

FIGURE 2.12 Our sketch for Example 2.5.

EVALUATE Our sketch shows that we've chosen $y = 0$ at the water, so $y_0 = 10$ m and Equation 2.11 gives

$$|v| = \sqrt{-2g(y - y_0)} = \sqrt{(-2)(9.8\text{ m/s}^2)(0\text{ m} - 10\text{ m})}$$
$$= 14\text{ m/s}$$

This is the magnitude of the velocity, hence the absolute value sign; the actual value is $v = -14$ m/s, with the minus sign indicating downward motion. Knowing the initial and final velocities, we use Equation 2.7 to find how long the dive takes. Solving that equation for t gives

$$t = \frac{v_0 - v}{g} = \frac{0\text{ m/s} - (-14\text{ m/s})}{9.8\text{ m/s}^2} = 1.4\text{ s}$$

Note the careful attention to signs here; we wrote v with its negative sign and used $a = -g$ in Equation 2.7 because we defined downward to be the negative direction in our coordinate system.

ASSESS Make sense? Our expression for v gives a higher speed with a greater acceleration or a greater distance $y - y_0$—both as expected. Our approach here isn't the only one possible; we could also have found the time by solving Equation 2.10 and then evaluating the speed using Equation 2.7. ■

In Example 2.5 the diver was moving downward, and the downward gravitational acceleration steadily increased his speed. But, as Conceptual Example 2.1 suggested, the acceleration of gravity is downward regardless of an object's motion. Throw a ball straight up, and it's accelerating *downward* even while moving *upward*. Since velocity and acceleration are in opposite directions, the ball slows until it reaches its peak, then pauses instantaneously, and then gains speed as it falls. All the while its acceleration is 9.8 m/s^2 downward.

EXAMPLE 2.6 Constant Acceleration due to Gravity: Tossing a Ball

You toss a ball straight up at 7.3 m/s; it leaves your hand at 1.5 m above the floor. Find when it hits the floor, the maximum height it reaches, and its speed when it passes your hand on the way down.

INTERPRET We have constant acceleration due to gravity, and here the object of interest is the ball. We want to find time, height, and speed.

DEVELOP The ball starts by going up, eventually comes to a stop, and then heads downward. Figure 2.13 is a sketch of the height versus time that we expect, showing what we know and the three quantities we're after. Equation 2.10, $y = y_0 + v_0t + \frac{1}{2}at^2$, determines position as a function of time, so our plan is to use that equation to find the time the ball hits the floor (again, we've replaced horizontal position x with height y in Equation 2.10). Then we can use Equation 2.11, $v^2 = v_0^2 + 2a(y - y_0)$, to find the height at which $v = 0$—that is, the peak height. Finally, Equation 2.11 will also give us the speed at any height, letting us answer the question about the speed when the ball passes the height of 1.5 m on its way down.

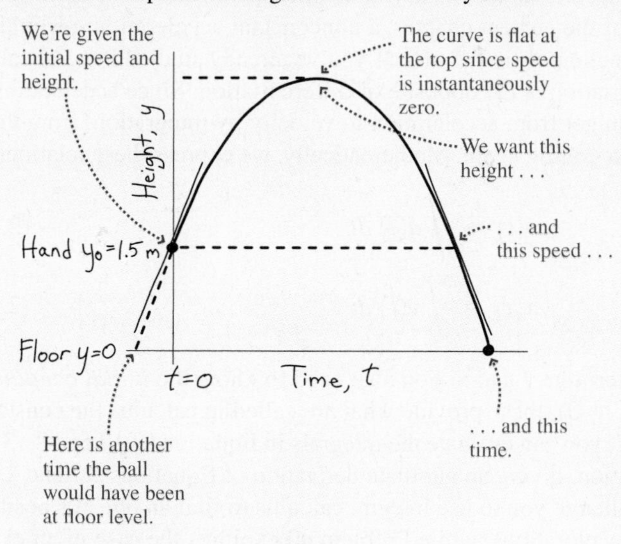

FIGURE 2.13 Our sketch for Example 2.6.

EVALUATE Our sketch shows that we've taken $y = 0$ at the floor; so when the ball is at the floor, Equation 2.10 becomes $0 = y_0 + v_0t - \frac{1}{2}gt^2$, which we can solve for t using the quadratic formula [Appendix A; $t = (v_0 \pm \sqrt{v_0^2 + 2y_0g})/g$]. Here v_0 is the initial velocity, 7.3 m/s; it's positive because the motion is initially upward. The initial position is the hand height, so $y_0 = 1.5$ m, and g of course is 9.8 m/s^2 (we accounted for the downward acceleration by putting $a = -g$ in Equation 2.10). Putting in these numbers gives $t = 1.7$ s or -0.18 s; the answer we want is 1.7 s. At the peak of its flight, the ball's velocity is instantaneously zero because it's moving neither up nor down. So we set $v^2 = 0$ in Equation 2.11 to get $0 = v_0^2 - 2g(y - y_0)$. Solving for y then gives the peak height:

$$y = y_0 + \frac{v_0^2}{2g} = 1.5\text{ m} + \frac{(7.3\text{ m/s})^2}{(2)(9.8\text{ m/s}^2)} = 4.2\text{ m}$$

To find the speed when the ball reaches 1.5 m on the way down, we set $y = y_0$ in Equation 2.11. The result is $v^2 = v_0^2$, so $v = \pm v_0$ or ± 7.3 m/s. Once again, there are two answers. The equation has given us *all* the velocities the ball has at 1.5 m—including the initial upward velocity and the later downward velocity. We've shown here that an upward-thrown object returns to its initial height with the same speed it had initially.

ASSESS Make sense? With no air resistance to sap the ball of its energy, it seems reasonable that the ball comes back down with the same speed—a fact we'll explore further when we introduce energy conservation in Chapter 7. But why are there two answers for time and velocity? Equation 2.10 doesn't "know" about your hand or the floor; it "assumes" the ball has always been undergoing downward acceleration g. We asked of Equation 2.10 when the ball would be at $y = 0$. The second answer, 1.7 s, was the one we wanted. But if the ball had always been in free fall, it would also have been on the floor 0.18 s earlier, heading upward. That's the meaning of the other answer, -0.18 s, as we've indicated on our sketch. Similarly, Equation 2.11 gave us all the velocities the ball had at a height of 1.5 m, including both the initial upward velocity and the later downward velocity. ■

✓TIP Multiple Answers

Frequently the mathematics of a problem gives more than one answer. Think about what each answer means before discarding it! Sometimes an answer isn't consistent with the physical assumptions of the problem, but other times all answers are meaningful even if they aren't all what you're looking for.

GOT IT? 2.5 Standing on a roof, you simultaneously throw one ball straight up and drop another from rest. Which hits the ground first? Which hits the ground moving faster?

APPLICATION **Keeping Time**

The NIST-F1 atomic clock, shown here with its developers, sets the U.S. standard of time. The clock is so accurate that it won't gain or lose more than a second in 100 million years! It gets its remarkable accuracy by monitoring a super-cold clump of freely falling cesium atoms for what is, in this context, a long time period of about 1 s. The atom clump is put in free fall by a more sophisticated version of the ball toss in Example 2.6. In the NIST-F1 clock, laser beams gently "toss" the ball of atoms upward with a speed that gives it an up-and-down travel time of about 1 s (see Problem 66). For this reason NIST-F1 is called an atomic fountain clock. In the photo you can see the clock's towerlike structure that accommodates this atomic fountain.

2.6 When Acceleration Isn't Constant

Sections 2.4 and 2.5 both dealt with *constant acceleration*. Fortunately, there are many important applications, such as situations involving gravity near Earth's surface, where acceleration *is* constant. But when it isn't, then the equations listed in Table 2.1 don't apply. In Chapter 3 you'll see that acceleration can vary in magnitude, direction, or both. In the one-dimensional situations of the current chapter, a nonconstant acceleration a would be specified by giving a as a function of time t: $a(t)$. If you've already studied integral calculus, then you know that integration is the opposite of differentiation. Since acceleration is the derivative of velocity, you get from acceleration to velocity by integration; from there you can get to position by integrating again. Mathematically, we express these relations as

$$v(t) = \int a(t)\, dt \tag{2.12}$$

$$x(t) = \int v(t)\, dt \tag{2.13}$$

These results don't fully determine v and x; you also need to know the *initial conditions* (usually, the values at time $t = 0$); these provide what are called in calculus the constants of integration. In Problem 87, you can evaluate the integrals in Equations 2.12 and 2.13 for the case of constant acceleration, giving an alternate derivation of Equations 2.7 and 2.10. Problems 82, 88, and 89 challenge you to use integral calculus to find an object's position in the case of nonconstant accelerations, while Problem 90 explores the case of an exponentially decreasing acceleration.

GOT IT? 2.6 The graph shows acceleration versus time for three different objects, all of which start at rest from the same position. Only object (b) undergoes constant acceleration. Which object is going fastest at the time t_1?

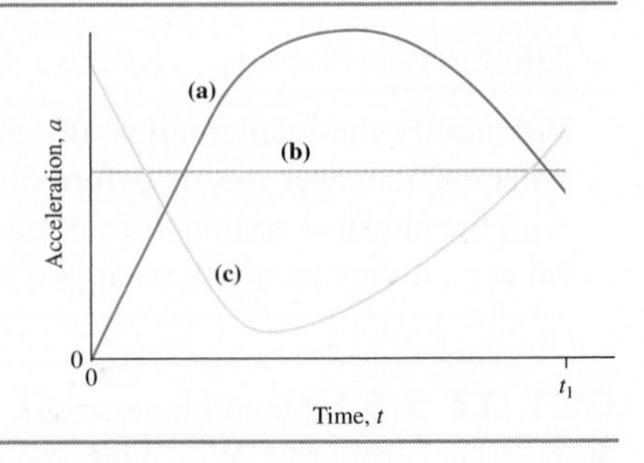

CHAPTER 2 SUMMARY

Big Idea

The big ideas here are those of **kinematics**—the study of motion without regard to its cause. **Position**, **velocity**, and **acceleration** are the quantities that characterize motion:

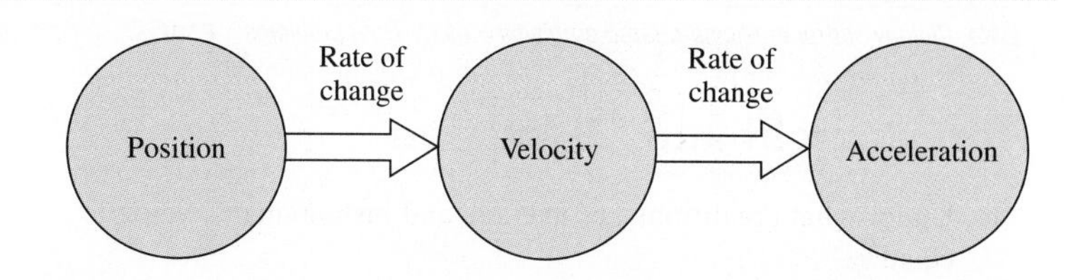

Key Concepts and Equations

Average velocity and acceleration involve changes in position and velocity, respectively, occurring over a time interval Δt:

$$\bar{v} = \frac{\Delta x}{\Delta t}$$

$$\bar{a} = \frac{\Delta v}{\Delta t}$$

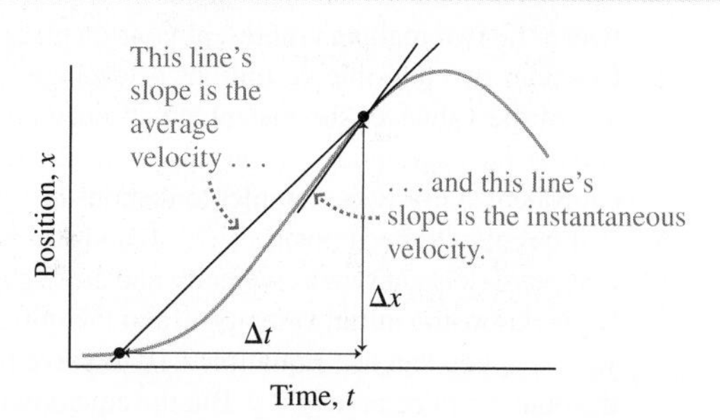

Here Δx is the **displacement**, or change in position, and Δv is the change in velocity.

Instantaneous values are the limits of infinitesimally small time intervals and are given by calculus as the time derivatives of position and velocity:

$$v = \frac{dx}{dt}$$

$$a = \frac{dv}{dt}$$

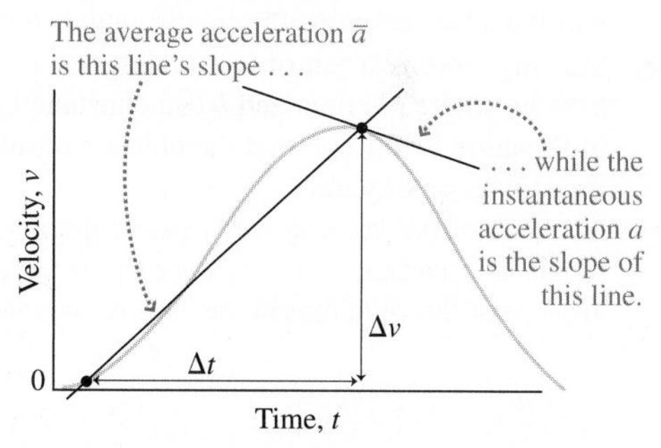

Applications

Constant acceleration is a special case that yields simple equations describing one-dimensional motion:

$$v = v_0 + at$$

$$x = x_0 + v_0 t + \tfrac{1}{2}at^2$$

$$v^2 = v_0^2 + 2a(x - x_0)$$

These equations apply only in the case of constant acceleration.

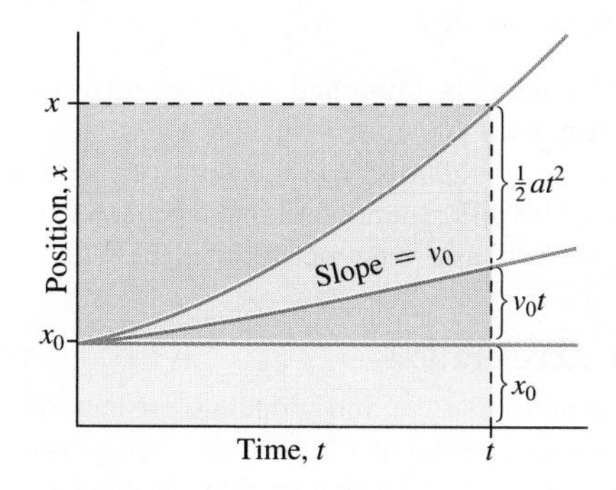

An important example is the acceleration of gravity, essentially constant near Earth's surface, with magnitude approximately $9.8\ \text{m/s}^2$.

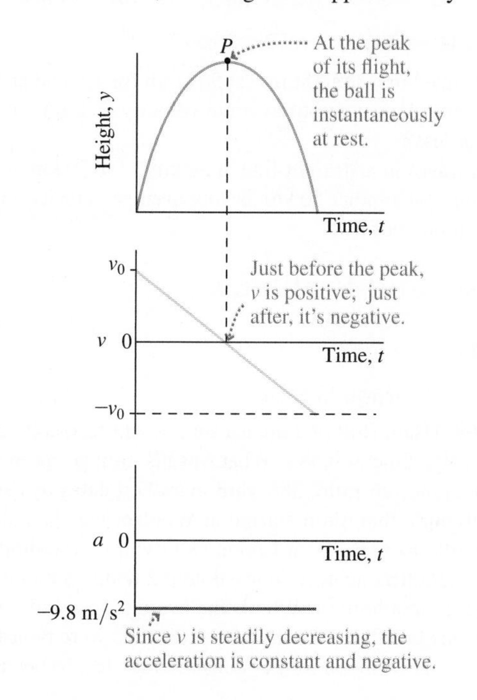

BIO *Biology and/or medicine-related problems* **DATA** *Data problems* **ENV** *Environmental problems* **CH** *Challenge problems* **COMP** *Computer problems*

For Thought and Discussion

1. Under what conditions are average and instantaneous velocity equal?
2. Does a speedometer measure speed or velocity?
3. You check your odometer at the beginning of a day's driving and again at the end. Under what conditions would the difference between the two readings represent your displacement?
4. Consider two possible definitions of average speed: (a) the average of the values of the instantaneous speed over a time interval and (b) the magnitude of the average velocity. Are these definitions equivalent? Give two examples to demonstrate your conclusion.
5. Is it possible to be at position $x = 0$ and still be moving?
6. Is it possible to have zero velocity and still be accelerating?
7. If you know the initial velocity v_0 and the initial and final heights y_0 and y, you can use Equation 2.10 to solve for the time t when the object will be at height y. But the equation is quadratic in t, so you'll get two answers. Physically, why is this?
8. Starting from rest, an object undergoes acceleration given by $a = bt$, where t is time and b is a constant. Can you use bt for a in Equation 2.10 to predict the object's position as a function of time? Why or why not?
9. In which of the velocity-versus-time graphs shown in Fig. 2.14 would the average velocity over the interval shown equal the average of the velocities at the ends of the interval?

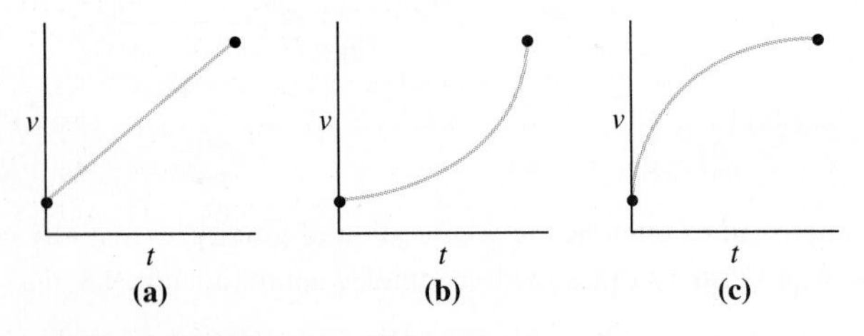

FIGURE 2.14 For Thought and Discussion 9

10. If you travel in a straight line at 50 km/h for 1 h and at 100 km/h for another hour, is your average velocity 75 km/h? If not, is it more or less?
11. If you travel in a straight line at 50 km/h for 50 km and then at 100 km/h for another 50 km, is your average velocity 75 km/h? If not, is it more or less?

Exercises and Problems

Exercises

Section 2.1 Average Motion

12. In 2009, Usain Bolt of Jamaica set a world record in the 100-m dash with a time of 9.58 s. What was his average speed?
13. The standard 26-mile, 385-yard marathon dates to 1908, when the Olympic marathon started at Windsor Castle and finished before the Royal Box at London's Olympic Stadium. Today's top marathoners achieve times around 2 hours, 3 minutes for the standard marathon. (a) What's the average speed of a marathon run in this time? (b) Marathons before 1908 were typically about 25 miles. How much longer does the race last today as a result of the extra mile and 385 yards, assuming it's run at part (a)'s average speed?
14. Starting from home, you bicycle 24 km north in 2.5 h and then turn around and pedal straight home in 1.5 h. What are your (a) displacement at the end of the first 2.5 h, (b) average velocity over the first 2.5 h, (c) average velocity for the homeward leg of the trip, (d) displacement for the entire trip, and (e) average velocity for the entire trip?
15. The Voyager 1 spacecraft is expected to continue broadcasting data until at least 2020, when it will be some 14 billion miles from Earth. How long will it take Voyager's radio signals, traveling at the speed of light, to reach Earth from this distance?
16. In 2008, Australian Emma Snowsill set an unofficial record in the women's Olympic triathlon, completing the 1.5-km swim, 40-km bicycle ride, and 10-km run in 1 h, 58 min, 27.66 s. What was her average speed?
17. Taking Earth's orbit to be a circle of radius 1.5×10^8 km, determine Earth's orbital speed in (a) meters per second and (b) miles per second.
18. What's the conversion factor from meters per second to miles per hour?

Section 2.2 Instantaneous Velocity

19. On a single graph, plot distance versus time for the first two trips from Houston to Des Moines described on page 16. For each trip, identify graphically the average velocity and, for each segment of the trip, the instantaneous velocity.
20. For the motion plotted in Fig. 2.15, estimate (a) the greatest velocity in the positive x-direction, (b) the greatest velocity in the negative x-direction, (c) any times when the object is instantaneously at rest, and (d) the average velocity over the interval shown.

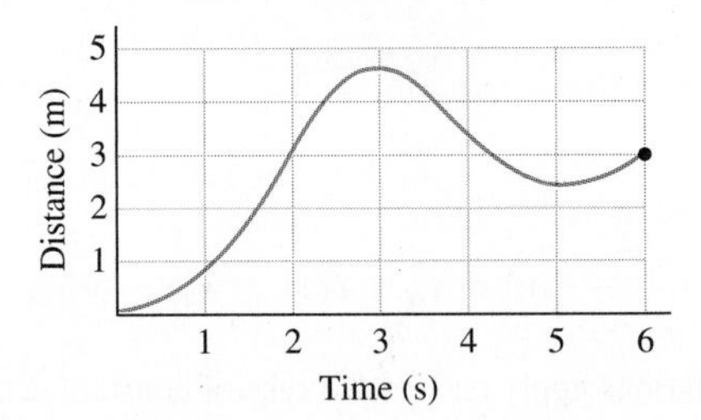

FIGURE 2.15 Exercise 20

21. A model rocket is launched straight upward. Its altitude y as a function of time is given by $y = bt - ct^2$, where $b = 82$ m/s, $c = 4.9$ m/s^2, t is the time in seconds, and y is in meters. (a) Use differentiation to find a general expression for the rocket's velocity as a function of time. (b) When is the velocity zero?

Section 2.3 Acceleration

22. A giant eruption on the Sun propels solar material from rest to 450 km/s over a period of 1 h. Find the average acceleration.
23. Starting from rest, a subway train first accelerates to 25 m/s, then brakes. Forty-eight seconds after starting, it's moving at 17 m/s. What's its average acceleration in this 48-s interval?

24. A space shuttle's main engines cut off 8.5 min after launch, at which time its speed is 7.6 km/s. What's the shuttle's average acceleration during this interval?
25. An egg drops from a second-story window, taking 1.12 s to fall and reaching 11.0 m/s just before hitting the ground. On contact, the egg stops completely in 0.131 s. Calculate the magnitudes of its average acceleration (a) while falling and (b) while stopping.
26. An airplane's takeoff speed is 320 km/h. If its average acceleration is 2.9 m/s^2, how much time is it accelerating down the runway before it lifts off?
27. ThrustSSC, the world's first supersonic car, accelerates from rest to 1000 km/h in 16 s. What's its acceleration?

Section 2.4 Constant Acceleration

28. You're driving at 70 km/h when you apply constant acceleration to pass another car. Six seconds later, you're doing 80 km/h. How far did you go in this time?
29. Differentiate both sides of Equation 2.10, and show that you get Equation 2.7.
30. An X-ray tube gives electrons constant acceleration over a distance of 15 cm. If their final speed is 1.2×10^7 m/s, what are (a) the electrons' acceleration and (b) the time they spend accelerating?
31. A rocket rises with constant acceleration to an altitude of 85 km, at which point its speed is 2.8 km/s. (a) What's its acceleration? (b) How long does the ascent take?
32. Starting from rest, a car accelerates at a constant rate, reaching 88 km/h in 12 s. Find (a) its acceleration and (b) how far it goes in this time.
33. A car moving initially at 50 mi/h begins slowing at a constant rate 100 ft short of a stoplight. If the car comes to a full stop just at the light, what is the magnitude of its acceleration?
34. **BIO** In a medical X-ray tube, electrons are accelerated to a velocity of 10^8 m/s and then slammed into a tungsten target. As they stop, the electrons' rapid acceleration produces X rays. If the time for an electron to stop is on the order of 10^{-9} s, approximately how far does it move while stopping?
35. California's Bay Area Rapid Transit System (BART) uses an automatic braking system triggered by earthquake warnings. The system is designed to prevent disastrous accidents involving trains traveling at a maximum of 112 km/h and carrying a total of some 45,000 passengers at rush hour. If it takes a train 24 s to brake to a stop, how much advance warning of an earthquake is needed to bring a 112-km/h train to a reasonably safe speed of 42 km/h when the earthquake strikes?
36. You're driving at speed v_0 when you spot a stationary moose on the road, a distance d ahead. Find an expression for the magnitude of the acceleration you need if you're to stop before hitting the moose.

Section 2.5 The Acceleration of Gravity

37. You drop a rock into a deep well and 4.4 s later hear a splash. How far down is the water? Neglect the travel time of sound.
38. Your friend is sitting 6.5 m above you on a tree branch. How fast should you throw an apple so it just reaches her?
39. A model rocket leaves the ground, heading straight up at 49 m/s. (a) What's its maximum altitude? Find its speed and altitude at (b) 1 s, (c) 4 s, and (d) 7 s.
40. A foul ball leaves the bat going straight up at 23 m/s. (a) How high does it rise? (b) How long is it in the air? Neglect the distance between bat and ground.
41. A Frisbee is lodged in a tree 6.5 m above the ground. A rock thrown from below must be going at least 3 m/s to dislodge the Frisbee. How fast must such a rock be thrown upward if it leaves the thrower's hand 1.3 m above the ground?
42. Space pirates kidnap an earthling and hold him on one of the solar system's planets. With nothing else to do, the prisoner amuses himself by dropping his watch from eye level (170 cm) to the floor. He observes that the watch takes 0.95 s to fall. On what planet is he being held? (*Hint*: Consult Appendix E.)

Problems

43. You allow 40 min to drive 25 mi to the airport, but you're caught in heavy traffic and average only 20 mi/h for the first 15 min. What must your average speed be on the rest of the trip if you're to make your flight?
44. A base runner can get from first to second base in 3.4 s. If he leaves first as the pitcher throws a 90 mi/h fastball the 61-ft distance to the catcher, and if the catcher takes 0.45 s to catch and rethrow the ball, how fast does the catcher have to throw the ball to second base to make an out? Home plate to second base is the diagonal of a square 90 ft on a side.
45. You can run 9.0 m/s, 20% faster than your brother. How much head start should you give him in order to have a tie race over 100 m?
46. A jetliner leaves San Francisco for New York, 4600 km away. With a strong tailwind, its speed is 1100 km/h. At the same time, a second jet leaves New York for San Francisco. Flying into the wind, it makes only 700 km/h. When and where do the two planes pass?
47. An object's position is given by $x = bt + ct^3$, where $b = 1.50$ m/s, $c = 0.640$ m/s^3, and t is time in seconds. To study the limiting process leading to the instantaneous velocity, calculate the object's average velocity over time intervals from (a) 1.00 s to 3.00 s, (b) 1.50 s to 2.50 s, and (c) 1.95 s to 2.05 s. (d) Find the instantaneous velocity as a function of time by differentiating, and compare its value at 2 s with your average velocities.
48. An object's position as a function of time t is given by $x = bt^4$, with b a constant. Find an expression for the instantaneous velocity, and show that the average velocity over the interval from $t = 0$ to any time t is one-fourth of the instantaneous velocity at t.
49. In a drag race, the position of a car as a function of time is given by $x = bt^2$, with $b = 2.000$ m/s^2. In an attempt to determine the car's velocity midway down a 400-m track, two observers stand at the 180-m and 220-m marks and note when the car passes. (a) What value do the two observers compute for the car's velocity over this 40-m stretch? Give your answer to four significant figures. (b) By what percentage does this observed value differ from the instantaneous value at $x = 200$ m?
50. Squaring Equation 2.7 gives an expression for v^2. Equation 2.11 also gives an expression for v^2. Equate the two expressions, and show that the resulting equation reduces to Equation 2.10.
51. During the complicated sequence that landed the rover *Curiosity* on Mars in 2012, the spacecraft reached an altitude of 142 m above the Martian surface, moving vertically downward at 32.0 m/s. It then entered a so-called constant deceleration (CD) phase, during which its velocity decreased steadily to 0.75 m/s while it dropped to an altitude of 23 m. What was the magnitude of the spacecraft's acceleration during this CD phase?

52. The position of a car in a drag race is measured each second, and
DATA the results are tabulated below.

Time t (s)	0	1	2	3	4	5
Position x (m)	0	1.7	6.2	17	24	40

Assuming the acceleration is approximately constant, plot position versus a quantity that should make the graph a straight line. Fit a line to the data, and from it determine the approximate acceleration.

53. A fireworks rocket explodes at a height of 82.0 m, producing fragments with velocities ranging from 7.68 m/s downward to 16.7 m/s upward. Over what time interval are fragments hitting the ground?
54. The muscles in a grasshopper's legs can propel the insect upward
BIO at 3.0 m/s. How high can the grasshopper jump?
55. On packed snow, computerized antilock brakes can reduce a car's stopping distance by 55%. By what percentage is the stopping time reduced?
56. A particle leaves its initial position x_0 at time $t = 0$, moving in the positive x-direction with speed v_0 but undergoing acceleration of magnitude a in the negative x-direction. Find expressions for (a) the time when it returns to x_0 and (b) its speed when it passes that point.
57. A hockey puck moving at 32 m/s slams through a wall of snow 35 cm thick. It emerges moving at 18 m/s. Assuming constant acceleration, find (a) the time the puck spends in the snow and (b) the thickness of a snow wall that would stop the puck entirely.
58. Amtrak's 20th-Century Limited is en route from Chicago to New York at 110 km/h when the engineer spots a cow on the track. The train brakes to a halt in 1.2 min, stopping just in front of the cow. (a) What is the magnitude of the train's acceleration? (b) What's the direction of the acceleration? (c) How far was the train from the cow when the engineer applied the brakes?
59. A jetliner touches down at 220 km/h and comes to a halt 29 s later. What's the shortest runway on which this aircraft can land?
60. A motorist suddenly notices a stalled car and slams on the brakes, negatively accelerating at 6.3 m/s^2. Unfortunately, this isn't enough, and a collision ensues. From the damage sustained, police estimate that the car was going 18 km/h at the time of the collision. They also measure skid marks 34 m long. (a) How fast was the motorist going when the brakes were first applied? (b) How much time elapsed from the initial braking to the collision?
61. A racing car undergoing constant acceleration covers 140 m in 3.6 s. (a) If it's moving at 53 m/s at the end of this interval, what was its speed at the beginning of the interval? (b) How far did it travel from rest to the end of the 140-m distance?
62. The maximum braking acceleration of a car on a dry road is about 8 m/s^2. If two cars move head-on toward each other at 88 km/h (55 mi/h), and their drivers brake when they're 85 m apart, will they collide? If so, at what relative speed? If not, how far apart will they be when they stop? Plot distance versus time for both cars on a single graph.
63. After 35 min of running, at the 9-km point in a 10-km race, you find yourself 100 m behind the leader and moving at the same speed. What should your acceleration be if you're to catch up by the finish line? Assume that the leader maintains constant speed.
64. You're speeding at 85 km/h when you notice that you're only 10 m behind the car in front of you, which is moving at the legal speed limit of 60 km/h. You slam on your brakes, and your car negatively accelerates at 4.2 m/s^2. Assuming the other car continues at constant speed, will you collide? If so, at what relative speed? If not, what will be the distance between the cars at their closest approach?
65. Airbags cushioned the Mars rover Spirit's landing, and the rover bounced some 15 m vertically after its first impact. Assuming no loss of speed at contact with the Martian surface, what was Spirit's impact speed?
66. Calculate the speed with which cesium atoms must be "tossed" in the NIST-F1 atomic clock so that their up-and-down travel time is 1.0 s. (See the Application on page 26.)
67. A falling object travels one-fourth of its total distance in the last
CH second of its fall. From what height was it dropped?
68. You're on a NASA team engineering a probe to land on Jupiter's moon Io, and your job is to specify the impact speed the probe can tolerate without damage. Rockets will bring the probe to a halt 100 m above the surface, after which it will fall freely. What speed do you specify? (Consult Appendix E.)
69. You're atop a building of height h, and a friend is poised to drop a ball from a window at $h/2$. Find an expression for the speed at which you should simultaneously throw a ball downward, so the two hit the ground at the same time.
70. A castle's defenders throw rocks down on their attackers from a 15-m-high wall, with initial speed 10 m/s. How much faster are the rocks moving when they hit the ground than if they were simply dropped?
71. Two divers jump from a 3.00-m platform. One jumps upward at
CH 1.80 m/s, and the second steps off the platform as the first passes it on the way down. (a) What are their speeds as they hit the water? (b) Which hits the water first and by how much?
72. A balloon is rising at 10 m/s when its passenger throws a ball straight up at 12 m/s relative to the balloon. How much later does the passenger catch the ball?
73. Landing on the Moon, a spacecraft fires its rockets and comes to a complete stop just 12 m above the lunar surface. It then drops freely to the surface. How long does it take to fall, and what's its impact speed? (*Hint*: Consult Appendix E.)
74. You're at mission control for a rocket launch, deciding whether to let the launch proceed. A band of clouds 5.3 km thick extends upward from 1.9 km altitude. The rocket will accelerate at 4.6 m/s^2, and it isn't allowed to be out of sight for more than 30 s. Should you allow the launch?
75. You're an investigator for the National Transportation Safety Board, examining a subway accident in which a train going at 80 km/h collided with a slower train traveling in the same direction at 25 km/h. Your job is to determine the relative speed of the collision, to help establish new crash standards. The faster train's "black box" shows that its brakes were applied and it began slowing at the rate of 2.1 m/s^2 when it was 50 m from the slower train, while the slower train continued at constant speed. What do you report?
76. You toss a book into your dorm room, just clearing a windowsill
CH 4.2 m above the ground. (a) If the book leaves your hand 1.5 m above the ground, how fast must it be going to clear the sill? (b) How long after it leaves your hand will it hit the floor, 0.87 m below the windowsill?
77. Consider an object traversing a distance L, part of the way at speed v_1 and the rest of the way at speed v_2. Find expressions for the object's average speed over the entire distance L when the object moves at each of the two speeds v_1 and v_2 for (a) half the total *time* and (b) half the total *distance*. (c) In which case is the average speed greater?
78. A particle's position as a function of time is given by $x = x_0 \sin\omega t$, where x_0 and ω are constants. (a) Find expressions for the velocity and acceleration. (b) What are the maximum values of velocity and acceleration? (*Hint*: Consult the table of derivatives in Appendix A.)

79. **CH** Ice skaters, ballet dancers, and basketball players executing vertical leaps often give the illusion of "hanging" almost motionless near the top of the leap. To see why this is, consider a leap to maximum height h. Of the total time spent in the air, what fraction is spent in the upper half (i.e., at $y > \frac{1}{2}h$)?
80. You're staring idly out your dorm window when you see a water balloon fall past. If the balloon takes 0.22 s to cross the 1.3-m-high window, from what height above the window was it dropped?
81. A police radar's effective range is 1.0 km, and your radar detector's range is 1.9 km. You're going 110 km/h in a 70 km/h zone when the radar detector beeps. At what rate must you negatively accelerate to avoid a speeding ticket?
82. **CH** An object starts moving in a straight line from position x_0, at time $t = 0$, with velocity v_0. Its acceleration is given by $a = a_0 + bt$, where a_0 and b are constants. Use integration to find expressions for (a) the instantaneous velocity and (b) the position, as functions of time.
83. You're a consultant on a movie set, and the producer wants a car to drop so that it crosses the camera's field of view in time Δt. The field of view has height h. Derive an expression for the height above the top of the field of view from which the car should be released.
84. **CH** (a) For the ball in Example 2.6, find its velocity just before it hits the floor. (b) Suppose you had tossed a second ball straight down at 7.3 m/s (from the same place 1.5 m above the floor). What would its velocity be just before it hits the floor? (c) When would the second ball hit the floor? (Interpret any multiple answers.)
85. Your roommate is an aspiring novelist and asks your opinion on a matter of physics. The novel's central character is kept awake at night by a leaky faucet. The sink is 19.6 cm below the faucet. At the instant one drop leaves the faucet, another strikes the sink below and two more are in between on the way down. How many drops per second are keeping the protagonist awake?
86. You and your roommate plot to drop water balloons on students entering your dorm. Your window is 20 m above the sidewalk. You plan to place an X on the sidewalk to mark the spot a student must be when you drop the balloon. You note that most students approach the dorm at about 2 m/s. How far from the impact point do you place the X?
87. **CH** Derive Equation 2.10 by integrating Equation 2.7 over time. You'll have to interpret the constant of integration.
88. **CH** An object's acceleration increases quadratically with time: $a(t) = bt^2$, where $b = 0.041\ \text{m/s}^4$. If the object starts from rest, how far does it travel in 6.3 s?
89. **CH** An object's acceleration is given by the expression $a(t) = -a_0 \cos \omega t$, where a_0 and ω are positive constants. Find expressions for the object's (a) velocity and (b) position as functions of time. Assume that at time $t = 0$ it starts from rest at its greatest positive displacement from the origin. (c) Determine the magnitudes of the object's maximum velocity and maximum displacement from the origin.
90. **CH** An object's acceleration decreases exponentially with time: $a(t) = a_0 e^{-bt}$, where a_0 and b are constants. (a) Assuming the object starts from rest, determine its velocity as a function of time. (b) Will its speed increase indefinitely? (c) Will it travel indefinitely far from its starting point?
91. **CH** A ball is dropped from rest at a height h_0 above the ground. At the same instant, a second ball is launched with speed v_0 straight up from the ground, at a point directly below where the other ball is dropped. (a) Find a condition on v_0 such that the two balls will collide in mid-air. (b) Find an expression for the height at which they collide.

Passage Problems

A wildlife biologist is studying the hunting patterns of tigers. She anesthetizes a tiger and attaches a GPS collar to track its movements. The collar transmits data on the tiger's position and velocity. Figure 2.16 shows the tiger's velocity as a function of time as it moves on a one-dimensional path.

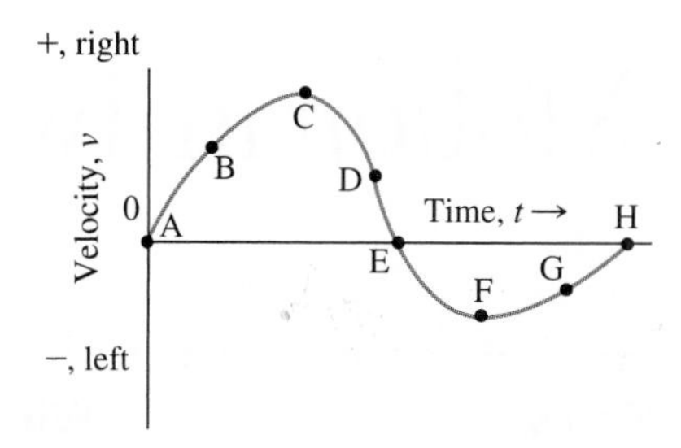

FIGURE 2.16 The tiger's velocity (Passage Problems 92–96)

92. At which marked point(s) is the tiger not moving?
    a. E only
    b. A, E, and H
    c. C and F
    d. none of the points (it's always moving)
93. At which marked point(s) is the tiger not accelerating?
    a. E only
    b. A, E, and H
    c. C and F
    d. all of the points (it's never accelerating)
94. At which point does the tiger have the greatest speed?
    a. B
    b. C
    c. D
    d. F
95. At which point does the tiger's acceleration have the greatest magnitude?
    a. B
    b. C
    c. D
    d. F
96. At which point is the tiger farthest from its starting position at $t = 0$?
    a. C
    b. E
    c. F
    d. H

Answers to Chapter Questions

Answer to Chapter Opening Question

Although the ball's velocity is zero at the top of its motion, its acceleration is $-9.8\ \text{m/s}^2$, as it is throughout the toss.

Answers to GOT IT? Questions

2.1 (a) and (b); average speed is greater for (c)
2.2 (b) moves with constant speed; (a) reverses; (d) speeds up
2.3 (b) downward
2.4 (a) halfway between the times; because its acceleration is constant, the police car's speed increases by equal amounts in equal times. So it gets from 0 to half its final velocity—which is twice the car's velocity—in half the total time.
2.5 The dropped ball hits first; the thrown ball hits moving faster.
2.6 (c)

Motion in Two and Three Dimensions

What You Know

- You understand basic motion concepts: position, velocity, and acceleration.
- You can interpret graphs of these quantities as functions of time.
- You know how to analyze motion in one dimension under constant acceleration, including the acceleration of gravity near Earth.

What You're Learning

- You'll learn to describe the richness of motion in two and three dimensions using the language of vectors.
- You'll develop vector expressions for position, velocity, and acceleration.
- You'll see how the analysis of multidimensional motion is based on the techniques of Chapter 2, now applied in mutually perpendicular directions.
- You'll learn about motion under the influence of gravity near Earth's surface.
- You'll see how circular motion is a special case of accelerated motion, and you'll see how to find the magnitude and direction of that acceleration.

How You'll Use It

- You'll study Newton's laws of motion in Chapters 4 and 5, and you'll see how acceleration—involving *change* in motion—is a key concept in Newtonian mechanics.
- Your understanding of accelerated motion developed here will set the stage for applying Newton's laws of motion in multidimensional situations.
- The language of vectors will serve you throughout the rest of this course because physical quantities ranging from forces to angular momentum to electric and magnetic fields are all vectors.

At what angle should this penguin leave the water to maximize the range of its jump?

What's the speed of an orbiting satellite? How should I leap to win the long-jump competition? How do I engineer a curve in the road for safe driving? These and many other questions involve motion in more than one dimension. In this chapter we extend the ideas of one-dimensional motion to these more complex—and more interesting—situations.

3.1 Vectors

We've seen that quantities describing motion have direction as well as magnitude. In Chapter 2, a simple plus or minus sign took care of direction. But now, in two or three dimensions, we need a way to account for all possible directions. We do this with mathematical quantities called **vectors**, which express both magnitude and direction. Vectors stand in contrast to **scalars**, which are quantities that have no direction.

Position and Displacement

The simplest vector quantity is position. Given an origin, we can characterize any position in space by drawing an arrow from the origin to that position. That arrow is a pictorial representation of a **position vector**, which we call $\vec{r}$. The arrow over the r indicates that this is a vector quantity, and it's crucial to include the arrow whenever you're dealing with vectors. Figure 3.1 shows a position vector in a two-dimensional coordinate system; this vector describes a point a distance of 2 m from the origin, in a direction 30° from the horizontal axis.

Suppose you walk from the origin straight to the point described by the vector $\vec{r}_1$ in Fig. 3.1, and then you turn right and walk another 1 m. Figure 3.2 shows how you can tell where you end up. Draw a second vector whose length represents 1 m and that points to the right; we'll call this vector $\Delta\vec{r}$ because it's a **displacement vector**, representing a *change* in position. Put the tail of $\Delta\vec{r}$ at the head of the vector $\vec{r}_1$; then the head of $\Delta\vec{r}$ shows your ending position. The result is the same as if you had walked straight from the origin to this position. So the new position is described by a third vector $\vec{r}_2$, as indicated in Fig. 3.2. What we've just described is **vector addition**. To add two vectors, put the second vector's tail at the head of the first; the sum is then the vector that extends from the tail of the first vector to the head of the second, as does $\vec{r}_2$ in Fig. 3.2.

A vector has both magnitude and direction—but because that's all the information it contains, it doesn't matter where it starts. So you're free to move a vector around to form vector sums. Figure 3.3 shows some examples of vector addition and also shows that vector addition obeys simple rules you know for regular arithmetic.

The vector $\vec{r}_1$ describes the position of this point.
O is the arbitrary origin.
$\vec{r}_1$
2.0 m
30°
O

FIGURE 3.1 A position vector $\vec{r}_1$.

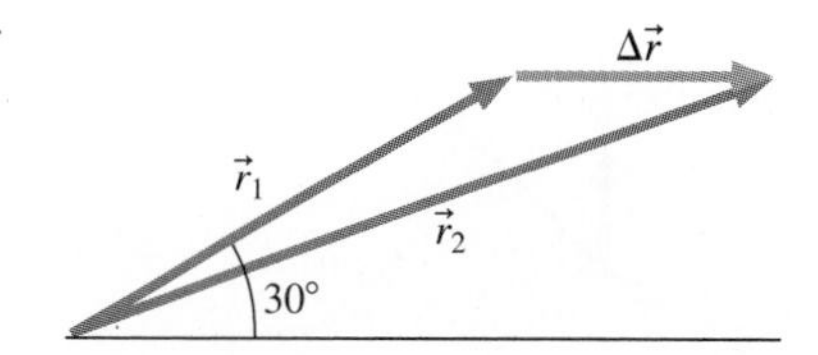

FIGURE 3.2 Vectors $\vec{r}_1$ and $\Delta\vec{r}$ sum to $\vec{r}_2$.

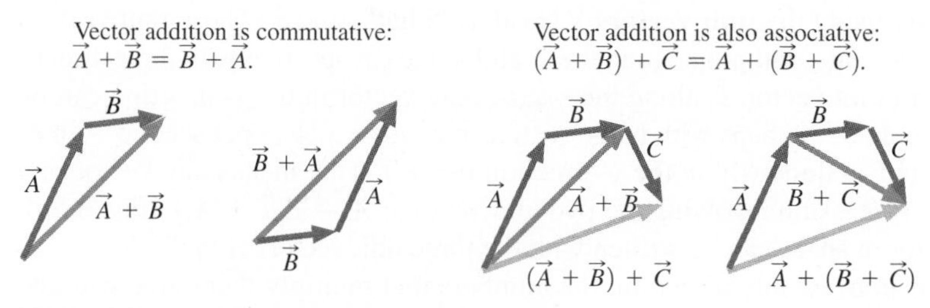

FIGURE 3.3 Vector addition is commutative and associative.

PhET: Vector Addition

Multiplication

You and I jog in the same direction, but you go twice as far. Your displacement vector, $\vec{B}$, is twice as long as my displacement vector, $\vec{A}$; mathematically, $\vec{B} = 2\vec{A}$. That's what it means to multiply a vector by a scalar; simply rescale the magnitude of the vector by that scalar. If the scalar is negative, then the vector direction reverses—and that provides a way to subtract vectors. In Fig. 3.2, for example, you can see that $\vec{r}_1 = \vec{r}_2 + (-1)\Delta\vec{r}$, or simply $\vec{r}_1 = \vec{r}_2 - \Delta\vec{r}$. Later, we'll see ways to multiply two vectors, but for now the only multiplication we consider is a vector multiplied by a scalar.

Vector Components

You can always add vectors graphically, as shown in Fig. 3.2, or you can use geometric relationships like the laws of sines and cosines to accomplish the same thing algebraically. In both these approaches, you specify a vector by giving its magnitude and direction. But often it's more convenient instead to describe vectors using their **components** in a given coordinate system.

A **coordinate system** is a framework for describing positions in space. It's a mathematical construct, and you're free to choose whatever coordinate system you want. You've already seen **Cartesian** or **rectangular coordinate systems**, in which a pair of numbers (x, y) represents each point in a plane. You could also think of each point as representing the head of a position vector, in which case the numbers x and y are the vector components. The components tell how much of the vector is in the x-direction and how much is in the y-direction. Not all vectors represent actual positions in space; for example, there are velocity, acceleration, and force vectors. The lengths of these vectors represent the magnitudes of the corresponding physical quantities. For an arbitrary vector quantity $\vec{A}$, we designate the components A_x and A_y (Fig. 3.4). Note that the components themselves aren't vectors but scalars.

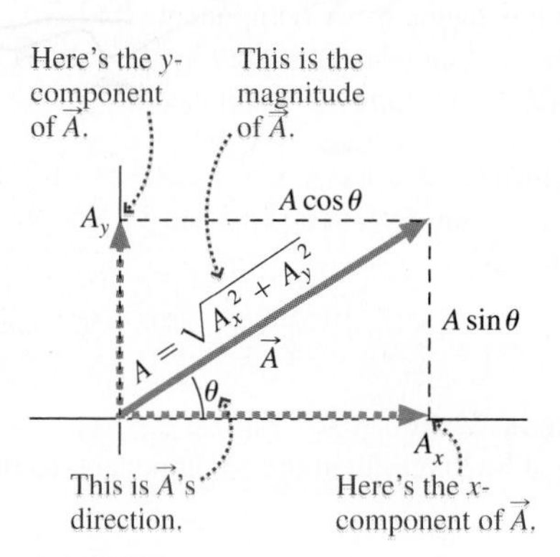

FIGURE 3.4 Magnitude/direction and component representations of vector $\vec{A}$.

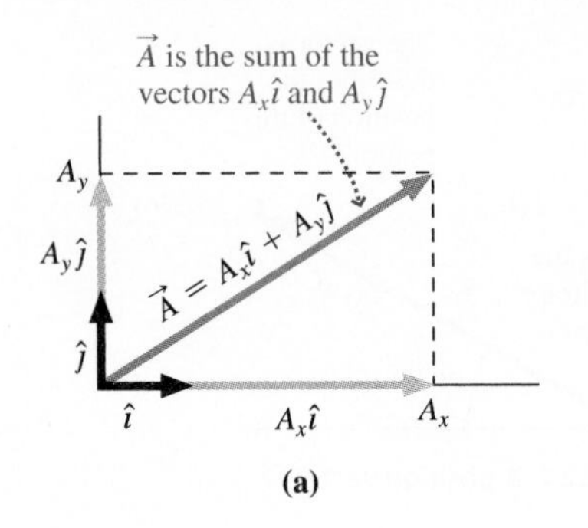

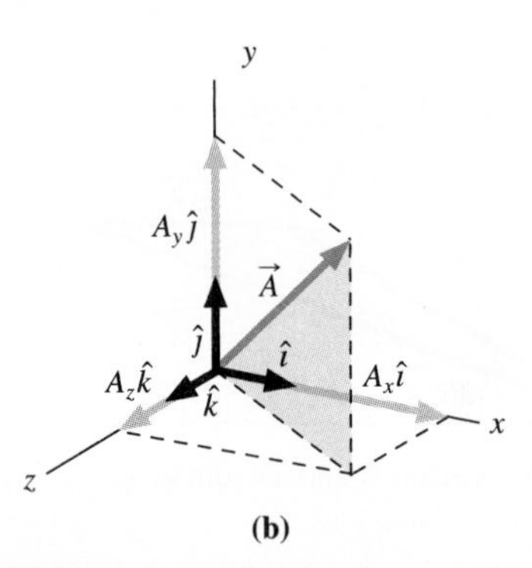

FIGURE 3.5 Vectors in (a) a plane and (b) space, expressed using unit vectors.

In two dimensions it takes two quantities to specify a vector—either its magnitude and direction or its components. They're related by the Pythagorean theorem and the definitions of the trig functions, as shown in Fig. 3.4:

$$A = \sqrt{A_x^2 + A_y^2} \quad \text{and} \quad \tan\theta = \frac{A_y}{A_x} \qquad \text{(vector magnitude and direction)} \quad (3.1)$$

Without the arrow above it, a vector's symbol stands for the vector's magnitude. Going the other way, we have

$$A_x = A\cos\theta \quad \text{and} \quad A_y = A\sin\theta \qquad \text{(vector components)} \quad (3.2)$$

If a vector $\vec{A}$ has zero magnitude, we write $\vec{A} = \vec{0}$, where the vector arrow on the zero indicates that both components must be zero.

Unit Vectors

It's cumbersome to say "a vector of magnitude 2 m at 30° to the x-axis" or, equivalently, "a vector whose x- and y-components are 1.73 m and 1.0 m, respectively." We can express this more succinctly using the **unit vectors** $\hat{\imath}$ (read as "i hat") and $\hat{\jmath}$. These unit vectors have magnitude 1, no units, and point along the x- and y-axes, respectively. In three dimensions we add a third unit vector, $\hat{k}$, along the z-axis. Any vector in the x-direction can be written as some number—perhaps with units, such as meters or meters per second—times the unit vector $\hat{\imath}$, and analogously in the y-direction using $\hat{\jmath}$. That means any vector in a plane can be written as a sum involving the two unit vectors: $\vec{A} = A_x\hat{\imath} + A_y\hat{\jmath}$ (Fig. 3.5*a*). Similarly, any vector in space can be written with the three unit vectors (Fig. 3.5*b*).

The unit vectors convey only direction; the numbers that multiply them give size and units. Together they provide compact representations of vectors, including units. The displacement vector $\vec{r}_1$ in Fig. 3.1, for example, is $\vec{r}_1 = 1.7\hat{\imath} + 1.0\hat{\jmath}$ m.

EXAMPLE 3.1 Unit Vectors: Taking a Drive

You drive to a city 160 km from home, going 35° north of east. Express your new position in unit vector notation, using an east–west/north–south coordinate system.

INTERPRET We interpret this as a problem about writing a vector in unit vector notation, given its magnitude and direction.

DEVELOP Unit vector notation multiplies a vector's x- and y-components by the unit vectors $\hat{\imath}$ and $\hat{\jmath}$ and sums the results; so we draw a sketch showing those components (Fig. 3.6). Our plan is to solve for the two components, multiply by the unit vectors, and then add. Equations 3.2 determine the components.

EVALUATE We have $x = r\cos\theta = (160\text{ km})(\cos 35°) = 131$ km and $y = r\sin\theta = (160\text{ km})(\sin 35°) = 92$ km. Then the position of the city is

$$\vec{r} = 131\hat{\imath} + 92\hat{\jmath}\text{ km}$$

ASSESS Make sense? Figure 3.6 suggests that the x-component should be longer than the y component, as our answer indicates. Our sketch shows the component values and the final answer. Note that we treat $131\hat{\imath} + 92\hat{\jmath}$ as a single vector quantity, labeling it at the end with the appropriate unit, km. ■

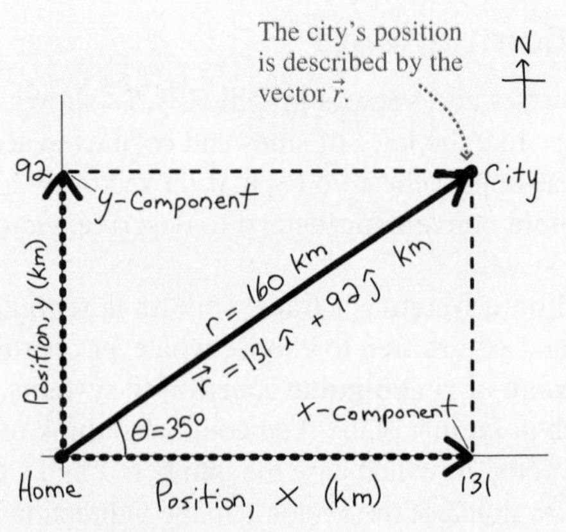

FIGURE 3.6 Our sketch for Example 3.1.

Vector Arithmetic with Unit Vectors

Vector addition is simple with unit vectors: Just add the corresponding components. If $\vec{A} = A_x\hat{\imath} + A_y\hat{\jmath}$ and $\vec{B} = B_x\hat{\imath} + B_y\hat{\jmath}$, for example, then their sum is

$$\vec{A} + \vec{B} = (A_x\hat{\imath} + A_y\hat{\jmath}) + (B_x\hat{\imath} + B_y\hat{\jmath}) = (A_x + B_x)\hat{\imath} + (A_y + B_y)\hat{\jmath}$$

Subtraction and multiplication by a scalar are similarly straightforward.

GOT IT? 3.1 Which vector describes a displacement of 10 units in a direction 30° below the positive x-axis? (a) $10\hat{\imath} - 10\hat{\jmath}$; (b) $5.0\hat{\imath} - 8.7\hat{\jmath}$; (c) $8.7\hat{\imath} - 5.0\hat{\jmath}$; (d) $10(\hat{\imath} + \hat{\jmath})$

3.2 Velocity and Acceleration Vectors

We defined velocity in one dimension as the rate of change of position. In two or three dimensions it's the same thing, except now the change in position—displacement—is a vector. So we write

$$\bar{\vec{v}} = \frac{\Delta\vec{r}}{\Delta t} \quad \text{(average velocity vector)} \qquad (3.3)$$

for the average velocity, in analogy with Equation 2.1. Here division by Δt simply means multiplying by $1/\Delta t$. As before, instantaneous velocity is given by a limiting process:

$$\vec{v} = \lim_{\Delta t \to 0} \frac{\Delta\vec{r}}{\Delta t} = \frac{d\vec{r}}{dt} \quad \text{(instantaneous velocity vector)} \qquad (3.4)$$

Again, that derivative $d\vec{r}/dt$ is shorthand for the result of the limiting process, taking ever smaller time intervals Δt and the corresponding displacements $\Delta\vec{r}$. Another way to look at Equation 3.4 is in terms of components. If $\vec{r} = x\hat{\imath} + y\hat{\jmath}$, then we can write

$$\vec{v} = \frac{d\vec{r}}{dt} = \frac{dx}{dt}\hat{\imath} + \frac{dy}{dt}\hat{\jmath} = v_x\hat{\imath} + v_y\hat{\jmath}$$

where the velocity components v_x and v_y are the derivatives of the position components.

Acceleration is the rate of change of velocity, so we write

$$\bar{\vec{a}} = \frac{\Delta\vec{v}}{\Delta t} \quad \text{(average acceleration vector)} \qquad (3.5)$$

for the average acceleration and

$$\vec{a} = \lim_{\Delta t \to 0} \frac{\Delta\vec{v}}{\Delta t} = \frac{d\vec{v}}{dt} \quad \text{(instantaneous acceleration vector)} \qquad (3.6)$$

for the instantaneous acceleration. We can also express instantaneous acceleration in components, as we did for velocity:

$$\vec{a} = \frac{d\vec{v}}{dt} = \frac{dv_x}{dt}\hat{\imath} + \frac{dv_y}{dt}\hat{\jmath} = a_x\hat{\imath} + a_y\hat{\jmath}$$

Velocity and Acceleration in Two Dimensions

Motion in a straight line may or may not involve acceleration, but motion on curved paths in two or three dimensions is *always* accelerated motion. Why? Because moving in multiple dimensions means *changing direction*—and *any* change in velocity, including direction, involves acceleration. Get used to thinking of acceleration as meaning more than "speeding up" or "slowing down." It can equally well mean "changing direction," whether or not speed is also changing. Whether acceleration results in a speed change, a direction change, or both depends on the relative orientation of the velocity and acceleration vectors.

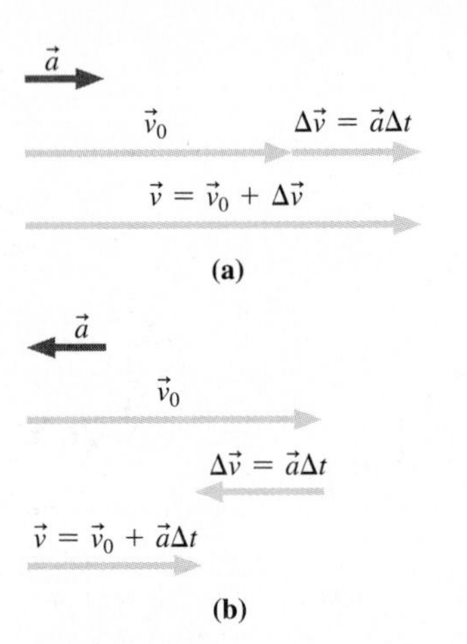

FIGURE 3.7 When $\vec{v}$ and $\vec{a}$ are colinear, only the speed changes.

Suppose you're driving down a straight road at speed v_0 when you step on the gas to give a constant acceleration $\vec{a}$ for a time Δt. Equation 3.5 shows that the change in your velocity is $\Delta\vec{v} = \vec{a}\Delta t$. In this case the acceleration is in the same direction as your velocity and, as Fig. 3.7*a* shows, the result is an increase in the magnitude of your velocity; that is, you speed up. Step on the brake, and your acceleration is opposite your velocity, and you slow down (Fig. 3.7*b*).

✓**TIP** Vectors Tell It All

Are you thinking there should be a minus sign in Fig. 3.7*b* because the speed is decreasing? Nope: Vectors have both magnitude and direction, and the vector addition $\vec{v} = \vec{v}_0 + \vec{a}\Delta t$ tells it all. In Fig. 3.7*b*, $\Delta\vec{v}$ points to the left, and that takes care of the "subtraction."

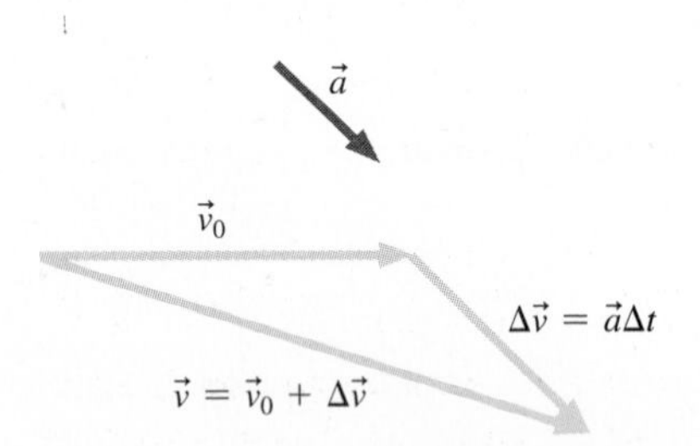

FIGURE 3.8 In general, acceleration changes both the magnitude and the direction of velocity.

In two dimensions acceleration and velocity can be at any angle. In general, acceleration then changes both the magnitude and the direction of the velocity (Fig. 3.8). Particularly interesting is the case when $\vec{a}$ is perpendicular to $\vec{v}$; then only the direction of motion changes. If acceleration is constant—in both magnitude and direction—then the two vectors won't stay perpendicular once the direction of $\vec{v}$ starts to change, and the magnitude will change, too. But in the special case where acceleration changes direction so it's always perpendicular to velocity, then it's strictly true that only the direction of motion changes. Figure 3.9 illustrates this point, which we'll soon explore quantitatively.

GOT IT? 3.2 An object is accelerating downward. Which, if any, of the following must be true? (a) the object cannot be moving upward; (b) the object cannot be moving in a straight line; (c) the object is moving directly downward; (d) if the object's motion is instantaneously horizontal, it can't continue to be so

3.3 Relative Motion

You stroll down the aisle of a plane, walking toward the front at a leisurely 4 km/h. Meanwhile the plane is moving relative to the ground at 1000 km/h. Therefore, you're moving at 1004 km/h relative to the ground. As this example suggests, velocity is meaningful only when we know the answer to the question, "Velocity relative to what?" That "what" is called a **frame of reference**. Often we know an object's velocity relative to one frame of reference—for example, your velocity relative to the plane—and we want to know its velocity relative to some other reference frame—in this case the ground. In this one-dimensional case, we can simply add the two velocities. If you had been walking toward the back of the plane, then the two velocities would have opposite signs and you would be going at 996 km/h relative to the ground.

The same idea works in two dimensions, but here we need to recognize that velocity is a vector. Suppose that airplane is flying with velocity $\vec{v}\,'$ relative to the air. If a wind is blowing, then the air is moving with some velocity $\vec{V}$ relative to the ground. The plane's velocity $\vec{v}$ relative to the ground is the vector sum of its velocity relative to the air and the air's velocity relative to the ground:

$$\vec{v} = \vec{v}\,' + \vec{V} \quad \text{(relative velocity)} \tag{3.7}$$

Here we use lowercase letters for the velocities of an object relative to two different reference frames; we distinguish the two with the prime on one of the velocities. The capital $\vec{V}$ is the relative velocity between the two frames. In general, Equation 3.7 lets us use the velocity of an object in one reference frame to find its velocity relative to another frame—provided we know that relative velocity $\vec{V}$. Example 3.2 illustrates the application of this idea to aircraft navigation.

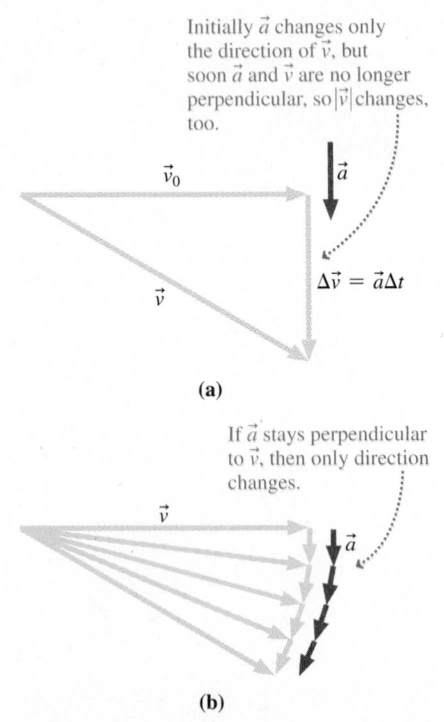

FIGURE 3.9 Acceleration that is always perpendicular to velocity changes only the direction.

EXAMPLE 3.2 Relative Velocity: Navigating a Jetliner

A jetliner flies at 960 km/h relative to the air. It's going from Houston to Omaha, 1290 km northward. At cruising altitude a wind is blowing eastward at 190 km/h. In what direction should the plane fly? How long will the trip take?

INTERPRET This is a problem involving relative velocities. We identify the given information: the plane's speed, but not its direction, in the reference frame of the air; the plane's direction, but not its speed, in the reference frame of the ground; and the wind velocity, both speed and direction.

DEVELOP Equation 3.7, $\vec{v} = \vec{v}' + \vec{V}$, applies, and we identify $\vec{v}$ as the plane's velocity relative to the ground, $\vec{v}'$ as its velocity relative to the air, and $\vec{V}$ as the wind velocity. Equation 3.7 shows that $\vec{v}'$ and $\vec{V}$ add vectorially to give $\vec{v}$ that, with the given information, helps us draw the situation (Fig. 3.10). Measuring the angle of $\vec{v}'$ and the length of $\vec{v}$ in the diagram would then give the answers. However, we'll work the problem algebraically using vector components. Since the plane is flying northward and the wind is blowing eastward, a suitable coordinate system has x-axis eastward and y-axis northward. Our plan is to work out the vector components in these coordinates and then apply Equation 3.7.

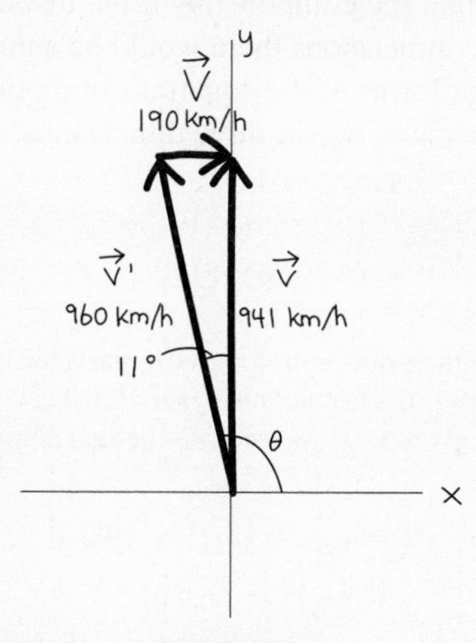

FIGURE 3.10 Our vector diagram for Example 3.2.

EVALUATE Using Equations 3.2 for the vector components, we can express the three vectors as

$$\vec{v}' = v' \cos\theta\hat{\imath} + v' \sin\theta\hat{\jmath}, \quad \vec{V} = V\hat{\imath}, \quad \text{and} \quad \vec{v} = v\hat{\jmath}$$

Here we know the magnitude v' of the velocity $\vec{v}'$, but we don't know the angle θ. We know the magnitude V of the wind velocity $\vec{V}$, and we also know its direction—toward the east. So $\vec{V}$ has only an x-component. Meanwhile we want the velocity $\vec{v}$ relative to the ground to be purely northward, so it has only a y-component—although we don't know its magnitude v. We're now ready to put the three velocities into Equation 3.7. Since two vectors are equal only if all their components are equal, we can express the vector Equation 3.7 as two separate scalar equations for the x- and y-components:

$$x\text{-component:} \quad v' \cos\theta + V = 0$$

$$y\text{-component:} \quad v' \sin\theta + 0 = v$$

The rest is math, evaluating the unknowns θ and v. Solving the x equation gives

$$\theta = \cos^{-1}\left(-\frac{V}{v'}\right) = \cos^{-1}\left(-\frac{190 \text{ km/h}}{960 \text{ km/h}}\right) = 101.4°$$

This angle is measured from the x-axis (eastward; see Fig. 3.10), so it amounts to a flight path 11° west of north. We can then evaluate v from the y equation:

$$v = v' \sin\theta = (960 \text{ km/h})(\sin 101.4°) = 941 \text{ km/h}$$

That's the plane's speed relative to the ground. Going 1290 km will then take $(1290 \text{ km})/(941 \text{ km/h}) = 1.4$ h.

ASSESS Make sense? The plane's heading of 11° west of north seems reasonable compensation for an eastward wind blowing at 190 km/h, given the plane's airspeed of 960 km/h. If there were no wind, the trip would take 1 h, 20 min (1290 km divided by 960 km/h), so our time of 1 h, 24 min with the wind makes sense. ■

GOT IT? 3.3 An airplane is making a 500-km trip directly north that is supposed to take exactly 1 h. For 100-km/h winds blowing in each of the directions (1), (2), and (3) shown, does the plane's speed relative to the air need to be (a) less than, (b) equal to, or (c) greater than 500 km/h?

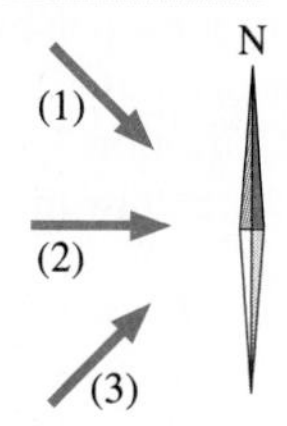

3.4 Constant Acceleration

When acceleration is constant, the individual components of the acceleration vector are themselves constant. Furthermore, the component of acceleration in one direction has no effect on the motion in a perpendicular direction (Fig. 3.11, next page). Then with constant acceleration, the separate components of the motion must obey the constant-acceleration formulas we developed in Chapter 2 for one-dimensional motion. Using vector notation, we can then generalize Equations 2.7 and 2.10 to read

$$\vec{v} = \vec{v}_0 + \vec{a}t \quad \text{(for constant acceleration only)} \tag{3.8}$$

$$\vec{r} = \vec{r}_0 + \vec{v}_0 t + \tfrac{1}{2}\vec{a}t^2 \quad \text{(for constant acceleration only)} \tag{3.9}$$

Video Tutor Demo | **Dropped and Thrown Balls**

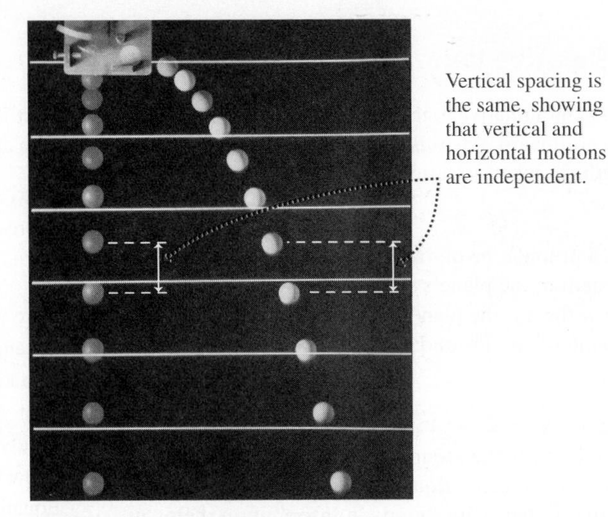

FIGURE 3.11 Two marbles, one dropped and the other projected horizontally.

where $\vec{r}$ is the position vector. In two dimensions, each of these vector equations represents a pair of scalar equations describing constant acceleration in two mutually perpendicular directions. Equation 3.9, for example, contains the pair $x = x_0 + v_{x0}t + \frac{1}{2}a_xt^2$ and $y = y_0 + v_{y0}t + \frac{1}{2}a_yt^2$. (Remember that the components of the displacement vector $\vec{r}$ are just the coordinates x and y.) In three dimensions there would be a third equation for the z-component. Starting with these vector forms of the equations of motion, you can apply Problem-Solving Strategy 2.1 to problems in two or three dimensions.

EXAMPLE 3.3 Acceleration in Two Dimensions: Windsurfing

You're windsurfing at 7.3 m/s when a gust hits, accelerating your sailboard at 0.82 m/s^2 at 60° to your original direction. If the gust lasts 8.7 s, what's the board's displacement during this time?

INTERPRET This is a problem involving constant acceleration in two dimensions. The key concept is that motion in perpendicular directions is independent, so we can treat the problem as involving two separate one-dimensional motions.

DEVELOP Equation 3.9, $\vec{r} = \vec{r}_0 + \vec{v}_0t + \frac{1}{2}\vec{a}t^2$, will give the board's displacement. We need a coordinate system, so we take the x-axis along the board's initial motion, with the origin at the point where the gust first hits. Our plan is to find the components of the acceleration vector and then apply the two components of Equation 3.9 to get the components of the displacement. In Fig. 3.12 we draw the acceleration vector to determine its components.

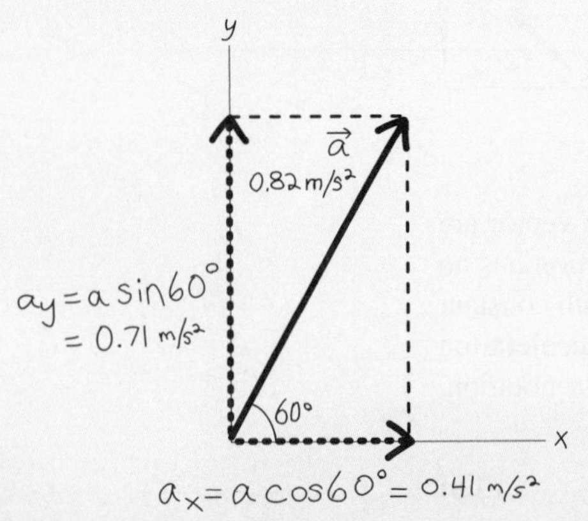

FIGURE 3.12 Our sketch of the sailboard's acceleration components.

EVALUATE With the x-direction along the initial velocity, $\vec{v}_0 = 7.3\hat{\imath}$ m/s. As Fig. 3.12 shows, the acceleration is $\vec{a} = 0.41\hat{\imath} + 0.71\hat{\jmath}$ m/s^2. Our choice of origin gives $x_0 = y_0 = 0$, so the two components of Equation 3.9 are

$$x = v_{x0}t + \tfrac{1}{2}a_xt^2 = 79.0 \text{ m}$$

$$y = \tfrac{1}{2}a_yt^2 = 26.9 \text{ m}$$

where we used the appropriate components of $\vec{a}$ and where $t = 8.7$ s. The new position vector is then $\vec{r} = x\hat{\imath} + y\hat{\jmath} = 79.0\hat{\imath} + 26.9\hat{\jmath}$ m, giving a net displacement of $r = \sqrt{x^2 + y^2} = 83$ m.

ASSESS Make sense? Figure 3.13 shows how the acceleration deflects the sailboard from its original path and also increases its speed somewhat. Since the acceleration makes a fairly large angle with the initial velocity, the change in direction is the greater effect. ■

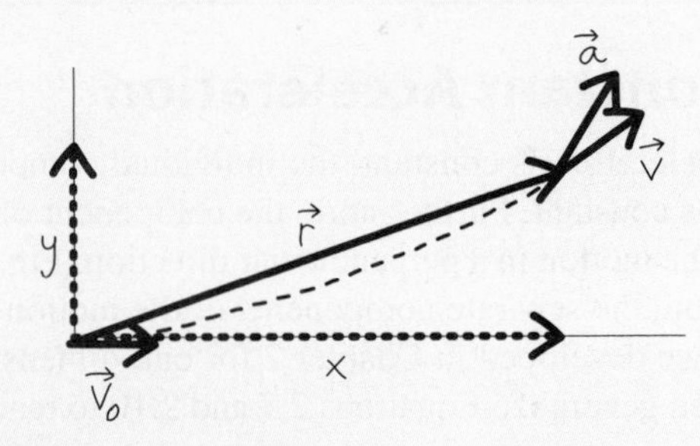

FIGURE 3.13 Our sketch of the displacement $\vec{r}$, velocity $\vec{v}$, and acceleration $\vec{a}$ at the end of the wind gust. The actual path of the sailboard during the gust is indicated by the dashed curve.

GOT IT? 3.4 An object is moving initially in the $+x$-direction. Which of the following accelerations, all acting for the same time interval, will cause the greatest change in its speed? In its direction? (a) $10\hat{\imath}$ m/s^2; (b) $10\hat{\jmath}$ m/s^2; (c) $10\hat{\imath} + 5\hat{\jmath}$ m/s^2; (d) $2\hat{\imath} - 8\hat{\jmath}$ m/s^2

Video Tutor Demo | **Ball Fired from Cart on Incline**

Video Tutor Demo | **Ball Fired Upward from Accelerating Cart**

PhET: Projectile Motion

3.5 Projectile Motion

A **projectile** is an object that's launched into the air and then moves predominantly under the influence of gravity. Examples are numerous; baseballs, jets of water (Fig. 3.14), fireworks, missiles, ejecta from volcanoes, drops of ink in an ink-jet printer, and leaping dolphins are all projectiles.

To treat projectile motion, we make two simplifying assumptions: (1) We neglect any variation in the direction or magnitude of the gravitational acceleration, and (2) we neglect air resistance. The first assumption is equivalent to neglecting Earth's curvature, and is valid for projectiles whose displacements are small compared with Earth's radius. Air resistance has a more variable effect; for dense, compact objects it's often negligible, but for objects whose ratio of surface area to mass is large—like ping-pong balls and parachutes—air resistance dramatically alters the motion.

To describe projectile motion, it's convenient to choose a coordinate system with the y-axis vertically upward and the x-axis horizontal. With the only acceleration provided by gravity, $a_x = 0$ and $a_y = -g$, so the components of Equations 3.8 and 3.9 become

$$v_x = v_{x0} \quad (3.10)$$
$$v_y = v_{y0} - gt \quad (3.11)$$
$$x = x_0 + v_{x0}t \quad (3.12)$$
$$y = y_0 + v_{y0}t - \tfrac{1}{2}gt^2 \quad (3.13)$$
(for constant gravitational acceleration)

We take g to be positive, and account for the downward direction using minus signs. Equations 3.10–3.13 tell us mathematically what Fig. 3.15 tells us physically: Projectile motion comprises two perpendicular and independent components—horizontal motion with constant velocity and vertical motion with constant acceleration.

FIGURE 3.14 Water droplets–each an individual projectile–combine to form graceful parabolic arcs in this fountain.

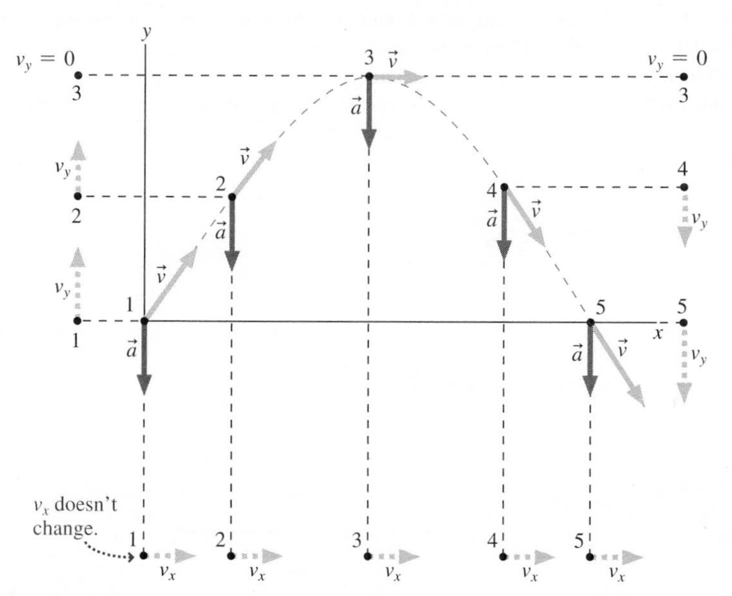

FIGURE 3.15 Velocity and acceleration at five points on a projectile's path. Also shown are horizontal and vertical components.

PROBLEM-SOLVING STRATEGY 3.1 Projectile Motion

INTERPRET Make sure that you have a problem involving the constant acceleration of gravity near Earth's surface, and that the motion involves both horizontal and vertical components. Identify the object or objects in question and whatever initial or final positions and velocities are given. Know what quantities you're being asked to find.

DEVELOP Establish a horizontal/vertical coordinate system, and write the separate components of the equations of motion (Equations 3.10–3.13). The equations for different components will be linked by a common variable—namely, time. Draw a sketch showing the initial motion and a rough trajectory.

EVALUATE Solve your individual equations simultaneously for the unknowns of the problem.

ASSESS Check that your answer makes sense. Consider special cases, like purely vertical or horizontal initial velocities. Because the equations of motion are quadratic in time, you may have two answers. One answer may be the one you want, but you gain more insight into physics if you consider the meaning of the second answer, too.

EXAMPLE 3.4 Finding the Horizontal Distance: Washout!

A raging flood has washed away a section of highway, creating a gash 1.7 m deep. A car moving at 31 m/s goes straight over the edge. How far from the edge of the washout does it land?

INTERPRET This is a problem involving projectile motion, and it asks for the horizontal distance the car moves after it leaves the road. We're given the car's initial speed and direction (horizontal) and the distance it falls.

DEVELOP Figure 3.16*a* shows the situation, and we've sketched the essentials in Fig. 3.16*b*. Since there's no horizontal acceleration, Equation 3.12, $x = x_0 + v_{x0}t$, would determine the unknown distance if we knew the time. But horizontal and vertical motions are independent, so we can find the time until the car hits the ground from the vertical motion alone, as determined by Equation 3.13, $y = y_0 + v_{y0}t - \frac{1}{2}gt^2$. So our plan is to get the time from Equation 3.13 and then use that time in Equation 3.12 to get the horizontal distance. If we choose the origin as the bottom of the washout, then $y_0 = 1.7$ m. Then we want the time when $y = 0$.

EVALUATE With $v_{y0} = 0$, we solve Equation 3.13 for t:

$$t = \sqrt{\frac{2y_0}{g}} = \sqrt{\frac{(2)(1.7\text{ m})}{(9.8\text{ m/s}^2)}} = 0.589\text{ s}$$

During this time the car continues to move horizontally at $v_{x0} = 31$ m/s, so Equation 3.12 gives $x = v_{x0}t = (31\text{ m/s})(0.589\text{ s}) = 18$ m.

Note that we carried three significant figures in our intermediate answer for the time t to avoid roundoff error in our final two-significant-figure answer. Alternatively, we could have kept the time in symbolic form, $t = \sqrt{2y_0/g}$. Often you can gain more physical insight from an answer that's expressed symbolically before you put in the numbers.

ASSESS Make sense? About half a second to drop 1.7 m or about 6 ft seems reasonable, and at 31 m/s an object will go somewhat farther than 15 m in this time. ■

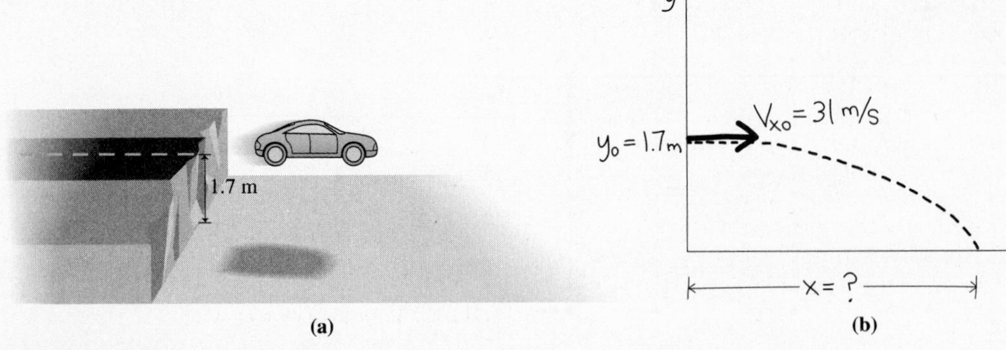

FIGURE 3.16 (a) The highway and car, and (b) our sketch.

✓TIP Multistep Problems

Example 3.4 asked for the horizontal distance the car traveled. For that we needed the time—which we weren't given. This is a common situation in all but the simplest physics problems. You need to work through several steps to get the answer—in this case solving first for the unknown time and then for the distance. In essence, we solved two problems in Example 3.4: the first involving vertical motion and the second horizontal motion.

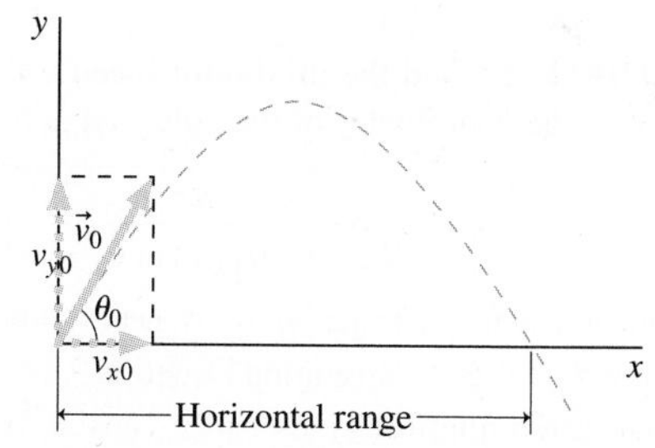

FIGURE 3.17 Parabolic trajectory of a projectile.

Projectile Trajectories

We're often interested in the path, or **trajectory**, of a projectile without the details of where it is at each instant of time. We can specify the trajectory by giving the height y as a function of the horizontal position x. Consider a projectile launched from the origin at some angle θ_0 to the horizontal, with initial speed v_0. As Fig. 3.17 suggests, the components of the initial velocity are $v_{x0} = v_0 \cos\theta_0$ and $v_{y0} = v_0 \sin\theta_0$. Then Equations 3.12 and 3.13 become

$$x = v_0 \cos\theta_0 t \quad \text{and} \quad y = v_0 \sin\theta_0 t - \tfrac{1}{2}gt^2$$

Solving the x equation for the time t gives

$$t = \frac{x}{v_0 \cos\theta_0}$$

Using this result in the y equation, we have

$$y = v_0 \sin\theta_0\left(\frac{x}{v_0\cos\theta_0}\right) - \tfrac{1}{2}g\left(\frac{x}{v_0\cos\theta_0}\right)^2$$

or

$$y = x\tan\theta_0 - \frac{g}{2v_0^2\cos^2\theta_0}x^2 \quad \text{(projectile trajectory)} \qquad (3.14)$$

Equation 3.14 gives a mathematical description of the projectile's trajectory. Since y is a quadratic function of x, the trajectory is a parabola.

APPLICATION **Pop Flies, Line Drives, and Hang Times**

Although air resistance significantly influences baseball trajectories, to a first approximation baseballs behave like projectiles. For a given speed off the bat, this means a pop fly's "hang time" is much greater than that of a nearly horizontal line drive, and that makes the fly ball much easier to catch (see photo).

EXAMPLE 3.5 Finding the Trajectory: Out of the Hole

A construction worker stands in a 2.6-m-deep hole, 3.1 m from the edge of the hole. He tosses a hammer to a companion outside the hole. If the hammer leaves his hand 1.0 m above the bottom of the hole at an angle of 35°, what's the minimum speed it needs to clear the edge of the hole? How far from the edge of the hole does it land?

INTERPRET We're concerned about *where* an object is but not *when*, so we interpret this as a problem about the trajectory—specifically, the minimum-speed trajectory that just grazes the edge of the hole.

DEVELOP We draw the situation in Fig. 3.18. Equation 3.14 determines the trajectory, so our plan is to find the speed that makes the trajectory pass just over the edge of the hole at $x = 3.1$ m, $y = 1.6$ m, where Fig. 3.18 shows that we've chosen a coordinate system with its origin at the worker's hand.

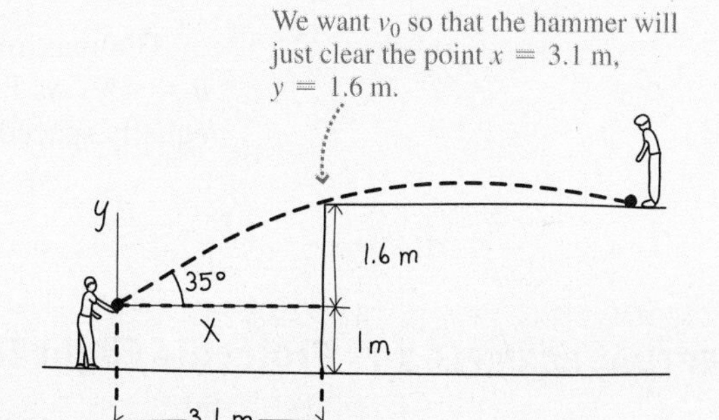

FIGURE 3.18 Our sketch for Example 3.5.

(continued)

EVALUATE To find the minimum speed we solve Equation 3.14 for v_0, using the coordinates of the hole's edge for x and y:

$$v_0 = \sqrt{\frac{gx^2}{2\cos^2\theta_0\,(x\tan\theta_0 - y)}} = 11\text{ m/s}$$

To find where the hammer lands, we need to know the horizontal position when $y = 1.6$ m. Rearranging Equation 3.14 into the standard form for a quadratic equation gives $(g/2v_0^2\cos^2\theta_0)x^2 - (\tan\theta_0)x + y = 0$. Applying the quadratic formula (Appendix A) gives $x = 3.1$ m and $x = 8.7$ m; the second value is the one we want. That 8.7 m is the distance from our origin at the worker's hand, and amounts to 8.7 m − 3.1 m = 5.6 m from the hole's edge.

ASSESS Make sense? The other answer to the quadratic, $x = 3.1$ m, is a clue that we did the problem correctly. That 3.1 m is the distance to the edge of the hole. The fact that we get this position when we ask for a vertical height of 1.6 m confirms that the trajectory does indeed just clear the edge of the hole. ■

The Range of a Projectile

How far will a soccer ball go if I kick it at 12 m/s at 50° to the horizontal? If I can throw a rock at 15 m/s, can I get it across a 30-m-wide pond? How far off vertical can a rocket's trajectory be and still land within 50 km of its launch point? As in these examples, we're frequently interested in the **horizontal range** of a projectile—that is, how far it moves horizontally over level ground.

For a projectile launched on level ground, we can determine when the projectile will return to the ground by setting $y = 0$ in Equation 3.14:

$$0 = x\tan\theta_0 - \frac{g}{2v_0^2\cos^2\theta_0}x^2 = x\left(\tan\theta_0 - \frac{gx}{2v_0^2\cos^2\theta_0}\right)$$

There are two solutions: $x = 0$, corresponding to the launch point, and

$$x = \frac{2v_0^2}{g}\cos^2\theta_0\tan\theta_0 = \frac{2v_0^2}{g}\sin\theta_0\cos\theta_0$$

But $\sin 2\theta_0 = 2\sin\theta_0\cos\theta_0$, so this becomes

$$x = \frac{v_0^2}{g}\sin 2\theta_0 \qquad \text{(horizontal range)} \qquad (3.15)$$

Here the particle returns to its starting height, so Equation 3.15 applies.

(a)

Here the particle lands at a different height, so Equation 3.15 doesn't apply.

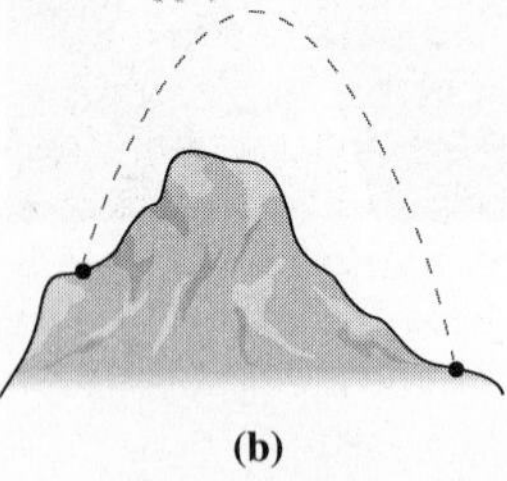

(b)

FIGURE 3.19 Equation 3.15 applies in (a) but not in (b).

✓TIP Know Your Limits

We emphasize that Equation 3.15 gives the *horizontal* range—the distance a projectile travels horizontally before returning *to its starting height*. From the way it was derived—setting $y = 0$—you can see that it does *not* give the horizontal distance when the projectile returns to a different height (Fig. 3.19).

The maximum range occurs when $\sin 2\theta = 1$ in Equation 3.15, which occurs when $\theta = 45°$. As Fig. 3.20 suggests, the range for a given launch speed v_0 is equal for angles equally spaced on either side of 45°—as you can prove in Problem 70.

CONCEPTUAL EXAMPLE 3.1 **Projectile Flight Times**

The ranges in Fig. 3.20 are equal for angles on either side of 45°. How do the flight times compare?

EVALUATE We're being asked about the times projectiles spend on the trajectories shown. Since horizontal and vertical motions are independent, flight time depends on how high the projectile goes. So we can argue from the vertical motions that the trajectory with the higher launch angle takes longer. We can also argue from horizontal motions: Horizontal distances of the paired trajectories are the same, but the lower trajectory has a greater horizontal velocity component, so again the lower trajectory takes less time.

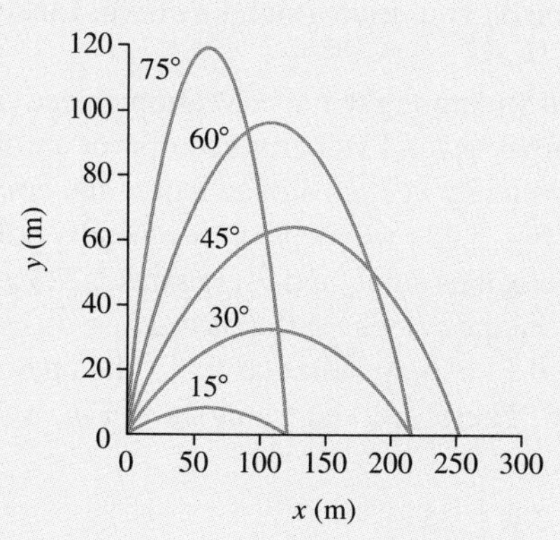

FIGURE 3.20 Trajectories for a projectile launched at 50 m/s.

ASSESS Consider the extreme cases of near-vertical and near-horizontal trajectories. The former goes nearly straight up and down, taking a relatively long time but returning essentially to its starting point. The latter hardly gets anywhere because it immediately hits the ground right at its starting point, so it takes just about no time!

MAKING THE CONNECTION Find the flight times for the 30° and 60° trajectories in Fig. 3.20.

EVALUATE The range of Equation 3.15 is also equal to the horizontal velocity v_x multiplied by the time: $v_x t = v_0^2 \sin 2\theta_0/g$. Using $v_{x0} = v_0 \cos \theta_0$ and solving for t gives $t = 2v_0 \sin \theta_0/g$. Using Fig. 3.20's $v_0 = 50$ m/s yields $t_{30} = 5.1$ s and $t_{60} = 8.8$ s. You can explore this time difference more generally in Problem 65.

✓TIP Know the Fundamentals

Equations 3.14 and 3.15 for a projectile's trajectory and range are useful, but they're not fundamental equations of physics. Both follow directly from the equations for constant acceleration. If you think that specialized results like Equations 3.14 and 3.15 are on an equal footing with more fundamental equations and principles, then you're seeing physics as a hodgepodge of equations and missing the big picture of a science with a few underlying principles from which all else follows.

Video Tutor Demo | **Range of a Gun at Two Firing Angles**

EXAMPLE 3.6 Projectile Range: Probing the Atmosphere

After a short engine firing, an atmosphere-probing rocket reaches 4.6 km/s. If the rocket must land within 50 km of its launch site, what's the maximum allowable deviation from a vertical trajectory?

INTERPRET Although we're asked about the launch angle, the 50-km criterion is a clue that we can interpret this as a problem about the horizontal range. That "short engine firing" means we can neglect the distance over which the rocket fires and consider it a projectile that leaves the ground at $v_0 = 4.6$ km/s.

DEVELOP Equation 3.15, $x = (v_0^2/g) \sin 2\theta_0$, determines the horizontal range, so our plan is to solve that equation for θ_0 with range $x = 50$ km.

EVALUATE We have $\sin 2\theta_0 = gx/v_0^2 = 0.0232$. There are two solutions, corresponding to $2\theta_0 = 1.33°$ and $2\theta_0 = 180° - 1.33°$. The second is the one we want, giving a launch angle $\theta_0 = 90° - 0.67°$. Therefore the launch angle must be within 0.67° of vertical.

ASSESS Make sense? At 4.6 km/s, this rocket goes quite high, so with even a small deviation from vertical it will land far from its launch point. Again we've got two solutions. The one we rejected is like the low trajectories of Fig. 3.20; although it gives a 50-km range, it isn't going to get our rocket high into the atmosphere. ■

GOT IT? 3.5 Two projectiles are launched simultaneously from the same point on a horizontal surface, one at 45° to the horizontal and the other at 60°. Their launch speeds are different and are chosen so that the two projectiles travel the same horizontal distance before landing. Which of the following statements is true? (a) A and B land at the same time; (b) B's launch speed is lower than A's and B lands sooner; (c) B's launch speed is lower than A's and B lands later; (d) B's launch speed is higher than A's and B lands sooner; or (e) B's launch speed is higher than A's and B lands later.

3.6 Uniform Circular Motion

An important case of accelerated motion in two dimensions is **uniform circular motion**—that of an object describing a circular path at constant speed. Although the speed is constant, the motion is accelerated because the *direction* of the velocity is changing.

PhET: Ladybug Motion 2D
PhET: Motion in 2D

Uniform circular motion is common. Many spacecraft are in circular orbits, and the orbits of the planets are approximately circular. Earth's daily rotation carries you around in

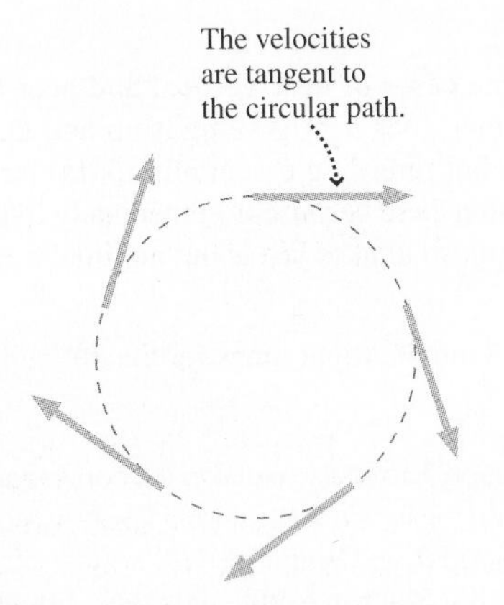

FIGURE 3.21 Velocity vectors in circular motion are tangent to the circular path.

uniform circular motion. Pieces of rotating machinery describe uniform circular motion, and you're temporarily in circular motion as you drive around a curve. Electrons undergo circular motion in magnetic fields.

Here we derive an important relationship among the acceleration, speed, and radius of uniform circular motion. Figure 3.21 shows several velocity vectors for an object moving with speed v around a circle of radius r. Velocity vectors are tangent to the circle, indicating the instantaneous direction of motion. In Fig. 3.22*a* we focus on two nearby points described by position vectors $\vec{r}_1$ and $\vec{r}_2$, where the velocities are $\vec{v}_1$ and $\vec{v}_2$. Figures 3.22*b* and *c* show the corresponding displacement $\Delta\vec{r} = \vec{r}_2 - \vec{r}_1$ and velocity difference $\Delta\vec{v} = \vec{v}_2 - \vec{v}_1$.

Because $\vec{v}_1$ is perpendicular to $\vec{r}_1$, and $\vec{v}_2$ is perpendicular to $\vec{r}_2$, the angles θ shown in all three parts of Fig. 3.22 are the same. Therefore, the triangles in Fig. 3.22*b* and *c* are similar, and we can write

$$\frac{\Delta v}{v} = \frac{\Delta r}{r}$$

Now suppose the angle θ is small, corresponding to a short time interval Δt for motion from position $\vec{r}_1$ to $\vec{r}_2$. Then the length of the vector $\Delta\vec{r}$ is approximately the length of the circular arc joining the endpoints of the position vectors, as suggested in Fig. 3.22*b*. The length of this arc is the distance the object travels in the time Δt, or $v\Delta t$, so $\Delta r \simeq v\Delta t$. Then the relation between similar triangles becomes

$$\frac{\Delta v}{v} \simeq \frac{v\,\Delta t}{r}$$

Rearranging this equation gives an approximate expression for the magnitude of the average acceleration:

$$\bar{a} = \frac{\Delta v}{\Delta t} \simeq \frac{v^2}{r}$$

Taking the limit $\Delta t \to 0$ gives the instantaneous acceleration; in this limit the angle θ approaches 0, the circular arc and $\Delta\vec{r}$ become indistinguishable, and the relation $\Delta r \simeq v\,\Delta t$ becomes exact. So we have

$$a = \frac{v^2}{r} \quad \text{(uniform circular motion)} \tag{3.16}$$

for the magnitude of the instantaneous acceleration of an object moving in a circle of radius r at constant speed v. What about its direction? As Fig. 3.22*c* suggests, $\Delta\vec{v}$ is very nearly perpendicular to both velocity vectors; in the limit $\Delta t \to 0$, $\Delta\vec{v}$ and the acceleration $\Delta\vec{v}/\Delta t$ become exactly perpendicular to the velocity. The direction of the acceleration vector is therefore toward the center of the circle.

Our geometric argument would work for any point on the circle, so we conclude that the acceleration has constant magnitude v^2/r and always points toward the center of the circle. Isaac Newton coined the term *centripetal* to describe this center-pointing acceleration. However, we'll use that term sparingly because we want to emphasize that centripetal acceleration is fundamentally no different from any other acceleration: It's simply a vector describing the rate of change of velocity.

Does Equation 3.16 make sense? Yes. An increase in speed v means the time Δt for a given change in direction of the velocity becomes shorter. Not only that, but the associated change $\Delta\vec{v}$ in velocity is larger. These two effects combine to give an acceleration that depends on the *square* of the speed. On the other hand, an increase in the radius with a fixed speed increases the time Δt associated with a given change in velocity, so the acceleration is inversely proportional to the radius.

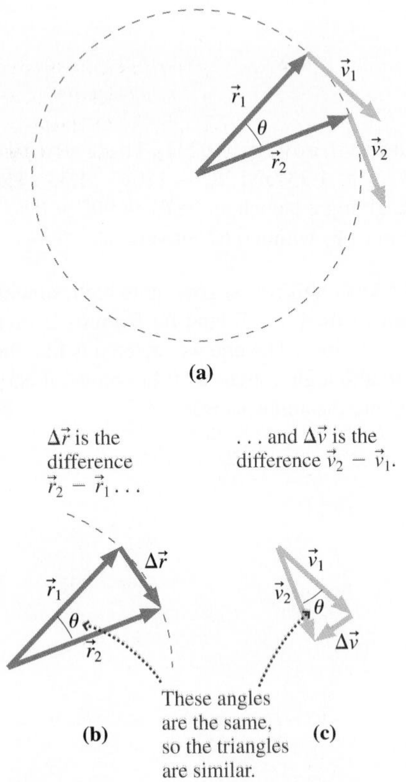

(a)

$\Delta\vec{r}$ is the difference $\vec{r}_2 - \vec{r}_1$ . . .

. . . and $\Delta\vec{v}$ is the difference $\vec{v}_2 - \vec{v}_1$.

These angles are the same, so the triangles are similar.

(b) (c)

FIGURE 3.22 Position and velocity vectors for two nearby points on the circular path.

✓TIP Circular Motion and Constant Acceleration

The direction toward the center changes as an object moves around a circular path, so the acceleration vector is *not constant*, even though its magnitude is. Uniform circular motion is *not* motion with constant acceleration, and our constant-acceleration equations *do not apply*. In fact, we know that constant acceleration in two dimensions implies a parabolic trajectory, not a circle.

EXAMPLE 3.7 Uniform Circular Motion: The International Space Station

Find the orbital period (the time to complete one orbit) of the International Space Station in its circular orbit at altitude 400 km, where the acceleration of gravity is 89% of its surface value.

INTERPRET This is a problem about uniform circular motion.

DEVELOP Given the radius and acceleration, we could use Equation 3.16, $a = v^2/r$, to determine the orbital speed. But we're given the altitude, not the orbital radius, and we want the period, not the speed. So our plan is to write the speed in terms of the period and use the result in Equation 3.16. The orbital altitude is the distance from Earth's surface, so we'll need to add Earth's radius to get the orbital radius r.

EVALUATE The speed v is the orbital circumference, $2\pi r$, divided by the period T. Using this in Equation 3.16 gives

$$a = \frac{v^2}{r} = \frac{(2\pi r/T)^2}{r} = \frac{4\pi^2 r}{T^2}$$

Appendix E lists Earth's radius as $R_E = 6.37$ Mm, giving an orbital radius $r = R_E + 400\text{ km} = 6.77$ Mm. Solving our acceleration expression for the period then gives $T = \sqrt{4\pi^2 r/a} = 5536\text{ s} = 92\text{ min}$, where we used $a = 0.89g$.

ASSESS Make sense? Astronauts orbit Earth in about an hour and a half, experiencing multiple sunrises and sunsets in a 24-hour day. Our answer of 92 min is certainly consistent with that. There's no choice here; for a given orbital radius, Earth's size and mass determine the period. Because astronauts' orbits are limited to a few hundred kilometers, a distance small compared with R_E, variations in g and T are minimal. Any such "low Earth orbit" has a period of approximately 90 min. At higher altitudes, gravity diminishes significantly and periods lengthen; the Moon, for example, orbits in 27 days. We'll discuss orbits more in Chapter 8. ■

EXAMPLE 3.8 Uniform Circular Motion: Engineering a Road

An engineer is designing a flat, horizontal road for an 80 km/h speed limit (that's 22.2 m/s). If the maximum acceleration of a vehicle on this road is 1.5 m/s^2, what's the minimum safe radius for curves in the road?

INTERPRET Even though a curve is only a portion of a circle, we can still interpret this problem as involving uniform circular motion.

DEVELOP Equation 3.16, $a = v^2/r$, gives the acceleration in terms of the speed and radius. Here we have the acceleration and speed, so our plan is to solve for the radius.

EVALUATE Using the given numbers, we have $r = v^2/a = (22.2\text{ m/s})^2/1.5\text{ m/s}^2 = 329$ m.

ASSESS Make sense? A speed of 80 km/h is pretty fast, so we need a wide curve to keep the required acceleration below its design value. If the curve is sharper, vehicles may slide off the road. We'll see more clearly in subsequent chapters how vehicles manage to negotiate high-speed curves. ■

Nonuniform Circular Motion

What if an object moves in a circular path but its speed changes? Then it has components of acceleration both perpendicular and parallel to its velocity. The former, the **radial acceleration** a_r, is what changes the direction to keep the object in circular motion. Its magnitude is still v^2/r, with v now the instantaneous speed. The parallel component of acceleration, also called **tangential acceleration** a_t because it's tangent to the circle, changes the speed but not the direction. Its magnitude is therefore the rate of change of speed, or dv/dt. Figure 3.23 shows these two acceleration components for a car rounding a curve. We'll explore these two components of acceleration further in Chapter 10, when we study rotational motion.

Finally, what if the radius of a curved path changes? At any point on a curve we can define a **radius of curvature**. Then the radial acceleration is still v^2/r, and it can vary if either v or r changes along the curve. The tangential acceleration is still tangent to the curve, and it still describes the rate of change of speed. So it's straightforward to generalize the ideas of uniform circular motion to cases where the motion is nonuniform either because the speed changes, or because the radius changes, or both.

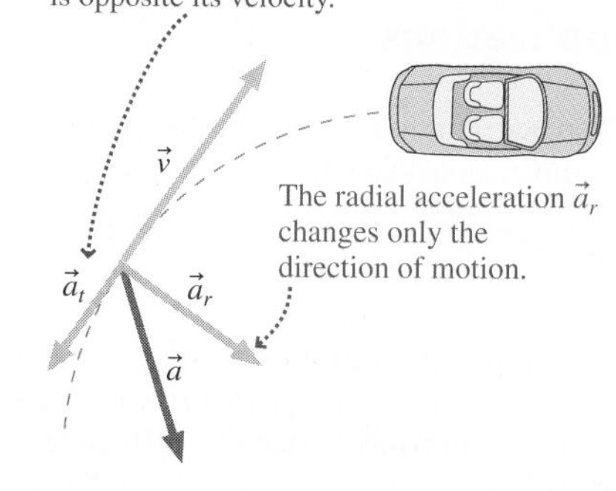

FIGURE 3.23 Acceleration of a car that slows as it rounds a curve.

GOT IT? 3.6 An object moves in a horizontal plane with constant speed on the path shown. At which marked point is the magnitude of its acceleration greatest?

A B C D E

CHAPTER 3 SUMMARY

Big Idea

Quantities characterizing motion in two and three dimensions have both **magnitude** and **direction** and are described by **vectors**. Position, velocity, and acceleration are all vector quantities, related as they are in one dimension:

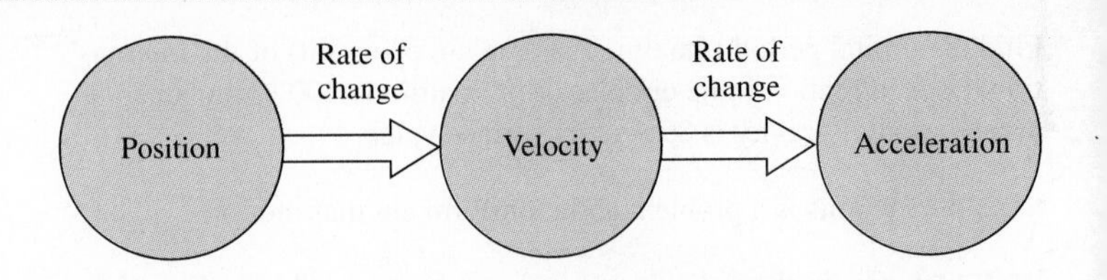

These vector quantities need not have the same direction. In particular, acceleration that's perpendicular to velocity changes the direction but not the magnitude of the velocity. Acceleration that's colinear changes only the magnitude of the velocity. In general, both change.

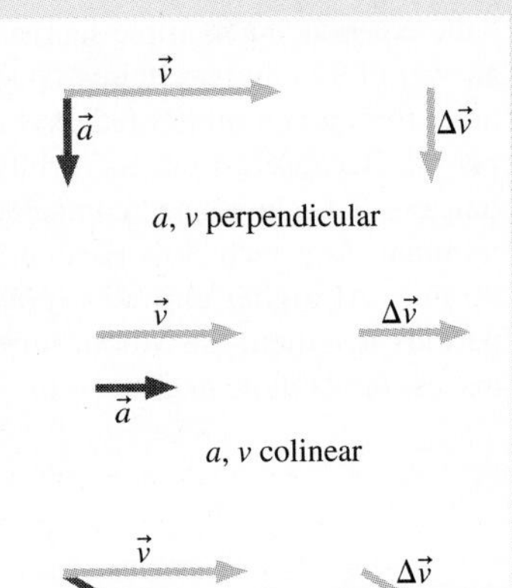

Components of motion in two perpendicular directions are independent. This reduces problems in two and three dimensions to sets of one-dimensional problems that can be solved with the methods of Chapter 2.

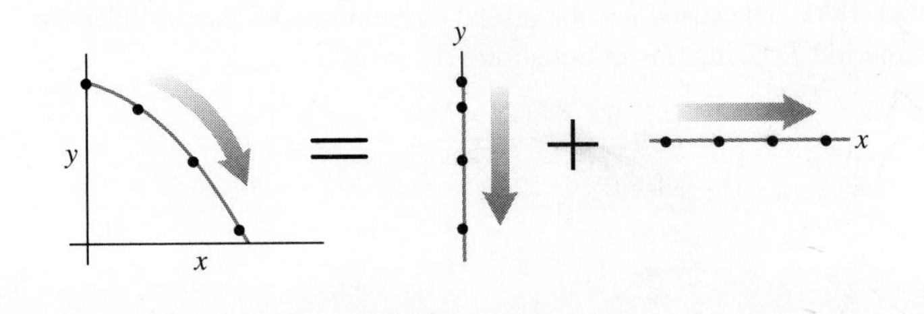

Key Concepts and Equations

Vectors can be described by magnitude and direction or by components. In two dimensions these representations are related by

$$A = \sqrt{A_x^2 + A_y^2} \quad \text{and} \quad \theta = \tan^{-1}\frac{A_y}{A_x}$$

$$A_x = A\cos\theta \quad \text{and} \quad A_y = A\sin\theta$$

A compact way to express vectors involves unit vectors that have magnitude 1, have no units, and point along the coordinate axes:

$$\vec{A} = A_x\hat{\imath} + A_y\hat{\jmath}$$

Velocity is the rate of change of the position vector $\vec{r}$:

$$\vec{v} = \frac{d\vec{r}}{dt}$$

Acceleration is the rate of change of velocity:

$$\vec{a} = \frac{d\vec{v}}{dt}$$

Applications

When acceleration is constant, motion is described by vector equations that generalize the one-dimensional equations of Chapter 2:

$$\vec{v} = \vec{v}_0 + \vec{a}t \qquad \vec{r} = \vec{r}_0 + \vec{v}_0 t + \tfrac{1}{2}\vec{a}t^2$$

An important application of constant-acceleration motion in two dimensions is **projectile motion** under the influence of gravity.

Projectile trajectory:

$$y = x\tan\theta_0 - \frac{g}{2v_0^2\cos^2\theta_0}x^2$$

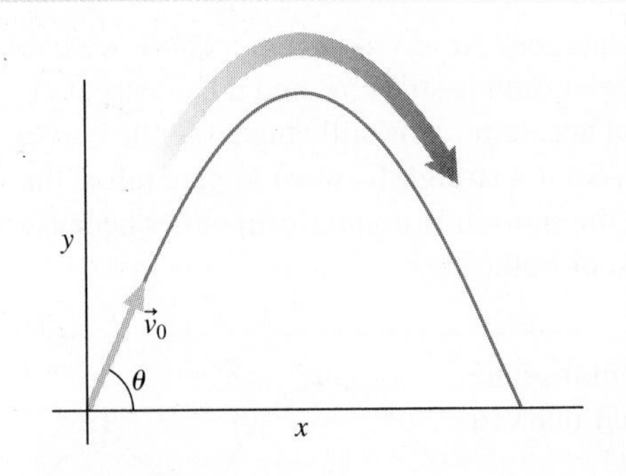

In **uniform circular motion** the magnitudes of velocity and acceleration remain constant, but their directions continually change. For an object moving in a circular path of radius r, the magnitudes of $\vec{a}$ and $\vec{v}$ are related by $a = v^2/r$.

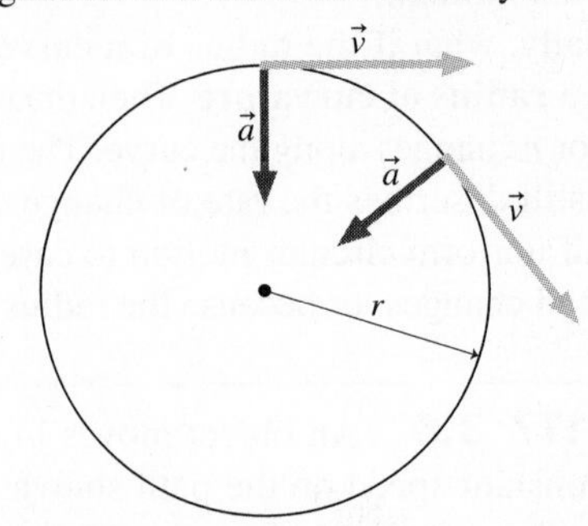

MP For homework assigned on MasteringPhysics, go to www.masteringphysics.com

BIO *Biology and/or medicine-related problems* **DATA** *Data problems* **ENV** *Environmental problems* **CH** *Challenge problems* **COMP** *Computer problems*

For Thought and Discussion

1. Under what conditions is the magnitude of the vector sum $\vec{A} + \vec{B}$ equal to the sum of the magnitudes of the two vectors?
2. Can two vectors of equal magnitude sum to zero? How about two vectors of unequal magnitude?
3. Repeat Question 2 for three vectors.
4. Can an object have a southward acceleration while moving northward? A westward acceleration while moving northward?
5. You're a passenger in a car rounding a curve. The driver claims the car isn't accelerating because the speedometer reading is unchanging. Explain why the driver is wrong.
6. In what sense is Equation 3.8 really two (or three) equations?
7. Is a projectile's speed constant throughout its parabolic trajectory?
8. Is there any point on a projectile's trajectory where velocity and acceleration are perpendicular?
9. How is it possible for an object to be moving in one direction but accelerating in another?
10. You're in a bus moving with constant velocity on a level road when you throw a ball straight up. When the ball returns, does it land ahead of you, behind you, or back at your hand? Explain.
11. Which of the following are legitimate mathematical equations? Explain. (a) $v = 5\hat{\imath}$ m/s; (b) $\vec{v} = 5$ m/s; (c) $\vec{a} = dv/dt$; (d) $\vec{a} = d\vec{v}/dt$; (e) $\vec{v} = 5\hat{\imath}$ m/s.
12. You would probably reject as unscientific any claim that Earth is flat. Yet the assumption of Section 3.5 that leads to parabolic projectile trajectories is tantamount to assuming a flat Earth. Explain.

Exercises and Problems

Exercises

Section 3.1 Vectors

13. You walk west 220 m, then north 150 m. What are the magnitude and direction of your displacement vector?
14. An ion in a mass spectrometer follows a semicircular path of radius 15.2 cm. What are (a) the distance it travels and (b) the magnitude of its displacement?
15. A migrating whale follows the west coast of Mexico and North America toward its summer home in Alaska. It first travels 360 km northwest to just off the coast of northern California, and then turns due north and travels 400 km toward its destination. Determine graphically the magnitude and direction of its displacement.
16. Vector $\vec{A}$ has magnitude 3.0 m and points to the right; vector $\vec{B}$ has magnitude 4.0 m and points vertically upward. Find the magnitude and direction of vector $\vec{C}$ such that $\vec{A} + \vec{B} + \vec{C} = \vec{0}$.
17. Use unit vectors to express a displacement of 120 km at 29° counterclockwise from the x-axis.
18. Find the magnitude of the vector $34\hat{\imath} + 13\hat{\jmath}$ m and determine its angle to the x-axis.
19. (a) What's the magnitude of $\hat{\imath} + \hat{\jmath}$? (b) What angle does it make with the x-axis?

Section 3.2 Velocity and Acceleration Vectors

20. You're leading an international effort to save Earth from an asteroid heading toward us at 15 km/s. Your team mounts a rocket on the asteroid and fires it for 10 min, after which the asteroid is moving at 19 km/s at 28° to its original path. In a news conference, what do you report for the magnitude of the acceleration imparted to the asteroid?
21. An object is moving at 18 m/s at 220° counterclockwise from the x-axis. Find the x- and y-components of its velocity.
22. A car drives north at 40 mi/h for 10 min, then turns east and goes 5.0 mi at 60 mi/h. Finally, it goes southwest at 30 mi/h for 6.0 min. Determine the car's (a) displacement and (b) average velocity for this trip.
23. An object's velocity is $\vec{v} = ct^3\hat{\imath} + d\hat{\jmath}$, where t is time and c and d are positive constants with appropriate units. What's the direction of the object's acceleration?
24. A car, initially going eastward, rounds a 90° curve and ends up heading southward. If the speedometer reading remains constant, what's the direction of the car's average acceleration vector?
25. What are (a) the average velocity and (b) the average acceleration of the tip of the 2.4-cm-long hour hand of a clock in the interval from noon to 6 PM? Use unit vector notation, with the x-axis pointing toward 3 and the y-axis toward noon.
26. An ice skater is gliding along at 2.4 m/s, when she undergoes an acceleration of magnitude 1.1 m/s^2 for 3.0 s. After that she's moving at 5.7 m/s. Find the angle between her acceleration vector and her initial velocity. *Hint*: You don't need to do a complicated calculation.
27. An object is moving in the x-direction at 1.3 m/s when it undergoes an acceleration $\vec{a} = 0.52\hat{\jmath}$ m/s^2. Find its velocity vector after 4.4 s.

Section 3.3 Relative Motion

28. You're a pilot beginning a 1500-km flight. Your plane's speed is 1000 km/h, and air traffic control says you'll have to head 15° west of south to maintain a southward course. If the flight takes 100 min, what's the wind velocity?
29. You wish to row straight across a 63-m-wide river. You can row at a steady 1.3 m/s relative to the water, and the river flows at 0.57 m/s. (a) What direction should you head? (b) How long will it take you to cross the river?
30. A plane with airspeed 370 km/h flies perpendicularly across the jet stream, its nose pointed into the jet stream at 32° from the perpendicular direction of its flight. Find the speed of the jet stream.
31. A flock of geese is attempting to migrate due south, but the wind is blowing from the west at 5.1 m/s. If the birds can fly at 7.5 m/s relative to the air, what direction should they head?

Section 3.4 Constant Acceleration

32. The position of an object as a function of time is given by $\vec{r} = (3.2t + 1.8t^2)\hat{\imath} + (1.7t - 2.4t^2)\hat{\jmath}$ m, with t in seconds. Find the object's acceleration vector.
33. You're sailboarding at 6.5 m/s when a wind gust hits, lasting 6.3 s accelerating your board at 0.48 m/s^2 at 35° to your original direction. Find the magnitude and direction of your displacement during the gust.

Section 3.5 Projectile Motion

34. You toss an apple horizontally at 8.7 m/s from a height of 2.6 m. Simultaneously, you drop a peach from the same height. How long does each take to reach the ground?

35. A carpenter tosses a shingle horizontally off an 8.8-m-high roof at 11 m/s. (a) How long does it take the shingle to reach the ground? (b) How far does it move horizontally?
36. An arrow fired horizontally at 41 m/s travels 23 m horizontally. From what height was it fired?
37. Droplets in an ink-jet printer are ejected horizontally at 12 m/s and travel a horizontal distance of 1.0 mm to the paper. How far do they fall in this interval?
38. Protons drop 1.2 μm over the 1.7-km length of a particle accelerator. What's their approximate average speed?
39. If you can hit a golf ball 180 m on Earth, how far can you hit it on the Moon? (Your answer will be an underestimate because it neglects air resistance on Earth.)

Section 3.6 Uniform Circular Motion

40. China's high-speed rail network calls for a minimum turn radius of 7.0 km for 350-km/h trains. What's the magnitude of a train's acceleration in this case?
41. The minute hand of a clock is 7.50 cm long. Find the magnitude of the acceleration of its tip.
42. How fast would a car have to round a 75-m-radius turn for its acceleration to be numerically equal to that of gravity?
43. Estimate the acceleration of the Moon, which completes a nearly circular orbit of 384.4 Mm radius in 27 days.
44. Global Positioning System (GPS) satellites circle Earth at altitudes of approximately 20,000 km, where the gravitational acceleration has 5.8% of its surface value. To the nearest hour, what's the orbital period of the GPS satellites?

Problems

45. Two vectors $\vec{A}$ and $\vec{B}$ have the same magnitude A and are at right angles. Find the magnitudes of (a) $\vec{A} + 2\vec{B}$ and (b) $3\vec{A} - \vec{B}$.
46. Vector $\vec{A}$ has magnitude 1.0 m and points 35° clockwise from the x-axis. Vector $\vec{B}$ has magnitude 1.8 m. Find the direction of $\vec{B}$ such that $\vec{A} + \vec{B}$ is in the y-direction.
47. Let $\vec{A} = 15\hat{\imath} - 40\hat{\jmath}$ and $\vec{B} = 31\hat{\jmath} + 18\hat{k}$. Find $\vec{C}$ such that $\vec{A} + \vec{B} + \vec{C} = \vec{0}$.
48. **BIO** A biologist looking through a microscope sees a bacterium at $\vec{r}_1 = 2.2\hat{\imath} + 3.7\hat{\jmath} - 1.2\hat{k}$ μm. After 6.2 s, it's located at $\vec{r}_2 = 4.6\hat{\imath} + 1.9\hat{k}$ μm. Find (a) its average velocity, expressed in unit vectors, and (b) its average speed.
49. A particle's position is $\vec{r} = (ct^2 - 2dt^3)\hat{\imath} + (2ct^2 - dt^3)\hat{\jmath}$, where c and d are positive constants. Find expressions for times $t > 0$ when the particle is moving in (a) the x-direction and (b) the y-direction.
50. For the particle in Problem 49, is there any time $t > 0$ when the particle is (a) at rest and (b) accelerating in the x-direction? If either answer is "yes," find the time(s).
51. You're designing a "cloverleaf" highway interchange. Vehicles will exit the highway and slow to a constant 70 km/h before negotiating a circular turn. If a vehicle's acceleration is not to exceed $0.40g$ (i.e., 40% of Earth's gravitational acceleration), then what's the minimum radius for the turn? Assume the road is flat, not banked (more on this in Chapter 5).
52. An object undergoes acceleration $2.3\hat{\imath} + 3.6\hat{\jmath}$ m/s² for 10 s. At the end of this time, its velocity is $33\hat{\imath} + 15\hat{\jmath}$ m/s. (a) What was its velocity at the beginning of the 10-s interval? (b) By how much did its speed change? (c) By how much did its direction change? (d) Show that the speed change is not given by the magnitude of the acceleration multiplied by the time. Why not?
53. The New York Wheel is the world's largest Ferris wheel. It's 183 meters in diameter and rotates once every 37.3 min. Find the magnitudes of (a) the average velocity and (b) the average acceleration at the wheel's rim, over a 5.00-min interval. (c) Compare your answer to (b) with the wheel's instantaneous accelerations.
54. A ferryboat sails between towns directly opposite each other on a river, moving at speed v' relative to the water. (a) Find an expression for the angle it should head at if the river flows at speed V. (b) What's the significance of your answer if $V > v'$?
55. The sum of two vectors, $\vec{A} + \vec{B}$, is perpendicular to their difference, $\vec{A} - \vec{B}$. How do the vectors' magnitudes compare?
56. Write an expression for a unit vector at 45° clockwise from the x-axis.
57. An object is initially moving in the x-direction at 4.5 m/s, when it undergoes an acceleration in the y-direction for a period of 18 s. If the object moves equal distances in the x- and y-directions during this time, what's the magnitude of its acceleration?
58. A particle leaves the origin with its initial velocity given by $\vec{v}_0 = 11\hat{\imath} + 14\hat{\jmath}$ m/s, undergoing constant acceleration $\vec{a} = -1.2\hat{\imath} + 0.26\hat{\jmath}$ m/s². (a) When does the particle cross the y-axis? (b) What's its y-coordinate at the time? (c) How fast is it moving, and in what direction?
59. A kid fires a squirt gun horizontally from 1.6 m above the ground. It hits another kid 2.1 m away square in the back, 0.93 m above the ground. What was the water's initial speed?
60. A projectile has horizontal range R on level ground and reaches maximum height h. Find an expression for its initial speed.
61. You throw a baseball at a 45° angle to the horizontal, aiming at a friend who's sitting in a tree a distance h above level ground. At the instant you throw your ball, your friend drops another ball. (a) Show that the two balls will collide, no matter what your ball's initial speed, provided it's greater than some minimum value. (b) Find an expression for that minimum speed.
62. In a chase scene, a movie stuntman runs horizontally off the flat roof of one building and lands on another roof 1.9 m lower. If the gap between the buildings is 4.5 m wide, how fast must he run to cross the gap?
63. Standing on the ground 3.0 m from a building, you want to throw a package from your 1.5-m shoulder level to someone in a window 4.2 m above the ground. At what speed and angle should you throw the package so it just barely clears the windowsill?
64. Derive a general formula for the horizontal distance covered by a projectile launched horizontally at speed v_0 from height h.
65. Consider two projectiles launched on level ground with the same speed, at angles 45° ± α. Show that the ratio of their flight times is $\tan(\alpha + 45°)$.
66. You toss a protein bar to your hiking companion located 8.6 m up a 39° slope, as shown in Fig. 3.24. Determine the initial velocity vector so that when the bar reaches your friend, it's moving horizontally.

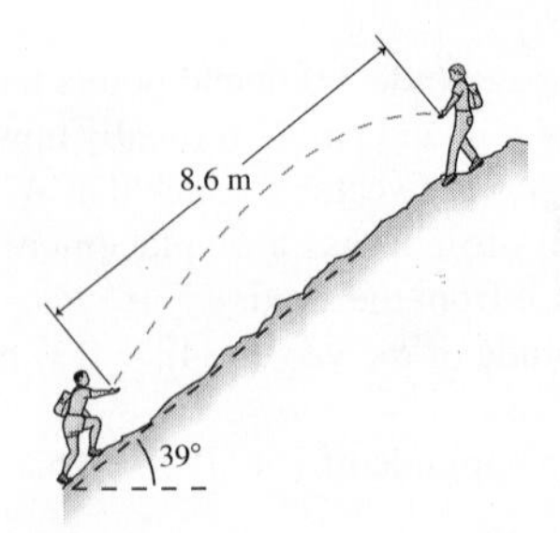

FIGURE 3.24 Problem 66

67. **DATA** The table below lists position versus time for an object moving in the x–y plane, which is horizontal in this case. Make a plot

of position y versus x to determine the nature of the object's path. Then determine the magnitudes of the object's velocity and acceleration.

Time, t (s)	x (m)	y (m)	Time, t (s)	x (m)	y (m)
0	0	0	0.70	2.41	3.15
0.10	0.65	0.09	0.80	2.17	3.75
0.20	1.25	0.33	0.90	1.77	4.27
0.30	1.77	0.73	1.00	1.25	4.67
0.40	2.17	1.25	1.10	0.65	4.91
0.50	2.41	1.85	1.20	0.00	5.00
0.60	2.50	2.50			

68. A projectile launched at angle θ to the horizontal reaches maximum height h. Show that its horizontal range is $4h/\tan\theta$.
69. As an expert witness, you're testifying in a case involving a motorcycle accident. A motorcyclist driving in a 60-km/h zone hit a stopped car on a level road. The motorcyclist was thrown from his bike and landed 39 m down the road. You're asked whether he was speeding. What's your answer?
70. Show that, for a given initial speed, the horizontal range of a projectile is the same for launch angles $45° + \alpha$ and $45° - \alpha$.
71. A basketball player is 15 ft horizontally from the center of the basket, which is 10 ft off the ground. At what angle should the player aim the ball from a height of 8.2 ft with a speed of 26 ft/s?
72. Two projectiles are launched simultaneously from the same point, with different launch speeds and angles. Show that no combination of speeds and angles will permit them to land simultaneously and at the same point.
73. CH Consider the two projectiles in GOT IT? 3.5. Suppose the 45° projectile is launched with speed v and that it's in the air for time t. Find expressions for (a) the launch speed and (b) the flight time of the 60° projectile, in terms of v and t.
74. The portion of a projectile's parabolic trajectory in the vicinity of the peak can be approximated as a circle. If the projectile's speed at the peak of the trajectory is v, formulate an argument to show that the curvature radius of the circle that approximates the parabola is $r = v^2/g$.
75. A jet is diving vertically downward at 1200 km/h. If the pilot can withstand a maximum acceleration of $5g$ (i.e., 5 times Earth's gravitational acceleration) before losing consciousness, at what height must the plane start a 90° circular turn, from vertical to horizontal, in order to pull out of the dive? See Fig. 3.25, assume the speed remains constant, and neglect gravity.

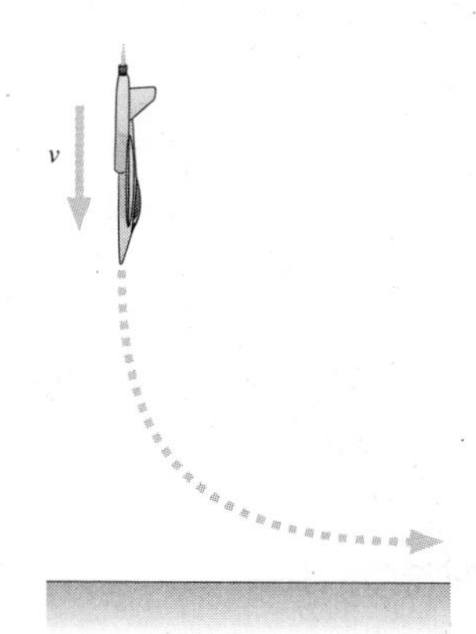

FIGURE 3.25

76. Your alpine rescue team is using a slingshot to send an emergency medical packet to climbers stranded on a ledge, as shown in Fig. 3.26; your job is to calculate the launch speed. What do you report?

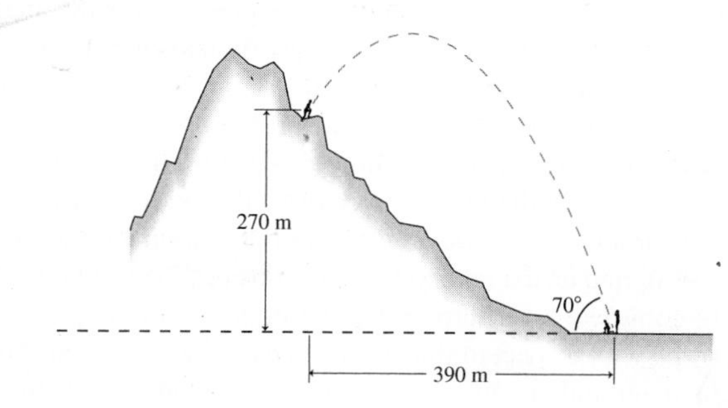

FIGURE 3.26 Problem 76

77. If you can throw a stone straight up to height h, what's the maximum horizontal distance you could throw it over level ground?
78. In a conversion from military to peacetime use, a missile with maximum horizontal range 180 km is being adapted for studying Earth's upper atmosphere. What is the maximum altitude it can achieve if launched vertically?
79. A soccer player can kick the ball 28 m on level ground, with its initial velocity at 40° to the horizontal. At the same initial speed and angle to the horizontal, what horizontal distance can the player kick the ball on a 15° upward slope?
80. A diver leaves a 3-m board on a trajectory that takes her 2.5 m above the board and then into the water 2.8 m horizontally from the end of the board. At what speed and angle did she leave the board?
81. Using calculus, you can find a function's maximum or minimum by differentiating and setting the result to zero. Do this for Equation 3.15, differentiating with respect to θ, and thus verify that the maximum range occurs for $\theta = 45°$.
82. CH You're a consulting engineer specializing in athletic facilities, and you've been asked to help design the Olympic ski jump pictured in Fig. 3.27. Skiers will leave the jump at 28 m/s and 9.5° below the horizontal, and land 55 m horizontally from the end of the jump. Your job is to specify the slope of the ground so skiers' trajectories make an angle of only 3.0° with the ground on landing, ensuring their safety. What slope do you specify?

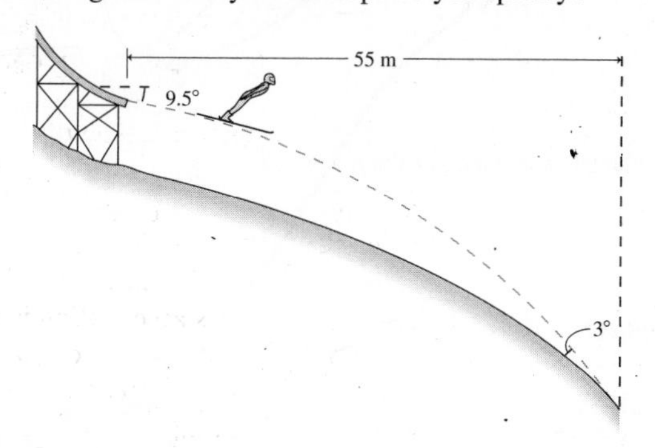

FIGURE 3.27 Problem 82

83. CH Differentiate the trajectory Equation 3.14 to find its slope, $\tan\theta = dy/dx$, and show that the slope is in the direction of the projectile's velocity, as given by Equations 3.10 and 3.11.
84. Your medieval history class is constructing a trebuchet, a catapult-like weapon for hurling stones at enemy castles. The plan is to launch stones off a 75-m-high cliff, with initial speed

36 m/s. Some members of the class think a 45° launch angle will give the maximum range, but others claim the cliff height makes a difference. What do you give for the angle that will maximize the range?

85. Generalize Problem 84 to find an expression for the angle that will maximize the range of a projectile launched with speed v_0 from height h above level ground.
CH

86. (a) Show that the position of a particle on a circle of radius R with its center at the origin is $\vec{r} = R(\cos\theta\hat{\imath} + \sin\theta\hat{\jmath})$, where θ is the angle the position vector makes with the x-axis. (b) If the particle moves with constant speed v starting on the x-axis at $t = 0$, find an expression for θ in terms of time t and the period T to complete a full circle. (c) Differentiate the position vector twice with respect to time to find the acceleration, and show that its magnitude is given by Equation 3.16 and its direction is toward the center of the circle.
CH

87. In dealing with nonuniform circular motion, as shown in Fig. 3.23, we should write Equation 3.16 as $a_r = v^2/r$, to show that this is only the radial component of the acceleration. Recognizing that v is the object's speed, which changes only in the presence of tangential acceleration, differentiate this equation with respect to time to find a relation between the magnitude of the tangential acceleration and the rate of change of the magnitude of the radial acceleration. Assume the radius stays constant.

88. Repeat Problem 87, now generalizing to the case where not only the speed but also the radius may be changing.
CH

Passage Problems

Alice (A), Bob (B), and Carrie (C) all start from their dorm and head for the library for an evening study session. Alice takes a straight path, while the paths Bob and Carrie follow are portions of circular arcs, as shown in Fig. 3.28. Each student walks at a constant speed. All three leave the dorm at the same time, and they arrive simultaneously at the library.

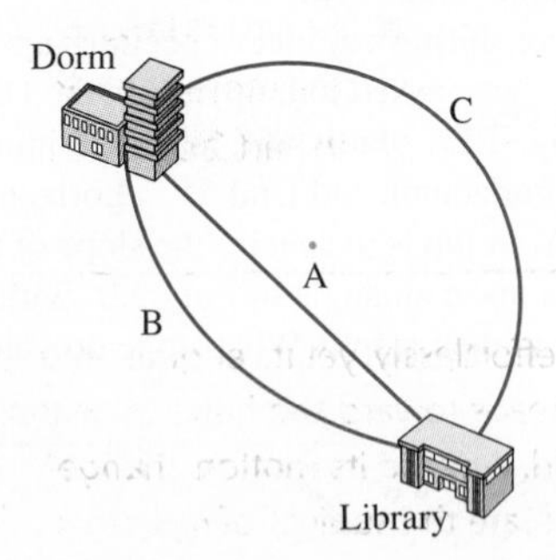

FIGURE 3.28 Passage Problems 89–92

89. Which statement characterizes the distances the students travel?
    a. They're equal.
    b. C > A > B
    c. C > B > A
    d. B > C > A

90. Which statement characterizes the students' displacements?
    a. They're equal.
    b. C > A > B
    c. C > B > A
    d. B > C > A

91. Which statement characterizes their average speeds?
    a. They're equal.
    b. C > A > B
    c. C > B > A
    d. B > C > A

92. Which statement characterizes their accelerations while walking (not starting and stopping)?
    a. They're equal.
    b. None accelerates.
    c. A > B > C
    d. C > B > A
    e. B > C > A
    f. There's not enough information to decide.

Answers to Chapter Questions

Answer to Chapter Opening Question

Assuming negligible air resistance, the penguin should leave the water at a 45° angle.

Answers to GOT IT? Questions

3.1 (c)
3.2 (d) only
3.3 (1) (c); (2) (c); (3) (a)
3.4 (c) gives the greatest change in speed; (b) gives the greatest change in direction
3.5 (e)
3.6 (c)

Force and Motion

What You Know

- You can express motion quantities as vectors in one, two, or three dimensions.
- You can describe motion quantitatively when acceleration is constant.
- You're especially familiar with projectile motion under the influence of gravity near Earth's surface.
- You understand that circular motion is a case of accelerated motion, and you can find the acceleration given speed and radius of the circle.

What You're Learning

- Here you'll learn to adopt the Newtonian paradigm that the laws governing motion are about *change* in motion.
- You'll explore Newton's three laws of motion and see how together they provide a consistent picture of motion from microscopic to cosmic scales.
- You'll learn the concept of *force* and its relation to acceleration.
- You'll be introduced to the fundamental forces of nature.
- You'll see how the force of gravity acts on objects near Earth's surface.
- You'll learn to distinguish *weight* from *apparent weight*.
- You'll see how springs provide a convenient means of measuring force.

How You'll Use It

- Chapter 4 is largely restricted to one-dimensional motion. But the principles you learn here work in Chapter 5, where you'll apply Newton's laws in multiple dimensions.
- In Chapters 6 and 7 you'll see how Newton's second law leads to the fundamental concepts of energy and energy conservation.
- You'll apply Newtonian ideas to gravity in Chapter 8.
- In Chapter 9 you'll see how the interplay of Newton's second and third laws underlies the behavior of complex systems.
- Newton's laws will continue to govern the more complex motions you'll study in Part 2 of the book and beyond.

What forces did engineers have to consider when they developed the Mars *Curiosity* rover's "sky crane" landing system?

An interplanetary spacecraft moves effortlessly, yet its engines shut down years ago. Why does it keep moving? A baseball heads toward the batter. The batter swings, and suddenly the ball is heading toward left field. Why did its motion change?

Questions about the "why" of motion are the subject of **dynamics**. Here we develop the basic laws that answer those questions. Isaac Newton first stated these laws more than 300 years ago, yet they remain a vital part of physics and engineering today, helping us guide spacecraft to distant planets, develop better cars, and manipulate the components of individual cells.

4.1 The Wrong Question

We began this chapter with two questions: one about why a spacecraft *moved* and the other about why a baseball's motion *changed*. For nearly 2000 years following the work of Aristotle (384–322 BCE), the first question—Why do things move?—was the crucial one. And the answer seemed obvious: It took a force—a push or a pull—to keep something moving. This idea makes sense: Stop exerting yourself when jogging, and you stop moving; take your foot off the gas pedal, and your car soon stops. Everyday experience seems to suggest that Aristotle was right, and most of us carry in our heads the Aristotelian idea that motion requires a cause—something that pushes or pulls on a moving object to keep it going.

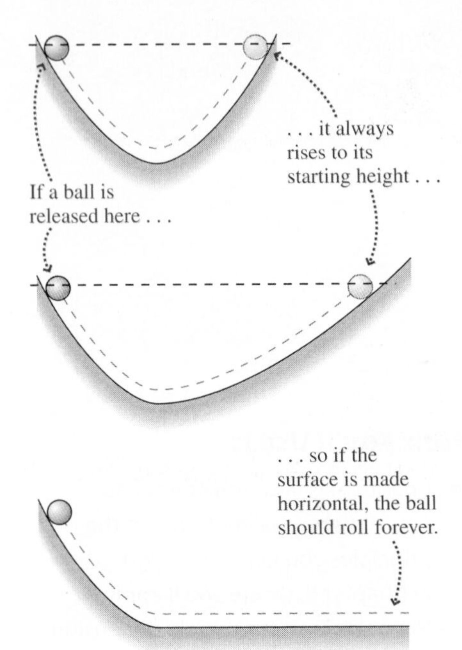

FIGURE 4.1 Galileo considered balls rolling on inclines and concluded that a ball on a horizontal surface should roll forever.

Actually, "What keeps things moving?" is the wrong question. In the early 1600s, Galileo Galilei did experiments that convinced him that a moving object has an intrinsic "quantity of motion" and needs no push to keep it moving (Fig. 4.1). Instead of answering "What keeps things moving?," Galileo declared that the question needs no answer. In so doing, he set the stage for centuries of progress in physics, beginning with the achievements of Issac Newton and culminating in the work of Albert Einstein.

The Right Question

Our first question—about why the spacecraft keeps moving—is the wrong question. So what's the right question? It's the second one, about why the baseball's motion *changed*. Dynamics isn't about what causes motion itself; it's about what causes *changes* in motion. Changes include starting and stopping, speeding up and slowing down, and changing direction. Any *change* in motion begs an explanation, but motion itself does not. Get used to this important idea and you'll have a much easier time with physics. But if you remain a "closet Aristotelian," secretly looking for causes of motion itself, you'll find it difficult to understand and apply the simple laws that actually govern motion.

Galileo identified the right question about motion. But it was Isaac Newton who formulated the quantitative laws describing how motion changes. We use those laws today for everything from designing antilock braking systems, to building skyscrapers, to guiding spacecraft.

4.2 Newton's First and Second Laws

What caused the baseball's motion to change? It was the bat's push. The term **force** describes a push or a pull. And the essence of dynamics is simply this:

> Force causes change in motion.

We'll soon quantify this idea, writing equations and solving numerical problems. But the essential point is in the simple sentence above. If you want to change an object's motion, you need to apply a force. If you see an object's motion change, you know there's a force acting. Contrary to Aristotle, and probably to your own intuitive sense, it does *not* take a force to keep something in unchanging motion; force is needed *only* to *change* an object's motion.

The Net Force

You can push a ball left or right, up or down. Your car's tires can push the car forward or backward, or make it round a curve. Force has direction and is a vector quantity. Furthermore, more than one force can act on an object. We call the individual forces on an object **interaction forces** because they always involve other objects interacting with the object in question. In Fig. 4.2*a*, for example, the interaction forces are exerted by the people pushing the car. In Fig. 4.2*b*, the interaction forces include the force of air on the plane, the engine force from the hot exhaust gases, and Earth's gravitational force.

We now explore in more detail the relation between force and change in motion. Experiment shows that what matters is the **net force**, meaning the vector sum of all individual interaction forces acting on an object. If the net force on an object isn't zero, then the object's motion must be changing—in direction or speed or both (Fig. 4.2*a*). If the net force on an object is zero—no matter what individual interaction forces contribute to the net force—then the object's motion is unchanging (Fig. 4.2*b*).

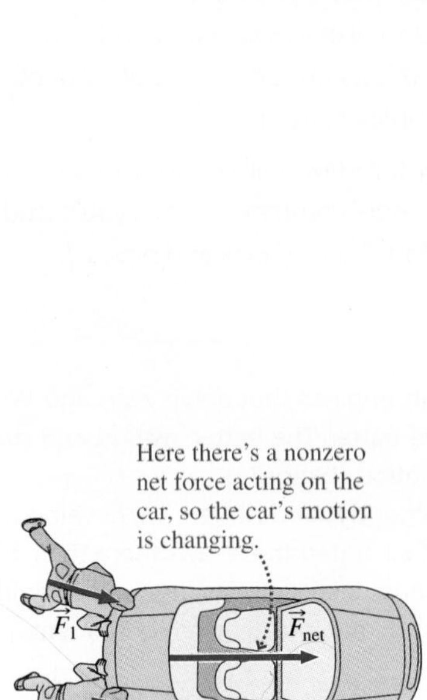

Here there's a nonzero net force acting on the car, so the car's motion is changing.

$\vec{F}_1$ $\vec{F}_2$ $\vec{F}_{net}$

(a)

The three forces sum to zero, so the plane moves in a straight line with constant speed.

$\vec{F}_{air}$ $\vec{F}_{net} = \vec{0}$ $\vec{F}_{engine}$ $\vec{F}_g$

(b)

FIGURE 4.2 The net force determines the change in an object's motion.

Newton's First Law

The basic idea that force causes change in motion is the essence of **Newton's first law**:

> **Newton's first law of motion:** A body in uniform motion remains in uniform motion, and a body at rest remains at rest, unless acted on by a nonzero net force.

The word "uniform" here is essential; **uniform motion** means unchanging motion—that is, motion in a straight line at constant speed. The phrase "a body at rest" isn't really necessary because rest is just the special case of uniform motion with zero speed, but we include it for consistency with Newton's original statement.

Video Tutor Demo | **Cart with Fan and Sail**

The first law says that uniform motion is a perfectly natural state, requiring no explanation. Again, the word "uniform" is crucial. The first law does *not* say that an object moving in a circle will continue to do so without a nonzero net force; in fact, it says that an object moving in a circle—or in any other curved path—*must* be subject to a nonzero net force because its motion is changing.

GOT IT? 4.1 A curved barrier lies on a horizontal tabletop, as shown. A ball rolls along the barrier, and the barrier exerts a force that guides the ball in its curved path. After the ball leaves the barrier, which of the dashed paths shown does it follow?

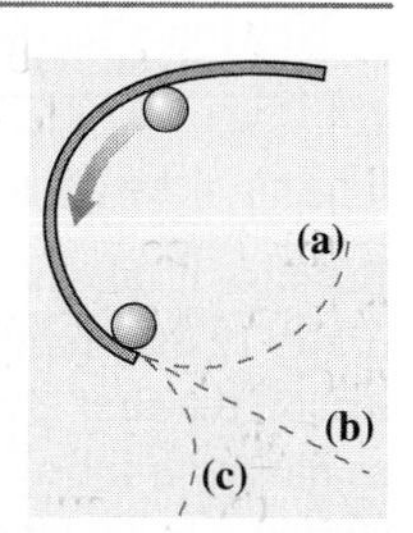

Video Tutor Demo | **Ball Leaves Circular Track**

Newton's first law is simplicity itself, but it's counter to our Aristotelian preconceptions; after all, your car soon stops when you take your foot off the gas. But because the motion changes, that just means—as the first law says—that there must be a nonzero net force acting. That force is often a "hidden" one, like friction, that isn't as obvious as the push or pull of muscle. Watch an ice show or hockey game, where frictional forces are minimal, and the first law becomes a lot clearer.

Newton's Second Law

Newton's second law quantifies the relation between force and change in motion. Newton reasoned that the product of mass and velocity was the best measure of an object's "quantity of motion." The modern term is **momentum**, and we write

$$\vec{p} = m\vec{v} \quad \text{(momentum)} \tag{4.1}$$

for the momentum of an object with mass m and velocity $\vec{v}$. As the product of a scalar (mass) and a vector (velocity), momentum is itself a vector quantity. Newton's second law relates the rate of change of an object's momentum to the net force acting on that object:

Newton's second law of motion: The rate at which a body's momentum changes is equal to the net force acting on the body:

$$\vec{F}_{\text{net}} = \frac{d\vec{p}}{dt} \quad \text{(Newton's 2}^{\text{nd}}\text{ law)} \tag{4.2}$$

When a body's mass remains constant, we can use the definition of momentum, $\vec{p} = m\vec{v}$, to write

$$\vec{F}_{\text{net}} = \frac{d\vec{p}}{dt} = \frac{d(m\vec{v})}{dt} = m\frac{d\vec{v}}{dt}$$

But $d\vec{v}/dt$ is the acceleration $\vec{a}$, so

$$\vec{F}_{\text{net}} = m\vec{a} \quad \text{(Newton's 2}^{\text{nd}}\text{ law, constant mass)} \tag{4.3}$$

We'll be using the form given in Equation 4.3 almost exclusively in the next few chapters. But keep in mind that Equation 4.2 is Newton's original expression of the second law, that it's more general than Equation 4.3, and that it embodies the fundamental concept of momentum. We'll return to Newton's law in the form of Equation 4.3, and elaborate on momentum, when we consider many-particle systems in Chapter 9.

Product of object's mass and its acceleration; not a force.

$$\vec{F}_{net} = m\vec{a}$$

Net force: the vector sum of all real, physical forces acting on an object

Equal sign indicates that the two sides are mathematically equal—but that doesn't mean they're the same physically. Only $\vec{F}_{net}$ involves physical forces.

FIGURE 4.3 Meaning of the terms in Newton's second law.

Newton's second law includes the first law as the special case $\vec{F}_{net} = \vec{0}$. In this case Equation 4.3 gives $\vec{a} = \vec{0}$, so an object's velocity doesn't change.

✓TIP Understanding Newton

To apply Newton's law successfully, you have to understand the terms, summarized in Fig. 4.3. On the left is the net force $\vec{F}_{net}$—the vector sum of all real, physical interaction forces acting on an object. On the right is $m\vec{a}$—not a force but the product of the object's mass and acceleration. The equal sign says that they have the same value, not that they're the same thing. So don't go adding an extra force $m\vec{a}$ when you're applying Newton's second law.

Mass, Inertia, and Force

Because it takes force to change an object's motion, the first law implies that objects naturally resist changes in motion. The term **inertia** describes this resistance, and for that reason the first law is also called the **law of inertia**. Just as we describe a sluggish person as having a lot of inertia, so an object that is hard to start moving—or hard to stop once started—has a lot of inertia. If we solve the second law for the acceleration $\vec{a}$, we find that $\vec{a} = \vec{F}/m$—showing that a given force is *less* effective in changing the motion of a *more massive* object (Fig. 4.4). The mass m that appears in Newton's laws is thus a measure of an object's inertia and determines the object's response to a given force.

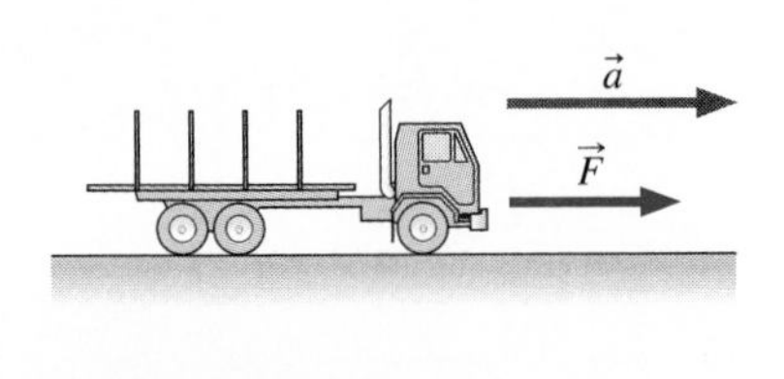

FIGURE 4.4 The loaded truck has greater mass—more inertia—so its acceleration is smaller when the same force is applied.

By comparing the acceleration of a known and an unknown mass in response to the same force, we can determine the unknown mass. From Newton's second law for a force of magnitude F,

$$F = m_{known}a_{known} \quad \text{and} \quad F = m_{unknown}a_{unknown}$$

where we're interested only in magnitudes so we don't use vectors. Equating these two expressions for the same force, we get

$$\frac{m_{unknown}}{m_{known}} = \frac{a_{known}}{a_{unknown}} \tag{4.4}$$

Equation 4.4 is an operational definition of mass; it shows how, given a known mass and force, we can determine other masses.

The force required to accelerate a 1-kg mass at the rate of 1 m/s^2 is defined to be 1 **newton** (N). Equation 4.3 shows that 1 N is equivalent to 1 kg · m/s^2. Other common force units are the English pound (lb, equal to 4.448 N) and the dyne, a metric unit equal to 10^{-5} N. A 1-N force is rather small; you can readily exert forces measuring hundreds of newtons with your own body.

Video Tutor Demo | **Suspended Balls: Which String Breaks?**

EXAMPLE 4.1 Force from Newton: A Car Accelerates

A 1200-kg car accelerates from rest to 20 m/s in 7.8 s, moving in a straight line with constant acceleration. (a) Find the net force acting on the car. (b) If the car then rounds a bend 85 m in radius at a steady 20 m/s, what net force acts on it?

INTERPRET In this problem we're asked to evaluate the net force on a car (a) when it undergoes constant acceleration and (b) when it rounds a turn. In both cases the net force is entirely horizontal, so we need to consider only the horizontal component of Newton's law.

DEVELOP Figure 4.5 shows the horizontal force acting on the car in each case; since this is the net force, it's equal to the car's mass multiplied by its acceleration. We aren't actually given the acceleration in this problem, but for (a) we know the change in speed and the time involved, so we can write $a = \Delta v/\Delta t$. For (b) we're given the speed and the radius of the turn; since the car is in uniform circular motion, Equation 3.16 applies, and we have $a = v^2/r$.

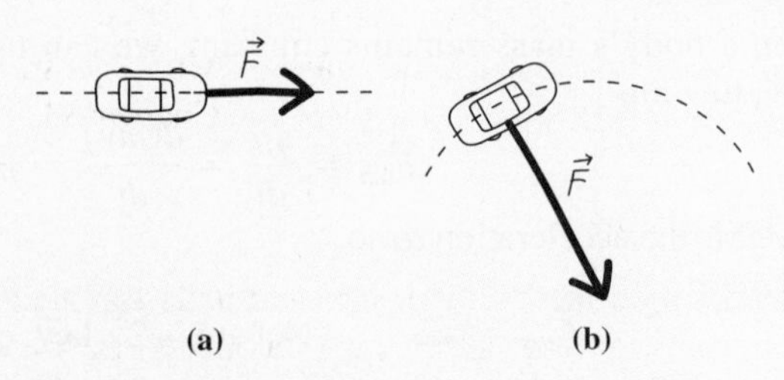

FIGURE 4.5 Our sketch of the net force on the car in Example 4.1.

EVALUATE We solve for the unknown acceleration and evaluate the numerical answers for both cases:

$$\text{(a)} \quad F_{\text{net}} = ma = m\frac{\Delta v}{\Delta t} = (1200\text{ kg})\left(\frac{20\text{ m/s}}{7.8\text{ s}}\right) = 3.1\text{ kN}$$

$$\text{(b)} \quad F_{\text{net}} = ma = m\frac{v^2}{r} = (1200\text{ kg})\frac{(20\text{ m/s})^2}{85\text{ m}} = 5.6\text{ kN}$$

ASSESS First, the units worked out; they were actually kg·m/s^2, but that defines the newton. The answers came out in thousands of N, but we moved the decimal point three places and changed to kilonewtons (kN) for convenience. And the numbers seem to make sense; we mentioned that 1 newton is a rather small force, so it's not surprising to find forces on cars measured in kN.

Note that Newton's law doesn't distinguish between forces that change an object's speed, as in (a), and forces that change its direction, as in (b). Newton's law relates force, mass, and acceleration in *all* cases. ■

GOT IT? 4.2 A nonzero net force acts on an object. Which of the following is true? (a) the object necessarily moves in the same direction as the net force; (b) under some circumstances the object could move in the same direction as the net force, but in other situations it might not; (c) the object cannot move in the same direction as the net force

Inertial Reference Frames

Why don't flight attendants serve beverages when an airplane is accelerating down the runway? For one thing, their beverage cart wouldn't stay put, but would accelerate toward the back of the plane even in the absence of a net force. So is Newton's first law wrong? No, but Newton's laws don't apply in an accelerating airplane. With respect to the ground, in fact, the beverage cart is doing just what Newton says it should: It remains in its original state of motion, while all around it plane and passengers accelerate toward takeoff.

In Section 3.3 we defined a reference frame as a system against which we measure velocities; more generally, a reference frame is the "background" in which we study physical reality. Our airplane example shows that Newton's laws don't work in all reference frames; in particular, they're not valid in accelerating frames. Where they are valid is in reference frames undergoing uniform motion—called **inertial reference frames** because only in these frames does the law of inertia hold. In a noninertial frame like an accelerating airplane, a car rounding a curve, or a whirling merry-go-round, an object at rest doesn't remain at rest, even when no force is acting. A good test for an inertial frame is to check whether Newton's first law is obeyed—that is, whether an object at rest remains at rest, and an object in uniform motion remains in uniform motion, when no force is acting on it.

Strictly speaking, our rotating Earth is not an inertial frame, and therefore Newton's laws aren't exactly valid on Earth. But Earth's rotation has an insignificant effect on most motions of interest, so we can usually treat Earth as an inertial reference frame. An important exception is the motion of oceans and atmosphere; here, scientists must take Earth's rotation into account.

If Earth isn't an inertial frame, what is? That's a surprisingly subtle question, and it pointed Einstein toward his general theory of relativity. The law of inertia is intimately related to questions of space, time, and gravity—questions whose answers lie in Einstein's theory. We'll look briefly at that theory in Chapter 33.

4.3 Forces

The most familiar forces are pushes and pulls you apply yourself, but passive objects can apply forces, too. A car collides with a parked truck and comes to a stop. Why? Because the truck exerts a force on it. The Moon circles Earth rather than moving in a straight line. Why? Because Earth exerts a gravitational force on it. You sit in a chair and don't fall to the floor. Why not? Because the chair exerts an upward force on you, countering gravity.

Some forces, like those you apply with your muscles, can have values that you choose. Other forces take on values determined by the situation. When you sit in the chair shown in Fig. 4.6, the downward force of gravity on you causes the chair to compress slightly.

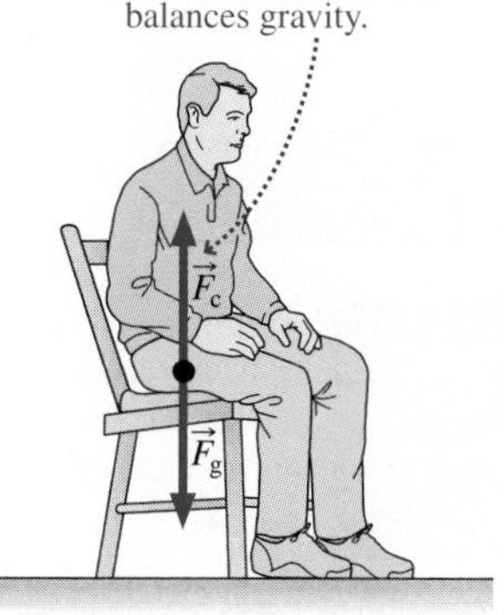

FIGURE 4.6 A compression force.

The chair acts like a spring and exerts an upward force. When the chair compresses enough that the upward force is equal in magnitude to the downward force of gravity, there's no net force and you sit without accelerating. The same thing happens with **tension forces** when objects are suspended from ropes or cables—the ropes stretch until the force they exert balances the force of gravity (Fig. 4.7).

FIGURE 4.7 The climbing rope exerts an upward tension force $\vec{T}$ that balances the force of gravity.

Forces like the pull you exert on your rolling luggage, the force of a chair on your body, and the force a baseball exerts on a bat are **contact forces** because the force is exerted through direct contact. Other forces, like gravity and electric and magnetic forces, are **action-at-a-distance forces** because they seemingly act between distant objects, like Earth and the Moon. Actually, the distinction isn't clear-cut; at the microscopic level, contact forces involve action-at-a-distance electric forces between molecules. And the action-at-a-distance concept itself is troubling. How can Earth "reach out" across empty space and pull on the Moon? Later we'll look at an approach to forces that avoids this quandary.

The Fundamental Forces

Gravity, tension forces, compression forces, contact forces, electric forces, friction forces—how many kinds of forces are there? At present, physicists identify three basic forces: the gravitational force, the electroweak force, and the strong force.

Gravity is the weakest of the fundamental forces, but because it acts attractively between all matter, gravity's effect is cumulative. That makes gravity the dominant force in the large-scale universe, determining the structure of planets, stars, galaxies, and the universe itself.

The **electroweak force** subsumes **electromagnetism** and the **weak nuclear force**. Virtually all the nongravitational forces we encounter in everyday life are electromagnetic, including contact forces, friction, tension and compression forces, and the forces that bind atoms into chemical compounds. The weak nuclear force is less obvious, but it's crucial in the Sun's energy production—providing the energy that powers life on Earth.

The **strong force** describes how particles called **quarks** bind together to form protons, neutrons, and a host of less-familiar particles. The force that joins protons and neutrons to make atomic nuclei is a residue of the strong force between their constituent quarks. Although the strong force isn't obvious in everyday life, it's ultimately responsible for the structure of matter. If its strength were slightly different, atoms more complex than helium couldn't exist, and the universe would be devoid of life!

Unifying the fundamental forces is a major goal of physics. Over the centuries we've come to understand seemingly disparate forces as manifestations of a more fundamental underlying force. Figure 4.8 suggests that the process continues, as physicists attempt first to unify the strong and electroweak forces, and then ultimately to add gravity to give a "Theory of Everything."

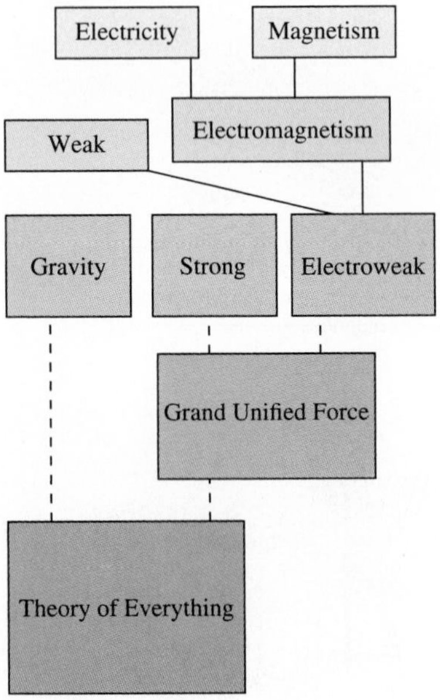

Electricity
Magnetism
Weak
Electromagnetism
Gravity
Strong
Electroweak
Grand Unified Force
Theory of Everything

FIGURE 4.8 Unification of forces is a major theme in physics.

4.4 The Force of Gravity

Newton's second law shows that mass is a measure of a body's resistance to changes in motion—its inertia. A body's mass is an intrinsic property; it doesn't depend on location. If my mass is 65 kg, it's 65 kg on Earth, in an orbiting spacecraft, or on the Moon. That means no matter where I am, a force of 65 N gives me an acceleration of 1 m/s^2.

We commonly use the term "weight" to mean the same thing as mass. In physics, though, **weight** is the *force* that gravity exerts on a body. Near Earth's surface, a freely falling body accelerates downward at 9.8 m/s^2; we designate this acceleration vector by $\vec{g}$. Newton's second law, $\vec{F} = m\vec{a}$, then says that the force of gravity on a body of mass m is $m\vec{g}$; this force is the body's weight:

$$\vec{w} = m\vec{g} \qquad \text{(weight)}. \tag{4.5}$$

With my 65-kg mass, my weight near Earth's surface is then $(65\text{ kg})(9.8\text{ m/s}^2)$ or 640 N. On the Moon, where the acceleration of gravity is only 1.6 m/s^2, I would weigh only 100 N. And in the remote reaches of intergalactic space, far from any gravitating object, my weight would be essentially zero.

EXAMPLE 4.2 Mass and Weight: Exploring Mars

The rover *Curiosity* that landed on Mars in 2012 weighed 8.82 kN on Earth. What were its mass and weight on Mars?

INTERPRET Here we're asked about the relation between mass and weight, and the object we're interested in is the *Curiosity* rover.

DEVELOP Equation 4.5 describes the relation between mass and weight. Writing this equation in scalar form because we're interested only in magnitudes, we have $w = mg$.

EVALUATE First we want to find mass from weight, so we solve for m using the Earth weight and Earth's gravity:

$$m = \frac{w}{g} = \frac{8.82 \text{ kN}}{9.81 \text{ m/s}^2} = 899 \text{ kg}$$

This mass is the same everywhere, so the weight on Mars is given by $w = mg_{\text{Mars}} = (899 \text{ kg})(3.71 \text{ m/s}^2) = 3.34 \text{ kN}$. Here we found the acceleration of gravity on Mars in Appendix E. We used $g = 9.81 \text{ m/s}^2$ in this calculation because we were given other information to three significant figures.

ASSESS Make sense? Sure: Mars's gravitational acceleration is lower than Earth's, and so is the spacecraft's weight on Mars. ■

One reason we confuse mass and weight is the common use of the SI unit kilogram to describe "weight." At the doctor's office you may be told that you "weigh" 55 kg. You don't; you have a mass of 55 kg, so your weight is $(55 \text{ kg})(9.8 \text{ m/s}^2)$ or 540 N. The unit of force in the English system is the pound, so giving your weight in pounds is correct.

That we confuse mass and weight at all results from the remarkable fact that the gravitational acceleration of all objects is the same. This makes a body's *weight*, a gravitational property, proportional to its *mass*, a measure of its inertia in terms that have nothing to do with gravity. First inferred by Galileo from his experiments with falling bodies, this relation between gravitation and inertia seemed a coincidence until the early 20th century. Finally Albert Einstein showed how that simple relation reflects the underlying geometry of space and time in a way that intimately links gravitation and acceleration.

Weightlessness

Aren't astronauts "weightless"? Not according to our definition. At the altitude of the International Space Station, the acceleration of gravity has about 89% of its value at Earth's surface, so the gravitational forces $m\vec{g}$ on the station and its occupants are almost as large as on Earth. But the astronauts *seem* weightless, and indeed they *feel* weightless (Fig. 4.9). What's going on?

Imagine yourself in an elevator whose cable has broken and is dropping freely downward with the gravitational acceleration g. In other words, the elevator and its occupant are in **free fall**, with only the force of gravity acting. If you let go of a book, it too falls freely with acceleration g. But so does everything else around it—and therefore the book stays

FIGURE 4.9 These astronauts only *seem* weightless.

In a freely falling elevator you and your book seem weightless because both fall with the same acceleration as the elevator.

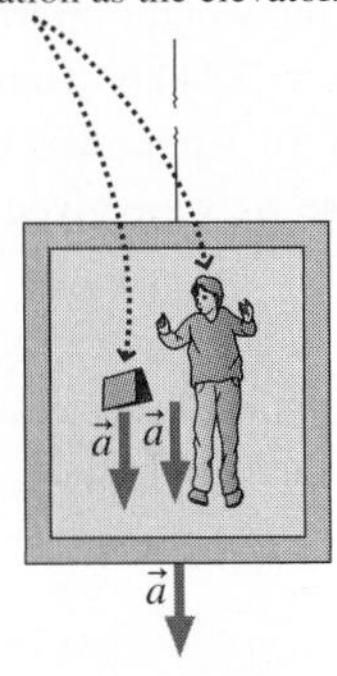

Earth

(a)

Like the elevator in (a), an orbiting spacecraft is falling toward Earth, and because its occupants also fall with the same acceleration, they experience apparent weightlessness.

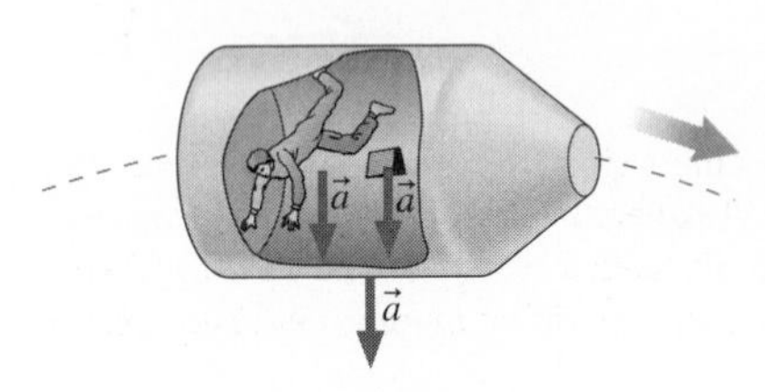

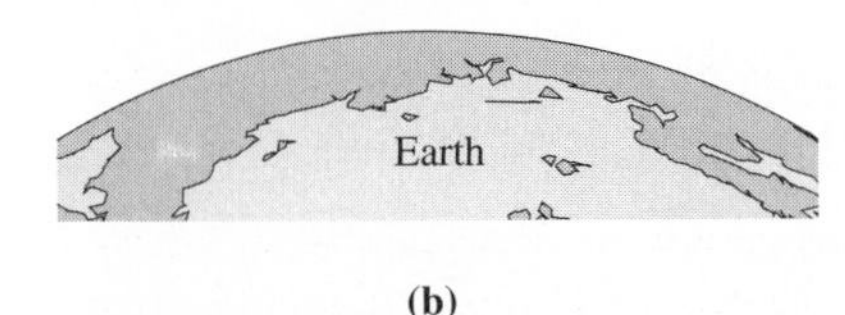

(b)

FIGURE 4.10 Objects in free fall appear weightless because they all experience the same acceleration.

put relative to you (Fig. 4.10*a*). To you, the book seems "weightless," since it doesn't seem to fall when you let go of it. And you're "weightless" too; if you jump off the elevator's floor, you float to the ceiling rather than falling back. You, the book, and the elevator are *all* falling, but because all have the same acceleration that isn't obvious to you. The gravitational force is still acting; it's making you fall. So you really do have weight, and your condition is best termed **apparent weightlessness**.

A falling elevator is a dangerous place; your state of apparent weightlessness would end with a deadly smash caused by nongravitational contact forces when you hit the ground. But apparent weightlessness occurs permanently in a state of free fall that doesn't intersect Earth—as in an orbiting spacecraft (Fig. 4.10*b*). It's not being in outer space that makes astronauts seem weightless; it's that they, like our hapless elevator occupant, are in free fall—moving under the influence of the gravitational force alone. The condition of apparent weightlessness in orbiting spacecraft is sometimes called "microgravity."

GOT IT? 4.3 A popular children's book explains the weightlessness astronauts experience by saying there's no gravity in space. If there were no gravity in space, what would be the motion of a space shuttle, a satellite, or, the Moon? (a) a circular orbit; (b) an elliptical orbit; (c) a straight line

---

4.5 Using Newton's Second Law

The interesting problems involving Newton's second law are those where more than one force acts on an object. To apply the second law, we then need the net force. For an object of constant mass, the second law relates the net force and the acceleration:

$$\vec{F}_{\text{net}} = m\vec{a}$$

Using Newton's second law with multiple forces is easier if we draw a **free-body diagram**, a simple diagram that shows only the object of interest and the forces acting on it.

TACTICS 4.1 Drawing a Free-Body Diagram

Drawing a free-body diagram, which shows the forces acting on an object, is the key to solving problems with Newton's laws. To make a free-body diagram:

1. Identify the object of interest and all the forces acting on it.
2. Represent the object as a dot.
3. Draw the vectors for *only* those forces acting *on* the object, with their tails all starting on the dot.

Figure 4.11 shows two examples where we reduce physical scenarios to free-body diagrams. We often add a coordinate system to the free-body diagram so that we can express force vectors in components.

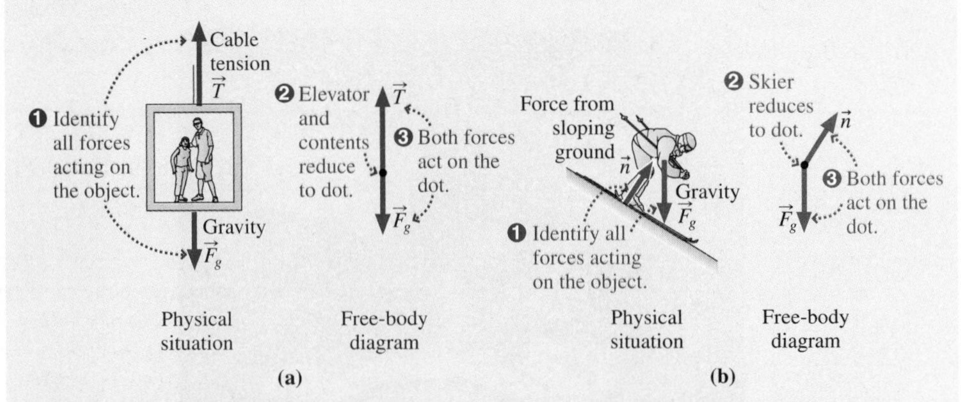

FIGURE 4.11 Free-body diagrams. (a) A one-dimensional situation like those we discuss in this chapter. (b) A two-dimensional situation. We'll deal with such cases in Chapter 5.

Our IDEA strategy applies to Newton's laws as it does to other physics problems. For the second law, we can elaborate on the four IDEA steps:

PROBLEM-SOLVING STRATEGY 4.1 Newton's Second Law

INTERPRET Interpret the problem to be sure that you know what it's asking and that Newton's second law is the relevant concept. Identify the object of interest and all the individual interaction forces acting on it.

DEVELOP Draw a free-body diagram as described in Tactics 4.1. Develop your solution plan by writing Newton's second law, $\vec{F}_{net} = m\vec{a}$, with $\vec{F}_{net}$ expressed as the sum of the forces you've identified. Then choose a coordinate system so you can express Newton's law in components.

EVALUATE At this point the physics is done, and you're ready to execute your plan by solving Newton's second law and evaluating the numerical answer(s), if called for. Even in the one-dimensional problems of this chapter, remember that Newton's law is a vector equation; that will help you get the signs right. You need to write the components of Newton's law in the coordinate system you chose, and then solve the resulting equation(s) for the quantity(ies) of interest.

ASSESS Assess your solution to see that it makes sense. Are the numbers reasonable? Do the units work out correctly? What happens in special cases—for example, when a mass, force, or acceleration becomes very small or very large, or an angle becomes 0° or 90°?

EXAMPLE 4.3 Newton's Second Law: In the Elevator

A 740-kg elevator accelerates upward at 1.1 m/s^2, pulled by a cable of negligible mass. Find the tension force in the cable.

INTERPRET In this problem we're asked to evaluate one of the forces on an object. First we identify the object of interest. Although the problem asks about the cable tension, it's the elevator on which that tension acts, so the elevator is the object of interest. Next, we identify the forces acting on the elevator. There are two: the downward force of gravity $\vec{F}_g$ and the upward cable tension $\vec{T}$.

DEVELOP Figure 4.12*a* shows the elevator accelerating upward; Fig. 4.12*b* is a free-body diagram representing the elevator as a dot with the two force vectors acting on it. The applicable equation is Newton's second law, $\vec{F}_{net} = m\vec{a}$, with $\vec{F}_{net}$ given by the sum of the forces we've identified:

$$\vec{F}_{net} = \vec{T} + \vec{F}_g = m\vec{a} \quad (4.6)$$

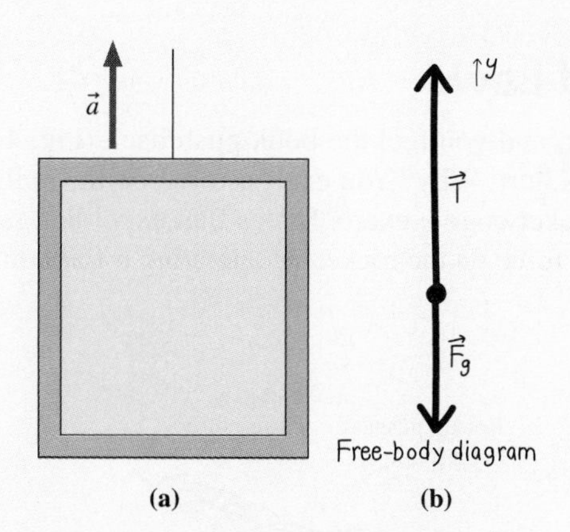

FIGURE 4.12 The forces on the elevator are the cable tension $\vec{T}$ and gravity $\vec{F}_g$.

✓TIP Vectors Tell it All

Are you tempted to put a minus sign in this equation because one force is downward? *Don't*! A vector contains all the information about its direction. You don't have to worry about signs until you write the components of a vector equation in the coordinate system you chose.

Now we need to choose a coordinate system. Here all the forces are vertical, so we'll choose our *y*-axis pointing upward.

EVALUATE Now we're ready to rewrite Newton's second law—Equation 4.6 in this case—in our coordinate system. Formally, we remove the vector signs and add coordinate subscripts—just *y* in this case:

$$T_y + F_{gy} = ma_y \quad (4.7)$$

Still no need to worry about signs. Now, what is T_y? Since the tension points upward and we've chosen that to be the positive direction, the component of tension in the *y*-direction is its magnitude *T*. What about F_{gy}? Gravity points downward, so this component is negative. Furthermore, we know that the magnitude of the gravitational force is mg. So $F_{gy} = -mg$. Then our Newton's law equation becomes

$$T - mg = ma_y$$

so

$$T = ma_y + mg = m(a_y + g) \quad (4.8)$$

For the numbers given, this equation yields

$$T = m(a_y + g) = (740\ \text{kg})(1.1\ \text{m/s}^2 + 9.8\ \text{m/s}^2) = 8.1\ \text{kN}$$

ASSESS We can see that this answer makes sense—and learn a lot more about physics—from the algebraic form of the answer in Equation 4.8. Consider some special cases: If the acceleration a_y were zero, then the net force on the elevator would have to be zero. In that case Equation 4.8 gives $T = mg$. Makes sense: The cable is then supporting the elevator's weight *mg* but not exerting any additional force to accelerate it.

(continued)

On the other hand, if the elevator is accelerating upward, then the cable has to provide an extra force in addition to the weight; that's why the tension becomes $ma_y + mg$. Numerically, our answer of 8.1 kN is *greater* than the elevator's weight—and the cable had better be strong enough to handle the extra force.

Finally, if the elevator is accelerating downward, then a_y is negative, and the cable tension is *less* than the weight. In free fall, $a_y = -g$, and the cable tension would be zero.

You might have reasoned out this problem in your head. But we did it very thoroughly because the strategy we followed will let you solve all problems involving Newton's second law, even if they're much more complicated. If you always follow this strategy and don't try to find shortcuts, you'll become confident in using Newton's second law. ■

GOT IT? 4.4 For each of the following situations, would the cable tension in Example 4.3 be (a) greater than, (b) less than, or (c) equal to the elevator's weight? (1) elevator starts moving upward, accelerating from rest; (2) elevator decelerates to a stop while moving upward; (3) elevator starts moving downward, accelerating from rest; (4) elevator slows to a stop while moving downward; (5) elevator is moving upward with constant speed

CONCEPTUAL EXAMPLE 4.1 At the Equator

When you stand on a scale, the scale pushes up to support you, and the scale reading shows the force with which it's pushing. If you stand on a scale at Earth's equator, is the reading greater or less than your weight?

EVALUATE The question asks about the force the scale exerts on you, in comparison to your weight (the gravitational force on you). Figure 4.13 is our sketch, showing the scale force upward and the gravitational force downward, toward Earth's center. You're in circular motion about Earth's center, so the direction of your acceleration is toward the center (downward). According to Newton's second law, the net force and acceleration are in the same direction. The only two forces acting on you are the downward force of gravity and the upward force of the scale. For them to sum to a net force that's downward, the force of gravity—your weight—must be larger. Therefore, the scale reading must be less than your weight.

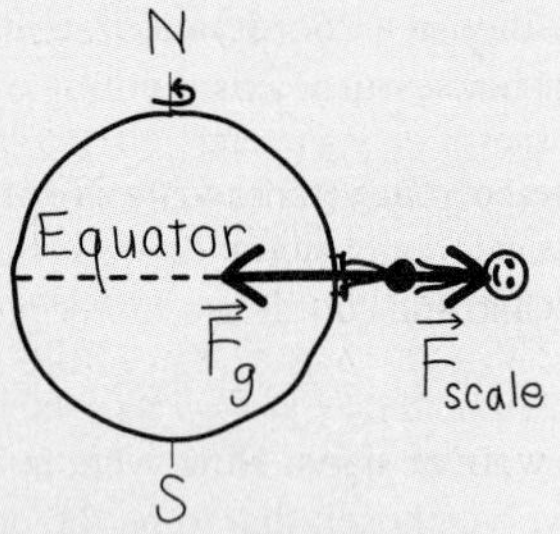

FIGURE 4.13 Our sketch for Conceptual Example 4.1.

ASSESS Make sense? Yes: If the two forces had equal magnitudes, the net force would be zero—inconsistent with the fact that you're accelerating. And if the scale force were greater, you'd be accelerating in the wrong direction! The same effect occurs everywhere except at the poles, but its analysis is more complicated because the acceleration is toward Earth's axis, not the center.

MAKING THE CONNECTION By what percentage is your apparent weight (the scale reading) at the equator less than your actual weight?

EVALUATE Using Earth's radius R_E from Appendix E, and its 24-hour rotation period, you can find your acceleration: From Equation 3.16, it's v^2/R_E. Following Problem-Solving Strategy 4.1 and working in a coordinate system with the vertical direction upward, you'll find that Newton's second law becomes $F_{scale} - mg = -m\frac{v^2}{R_E}$, or $F_{scale} = mg - mv^2/R_E$. So the scale reading differs from your weight mg by mv^2/R_E. Working the numbers shows that's a difference of only 0.34%. Note that this result doesn't depend on your mass m.

4.6 Newton's Third Law

Push your book across your desk, and you feel the book push back (Fig. 4.14*a*). Kick a ball with bare feet, and your toes hurt. Why? You exert a force on the ball, and the ball exerts a force back on you. A rocket engine exerts forces that expel hot gases out of its nozzle—and the hot gases exert a force on the rocket, accelerating it forward (Fig. 4.14*b*).

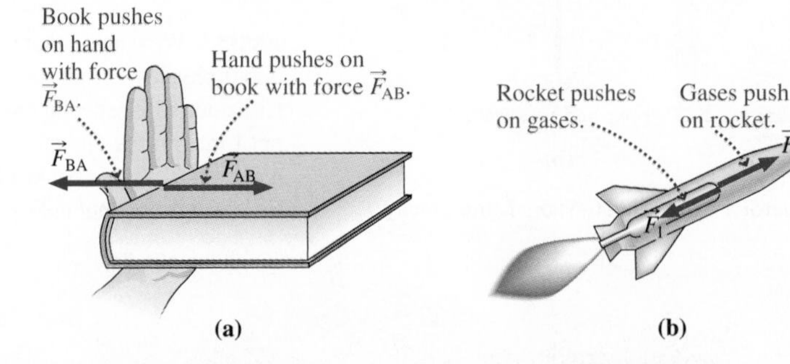

FIGURE 4.14 Newton's third law says that forces always come in pairs. With objects in contact, both forces act at the contact point. To emphasize that the two forces act on *different* objects, we draw them slightly displaced.

Whenever one object exerts a force on a second object, the second object also exerts a force on the first. The two forces are in opposite directions, but they have equal magnitudes. This fact constitutes **Newton's third law** of motion. The familiar expression "for every action there is an equal and opposite reaction" is Newton's 17th-century language. But there's really no distinction between "action" and "reaction"; both are always present. In modern language, the third law states:

> **Newton's third law of motion:** If object A exerts a force on object B, then object B exerts an oppositely directed force of equal magnitude on A.

Newton's third law is about forces between objects. It says that such forces always occur in pairs—that it's not possible for object A to exert a force on object B without B exerting a force back on A. You can now see why we coined the term "interaction forces"—when there's force between two objects, it's always a true *inter*action, with both objects exerting forces and both experiencing forces. We'll use the terms **interaction force pair** and **third-law pair** for the two forces described by Newton's third law.

It's crucial to recognize that the forces of a third-law pair act on *different* objects; the force $\vec{F}_{AB}$ of object A acts on object B, and the force $\vec{F}_{BA}$ of B acts on A. The forces have equal magnitudes and opposite directions, but they don't cancel to give zero net force *because they don't act on the same object*. In Fig. 4.14*a*, for example, $\vec{F}_{AB}$ is the force the hand exerts on the book. There's no other horizontal force acting on the book, so the net force on the book is nonzero and the book accelerates. Failure to recognize that the two forces of a third-law pair act on different objects leads to a contradiction, embodied in the famous horse-and-cart dilemma illustrated in Fig. 4.15.

These forces constitute an equal but opposite pair, but they don't act on the same object so they don't cancel.

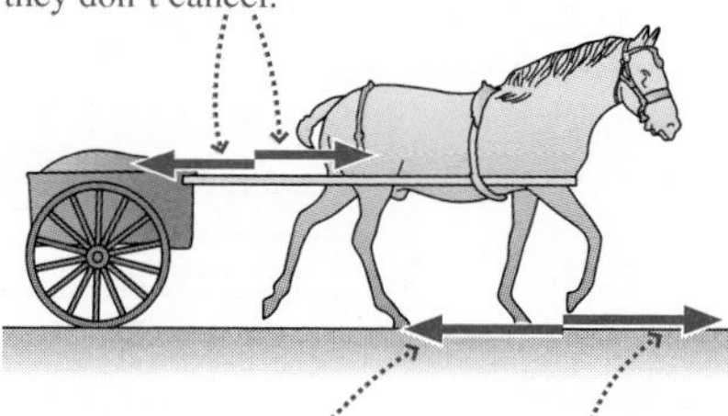

The force on the horse arises as a reaction to the horse pushing back on the road.

The forward force from the road is greater than the backward force from the cart so the net force is forward.

FIGURE 4.15 The horse-and-cart dilemma: The horse pulls on the cart, and the cart pulls back on the horse with a force of equal magnitude. So how can the pair ever get moving? No problem: The *net* force on the horse involves forces from *different* third-law pairs. Their magnitudes aren't equal and the horse experiences a net force in the forward direction.

APPLICATION Hollywood Goes Weightless

The film *Apollo 13* shows Tom Hanks and his fellow actors floating weightlessly around the cabin of their movie-set spacecraft. What special effects did Hollywood use here? None. The actors' apparent weightlessness was the real thing. But even Hollywood's budget wasn't enough to buy a space-shuttle flight. So the producers rented NASA's weightlessness training aircraft, aptly dubbed the "vomit comet." This airplane executes parabolic trajectories that mimic the free-fall motion of a projectile, so its occupants experience apparent weightlessness.

Movie critics marveled at how *Apollo 13* "simulated the weightlessness of outer space." Nonsense! The actors were in free fall just like the real astronauts on board the real *Apollo 13*, and they experienced exactly the same physical phenomenon—apparent weightlessness when moving under the influence of gravity alone.

In contrast to *Apollo 13*, scenes of apparent weightlessness in the 2013 film *Gravity* were done with special effects. That's one reason the film's star, Sandra Bullock, wears her hair short; it would be too difficult to simulate individual free-floating strands of long hair.

EXAMPLE 4.4 Newton's Third Law: Pushing Books

On a frictionless horizontal surface, you push with force $\vec{F}$ on a book of mass m_1 that in turn pushes on a book of mass m_2 (Fig. 4.16*a*). What force does the second book exert on the first?

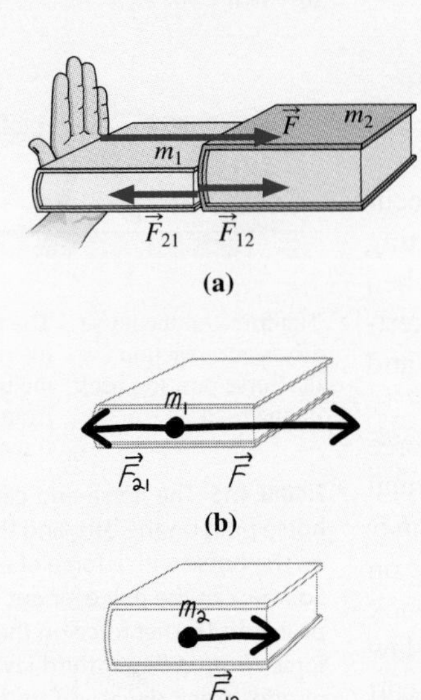

FIGURE 4.16 Horizontal forces on the books of Example 4.4. Not shown are the vertical forces of gravity and the normal force from the surface supporting the books.

INTERPRET This problem is about the interaction between two objects, so we identify both books as objects of interest.

DEVELOP In a problem with multiple objects, it's a good idea to draw a separate free-body diagram for each object. We've done that in Fig. 4.16*b* and *c*, keeping very light images of the books themselves. Now, we're asked about the force the second book exerts on the first. Newton's third law would give us that force if we knew the force the first book exerts on the second. Since that's the only horizontal force acting on book 2, we could get it from Newton's *second* law if we knew the acceleration of book 2. So here's our plan: (1) Find the acceleration of book 2; (2) use Newton's second law to find the net force on book 2, which in this case is the single force $\vec{F}_{12}$; and (3) apply Newton's third law to get $\vec{F}_{21}$, which is what we're looking for.

EVALUATE (1) The total mass of the two books is $m_1 + m_2$, and the net force applied to the combination is $\vec{F}$. Newton's second law, $\vec{F} = m\vec{a}$, gives

$$\vec{a} = \frac{\vec{F}}{m} = \frac{\vec{F}}{m_1 + m_2}$$

for the acceleration of both books, including book 2. (2) Now that we know book 2's acceleration, we use Newton's second law to find $\vec{F}_{12}$, which, since it's the only horizontal force on book 2, is the net force on that book:

$$\vec{F}_{12} = m_2\vec{a} = m_2\frac{\vec{F}}{m_1 + m_2} = \frac{m_2}{m_1 + m_2}\vec{F}$$

(3) Finally, the forces the books exert on each other constitute a third-law pair, so we have

$$\vec{F}_{21} = -\vec{F}_{12} = -\frac{m_2}{m_1 + m_2}\vec{F}$$

ASSESS You can see that this result makes sense by considering the first book. It too undergoes acceleration $\vec{a} = \vec{F}/(m_1 + m_2)$, but there are *two* forces acting on it: the applied force $\vec{F}$ and the force $\vec{F}_{21}$ from the second book. So the net force on the first book is

$$\vec{F} + \vec{F}_{21} = \vec{F} - \frac{m_2}{m_1 + m_2}\vec{F} = \frac{m_1}{m_1 + m_2}\vec{F} = m_1\vec{a}$$

consistent with Newton's second law. Our result shows that Newton's second and third laws are both necessary for a fully consistent description of the motion. ■

GOT IT? 4.5 The figure shows two blocks with two forces acting on the pair. Is the net force on the larger block (a) greater than 2 N, (b) equal to 2 N, or (c) less than 2 N?

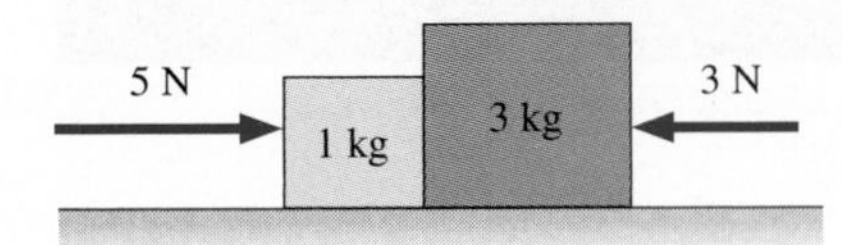

A contact force such as the force between the books in Example 4.4 is called a **normal force** (symbol $\vec{n}$) because it acts at right angles to the surfaces in contact. Other examples of normal forces are the upward force that a table or bridge exerts on objects it supports, and the force perpendicular to a sloping surface supporting an object (Fig. 4.17).

Video Tutor Demo | **Weighing a Hovering Magnet**

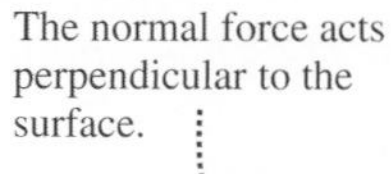
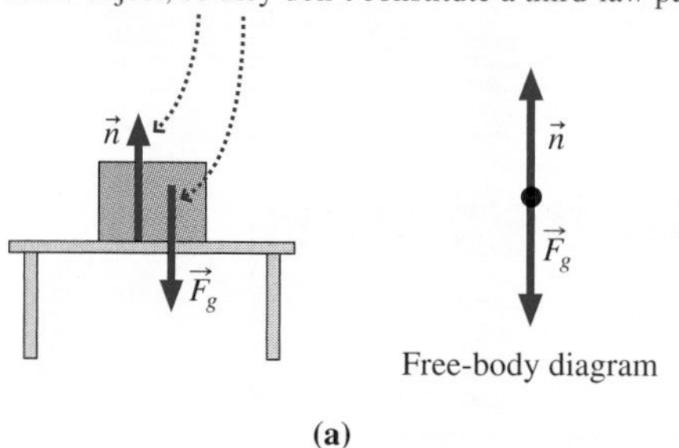

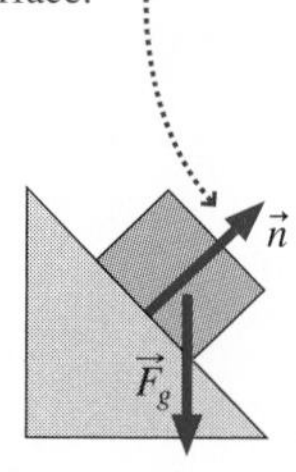

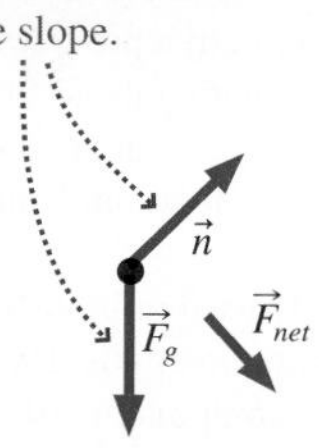

FIGURE 4.17 Normal forces. Also shown in each case is the gravitational force.

Newton's third law also applies to forces like gravity that don't involve direct contact. Since Earth exerts a downward force on you, the third law says that you exert an equal upward force on Earth (Fig. 4.18). If you're in free fall, then Earth's gravity causes you to accelerate toward Earth. Earth, too, accelerates toward you—but it's so massive that this acceleration is negligible.

FIGURE 4.18 Gravitational forces on you and on Earth form a third-law pair. Figure is obviously not to scale!

Measuring Force

Newton's third law provides a convenient way to measure forces using the tension or compression force in a spring. A spring stretches or compresses in proportion to the force exerted on it. By Newton's third law, the force *on* the spring is equal and opposite to the force the spring exerts on whatever is stretching or compressing it (Fig. 4.19). The spring's stretch or compression thus provides a measure of the force on whatever object is attached to the spring.

In an **ideal spring**, the stretch or compression is directly proportional to the force exerted by the spring. **Hooke's law** expresses this proportionality mathematically:

$$F_s = -kx \qquad \text{(Hooke's law, ideal spring)} \qquad (4.9)$$

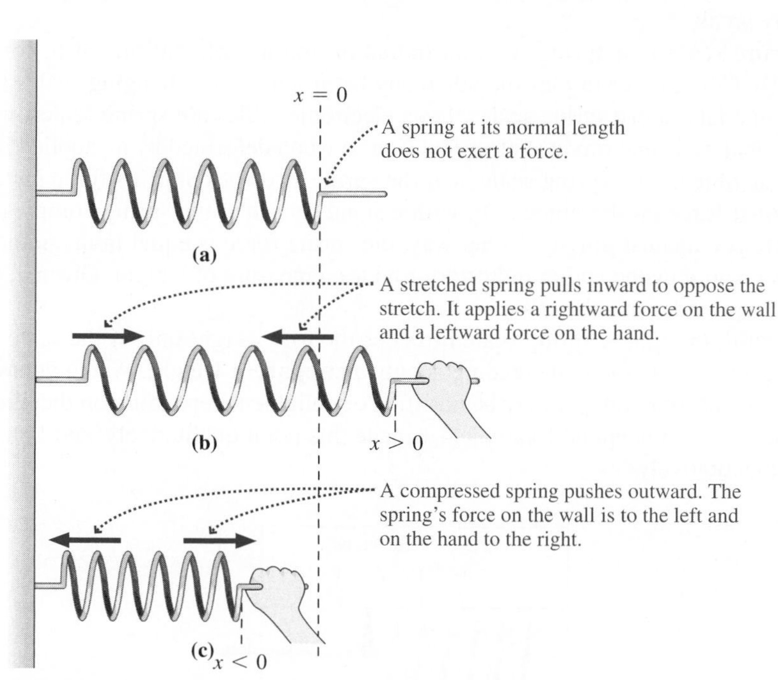

FIGURE 4.19 A spring responds to stretching or compression with an oppositely directed force.

APPLICATION Accelerometers, MEMS, Airbags, and Smartphones

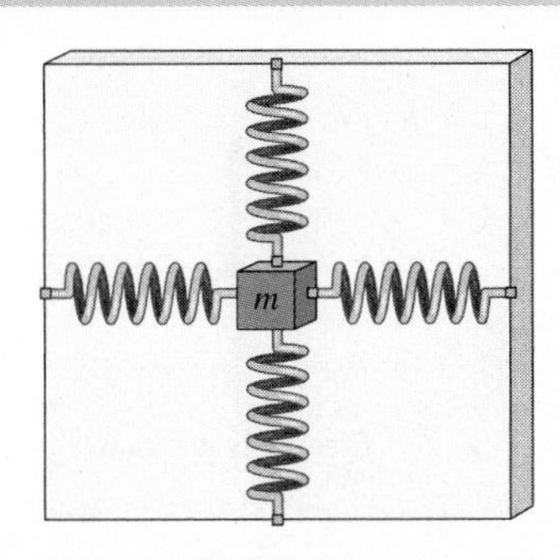

Hook one end of a spring to any part of an accelerating car, airplane, rocket, or whatever, and attach a mass m to the other end. The spring stretches until it provides the force needed to bring the mass along with the accelerating vehicle, sharing its acceleration. If you measure the spring's stretch and know its spring constant, you can get the force. If you know the mass m, you can then use $F = ma$ to get the acceleration. You've made an **accelerometer**!

Accelerometers based on this simple principle are widely used in industrial, transportation, robotics, and scientific applications. Often they're three-axis devices, with three mutually perpendicular springs to measure all three components of the acceleration vector. The drawing shows a simplified two-axis accelerometer for measuring accelerations in a horizontal plane.

Today's accelerometers are miniature devices based on technology called MEMS, for microelectromechanical systems. They're etched out of a tiny silicon chip that includes electronics for measuring stretch and determining acceleration. Your car employs a number of these accelerometers, including those that sense when to deploy the airbags.

Your smartphone contains a three-axis MEMS accelerometer (see photo, which is magnified some 700 times) that determines the phone's acceleration in three mutually perpendicular directions. The components in a smartphone accelerometer are only a fraction of a millimeter across. Apps are available to record accelerometer data, making your smartphone a useful device for physics experiments. Problem 64 explores smartphone accelerometer data.

Here F_s is the spring force, x is the distance the spring has been stretched or compressed from its normal length, and k is the **spring constant**, which measures the "stiffness" of the spring. Its units are N/m. The minus sign shows that the spring force is *opposite* the distortion of the spring: Stretch it, and the spring responds with a force *opposite* the stretching force; compress it, and the spring pushes back against the compressing force. Real springs obey Hooke's law only up to a point; stretch it too much, and a spring will deform and eventually break.

A **spring scale** is a spring with an indicator and a scale calibrated in force units (Fig. 4.20). Common examples include many bathroom scales, hanging scales in supermarkets, and laboratory spring scales. Even electronic scales are spring scales, with their "springs" materials that produce electrical signals when deformed by an applied force.

Hang an object on a spring scale, and the spring stretches until its force counters the gravitational force on the object. Or, with a stand-on scale, the spring compresses until it supports you against gravity. Either way, the spring force is equal in magnitude to the weight mg, and thus the spring indicator provides a measure of weight. Given g, this procedure also provides the object's mass.

Be careful, though: A spring scale provides the true weight only if the scale isn't accelerating; otherwise, the scale reading is only an **apparent weight**. Weigh yourself in an accelerating elevator and you may be horrified or delighted, depending on the direction of the acceleration. Conceptual Example 4.1 made this point qualitatively, and Example 4.5 does so quantitatively.

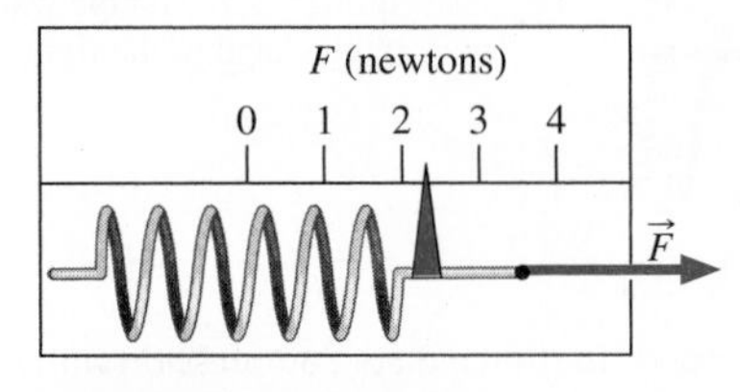

FIGURE 4.20 A spring scale.

EXAMPLE 4.5 True and Apparent Weight: A Helicopter Ride

A helicopter rises vertically, carrying concrete for a ski-lift foundation. A 35-kg bag of concrete sits in the helicopter on a spring scale whose spring constant is 3.4 kN/m. By how much does the spring compress (a) when the helicopter is at rest and (b) when it's accelerating upward at 1.9 m/s^2?

INTERPRET This problem is about concrete, a spring scale, and a helicopter. Ultimately, that means it's about mass, force, and acceleration—the content of Newton's laws. We're interested in the spring and the concrete mass resting on it, which share the motion of the helicopter. We identify two forces acting on the concrete: gravity and the spring force $\vec{F}_s$.

DEVELOP As with any Newton's law problem, we start with a free-body diagram (Fig. 4.21). We then write Newton's second law in its vector form

$$\vec{F}_{\text{net}} = \vec{F}_s + \vec{F}_g = m\vec{a}$$

Vectors tell it all; don't worry about signs at this point. Our equation expresses all the physics of the situation, but before we can move on to the solution, we need to choose a coordinate system. Here it's convenient to take the y-axis vertically upward.

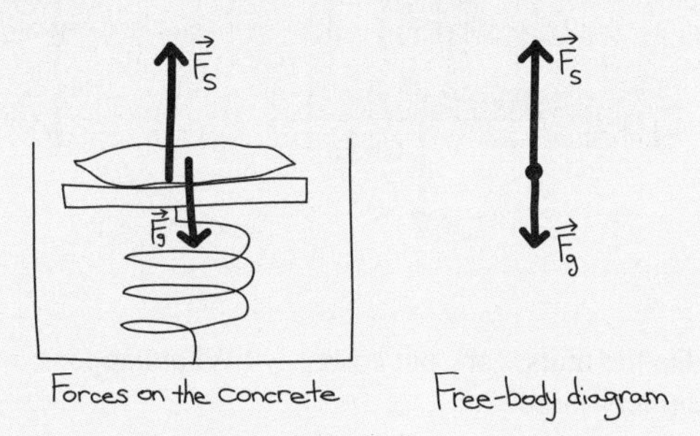

FIGURE 4.21 Our drawings for Example 4.5.

EVALUATE The forces are in the vertical direction, so we're concerned with only the y-component of Newton's law: $F_{sy} + F_{gy} = ma_y$. The spring force is upward and, from Hooke's law, it has magnitude kx, so $F_{sy} = kx$. Gravity is downward with magnitude mg, so $F_{gy} = -mg$. The y-component of Newton's law then becomes $kx - mg = ma_y$, which we solve to get

$$x = \frac{m(a_y + g)}{k}$$

Putting in the numbers (a) with the helicopter at rest $(a_y = 0)$ and (b) with $a_y = 1.9\text{ m/s}^2$ gives

$$\text{(a)} \quad x = \frac{m(a_y + g)}{k} = \frac{(35\text{ kg})\,(0 + 9.8\text{ m/s}^2)}{3400\text{ N/m}} = 10\text{ cm}$$

$$\text{(b)} \quad x = \frac{(35\text{ kg})(1.9\text{ m/s}^2 + 9.8\text{ m/s}^2)}{3400\text{ N/m}} = 12\text{ cm}$$

ASSESS Why is the answer to (b) larger? Because, just as with the cable in Example 4.3, the spring needs to provide an additional force to accelerate the concrete upward. ■

GOT IT? 4.6 (1) Would the answer to (a) in Example 4.5 change if the helicopter were not at rest but moving upward at constant speed? (2) Would the answer to (b) change if the helicopter were moving *downward* but still accelerating *upward*?

CHAPTER 4 SUMMARY

Big Idea

The big idea of this chapter—and of all Newtonian mechanics—is that **force** causes *change* in motion, not motion itself. Uniform motion—straight line, constant speed—needs no cause or explanation. Any deviation, in speed or direction, requires a **net force**. This idea is the essence of Newton's first and second laws. Combined with Newton's third law, these laws provide a consistent description of motion.

Newton's First Law

A body in uniform motion remains in uniform motion, and a body at rest remains at rest, unless acted on by a nonzero net force.

This law is implicit in Newton's second law.

Newton's Second Law

The rate at which a body's momentum changes is equal to the net force acting on the body.

Here **momentum** is the "quantity of motion," the product of mass and velocity.

Newton's Third Law

If object A exerts a force on object B, then object B exerts an oppositely directed force of equal magnitude on A.

Newton's third law says that forces come in pairs.

Solving Problems with Newton's Laws

INTERPRET Interpret the problem to be sure that you know what it's asking and that Newton's second law is the relevant concept. Identify the object of interest and all the individual **interaction forces** acting on it.

DEVELOP Draw a **free-body diagram** as described in Tactics 4.1. Develop your solution plan by writing Newton's second law, $\vec{F}_{net} = m\vec{a}$, with $\vec{F}_{net}$ expressed as the sum of the forces you've identified. Then choose a coordinate system so you can express Newton's law in components.

EVALUATE At this point the physics is done, and you're ready to execute your plan by solving Newton's second law and evaluating the numerical answer(s), if called for. Remember that even in the one-dimensional problems of this chapter, Newton's law is a vector equation; that will help you get the signs right. You need to write the components of Newton's law in the coordinate system you chose, and then solve the resulting equation(s) for the quantity(ies) of interest.

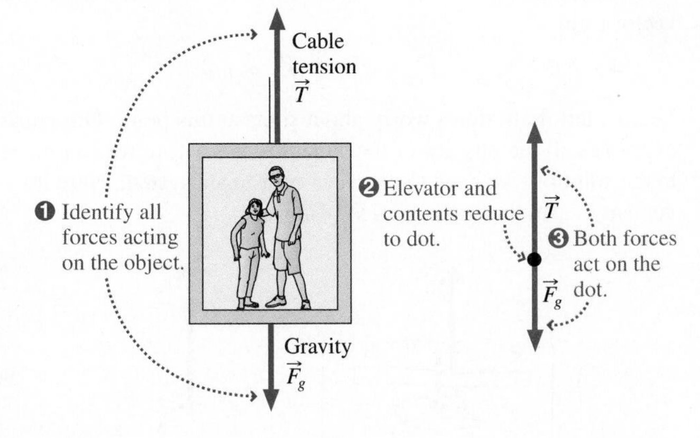

ASSESS Assess your solution to see that it makes sense. Are the numbers reasonable? Do the units work out correctly? What happens in special cases—for example, when a mass, a force, an acceleration, or an angle gets very small or very large?

Key Concepts and Equations

Mathematically, Newton's second law is $\vec{F}_{net} = d\vec{p}/dt$, where $\vec{p} = m\vec{v}$ is an object's momentum, and $\vec{F}_{net}$ is the sum of all the individual forces acting on the object. When an object has constant mass, the second law takes the familiar form

$$\vec{F}_{net} = m\vec{a} \qquad \text{(Newton's second law)}$$

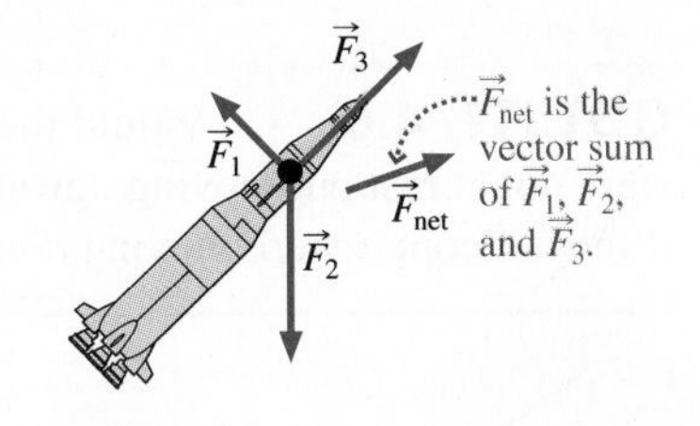

Newton's second law is a *vector* equation. To use it correctly, you must write the components of the equation in a chosen coördinate system. In one-dimensional problems the result is a single equation.

Applications

The force of gravity on an object is its **weight**. Since all objects at a given location experience the same gravitational acceleration, weight is proportional to mass:

$$\vec{w} = m\vec{g} \qquad \text{(weight on Earth)}$$

In an accelerated reference frame, an object's **apparent weight** differs from its actual weight; in particular, an object in free fall experiences **apparent weightlessness**.

Springs are convenient force-measuring devices, stretching or compressing in response to the applied force. For an ideal spring, the stretch or compression is directly proportional to the force:

$$F_s = -kx \qquad \text{(Hooke's law)}$$

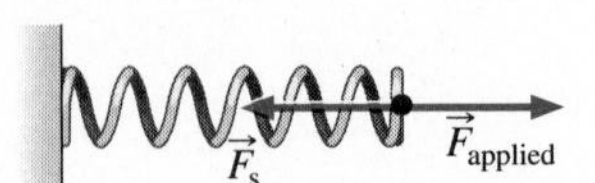

where k is the **spring constant,** with units of N/m.

MP For homework assigned on MasteringPhysics, go to www.masteringphysics.com

BIO *Biology and/or medicine-related problems* **DATA** *Data problems* **ENV** *Environmental problems* **CH** *Challenge problems* **COMP** *Computer problems*

For Thought and Discussion

1. Distinguish the Aristotelian and Galilean/Newtonian views of the natural state of motion.
2. A ball bounces off a wall with the same speed it had before it hit the wall. Has its momentum changed? Has a force acted on the ball? Has a force acted on the wall? Relate your answers to Newton's laws of motion.
3. We often use the term "inertia" to describe human sluggishness. How is this usage related to the meaning of "inertia" in physics?
4. Does a body necessarily move in the direction of the net force acting on it?
5. A truck crashes into a stalled car. A student trying to explain the physics of this event claims that no forces are involved; the car was just "in the way" so it got hit. Comment.
6. A barefoot astronaut kicks a ball, hard, across a space station. Does the ball's apparent weightlessness mean the astronaut's toes don't hurt? Explain.
7. The surface gravity on Jupiter's moon Io is one-fifth that on Earth. What would happen to your weight and to your mass if you were on Io?
8. In paddling a canoe, you push water backward with your paddle. What force actually propels the canoe forward?
9. Is it possible for a nonzero net force to act on an object without the object's speed changing? Explain.
10. As your plane accelerates down the runway, you take your keys from your pocket and suspend them by a thread. Do they hang vertically? Explain.
11. A driver tells passengers to buckle their seatbelts, invoking the law of inertia. What's that got to do with seatbelts?
12. If you cut a spring in half, is the spring constant of each new spring less than, equal to, or greater than the spring constant of the original spring? (See Problem 50 for a quantitative look at this question.)
13. As you're sitting on a chair, there's a gravitational force downward on you, and an upward normal force from the chair on you. Do these forces constitute a third-law pair? If not, what forces are paired with each of these?

Exercises and Problems

Exercises

Section 4.2 Newton's First and Second Laws

14. A subway train's mass is 1.5×10^6 kg. What force is required to accelerate the train at 2.5 m/s^2?
15. A 61-Mg railroad locomotive can exert a 0.12-MN force. At what rate can it accelerate (a) by itself and (b) when pulling a 1.4-Gg train?
16. A small plane accelerates down the runway at 7.2 m/s^2. If its propeller provides an 11-kN force, what's the plane's mass?
17. A car leaves the road traveling at 110 km/h and hits a tree, coming to a stop in 0.14 s. What average force does a seatbelt exert on a 60-kg passenger during this collision?
18. By how much does the force required to stop a car increase if the initial speed is doubled while the stopping distance remains the same?
19. **BIO** Kinesin is a "motor protein" responsible for moving materials within living cells. If it exerts a 6.0-pN force, what acceleration will it give a molecular complex with mass 3.0×10^{-18} kg?
20. Starting from rest and undergoing constant acceleration, a 940-kg racing car covers 400 m in 4.95 s. Find the force on the car.
21. In an egg-dropping contest, a student encases an 85-g egg in a large Styrofoam block. If the force on the egg can't exceed 28 N, and if the block hits the ground at 12 m/s, by how much must the Styrofoam compress on impact? Note: The acceleration associated with stopping the egg is so great that you can neglect gravity while the Styrofoam block is slowing due to contact with the ground.
22. In a front-end collision, a 1300-kg car with shock-absorbing bumpers can withstand a maximum force of 65 kN before damage occurs. If the maximum speed for a nondamaging collision is 10 km/h, by how much must the bumper be able to move relative to the car?

Section 4.4 The Force of Gravity

23. Show that the units of acceleration can be written as N/kg. Why does it make sense to give g as 9.8 N/kg when talking about mass and weight?
24. Your spaceship crashes on one of the Sun's planets. Fortunately, the ship's scales are intact and show that your weight is 532 N. If your mass is 60 kg, where are you? (*Hint:* Consult Appendix E.)
25. Your friend can barely lift a 35-kg concrete block on Earth. How massive a block could she lift on the Moon?
26. A cereal box says "net weight 340 grams." What's the actual weight (a) in SI units and (b) in ounces?
27. You're a safety engineer for a bridge spanning the U.S.–Canadian border. U.S. specifications permit a maximum load of 10 tons. What load limit should you specify on the Canadian side, where "weight" is given in kilograms?
28. The gravitational acceleration at the International Space Station's altitude is about 89% of its surface value. What's the weight of a 68-kg astronaut at this altitude?

Section 4.5 Using Newton's Second Law

29. A 50-kg parachutist descends at a steady 40 km/h. What force does air exert on the parachute?
30. A 930-kg motorboat accelerates away from a dock at 2.3 m/s^2. Its propeller provides a 3.9-kN thrust force. What drag force does the water exert on the boat?
31. An elevator accelerates downward at 2.4 m/s^2. What force does the elevator's floor exert on a 52-kg passenger?
32. At 560 metric tons, the Airbus A-380 is the world's largest airliner. What's the upward force on an A-380 when the plane is (a) flying at constant altitude and (b) accelerating upward at 1.1 m/s^2?
33. You're an engineer working on Ares I, NASA's replacement for the space shuttles. Performance specs call for a first-stage rocket capable of accelerating a total mass of 630 Mg vertically from rest to 7200 km/h in 2.0 min. You're asked to determine the required engine thrust (force) and the force exerted on a 75-kg astronaut during liftoff. What do you report?
34. You step into an elevator, and it accelerates to a downward speed of 9.2 m/s in 2.1 s. Quantitatively compare your apparent weight during this time with your actual weight.

Section 4.6 Newton's Third Law

35. What upward gravitational force does a 5600-kg elephant exert on Earth?
36. Your friend's mass is 65 kg. If she jumps off a 120-cm-high table, how far does Earth move toward her as she falls?

37. What force is necessary to stretch a spring 48 cm, if its spring constant is 270 N/m?
38. A 35-N force is applied to a spring with spring constant $k = 220$ N/m. How much does the spring stretch?
39. A spring with spring constant $k = 340$ N/m is used to weigh a 6.7-kg fish. How far does the spring stretch?

Problems

40. A 1.25-kg object is moving in the x-direction at 17.4 m/s. Just 3.41 s later, it's moving at 26.8 m/s at 34.0° to the x-axis. Find the magnitude and direction of the force applied during this time.
41. An airplane encounters sudden turbulence, and you feel momentarily lighter. If your apparent weight seems to be about 70% of your normal weight, what are the magnitude and direction of the plane's acceleration?
42. A 74-kg tree surgeon rides a "cherry picker" lift to reach the upper branches of a tree. What force does the lift exert on the surgeon when it's (a) at rest; (b) moving upward at a steady 2.4 m/s; (c) moving downward at a steady 2.4 m/s; (d) accelerating upward at 1.7 m/s^2; (e) accelerating downward at 1.7 m/s^2?
43. A dancer executes a vertical jump during which the floor pushes up on his feet with a force 50% greater than his weight. What's his upward acceleration?
44. Find expressions for the force needed to bring an object of mass m from rest to speed v (a) in time Δt and (b) over distance Δx.
45. An elevator moves upward at 5.2 m/s. What's its minimum stopping time if the passengers are to remain on the floor?
46. A 2.50-kg object is moving along the x-axis at 1.60 m/s. As it passes the origin, two forces $\vec{F}_1$ and $\vec{F}_2$ are applied, both in the y-direction (plus or minus). The forces are applied for 3.00 s, after which the object is at $x = 4.80$ m, $y = 10.8$ m. If $\vec{F}_1 = 15.0$ N, what's $\vec{F}_2$?
47. Blocks of 1.0, 2.0, and 3.0 kg are lined up on a frictionless table, as shown in Fig. 4.22, with a 12-N force applied to the leftmost block. What's the magnitude of the force that the rightmost block exerts on the middle one?

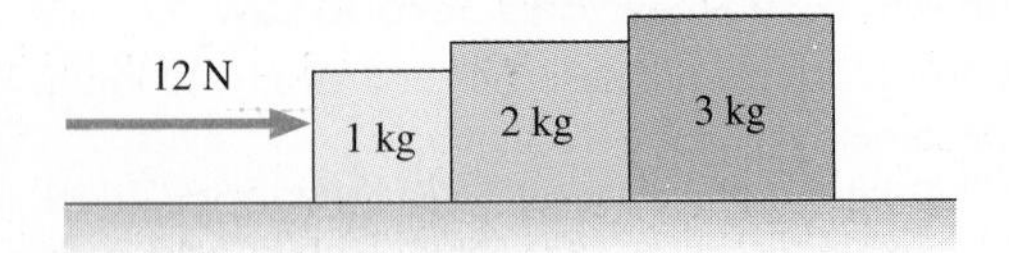

FIGURE 4.22 Problem 47

48. A child pulls an 11-kg wagon with a horizontal handle whose mass is 1.8 kg, accelerating the wagon and handle at 2.3 m/s^2. Find the tension forces at each end of the handle. Why are they different?
49. BIO Biophysicists use an arrangement of laser beams called *optical tweezers* to manipulate microscopic objects. In a particular experiment, optical tweezers exerting a force of 0.373 pN were used to stretch a DNA molecule by 2.30 μm. What was the spring constant of the DNA?
50. A force F is applied to a spring of spring constant k_0, stretching it a distance x. Consider the spring to be made up of two smaller springs of equal length, with the same force F still applied. Use $F = -kx$ to find the spring constant k_1 of each of the smaller springs. Your result is a quantitative answer to Question 12.
51. A 2200-kg airplane pulls two gliders, the first of mass 310 kg and the second of mass 260 kg, down the runway with acceleration 1.9 m/s^2 (Fig. 4.23). Neglecting the mass of the two ropes and any frictional forces, determine the magnitudes of (a) the horizontal thrust of the plane's propeller; (b) the tension force in the first rope; (c) the tension force in the second rope; and (d) the net force on the first glider.

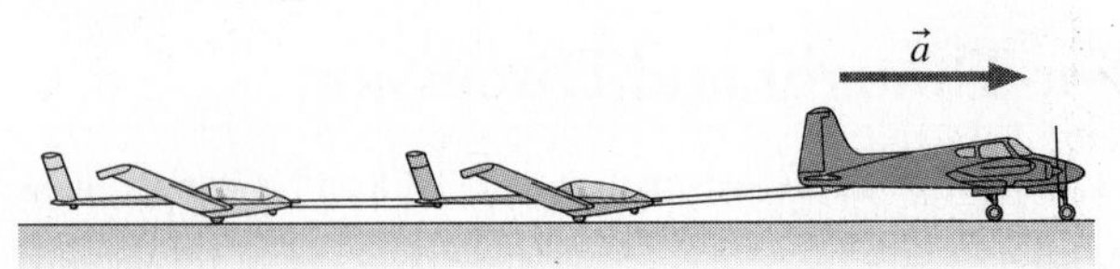

FIGURE 4.23 Problem 51

52. BIO A biologist is studying the growth of rats on the Space Station. To determine a rat's mass, she puts it in a 320-g cage, attaches a spring scale, and pulls so that the scale reads 0.46 N. If rat and cage accelerate at 0.40 m/s^2, what's the rat's mass?
53. An elastic towrope has spring constant 1300 N/m. It's connected between a truck and a 1900-kg car. As the truck tows the car, the rope stretches 55 cm. Starting from rest, how far do the truck and the car move in 1 min? Assume the car experiences negligible friction.
54. A 2.0-kg mass and a 3.0-kg mass are on a horizontal frictionless surface, connected by a massless spring with spring constant $k = 140$ N/m. A 15-N force is applied to the larger mass, as shown in Fig. 4.24. How much does the spring stretch from its equilibrium length?

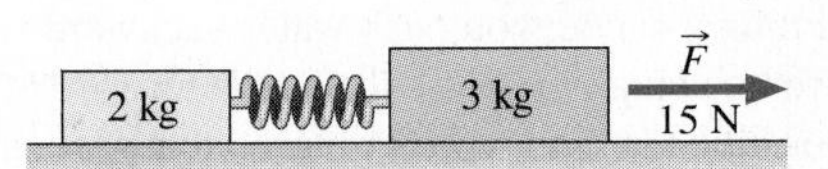

FIGURE 4.24 Problem 54

55. You're an automotive engineer designing the "crumple zone" of a new car—the region that compresses as the car comes to a stop in a head-on collision. If the maximum allowable force on a passenger in a 70-km/h collision is 20 times the passenger's weight, what do you specify for the amount of compression in the crumple zone?
56. BIO Frogs' tongues dart out to catch insects, with maximum tongue accelerations of about 250 m/s^2. What force is needed to give a 500-mg tongue such an acceleration?
57. Two large crates, with masses 640 kg and 490 kg, are connected by a stiff, massless spring ($k = 8.1$ kN/m) and propelled along an essentially frictionless factory floor by a horizontal force applied to the more massive crate. If the spring compresses 5.1 cm, what's the applied force?
58. What force do the blades of a 4300-kg helicopter exert on the air when the helicopter is (a) hovering at constant altitude; (b) dropping at 21 m/s with speed decreasing at 3.2 m/s^2; (c) rising at 17 m/s with speed increasing at 3.2 m/s^2; (d) rising at a steady 15 m/s; (e) rising at 15 m/s with speed decreasing at 3.2 m/s^2?
59. What engine thrust (force) is needed to accelerate a rocket of mass m (a) downward at $1.40g$ near Earth's surface; (b) upward at $1.40g$ near Earth's surface; (c) at $1.40g$ in interstellar space, far from any star or planet?
60. Your engineering firm is asked to specify the maximum load for the elevators in a new building. Each elevator has mass 490 kg when empty and maximum acceleration 2.24 m/s^2. The elevator cables can withstand a maximum tension of 19.5 kN before breaking. For safety, you need to ensure that the tension never exceeds two-thirds of that value. What do you specify for the maximum load? How many 70-kg people is that?

61. With its fuel tanks half full, an F-35A jet fighter has mass 18 Mg and engine thrust 191 kN. An Airbus A-380 has mass 560 Mg and total engine thrust 1.5 MN. Could either aircraft climb vertically with no lift from its wings? If so, what vertical acceleration could it achieve?

62. Two springs have the same unstretched length but different spring constants, k_1 and k_2. (a) If they're connected side by side and stretched a distance x, as shown in Fig. 4.25a, show that the force exerted by the combination is $(k_1 + k_2)x$. (b) If they're connected end to end (Fig. 4.25b) and the combination is stretched a distance x, show that they exert a force $k_1k_2x/(k_1 + k_2)$.

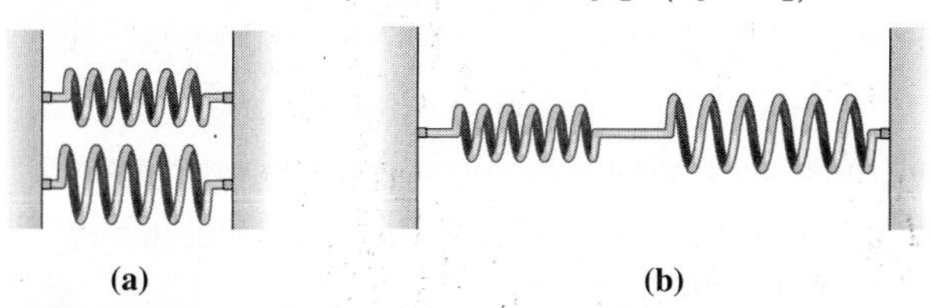

FIGURE 4.25 Problem 62

63. Although we usually write Newton's second law for one-dimensional motion in the form $F = ma$, which holds when mass is constant, a more fundamental version is $F = \dfrac{d(mv)}{dt}$. Consider an object whose mass is changing, and use the product rule for derivatives to show that Newton's law then takes the form $F = ma + v\dfrac{dm}{dt}$.

64. A railroad car is being pulled beneath a grain elevator that dumps grain at the rate of 450 kg/s. Use the result of Problem 63 to find the force needed to keep the car moving at a constant 2.0 m/s.

65. A block 20% more massive than you hangs from a rope that goes over a frictionless, massless pulley. With what acceleration must you climb the other end of the rope to keep the block from falling?

66. You're asked to calibrate a device used to measure vertical acceleration in helicopters. The device consists of a mass m hanging from a massless spring of constant k. Your job is to express the acceleration as a function of the position y of the mass relative to where it is when there's no acceleration. Take the positive y-axis to point upward.

67. **CH** A spider of mass m_s drapes a silk thread of negligible mass over a stick with its far end a distance h off the ground, as shown in Fig. 4.26. A drop of dew lubricates the stick, making friction negligible. The spider waits on the ground until a fly of mass m_f ($m_f > m_s$) lands on the other end of the silk and sticks to it. The spider immediately begins to climb her end of the silk. (a) With what acceleration must she climb to keep the fly from falling? If she climbs with acceleration a_s, at what height y will she encounter the fly?

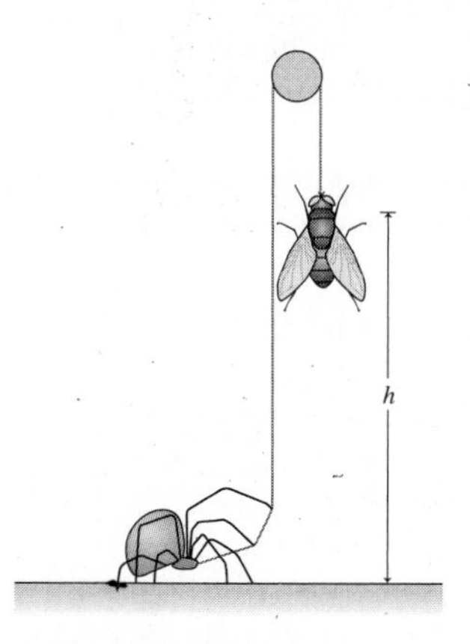

FIGURE 4.26 Problem 67

68. **DATA** Figure 4.27 shows vertical accelerometer data from an iPhone that was dropped onto a pillow. The phone's accelerometer, like all accelerometers, can't distinguish gravity from acceleration, so it reads $1g$ when it's not accelerating and $0g$ when it's in free fall. Interpret the graph to determine (a) how long the phone was in free fall and therefore how far it fell, (b) how many times it bounced, (c) the maximum force the phone experienced, expressed in terms of its weight w, and (d) when it finally came completely to rest. (Note: The phone was held flat when dropped, with the screen up for protection. In that orientation, it recorded negative values for acceleration; the graph shows the corresponding positive values that would have been recorded had it fallen screen side down.)

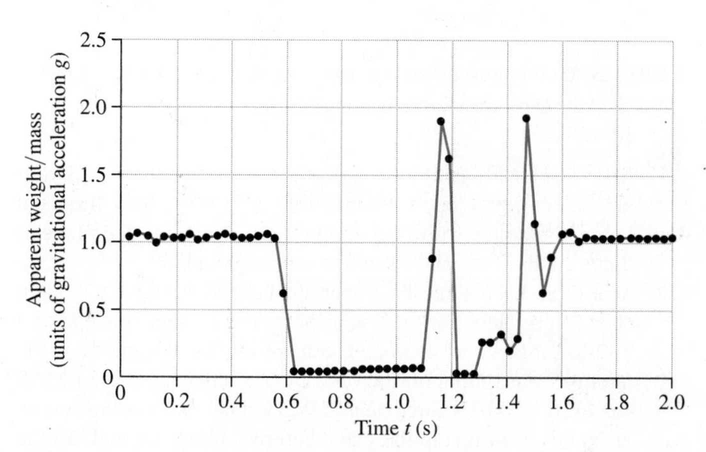

FIGURE 4.27 Accelerometer data for Problem 68, taken with an iPhone. The accelerometer can't distinguish gravity from acceleration, so what it actually measures is apparent weight divided by mass, expressed in units of g.

69. **CH** A hockey stick is in contact with a 165-g puck for 22.4 ms; during this time, the force on the puck is given approximately by $F(t) = a + bt + ct^2$, where $a = -25.0$ N, $b = 1.25\times10^5$ N/s, and $c = -5.58\times10^6$ N/s^2. Determine (a) the speed of the puck after it leaves the stick and (b) how far the puck travels while it's in contact with the stick.

70. **DATA** After parachuting through the Martian atmosphere, the Mars Science Laboratory executed a complex series of maneuvers that successfully placed the rover *Curiosity* on the surface of Mars in 2012. The final ~22 s of the landing involved, in this order, firing rockets (1) to maintain a constant downward velocity of 32 m/s, (2) to achieve a constant deceleration that brought the downward speed to 0.75 m/s, and (3) to hold that constant velocity while the rover was lowered on cables from the rest of the spacecraft (see this chapter's opening image). The rover's touchdown was indicated by a sudden decrease in the rocket thrust needed to maintain constant velocity. Figure 4.28 shows the rocket thrust (upward force) as a function of time during these final 22 s of the flight and the first few seconds after touchdown. (a) Identify the two constant-velocity phases, the constant-deceleration phase, and the post-touchdown phase. (b) Find the magnitude of the spacecraft's acceleration during the constant-deceleration phase. Finally, determine (c) the mass of the so-called powered descent vehicle (PDF), meaning the spacecraft with the rover attached and (d) the mass of the rover

alone. Remember that all this happened at Mars, so you'll need to consult Appendix E.

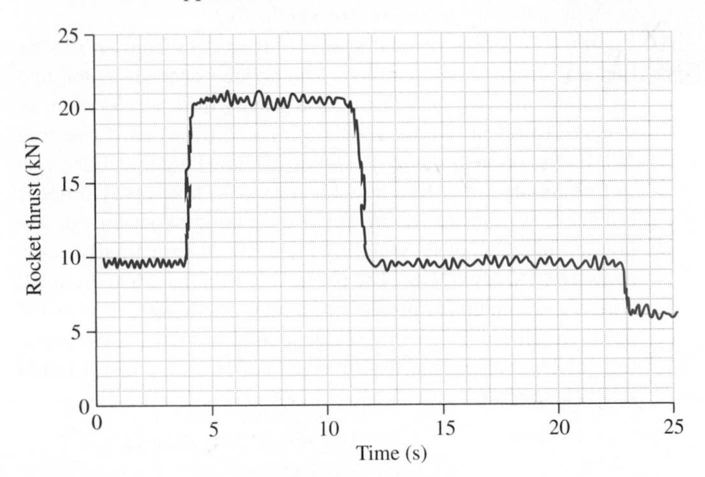

FIGURE 4.28 Rocket thrust (upward force of rocket engines) during the final descent of the Mars rover Curiosity (Problem 70).

71. Your airplane is caught in a brief, violent downdraft. To your amazement, pretzels rise vertically off your seatback tray, and you estimate their upward acceleration relative to the plane at 2 m/s^2. What's the plane's downward acceleration?
72. You're assessing the Engineered Material Arresting System (EMAS) at New York's JFK airport. The system consists of a 132-m-long bed of crushable cement blocks, designed to stop aircraft from sliding off the runway in emergencies. The EMAS can exert a 300-kN force on a 55-Mg jetliner that hits the system at 36 m/s. Can it stop the plane before it plows through all the blocks?
73. Two masses are joined by a massless string. A 30-N force applied vertically to the upper mass gives the system a constant upward acceleration of 3.2 m/s^2. If the string tension is 18 N, what are the two masses?
74. A mass M hangs from a uniform rope of length L and mass m. Find an expression for the rope tension as a function of the distance y measured downward from the top of the rope.
75. CH "Jerk" is the rate of change of acceleration, and it's what can make you sick on an amusement park ride. In a particular ride, a car and passengers with total mass M are subject to a force given by $F = F_0 \sin \omega t$, where F_0 and ω are constants. Find an expression for the maximum jerk.

Passage Problems

Laptop computers are equipped with accelerometers that sense when the device is dropped and then put the hard drive into a protective mode. Your computer geek friend has written a program that reads the accelerometer and calculates the laptop's apparent weight. You're amusing yourself with this program on a long plane flight. Your laptop weighs just 5 pounds, and for a long time that's what the program reports. But then the "Fasten Seatbelt" light comes on as the plane encounters turbulence. For the next 12 seconds, your laptop reports rapid changes in apparent weight, as shown in Fig. 4.29.

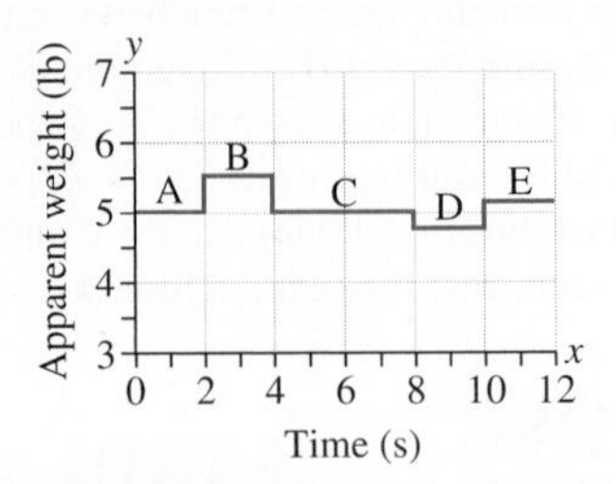

FIGURE 4.29 The laptop's apparent weight (Passage Problems 76–79).

76. At the first sign of turbulence, the plane's acceleration
    a. is upward.
    b. is downward.
    c. is impossible to tell from the graph.
77. The plane's vertical acceleration has its greatest magnitude
    a. during interval B.
    b. during interval C.
    c. during interval D.
78. During interval C, you can conclude for certain that the plane is
    a. at rest.
    b. accelerating upward.
    c. accelerating downward.
    d. moving with constant vertical velocity.
79. The magnitude of the greatest vertical acceleration the plane undergoes during the time shown on the graph is approximately
    a. 0.5 m/s^2.
    b. 1 m/s^2.
    c. 5 m/s^2.
    d. 10 m/s^2.

Answers to Chapter Questions

Answer to Chapter Opening Question

The engineers needed to consider Martian gravity, the upward thrust of the sky crane's rockets, and the tension in the cables used to lower the rover from the sky crane.

Answers to GOT IT? Questions

4.1 (b)
4.2 (b) (Look at Fig. 4.4.)
4.3 (c) All would move in straight lines.
4.4 (1) (a); (2) (b); (3) (b); (4) (a); (5) (c)
4.5 (c) less than 2 N
4.6 (1) No, because acceleration is still zero; (2) No, because the direction of the velocity is irrelevant to the acceleration

Using Newton's Laws

What You Know

- You're a Newtonian, meaning you understand that forces cause *change* in motion.
- You recognize how Newton's second law, $\vec{F}_{\text{net}} = m\vec{a}$, embodies a vector relation between force and change in motion.
- You can work one-dimensional problems involving Newton's second law, following a clear problem-solving strategy.
- You've seen how Newton's third law describes the forces between interacting objects as always coming in pairs.

What You're Learning

- You'll learn to apply Newton's second law in two dimensions, expanding the strategy you learned in Chapter 4.
- You'll practice choosing coordinate systems and using trigonometry to break force vectors into components.
- You'll learn about the force of friction and how to incorporate friction into problems involving Newton's laws.
- You'll come out of the chapter able to analyze the richness of motion as it occurs throughout our three-dimensional universe.

How You'll Use It

- Newton's second law will help in developing energy concepts in Chapters 6 and 7.
- In Chapter 8, you'll see how Newton's laws describe motion in response to gravity as it applies to space flight and the cosmos.
- In Chapter 9, you'll learn how Newton's laws apply to systems of particles.
- In Chapters 10 and 11, rotational analogs of Newton's laws will help you understand rotational motion.
- Newton's laws will help you understand the motion of waves and fluids in Part 2, and the molecular motion that gives rise to temperature and pressure in Part 3.

Why does an airplane tip when it's turning?

Chapter 4 introduced Newton's three laws of motion and used them in one-dimensional situations. Now we apply Newton's laws in two dimensions. This material is at the heart of Newtonian physics, from textbook problems to systems that guide spacecraft to distant planets. The chapter consists largely of examples, to help you learn to apply Newton's laws and also to appreciate their wide range of applicability. We also introduce frictional forces and elaborate on circular motion. As you study the diverse examples, keep in mind that they all follow from the underlying principles embodied in Newton's laws.

5.1 Using Newton's Second Law

Newton's second law, $\vec{F}_{\text{net}} = m\vec{a}$, is the cornerstone of mechanics. We can use it to develop faster skis, engineer skyscrapers, design safer roads, compute a rocket's thrust, and solve myriad other practical problems.

We'll work Example 5.1 in great detail, applying Problem-Solving Strategy 4.1. Follow this example closely, and try to understand how our strategy is grounded in Newton's basic statement that the net force on an object determines that object's acceleration.

EXAMPLE 5.1 Newton's Law in Two Dimensions: Skiing

A skier of mass $m = 65$ kg glides down a slope at angle $\theta = 32°$, as shown in Fig. 5.1. Find (a) the skier's acceleration and (b) the force the snow exerts on the skier. The snow is so slippery that you can neglect friction.

FIGURE 5.1 What's the skier's acceleration?

INTERPRET This problem is about the skier's motion, so we identify the skier as the object of interest. Next, we identify the forces acting on the object. In this case there are just two: the downward force of gravity and the normal force the ground exerts on the skier. As always, the normal force is perpendicular to the surfaces in contact—in this case perpendicular to the slope.

DEVELOP Our strategy for using Newton's second law calls for drawing a free-body diagram that shows only the object and the forces acting on it; that's Fig. 5.2. Determining the relevant equation is straightforward here: It's Newton's second law, $\vec{F}_{net} = m\vec{a}$. We write Newton's law explicitly for the forces we've identified:

$$\vec{F}_{net} = \vec{n} + \vec{F}_g = m\vec{a}$$

To apply Newton's law in two dimensions, we need to choose a coordinate system so that we can write this vector equation in components. Since the coordinate system is just a mathematical construct, you're free to choose any coordinate system you like—but a smart choice can make the problem a lot easier. In this example, the normal force is perpendicular to the slope and the skier's acceleration is along the slope. If you choose a coordinate system with axes perpendicular and parallel to the slope, then these two vectors will lie along the coordinate axes, and you'll have only one vector—the gravitational force—that you'll need to break into components. So a tilted coordinate system makes this problem easier, and we've sketched this system on the free-body diagram in Fig. 5.2. But, again, any coordinate system will do. In Problem 34, you can rework this example in a horizontal/vertical coordinate system—getting the same answer at the expense of a lot more algebra.

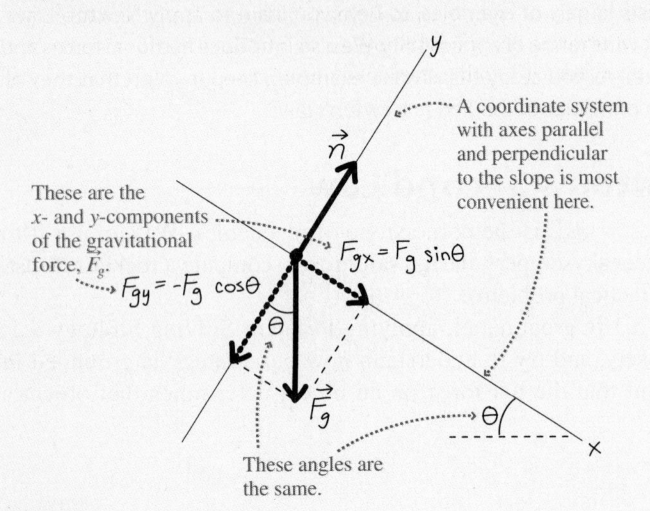

FIGURE 5.2 Our free-body diagram for the skier.

EVALUATE The rest is math. First, we write the components of Newton's law in our coordinate system. That means writing a version of the equation for each coordinate direction by removing the arrows indicating vector quantities and adding subscripts for the coordinate directions:

$$x\text{-component:} \quad n_x + F_{gx} = ma_x$$

$$y\text{-component:} \quad n_y + F_{gy} = ma_y$$

Don't worry about signs until the next step, when we actually evaluate the individual terms in these equations. Let's begin with the x equation. With the x-axis parallel and the y-axis perpendicular to the slope, the normal force has only a y-component, so $n_x = 0$. Meanwhile, the acceleration points downslope—that's the positive x-direction—so $a_x = a$, the magnitude of the acceleration. Only gravity has two nonzero components and, as Fig. 5.2 shows, trigonometry gives $F_{gx} = F_g \sin\theta$. But F_g, the magnitude of the gravitational force, is just mg, so $F_{gx} = mg \sin\theta$. This component has a positive sign because our x-axis slopes downward. Then, with $n_x = 0$, the x equation becomes

$$x\text{-component:} \quad mg \sin\theta = ma$$

On to the y equation. The normal force points in the positive y-direction, so $n_y = n$, the magnitude of the normal force. The acceleration has no component perpendicular to the slope, so $a_y = 0$. Figure 5.2 shows that $F_{gy} = -F_g \cos\theta = -mg \cos\theta$, so the y equation is

$$y\text{-component:} \quad n - mg \cos\theta = 0$$

Now we can evaluate to get the answers. The x equation solves directly to give

$$a = g \sin\theta = (9.8 \text{ m/s}^2)(\sin 32°) = 5.2 \text{ m/s}^2$$

which is the acceleration we were asked to find in (a). Next, we solve the y equation to get $n = mg \cos\theta$. Putting in the numbers gives $n = 540$ N. This is the answer to (b), the force the snow exerts on the skier.

ASSESS A look at two special cases shows that these results make sense. First, suppose $\theta = 0°$, so the surface is horizontal. Then the x equation gives $a = 0$, as expected. The y equation gives $n = mg$, showing that a horizontal surface exerts a force that just balances the skier's weight. At the other extreme, consider $\theta = 90°$, so the slope is a vertical cliff. Then the skier falls freely with acceleration g, as expected. In this case $n = 0$ because there's no contact between skier and slope. At intermediate angles, the slope's normal force lessens the effect of gravity, resulting in a lower acceleration. As the x equation shows, that acceleration is independent of the skier's mass—just as in the case of a vertical fall. The force exerted by the snow—here $mg \cos\theta$, or 540 N—is less than the skier's weight mg because the slope has to balance only the perpendicular component of the gravitational force.

If you understand this example, you should be able to apply Newton's second law confidently in other problems involving motion with forces in two dimensions. ■

Sometimes we're interested in finding the conditions under which an object won't accelerate. Examples are engineering problems, such as ensuring that bridges and buildings don't fall down, and physiology problems involving muscles and bones. Next we give a wilder example.

PhET: The Ramp

EXAMPLE 5.2 Objects at Rest: Bear Precautions

To protect her 17-kg pack from bears, a camper hangs it from ropes between two trees (Fig. 5.3). What's the tension in each rope?

FIGURE 5.3 Bear precautions.

INTERPRET Here the pack is the object of interest. The only forces acting on it are gravity and tension forces in the two halves of the rope. To keep the pack from accelerating, they must sum to zero net force.

DEVELOP Figure 5.4 is our free-body diagram for the pack. The relevant equation is again Newton's second law, $\vec{F}_{\text{net}} = m\vec{a}$—this time with $\vec{a} = \vec{0}$. For the three forces acting on the pack, Newton's law is then $\vec{T}_1 + \vec{T}_2 + \vec{F}_g = \vec{0}$. Next, we need a coordinate system. The two rope tensions point in different directions that aren't perpendicular, so it doesn't make sense to align a coordinate axis with either of them. Instead, a horizontal/vertical system is simplest.

EVALUATE First we need to write Newton's law in components. Formally, we have $T_{1x} + T_{2x} + F_{gx} = 0$ and $T_{1y} + T_{2y} + F_{gy} = 0$ for the component equations. Figure 5.4 shows the components of the tension forces, and we see that $F_{gx} = 0$ and $F_{gy} = -F_g = -mg$. So our component equations become

$$x\text{-component:} \quad T_1 \cos\theta - T_2 \cos\theta = 0$$

$$y\text{-component:} \quad T_1 \sin\theta + T_2 \sin\theta - mg = 0$$

The x equation tells us something that's apparent from the symmetry of the situation: Since the angle θ is the same for both halves

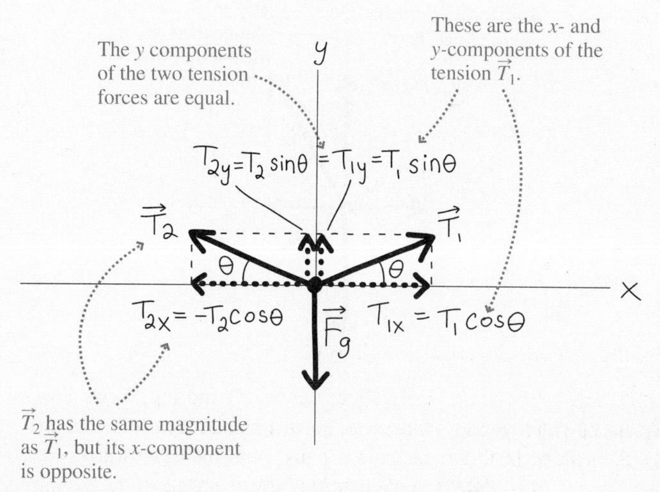

FIGURE 5.4 Our free-body diagram for the pack.

of the rope, the magnitudes T_1 and T_2 of the tension forces are the same. Let's just call the magnitude T: $T_1 = T_2 = T$. Then the terms $T_1 \sin\theta$ and $T_2 \sin\theta$ in the y equation are equal, and the equation becomes $2T \sin\theta - mg = 0$, which gives

$$T = \frac{mg}{2 \sin\theta} = \frac{(17\text{ kg})(9.8\text{ m/s}^2)}{2 \sin 22°} = 220\text{ N}$$

ASSESS Make sense? Let's look at some special cases. With $\theta = 90°$, the rope hangs vertically, $\sin\theta = 1$, and the tension in each half of the rope is $\frac{1}{2}mg$. That makes sense, because each piece of the rope supports half the pack's weight. But as θ gets smaller, the ropes become more horizontal and the tension increases. That's because the vertical tension components together still have to support the pack's weight—but now there's a horizontal component as well, increasing the overall tension. Ropes break if the tension becomes too great, and in this example that means the rope's so-called breaking tension must be considerably greater than the pack's weight. If $\theta = 0$, in fact, the tension would become infinite—demonstrating that it's impossible to support a weight with a purely horizontal rope. ■

EXAMPLE 5.3 Objects at Rest: Restraining a Ski Racer

A starting gate acts horizontally to restrain a 62-kg ski racer on a frictionless 30° slope (Fig. 5.5). What horizontal force does the starting gate apply to the skier?

INTERPRET Again, we want the skier to remain unaccelerated. The skier is the object of interest, and we identify three forces acting: gravity, the normal force from the slope, and a horizontal restraining force $\vec{F}_{\text{h}}$ that we're asked to find.

DEVELOP Figure 5.6 is our free-body diagram. The applicable equation is Newton's second law. Again, we want $\vec{a} = \vec{0}$, so with

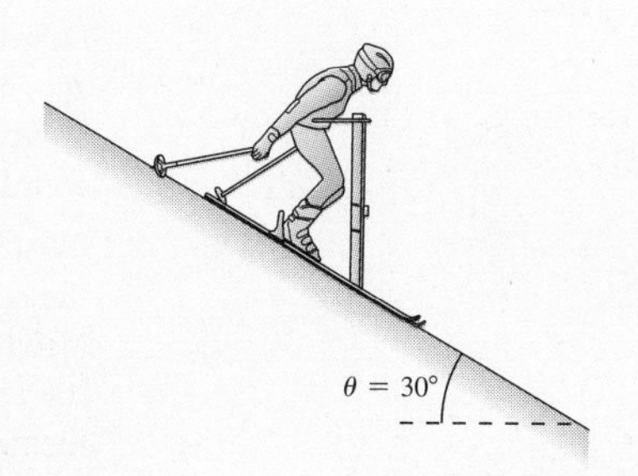

FIGURE 5.5 Restraining a skier.

(continued)

the forces we identified, $\vec{F}_{\text{net}} = m\vec{a}$ becomes $\vec{F}_{\text{h}} + \vec{n} + \vec{F}_g = \vec{0}$. Developing our solution strategy, we choose a coordinate system. With two forces now either horizontal or vertical, a horizontal/vertical system makes the most sense; we've shown this coordinate system in Fig. 5.6.

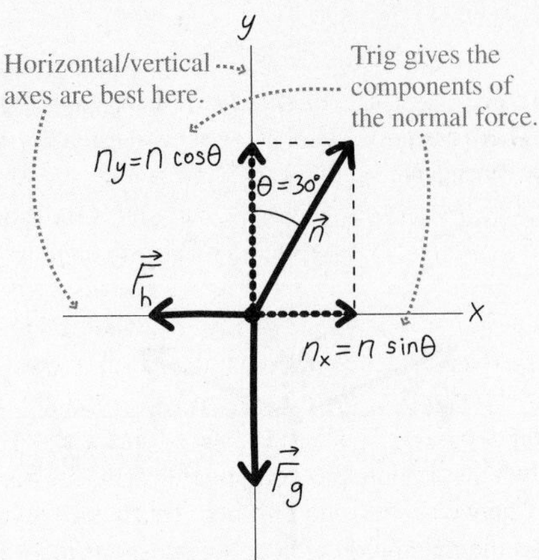

FIGURE 5.6 Our free-body diagram for the restrained skier.

EVALUATE As usual, the component equations follow directly from the vector form of Newton's law: $F_{\text{h}x} + n_x + F_{gx} = 0$ and $F_{\text{h}y} + n_y + F_{gy} = 0$. Figure 5.6 gives the components of the normal force and shows that $F_{\text{h}x} = -F_{\text{h}}$, $F_{gy} = -F_g = -mg$, and $F_{gx} = F_{\text{h}y} = 0$. Then the component equations become

$$x\text{:}\quad -F_{\text{h}} + n\sin\theta = 0 \qquad y\text{:}\quad n\cos\theta - mg = 0$$

There are two unknowns here—namely, the horizontal force F_{h} that we're looking for and the normal force n. We can solve the y equation to get $n = mg/\cos\theta$. Using this expression in the x equation and solving for F_{h} then give the answer:

$$F_{\text{h}} = \frac{mg}{\cos\theta}\sin\theta = mg\tan\theta = (62\text{ kg})(9.8\text{ m/s}^2)(\tan 30°) = 350\text{ N}$$

ASSESS Again, let's look at the extreme cases. With $\theta = 0$, we have $F_{\text{h}} = 0$, showing that it doesn't take any force to restrain a skier on flat ground. But as the slope becomes more vertical, $\tan\theta \to \infty$, and in the vertical limit, it becomes impossible to restrain the skier with a purely horizontal force. ■

GOT IT? 5.1 A roofer's toolbox rests on an essentially frictionless metal roof with a 45° slope, secured by a horizontal rope as shown. Is the rope tension (a) greater than, (b) less than, or (c) equal to the box's weight?

How does the rope tension compare with the toolbox weight?

45°

5.2 Multiple Objects

In the preceding examples there was a single object of interest. But often we have several objects whose motion is linked. Our Newton's law strategy still applies, with extensions to handle multiple objects.

PROBLEM-SOLVING STRATEGY 5.1 Newton's Second Law and Multiple Objects

INTERPRET Interpret the problem to be sure that you know what it's asking and that Newton's second law is the relevant concept. Identify the *multiple* objects of interest and all the individual interaction forces acting on *each* object. Finally, identify *connections* between the objects and the resulting *constraints* on their motions.

DEVELOP Draw a *separate* free-body diagram showing all the forces acting on *each* object. Develop your solution plan by writing Newton's law, $\vec{F}_{\text{net}} = m\vec{a}$, separately for each object, with $\vec{F}_{\text{net}}$ expressed as the sum of the forces acting on that object. Then choose a coordinate system appropriate to each object, so you can express each Newton's law equation in components. The coordinate systems for different objects don't need to have the same orientation.

EVALUATE At this point the physics is done, and you're ready to execute your plan by solving the equations and evaluating the numerical answer(s), if called for. Write the components of Newton's law for each object in the coordinate system you chose for each. You can then solve the resulting equations for the quantity(ies) you're interested in, using the connections you identified to relate the quantities that appear in the equations for the different objects.

ASSESS Assess your solution to see whether it makes sense. Are the numbers reasonable? Do the units work out correctly? What happens in special cases—for example, when a mass, a force, an acceleration, or an angle gets very small or very large?

EXAMPLE 5.4 Multiple Objects: Rescuing a Climber

A 73-kg climber finds himself dangling over the edge of an ice cliff, as shown in Fig. 5.7. Fortunately, he's roped to a 940-kg rock located 51 m from the edge of the cliff. Unfortunately, the ice is frictionless, and the climber accelerates downward. What's his acceleration, and how much time does he have before the rock goes over the edge? Neglect the rope's mass.

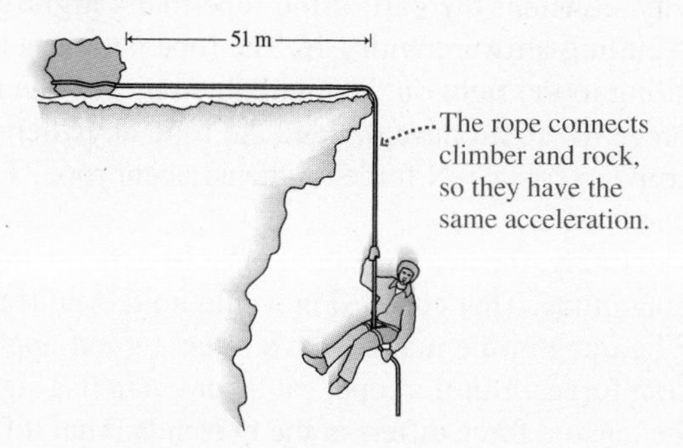

FIGURE 5.7 A climber in trouble.

INTERPRET We need to find the climber's acceleration, and from that we can get the time before the rock goes over the edge. We identify two objects of interest, the climber and the rock, and we note that the rope connects them. There are two forces on the climber: gravity and the upward rope tension. There are three forces on the rock: gravity, the normal force from the surface, and the rightward-pointing rope tension.

DEVELOP Figure 5.8 shows a free-body diagram for each object. Newton's law applies to each, so we write two vector equations:

$$\text{climber:} \quad \vec{T}_c + \vec{F}_{gc} = m_c\vec{a}_c$$
$$\text{rock:} \quad \vec{T}_r + \vec{F}_{gr} + \vec{n} = m_r\vec{a}_r$$

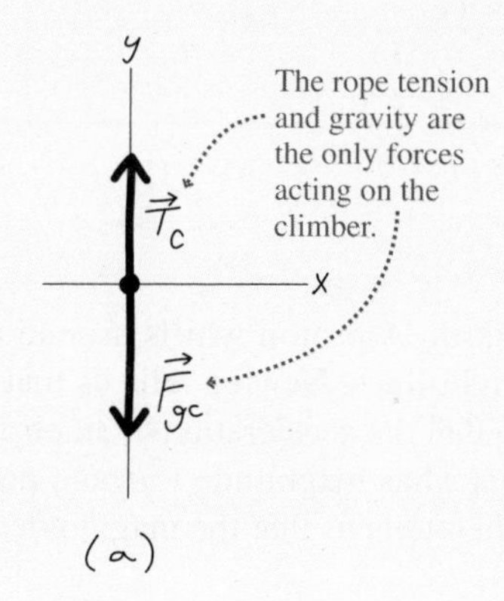

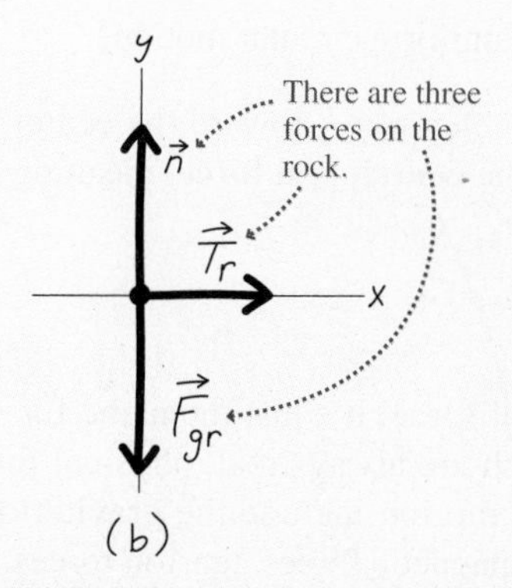

FIGURE 5.8 Our free-body diagrams for (a) the climber and (b) the rock.

where the subscripts c and r stand for climber and rock, respectively. All forces are either horizontal or vertical, so we can use the same horizontal/vertical coordinate system for both objects, as shown in Fig. 5.8.

EVALUATE Again, the component equations follow directly from the vector forms. There are no horizontal forces on the climber, so only the y equation is significant. We're skilled enough now to skip the intermediate step of writing the components without their actual expressions, and we see from Fig. 5.8*a* that the y-component of Newton's law for the climber becomes $T_c - m_c g = m_c a_c$. For the rock, the only horizontal force is the tension, pointing to the right or positive x-direction, so the rock's x equation is $T_r = m_r a_r$. Since it's on a horizontal surface, the rock has no vertical acceleration, so its y equation is $n - m_r g = 0$. In writing these equations, we haven't added the subscripts x and y because each vector has only a single nonzero component. Now we need to consider the connection between rock and climber. That's the rope, and its presence means that the magnitude of both accelerations is the same. Calling that magnitude a, we can see from Fig. 5.8 that $a_r = a$ and $a_c = -a$. The value for the rock is positive because $\vec{T}_r$ points to the right, which we defined as the positive x-direction; the value for the climber is negative because he's accelerating downward, which we defined as the negative y-direction. The rope, furthermore, has negligible mass, so the tension throughout it must be the same (more on this point just after the example). Therefore, the tension forces on rock and climber have equal magnitude T, so $T_c = T_r = T$. Putting this all together gives us three equations:

$$\text{climber, } y\text{:} \quad T - m_c g = -m_c a$$
$$\text{rock, } x\text{:} \quad T = m_r a$$
$$\text{rock, } y\text{:} \quad n - m_r g = 0$$

The rock's x equation gives the tension, which we can substitute into the climber's equation to get $m_r a - m_c g = -m_c a$. Solving for a then gives the answer:

$$a = \frac{m_c g}{m_c + m_r} = \frac{(73\text{ kg})(9.8\text{ m/s}^2)}{(73\text{ kg} + 940\text{ kg})} = 0.71\text{ m/s}^2$$

We didn't need the rock's y equation, which just says that the normal force supports the rock's weight.

ASSESS Again, let's look at special cases. Suppose the rock's mass is zero; then our expression gives $a = g$. In this case there's no rope tension and the climber plummets in free fall. Also, acceleration decreases as the rock's mass increases, so with an infinitely massive rock, the climber would dangle without accelerating. You can see physically why our expression for acceleration makes sense. The gravitational force $m_c g$ acting on the climber has to accelerate both rock and climber—whose combined mass is $m_c + m_r$. The result is an acceleration of $m_c g/(m_c + m_r)$.

We're not quite done because we were also asked for the time until the rock goes over the cliff, putting the climber in real trouble. We interpret this as a problem in one-dimensional motion from Chapter 2, and we determine that Equation 2.10, $x = x_0 + v_0 t + \frac{1}{2}at^2$, applies. With $x_0 = 0$ and $v_0 = 0$, we have $x = \frac{1}{2}at^2$. We evaluate by solving for t and using the acceleration we found along with $x = 51$ m for the distance from the rock to the cliff edge:

$$t = \sqrt{\frac{2x}{a}} = \sqrt{\frac{(2)(51\text{ m})}{0.71\text{ m/s}^2}} = 12\text{ s}$$

■

✓TIP Ropes and Tension Forces

Tension forces can be confusing. In Example 5.4, the rock pulls on one end of the rope and the climber pulls on the other. So why isn't the rope tension the sum of these forces? And why is it important to neglect the rope's mass? The answers lie in the meaning of tension.

Figure 5.9 shows a situation similar to Example 5.4, with two people pulling on opposite ends of a rope with forces of 1 N each. You might think the rope tension is then 2 N, but it's not. To see why, consider the part of the rope that's highlighted in Fig. 5.9*b*. To the left is the hand pulling leftward with 1 N. The rope isn't accelerating, so there must be a 1-N force pulling to the right on the highlighted piece. The remainder of the rope provides that force. We could have divided the rope anywhere, so we conclude that every part of the rope exerts a 1-N force on the adjacent rope. That 1-N force is what we mean by the rope tension.

As long as the rope isn't accelerating, the net force on it must be zero, so the forces at the two ends have the same magnitude. That conclusion would hold even if the rope were accelerating—provided it had negligible mass. That's often a good approximation in situations involving tension forces. But if a rope, cable, or chain has significant mass and is accelerating, then the tension force differs at the two ends. That difference, according to Newton's second law, is the net force that accelerates the rope.

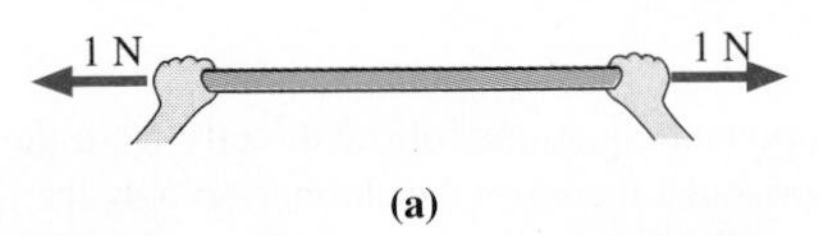

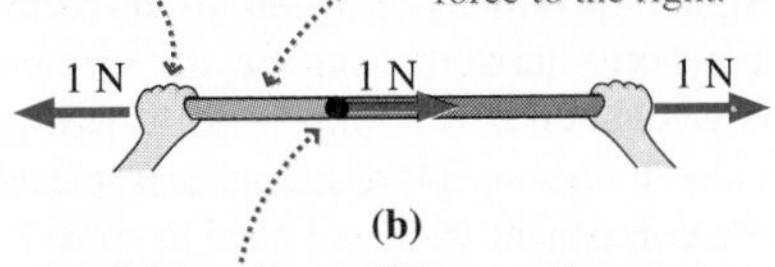

FIGURE 5.9 Understanding tension forces.

Video Tutor Demo | **Tension in String between Hanging Weights**

GOT IT? 5.2 In the figure below we've replaced one of the hands from Fig. 5.9 with a hook attaching the rope to a wall. On the right, the hand still pulls with a 1-N force. How do the forces now differ from what they were in Fig. 5.9? (a) there's no difference; (b) the force exerted by the hook is zero; (c) the rope tension is now 0.5 N

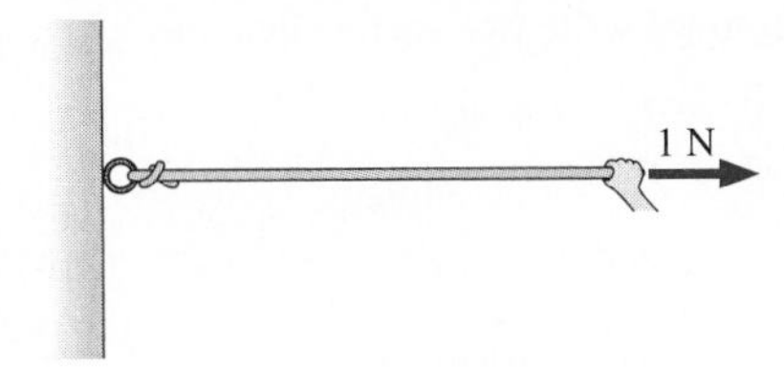

5.3 Circular Motion

A car rounds a curve. A satellite circles Earth. A proton whirls around a giant particle accelerator. Since they're not going in straight lines, Newton tells us that a force acts on each (Fig. 5.10). We know from Section 3.6 that the acceleration of an object moving with constant speed v in a circular path of radius r has magnitude v^2/r and points toward the center of the circle. Newton's second law then tells us that the magnitude of the net force on an object of mass m in circular motion is

$$F_{\text{net}} = ma = \frac{mv^2}{r} \quad \text{(uniform circular motion)} \tag{5.1}$$

The force is in the same direction as the acceleration—toward the center of the circular path. For that reason it's sometimes called the **centripetal force**, meaning center-seeking (from the Latin *centrum*, "center," and *petere*, "to seek").

FIGURE 5.10 A car rounds a turn on the Trans-Sahara highway.

✓TIP Look for Real Forces

Centripetal force is *not* some new kind of force. It's just the name for *any* forces that keep an object in circular motion—which are always real, physical forces. Common examples of forces involved in circular motion include the gravitational force on a satellite, friction between tires and road, magnetic forces, tension forces, normal forces, and combinations of these and other forces.

Newton's second law describes circular motion exactly as it does any other motion: by relating net force, mass, and acceleration. Therefore, we can analyze circular motion with the same strategy we've used in other Newton's law problems.

EXAMPLE 5.5 Circular Motion: Whirling a Ball on a String

A ball of mass m whirls around in a horizontal circle at the end of a massless string of length L (Fig. 5.11). The string makes an angle θ with the horizontal. Find the ball's speed and the string tension.

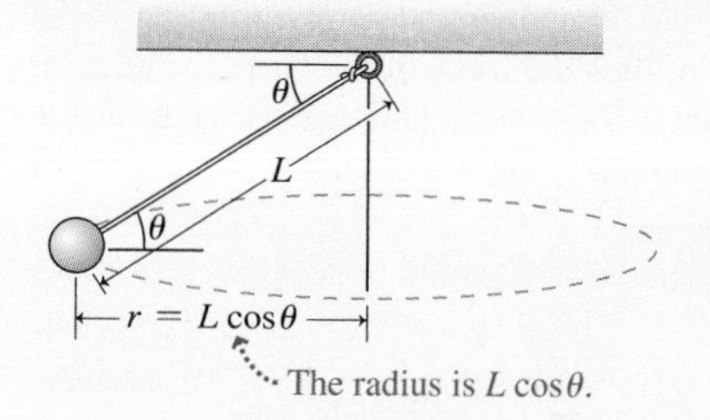

FIGURE 5.11 A ball whirling on a string.

INTERPRET This problem is similar to other Newton's law problems we've worked involving force and acceleration. The object of interest is the ball, and only two forces are acting on it: gravity and the string tension.

DEVELOP Figure 5.12 is our free-body diagram showing the two forces we've identified. The relevant equation is Newton's second law, which becomes

$$\vec{T} + \vec{F}_g = m\vec{a}$$

The ball's path is in a horizontal plane, so its acceleration is horizontal. Then two of the three vectors in our problem—$\vec{F}_g$ and $\vec{a}$—are horizontal or vertical, so in developing our strategy, we choose a horizontal/vertical coordinate system.

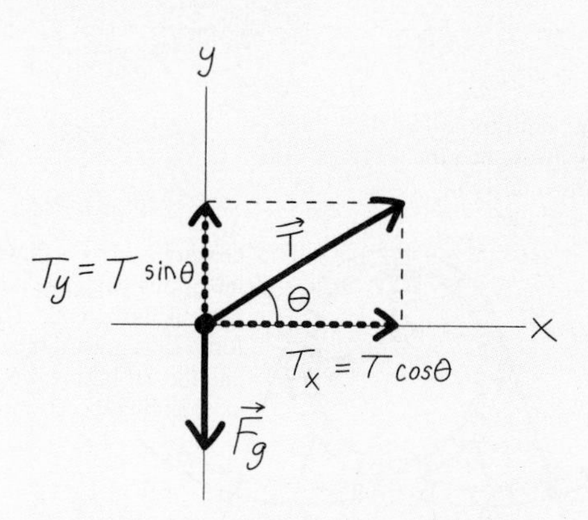

FIGURE 5.12 Our free-body diagram for the whirling ball.

✓TIP Real Forces Only!

Were you tempted to draw a third force in Fig. 5.12, perhaps pointing outward to balance the other two? *Don't!* Because the ball is accelerating, the net force is nonzero and the individual forces *do not balance*. Or maybe you were tempted to draw an inward-pointing force, mv^2/r. *Don't!* The quantity mv^2/r is *not* another force; it's just the product of mass and acceleration that appears in Newton's law (recall Fig. 4.3 and the associated tip). Students often complicate problems by introducing forces that aren't there. That makes physics seem harder than it is!

EVALUATE We now need the x- and y-components of Newton's law. Figure 5.12 shows that $F_{gy} = -F_g = -mg$ and also gives tension components in terms of trig functions. The acceleration is purely horizontal, so $a_y = 0$, and since the ball is in circular motion, $a_x = v^2/r$. But what's r? It's the radius of the circular path and, as Fig. 5.11 shows, that's not the string length L but $L\cos\theta$. With all these expressions, the components of Newton's law become

$$x: \quad T\cos\theta = \frac{mv^2}{L\cos\theta} \qquad y: \quad T\sin\theta - mg = 0$$

We can get the tension directly from the y equation: $T = mg/\sin\theta$. Using this result in the x equation lets us solve for the speed v:

$$v = \sqrt{\frac{TL\cos^2\theta}{m}} = \sqrt{\frac{(mg/\sin\theta)L\cos^2\theta}{m}} = \sqrt{\frac{gL\cos^2\theta}{\sin\theta}}$$

ASSESS In the special case $\theta = 90°$, the string hangs vertically; here $\cos\theta = 0$, so $v = 0$. There's no motion, and the string tension equals the ball's weight. But as the string becomes increasingly horizontal, both speed and tension increase. And, just as in Example 5.2, the tension becomes very great as the string approaches horizontal. Here the string tension has two jobs to do: Its vertical component supports the ball against gravity, while its horizontal component keeps the ball in its circular path. The vertical component is always equal to mg, but as the string approaches horizontal, that becomes an insignificant part of the overall tension—and thus the tension and speed grow very large. ■

EXAMPLE 5.6 Circular Motion: Engineering a Road

Roads designed for high-speed travel have banked curves to give the normal force a component toward the center of the curve. That lets cars turn without relying on friction between tires and road. At what angle should a road with 350-m curvature radius be banked for travel at 90 km/h (25 m/s)?

INTERPRET This is another example involving circular motion and Newton's second law. Although we're asked about the road, a car on the road is the object we're interested in, and we need to design the road so the car can round the curve without needing a frictional force. That means the only forces on the car are gravity and the normal force.

DEVELOP Figure 5.13 shows the physical situation, and Fig. 5.14 is our free-body diagram for the car. Newton's second law is the applicable equation, and here it becomes $\vec{n} + \vec{F}_g = m\vec{a}$. Unlike the skier of Example 5.1, the car isn't accelerating down the slope, so a horizontal/vertical coordinate system makes the most sense.

EVALUATE First we write Newton's law in components. Gravity has only a vertical component, $F_{gy} = -mg$ in our coordinate system, and Fig. 5.14 shows the two components of the normal force. The acceleration is purely horizontal and points toward the center of the curve; in our coordinate system that's the positive x-direction. Since the car is in circular motion, the magnitude of the acceleration is v^2/r. So the components of Newton's law become

$$x: \quad n \sin\theta = \frac{mv^2}{r} \qquad y: \quad n\cos\theta - mg = 0$$

where the 0 on the right-hand side of the y equation reflects the fact that we don't want the car to accelerate in the vertical direction. Solving the y equation gives $n = mg/\cos\theta$. Then using this result in the x equation gives $mg \sin\theta/\cos\theta = mv^2/r$, or $g\tan\theta = v^2/r$. The mass canceled, which is good news because it means our banked road will work for a vehicle of any mass. Now we can solve for the banking angle:

$$\theta = \tan^{-1}\left(\frac{v^2}{gr}\right) = \tan^{-1}\left(\frac{(25\text{ m/s})^2}{(9.8\text{ m/s}^2)(350\text{ m})}\right) = 10°$$

ASSESS Make sense? At low speed v or large radius r, the car's motion changes gently and it doesn't take a large force to keep it on its circular path. But as v increases or r decreases, the required force increases and so does the banking angle. That's because the horizontal component of the normal force is what keeps the car in circular motion, and the steeper the angle, the greater that component. A similar thing happens when an airplane banks to turn; then the force of the air perpendicular to the wings acquires a horizontal component, and that's what turns the plane (see this chapter's opening photo and Problem 43). ■

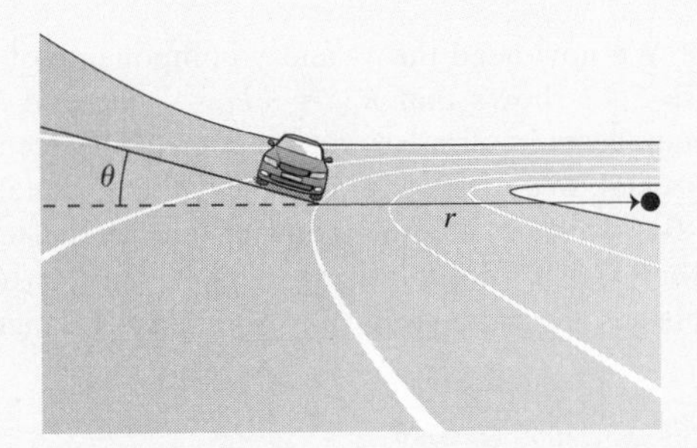

FIGURE 5.13 Car on a banked curve.

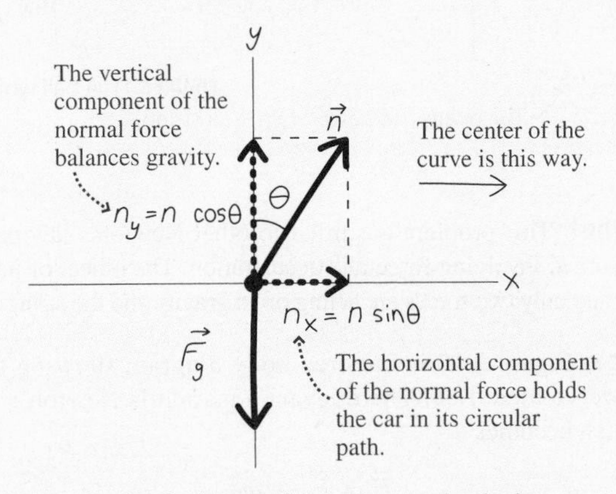

FIGURE 5.14 Our free-body diagram for the car on a banked curve.

EXAMPLE 5.7 Circular Motion: Looping the Loop

The "Great American Revolution" roller coaster at Valencia, California, includes a loop-the-loop section whose radius is 6.3 m at the top. What's the minimum speed for a roller-coaster car at the top of the loop if it's to stay on the track?

INTERPRET Again, we have circular motion described by Newton's second law. We're asked about the minimum speed for the car to stay on the track. What does it mean to stay on the track? It means there must be a normal force between car and track; otherwise, the two aren't in contact. So we can identify two forces acting on the car: gravity and the normal force from the track.

DEVELOP Figure 5.15 shows the physical situation. Things are especially simple at the top of the track, where both forces point in the same direction. We show this in our free-body diagram, Fig. 5.16 (next page). Since that common direction is downward, it makes sense to choose a coordinate system with the positive y-axis *downward*. The applicable equation is Newton's second law, and with the two forces we've identified, that becomes $\vec{n} + \vec{F}_g = m\vec{a}$.

EVALUATE With both forces in the same direction, we need only the y-component of Newton's law. With the downward direction positive, $n_y = n$ and $F_{gy} = mg$. At the top of the loop, the car is in circular motion, so its acceleration is toward the center—downward—and has magnitude v^2/r. So $a_y = v^2/r$, and the y-component of Newton's law becomes

$$n + mg = \frac{mv^2}{r}$$

Solving for the speed gives $v = \sqrt{(nr/m) + gr}$. Now, the minimum possible speed for contact with the track occurs when n gets arbitrarily

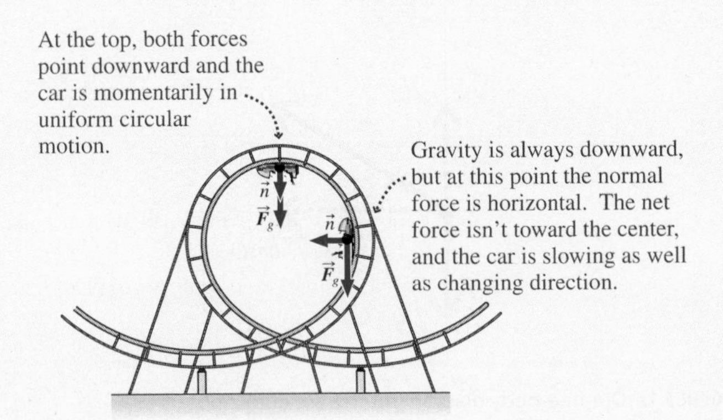

FIGURE 5.15 Forces on the roller-coaster car.

small right at the top of the track, so we find this minimum limit by setting $n = 0$. Then the answer is

$$v_{min} = \sqrt{gr} = \sqrt{(9.8\ m/s^2)(6.3\ m)} = 7.9\ m/s$$

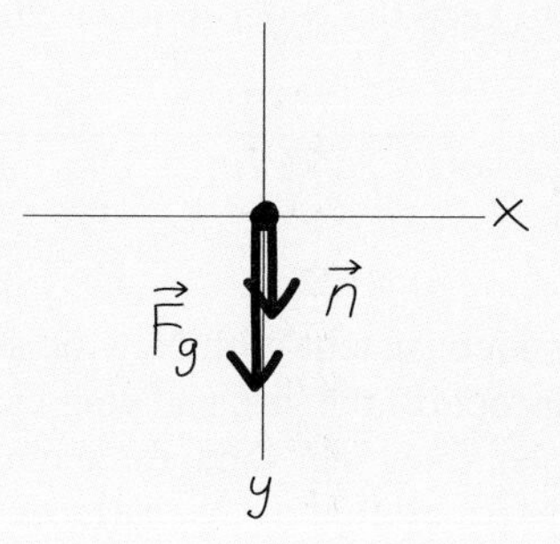

FIGURE 5.16 Our free-body diagram at the top of the loop.

ASSESS Do you see what's happening here? With the minimum speed, the normal force vanishes at the top of the loop, and gravity alone provides the force that keeps the object in its circular path. Since the motion is circular, that force must have magnitude mv^2/r. But the force of gravity alone is mg, and $v_{min} = \sqrt{gr}$ follows directly from equating those two quantities. A car moving any slower than v_{min} would lose contact with the track and go into the parabolic trajectory of a projectile. For a car moving faster, there would be a nonzero normal force contributing to the downward acceleration at the top of the loop. In the "Great American Revolution," the actual speed at the loop's top is 9.7 m/s to provide a margin of safety. As with most problems involving gravity, the mass cancels. That's a good thing because it means the safe speed doesn't depend on the number or mass of the riders. ■

✓TIP Force and Motion

We've said this before, but it's worth noting again: Force doesn't cause motion but rather *change* in motion. The direction of an object's motion need not be the direction of the force on the object. That's true in Example 5.7, where the car is moving horizontally at the top of the loop while subject to a downward force. What *is* in the same direction as the force is the *change* in motion, here embodied in the center-directed acceleration of circular motion.

CONCEPTUAL EXAMPLE 5.1 Bad Hair Day

What's wrong with this cartoon showing riders on a loop-the-loop roller coaster (Fig. 5.17)?

FIGURE 5.17 Conceptual Example 5.1.

EVALUATE Our objects of interest are the riders near the top of the roller coaster. We need to know the forces on them; one is obviously gravity. If the roller coaster is moving faster than Example 5.7's minimum speed—and it better be, for safety—then there are also normal forces from the seats as well as internal forces acting to accelerate parts of the riders' bodies.

Newton's law relates net force and acceleration: $\vec{F} = m\vec{a}$. This equation implies that the net force and acceleration must be in the same direction. At the top of the loop that direction is downward. Every part of the riders' bodies must therefore experience a net downward force. Again, Example 5.7 shows that the minimum force is that of gravity alone; for safety, there must be additional downward forces.

Now focus on the riders' hair, shown hanging downward. Forces on an individual hair are gravity and tension, and our safety argument shows that they should both point in the same direction—namely, downward—to provide a downward force stronger than gravity alone. How, then, can the riders' hair hang downward? That implies an *upward* tension force, inconsistent with our argument. The artist should have drawn the hair "hanging" upward.

ASSESS Make sense? Yes: To the riders, it feels like up is down! They feel the normal force of the seat pushing down, and their hairs experience a downward-pointing tension force. Even though the riders wear seatbelts, they don't need them: If the speed exceeds Example 5.7's minimum, then they feel tightly bound to their seats. Is there some mysterious new force that pushes them against their seats and that pulls their hair up? No! Newton's second law says the net force on the riders is in the direction of their acceleration—namely, downward. And for safety, that net force must be greater than gravity. It's those additional downward forces—the normal force from the seat and the tension force in the hair—that make up feel like down.

MAKING THE CONNECTION Suppose the riders feel like they weigh 50% of what they weigh at rest on the ground. How does the roller coaster's speed compare with Example 5.7's minimum?

EVALUATE In Example 5.7, we found the speed in terms of the normal force n and other quantities: $v = \sqrt{(nr/m) + gr}$. An apparent weight 50% of normal implies that $n = mg/2$. Then $v = \sqrt{(gr/2) + gr} = \sqrt{3/2}\sqrt{gr}$. Example 5.7 shows that the minimum speed is $\sqrt{gr}$, so our result is $\sqrt{3/2} \simeq 1.22$ times the minimum speed. And that 50% apparent weight the riders feel is *upward*!

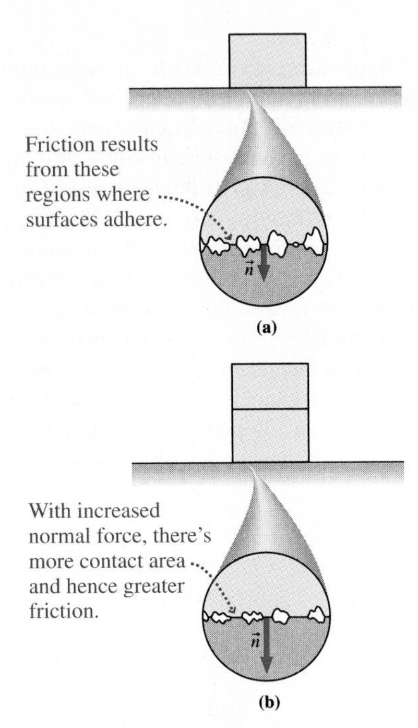

FIGURE 5.18 Friction originates in the contact between two surfaces.

PhET: Forces in 1 Dimension
PhET: Friction

GOT IT? 5.3 You whirl a bucket of water around in a vertical circle and the water doesn't fall out. A Newtonian explanation of why the water doesn't fall out is that (a) the centripetal force mv^2/r balances the gravitational force; (b) there's a centrifugal force pushing the water upward; (c) the normal force plus the gravitational force together provide the downward acceleration needed to keep the water in its circular path; or (d) an upward normal force balances gravity.

5.4 Friction

Your everyday experience of motion seems inconsistent with Newton's first law. Slide a book across the table, and it stops. Take your foot off the gas, and your car coasts to a stop. But Newton's law is correct, so these examples show that some force must be acting. That force is **friction**, a force that opposes the relative motion of two surfaces in contact.

On Earth, we can rarely ignore friction. Some 20% of the gasoline burned in your car is used to overcome friction inside the engine. Friction causes wear and tear on machinery and clothing. But friction is also useful; without it, you couldn't drive or walk.

The Nature of Friction

Friction is ultimately an electrical force between molecules in different surfaces. When two surfaces are in contact, microscopic irregularities adhere, as shown in Fig. 5.18*a*. At the macroscopic level, the result is a force that opposes any relative movement of the surfaces.

Experiments show that the magnitude of the frictional force depends on the normal force between surfaces in contact. Figure 5.18*b* shows why this makes sense: As the normal forces push the surfaces together, the actual contact area increases. There's more adherence, and this increases the frictional force.

At the microscopic level, friction is complicated. The simple equations we'll develop here provide approximate descriptions of frictional forces. Friction is important in everyday life, but it's not one of the fundamental physical interactions.

Frictional Forces

Try pushing a heavy trunk across the floor. At first nothing happens. Push harder; still nothing. Finally, as you push even harder, the trunk starts to slide—and you may notice that once it gets going, you don't have to push quite so hard. Why is that?

With the trunk at rest, microscopic contacts between trunk and floor solidify into relatively strong bonds. As you start pushing, you distort those bonds without breaking them; they respond with a force that opposes your applied force. This is the force of **static friction**, $\vec{f}_s$. As you increase the applied force, static friction increases equally, as shown in Fig. 5.19, and the trunk remains at rest. Experimentally, we find that the maximum static-friction force is proportional to the normal force between surfaces, and we write

$$f_s \le \mu_s n \qquad \text{(static friction)} \qquad (5.2)$$

Here the proportionality constant μ_s (lowercase Greek mu, with the subscript s for "static") is the **coefficient of static friction**, a quantity that depends on the two surfaces. The $\le$ sign indicates that the force of static friction ranges from zero up to the maximum value on the right-hand side.

Eventually you push hard enough to break the bonds between trunk and floor, and the trunk begins to move; this is the point in Fig. 5.19 where the frictional force suddenly drops. Now the microscopic bonds don't have time to strengthen, so the force needed to overcome them isn't so great. In Fig. 5.19 we're assuming you then push with just enough force to overcome friction, so the trunk now moves with constant speed.

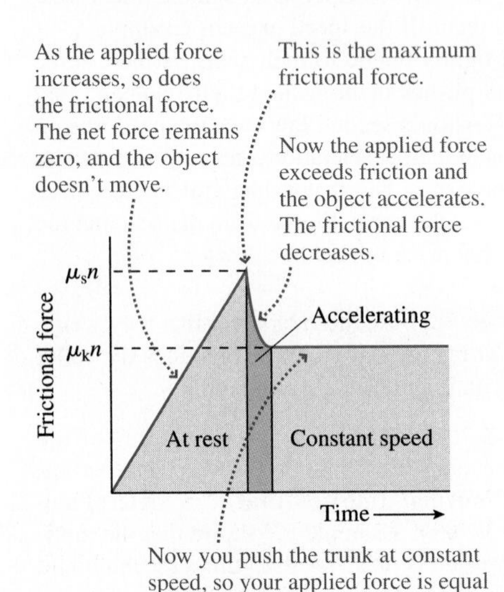

FIGURE 5.19 Behavior of frictional forces.

The weaker frictional force between surfaces in relative motion is the force of **kinetic friction**, $\vec{f}_k$. Again, it's proportional to the normal force between the surfaces:

$$f_k = \mu_k n \qquad \text{(kinetic friction)} \qquad (5.3)$$

where now the proportionality constant is μ_k, the **coefficient of kinetic friction**. Because kinetic friction is weaker, the coefficient of kinetic friction for a given pair of surfaces is less than the coefficient of static friction. Cross-country skiers exploit that fact by using waxes that provide a high coefficient of static friction for pushing against the snow and for climbing hills, while the lower kinetic friction permits effortless gliding.

Equations 5.2 and 5.3 give only the magnitudes of the frictional forces. The direction of the frictional force is parallel to the two surfaces, in the direction that opposes any applied force (Fig. 5.20*a*) or the surfaces' relative motion (Fig. 5.20*b*).

Since they describe proportionality between the magnitudes of two forces, the coefficients of friction are dimensionless. Typical values of μ_k range from less than 0.01 for smooth or lubricated surfaces to about 1.5 for very rough ones. Rubber on dry concrete—vital in driving an automobile—has μ_k about 0.8 and μ_s can exceed 1. A waxed ski on dry snow has $\mu_k \approx 0.04$, while the synovial fluid that lubricates your body's joints reduces μ_k to a low 0.003.

If you push a moving object with a force equal to the opposing force of kinetic friction, then the net force is zero and, according to Newton, the object moves at constant speed. Since friction is nearly always present, but not as obvious as the push of a hand or the pull of a rope, you can see why it's so easy to believe that force is needed to make things move—rather than, as Newton recognized, to make them accelerate.

We emphasize that the equations describing friction are empirical expressions that approximate the effects of complicated but more basic interactions at the microscopic level. Our friction equations have neither the precision nor the fundamental character of Newton's laws.

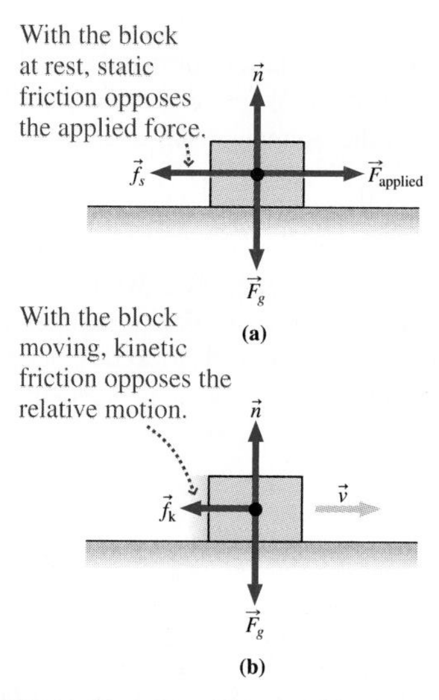

FIGURE 5.20 Direction of frictional forces.

Applications of Friction

Static friction plays a vital role in everyday activities such as walking and driving. As you walk, your foot contacting the ground is momentarily at rest, pushing back against the ground. By Newton's third law, the ground pushes forward, accelerating you forward (Fig. 5.21). Both forces of the third-law pair arise from static friction between foot and ground. On a frictionless surface, walking is impossible.

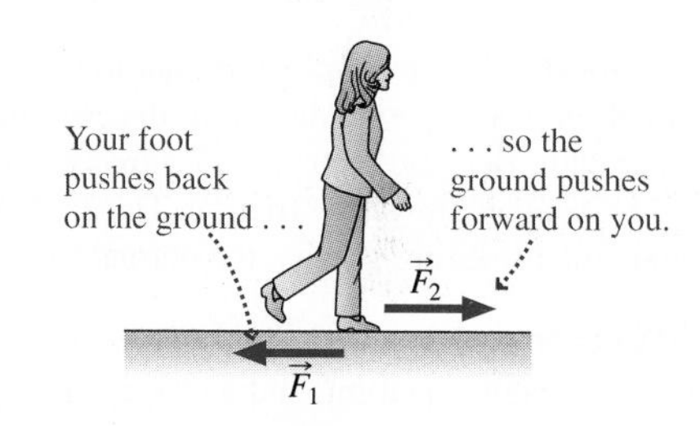

FIGURE 5.21 Walking.

Similarly, the tires of an accelerating car push back on the road. If they aren't slipping, the bottom of each tire is momentarily at rest (more on this in Chapter 10). Therefore the force is static friction. The third law then requires a frictional force of the road pushing forward on the tires; that's what accelerates the car. Braking is the opposite: The tires push forward, and the road pushes back to decelerate the car (Fig. 5.22). The brakes affect only the wheels; it's friction between tires and road that stops the car. You know this if you've applied your brakes on an icy road!

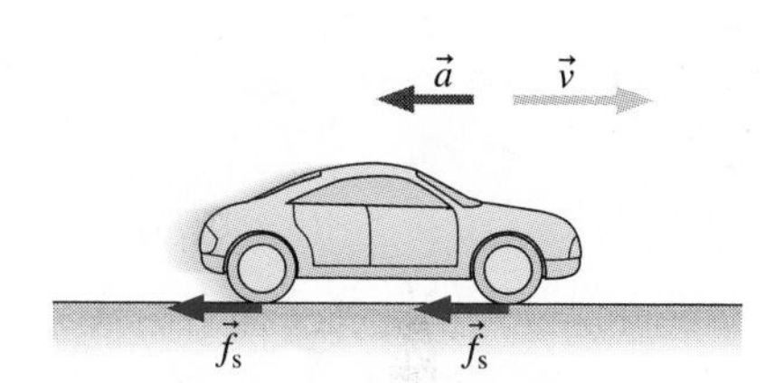

FIGURE 5.22 Friction stops the car.

EXAMPLE 5.8 Frictional Forces: Stopping a Car

The kinetic- and static-friction coefficients between a car's tires and a dry road are 0.61 and 0.89, respectively. The car is initially traveling at 90 km/h (25 m/s) on a level road. Determine (a) the minimum stopping distance, which occurs when the brakes are applied so that the wheels keep rolling as they slow and therefore static friction applies, and (b) the stopping distance with the wheels fully locked and the car skidding.

INTERPRET Since we're asked about the stopping distance, this is ultimately a question about accelerated motion in one dimension—the subject of Chapter 2. But here friction causes that acceleration, so we have a Newton's law problem. The car is the object of interest, and we identify three forces: gravity, the normal force, and friction.

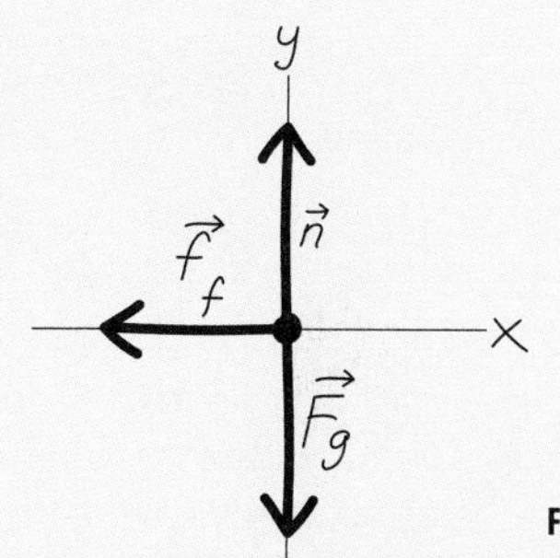

FIGURE 5.23 Our free-body diagram for the braking car.

DEVELOP Figure 5.23 is our free-body diagram. We have a two-part problem here: First, we need to use Newton's second law to find the acceleration, and then we can use Equation 2.11, $v^2 = v_0^2 + 2a\,\Delta x$, to relate distance and acceleration. With the three forces acting on the car, Newton's law becomes $\vec{F}_g + \vec{n} + \vec{f}_f = m\vec{a}$. A horizontal/vertical coordinate system is most appropriate for the components of Newton's law.

EVALUATE The only horizontal force is friction, which points in the $-x$-direction and has magnitude μn, where μ can be either the kinetic- or the static-friction coefficient. The normal force and gravity act in the vertical direction, so the component equations are

$$x:\quad -\mu n = ma_x \qquad y:\quad -mg + n = 0$$

(continued)

Solving the y equation for n and substituting in the x equation gives the acceleration: $a_x = -\mu g$. We then use this result in Equation 2.11 and solve for the stopping distance Δx. With final speed $v = 0$, this gives

$$\Delta x = \frac{v_0^2}{-2a_x} = \frac{v_0^2}{2\mu g}$$

Using the numbers given, we get (a) $\Delta x = 36$ m for the minimum stopping distance (no skid; static friction) and (b) 52 m for the car skidding with its wheels locked (kinetic friction). The difference could well be enough to prevent an accident.

ASSESS Our result $a_x = -\mu g$ shows that a higher friction coefficient leads to a larger acceleration; this makes sense because friction is what causes the acceleration. What happened to the car's mass? A more massive car requires a larger frictional stopping force for the same acceleration—but friction depends on the normal force, and the latter is greater in proportion to the car's mass. Thus the stopping distance doesn't depend on mass.

This example shows that stopping distance increases as the *square* of the speed. That's one reason high speeds are dangerous: Doubling your speed quadruples your stopping distance! ■

EXAMPLE 5.9 Frictional Forces: Steering

A level road makes a 90° turn with radius 73 m. What's the maximum speed for a car to negotiate this turn when the road is dry ($\mu_s = 0.88$) and when the road is snow covered ($\mu_s = 0.21$)?

INTERPRET This example is similar to Example 5.8, but now the frictional force acts perpendicular to the car's motion, keeping it in a circular path. Because the car isn't moving in the direction of the force, we're dealing with *static* friction. The car is the object of interest, and again the forces are gravity, the normal force, and friction.

DEVELOP Figure 5.24 is our free-body diagram. Newton's law is the applicable equation, and we're dealing with the acceleration v^2/r that occurs in circular motion. With the three forces acting on the car, Newton's law is $\vec{F}_g + \vec{n} + \vec{f}_s = m\vec{a}$. A horizontal/vertical coordinate system is most appropriate, and now it's most convenient to take the x-axis in the direction of the acceleration—namely, toward the center of the curve.

y
$\vec{n}$
The dot represents the car, whose direction of motion is out of the page.
The frictional force points toward the curve's center.
$\vec{f}_s$
x
$\vec{F}_g$

FIGURE 5.24 Our free-body diagram for the cornering car.

EVALUATE Again, the only horizontal force is friction, with magnitude $\mu_s n$. Here it points in the positive x-direction, as does the acceleration of magnitude v^2/r. So the x-component of Newton's law is $\mu_s n = mv^2/r$. There's no vertical acceleration, so the y-component is $-mg + n = 0$. Solving for n and using the result in the x equation give $\mu_s mg = mv^2/r$. Again the mass cancels, and we solve for v to get

$$v = \sqrt{\mu_s g r}$$

Putting in the numbers, we get $v = 25$ m/s (90 km/h) for the dry road and 12 m/s (44 km/h) for the snowy road. Exceed these speeds, and your car inevitably moves in a path with a larger radius—and that means going off the road!

ASSESS Once again, it makes sense that the car's mass doesn't matter. A more massive car needs a larger frictional force, and it gets what it needs because its larger mass results in a larger normal force. The safe speed increases with the curve radius r, and that, too, makes sense: A larger radius means a gentler turn, with less acceleration at a given speed. So less frictional force is needed. ■

APPLICATION Antilock Brakes

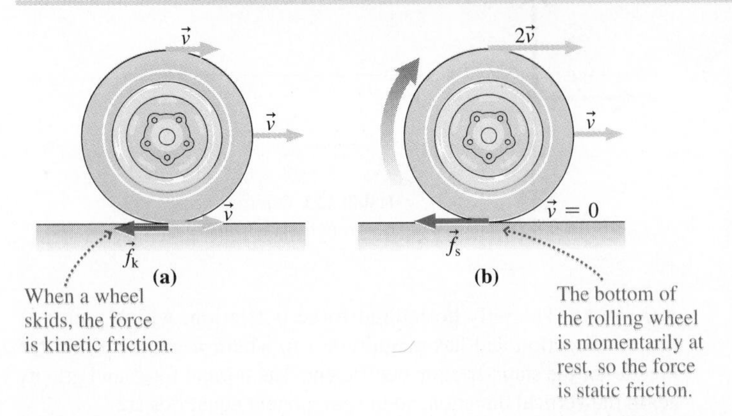

Today's cars have computer-controlled antilock braking systems (ABS). These systems exploit the fact that static friction is greater than kinetic friction. Slam on the brakes of a non-ABS car and the wheels lock and skid without turning. The force between tires and road is then *kinetic* friction (part a in the figure). But if you pump the brakes to keep the wheels from skidding, then it's the greater force of *static* friction (part b).

ABS improves on this brake-pumping strategy with a computer that independently controls the brakes at each wheel, keeping each just on the verge of slipping. Drivers of ABS cars should slam the brakes hard in an emergency; the ensuing clatter indicates the ABS is working.

Although ABS can reduce the stopping distance, its real significance is in preventing vehicles from skidding out of control as can happen when you apply the brakes with some wheels on ice and others on pavement. Increasingly, today's cars incorporate their computer-controlled brakes into sophisticated systems that enhance stability during emergency maneuvers.

EXAMPLE 5.10 Friction on a Slope: Avalanche!

A storm dumps new snow on a ski slope. The coefficient of static friction between the new snow and the older snow underneath is 0.46. What's the maximum slope angle to which the new snow can adhere?

INTERPRET The problem asks about an angle, but it's friction that holds the new snow to the old, so this is really a problem about the maximum possible static friction. We aren't given an object, but we can model the new snow as a slab of mass m resting on a slope of unknown angle θ. The forces on the slab are gravity, the normal force, and static friction $\vec{f}_s$.

DEVELOP Figure 5.25 shows the model, and Fig. 5.26 is our free-body diagram. Newton's second law is the applicable equation, here with $\vec{a} = \vec{0}$, giving $\vec{F}_g + \vec{n} + \vec{f}_s = \vec{0}$. We also need the maximum static-friction force, given in Equation 5.2, $f_{s\,max} = \mu_s n$. As in Example 5.1, a tilted coordinate system is simplest and is shown in Fig. 5.26.

EVALUATE With the positive x-direction downslope, Fig. 5.26 shows that the x-component of gravity is $F_g \sin\theta = mg\sin\theta$, while the frictional force acts upslope ($-x$-direction) and has maximum magnitude $\mu_s n$; therefore, $f_{sx} = -\mu_s n$. So the x-component of Newton's law is $mg\sin\theta - \mu_s n = 0$. We can read the y-component from Fig. 5.26: $-mg\cos\theta + n = 0$. Solving the y equation gives $n = mg\cos\theta$. Using this result in the x equation then yields $mg\sin\theta - \mu_s mg\cos\theta = 0$. Both m and g cancel, and we have $\sin\theta = \mu_s\cos\theta$ or, since $\tan\theta = \sin\theta/\cos\theta$,

$$\tan\theta = \mu_s$$

For the numbers given in this example, the result becomes $\theta = \tan^{-1}\mu_s = \tan^{-1}(0.46) = 25°$.

FIGURE 5.25 A layer of snow, modeled as a slab on a sloping surface.

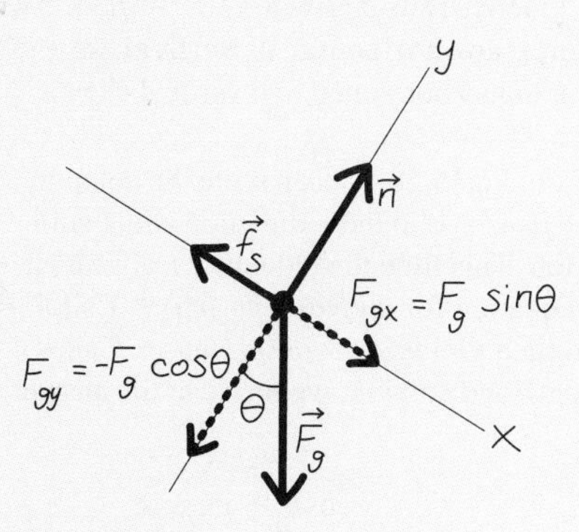

FIGURE 5.26 Our free-body diagram for the snow slab.

ASSESS Make sense? Sure: The steeper the slope, the greater the friction needed to keep the snow from sliding. Two effects are at work here: First, as the slope steepens, so does the component of gravity along the slope. Second, as the slope steepens, the normal force gets smaller, and that reduces the frictional force for a given friction coefficient. Note here that the normal force is not simply the weight mg of the snow; again, that's because of the sloping surface.

The real avalanche danger comes at angles slightly smaller than our answer $\tan\theta = \mu_s$, where a thick snowpack can build up. Changes in the snow's composition with temperature may decrease the friction coefficient and unleash an avalanche. ■

EXAMPLE 5.11 Friction: Dragging a Trunk

You drag a trunk of mass m across a level floor using a massless rope that makes an angle θ with the horizontal (Fig. 5.27). Given a kinetic-friction coefficient μ_k, what rope tension is required to move the trunk at constant speed?

FIGURE 5.27 Dragging a trunk.

INTERPRET Even though the trunk is moving, it isn't accelerating, so here's another problem involving Newton's law with zero acceleration. The object is the trunk, and now four forces act: gravity, the normal force, friction, and the rope tension.

DEVELOP Figure 5.28 is our free-body diagram showing all four forces acting on the trunk. The relevant equation is Newton's law.

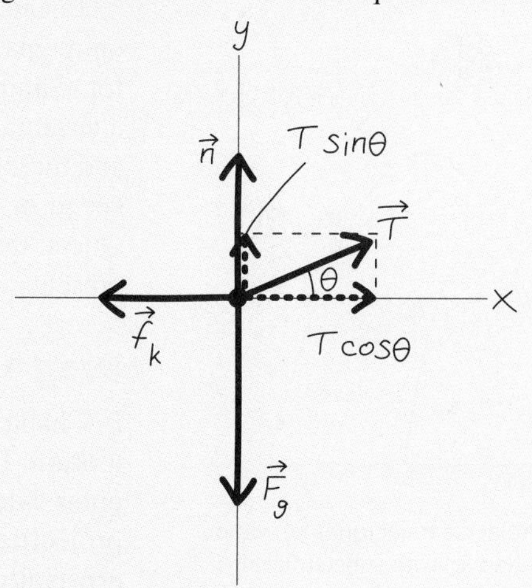

FIGURE 5.28 Our free-body diagram for the trunk.

With no acceleration, it reads $\vec{F}_g + \vec{n} + \vec{f}_k + \vec{T} = \vec{0}$, with the magnitude of kinetic friction given by $f_k = \mu_k n$. All vectors except the tension force are horizontal or vertical, so the most sensible coordinate system has horizontal and vertical axes.

EVALUATE From Fig. 5.28, we can write the components of Newton's law: $T\cos\theta - \mu_k n = 0$ in the x-direction and $T\sin\theta - mg + n = 0$ in the y-direction. This time the unknown T appears in both equations. Solving the y equation for n gives $n = mg - T\sin\theta$. Putting this n in the x equation then yields $T\cos\theta - \mu_k(mg - T\sin\theta) = 0$. Factoring terms involving T and solving, we arrive at the answer:

$$T = \frac{\mu_k mg}{\cos\theta + \mu_k \sin\theta}$$

ASSESS Make sense? Without friction, we wouldn't need any force to move the trunk at constant speed, and indeed our expression gives $T = 0$ in this case. On the other hand, if there is friction but $\theta = 0$, then $\sin\theta = 0$ and we get $T = \mu_k mg$. In this case the normal force equals the weight, so the frictional force is $\mu_k mg$. Since the frictional force is horizontal and with $\theta = 0$ we're pulling horizontally, this is also the magnitude of the tension force. At intermediate angles, two effects come into play: First, the upward component of tension helps support the trunk's weight, and that means less normal force is needed. With less normal force, there's less friction—making the trunk easier to pull. But as the angle increases, less of the tension is horizontal, and that means a larger tension force is needed to overcome friction. In combination, these two effects mean there's an optimum angle at which the rope tension is a minimum. Problem 68 explores this point further. ■

GOT IT? 5.4 The figure shows a logging vehicle pulling a redwood log. Is the frictional force in this case (a) less than, (b) equal to, or (c) greater than the weight multiplied by the coefficient of friction?

5.5 Drag Forces

Friction isn't the only "hidden" force that robs objects of their motion and obscures Newton's first law. Objects moving through fluids like water or air experience **drag forces** that oppose the relative motion of object and fluid. Ultimately, drag results from collisions between fluid molecules and the object. The drag force depends on several factors, including fluid density and the object's cross-sectional area and speed.

Terminal Speed

When an object falls from rest, its speed is initially low and so is the velocity-dependent drag force. It therefore accelerates downward with nearly the gravitational acceleration g. But as the object gains speed, the drag force increases—until eventually the drag force and gravity have equal magnitudes. At that point the net force on the object is zero, and it falls with constant speed, called its **terminal speed.**

Because the drag force depends on an object's area and the gravitational force depends on its mass, the terminal speed is lower for lighter objects with large areas. A parachute, for example, is designed specifically to have a large surface area that results, typically, in a terminal speed around 5 m/s. A ping-pong ball and a golf ball have about the same size and therefore the same area, but the ping-pong ball's much lower mass leads to a terminal speed of about 10 m/s compared with the golf ball's 50 m/s. For an irregularly shaped object, the drag and thus the terminal speed depend on how large a surface area the object presents to the air. Skydivers exploit this effect to vary their rates of fall.

Drag and Projectile Motion

In Chapter 3, we consistently neglected air resistance—the drag force of air—in projectile motion. Determining drag effects on projectiles is not trivial and usually requires computer calculations. The net effect, though, is that air resistance decreases the range of a projectile (Fig. 5.29). Despite the physicist's need for computer calculations, others—especially athletes—have a feel for drag forces that lets them play their sports by judging correctly the trajectory of a projectile under the influence of drag forces. You can explore drag forces further in Problems 70 and 71.

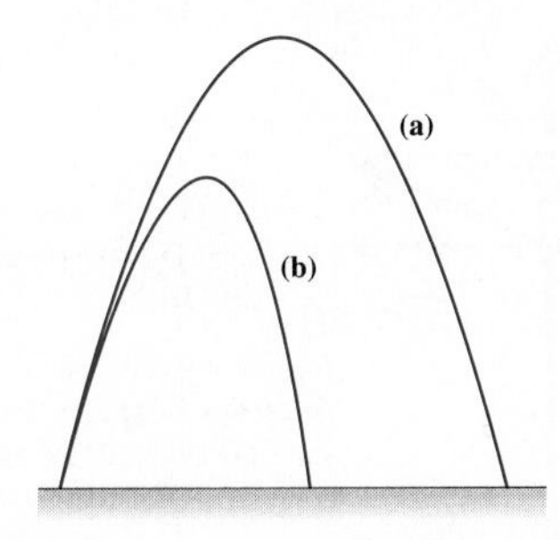

FIGURE 5.29 Projectile trajectories (a) without air resistance and (b) with substantial air resistance. Note that (b) not only achieves less height and range but that the trajectory is no longer a symmetric parabola.

CHAPTER 5 SUMMARY

Big Idea

The big idea here is the same as in Chapter 4—namely, that Newton's laws are a universal description of motion, in which force causes not motion itself but change in motion. Here we focus on Newton's second law, extended to the richer and more complex examples of motion in two dimensions. To use Newton's law, we now sum forces that may point in different directions, but the result is the same: The net force determines an object's acceleration.

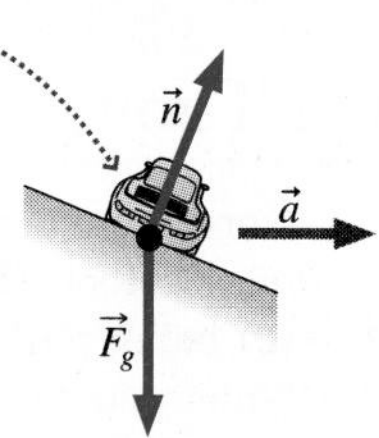

Here's a car on a banked turn. The forces on it don't sum to zero because the car is accelerating toward the center of the turn.

Common forces include gravity, the normal force from surfaces, tension forces, and a force introduced here: friction. Important examples are those where an object is accelerating, including in circular motion, and those where there's no acceleration and therefore the net force is zero.

A block sits at rest on a slope. The three forces—gravity, normal force, and friction—sum to zero.

$\vec{n}$ $\vec{f}_f$ $\vec{F}_g$

Solving Problems with Newton's Laws

The problem-solving strategy in this chapter is exactly the same as in Chapter 4, except that in two dimensions the choice of coordinate system and the division of forces into components become crucial steps. You usually need both component equations to solve a problem.

A skier on a frictionless slope

Free-body diagram showing the two forces acting

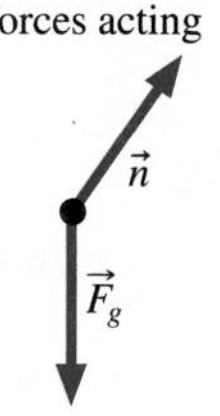

Coordinate system and vector components

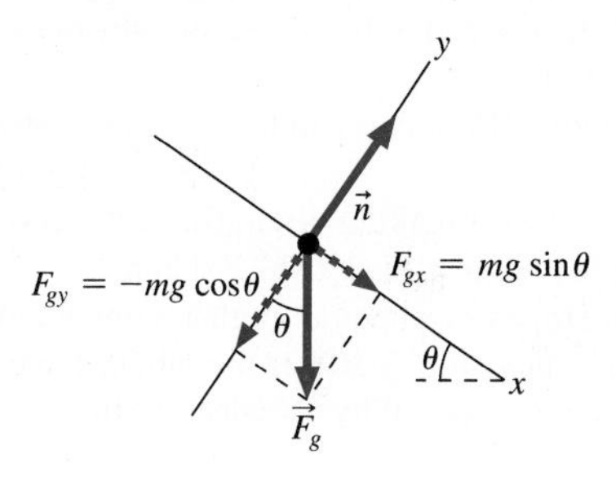

$$\vec{F} = m\vec{a} \rightarrow \vec{n} + \vec{F}_g = m\vec{a} \rightarrow \begin{cases} n_x + F_{gx} = ma_x \\ n_y + F_{gy} = ma_y \end{cases} \rightarrow \begin{cases} mg\sin\theta = ma_x \\ n - mg\cos\theta = 0 \end{cases}$$

Key Concepts and Equations

Newton's second law, $\vec{F}_{\text{net}} = m\vec{a}$, is the key equation in this chapter. It's crucial to remember that it's a *vector* equation, representing a pair of scalar equations for its two components in two dimensions.

Applications

Friction acts between surfaces to oppose their relative motion, and its strength depends on the normal force $\vec{n}$ acting perpendicular to them. When surfaces aren't actually in relative motion, the force is **static friction**, whose value ranges from zero to a maximum value $\mu_s n$ as needed to oppose any applied force: $f_s \leq \mu_s n$. Here μ_s is the **coefficient of static friction**, which depends on the nature of the two surfaces. For surfaces in relative motion, the force is **kinetic friction**, given by $f_k = \mu_k n$, where the **coefficient of kinetic friction** is less than the coefficient of static friction.

A block moving to the right experiences a frictional force f to the left.

The magnitude of the frictional force depends on the normal force: $f = \mu n$.

$\vec{n}$ $\vec{f}$ $\vec{v}$ $\vec{F}_g$

Here the frictional force is a little less than the normal force, so μ is a little less than 1.

MP For homework assigned on MasteringPhysics, go to www.masteringphysics.com

BIO *Biology and/or medicine-related problems* **DATA** *Data problems* **ENV** *Environmental problems* **CH** *Challenge problems* **COMP** *Computer problems*

For Thought and Discussion

1. Compare the net force on a heavy trunk when it's (a) at rest on the floor; (b) being slid across the floor at constant speed; (c) being pulled upward in an elevator whose cable tension equals the combined weight of the elevator and trunk; and (d) sliding down a frictionless ramp.
2. The force of static friction acts only between surfaces at rest. Yet that force is essential in walking and in accelerating or braking a car. Explain.
3. A jet plane flies at constant speed in a vertical circular loop. At what point in the loop does the seat exert the greatest force on the pilot? The least force?
4. In cross-country skiing, skis should easily glide forward but should remain at rest when the skier pushes back against the snow. What frictional properties should the ski wax have to achieve this goal?
5. Why do airplanes bank when turning?
6. Why is it easier for a child to stand nearer the inside of a rotating merry-go-round?
7. Gravity pulls a satellite toward Earth's center. So why doesn't the satellite actually fall to Earth?
8. Explain why a car with ABS brakes can have a shorter stopping distance.
9. A fishing line has a 20-lb breaking strength. Is it possible to break the line while reeling in a 15-lb fish? Explain.
10. Two blocks rest on slopes of unequal angles, connected by a rope passing over a pulley (Fig. 5.30). If the blocks have equal masses, will they remain at rest? Why? Neglect friction.

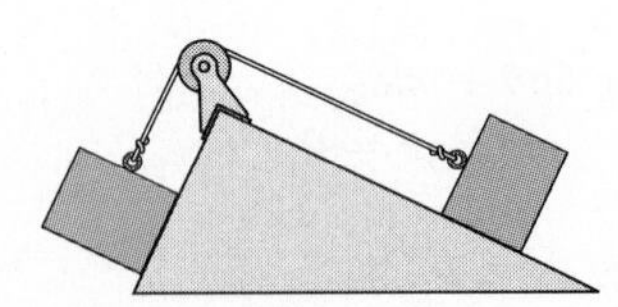

FIGURE 5.30 For Thought and Discussion 10; Exercises 21 and 22

11. You're on a plane undergoing a banked turn, so steep that out the window you see the ground below. Yet your pretzels stay put on the seatback tray, rather than sliding downward. Why?
12. A backcountry skier weighing 700 N skis down a steep slope, unknowingly crossing a snow bridge that spans a deep, hidden crevasse. If the bridge can support 580 N—meaning that's the maximum normal force it can sustain without collapsing—is there any chance the mountaineer can cross safely? Explain.

Exercises and Problems

Exercises

Section 5.1 Using Newton's Second Law

13. Two forces, both in the x-y plane, act on a 3.25-kg mass that accelerates at 5.48 m/s^2 in a direction 38.0° counterclockwise from the x-axis. One force has magnitude 8.63 N and points in the $+x$-direction. Find the other force.
14. Two forces act on a 3.1-kg mass that undergoes acceleration $\vec{a} = 0.91\hat{\imath} - 0.27\hat{\jmath}$ m/s^2. If one force is $-1.2\hat{\imath} - 2.5\hat{\jmath}$ N, what's the other?
15. At what angle should you tilt an air table to simulate free fall at the surface of Mars, where $g = 3.71$ m/s^2?
16. A skier starts from rest at the top of a 24° slope 1.3 km long. Neglecting friction, how long does it take to reach the bottom?
17. A tow truck is connected to a 1400-kg car by a cable that makes a 25° angle to the horizontal. If the truck accelerates at 0.57 m/s^2, what's the magnitude of the cable tension? Neglect friction and the cable's mass.
18. **BIO** Studies of gymnasts show that their high rate of injuries to the Achilles tendon is due to tensions in the tendon that typically reach 10 times body weight. That force is provided by a pair of muscles, each exerting a force at 25° to the vertical, with their horizontal components opposite. For a 55-kg gymnast, find the force in each of these muscles.

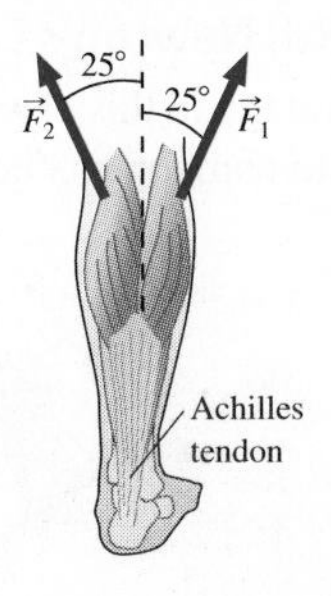

FIGURE 5.31

19. Find the minimum slope angle for which the skier in Question 12 can safely traverse the snow bridge.

Section 5.2 Multiple Objects

20. Your 12-kg baby sister pulls on the bottom of the tablecloth with all her weight. On the table, 60 cm from the edge, is a 6.8-kg roast turkey. (a) What's the turkey's acceleration? (b) From the time your sister starts pulling, how long do you have to intervene before the turkey goes over the edge? Neglect friction.
21. If the left-hand slope in Fig. 5.30 makes a 60° angle with the horizontal, and the right-hand slope makes a 20° angle, how should the masses compare if the objects are not to slide along the frictionless slopes?
22. Suppose the angles shown in Fig. 5.30 are 60° and 20°. If the left-hand mass is 2.1 kg, what should the right-hand mass be so that it accelerates (a) downslope at 0.64 m/s^2 and (b) upslope at 0.76 m/s^2?
23. Two unfortunate climbers, roped together, are sliding freely down an icy mountainside. The upper climber (mass 75 kg) is on a slope at 12° to the horizontal, but the lower climber (mass 63 kg) has gone over the edge to a steeper slope at 38°. (a) Assuming frictionless ice and a massless rope, what's the acceleration of the pair? (b) The upper climber manages to stop the slide with an ice ax. After the climbers have come to a complete stop, what force must the ax exert against the ice?

Section 5.3 Circular Motion

24. Suppose the Moon were held in its orbit not by gravity but by tension in a massless cable. Estimate the magnitude of the cable tension. (*Hint:* See Appendix E.)
25. Show that the force needed to keep a mass m in a circular path of radius r with period T is $4\pi^2 mr/T^2$.

26. A 940-g rock is whirled in a horizontal circle at the end of a 1.30-m-long string. (a) If the breaking strength of the string is 120 N, what's the minimum angle the string can make with the horizontal? (b) At this minimum angle, what's the rock's speed?
27. You're investigating a subway accident in which a train derailed while rounding an unbanked curve of radius 150 m, and you're asked to estimate whether the train exceeded the 35-km/h speed limit for this curve. You interview a passenger who had been standing and holding onto a strap; she noticed that an unused strap was hanging at about a 15° angle to the vertical just before the accident. What do you conclude?
28. A tetherball on a 1.55-m rope is struck so that it goes into circular motion in a horizontal plane, with the rope making a 12.0° angle to the horizontal. What's the ball's speed?
29. An airplane goes into a turn 3.6 km in radius. If the banking angle required is 28° from the horizontal, what's the plane's speed?

Section 5.4 Friction

30. Movers slide a 73-kg file cabinet along a floor where the coefficient of kinetic friction is 0.81. What's the frictional force on the cabinet?
31. A hockey puck is given an initial speed of 14 m/s. If it comes to rest in 56 m, what's the coefficient of kinetic friction?
32. Starting from rest, a skier slides 100 m down a 28° slope. How much longer does the run take if the coefficient of kinetic friction is 0.17 instead of 0?
33. A car moving at 40 km/h negotiates a 130-m-radius banked turn designed for 60 km/h. What coefficient of friction is needed to keep the car on the road?

Problems

34. Repeat Example 5.1, this time using a horizontal/vertical coordinate system.
35. A block is launched with initial speed 2.2 m/s up a 35° frictionless ramp. How far up the ramp does it slide?
36. **BIO** In the process of mitosis (cell division), two motor proteins pull on a spindle pole, each with a 7.3-pN force. The two force vectors make a 65° angle. What's the magnitude of the force the two motor proteins exert on the spindle pole?
37. A 14.6-kg monkey hangs from the middle of a massless rope, each half of which makes an 11.0° angle with the horizontal. What's the rope tension? Compare with the monkey's weight.
38. A camper hangs a 26-kg pack between two trees using separate ropes of different lengths, as shown in Fig. 5.32. Find the tension in each rope.

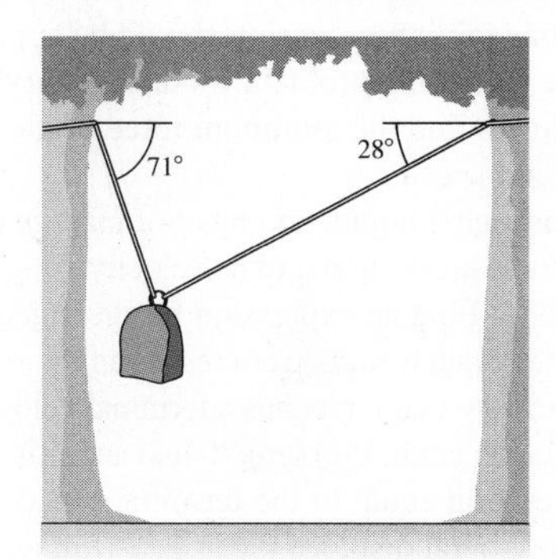

FIGURE 5.32 Problem 38

39. A mass m_1 undergoes circular motion of radius R on a horizontal frictionless table, connected by a massless string through a hole in the table to a second mass m_2 (Fig. 5.33). If m_2 is stationary, find expressions for (a) the string tension and (b) the period of the circular motion.

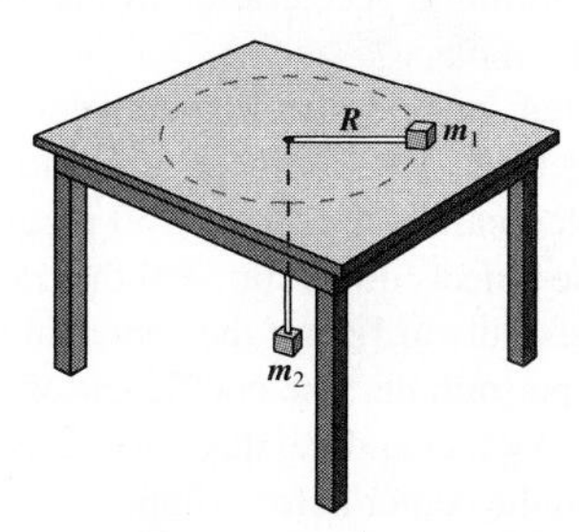

FIGURE 5.33 Problem 39

40. **BIO** Patients with severe leg breaks are often placed in *traction*, with an external force countering muscles that would pull too hard on the broken bones. In the arrangement shown in Fig. 5.34, the mass m is 4.8 kg, and the pulleys can be considered massless and frictionless. Find the horizontal traction force applied to the leg.

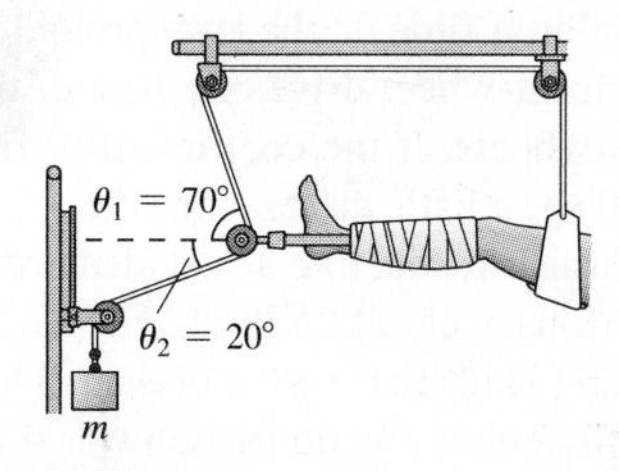

FIGURE 5.34 Problem 40

41. Riders on the "Great American Revolution" loop-the-loop roller coaster of Example 5.7 wear seatbelts as the roller coaster negotiates its 6.3-m-radius loop at 9.7 m/s. At the top of the loop, what are the magnitude and direction of the force exerted on a 60-kg rider (a) by the roller-coaster seat and (b) by the seatbelt? (c) What would happen if the rider unbuckled at this point?
42. A 45-kg skater rounds a 5.0-m-radius turn at 6.3 m/s. (a) What are the horizontal and vertical components of the force the ice exerts on her skate blades? (b) At what angle can she lean without falling over?
43. When a plane turns, it banks as shown in Fig. 5.35 to give the wings' lifting force $\vec{F}_w$ a horizontal component that turns the plane. If a plane is flying level at 950 km/h and the banking angle θ is not to exceed 40°, what's the minimum curvature radius for the turn?

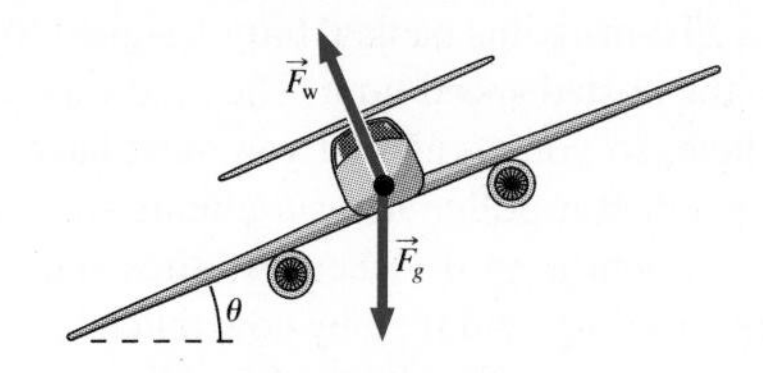

FIGURE 5.35 Problem 43

44. You whirl a bucket of water in a vertical circle of radius 85 cm. What's the minimum speed that will keep the water from falling out?
45. A child sleds down an 8.5° slope at constant speed. What's the frictional coefficient between slope and sled?
46. The handle of a 22-kg lawnmower makes a 35° angle with the horizontal. If the coefficient of friction between lawnmower and ground is 0.68, what magnitude of force, applied in the direction of the handle, is required to push the mower at constant velocity? Compare with the mower's weight.
47. Repeat Example 5.4, now assuming that the coefficient of kinetic friction between rock and ice is 0.057.

48. A bat crashes into the vertical front of an accelerating subway train. If the frictional coefficient between bat and train is 0.86, what's the minimum acceleration of the train that will allow the bat to remain in place?
49. The coefficient of static friction between steel train wheels and steel rails is 0.58. The engineer of a train moving at 140 km/h spots a stalled car on the tracks 150 m ahead. If he applies the brakes so the wheels don't slip, will the train stop in time?
50. A bug crawls outward from the center of a CD spinning at 200 revolutions per minute. The coefficient of static friction between the bug's sticky feet and the disc surface is 1.2. How far does the bug get from the center before slipping?
51. CH A 310-g paperback book rests on a 1.2-kg textbook. A force is applied to the textbook, and the two books accelerate together from rest to 96 cm/s in 0.42 s. The textbook is then brought to a stop in 0.33 s, during which time the paperback slides off. Within what range does the coefficient of static friction between the two books lie?
52. Children sled down a 41-m-long hill inclined at 25°. At the bottom, the slope levels out. If the coefficient of friction is 0.12, how far do the children slide on the level ground?
53. In a typical front-wheel-drive car, 70% of the car's weight rides on the front wheels. If the coefficient of friction between tires and road is 0.61, what's the car's maximum acceleration?
54. A police officer investigating an accident estimates that a moving car hit a stationary car at 25 km/h. Before the collision, the car left 47-m-long skid marks as it braked. The officer determines that the coefficient of kinetic friction was 0.71. What was the initial speed of the moving car?
55. A slide inclined at 35° takes bathers into a swimming pool. With water sprayed onto the slide to make it essentially frictionless, a bather spends only one-third as much time on the slide as when it's dry. What's the coefficient of friction on the dry slide?
56. You try to move a heavy trunk, pushing down and forward at an angle of 50° below the horizontal. Show that, no matter how hard you push, it's impossible to budge the trunk if the coefficient of static friction exceeds 0.84.
57. A block is shoved up a 22° slope with an initial speed of 1.4 m/s. The coefficient of kinetic friction is 0.70. (a) How far up the slope will the block get? (b) Once stopped, will it slide back down?
58. At the end of a factory production line, boxes start from rest and slide down a 30° ramp 5.4 m long. If the slide can take no more than 3.3 s, what's the maximum allowed frictional coefficient?
59. You're in traffic court, arguing against a speeding citation. You entered a 210-m-radius banked turn designed for 80 km/h, which was also the posted speed limit. The road was icy, yet you stayed in your lane, so you argue that you must have been going at the design speed. But police measurements show there was a frictional coefficient $\mu = 0.15$ between tires and road. Is it possible you were speeding, and if so by how much?
60. A space station is in the shape of a hollow ring, 450 m in diameter (Fig. 5.36). At how many revolutions per minute should it rotate in order to simulate Earth's gravity—that is, so the normal force on an astronaut at the outer edge would equal the astronaut's weight on Earth?

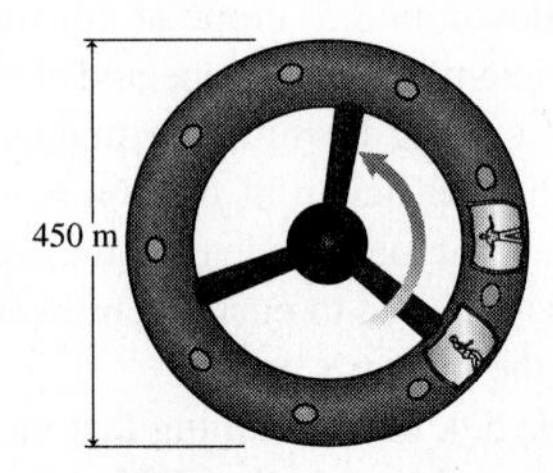

FIGURE 5.36 Problem 60

61. In a loop-the-loop roller coaster, show that a car moving too slowly would leave the track at an angle ϕ given by $\cos\phi = v^2/rg$, where ϕ is the angle made by a vertical line through the center of the circular track and a line from the center to the point where the car leaves the track.
62. CH Find an expression for the minimum frictional coefficient needed to keep a car with speed v on a banked turn of radius R designed for speed v_0.
63. An astronaut is training in an earthbound centrifuge that consists of a small chamber whirled horizontally at the end of a 5.1-m-long shaft. The astronaut places a notebook on the vertical wall of the chamber and it stays in place. If the coefficient of static friction is 0.62, what's the minimum rate at which the centrifuge must be revolving?
64. You stand on a spring scale at the north pole and again at the equator. Which scale reading will be lower, and by what percentage will it be lower than the higher reading? Assume g has the same value at pole and equator.
65. Driving in thick fog on a horizontal road, you spot a tractor-trailer truck jackknifed across the road. To avert a collision, you could brake to a stop or swerve in a circular arc, as suggested in Fig. 5.37. Which option offers the greater margin of safety? Assume that there is the same coefficient of static friction in both cases, and that you maintain constant speed if you swerve.

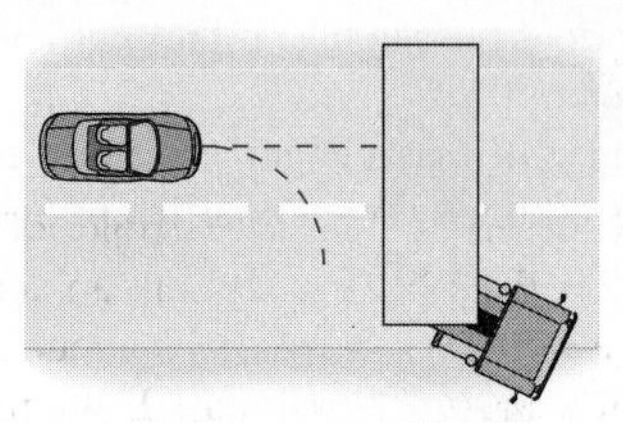

FIGURE 5.37 Problem 65

66. CH A block is projected up an incline at angle θ. It returns to its initial position with half its initial speed. Show that the coefficient of kinetic friction is $\mu_k = \frac{3}{5}\tan\theta$.
67. A 2.1-kg mass is connected to a spring with spring constant $k = 150$ N/m and unstretched length 18 cm. The two are mounted on a frictionless air table, with the free end of the spring attached to a frictionless pivot. The mass is set into circular motion at 1.4 m/s. Find the radius of its path.
68. Take $\mu_k = 0.75$ in Example 5.11, and plot the tension force in units of the trunk's weight, as a function of the rope angle θ (that is, plot T/mg versus θ). Use your plot to determine (a) the minimum tension necessary to move the trunk and (b) the angle at which this minimum tension should be applied.
69. CH Repeat the preceding problem for an arbitrary value of μ_k, by using calculus to find the minimum force needed to move the trunk with constant speed.
70. CH Moving through a liquid, an object of mass m experiences a resistive drag force proportional to its velocity, $F_{drag} = -bv$, where b is a constant. (a) Find an expression for the object's speed as a function of time, when it starts from rest and falls vertically through the liquid. (b) Show that it reaches a terminal velocity mg/b.
71. CH Suppose the object in Problem 70 had an initial velocity in the horizontal direction equal to the terminal speed, $v_{x0} = mg/b$. Show that the horizontal distance it can go is limited to $x_{max} = mv_{x0}/b$, and find an expression for its trajectory (y as a function of x).
72. CH A block is launched with speed v_0 up a slope making an angle θ with the horizontal; the coefficient of kinetic friction is μ_k. (a) Find an expression for the distance d the block travels along the slope. (b) Use calculus to determine the angle that minimizes d.

73. A florist asks you to make a window display with two hanging pots as shown in Fig. 5.38. The florist is adamant that the strings be as invisible as possible, so you decide to use fishing line but want to use the thinnest line you can. Will fishing line that can withstand 100 N of tension work?

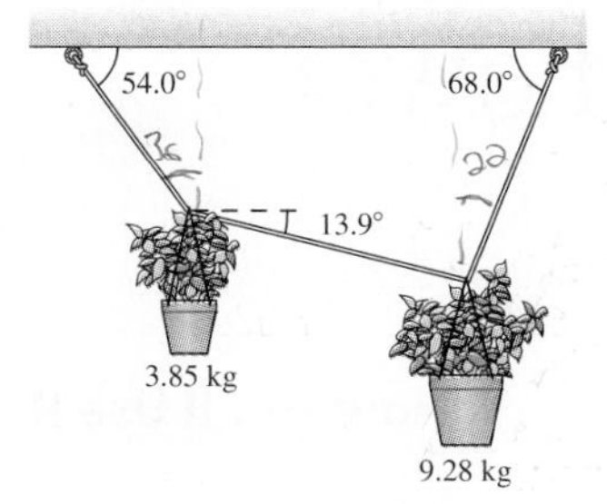

FIGURE 5.38 Problem 73

74. You're at the state fair. A sideshow barker claims that the star of the show can throw a 7.3-kg Olympic-style hammer "faster than a speeding bullet." You recall that bullets travel at several hundreds of meters per second. The burly hammer thrower whirls the hammer in a circle that you estimate to be 2.4 m in diameter. You guess the chain holding the hammer makes an angle of 10° with the horizontal. When the hammer flies off, is it really moving faster than a bullet?
75. One of the limiting factors in high-performance aircraft is the acceleration to which the pilot can be subjected without blacking out; it's measured in "gees," or multiples of the gravitational acceleration. The F-22 Raptor fighter can achieve Mach 1.8 (1.8 times the speed of sound, which is about 340 m/s). Suppose a pilot dives in a circle and pulls up. If the pilot can't exceed $6g$, what's the tightest circle (smallest radius) in which the plane can turn?
76. **DATA** Figure 5.39 shows an apparatus used to verify Newton's second law. A "pulling mass" m_1 hangs vertically from a string of negligible mass that passes over a pulley, also of negligible mass and with nearly frictionless bearings. The other end of the string is attached to a glider of mass m_2 riding on an essentially frictionless, horizontal air track. Both m_1 and m_2 may be varied by placing additional masses on the pulling mass and glider. The experiment consists of starting the glider from rest and letting the pulling mass accelerate it down the track. Three photogates are used to time the glider over two distance intervals, and an experimental value for its acceleration is determined from these data, using constant-acceleration equations from Chapter 2. The table in the next column lists the measured acceleration for a number of mass combinations. (a) Determine a quantity that, when plotted on the horizontal axis of a graph, should result in a straight line of slope g when acceleration is plotted on the vertical axis. (b) Make your plot, fit a line to the plotted data, and report the experimentally determined value of g.

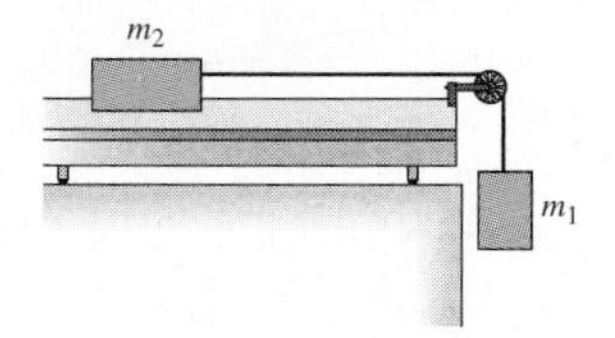

FIGURE 5.39 Problem 76

m_1 (g)	m_2 (g)	a (m/s^2)
10.0	170	0.521
10.0	270	0.376
10.0	370	0.274
20.0	170	1.06
20.0	270	0.652
20.0	370	0.534

Passage Problems

A *spiral* is an ice-skating position in which the skater glides on one foot with the other foot held above hip level. It's a required element in women's singles figure skating competition and is related to the arabesque performed in ballet. Figure 5.40 shows skater Sarah Hughes executing a spiral during her gold-medal performance at the Winter Olympics in Salt Lake City.

FIGURE 5.40 Passage Problems 77–80

77. From the photo, you can conclude that the skater is
    a. executing a turn to her left.
    b. executing a turn to her right.
    c. moving in a straight line out of the page.
78. The net force on the skater
    a. points to her left.
    b. points to her right.
    c. is zero.
79. If the skater were to execute the same maneuver but at higher speed, the tilt evident in the photo would be
    a. less.
    b. greater.
    c. unchanged.
80. The tilt angle θ that the skater's body makes with the vertical is given approximately by $\theta = \tan^{-1}(0.5)$. From this you can conclude that the skater's centripetal acceleration has approximate magnitude
    a. 0.
    b. 0.5 m/s^2.
    c. 5 m/s^2.
    d. can't be determined without knowing the skater's speed

Answers to Chapter Questions

Answer to Chapter Opening Question

The airplane tips, or *banks*, so there's a horizontal component of the aerodynamic force on the wings. That component provides the mv^2/r force that keeps the plane in its circular path. The vertical component of the aerodynamic force is what balances the gravitational force, keeping the plane aloft.

Answers to GOT IT? Questions

5.1 (c) Equal—but only because of the 45° slope. At larger angles, the tension would be greater than the weight; at smaller angles, less.
5.2 (a) The left hand in Fig. 5.9 and the hook in this figure play exactly the same role, balancing the 1-N tension force in the rope.
5.3 (c)
5.4 (c) Greater because the chain is pulling downward, making the normal force greater than the log's weight.

Energy, Work, and Power

What You Know

- You understand Newton's second law and how it relates force to change in motion.
- You're familiar with the terms *energy* and *work*, although not necessarily in their precise physics sense.
- Depending on your calculus background, you may or may not have seen integration—a calculus procedure that's the inverse of differentiation.

What You're Learning

- Here you're introduced to the fundamental concept of *energy*—ultimately, one of the two kinds of "stuff" our universe is made of.
- You'll learn how *work* means the transfer of energy by exerting forces.
- You'll see how to calculate work in both simple and complex situations.
- You'll have a quick introduction to integral calculus.
- You'll learn the vector dot product—one of two ways to multiply vectors.
- You'll see how applying Newton's second law in the context of work leads to the concept of kinetic energy.
- You'll come to appreciate the distinction between *power* and energy.

How You'll Use It

- In Chapter 7 you'll use the ideas developed here to formulate a statement about the conservation of energy.
- You'll see how energy conservation provides a "shortcut" to solving problems that would otherwise be very difficult.
- Energy will continue to play an important role as you explore Newtonian physics further in Parts 1 and 2, and as you expand your physics knowledge to encompass thermodynamics in Part 3 and electromagnetism in Part 4.
- You'll see how energy remains at the heart of your understanding of physical reality even when you move past Newtonian physics and into the realms of relativity and quantum physics.

Climbing a mountain, these cyclists do work against gravity. Does that work depend on the route chosen?

Figure 6.1*a* shows a skier starting from rest at the top of a uniform slope. What's the skier's speed at the bottom? You can solve this problem by applying Newton's second law to find the skier's constant acceleration and then the speed. But what about the skier in Fig. 6.1*b*? Here the slope is continuously changing and so is the acceleration. Constant-acceleration equations don't apply, so solving for the details of the skier's motion would be difficult.

There are many cases where motion involves changing forces and accelerations. In this chapter, we introduce the important physical concepts of **work** and **energy**. These powerful concepts enable us to "shortcut" the detailed application of Newton's law to analyze these

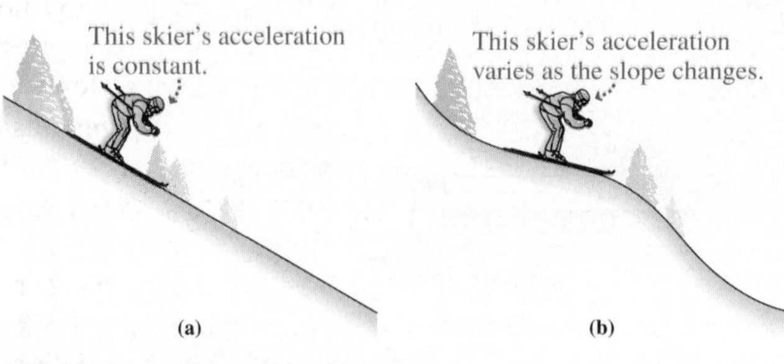

FIGURE 6.1 Two skiers.

more complex situations. But these new concepts have significance far beyond their practical applications in problem solving. Energy, in particular, is a fundamental aspect of the universe—a "substance" akin to, and every bit as real as, matter itself. In fact, as you'll see when you explore relativity in Chapter 33, energy and matter are really both aspects of a single "substance," linked by Einstein's equation $E = mc^2$.

6.1 Energy

"Energy" is a word you hear every day. You buy energy when you fill your car's gas tank. You use energy to heat your home and cook your food. You experience the awesome energy of a hurricane, a tornado, or an explosion. You sense the energy inherent in a truck barreling down the highway, or the energy generated by an airplane's engines as it surges down the runway. A power line crosses the countryside and, even though you can't see anything but the wires, you know that the line is carrying energy to a distant city. Your cellphone dies, its battery discharged, and you know that it needs to be replenished with energy. Your own body produces energy, which you sense as you climb a mountain, cycle, walk, or even think. You may have helped insulate or weatherize a home to reduce its energy loss. And the colossal rate at which humankind consumes energy is much on our minds as we become increasingly aware of the impact energy consumption has on our planet.

Actually, words like "consume," "generate," "produce," and "loss," although widely used in the context of energy, are misleading. That's because energy is *conserved*—meaning that it can change forms but cannot be created or destroyed. Much of your study of energy will involve ways to transform energy from one form to another or transfer it from place to place—all the while conserving the total amount of energy. Conservation of energy is a profound idea in physics, one whose richness we'll explore throughout the rest of this book.

Here in Part 1 of the book, we'll focus on *mechanical energy*, associated with motions and configurations of macroscopic objects such as cars, planets, baseballs, people, and springs. This chapter introduces *kinetic energy*, the energy of motion, as applied to such macroscopic objects. You'll see how the act of doing *work* is one way to transfer energy to an object. Chapter 7 will add the idea of *potential energy* and will develop a statement of energy conservation as it applies to mechanical energy. We'll also need to consider so-called *internal energy*, associated with random motions and configuration changes at the molecular level. In Part 3 (thermodynamics), we'll explore internal energy and show how it's incorporated into a broader statement of energy conservation. In Part 3, you'll also see how *heat* describes another way of transferring energy. In Part 4, on electricity and magnetism, we'll introduce forms of *electromagnetic energy* and associated energy-transfer processes. In Part 6, you'll see that energy concepts survive even into the realm of quantum physics—no mean feat given that many other ideas from classical physics become meaningless or even nonsensical in the quantum realm.

Energy and Systems

What's got energy? A moving car does. So does a whole highway full of cars. A warm house has energy. So does a stretched spring. A hurricane has energy, and so does our whole planet. When we're accounting for energy and studying energy flows and transformations, we need to have in mind a **system** whose energy we're interested in. Typically the system contains one or more objects, and it's defined by a closed boundary. Everything within the boundary is part of the system, whereas everything outside comprises the environment that surrounds the system. Like coordinate axes, a system is something you define for your convenience. Once you've defined a system, then you can talk about the system's energy and what forms it takes; about energy transformations within the system; and about any transfers of energy into or out of the system. Figure 6.2 shows conceptually how to think about energy in the context of a system, while the Application shows how the idea behind Fig. 6.2 is applied in the important case of climate modeling.

The boundary separates the system from its surroundings.

Energy flows within the system, and transforms from one form to another. Absent inflows and outflows, the total energy in the system wouldn't change.

System

Energy in

Energy out

K U E_{int} E_{EM}

Energy can flow into and out of the system, as mechanical energy, as heat, or as electromagnetic energy.

FIGURE 6.2 Diagram showing energy flows both within a system and across its boundaries, as well as transformations between different types of energy within the system. We show four common forms of energy: Kinetic energy (K) and potential energy (U), subjects of this chapter and the next; internal or thermal energy (E_{int}), a subject of Part 3; and electromagnetic energy, covered in Part 4.

APPLICATION Climate Modeling

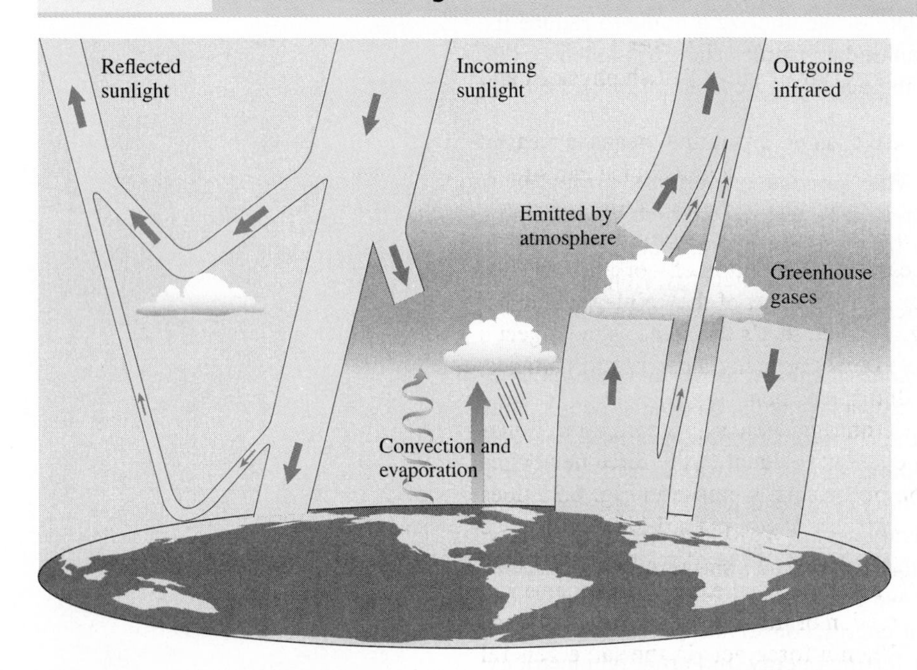

Figure 6.2 isn't just a pedagogical aid to understanding energy and systems concepts; it's a framework for realistic models used to characterize energy-related systems ranging from biological organisms to nuclear power plants to Earth's climate. The figure here shows the energy–systems concept as climate scientists use it. The system comprises Earth and its atmosphere. Many of the arrows represent energy flows and transformations between Earth and atmosphere—that is, within the system. These involve electromagnetic energy emitted in the form of infrared radiation, as well as energy associated with warm air and water vapor rising into the atmosphere. The three arrows at the top show energy exchanges with the planet's surroundings—that is, energy crossing the system boundary. As you'll see in Chapter 16, the incoming and outgoing flows must balance for climate to remain stable.

In this chapter, we'll often choose our system to coincide with a single object, and in that case we'll use the term "object" interchangeably with "system." But in Chapter 7, we'll need to consider systems comprising at least two interacting objects, and when we get to Chapter 9, we'll be dealing with systems of many particles. In these more complex situations we'll have to decide carefully what's in our system and what's outside it, and we'll need to make more use of systems terminology.

6.2 Work

PhET: The Ramp

One way to transfer energy to a system is to act on the system with an external force—a force applied by an entity that isn't part of the system. In this case we say that the force does **work** on the system. Doing work is an inherently mechanical process, involving the concept of force that you're already familiar with from Newtonian mechanics.

Imagine carrying a piece of furniture upstairs. You have to apply an upward force on the furniture as you climb the stairs. Define the furniture as your system, and the force you apply is an external force that does work on the system. Thus you transfer energy from your body to the furniture by doing mechanical work.

You've already got an intuitive sense of work and how it's quantified. Make that furniture heavier, or the stairs higher, and you do more work. Or try pushing a stalled car: The harder you push, or the farther you push, the more work you do. The precise definition of work reflects your intuition:

For an object moving in one dimension, the work W done on the object by constant applied force $\vec{F}$ is

$$W = F_x \Delta x \qquad (6.1)$$

where F_x is the component of the force in the direction of the object's motion and Δx is the object's displacement.

The force $\vec{F}$ need not be the net force. If you're interested, for example, in how much work *you* must do to drag a heavy box across the floor, then $\vec{F}$ is the force *you apply* and W is the work *you do*.

Equation 6.1 shows that the SI unit of work is the newton-meter (N·m). One newton-meter is given the name **joule**, in honor of the 19th-century British physicist and brewer James Joule.

Our definition of work involves an object's displacement. What that means is clear in the case of a rigid object like a block or a ball. But what about a spring, which can stretch or compress in response to an applied force? Or your own body, whose configuration alters as you lift, run, jump, dance, or swim? Or a more complex system of many parts, with a force applied to just one part? In all these cases our definition of work still applies, provided we interpret "displacement" to mean the displacement of the point at which the force is applied. For a system consisting of a rigid object, that's the same as the object's displacement. For a system consisting of a flexible object or many independent particles, it's not necessarily the same as the system's overall displacement.

Figure 6.3 considers several cases of work done on rigid objects. According to Equation 6.1, the person pushing the car in Fig. 6.3*a* does work equal to the force he applies times the distance the car moves. But the person pulling the suitcase in Fig. 6.3*b* does work equal to only the horizontal component of the force she applies times the distance the suitcase moves. Furthermore, by our definition, the waiter of Fig. 6.3*c* does no work on the tray. Why not? Because the force on the tray is vertical while the tray's displacement is horizontal; there's no component of force in the direction of the tray's motion.

Work can be positive or negative (Fig. 6.4). When a force acts in the same general direction as the motion, it does positive work. A force acting at 90° to the motion does no work. And when a force acts to oppose motion, it does negative work.

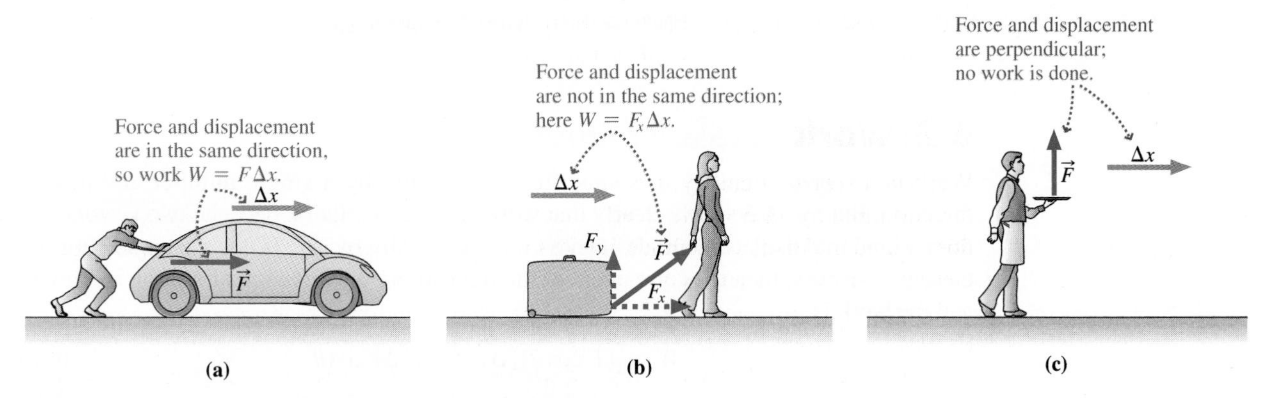

FIGURE 6.3 Work depends on the orientation of force and displacement.

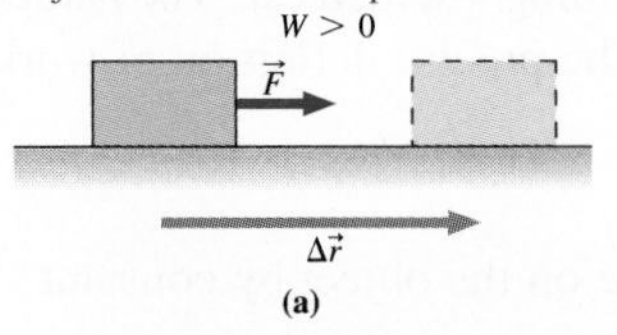

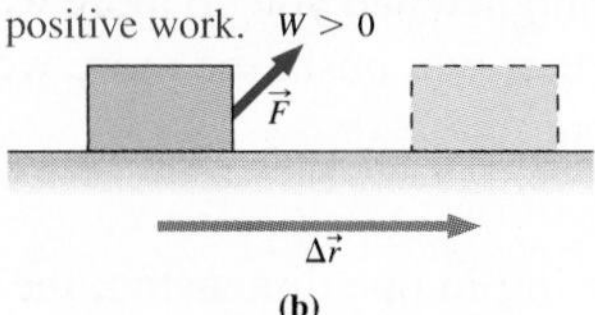

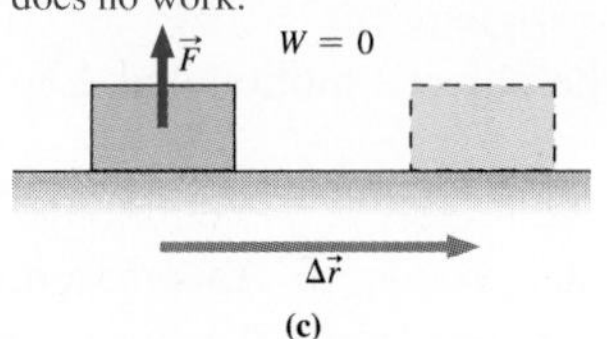

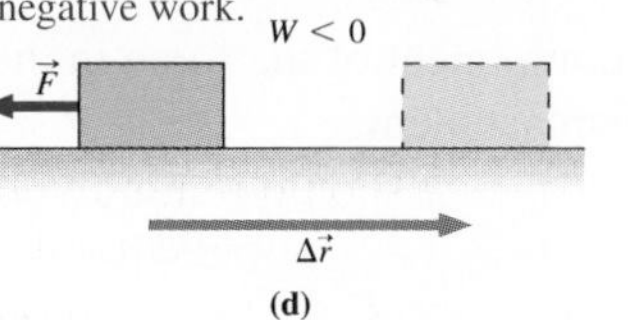

FIGURE 6.4 The sign of the work depends on the relative directions of force and motion. We use $\Delta\vec{r}$ here to indicate that the displacement can be any vector.

EXAMPLE 6.1 Calculating Work: Pushing a Car

The person in Fig. 6.3*a* pushes with a force of 650 N, moving the car a distance of 4.3 m. How much work does he do?

INTERPRET This problem is about work. We identify the car as the object *on* which the work is done and the person as the agent *doing* the work.

DEVELOP Figure 6.3*a* is our drawing. Equation 6.1, $W = F_x\,\Delta x$, is the relevant equation, so our plan is to apply that equation. The force is in the same direction as the displacement, so 650 N is the component we need.

EVALUATE We apply Equation 6.1 to get

$$W = F_x\,\Delta x = (650\text{ N})(4.3\text{ m}) = 2.8\text{ kJ}$$

ASSESS Make sense? The units work out, with newtons times meters giving joules—here expressed in kilojoules for convenience. ■

EXAMPLE 6.2 Calculating Work: Pulling a Suitcase

The airline passenger in Fig. 6.3*b* exerts a 60-N force on her suitcase, pulling at 35° to the horizontal. How much work does she do in pulling the suitcase 45 m on a level floor?

INTERPRET Again, this example is about work—here done *by* the passenger *on* the suitcase.

DEVELOP Equation 6.1, $W = F_x\,\Delta x$, applies here, but because the displacement is horizontal while the force isn't, we need to find the horizontal force component. We've redrawn the force vector in Fig. 6.5 to determine F_x.

EVALUATE Applying Equation 6.1 to the *x*-component from Fig. 6.5, we get

$$W = F_x\,\Delta x = [(60\text{ N})(\cos 35°)](45\text{ m}) = 2.2\text{ kJ}$$

ASSESS The answer of 2.2 kJ is less than the product of 60 N and 45 m, and that makes sense because only the *x*-component of that 60-N force contributes to the work. ■

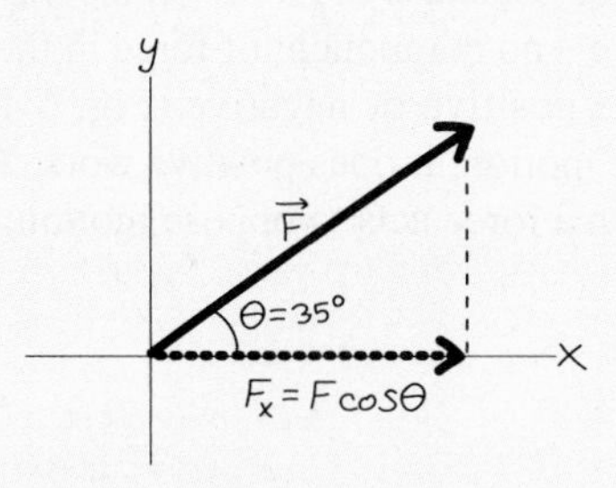

FIGURE 6.5 Our sketch for Example 6.2.

Work and the Scalar Product

Work is a *scalar* quantity; it's specified completely by a single number and has no direction. But Fig. 6.3 shows clearly that work involves a relation between two *vectors*: the force $\vec{F}$ and the displacement, designated more generally by $\Delta\vec{r}$. If θ is the angle between these two vectors, then the component of the force along the direction of motion is $F\cos\theta$, and the work is

$$W = (F\cos\theta)(\Delta r) = F\,\Delta r\cos\theta \tag{6.2}$$

This equation is a generalization of our definition 6.1. If we choose the *x*-axis along $\Delta\vec{r}$, then $\Delta r = \Delta x$ and $F\cos\theta = F_x$, so we recover Equation 6.1.

Equation 6.2 shows that work is the product of the magnitudes of the vectors $\vec{F}$ and $\Delta\vec{r}$ and the cosine of the angle between them. This combination occurs so often that it's given a special name: the **scalar product** of two vectors.

The scalar product of any two vectors $\vec{A}$ and $\vec{B}$ is defined as

$$\vec{A}\cdot\vec{B} = AB\cos\theta \qquad (6.3)$$

where A and B are the magnitudes of the vectors and θ is the angle between them.

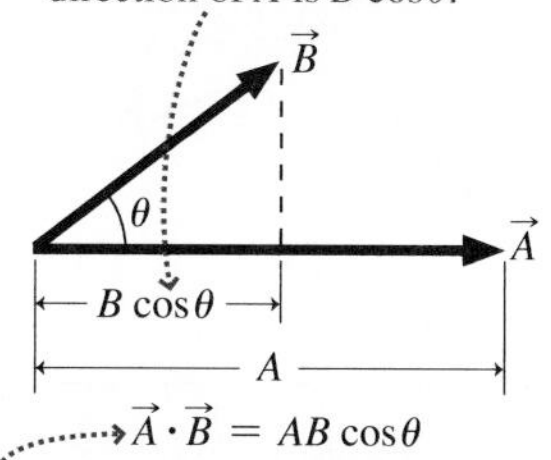

FIGURE 6.6 Geometric interpretation of the scalar product.

The term *scalar product* should remind you that $\vec{A}\cdot\vec{B}$ is itself a *scalar*, even though it's formed from two vectors. A centered dot designates the scalar product; for this reason, it's also called the **dot product**. Figure 6.6 gives a geometric interpretation.

The scalar product is commutative: $\vec{A}\cdot\vec{B} = \vec{B}\cdot\vec{A}$, and it's also distributive: $\vec{A}\cdot(\vec{B}+\vec{C}) = \vec{A}\cdot\vec{B} + \vec{A}\cdot\vec{C}$. With vectors expressed in unit vector notation, Problem 48 shows how the distributive law gives a simple form for the scalar product. If $\vec{A} = A_x\hat{\imath} + A_y\hat{\jmath} + A_z\hat{k}$ and $\vec{B} = B_x\hat{\imath} + B_y\hat{\jmath} + B_z\hat{k}$, then

$$\vec{A}\cdot\vec{B} = A_xB_x + A_yB_y + A_zB_z \qquad (6.4)$$

Comparing Equation 6.2 with Equation 6.3 shows that the work done by a constant force $\vec{F}$ moving an object through a straight-line displacement $\Delta\vec{r}$ can be expressed using the dot product:

$$W = \vec{F}\cdot\Delta\vec{r} \qquad (6.5)$$

As the examples below show, either Equation 6.3 or Equation 6.4 can be used in evaluating the dot product in this expression for work.

EXAMPLE 6.3 Work and the Scalar Product: A Tugboat

A tugboat pushes a cruise ship with force $\vec{F} = 1.2\hat{\imath} + 2.3\hat{\jmath}$ MN, moving the ship along a straight path with displacement $\Delta\vec{r} = 380\hat{\imath} + 460\hat{\jmath}$ m. Find (a) the work done by the tugboat and (b) the angle between the force and displacement.

INTERPRET Part (a) is about calculating work given force and displacement in unit vector notation. Part (b) is less obvious, but knowing that work involves the angle between force and displacement provides a clue, suggesting that the answer to (a) may lead us to (b).

DEVELOP Figure 6.7 is a sketch of the two vectors, which will serve as a check on our final answer. For (a), we want to use Equation 6.5, $W = \vec{F}\cdot\Delta\vec{r}$, with the scalar product in unit vector notation given by Equation 6.4. That will give us the work W. We also have the vectors $\vec{F}$ and $\Delta\vec{r}$, so we can find their magnitudes. That suggests a strategy for (b): Given the work and the vector magnitudes, we can write Equation 6.3 with a single unknown, the angle θ that we're asked to find.

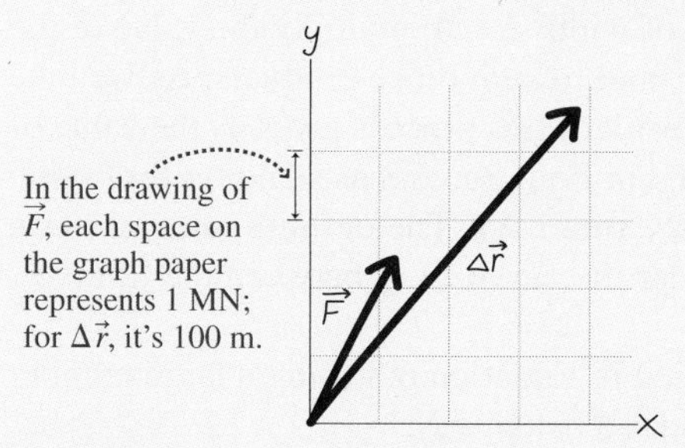

FIGURE 6.7 Our sketch of the vectors in Example 6.3.

EVALUATE For (a), we use Equations 6.5 and 6.4, respectively, to write

$$W = \vec{F}\cdot\Delta\vec{r} = F_x\,\Delta x + F_y\,\Delta y$$
$$= (1.2\text{ MN})(380\text{ m}) + (2.3\text{ MN})(460\text{ m}) = 1510\text{ MJ}$$

The first equality is from Equation 6.5; the second gives the scalar product in unit vector form from Equation 6.4. Δx and Δy are the components of the displacement $\Delta\vec{r}$. Now that we have the work, we can get the angle. The magnitude of a vector comes from the Pythagorean theorem, as expressed in Equation 3.1. So we have $F = \sqrt{F_x^2 + F_y^2} = \sqrt{(1.2\text{ MN})^2 + (2.3\text{ MN})^2} = 2.59$ MN; a similar calculation gives $\Delta r = 597$ m. Now we solve Equation 6.3 for θ:

$$\theta = \cos^{-1}\left(\frac{W}{F\,\Delta r}\right) = \cos^{-1}\left(\frac{1510\text{ MJ}}{(2.59\text{ MN})(597\text{ m})}\right) = 12°$$

ASSESS This small angle is consistent with our sketch in Fig. 6.7. And it makes good physical sense: A tugboat is most efficient when pushing in the direction the ship is supposed to go. Note how the units work out in that last calculation: MJ in the numerator and MN·m in the denominator. But 1 N·m is 1 J, so that's MJ in the denominator, too, giving the dimensionless cosine. ■

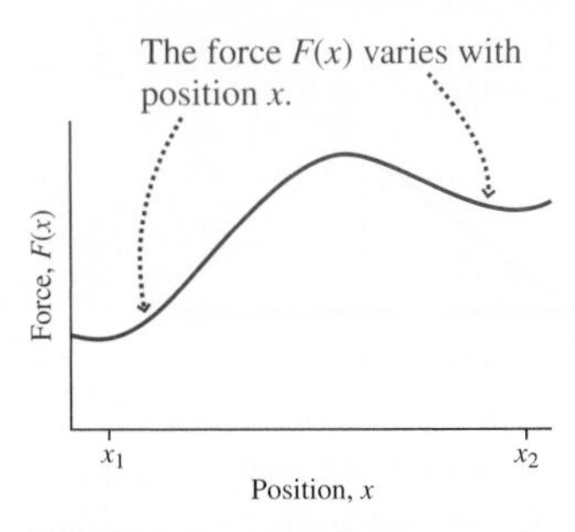

FIGURE 6.8 A varying force.

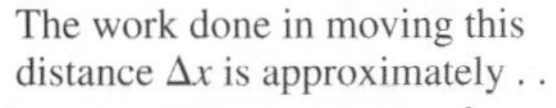

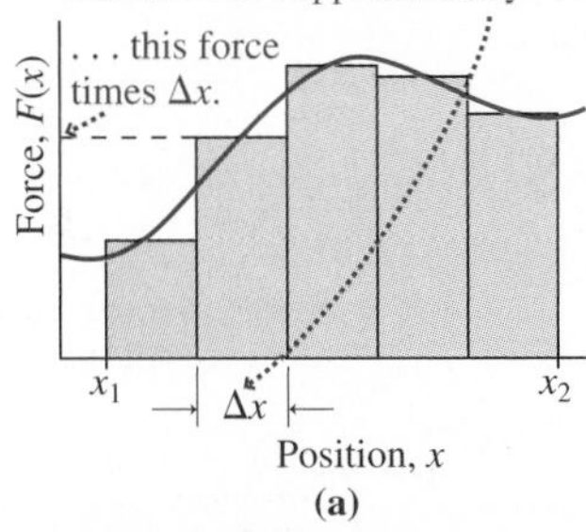

Making the rectangles smaller makes the approximation more accurate.

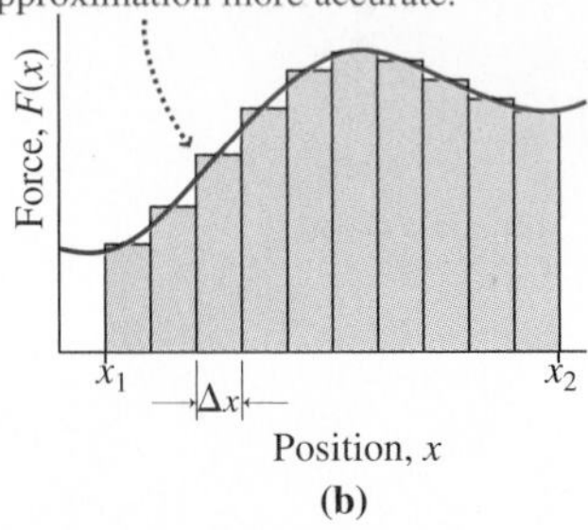

The exact value for the work is the area under the force-versus-position curve.

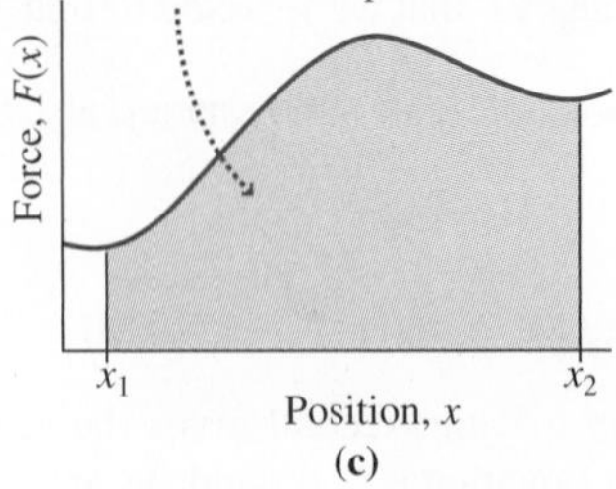

FIGURE 6.9 Work done by a varying force.

PhET: The Ramp

GOT IT? 6.1 Two objects are each displaced the same distance, one by a force F pushing in the direction of motion and the other by a force $2F$ pushing at 45° to the direction of motion. Which force does more work? (a) F; (b) $2F$; (c) they do equal work

---

6.3 Forces That Vary

Often the force applied to an object varies with position. Important examples include electric and gravitational forces, which vary with the distance between interacting objects. The force of a spring that we encountered in Chapter 4 provides another example; as the spring stretches, the force increases.

Figure 6.8 is a plot of a force F that varies with position x. We want to find the work done as an object moves from x_1 to x_2. We can't simply write $F(x_2 - x_1)$; since the force varies, there's no single value for F. What we can do, though, is divide the region into rectangles of width Δx, as shown in Fig. 6.9*a*. If we make Δx small enough, the force will be nearly constant over the width of each rectangle (Fig. 6.9*b*). Then the work ΔW done in moving the width Δx of one such rectangle is approximately $F(x)\,\Delta x$, where $F(x)$ is the force at the midpoint x of that rectangle. We write $F(x)$ to show explicitly that the force is a function of position. Note that the quantity $F(x)\,\Delta x$ is the area of the rectangle expressed in the appropriate units (N·m, or, equivalently, J).

Suppose there are N rectangles. Let x_i be the midpoint of the ith rectangle. Then the total work done in moving from x_1 to x_2 is given approximately by the sum of the individual amounts of work ΔW_i associated with each rectangle, or

$$W \simeq \sum_{i=1}^{N} \Delta W_i = \sum_{i=1}^{N} F(x_i)\,\Delta x \tag{6.6}$$

How good is this approximation? That depends on how small we make the rectangles. Suppose we let them get arbitrarily small. Then the number of rectangles must grow arbitrarily large. In the limit of infinitely many infinitesimally small rectangles, the approximation in Equation 6.6 becomes exact (Fig. 6.9*c*). Then we have

$$W = \lim_{\Delta x \to 0} \sum_i F(x_i)\,\Delta x \tag{6.7}$$

where the sum is over all the infinitesimal rectangles between x_1 and x_2. The quantity on the right-hand side of Equation 6.7 is the **definite integral** of the function $F(x)$ over the interval from x_1 to x_2. We introduce special symbolism for the limiting process of Equation 6.7:

$$W = \int_{x_1}^{x_2} F(x)\,dx \qquad \left(\begin{array}{c}\text{work done by a varying}\\ \text{force in one dimension}\end{array}\right) \tag{6.8}$$

Equation 6.8 means exactly the same thing as Equation 6.7: It tells us to divide the interval from x_1 to x_2 into many small rectangles of width Δx, to multiply the value of the function $F(x)$ at each rectangle by the width Δx, and to sum those products. As we take arbitrarily many arbitrarily small rectangles, the result of this process gives us the value of the definite integral. You can think of the symbol $\int$ in Equation 6.8 as standing for "sum" and the symbol dx as a limiting case of arbitrarily small Δx. The definite integral has a simple geometric interpretation: It's the area under the curve $F(x)$ between the limits x_1 and x_2 (Fig. 6.9*c*).

Computers approximate the infinite sum implied in Equation 6.8 using a large number of very small rectangles. But calculus often provides a better way.

TACTICS 6.1 Integrating

In your calculus course you've learned, or will soon learn, that integrals and derivatives are inverses. In Section 2.2, you saw that the derivative of x^n is nx^{n-1}; therefore, the integral of x^n is $(x^{n+1})/(n+1)$, as you can verify by differentiating. We determine the value of a definite integral by evaluating this expression at upper and lower limits and subtracting:

$$\int_{x_1}^{x_2} x^n\,dx = \left.\frac{x^{n+1}}{n+1}\right|_{x_1}^{x_2} = \frac{x_2^{n+1}}{n+1} - \frac{x_1^{n+1}}{n+1} \qquad (6.9)$$

where the middle term, with the vertical bar and the upper and lower limits, is a shorthand notation for the difference given in the rightmost term. Appendix A includes a review of integration and a table of common integrals.

Stretching a Spring

A spring provides an important example of a force that varies with position. We've seen that an ideal spring exerts a force proportional to its displacement from equilibrium: $F = -kx$, where k is the spring constant and the minus sign shows that the spring force is opposite the direction of the displacement. It's not just coiled springs that we're interested in here; many physical systems, from molecules to skyscrapers to stars, behave as though they contain springs. The work and energy considerations we develop here apply to those systems as well.

The force exerted *by* a stretched spring is $-kx$, so the force exerted *on* the spring *by* the external stretching force is $+kx$. If we let $x = 0$ be one end of the spring at equilibrium and if we hold the other end fixed and pull the spring until its free end is at a new position x, as shown in Fig. 6.10, then Equation 6.8 shows that the work done on the spring by the external force is

$$W = \int_0^x F(x)\,dx = \int_0^x kx\,dx = \left.\tfrac{1}{2}kx^2\right|_0^x = \tfrac{1}{2}kx^2 - \tfrac{1}{2}k(0)^2 = \tfrac{1}{2}kx^2 \qquad (6.10)$$

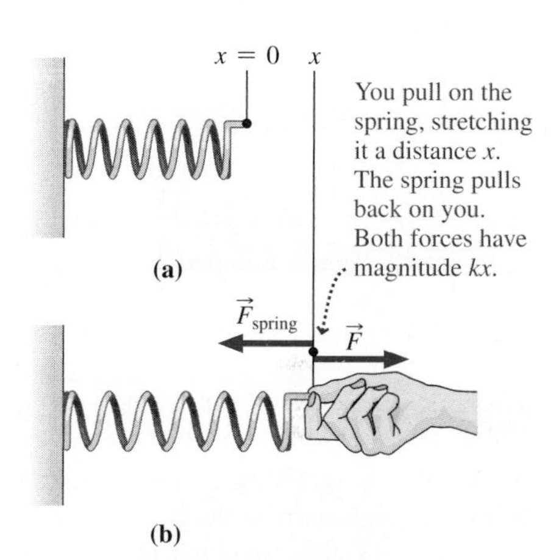

FIGURE 6.10 Stretching a spring.

where we used Equation 6.9 to evaluate the integral. The more we stretch the spring, the greater the force we must apply—and that means we must do more work for a given amount of additional stretch. Figure 6.11 shows graphically why the work depends quadratically on the displacement. Although we used the word *stretch* in developing Equation 6.10, the result applies equally to compressing a spring a distance x from equilibrium. Note here that we're explicitly using the displacement of the force application point—the end of the spring—which in this case of this flexible system isn't the same as the displacement of the whole spring.

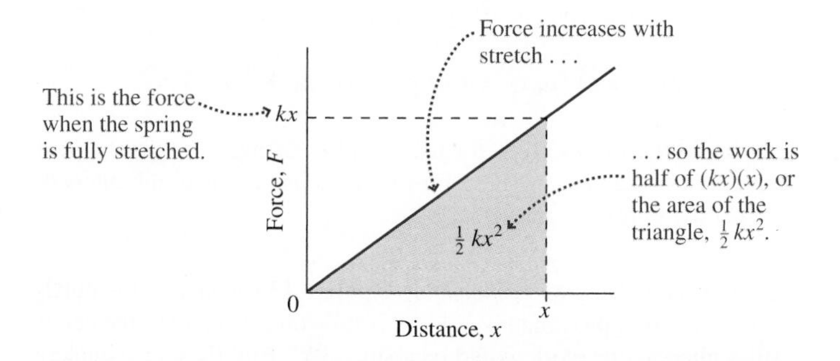

FIGURE 6.11 Work done in stretching a spring.

EXAMPLE 6.4 The Spring Force: Bungee Jumping

An elastic cord used in bungee jumping is normally 11 m long and has spring constant $k = 250$ N/m. At the lowest point in a jump, the cord length has doubled. How much work has been done on the cord?

INTERPRET The bungee cord behaves like a spring—as we can tell because we're given its spring constant. So this example is about the work done in stretching a spring. We're told the 11-m-long cord length doubles in length, so it's stretched another 11 m.

DEVELOP Equation 6.10 gives the work done in stretching the cord a distance x from its unstretched configuration.

EVALUATE Applying Equation 6.10 gives

$$W = \tfrac{1}{2}kx^2 = (\tfrac{1}{2})(250\text{ N/m})(11\text{ m})^2 = 15\text{ kN·m} = 15\text{ kJ}$$

ASSESS As you'll see shortly, that's just about equal to the work done by gravity on a 70-kg person dropping the 22-m distance from the attachment point of the cord to its full stretched extent. You'll see in the next chapter why this is no coincidence. ■

CONCEPTUAL EXAMPLE 6.1 Bungee Details

In Example 6.4, is the work done as the cord stretches its final meter greater than, less than, or equal to the work done in the first meter of stretch?

EVALUATE We're asked to compare the work done during the beginning and end of the bungee cord's stretch. We know that work is the area under the force–distance curve. We've sketched the force–distance curve in Fig. 6.12, highlighting the first and last meters. The figure makes it clear that the area associated with the last meter of stretch is much larger. Therefore, the work is greater.

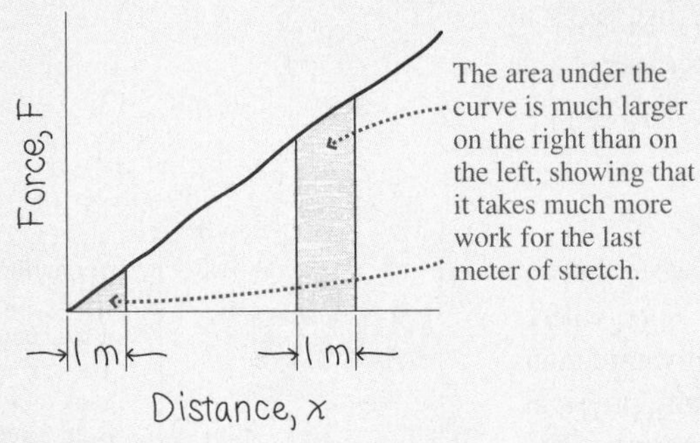

FIGURE 6.12 Conceptual Example 6.1.

ASSESS Makes sense! Once the cord has stretched 10 m, it exerts a large force. That makes it much harder to stretch farther—and thus the final meter requires a lot of work. The first meter takes much less work because at first the cord exerts very little force.

MAKING THE CONNECTION Find the work involved in stretching during the first and last meters, and compare.

EVALUATE We can use Equation 6.10, but instead of the limits 0 and x, we'll use 0 and 1 m for the first meter of stretch, and 10 m and 11 m for the last meter. The results are 125 J and 2.6 kJ. Stretching the final meter takes more than 20 times the work required for the first meter!

EXAMPLE 6.5 A Varying Friction Force: Rough Sliding

Workers pushing a 180-kg trunk across a level floor encounter a 10-m-long region where the floor becomes increasingly rough. The coefficient of kinetic friction here is given by $\mu_k = \mu_0 + ax^2$, where $\mu_0 = 0.17$, $a = 0.0062\ \text{m}^{-2}$, and x is the distance from the beginning of the rough region. How much work does it take to push the trunk across the region?

INTERPRET This example asks for the work needed to push the trunk. To move the trunk at constant speed, the workers must apply a force equal in magnitude to the frictional force. That force varies with position, so we're dealing with a varying force.

DEVELOP Our drawing, the force–position curve in Fig. 6.13, emphasizes that we have a varying force. Therefore, we have to integrate using Equation 6.8, $W = \int_{x_1}^{x_2} F(x)\,dx$. And we need to know the frictional force, which is given by Equation 5.3: $f_k = \mu_k n$. On a level floor, the normal force is equal in magnitude to the weight, mg, so Equation 6.8 becomes $W = \int_{x_1}^{x_2} \mu_k mg\,dx = \int_{x_1}^{x_2} mg(\mu_0 + ax^2)\,dx$.

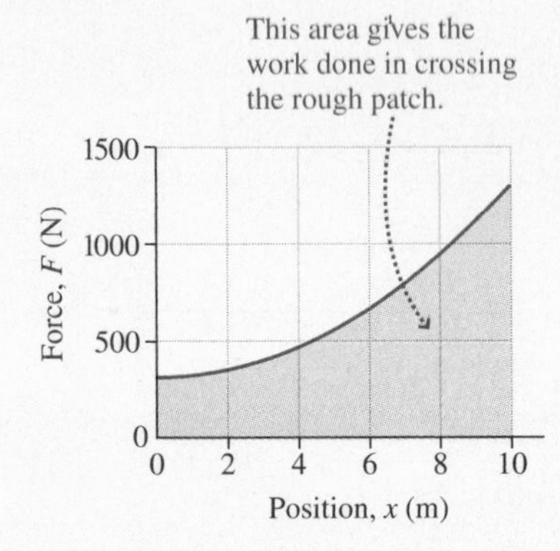

FIGURE 6.13 Force versus position for Example 6.5.

EVALUATE We evaluate the integral using Equation 6.9. Actually, we have two integrals here: one of dx alone and the other of $x^2\,dx$. According to Equation 6.9, the former gives x and the latter $x^3/3$. So the result is

$$W = \int_{x_1}^{x_2} mg(\mu_0 + ax^2)\,dx = mg(\mu_0 x + \tfrac{1}{3}ax^3)\Big|_{x_1}^{x_2}$$

$$= mg\left[(\mu_0 x_2 + \tfrac{1}{3}ax_2^3) - (\mu_0 x_1 + \tfrac{1}{3}ax_1^3)\right]$$

Putting in the values given for μ_0, a, and m, using $g = 9.8\ \text{m/s}^2$, and taking $x_1 = 0$ and $x_2 = 10$ m for the endpoints of the rough interval, we get 6.6 kJ for our answer.

ASSESS Is this answer reasonable? Figure 6.13 shows that the maximum force is approximately 1.3 kN. If this force acted over the entire 10-m interval, the work would be about 13 kJ. But it's approximately half that because the coefficient of kinetic friction and therefore the force start out quite low. You can see that the area under the curve in Fig. 6.13 is about half the area of the full rectangle, so our answer of 6.6 kJ makes sense. ■

✓TIP Don't Just Multiply!

When force depends on position, there's no single value for the force, so you can't just multiply force by distance to get work. You need either to integrate, as in Example 6.5, or to use a result that's been derived by integration, as with the equation $W = \frac{1}{2}kx^2$ used in Example 6.4.

Force and Work in Two and Three Dimensions

Sometimes a force varies in both magnitude and direction or an object moves on a curved path; either way, the angle between force and motion may vary. Then we have to take the scalar products of the force $\vec{F}$ with small displacements $\Delta\vec{r}$, writing $\Delta W = \vec{F}\cdot\Delta\vec{r}$ for the work involved in one such small displacement. Adding them all gives the total work, which in the limit of very small displacements becomes a **line integral**:

$$W = \int_{\vec{r}_1}^{\vec{r}_2} \vec{F}\cdot d\vec{r} \tag{6.11}$$

where the integral is taken over a specific path between positions $\vec{r}_1$ and $\vec{r}_2$. We won't pursue line integrals further here, but they'll be useful in later chapters.

Work Done against Gravity

When an object moves upward or downward on an arbitrary path, the angle between its displacement and the gravitational force varies. But here we don't really need the line integral of Equation 6.11 because we can consider any path as consisting of small horizontal and vertical steps (Fig. 6.14). Only the vertical steps contribute to the work, which then becomes simply $W = mgh$, where h is the total height the object rises—a result that's independent of the particular path taken. (As in our earlier work with gravity, this result holds only near Earth's surface, where we can neglect the variation in gravity with height.)

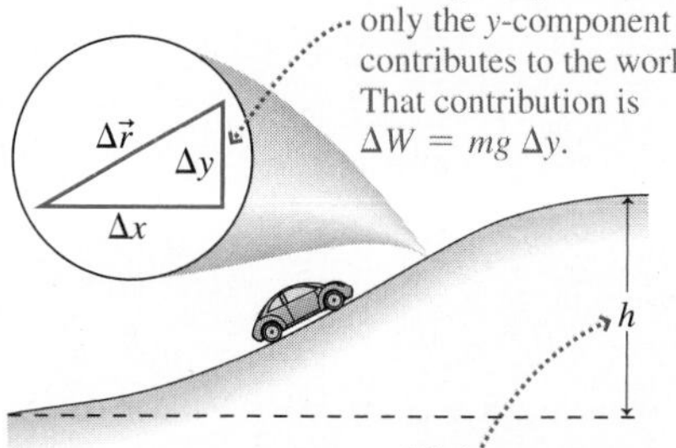

FIGURE 6.14 A car climbs a hill with varying slope.

GOT IT? 6.2 Three forces have magnitudes in newtons that are numerically equal to these quantities: (a) x, (b) x^2, and (c) $\sqrt{x}$, where x is the position in meters. Each force acts on an object as it moves from $x = 0$ to $x = 1$ m. Notice that all three forces have the same values at the two endpoints—namely, 0 N and 1 N. Which of the forces (a), (b), or (c) does the most work? Which does the least?

6.4 Kinetic Energy

Doing work on a system by applying a force is the mechanical way to transfer energy to the system. How does that energy manifest itself? Under some conditions it shows up as kinetic energy—energy of the system's motion. Here we develop a relation between the *net work* done by all forces acting on a system that consists of a single rigid object and the resulting change in the object's kinetic energy. In the process we'll develop a simple formula for kinetic energy.

We'll start by evaluating the net work done on the object and then apply Newton's second law. With our single object, the net work is the work done by the sum of all forces acting on the object—that is, by the net force. So we'll use the net force in our expression for work. (That wouldn't do for a more complicated system, with different forces acting on different points that might undergo different displacements; there we'd have to calculate the work associated with each force, and then sum.) We'll consider the simple case of one-dimensional motion, with force and displacement along the same line. In that case, Equation 6.8 gives the net work:

$$W_{\text{net}} = \int F_{\text{net}}\,dx$$

But the net force can be written in terms of Newton's second law: $F_{\text{net}} = ma$, or $F_{\text{net}} = m\,dv/dt$, so

$$W_{\text{net}} = \int m\frac{dv}{dt}\,dx$$

The quantities dv, dt, and dx arose as the limits of small numbers Δv, Δt, and Δx. In calculus, you've seen that the limit of a product or quotient is the product or quotient of the individual terms involved. For these reasons, we can rearrange the symbols dv, dt, and dx to rewrite our expression in the form

$$W_{net} = \int m\, dv \frac{dx}{dt}$$

But $dx/dt = v$, so we have

$$W_{net} = \int mv\, dv$$

The integral here is like $\int x\,dx$, which we evaluate by raising the exponent and dividing by the new exponent. What about the limits? Suppose our object starts at some speed v_1 and ends at v_2. Then we have

$$W_{net} = \int_{v_1}^{v_2} mv\, dv = \tfrac{1}{2} mv^2 \Big|_{v_1}^{v_2} = \tfrac{1}{2} mv_2^2 - \tfrac{1}{2} mv_1^2 \quad (6.12)$$

Equation 6.12 shows that an object has associated with it a quantity $\frac{1}{2}mv^2$ that changes when, and only when, net work is done on the object. This quantity is the object's **kinetic energy**:

The kinetic energy K of an object of mass m moving at speed v is

$$K = \tfrac{1}{2} mv^2 \quad (6.13)$$

Like velocity, kinetic energy is a relative term; its value depends on the reference frame in which it's measured. But unlike velocity, kinetic energy is a *scalar*. And since it depends on the *square* of the velocity, kinetic energy is never negative. All moving objects possess kinetic energy.

Equation 6.12 equates the change in an object's kinetic energy with the net work done on the object, a result known as the **work–kinetic energy theorem**:

Work–kinetic energy theorem: The change in an object's kinetic energy is equal to the net work done on the object:

$$\Delta K = W_{net} \quad (6.14)$$

Equations 6.12 and 6.14 are equivalent statements of the work–kinetic energy theorem.

We've seen that work can be positive or negative; the work-kinetic energy theorem (Equation 6.14) therefore shows that changes in kinetic energy are correspondingly positive or negative. If I stop a moving object, for example, I reduce its kinetic energy from $\frac{1}{2}mv^2$ to zero—a change $\Delta K = -\frac{1}{2}mv^2$. So I do negative work by applying a force directed opposite to the motion. By Newton's third law, the object exerts an equal but oppositely directed force on me, therefore doing positive work $\frac{1}{2}mv^2$ *on me*. So an object of mass m moving at speed v can do work equal to its initial kinetic energy, $\frac{1}{2}mv^2$, if it's brought to rest.

EXAMPLE 6.6 Work and Kinetic Energy: Passing Zone

A 1400-kg car enters a passing zone and accelerates from 70 to 95 km/h. (a) How much work is done on the car? (b) If the car then brakes to a stop, how much work is done on it?

INTERPRET Here we're asked about work, but we aren't given any forces as we were in previous examples. However, we now know the work–kinetic energy theorem. Kinetic energy depends on speed, which we're given. So this is a problem involving the work–kinetic energy theorem.

DEVELOP The relevant equation is Equation 6.14 or its more explicit form, Equation 6.12. Since we're given speeds, it's easiest to work with Equation 6.12.

EVALUATE For (a), Equation 6.12 gives

$$W_{net} = \tfrac{1}{2}mv_2^2 - \tfrac{1}{2}mv_1^2 = \tfrac{1}{2}m(v_2^2 - v_1^2)$$
$$= (\tfrac{1}{2})(1400\ \text{kg})[(26.4\ \text{m/s})^2 - (19.4\ \text{m/s})^2] = 220\ \text{kJ}$$

where we converted the speeds to meters per second before doing the calculation. The work–kinetic energy theorem applies equally to the braking car in (b), for which $v_1 = 26.4$ m/s and $v_2 = 0$. Here we have

$$W_{net} = \tfrac{1}{2}m(v_2^2 - v_1^2) = (\tfrac{1}{2})(1400\ \text{kg})[0^2 - (26.4\ \text{m/s})^2]$$
$$= -490\ \text{kJ}$$

ASSESS Make sense? Yes: There's a greater change in speed and thus in kinetic energy in the braking case, so the magnitude of the work involved is greater. Our second answer is negative because stopping the car means applying a force that *opposes* its motion—and that means negative work is done on the car. ■

GOT IT? 6.3 For each situation, tell whether the net work done on a soccer ball is (a) positive, (b) negative, or (c) 0. (1) You carry the ball out to the field, walking at constant speed. (2) You kick the stationary ball, starting it flying through the air. (3) The ball rolls along the field, gradually coming to a halt.

Energy Units

Since work is equal to the change in kinetic energy, the units of energy are the same as those of work. In SI, the unit of energy is therefore the joule, equal to 1 newton-meter. In science, engineering, and everyday life, though, you'll encounter other energy units. Scientific units include the **erg**, used in the centimeter-gram-second system of units and equal to 10^{-7} J; the **electronvolt**, used in nuclear, atomic, and molecular physics; and the **calorie**, used in thermodynamics and to describe the energies of chemical reactions. English units include the **foot-pound** and the **British thermal unit** (Btu); the latter is commonly used in engineering of heating and cooling systems. Your electric company charges you for energy use in **kilowatt-hours** (kW·h); we'll see in the next section how this unit relates to the SI joule. Appendix C contains an extensive table of energy units and conversion factors as well as the energy contents of common fuels.

6.5 Power

Climbing a flight of stairs requires the same amount of work no matter how fast you go. But it's harder to *run* up the stairs than to walk. Harder in what sense? In the sense that you do the same work in a shorter time; the *rate* at which you do the work is greater. We define **power** as the rate of doing work:

If an amount of work ΔW is done in time Δt, then the average power $\overline{P}$ is

$$\overline{P} = \frac{\Delta W}{\Delta t} \quad \text{(average power)} \qquad (6.15)$$

Often the rate of doing work varies with time. Then we define the **instantaneous power** as the average power taken in the limit of an arbitrarily small time interval Δt:

$$P = \lim_{\Delta t \to 0} \frac{\Delta W}{\Delta t} = \frac{dW}{dt} \quad \text{(instantaneous power)} \qquad (6.16)$$

Equations 6.15 and 6.16 both show that the units of power are joules/second. One J/s is given the name **watt** (W) in honor of James Watt, a Scottish engineer and inventor who was instrumental in developing the steam engine as a practical power source. Watt himself defined another unit, the horsepower. One horsepower (hp) is about 746 J/s or 746 W.

EXAMPLE 6.7 Power: Climbing Mount Washington

A 55-kg hiker ascends New Hampshire's Mount Washington, making the vertical rise of 1300 m in 2 h. A 1500-kg car drives up the Mount Washington Auto Road, taking half an hour. Neglecting energy lost to friction, what's the average power output for each?

INTERPRET This example is about power, which we identify as the *rate* at which hiker and car expend energy. So we need to know the work done by each and the corresponding time.

DEVELOP Equation 6.15, $\overline{P} = \Delta W/\Delta t$, is relevant, since we want the *average* power. To use this equation we'll need to find the work done in climbing the mountain. As you learned in Section 6.2, work done against gravity is independent of the path taken and is given by mgh, where h is the total height of the climb.

EVALUATE We apply Equation 6.15 in the two cases:

$$\overline{P}_{\text{hiker}} = \frac{\Delta W}{\Delta t} = \frac{(55\text{ kg})(9.8\text{ m/s}^2)(1300\text{ m})}{(2.0\text{ h})(3600\text{ s/h})} = 97\text{ W}$$

$$\overline{P}_{\text{car}} = \frac{\Delta W}{\Delta t} = \frac{(1500\text{ kg})(9.8\text{ m/s}^2)(1300\text{ m})}{(0.50\text{ h})(3600\text{ s/h})} = 11\text{ kW}$$

ASSESS Do these values make sense? A power of 97 W is typical of the sustained long-term output of the human body, as you can confirm by considering a typical daily diet of 2000 "calories" (actually kilocalories; see Exercise 31). The car's output amounts to 14 hp, which you may find low, given that the car's engine is probably rated at several hundred horsepower. But cars are notoriously inefficient machines, with only a small fraction of the rated horsepower available to do useful work. Most of the rest is lost to friction and heating. ■

When power is constant, so the average power and instantaneous power are the same, Equation 6.15 shows that the amount of work W done in time Δt is

$$W = P\,\Delta t \qquad (6.17)$$

When power isn't constant, we can consider small amounts of work ΔW, each taken over so small a time interval Δt that the power is nearly constant. Adding all these amounts of work and taking the limit as Δt becomes arbitrarily small, we have

$$W = \lim_{\Delta t \to 0} \sum P\,\Delta t = \int_{t_1}^{t_2} P\,dt \qquad (6.18)$$

where t_1 and t_2 are the beginning and end of the time interval over which we calculate the work.

EXAMPLE 6.8 Energy and Power: Yankee Stadium

Each of the 884 floodlights at Yankee Stadium uses electrical energy at the rate of 1650 W. How much does it cost to run these lights during a 5-h night game, if electricity costs 21¢/kW·h?

INTERPRET We're given a single floodlight's power consumption and the cost of electricity per kilowatt-hour, a unit of energy. So this problem is about calculating energy given power and time, with a little economics thrown in.

DEVELOP Since the power is constant, we can calculate the energy used over time with Equation 6.17, $W = P\,\Delta t$.

EVALUATE At 1.65 kW each, all 884 floodlights use energy at the rate $(884)(1.65\text{ kW}) = 1459\text{ kW}$. Then the total for a 5-h game is

$$W = P\,\Delta t = (1459\text{ kW})(5\text{ h}) = 7295\text{ kW·h}$$

The cost is then $(7295\text{ kW·h})(\$0.21/\text{kW·h}) = \$1532$.

ASSESS Do we have the right units here? Yes: With power in kilowatts and time in hours, the energy comes out immediately in kilowatt-hours. ■

APPLICATION Energy and Society

Humankind's rate of energy consumption is a matter of concern, especially given our dependence on fossil fuels whose carbon dioxide emissions threaten global climate change. Just how rapidly are we using energy?

Example 6.7 suggests that the average power output of the human body is approximately 100 W. Before our species harnessed fire and domesticated animals, that was all the power available to each of us. But in today's high-energy societies, we use energy at a much greater rate. For the average citizen of the United States in the early 21st century, for example, the rate of energy consumption is about 11 kW—the equivalent of more than a hundred human bodies. The rate is lower in most other industrialized countries, but it still amounts to many tens of human bodies' worth.

What do we do with all that energy? And where does it come from? The first pie chart shows that most of the United States's energy consumption goes for industry and transportation, with lesser amounts used in the residential and commercial sectors. The second chart is a stark reminder that our energy supply is neither diversified nor renewable, with some 81% coming from the fossil fuels coal, oil, and natural gas. That's going to have to change in the coming decades, as a result of both limited fossil-fuel resources and the environmental consequences of fossil-fuel combustion—especially climate change. Much of what you learn in an introductory physics course has direct relevance to the energy challenges we face today.

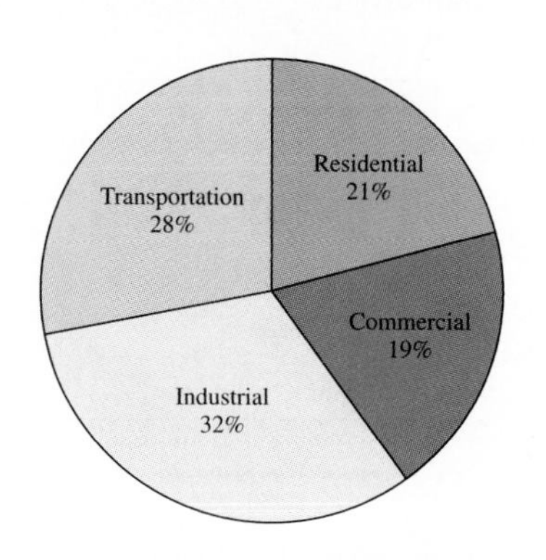

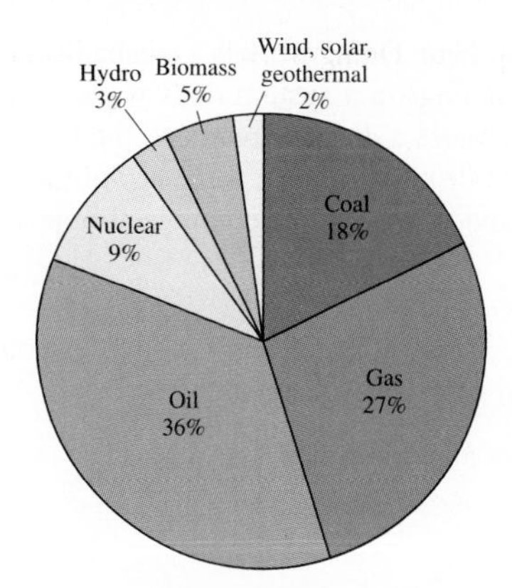

✓TIP Don't Confuse Energy and Power

Is that 11-kW per-capita energy consumption per year, per day, or what? That question reflects a common confusion of energy and power. Power is the *rate* of energy use; it doesn't need any "per time" attached to it. It's just 11 kW, period.

Power and Velocity

We can derive an expression relating power, applied force, and velocity by noting that the work dW done by a force $\vec{F}$ acting on an object that undergoes an infinitesimal displacement $d\vec{r}$ follows from Equation 6.5:

$$dW = \vec{F} \cdot d\vec{r}$$

Dividing both sides by the associated time interval dt gives the power:

$$P = \frac{dW}{dt} = \vec{F} \cdot \frac{d\vec{r}}{dt}$$

But $d\vec{r}/dt$ is the velocity $\vec{v}$, so

$$P = \vec{F} \cdot \vec{v}$$

EXAMPLE 6.9 Power and Velocity: Bicycling

Riding your 9.0-kg bicycle at a steady 16 km/h (4.4 m/s), you experience an 8.2-N force from air resistance. If your mass is 66 kg, what power must you supply on level ground and going up a 5° incline?

INTERPRET This example asks about power in two different situations: one with air resistance alone and the other when climbing. We identify the forces involved as air resistance and gravity. You need to exert forces of equal magnitude to overcome them.

DEVELOP Given that we have force and velocity, Equation 6.19, $P = \vec{F} \cdot \vec{v}$, applies. The force you apply to propel the bicycle is in the same direction as its motion, so $\vec{F} \cdot \vec{v}$ in that equation becomes just Fv.

EVALUATE On level ground, we have $P = Fv = (8.2\text{ N})(4.4\text{ m/s}) = 36\text{ W}$. Climbing the hill, you have to exert an additional force to overcome the downslope component of gravity, which in Example 5.1 we found to be $mg \sin\theta$. So here we have

$$P = Fv = (F_{\text{air}} + mg\sin\theta)v$$
$$= [8.2\text{ N} + (75\text{ kg})(9.8\text{ m/s}^2)(\sin 5°)](4.4\text{ m/s}) = 320\text{ W}$$

where we used your combined mass, body plus bicycle.

ASSESS Both numbers make sense. The values go from considerably less than to a lot more than your body's average power output of around 100 W, and as you've surely experienced, even a modest slope takes much more cycling effort than level ground. Top cyclists on mountain sections of the Tour de France can sustain power outputs of close to 500 W for extended periods. ■

GOT IT? 6.4 A newspaper reports that a new power plant will produce "50 megawatts per hour." What's wrong with this statement?

CHAPTER 6 SUMMARY

Big Idea

Energy and **work** are the big ideas here. Doing work is a mechanical means of transferring energy. A force acting on a system does work when the system (here a single object) undergoes a displacement and the force has a component in the direction of that displacement. A force at right angles to the displacement does no work, and a force with a component opposite the displacement does negative work.

Kinetic energy is the energy associated with an object's motion. An object's kinetic energy changes only when net work is done on the object.

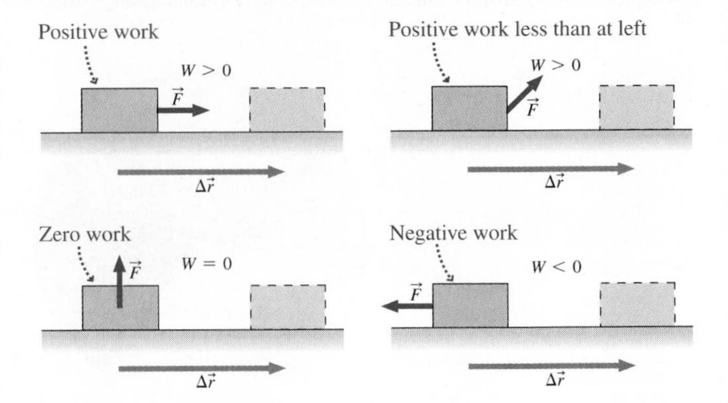

Key Concepts and Equations

Work is the product of force and displacement, but only the component of force in the direction of displacement counts toward the work.

For a constant force and displacement in the x-direction,

$$W = F_x \Delta x \qquad \text{(constant force only)}$$

More generally, for a constant force $\vec{F}$ and arbitrary displacement $\Delta\vec{r}$, the work is

$$W = \vec{F} \cdot \Delta\vec{r} = F\,\Delta r \cos\theta \qquad \text{(constant force only)}$$

Here F and Δr are the magnitudes of the force and displacement vectors, and θ is the angle between them. We've written work here using the shorthand notation of the scalar product, defined for any two vectors $\vec{A}$ and $\vec{B}$ as the product of their magnitudes and the cosine of the angle between them:

$$\vec{A} \cdot \vec{B} = AB\cos\theta \qquad \text{(scalar product)}$$

When force varies with position, calculating the work involves integrating. In one dimension:

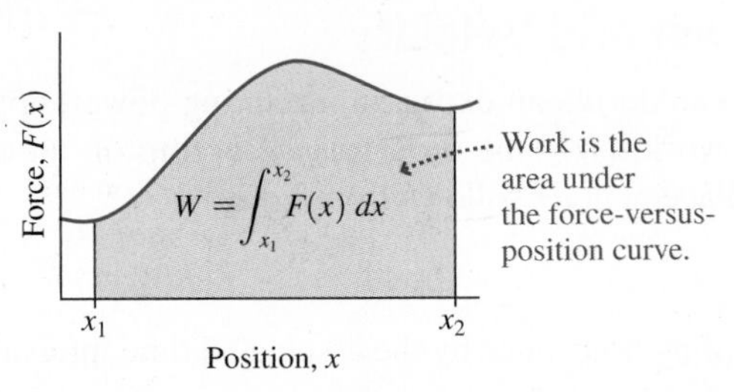

Most generally, work is the **line integral** of a varying force over an arbitrary path: $W = \int \vec{F} \cdot d\vec{r}$

Kinetic energy is a scalar quantity that depends on an object's mass and speed:

$$K = \tfrac{1}{2}mv^2$$

The **work–kinetic energy theorem** states that the change in an object's kinetic energy is equal to the net work done on it:

$$\Delta K = W_{\text{net}} \qquad \text{(work–kinetic energy theorem)}$$

The unit of energy and work is the joule (J), equal to 1 newton-meter.

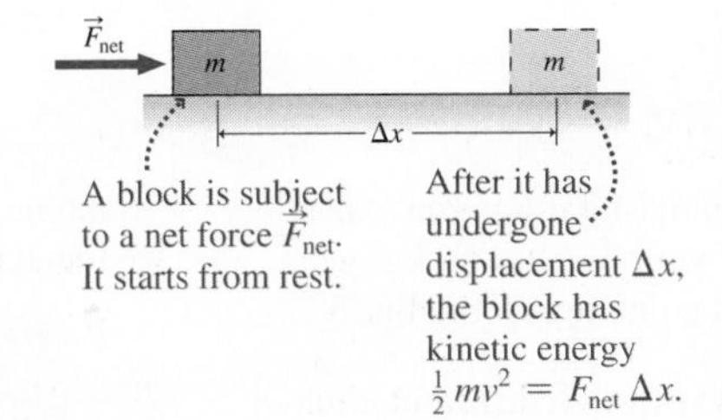

Power is the rate at which work is done or energy is used. The unit of power is the **watt** (W), equal to 1 joule/second.

$$P = \frac{dW}{dt} = \vec{F} \cdot \vec{v}$$

Applications

Common applications of work done against everyday forces are the work mgh needed to raise an object of mass m a distance h against gravity, and the work $\frac{1}{2}kx^2$ needed to stretch or compress a spring of spring constant k a distance x from its equilibrium length.

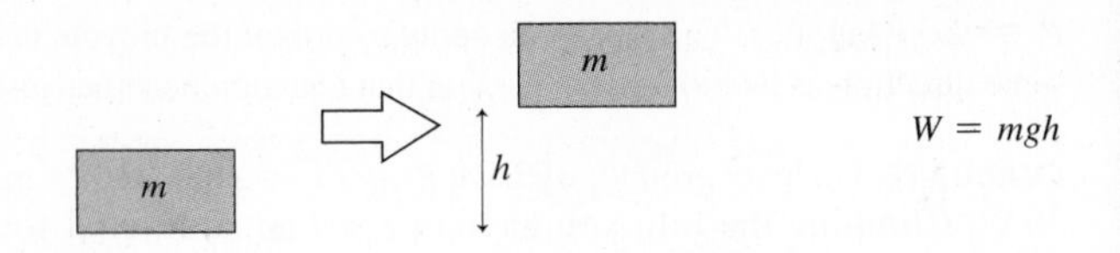

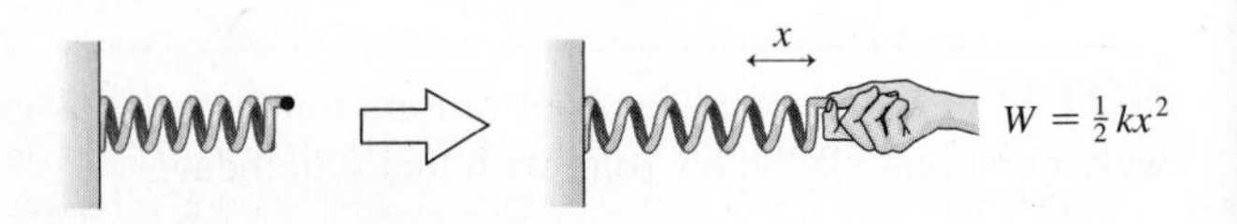

MP For homework assigned on MasteringPhysics, go to www.masteringphysics.com

BIO *Biology and/or medicine-related problems* **DATA** *Data problems* **ENV** *Environmental problems* **CH** *Challenge problems* **COMP** *Computer problems*

For Thought and Discussion

1. Give two examples of situations in which you might think you're doing work but in which, in the technical sense, you do no work.
2. If the scalar product of two nonzero vectors is zero, what can you conclude about their relative directions?
3. Must you do work to whirl a ball around on the end of a string? Explain.
4. If you pick up a suitcase and put it down, how much total work have you done on the suitcase? Does your answer change if you pick up the suitcase and drop it?
5. You want to raise a piano a given height using a ramp. With a fixed, nonzero coefficient of friction, will you have to do more work if the ramp is steeper or more gradual? Explain.
6. Does the gravitational force of the Sun do work on a planet in a circular orbit? On a comet in an elliptical orbit? Explain.
7. A pendulum bob swings back and forth on the end of a string, describing a circular arc. Does the tension force in the string do any work?
8. Does your car's kinetic energy change if you drive at constant speed for 1 hour?
9. A watt-second is a unit of what quantity? Relate it to a more standard SI unit.
10. A truck is moving northward at 55 mi/h. Later, it's moving eastward at the same speed. Has its kinetic energy changed? Has work been done on the truck? Has a force acted on the truck? Explain.
11. A news article reports that a new solar farm will produce 143 kilowatt-hours of electricity. Criticize this statement. What did the writer probably mean?
12. Is it possible for *you* to do work on an object without changing the object's kinetic energy? Explain.

Exercises and Problems

Exercises

Sections 6.1 and 6.2 Energy and Work

13. How much work do you do as you exert a 75-N force to push a shopping cart through a 12-m-long supermarket aisle?
14. If the coefficient of kinetic friction is 0.21, how much work do you do when you slide a 50-kg box at constant speed across a 4.8-m-wide room?
15. A crane lifts a 650-kg beam vertically upward 23 m and then swings it eastward 18 m. How much work does the crane do? Neglect friction, and assume the beam moves with constant speed.
16. The world's highest waterfall, the Cherun-Meru in Venezuela, has a total drop of 980 m. How much work does gravity do on a cubic meter of water dropping down the Cherun-Meru?
17. A meteorite plunges to Earth, embedding itself 75 cm in the ground. If it does 140 MJ of work in the process, what average force does the meteorite exert on the ground?
18. An elevator of mass m rises a vertical distance h with upward acceleration equal to one-tenth g. Find an expression for the work the elevator cable does on the elevator.
19. Show that the scalar product obeys the distributive law: $\vec{A}\cdot(\vec{B}+\vec{C}) = \vec{A}\cdot\vec{B} + \vec{A}\cdot\vec{C}$.
20. Find the work done by a force $\vec{F} = 1.8\,\hat{\imath} + 2.2\,\hat{\jmath}$ N as it acts on an object moving from the origin to the point $56\,\hat{\imath} + 31\,\hat{\jmath}$ m.
21. To push a stalled car, you apply a 470-N force at 17° to the car's motion, doing 860 J of work in the process. How far do you push the car?

Section 6.3 Forces That Vary

22. Find the total work done by the force shown in Fig. 6.15 as the object on which it acts moves (a) from $x = 0$ to $x = 3$ km and (b) from $x = 3$ km to $x = 4$ km.

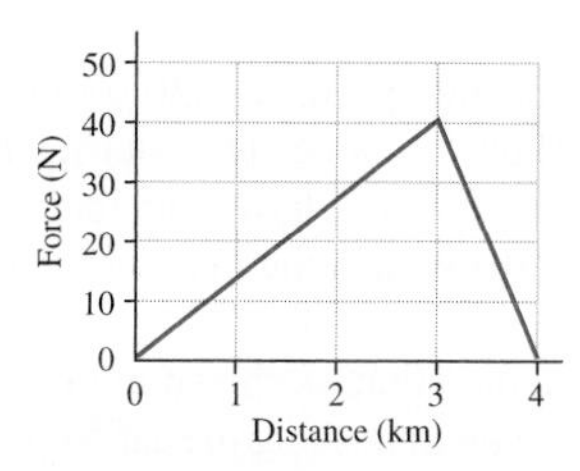

FIGURE 6.15 Exercise 22

23. How much work does it take to stretch a spring with $k = 200$ N/m (a) 10 cm from equilibrium and (b) from 10 cm to 20 cm from equilibrium?
24. Uncompressed, the spring for an automobile suspension is 45 cm long. It needs to be fitted into a space 32 cm long. If the spring constant is 3.8 kN/m, how much work does a mechanic have to do to fit the spring?
25. You do 8.5 J of work to stretch a spring with $k = 190$ N/m, starting with the spring unstretched. How far does the spring stretch?
26. **BIO** Spider silk is a remarkable elastic material. A particular strand has spring constant 70 mN/m, and it stretches 9.6 cm when a fly hits it. How much work did the fly's impact do on the silk strand?

Section 6.4 Kinetic Energy

27. What's the kinetic energy of a 2.4×10^5-kg airplane cruising at 900 km/h?
28. A cyclotron accelerates protons from rest to 21 Mm/s. How much work does it do on each proton?
29. At what speed must a 950-kg subcompact car be moving to have the same kinetic energy as a 3.2×10^4-kg truck going 20 km/h?
30. A 60-kg skateboarder comes over the top of a hill at 5.0 m/s and reaches 10 m/s at the bottom. Find the total work done on the skateboarder between the top and bottom of the hill.
31. After a tornado, a 0.50-g drinking straw was found embedded 4.5 cm in a tree. Subsequent measurements showed that the tree exerted a stopping force of 70 N on the straw. What was the straw's speed?
32. From what height would you have to drop a car for its impact to be equivalent to a 20-mi/h collision?

Section 6.5 Power

33. A typical human diet is "2000 calories" per day, where the "calorie" describing food energy is actually 1 kilocalorie. Express 2000 kcal/day in watts.
34. A horse plows a 200-m-long furrow in 5.0 min, exerting a 750-N force. Find its power output, measured in watts and in horsepower.
35. A typical car battery stores about 1 kW·h of energy. What's its power output if it drains completely in (a) 1 minute, (b) 1 hour, and (c) 1 day?
36. A sprinter completes a 100-m dash in 10.6 s, doing 22.4 kJ of work. What's her average power output?
37. How much work can a 3.5-hp lawnmower engine do in 1 h?
38. A 75-kg long-jumper takes 3.1 s to reach a prejump speed of 10 m/s. What's his power output?
39. Estimate your power output as you do deep knee bends at the rate of one per second.

40. In midday sunshine, solar energy strikes Earth at the rate of about 1 kW/m^2. How long would it take a perfectly efficient solar collector of 15-m^2 area to collect 40 kW·h of energy? (*Note:* This is roughly the energy content of a gallon of gasoline.) ENV
41. It takes about 20 kJ to melt an ice cube. A typical microwave oven produces 900 W of microwave power. How long will it take a typical microwave to melt the ice cube?
42. Which consumes more energy, a 1.2-kW hair dryer used for 10 min or a 7-W night-light left on for 24 h?

Problems

43. You slide a box of books at constant speed up a 30° ramp, applying a force of 200 N directed up the slope. The coefficient of sliding friction is 0.18. (a) How much work have you done when the box has risen 1 m vertically? (b) What's the mass of the box?
44. Two people push a stalled car at its front doors, each applying a 280-N force at 25° to the forward direction, as shown in Fig. 6.16. How much work does each person do in pushing the car 5.6 m?

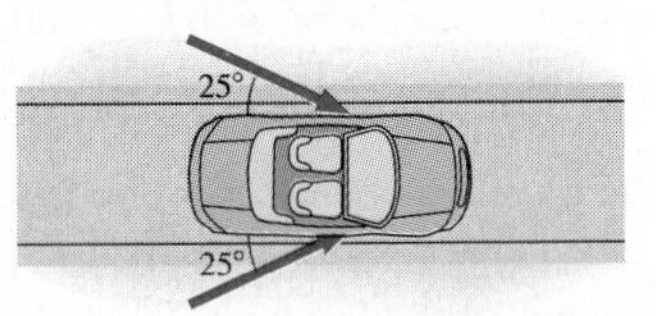

FIGURE 6.16 Problem 44

45. You're at the gym, doing arm raises. With each rep, you lift a 20-N weight 55 cm. (a) How many raises must you do before you've expended 200 kcal of work (see Exercise 33)? (b) If your workout takes 1.0 min, what's your average power output? BIO
46. A locomotive does 7.9×10^{11} J of work in pulling a 3.4×10^6-kg train 180 km. Find the average force in the coupling between the locomotive and the rest of the train.
47. You pull a box 23 m horizontally, using the rope shown in Fig. 6.17. If the rope tension is 120 N, and if the rope does 2500 J of work on the box, what angle θ does the rope make with the horizontal?

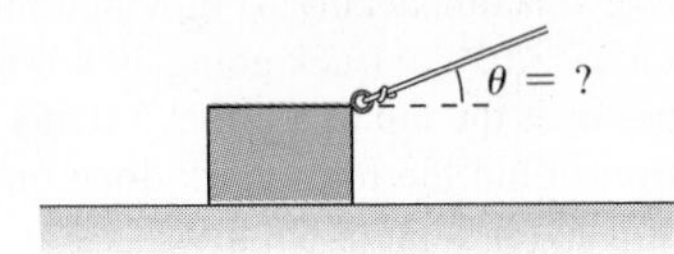

FIGURE 6.17 Problem 47

48. (a) Find the scalar products $\hat{\imath}\cdot\hat{\imath}$, $\hat{\jmath}\cdot\hat{\jmath}$, and $\hat{k}\cdot\hat{k}$. (b) Find $\hat{\imath}\cdot\hat{\jmath}$, $\hat{\jmath}\cdot\hat{k}$, and $\hat{k}\cdot\hat{\imath}$. (c) Use the distributive law to multiply out the scalar product of two arbitrary vectors $\vec{A} = A_x\hat{\imath} + A_y\hat{\jmath} + A_z\hat{k}$ and $\vec{B} = B_x\hat{\imath} + B_y\hat{\jmath} + B_z\hat{k}$, and use the results of (a) and (b) to verify Equation 6.4.
49. (a) Find the scalar product of the vectors $a\hat{\imath} + b\hat{\jmath}$ and $b\hat{\imath} - a\hat{\jmath}$, where a and b are arbitrary constants. (b) What's the angle between the two vectors?
50. Looking to cut costs, the airline you work for asks you to investigate the efficiency of the tractors that push aircraft away from the gates. One model is supposed to do no more than 10 MJ of work in pushing a 747 aircraft 25 m. If the tractor exerts a 0.42-MN force, does it meet its specifications?
51. How much work does a force $\vec{F} = 67\hat{\imath} + 23\hat{\jmath} + 55\hat{k}$ N do as it acts on a body moving in a straight line from $\vec{r}_1 = 16\hat{\imath} + 31\hat{\jmath}$m to $\vec{r}_2 = 21\hat{\imath} + 10\hat{\jmath} + 14\hat{k}$ m?
52. A force $\vec{F}$ acts in the x-direction, its magnitude given by $F = ax^2$, where x is in meters and $a = 5.0$ N/m^2. Find the work done by this force as it acts on a particle moving from $x = 0$ to $x = 6.0$ m.
53. A certain amount of work is required to stretch spring A a certain distance. Twice as much work is required to stretch spring B half that distance. Compare the spring constants of the two.
54. A force with magnitude $F = a\sqrt{x}$ acts in the x-direction, where $a = 9.5$ N/m$^{1/2}$. Calculate the work this force does as it acts on an object moving from (a) $x = 0$ to $x = 3.0$ m; (b) 3.0 m to 6.0 m; and (c) 6.0 m to 9.0 m.
55. The force exerted by a rubber band is given approximately by CH

$$F = F_0\left[\frac{L_0 - x}{L_0} - \frac{L_0^2}{(L_0 + x)^2}\right]$$

where L_0 is the unstretched length, x is the stretch, and F_0 is a constant. Find an expression for the work needed to stretch the rubber band a distance x.
56. You put your little sister (mass m) on a swing whose chains have length L and pull slowly back until the swing makes an angle ϕ with the vertical. Show that the work you do is $mgL(1 - \cos\phi)$.
57. Two unknown elementary particles pass through a detection chamber. If they have the same kinetic energy and their mass ratio is 4:1, what's the ratio of their speeds?
58. A tractor tows a plane from its airport gate, doing 8.7 MJ of work. The link from the plane to the tractor makes a 22° angle with the plane's motion, and the tension in the link is 0.41 MN. How far does the tractor move the plane?
59. *E. coli* bacteria swim by means of flagella that rotate about 100 times per second. A typical *E. coli* bacterium swims at 22 μm/s, its flagella exerting a force of 0.57 pN to overcome the resistance due to its liquid environment. (a) What's the bacterium's power output? (b) How much work would it do in traversing the 25-mm width of a microscope slide? BIO
60. On February 15, 2013, an asteroid moving at 19 km/s entered Earth's atmosphere over Chelyabinsk, Russia, and exploded at an altitude of more than 20 km. This was the largest object known to have entered the atmosphere in over a century. The asteroid's kinetic energy just before entering the atmosphere was estimated as the energy equivalent of 500 kilotons of the explosive TNT. (Kilotons [kt] and megatons [Mt] are energy units used to describe the explosive yields of nuclear weapons, and you'll find the energy equivalent of 1 Mt in Appendix C).What was the approximate mass of the Chelyabinsk asteroid?
61. An elevator ascends from the ground floor to the 10th floor, a height of 41 m, in 35 s. If the mass of the elevator and passengers is 840 kg, what's the power necessary to lift the elevator? (Your answer is greater than the actual power needed because elevators are counterweighted, thus reducing the work the motor needs to do.)
62. You're asked to assess the reliability of a nuclear power plant, as measured by the *capacity factor*—the ratio of the energy it actually produces to what it could produce if it operated all the time. The plant is rated at 840 MW of electrical power output, and in a full year it produces 6.8×10^9 kW·h of electrical energy. What's its capacity factor?
63. A force pointing in the x-direction is given by $F = F_0(x/x_0)^2$, where F_0 and x_0 are constants and x is position. Find an expression for the work done by this force as it acts on an object moving from $x = 0$ to $x = x_0$.
64. A force pointing in the x-direction is given by $F = ax^{3/2}$, where $a = 0.75$ N/m$^{3/2}$. Find the work done by this force as it acts on an object moving from $x = 0$ to $x = 14$ m.
65. Two vectors have equal magnitude, and their scalar product is one-third the square of their magnitude. Find the angle between them.
66. At what rate can a half-horsepower well pump deliver water to a tank 60 m above the water level in the well? Give your answer in kg/s and gal/min.

67. The rate at which the United States imports oil, expressed in terms of the energy content of the imported oil, is about 600 GW. Using the "Energy Content of Fuels" table in Appendix C, convert this figure to gallons per day. ENV

68. By measuring oxygen uptake, sports physiologists have found that long-distance runners' power output is given approximately by $P = m(bv - c)$, where m and v are the runner's mass and speed, and b and c are constants given by $b = 4.27$ J/kg·m and $c = 1.83$ W/kg. Determine the work done by a 54-kg runner who runs a 10-km race at 5.2 m/s. BIO

69. You're writing performance specifications for a new car model. The 1750-kg car delivers energy to its drive wheels at the rate of 35 kW. Neglecting air resistance, what do you list for the greatest speed at which it can climb a 4.5° slope?

70. A 1400-kg car ascends a mountain road at a steady 60 km/h, against a 450-N force of air resistance. If the engine supplies energy to the drive wheels at the rate of 38 kW, what's the slope angle of the road? CH

71. You do 2.2 kJ of work pushing a 78-kg trunk at constant speed 3.1 m along a ramp inclined upward at 22°. What's the frictional coefficient between trunk and ramp?

72. (a) Find the work done in lifting 1 L of blood (mass 1 kg) from the foot to the head of a 1.7-m-tall person. (b) If blood circulates through the body at the rate of 5.0 L/min, estimate the heart's power output. (Your answer underestimates the power by a factor of about 5 because it neglects fluid friction and other factors.) BIO

73. (a) What power is needed to push a 95-kg crate at 0.62 m/s along a horizontal floor where the coefficient of friction is 0.78? (b) How much work is done in pushing the crate 11 m?

74. You mix flour into bread dough, exerting a 45-N force on the spoon, which you move at 0.29 m/s. (a) What power do you supply? (b) How much work do you do if you stir for 1.0 min?

75. A machine does work at a rate given by $P = ct^2$, where $c = 18\ \text{W/s}^2$ and t is time. Find the work done between $t = 10$ s and $t = 20$ s.

76. A typical bumblebee has mass 0.25 mg. It beats its wings 100 times per second, and the wings undergo an average displacement of about 1.5 mm. When the bee is hovering over a flower, the average force between wings and air must support the bee's weight. Estimate the average power the bee expends in hovering. BIO

77. You're trying to decide whether to buy an energy-efficient 225-W refrigerator for $1150 or a standard 425-W model for $850. The standard model will run 20% of the time, but better insulation means the energy-efficient model will run 11% of the time. If electricity costs 9.5¢/kW·h, how long would you have to own the energy-efficient model to make up the difference in cost? Neglect interest you might earn on your money. ENV

78. Your friend does five reps with a barbell, on each rep lifting 45 kg 0.50 m. She claims the work done is enough to "burn off" a chocolate bar with energy content 230 kcal (see Exercise 33). Is that true? If not, how many lifts would it take?

79. A machine delivers power at a decreasing rate $P = P_0t_0^2/(t + t_0)^2$, where P_0 and t_0 are constants. The machine starts at $t = 0$ and runs forever. Show that it nevertheless does only a finite amount of work, equal to P_0t_0. CH

80. A locomotive accelerates a freight train of total mass M from rest, applying constant power P. Determine the speed and position of the train as functions of time, assuming all the power goes to increasing the train's kinetic energy. CH

81. A force given by $F = b/\sqrt{x}$ acts in the x-direction, where b is a constant with the units $\text{N·m}^{1/2}$. Show that even though the force becomes arbitrarily large as x approaches zero, the work done in moving from x_1 to x_2 remains finite even as x_1 approaches zero. Find an expression for that work in the limit $x_1 \to 0$.

82. You're assisting a cardiologist in planning a stress test for a 75-kg patient. The test involves rapid walking on an inclined treadmill, and the patient is to reach a peak power output of 350 W. If the patient's maximum walking speed is 8.0 km/h, what should be the treadmill's inclination angle? BIO

83. You're an engineer for a company that makes bungee-jump cords, and you're asked to develop a formula for the work involved in stretching cords to double their length. Your cords have force-distance relations described by $F = -(kx + bx^2 + cx^3 + dx^4)$, where k, b, c, and d are constants. (a) Given a cord with unstretched length L_0, what's your formula? (b) Evaluate the work done in doubling the stretch of a 10-m cord with $k = 420$ N/m, $b = -86\ \text{N/m}^2$, $c = 12\ \text{N/m}^3$, and $d = -0.50\ \text{N/m}^4$.

84. You push an object of mass m slowly, partway up a loop-the-loop track of radius R, starting from the bottom, and ending at a height $h < R$ above the bottom. The coefficient of friction between the object and the track is a constant μ. Show that the work you do against friction is $\mu mg\sqrt{2hR - h^2}$. CH

85. A particle moves from the origin to the point $x = 3$ m, $y = 6$ m along the curve $y = ax^2 - bx$, where $a = 2\ \text{m}^{-1}$ and $b = 4$. It's subject to a force $cxy\hat{\imath} + d\hat{\jmath}$, where $c = 10\ \text{N/m}^2$ and $d = 15$ N. Calculate the work done by the force. CH

86. Repeat Problem 85 for the following cases: (a) the particle moves first along the x-axis from the origin to the point (3 m, 0) and then parallel to the y-axis until it reaches (3 m, 6 m); (b) it moves first along the y-axis from the origin to the point (0, 6 m) and then parallel to the x-axis until it reaches (3 m, 6 m).

87. The world's fastest elevator, in Taiwan's Taipei 101 skyscraper (Fig. 6.18), ascends at the rate of 1010 m/min. Counterweights balance the weight of the elevator car, so the motor doesn't have to lift the car's weight. If the motor produces 330 kW of power, what's the maximum number of 67-kg people the elevator can accommodate? (Your answer somewhat overestimates the actual rated load of 24 people.)

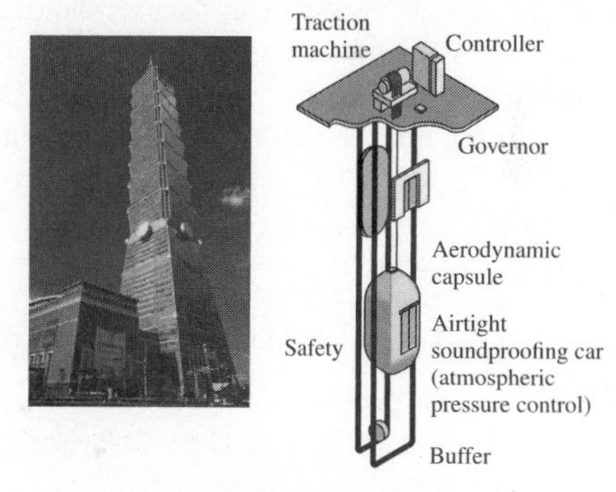

FIGURE 6.18 Problem 87

88. An experimental measurement of the force required to stretch a slingshot is given in the table below. Plot the force–distance curve for this slingshot and use graphical integration to determine the work done in stretching the slingshot the full 40-cm distance. DATA

Stretch (cm)	Force (N)
0	0
5.00	0.885
10.0	1.89
15.0	3.05
20.0	4.48
25.0	6.44
30.0	8.22
35.0	9.95
40.0	12.7

89. BIO You're an expert witness in a medical malpractice lawsuit. A hospital patient's leg slipped off a stretcher and his heel hit the floor. The defense attorney for the hospital claims the leg, with mass 8 kg, hit the floor with a force equal to the weight of the leg—about 80 N—and any damage was due to a prior injury. You argue that the leg and heel dropped freely for 0.7 m, then hit the floor and stopped in 2 cm. What do you tell the jury about the force on the heel?

Passage Problems

The energy in a batted baseball comes from the power delivered while the bat is in contact with the ball. The most powerful hitters can supply some 10 horsepower during the brief contact time, propelling the ball to over 100 miles per hour. Figure 6.19 shows data taken from a particular hit, giving the power the bat delivers to the ball as a function of time.

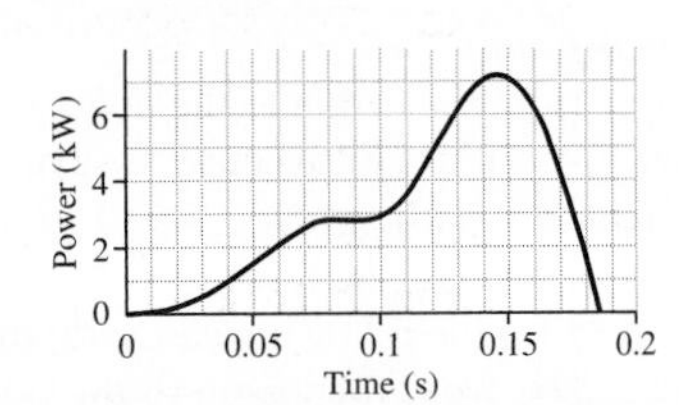

FIGURE 6.19 Passage Problems 90–93

90. Which of the following is greatest at the peak of the curve?
    a. the ball's kinetic energy
    b. the ball's speed
    c. the rate at which the bat supplies energy to the ball
    d. the total work the bat has done on the ball
91. The ball has its maximum speed at about
    a. 85 ms.
    b. 145 ms.
    c. 185 ms.
    d. whenever the force is greatest.
92. As a result of being hit, the ball's kinetic energy increases by about
    a. 550 J.
    b. 1.3 kJ.
    c. 7.0 kJ.
    d. You can't tell because you don't know its speed coming from the pitcher.
93. The force on the ball is greatest approximately
    a. at 185 ms.
    b. at the peak in Fig. 6.19.
    c. before the peak in Fig. 6.19.
    d. after the peak in Fig. 6.19 but before 185 ms.

Answers to Chapter Questions

Answer to Chapter Opening Question

No. The work done against gravity in climbing a particular height is independent of the path. A rider on a bicycle with a combined mass of 80 kg does roughly 400 kilojoules or 100 kilocalories of work against gravity regardless of the path up a 500-m mountain. To climb such a mountain in 20 minutes, the rider's power output must exceed 300 watts.

Answers to GOT IT? Questions

6.1 (b) $2F$ does $\sqrt{2}$ more work than F does. That's because $2F$'s component along the direction of motion is $2F\cos 45°$, or $2F\sqrt{2}/2 = F\sqrt{2}$.

6.2 (c) $\sqrt{x}$ does the most work. (b) x^2 does the least. You can see this by plotting these two functions from $x = 0$ to $x = 1$ and comparing the areas under each. The case of x is intermediate.

6.3 (1) (c): Kinetic energy doesn't change, so the net work done on the ball is zero. (2) (a): Kinetic energy increases, so the net work is positive. (3) (b): Kinetic energy decreases, so the net work is negative.

6.4 The megawatt is a unit of power; the "per time" is already built in. A correct statement would be that the power plant will produce "energy at the rate of 50 megawatts."

Conservation of Energy

What You Know

- You understand the concept of energy.
- You know how to define a system so you can consider energy flows into and out of the system.
- You've seen how work involves a transfer of mechanical energy.
- You recognize kinetic energy and understand how the work–kinetic energy theorem relates the net work done on a system to the change in its kinetic energy.

What You're Learning

- You'll see how forces come in two types: *conservative forces* and *nonconservative forces*.
- You'll learn that energy transferred as a result of work done against conservative forces ends up stored as *potential energy*.
- You'll learn expressions for potential energy associated with gravitational and elastic forces.
- You'll learn to treat the sum of kinetic energy and potential energy as a system's *total mechanical energy*.
- You'll see that mechanical energy is conserved in the absence of nonconservative forces.
- You'll learn to use the conservation-of-mechanical-energy principle to solve problems that would otherwise be difficult because they involve varying acceleration.
- You'll see how to evaluate situations where nonconservative forces result in a loss of mechanical energy.
- You'll see how *potential energy curves* describe a wide variety of systems, including molecules.

How You'll Use It

- In Chapter 8, you'll use the calculus expression for potential energy to explore the potential energy associated with the gravitational force over distances significant in space flight and astronomy.
- You'll also see how conservation of energy leads you to understand the physics of simple orbits and how energy conservation leads to the concept of *escape speed*.
- In Chapter 9, you'll apply conservation of mechanical energy to so-called *elastic collisions*, which conserve mechanical energy.
- Energy will continue to play an important role as you explore Newtonian physics further in Parts 1 and 2.
- In Part 3, you'll see how to extend the conservation-of-energy principle to account for heat as well as mechanical work.
- In Part 4, you'll see that electric and magnetic fields are repositories of potential energy.

How many different energy conversions take place as the Yellowstone River plunges over Yellowstone Falls?

(a)

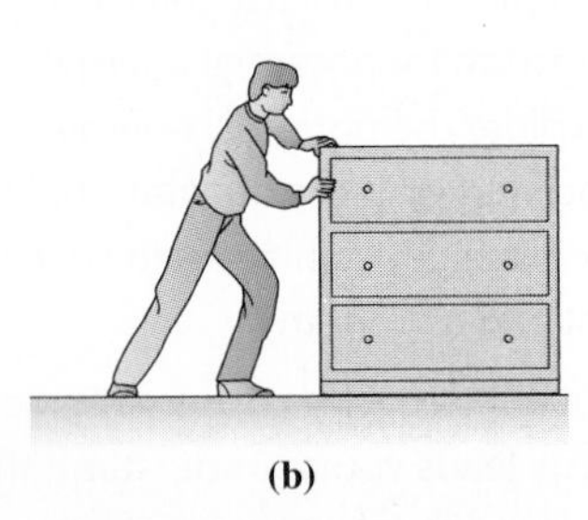

(b)

FIGURE 7.1 Both the rock climber and the mover do work, but only the climber can recover that work as kinetic energy.

The rock climber of Fig. 7.1*a* does work as she ascends the vertical cliff. So does the mover of Fig. 7.1*b*, as he pushes a heavy chest across the floor. But there's a difference. If the rock climber lets go, down she goes, gaining kinetic energy as she falls. If the mover lets go of the chest, though, he and the chest stay right where they are.

This contrast highlights a distinction between two types of forces, called *conservative* and *nonconservative*. That distinction will help us develop one of the most important principles in physics: **conservation of energy**. The introduction to Chapter 6 briefly mentioned three forms of energy: kinetic energy, potential energy, and internal energy—although there we worked quantitatively only with kinetic energy. Here we'll develop the concept of potential energy and show how it's associated with conservative forces. Nonconservative forces, in contrast, are associated with irreversible transformations of mechanical energy into internal energy. We'll take a brief look at such transformations here and formulate a broad statement of energy conservation. In Chapters 16–19 we'll elaborate on internal energy and see how it's related to temperature, and we'll expand our statement of energy conservation to include not only work but also heat as means of energy transfer.

7.1 Conservative and Nonconservative Forces

Both the climber and the mover in Fig. 7.1 are doing work against forces—gravity for the climber and friction for the mover. The difference is this: If the climber lets go, the gravitational force "gives back" the energy she supplied by doing work, which then manifests itself as the kinetic energy of her fall. But the frictional force doesn't "give back" the energy supplied by the mover, in the sense that this energy can't be recovered as kinetic energy.

A **conservative force** is a force like gravity or a spring that "gives back" energy that was transferred by doing work. A more precise description of what it means for a force to be conservative follows from considering the work involved as an object moves over a closed path—one that ends where it started. Suppose our rock climber ascends a cliff of height h and then descends to her starting point. As she climbs, the gravitational force is directed opposite to her motion, so gravity does negative work $-mgh$ (recall Fig. 6.4). When she descends, the gravitational force is in the same direction as her motion, so the gravitational work is $+mgh$. The total work that gravity does on the climber as she traverses the closed path up and down the cliff is therefore zero.

Now consider the mover in Fig. 7.1*b*. Suppose he pushes the chest across a room, discovers it's the wrong room, and pushes it back to the door. Like the climber, the mover and chest describe a closed path. But the frictional force always acts to oppose the motion of the chest. The mover needs to apply a force to oppose friction—that is, in the same direction as the chest's motion—so he ends up doing positive work as he crosses the room in both directions. Therefore, the total work he does is positive even when he moves the chest over a closed path. That's the nature of the frictional force, and, in contrast to the conservative gravitational force the climber had to deal with, this makes the frictional force **nonconservative**.

You'll notice that we didn't talk here about the work done by friction but rather the work done by the mover in opposing friction. That's because frictional work is a rather subtle concept, which we'll touch on later in the chapter. Nevertheless, our two examples make the distinction between conservative and nonconservative forces quite clear: Only for *conservative* forces is the work done in moving over a closed path equal to zero. That property provides a precise mathematical definition of a conservative force:

> When the total work done by a force $\vec{F}$ acting as an object moves over any closed path is zero, the force is conservative. Mathematically,
>
> $$\oint \vec{F} \cdot d\vec{r} = 0 \qquad \text{(conservative force)} \qquad (7.1)$$

The expression given in Equation 7.1 comes from the most general formula for work, Equation 6.11: $W = \int_{\vec{r}_1}^{\vec{r}_2} \vec{F} \cdot d\vec{r}$. The circle on the integral sign indicates that the integral is to be taken over a *closed* path.

Equation 7.1 suggests a related property of conservative forces. Suppose a conservative force acts on an object in the region shown in Fig. 7.2. Move the object along the straight path from point A to point B, and designate the work done by the conservative force as W_{AB}. Since the work done over any closed path is zero, the work W_{BA} done in moving back from B to A must be $-W_{AB}$, whether we return along the straight path, the curved path, or any other path. So, going from A to B involves work W_{AB}, regardless of the path taken. In other words:

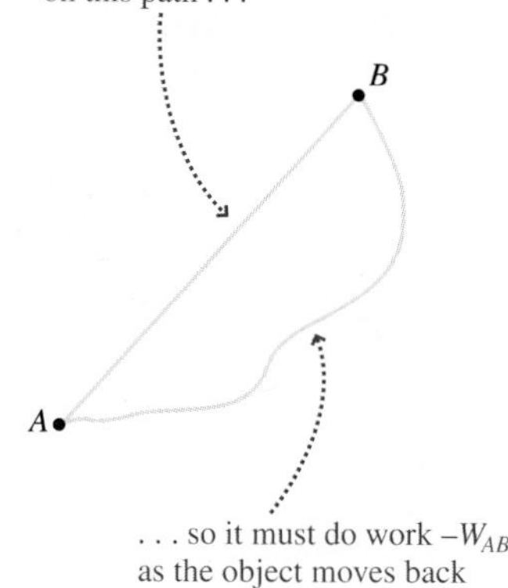

FIGURE 7.2 The work done by a conservative force is independent of path.

> The work done by a conservative force in moving between two points is independent of the path taken; mathematically, $\int_A^B \vec{F} \cdot d\vec{r}$ depends only on the endpoints A and B, not on the path between them.

Important examples of conservative forces include gravity and the static electric force. The force of an ideal spring—fundamentally an electric force—is also conservative. Nonconservative forces include friction, drag forces, and the electric force in the presence of time-varying magnetic effects, which we'll encounter in Chapter 27.

GOT IT? 7.1 Suppose it takes the same amount of work to push a trunk straight across a rough floor as it does to lift a weight the same distance straight upward. If both trunk and weight are moved instead on identically shaped curved paths between the same two points as before, is the work (a) still the same for both, (b) greater for the weight, or (c) greater for the trunk?

7.2 Potential Energy

The climber in Fig. 7.1*a* did work ascending the cliff, and the energy transferred as she did that work was somehow stored, in that she could get it back in the form of kinetic energy. She's acutely aware of that stored energy, since it gives her the potential for a dangerous fall. *Potential* is an appropriate word here: The stored energy is **potential energy**, in the sense that it has the potential to be converted into kinetic energy.

We'll give potential energy the symbol U, and we begin by defining *changes* in potential energy. Specifically:

> The change ΔU_{AB} in potential energy associated with a conservative force is the negative of the work done by that force as it acts over any path from point A to point B:
>
> $$\Delta U_{AB} = -\int_A^B \vec{F} \cdot d\vec{r} \qquad \text{(potential energy)} \qquad (7.2)$$

Here $\int_A^B \vec{F} \cdot d\vec{r}$ is the work done by the force $\vec{F}$, as defined in Equation 6.11. But why the minus sign? Because, if a conservative force does *negative* work (as does gravity on a weight being lifted), then energy is stored and ΔU must be *positive*. Another way to think about this is to consider the work *you* would have to do in order to just counter a conservative force like gravity. If $\vec{F}$ is the conservative force (e.g., gravity, pointing down), then you'd have to apply a force $-\vec{F}$ (e.g., upward), and the work you do would be $\int_A^B (-\vec{F}) \cdot d\vec{r}$ or $-\int_A^B \vec{F} \cdot d\vec{r}$, which is the right-hand side of Equation 7.2. Your work represents a transfer of energy, which here ends up stored as potential energy. So another way of interpreting Equation 7.2 is to say that the change in potential energy is equal to the work an external agent would have to do in just countering a conservative force.

Changes in potential energy are all that ever matter physically; the actual value of potential energy is meaningless. Often, though, it's convenient to establish a reference point at which the potential energy is defined to be zero. When we say "the potential energy U," we really mean the potential-energy difference ΔU between that reference point and whatever other point we're considering. Our rock climber, for example, might find it convenient to take the

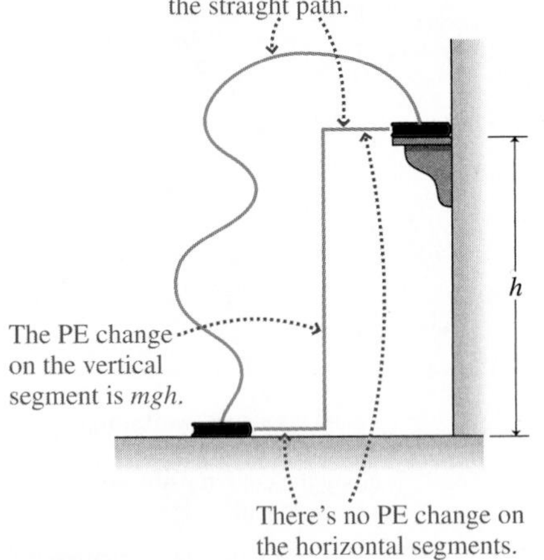

FIGURE 7.3 A good choice of path makes it easier to calculate the potential-energy change.

zero of potential energy at the base of the cliff. But the choice is purely for convenience; only potential-energy *differences* really matter. We'll often drop the subscript AB and write simply ΔU for a potential-energy difference. Keeping the subscript is important, though, when we need to be clear about whether we're going from A to B or from B to A.

Equation 7.2 is a completely general definition of potential energy, applicable in all circumstances. Often, though, we can consider a path where force and displacement are parallel (or antiparallel). Then Equation 7.2 simplifies to

$$\Delta U = -\int_{x_1}^{x_2} F(x)\, dx \tag{7.2a}$$

where x_1 and x_2 are the starting and ending points on the x-axis, taken to coincide with the path. When the force is constant, this equation simplifies further to

$$\Delta U = -F(x_2 - x_1) \tag{7.2b}$$

✓TIP Understand Your Equations

Equation 7.2b provides a very simple expression for potential-energy changes, but it applies *only* when the force is constant. Equation 7.2b is a special case of Equation 7.2a that follows because a constant force can be taken outside the integral.

Gravitational Potential Energy

We're frequently moving things up and down, causing changes in potential energy. Figure 7.3 shows two possible paths for a book that's lifted from the floor to a shelf of height h. Since the gravitational force is conservative, we can use either path to calculate the potential-energy change. It's easiest to use the path consisting of straight segments. No work or potential-energy change is associated with the horizontal motion, since the gravitational force is perpendicular to the motion. For the vertical lift, the force of gravity is constant and Equation 7.2b gives immediately $\Delta U = mgh$, where the minus sign in Equation 7.2b cancels with the minus sign associated with the *downward* direction of gravity. This result is quite general: When a mass m undergoes a vertical displacement Δy near Earth's surface, gravitational potential energy changes by

$$\Delta U = mg\,\Delta y \qquad \text{(gravitational potential energy)} \tag{7.3}$$

The quantity Δy can be positive or negative, depending on whether the object moves up or down; correspondingly, the potential energy can either increase or decrease. We emphasize that Equation 7.3 applies *near Earth's surface*—that is, for distances small compared with Earth's radius. That assumption allows us to treat the gravitational force as constant over the path. We'll explore the more general case in Chapter 8.

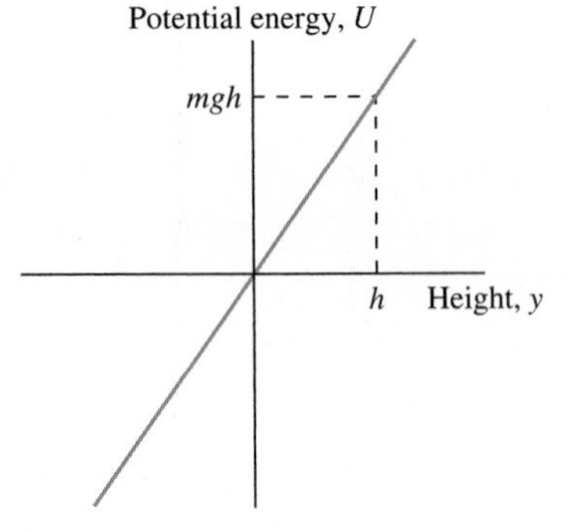

FIGURE 7.4 Gravitational force is constant, so potential energy increases linearly with height.

We've found the *change* in potential energy associated with raising the book, but what about the potential energy itself? That depends on where we define the zero of potential energy. If we choose $U = 0$ at the floor, then $U = mgh$ on the shelf. But we could just as well take $U = 0$ at the shelf; then potential energy when the book is on the floor would be $-mgh$. Negative potential energies arise frequently, and that's OK because only *differences* in potential energy really matter. Figure 7.4 shows a plot of potential energy versus height with $U = 0$ taken at the floor. The *linear* increase in potential energy with height reflects the *constant* gravitational force.

EXAMPLE 7.1 Gravitational Potential Energy: Riding the Elevator

A 55-kg engineer leaves her office on the 33rd floor of a skyscraper and takes an elevator up to the 59th floor. Later she descends to street level. If the engineer chooses the zero of potential energy at her office and if the distance from one floor to the next is 3.5 m, what's the potential energy when the engineer is (a) in her office, (b) on the 59th floor, and (c) at street level?

INTERPRET This is a problem about gravitational potential energy relative to a specified point of zero energy—namely, the engineer's office.

DEVELOP Equation 7.3, $\Delta U = mg\,\Delta y$, gives the change in gravitational energy associated with a change Δy in vertical position. We're given positions in floors, not meters, so we need to convert using the given factor 3.5 m per floor.

EVALUATE (a) When the engineer is in her office, the potential energy is zero, since she defined it that way. (b) The 59th floor is $59 - 33 = 26$ floors higher, so the potential energy when she's there is

$$U_{59} = mg\,\Delta y = (55\text{ kg})(9.8\text{ m/s}^2)(26\text{ floors})(3.5\text{ m/floor}) = 49\text{ kJ}$$

Here we can write U rather than ΔU because we're calculating the potential-energy *change* from the place where $U = 0$. (c) The street level is 32 floors *below* the engineer's office, so

$$U_{\text{street}} = mg\,\Delta y = (55\text{ kg})(9.8\text{ m/s}^2)(-32\text{ floors})(3.5\text{ m/floor}) = -60\text{ kJ}$$

ASSESS Makes sense: When the engineer goes *up*, the potential energy relative to her office is positive; when she goes *down*, it's negative. And the distance down is a bit farther, so the magnitude of the change is greater going down. ■

APPLICATION Pumped Storage

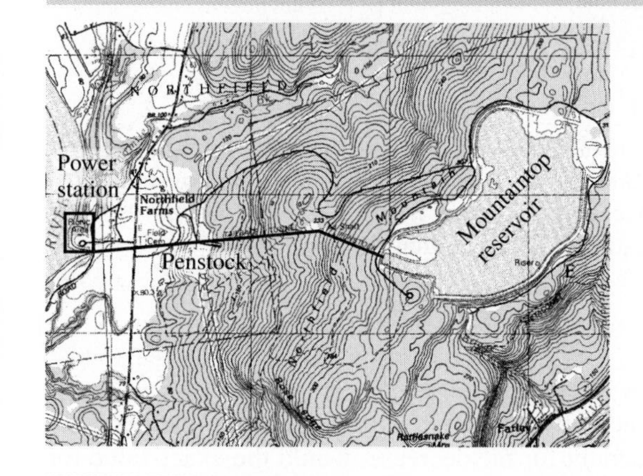

Electricity is a wonderfully versatile form of energy, but it's not easy to store. Large electric power plants are most efficient when operated continuously, yet the demand for power fluctuates. Renewable energy sources like wind and solar vary, not necessarily with demand. Energy storage can help in both cases. Today, the only practical way to store large amounts of excess electrical energy is to convert it to gravitational potential energy. In so-called pumped-storage facilities, surplus electric power pumps water from a lower reservoir to a higher one, thereby increasing gravitational potential energy. When power demand is high, water runs back down, turning the pump motors into generators that produce electricity. The map here shows the Northfield Mountain Pumped Storage Project in Massachusetts, including the mountaintop reservoir, the location of the power station 214 m below on the Deerfield River, and the *penstock*, the pipe that conveys water in both directions between the power station and the reservoir. You can explore this facility quantitatively in Problem 29.

Elastic Potential Energy

When you stretch or compress a spring or other elastic object, you do work against the spring force, and that work ends up stored as **elastic potential energy**. For an ideal spring, the force is $F = -kx$, where x is the distance the spring is stretched from equilibrium, and the minus sign shows that the force opposes the stretching or compression. Since the force varies with position, we use Equation 7.2a to evaluate the potential energy:

$$\Delta U = -\int_{x_1}^{x_2} F(x)\,dx = -\int_{x_1}^{x_2} (-kx)\,dx = \tfrac{1}{2}kx_2^2 - \tfrac{1}{2}kx_1^2$$

where x_1 and x_2 are the initial and final values of the stretch. If we take $U = 0$ when $x = 0$ (that is, when the spring is neither stretched nor compressed) then we can use this result to write the potential energy at an arbitrary stretch (or compression) x as

$$U = \tfrac{1}{2}kx^2 \quad \text{(elastic potential energy)} \qquad (7.4)$$

Comparison with Equation 6.10, $W = \frac{1}{2}kx^2$, shows that this is equal to the work done in stretching the spring. Thus the energy transferred by doing work gets stored as potential energy. Figure 7.5 shows potential energy as a function of the stretch or compression of a spring. The *parabolic* shape of the potential-energy curve reflects the *linear* change of the spring force with stretch or compression.

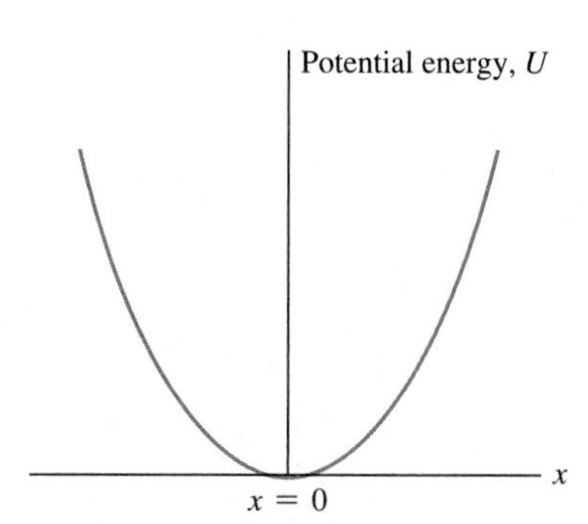

FIGURE 7.5 The potential-energy curve for a spring is a parabola.

EXAMPLE 7.2 Energy Storage: Springs versus Gasoline

A car's suspension consists of springs with an overall effective spring constant of 120 kN/m. How much would you have to compress the springs to store the same amount of energy as in 1 gram of gasoline?

INTERPRET This problem is about the energy stored in a spring, as compared with the chemical energy of gasoline.

DEVELOP Equation 7.4, $U = \frac{1}{2}kx^2$, gives a spring's stored energy when it's been compressed a distance x. Here we want that energy to equal the energy in 1 gram of gasoline. We can get that value from the "Energy Content of Fuels" table in Appendix C, which lists 44 MJ/kg for gasoline.

EVALUATE At 44 MJ/kg, the energy in 1 g of gasoline is 44 kJ. Setting this equal to the spring energy $\frac{1}{2}kx^2$ and solving for x, we get

$$x = \sqrt{\frac{2U}{k}} = \sqrt{\frac{(2)(44\text{ kJ})}{120\text{ kN/m}}} = 86\text{ cm}$$

ASSESS This answer is absurd. A car's springs couldn't compress anywhere near that far before the underside of the car hit the ground. And 1 g isn't much gasoline. This example shows that springs, though useful energy-storage devices, can't possibly compete with chemical fuels. ■

EXAMPLE 7.3 Elastic Potential Energy: A Climbing Rope

Ropes used in rock climbing are "springy" so that they cushion a fall. A particular rope exerts a force $F = -kx + bx^2$, where $k = 223$ N/m, $b = 4.10$ N/m^2, and x is the stretch. Find the potential energy stored in this rope when it's been stretched 2.62 m, taking $U = 0$ at $x = 0$.

INTERPRET Like Example 7.2, this one is about elastic potential energy. But this one isn't so easy because the rope isn't a simple $F = -kx$ spring for which we already have a potential-energy formula.

DEVELOP Because the rope force varies with stretch, we'll have to integrate. Since force and displacement are in the same direction, we can use Equation 7.2a, $\Delta U = -\int_{x_1}^{x_2} F(x)\,dx$. But that's not so much a formula as a strategy for deriving one.

EVALUATE Applying Equation 7.2 to this particular rope, we have

$$\begin{aligned} U &= -\int_{x_1}^{x_2} F(x)\,dx = -\int_0^x (-kx + bx^2)\,dx = \tfrac{1}{2}kx^2 - \tfrac{1}{3}bx^3\Big|_0^x \\ &= \tfrac{1}{2}kx^2 - \tfrac{1}{3}bx^3 \\ &= (\tfrac{1}{2})(223\text{ N/m})(2.62\text{ m})^2 - (\tfrac{1}{3})(4.1\text{ N/m}^2)(2.62\text{ m})^3 \\ &= 741\text{ J} \end{aligned}$$

ASSESS This result is about 3% less than the potential energy $U = \frac{1}{2}kx^2$ of an ideal spring with the same spring constant. This shows the effect of the extra term $+bx^2$, whose positive sign reduces the restoring force and thus the work needed to stretch the spring. ■

GOT IT? 7.2 Gravitational force actually decreases with height, but that decrease is negligible near Earth's surface. To account for the decrease, would the exact value for the potential-energy change associated with a height change h be (a) greater than, (b) less than, or (c) equal to mgh, where g is the gravitational acceleration at Earth's surface?

Where's the Stored Energy and What's the System?

In discussing the climber of Fig. 7.1*a*, the book of Fig. 7.3, and the engineer of Example 7.1, we were careful not to use phrases like "the climber's potential energy," "the potential energy of the book," or "the engineer's potential energy." After all, the climber herself hasn't changed in going from the bottom to the top of the cliff; nor is the book any different after you've returned it to the shelf. So it doesn't make a lot of sense to say that potential energy is somehow a property of these objects. Indeed, the idea of potential energy requires that two (or more) objects interact via a force. In the examples of the climber, the book, and the engineer, that force is gravity—and the pairs of interacting objects are, correspondingly, the climber and Earth, the book and Earth, and the engineer and Earth. So to characterize potential energy, we need in each case to consider a system consisting of at least two objects. In each example the *configuration* of that system changes, because the relative positions of the objects making up the system are altered. In each case, one member of the system—climber, book, or engineer—has moved relative to Earth. So potential energy is energy associated with the *configuration of a system.* It really makes no sense to talk about the potential energy of a single, structureless object. That's in contrast with kinetic energy, which is associated with the motion of a system that might be as simple as a single object.

So where is potential energy stored? In the system of interacting objects. Potential energy is inherently a property of a system and can't be assigned to individual objects.

In the case of gravity, we can go further and say that the energy is stored in the *gravitational field*—a concept that we'll introduce in the next chapter. It's the gravitational field that changes, not the individual objects, when we change the configuration of a system whose components interact via gravity.

What about a spring? We *can* talk about "the potential energy of a spring" because any flexible object, including a spring, necessarily comprises a system of interacting parts. In the case of a spring, the individual molecules in the spring ultimately interact via electric forces, and the associated *electric field* is what changes as the spring stretches or compresses. And, as we'll see quantitatively in Chapter 23, it's in the electric field that the potential energy resides. When we talk about "elastic potential energy" we're really describing potential energy stored in molecular electric fields.

7.3 Conservation of Mechanical Energy

The work–kinetic energy theorem, developed in Section 6.3, shows that the change ΔK in an object's kinetic energy is equal to the net work done on it:

$$\Delta K = W_{\text{net}}$$

Here we'll consider the case where the only forces acting are conservative; then, as our interpretation of Equation 7.2 shows, the work done is the negative of the potential-energy change: $W_{\text{net}} = -\Delta U$. As a result, we have $\Delta K = -\Delta U$, or

$$\Delta K + \Delta U = 0$$

What does this equation tell us? It says that any change ΔK in kinetic energy K must be compensated by an opposite change ΔU in potential energy U in order that the two changes sum to zero. If kinetic energy goes up, then potential energy goes down by the same amount, and vice versa. In other words, the total **mechanical energy**, defined as the sum of kinetic and potential energy, does not change.

Remember that at this point we're considering the case where only conservative forces act. For that case, we've just shown that mechanical energy is conserved. This principle, called **conservation of mechanical energy**, is expressed mathematically in the two equivalent ways we've just discussed:

$$\Delta K + \Delta U = 0 \qquad (7.5)$$

and, equivalently, (conservation of mechanical energy)

$$K + U = \text{constant} = K_0 + U_0 \qquad (7.6)$$

Here K_0 and U_0 are the kinetic and potential energy when an object is at some point, and K and U are their values when it's at any other point. Equations 7.5 and 7.6 both say the same thing: In the absence of nonconservative forces, the total mechanical energy $K + U$ doesn't change. Individually, K and U can change, as energy is transformed from kinetic to potential and vice versa—but when only conservative forces are acting, then the total mechanical energy remains unchanged.

The work–kinetic energy theorem—which itself follows from Newton's second law—is what lies behind the principle of mechanical energy conservation. Although we derived the work–kinetic energy theorem by considering a single object, the principle of mechanical energy conservation holds for any isolated system of macroscopic objects, no matter how complex, as long as its constituents interact only via conservative forces. Individual constituents of a complex system may exchange kinetic energy as, for example, they undergo collisions. Furthermore, the system's potential energy may change as the configuration of the system changes—but add all the constituents' kinetic energies and the potential energy contained in the entire system, and you'll find that the sum remains unchanged.

Keep in mind that we're considering here only isolated systems. If energy is transferred to the system from outside, by external forces doing work, then the system's mechanical energy increases. And if the system does work on its environment, then its mechanical energy decreases. Ultimately, however, energy is always conserved, and if you make the system large enough to encompass all interacting objects, and if those objects interact only via conservative forces, then the system's mechanical energy will be strictly conserved.

Conservation of mechanical energy is a powerful principle. Throughout physics, from the subatomic realm through practical problems in engineering and on to astrophysics, the principle of energy conservation is widely used in solving problems that would be intractable without it. Here we consider its use in macroscopic systems subject only to conservative forces; later we'll expand the principle to more general cases.

PROBLEM-SOLVING STRATEGY 7.1 Conservation of Mechanical Energy

When you're using energy conservation to solve problems, Equation 7.6 basically tells it all. Our IDEA problem-solving strategy adapts well to such problems.

INTERPRET First, interpret the problem to be sure that conservation of mechanical energy applies. Are all the forces conservative? If so, mechanical energy is conserved. Next, identify a point at which you know both the kinetic and the potential energy; then you know the total mechanical energy, which is what's conserved. If the problem doesn't do so and it's not implicit in the equations you use, you may need to identify the zero of potential energy—although that's your own arbitrary choice. You also need to identify the quantity the problem is asking for, and the situation in which it has the value you're after. The quantity may be the energy itself or a related quantity like height, speed, or spring compression. In some situations, you may have to deal with several types of potential energy—such as gravitational and elastic potential energy—appearing in the same problem.

DEVELOP Draw your object first in the situation where you know the energies and then in the situation that contains the unknown. It's helpful to draw simple bar charts suggesting the relative sizes of the potential- and kinetic-energy terms; we'll show you how in several examples. Then you're ready to set up the quantitative statement of mechanical energy conservation, Equation 7.6: $K + U = K_0 + U_0$. Consider which of the four terms you know or can calculate from the given information. You'll probably need secondary equations like the expressions for kinetic energy and for various forms of potential energy. Consider how the quantity you're trying to find is related to an energy.

EVALUATE Write Equation 7.6 for your specific problem, including expressions for kinetic or potential energy that contain the quantity you're after. Solving is then a matter of algebra.

ASSESS As usual, ask whether your answer makes physical sense. Does it have the right units? Are the numbers reasonable? Do the signs make sense? Is your answer consistent with the bar charts in your drawing?

EXAMPLE 7.4 Energy Conservation: Tranquilizing an Elephant

A biologist uses a spring-loaded gun to shoot tranquilizer darts into an elephant. The gun's spring has $k = 940$ N/m and is compressed a distance $x_0 = 25$ cm before firing a 38-g dart. Assuming the gun is pointed horizontally, at what speed does the dart leave the gun?

INTERPRET We're dealing with a spring, assumed ideal, so conservation of mechanical energy applies. We identify the initial state—dart at rest, spring fully compressed—as the point where we know both kinetic and potential energy. The state we're then interested in is when the dart just leaves the gun, when potential energy has been converted to kinetic energy and before gravity has changed its vertical position.

DEVELOP In Fig. 7.6 we've sketched the two states, giving the potential and kinetic energy for each. We've also sketched bar graphs showing the relative sizes of the energies. To use the statement of energy conservation, Equation 7.6, we also need expressions for the kinetic energy ($\frac{1}{2}mv^2$) and the spring potential energy ($\frac{1}{2}kx^2$; Equation 7.4). Incidentally, using Equation 7.4 implicitly sets the zero of elastic potential energy when the spring

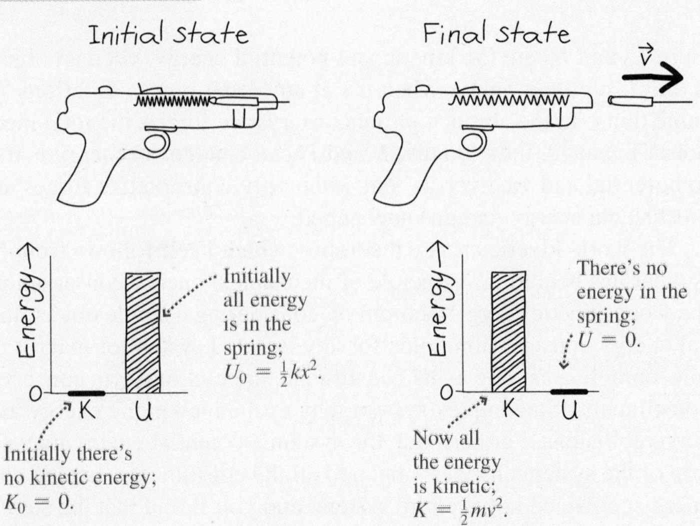

FIGURE 7.6 Our sketches for Example 7.4, showing bar charts for the initial and final states.

is in its equilibrium position. We might as well set the zero of gravitational energy at the height of the gun, since there's no change in the dart's vertical position between our initial and final states.

EVALUATE We're now ready to write Equation 7.6, $K + U = K_0 + U_0$. We know three of the terms in this equation: The initial kinetic energy K_0 is 0, since the dart is initially at rest. The initial potential energy is that of the compressed spring, $U_0 = \frac{1}{2}kx_0^2$. The final potential energy is $U = 0$ because the spring is now in its equilibrium position and we've taken the gravitational potential energy to be zero. What we don't know is the final kinetic energy, but we do know that it's given by $K = \frac{1}{2}mv^2$. So Equation 7.6 becomes $\frac{1}{2}mv^2 + 0 = 0 + \frac{1}{2}kx^2$, which solves to give

$$v = \sqrt{\frac{k}{m}}x_0 = \left(\sqrt{\frac{940\text{ N/m}}{0.038\text{ kg}}}\right)(0.25\text{ m}) = 39\text{ m/s}$$

ASSESS Take a look at the answer in algebraic form; it says that a stiffer spring or a greater compression will give a higher dart speed. Increasing the dart mass, on the other hand, will decrease the speed. All this makes good physical sense. And the outcome shows quantitatively what our bar charts suggest—that the dart's energy starts out all potential and ends up all kinetic. ■

Example 7.4 shows the power of the conservation-of-energy principle. If you had tried to find the answer using Newton's law, you would have been stymied by the fact that the spring force and thus the acceleration of the dart vary continuously. But you don't need to worry about those details; all you want is the final speed, and energy conservation gets you there, shortcutting the detailed application of $\vec{F} = m\vec{a}$.

EXAMPLE 7.5 Conservation of Energy: A Spring and Gravity

The spring in Fig. 7.7 has $k = 140$ N/m. A 50-g block is placed against the spring, which is compressed 11 cm. When the block is released, how high up the slope does it rise? Neglect friction.

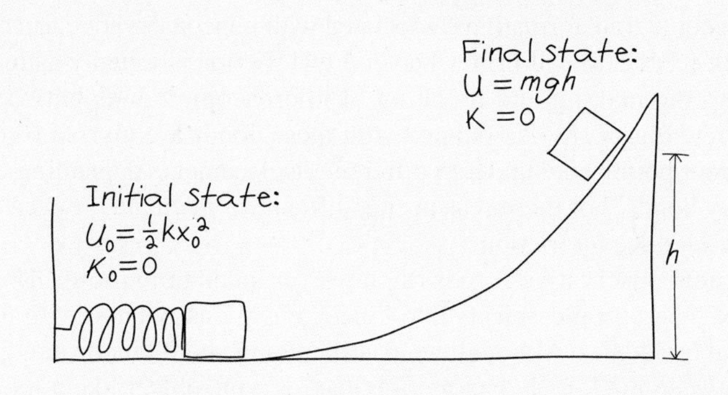

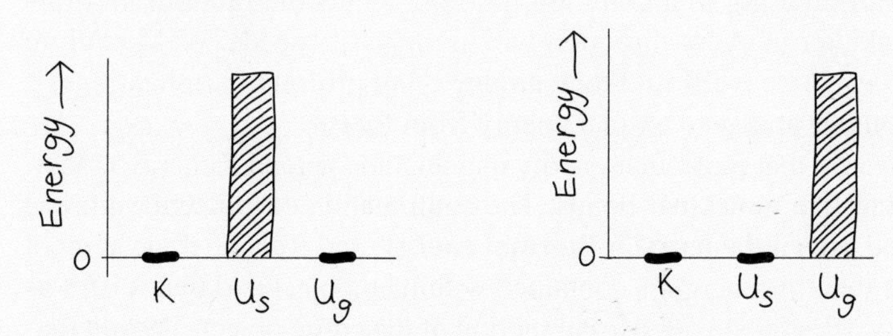

FIGURE 7.7 Our sketches for Example 7.5.

INTERPRET This example is similar to Example 7.4, but now we have changes in both elastic and gravitational potential energy. Since friction is negligible, we can consider that only conservative forces act, in which case we can apply conservation of mechanical energy. We identify the initial state as the block at rest against the compressed spring; the final state is the block momentarily at rest at its topmost point on the slope. We'll take the zero of gravitational potential energy at the bottom.

DEVELOP Figure 7.7 shows the initial and final states, along with bar charts for each. We've drawn separate bars for the spring and gravitational potential energies, U_s and U_g. Now apply Equation 7.6, $K + U = K_0 + U_0$.

EVALUATE In both states the block is at rest, so kinetic energy is zero. In the initial state we know the potential energy U_0: It's the spring energy $\frac{1}{2}kx^2$. We don't know the final-state potential energy, but we do know that it's gravitational energy—and with the zero of potential energy at the bottom, it's $U = mgh$. With $K = K_0 = 0$, $U_0 = \frac{1}{2}kx^2$, and $U = mgh$, Equation 7.6 reads $0 + mgh = 0 + \frac{1}{2}kx^2$. We then solve for the unknown h to get

$$h = \frac{kx^2}{2mg} = \frac{(140\text{ N/m})(0.11\text{ m})^2}{(2)(0.050\text{ kg})(9.8\text{ m/s}^2)} = 1.7\text{ m}$$

ASSESS Again, the answer in algebraic form makes sense; the stiffer the spring or the more it's compressed, the higher the block will go. But if the block is more massive or gravity is stronger, then the block won't get as far.

✓TIP Save Steps

You might be tempted to solve first for the block's speed when it leaves the spring and then equate $\frac{1}{2}mv^2$ to mgh to find the height. You could—but conservation of mechanical energy shortcuts all the details, getting you right from the initial to the final state. As long as energy is conserved, you don't need to worry about what happens in between.

■

Video Tutor Demo | **Chin Basher?**

GOT IT? 7.3 A bowling ball is tied to the end of a long rope and suspended from the ceiling. A student stands at one side of the room and holds the ball to her nose, then releases it from rest. Should she duck as it swings back? Explain.

7.4 Nonconservative Forces

In the examples in Section 7.3, we assumed that mechanical energy was strictly conserved. In the everyday world of friction and other nonconservative forces, however, conservation of mechanical energy is sometimes a reasonable approximation and sometimes not. When it's not, we have to consider energy transformations associated with nonconservative forces.

Friction is a nonconservative force. Recall from Chapter 5 that friction is actually a complex phenomenon, involving the making and breaking of microscopic bonds between two surfaces in contact (review Fig. 5.18). Associated with these bonds are myriad force application points, and different points may undergo different displacements depending on the strengths of the temporary bonds. For these reasons it's difficult to calculate, or even to define unambiguously, the work done by friction.

What friction and other nonconservative forces do, however, is unambiguous: They convert the kinetic energy of macroscopic objects into kinetic energy associated with the random motions of individual molecules. Although we're still talking about kinetic energy, there's a huge difference between the kinetic energy of a macroscopic object like a moving car, with all its parts participating in a common motion, versus the random motions of molecules going helter-skelter in every direction with a range of speeds. We'll explore that difference in Chapter 19, where we'll find that, among other profound implications, it places serious constraints on our ability to extract energy from fuels.

You'll also see, in Chapter 18, that molecular energy may include potential energy associated with stretching of spring-like molecular bonds. The combination of molecular kinetic and potential energy is called **internal energy** or **thermal energy**, and we give it the symbol E_{int}. Here "internal" implies that this energy is contained within an object and that it isn't as obvious as the kinetic energy associated with overall motion of the entire object. The alternative term "thermal" hints that internal energy is associated with temperature, heat, and related phenomena. We'll see in Chapters 16–19 that temperature is a measure of the internal energy per molecule, and that what you probably think of as "heat" is actually internal energy. In physics, "heat" has a very specific meaning: It designates another way of transferring energy to a system, in addition to the mechanical work we've considered in Chapters 6 and 7.

So friction and other nonconservative forces convert mechanical energy into internal energy. How much internal energy? Both theory and experiment give a simple answer: The amount of mechanical energy converted to internal energy is given by the product of the nonconservative force with the distance over which it acts. With friction, that means $\Delta E_{\text{int}} = f_k d$, where d is the distance over which the frictional force acts. (Here we write *kinetic* friction f_k explicitly because *static* friction f_s does not convert mechanical energy to internal energy because there's no relative motion involved.) Since the increase in internal energy comes at the expense of mechanical energy $K + U$, we can write

$$\Delta K + \Delta U = -\Delta E_{\text{int}} = -f_k d \qquad (7.7)$$

Example 7.6 describes a system in which friction converts mechanical energy to internal energy.

EXAMPLE 7.6 Nonconservative Forces: A Sliding Block

A block of mass m is launched from a spring of constant k that's initially compressed a distance x_0. After leaving the spring, the block slides on a horizontal surface with frictional coefficient μ. Find an expression for the distance the block slides before coming to rest.

INTERPRET The presence of friction means that mechanical energy isn't conserved. But we can still identify the kinetic and potential energy in the initial state: The kinetic energy is zero and the potential energy is that of the spring. In the final state, there's no mechanical energy at all. The nonconservative frictional force converts the block's mechanical energy into internal energy of the block and the surface it's sliding on. The block comes to rest when all its mechanical energy has been converted.

DEVELOP Figure 7.8 shows the situation. With $K_0 = 0$, we determine the total initial energy from Equation 7.4, $U_0 = \frac{1}{2}kx_0^2$. As the block slides a distance d, Equation 7.7 shows that the frictional force converts mechanical energy equal to $f_k d$ into internal energy. All the mechanical energy will be gone, therefore, when $f_k d = \frac{1}{2}kx_0^2$. Here the frictional force has magnitude $f_k = \mu n = \mu mg$, where in this case of a horizontal surface the normal force n has the same magnitude as the weight mg. So our statement that all the mechanical energy gets converted to internal energy becomes $\frac{1}{2}kx_0^2 = \mu mgd$.

EVALUATE We solve this equation for the unknown distance d to get $d = kx_0^2/2\mu mg$. Since we weren't given numbers, there's nothing further to evaluate.

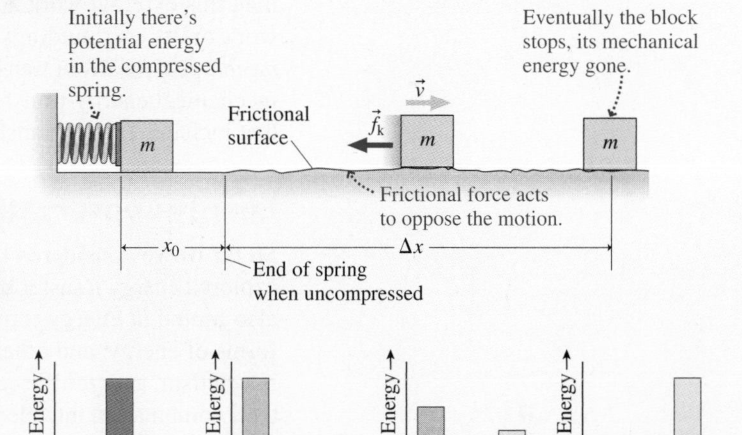

FIGURE 7.8 Intermediate bar charts show gradual conversion of mechanical energy into internal energy.

ASSESS Make sense? The stiffer the spring or the more it's compressed, the farther the block goes. The greater the friction or the normal force mg, the sooner the block stops. If $\mu = 0$, mechanical energy is once again conserved; then our result shows that the block would slide forever. ■

GOT IT? 7.4 For which of the following systems is (1) mechanical energy conserved and (2) total energy conserved? (a) the system is isolated, and all forces among its constituents are conservative; (b) the system is not isolated, and work is done on it by external forces; (c) the system is isolated, and some forces among its constituents are not conservative

---

7.5 Conservation of Energy

We often speak of energy being "lost" due to friction, or to air resistance, or to electrical resistance in power transmission. But that energy isn't really lost; instead, as we've just seen for friction, it's converted to internal energy. Physically, the internal energy manifests itself by warming the system. So the energy really is still there; it's just that we can't get it back as the kinetic energy of macroscopic objects.

Accounting for internal energy leads to a broader statement of energy conservation. Rearranging the first equality of Equation 7.7 lets us write

$$\Delta K + \Delta U + \Delta E_{\text{int}} = 0$$

This equation shows that the sum of the kinetic, potential, and internal energy of an isolated system doesn't change even though energy may be converted among these three different forms. You can see this conservation of energy graphically in Fig. 7.8, which plots all three forms of energy for the situation of Example 7.6.

So far we've considered only isolated systems, in which all forces are internal to the system. For Example 7.6 to be about an isolated system, for instance, that system had to include the spring, the block, and the surface on which the block slides. What if a system isn't isolated? Then external forces may do work on it, increasing its energy. Or the system may do work on its environment, decreasing its energy. In that case we can generalize Equation 7.7 to read

$$\Delta K + \Delta U + \Delta E_{\text{int}} = W_{\text{ext}} \tag{7.8}$$

where W_{ext} is the work done on the system by forces acting from outside. If W_{ext} is positive, then this external work adds energy to the system; if it's negative, then the system does work on its surroundings, and its total energy decreases. Recall that doing work is the *mechanical* means of transferring energy; in Chapters 16–18 we'll introduce *heat* as a nonmechanical energy-transfer mechanism, and we'll develop a statement like Equation 7.8 that includes energy transfers by both work and heat.

Energy Conservation: The Big Picture

So far we've considered kinetic energy, potential energy, and internal energy, and we've explored energy transfer by mechanical work and by dissipative forces like friction. We've also hinted at energy transfer by heat, to be defined in Chapter 16. But there are other forms of energy, and other energy-transfer mechanisms. In Part 3, you'll explore electromagnetism, and you'll see how energy can be stored in both electric and magnetic fields; their combination into electromagnetic waves results in energy transfer by *electromagnetic radiation*—the process that delivers life-sustaining energy from Sun to Earth and that also carries your cell phone conversations and data. Electromagnetic fields interact with matter, so energy transfers among electromagnetic, mechanical, and internal energy are important processes in the everyday physics of both natural and technological systems. But again, for any isolated system, such transfers only interchange *types* of energy and don't change the total *amount* of energy. Energy, it seems, is strictly conserved.

In Newtonian physics, conservation of energy stands alongside the equally fundamental principle of conservation of mass (the statement that the total mass of an isolated system can't change). A closer look, however, shows that neither principle stands by itself. If you measure precisely enough the mass of a system before it emits energy, and again afterward, you'll find that the mass has decreased. Einstein's equation $E = mc^2$ describes this effect, which ultimately shows that mass and energy are interchangeable. So Einstein replaces the separate conservation laws for mass and energy with a single statement: **conservation of mass–energy**. You'll see how mass–energy interchangeability arises when we study relativity in Chapter 33. Until then, we'll be dealing in the realm of Newtonian physics, where it's an excellent approximation to assume that energy and mass are separately conserved.

GOT IT? 7.5 Consider Earth and its atmosphere as a system. Which of the following processes conserves the total energy of this system? (a) a volcano erupts, spewing hot gases and particulate matter high into the atmosphere; (b) a small asteroid plunges into Earth's atmosphere, heating and exploding high over the planet; (c) over geologic time, two continents collide, and the one that is subducted under the other heats up and undergoes melting; (d) a solar flare delivers high-energy particles to Earth's upper atmosphere, lighting the atmosphere with colorful auroras; (e) a hurricane revs up its winds, extracting energy from water vapor evaporated from warm tropical seas; (f) coal burns in numerous power plants, and uranium fissions in nuclear reactors, with both processes sending electrical energy into the world's power grids and dumping warmed water into the environment

---

7.6 Potential-Energy Curves

MP

PhET: Calculus Grapher
PhET: Energy Skate Park

Figure 7.9 shows a frictionless roller-coaster track. How fast must a car be coasting at point A if it's to reach point D? Conservation of mechanical energy provides the answer. To get to D, the car must clear peak C. Clearing C requires that the total energy exceed the potential energy at C; that is, $\frac{1}{2}mv_A^2 + mgh_A > mgh_C$, where we've taken the zero of potential energy with the car at the bottom of the track. Solving for v_A gives $v_A > \sqrt{2g(h_C - h_A)}$. If v_A satisfies this inequality, the car will reach C with some kinetic energy remaining and will coast over the peak.

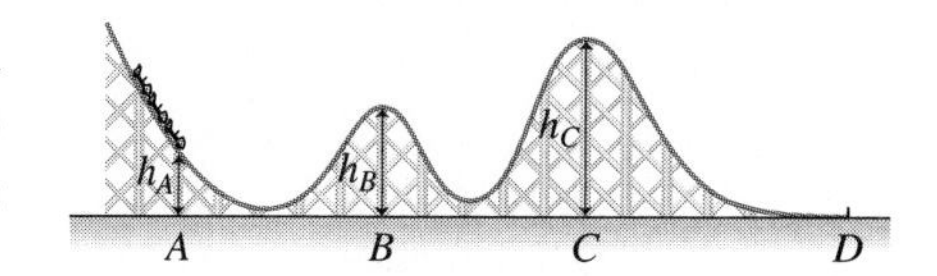

FIGURE 7.9 A roller-coaster track.

Figure 7.9 is a drawing of the actual roller-coaster track. But because gravitational potential energy is directly proportional to height, it's also a plot of potential energy versus position: a **potential-energy curve**. Conceptual Example 7.1 shows how we can study the car's motion by plotting total energy on the same graph as the potential-energy curve.

CONCEPTUAL EXAMPLE 7.1 Potential-Energy Curves

Figure 7.10 plots potential energy for our roller-coaster system, along with three possible values for the total mechanical energy. Since mechanical energy is conserved in the absence of nonconservative forces, the total-energy curve is a horizontal line. Use these graphs to describe the motion of a roller-coaster car, initially at point A and moving to the right.

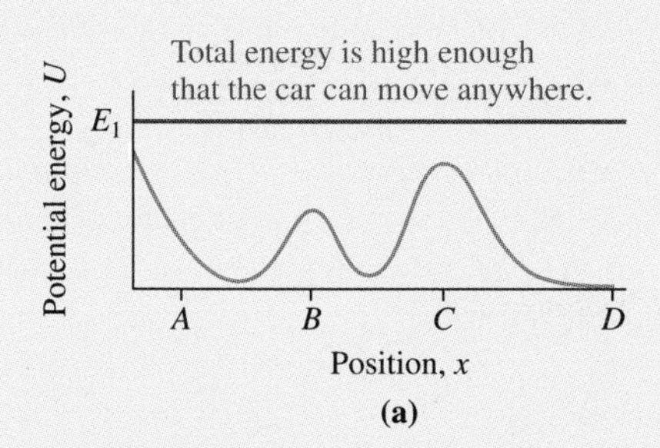

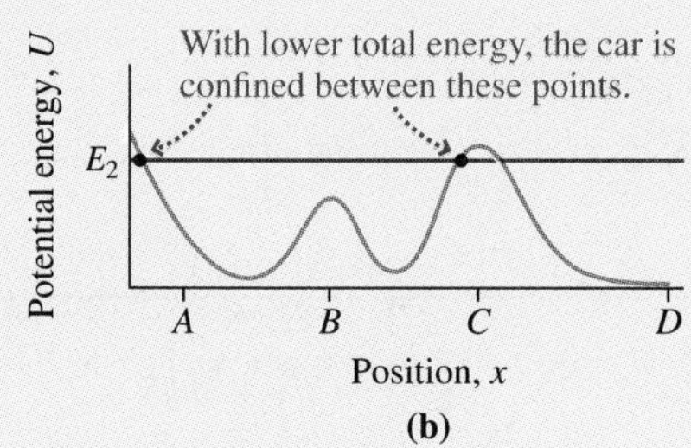

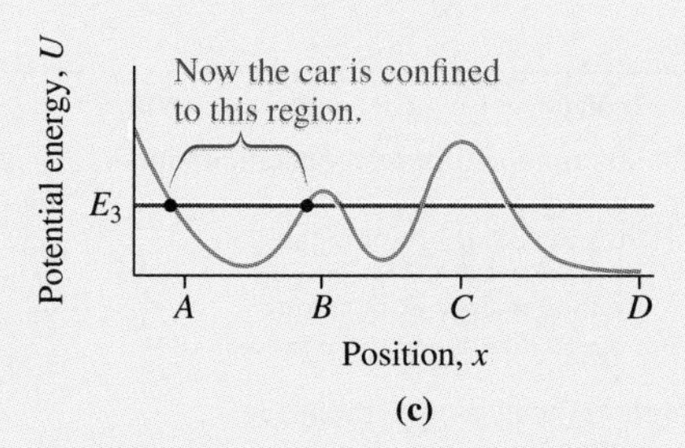

FIGURE 7.10 Potential and total energy for a roller coaster.

EVALUATE We're assuming there are no nonconservative forces (an approximation for a real roller coaster), so mechanical energy is conserved. In each figure, the sum of kinetic and potential energy therefore remains equal to the value set by the line indicating the total energy. When the roller-coaster car rises, potential energy increases and kinetic energy consequently decreases. But as long as potential energy remains below the total energy, the car still has kinetic energy and is still moving. Anywhere potential energy equals the total energy, the car has no kinetic energy and is momentarily at rest.

In Fig. 7.10*a* the car's total energy exceeds the maximum potential energy. Therefore, it can move anywhere from its initial position at A. Since it's initially moving to the right, it will clear peaks B and C and will end up at D still moving to the right—and, since D is lower than A, it will be moving faster than it was at A.

In Fig. 7.10*b* the highest peak in the potential-energy curve exceeds the total energy; so does the very leftmost portion of the curve. Therefore, the car will move rightward from A, clearing peak B, but will come to a stop just before peak C, a so-called **turning point** where potential energy equals the total energy. Then it will roll back down to the left, again clearing peak B and climbing to another turning point where the potential-energy curve and total-energy line again intersect. Absent friction, it will run back and forth between the two turning points.

In Fig. 7.10*c* the total energy is lower, and the car can't clear peak B. So now it will run back and forth between the two turning points we've marked.

ASSESS Make sense? Yes: The higher the total energy, the larger the extent of the car's allowed motion. That's because, for a given potential energy, the car it has more energy available in the form of kinetic energy.

MAKING THE CONNECTION Find a condition on the speed at A that will allow the car to move beyond peak B.

EVALUATE With total energy equal to U_B, the car could just barely clear peak B. The initial energy is $\frac{1}{2}mv_A^2 + mgh_A$, where v_A and h_A are the car's speed and height at A, and where we've taken the zero of potential energy at the bottom of the curve. Requiring that this quantity exceed $U_B = mgh_B$ then gives $v_A > \sqrt{2g(h_B - h_A)}$.

Even though the car in Figs. 7.10*b* and *c* can't get to D, the total energy still exceeds the potential energy at D. But the car is blocked from reaching D by the **potential barrier** of peak C. We say that it's **trapped** in a **potential well** between its turning points.

Potential-energy curves are useful even with nongravitational forces where there's no direct correspondence with hills and valleys. The terminology used here—potential barriers, wells, and trapping—remains appropriate in such cases and indeed is widely used throughout physics.

Figure 7.11 shows the potential energy of a system comprising a pair of hydrogen atoms, as a function of their separation. This energy is associated with attractive and repulsive electrical forces involving the electrons and the nuclei of the two atoms. The potential-energy curve exhibits a potential well, showing that the atoms can form a **bound system** in which they're unable to separate fully. That bound system is a hydrogen molecule (H_2). The minimum energy, -7.6×10^{-19} J, corresponds to the molecule's equilibrium separation of 0.074 nm. It's convenient to define the zero of potential energy when the atoms are infinitely far apart; Fig. 7.11 then shows that any total energy less than zero results in a bound system. But if the total energy is greater than zero, the atoms are free to move arbitrarily far apart, so they don't form a molecule.

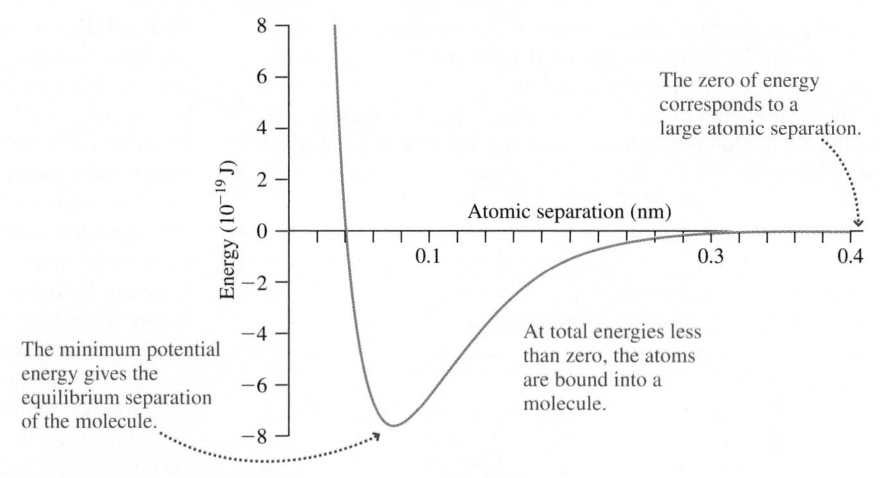

FIGURE 7.11 Potential-energy curve for two hydrogen atoms.

EXAMPLE 7.7 Molecular Energy: Finding Atomic Separation

Very near the bottom of the potential well in Fig. 7.11, the potential energy of the two-atom system given approximately by $U = U_0 + a(x - x_0)^2$, where $U_0 = -0.760$ aJ, $a = 286$ aJ/nm^2, and $x_0 = 0.0741$ nm is the equilibrium separation. What range of atomic separations is allowed if the total energy is -0.717 aJ?

INTERPRET This problem sounds complicated, with strange units and talk of molecular energies. But it's about just what's shown in Figs. 7.10 and 7.11. Specifically, we're given the total energy and asked to find the turning points—the points where the line representing total energy intersects the potential-energy curve. If the units look strange, remember the SI prefixes (there's a table inside the front cover), which we use to avoid writing large powers of 10. Here 1 aJ $= 10^{-18}$ J and 1 nm $= 10^{-9}$ m.

DEVELOP Figure 7.12 is a plot of the potential-energy curve from the function we've been given. The straight line represents the total energy E. The turning points are the values of atomic separation where the two curves intersect. We could read them off the graph, or we can solve algebraically by setting the total energy equal to the potential energy.

EVALUATE With the potential energy given by $U = U_0 + a(x - x_0)^2$ and the total energy E, the two turning points occur when $E = U_0 + a(x - x_0)^2$. We could solve directly for x, but then we'd have to use the quadratic formula. Solving for $x - x_0$ is easier:

$$x - x_0 = \pm\sqrt{\frac{E - U_0}{a}} = \pm\sqrt{\frac{-0.717\text{ aJ} - (-0.760\text{ aJ})}{286\text{ aJ/nm}^2}}$$
$$= \pm 0.0123\text{ nm}$$

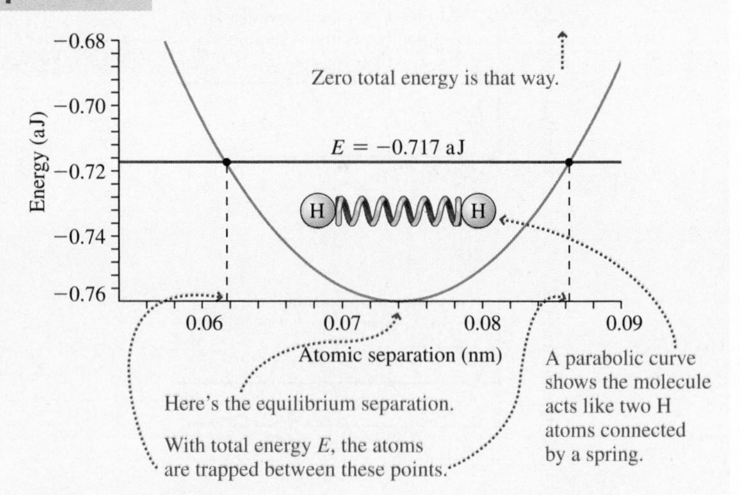

FIGURE 7.12 Analyzing the hydrogen molecule.

Then the turning points are at $x_0 \pm 0.0123$ nm—namely, 0.0864 nm and 0.0618 nm.

ASSESS Make sense? A look at Fig. 7.12 shows that we've correctly located the turning points. The fact that its potential-energy curve is parabolic (like a spring's $U = \frac{1}{2}kx^2$) shows that the molecule can be modeled approximately as two atoms joined by a spring. Chemists frequently use such models and even talk of the "spring constant" of the bond joining atoms into a molecule. ■

Force and Potential Energy

The roller-coaster track in Fig. 7.9 traces the potential-energy curve for a car on the track. But it also shows the force acting to accelerate the car: Where the graph is steep—that is, where the potential energy is changing rapidly—the force is greatest. At the peaks and valleys, the force is zero. So it's the *slope* of the potential-energy curve that tells us about the force (Fig. 7.13).

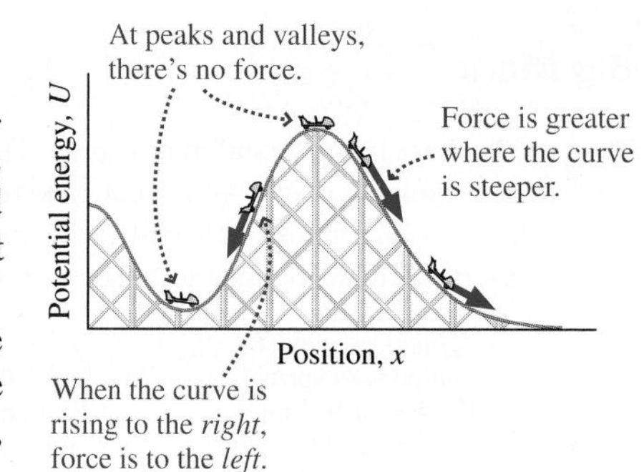

FIGURE 7.13 Force depends on the *slope* of the potential-energy curve.

Just how strong is this force? Consider a small change Δx, so small that the force is essentially constant over this distance. Then we can use Equation 7.2b to write $\Delta U = -F_x \Delta x$, or $F_x = -\Delta U/\Delta x$. In the limit $\Delta x \rightarrow 0$, $\Delta U/\Delta x$ becomes the derivative, and we have

$$F_x = -\frac{dU}{dx} \qquad (7.9)$$

This equation makes mathematical as well as physical sense. We've already written potential energy as the *integral* of force over distance, so it's no surprise that force is the *derivative* of potential energy. Equation 7.9 gives the force component in the *x*-direction only. In a three-dimensional situation, we'd have to take derivatives of potential energy with respect to *y* and *z* to find the full force vector.

Why the minus sign in Equation 7.9? You can see the answer in the molecular energy curve of Fig. 7.11, where pushing the atoms too close together—moving to the *left* of equilibrium—results in a repulsive force to the *right*, and pulling them apart—moving to the *right*—gives an attractive force to the *left*. You can see the same thing for the roller coaster in Fig. 7.13. In both cases the forces tend to drive the system back toward a minimum-energy state. We'll explore such minimum-energy equilibrium states further in Chapter 12.

GOT IT? 7.6 The figure shows the potential energy associated with an electron in a microelectronic device. From among the labeled points, find (1) the point where the force on the electron is greatest, (2) the rightmost position possible if the electron has total energy E_1, (3) the leftmost position possible if the electron has total energy E_2 and starts out to the right of *D*, (4) a point where the force on the electron is zero, and (5) a point where the force on the electron points to the left. In some cases there may be more than one answer.

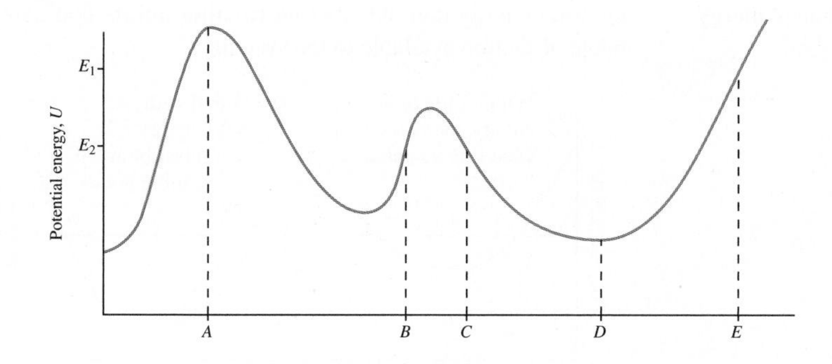

CHAPTER 7 SUMMARY

Big Ideas

The big idea here is conservation of energy. This chapter emphasizes the special case of systems subject only to conservative forces, in which case the total mechanical energy—the sum of kinetic and potential energy—cannot change. Energy may change from kinetic to potential, and vice versa, but the total remains constant. Applying conservation of mechanical energy requires the concept of potential energy—energy stored in a system as a result of work done against conservative forces.

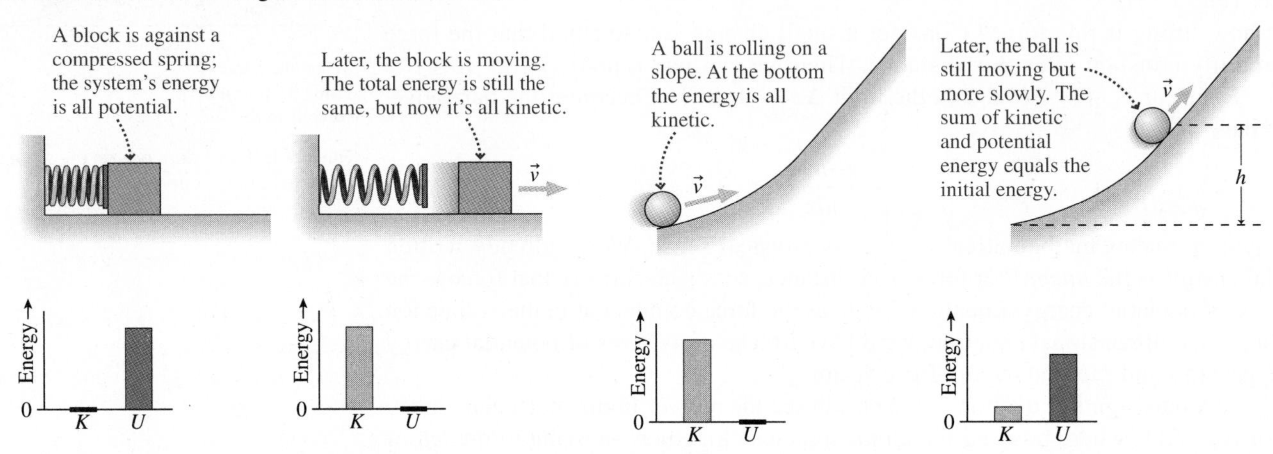

If nonconservative forces act in a system, then mechanical energy isn't conserved; instead, mechanical energy gets converted to internal energy.

Key Concepts and Equations

The important new concept here is potential energy, defined as the negative of the work done by a conservative force. Only the change ΔU has physical significance. Expressions for potential energy include:

$$\Delta U_{AB} = -\int_A^B \vec{F} \cdot d\vec{r}$$

This one is the most general, but it's mathematically involved. The force can vary over an arbitrary path between points A and B.

$$\Delta U = -\int_{x_1}^{x_2} F(x)dx$$

This is a special case, when force and displacement are in the same direction and force may vary with position.

$$\Delta U = -F(x_2 - x_1)$$

This is the most specialized case, where the force is constant.

Given the concept of potential energy, the principle of conservation of mechanical energy follows from the work–kinetic energy theorem of Chapter 6. Here's the mathematical statement of mechanical energy conservation:

$$K + U = K_0 + U_0$$

K and U are the kinetic and potential energy at some point where we don't know one of these quantities.

The total mechanical energy is conserved, as indicated by the equal sign.

K_0 and U_0 are the kinetic and potential energy at some point where both are known. $K_0 + U_0$ is the *total mechanical energy*.

We can describe a wide range of systems—from molecules to roller coasters to planets—in terms of **potential-energy curves**. Knowing the total energy then lets us find **turning points** that determine the range of motion available to the system.

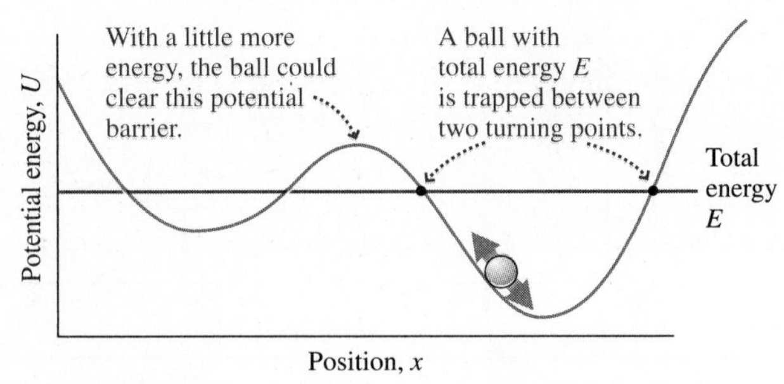

Applications

Two important cases of potential energy are the elastic potential energy of a spring, $U = \frac{1}{2}kx^2$, and the gravitational potential energy change, $\Delta U = mgh$, associated with lifting an object of mass m through a height h.

The former is limited to ideal springs for which $F = -kx$, the latter to the proximity of Earth's surface, where the variation of gravity with height is negligible.

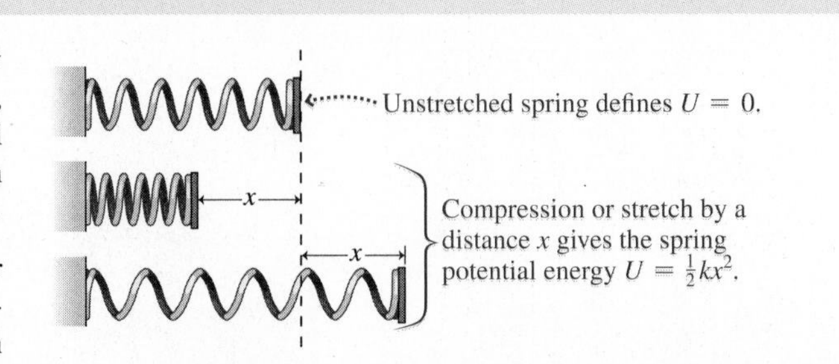

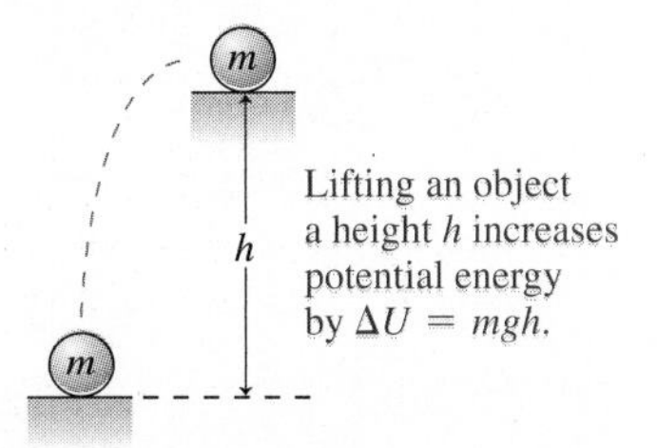

MP For homework assigned on MasteringPhysics, go to www.masteringphysics.com

BIO *Biology and/or medicine-related problems* **DATA** *Data problems* **ENV** *Environmental problems* **CH** *Challenge problems* **COMP** *Computer problems*

For Thought and Discussion

1. Figure 7.14 shows force vectors at different points in space for two forces. Which is conservative and which nonconservative? Explain.

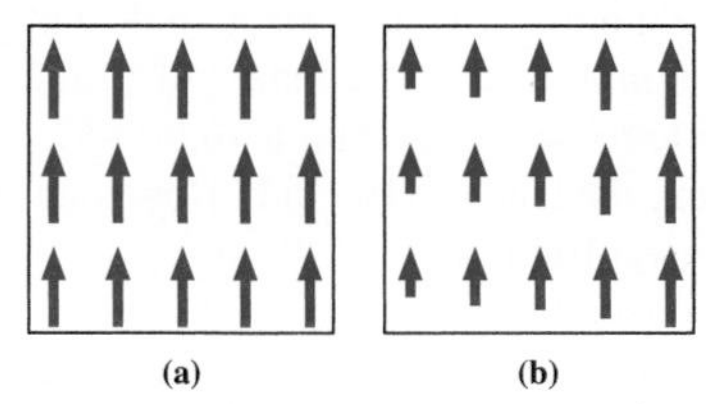

FIGURE 7.14 For Thought and Discussion 1; Problem 30

2. Is the conservation-of-mechanical-energy principle related to Newton's laws, or is it an entirely separate physical principle? Discuss.
3. Why can't we define a potential energy associated with friction?
4. Can potential energy be negative? Can kinetic energy? Can total mechanical energy? Explain.
5. If the potential energy is zero at a given point, must the force also be zero at that point? Give an example.
6. If the force is zero at a given point, must the potential energy also be zero at that point? Give an example.
7. If the difference in potential energy between two points is zero, does that necessarily mean that an object moving between those points experiences no force?
8. A tightrope walker follows an essentially horizontal rope between two mountain peaks of equal altitude. A climber descends from one peak and climbs the other. Compare the work done by the gravitational force on the tightrope walker and the climber.
9. If conservation of energy is a law of nature, why do we have programs—like mileage requirements for cars or insulation standards for buildings—designed to encourage energy conservation?

Exercises and Problems

Exercises

Section 7.1 Conservative and Nonconservative Forces

10. Determine the work you would have to do to move a block of mass m from point 1 to point 2 at constant speed over the two paths shown in Fig. 7.15. The coefficient of friction has the constant value μ over the surface. *Note:* The diagram lies in a horizontal plane.

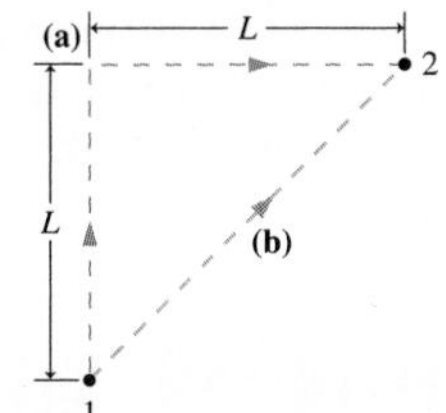

FIGURE 7.15 Exercises 10 and 11

11. Now take Fig. 7.15 to lie in a vertical plane, and find the work done by the gravitational force as an object moves from point 1 to point 2 over each of the paths shown.

Section 7.2 Potential Energy

12. Rework Example 7.1, now taking the zero of potential energy at street level.
13. Find the potential energy associated with a 70-kg hiker (a) atop New Hampshire's Mount Washington, 1900 m above sea level, and (b) in Death Valley, California, 86 m below sea level. Take the zero of potential energy at sea level.
14. You fly from Boston's Logan Airport, at sea level, to Denver, altitude 1.6 km. Taking your mass as 65 kg and the zero of potential energy at Boston, what's the gravitational potential energy when you're (a) at the plane's 11-km cruising altitude and (b) in Denver?
15. The potential energy associated with a 60-kg hiker ascending 1250-m-high Camel's Hump mountain in Vermont is -240 kJ; the zero of potential energy is taken at the mountaintop. What's her altitude?
16. How much energy can be stored in a spring with $k = 320$ N/m if the maximum allowed stretch is 18 cm?
17. How far would you have to stretch a spring with $k = 1.4$ kN/m for it to store 210 J of energy?
18. **BIO** A biophysicist grabs the ends of a DNA strand with optical tweezers and stretches it 26 μm. How much energy is stored in the stretched molecule if its spring constant is 0.046 pN/μm?

Section 7.3 Conservation of Mechanical Energy

19. A skier starts down a frictionless 32° slope. After a vertical drop of 25 m, the slope temporarily levels out and then slopes down at 20°, dropping an additional 38 m vertically before leveling out again. Find the skier's speed on the two level stretches.
20. A 10,000-kg Navy jet lands on an aircraft carrier and snags a cable to slow it down. The cable is attached to a spring with $k = 40$ kN/m. If the spring stretches 25 m to stop the plane, what was its landing speed?
21. A 120-g arrow is shot vertically from a bow whose effective spring constant is 430 N/m. If the bow is drawn 71 cm before shooting, to what height does the arrow rise?
22. In a railroad yard, a 35,000-kg boxcar moving at 7.5 m/s is stopped by a spring-loaded bumper mounted at the end of the level track. If $k = 2.8$ MN/m, how far does the spring compress in stopping the boxcar?
23. You work for a toy company, and you're designing a spring-launched model rocket. The launching apparatus has room for a spring that can be compressed 14 cm, and the rocket's mass is 65 g. If the rocket is to reach an altitude of 35 m, what should you specify for the spring constant?

Section 7.4 Nonconservative Forces

24. A 54-kg ice skater pushes off the wall of the rink, giving herself an initial speed of 3.2 m/s. She then coasts with no further effort. If the frictional coefficient between skates and ice is 0.023, how far does she go?
25. You push a 33-kg table across a 6.2-m-wide room. In the process, 1.5 kJ of mechanical energy gets converted to internal energy of the table/floor system. What's the coefficient of kinetic friction between table and floor?

Section 7.6 Potential-Energy Curves

26. A particle slides along the frictionless track shown in Fig. 7.16, starting at rest from point A. Find (a) its speed at B, (b) its speed at C, and (c) the approximate location of its right-hand turning point.

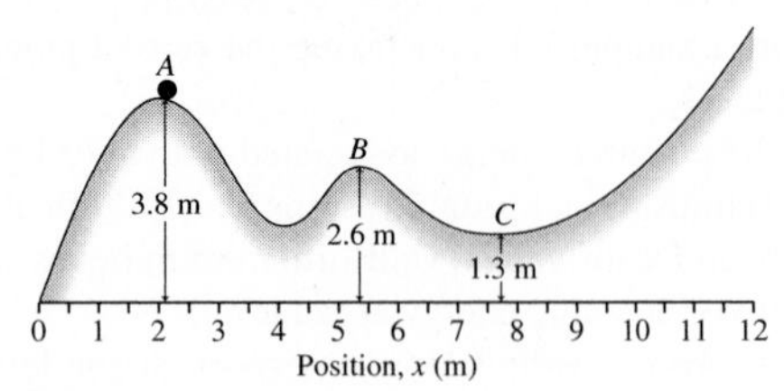

FIGURE 7.16 Exercise 26

27. A particle slides back and forth on a frictionless track whose height as a function of horizontal position x is $y = ax^2$, where $a = 0.92\ \text{m}^{-1}$. If the particle's maximum speed is 8.5 m/s, find its turning points.
28. A particle is trapped in a potential well described by $U(x) = 16x^2 - b$, with U in joules, x in meters, and $b = 4.0$ J. Find the force on the particle when it's at (a) $x = 2.1$ m, (b) $x = 0$, and (c) $x = -1.4$ m.

Problems

29. The reservoir at Northfield Mountain Pumped Storage Project is 214 m above the pump/generators and holds 2.1×10^{10} kg of water (see Application on p. 113). The generators can produce electrical energy at the rate of 1.08 GW. Find (a) the gravitational potential energy stored, taking zero potential energy at the generators, and (b) the length of time the station can generate power before the reservoir is drained. **ENV**
30. The force in Fig. 7.14*a* is given by $\vec{F}_a = F_0\,\hat{j}$, where F_0 is a constant. The force in Fig. 7.14*b* is given by $\vec{F}_b = F_0(x/a)\,\hat{j}$, where the origin is at the lower left corner of the box, a is the width of the square box, and x increases horizontally to the right. Determine the work you would have to do to move an object around the perimeter of each each box, going clockwise at constant speed, starting at the lower left corner.
31. A 1.50-kg brick measures 20.0 cm × 8.00 cm × 5.50 cm. Taking the zero of potential energy when the brick lies on its broadest face, what's the potential energy (a) when the brick is standing on end and (b) when it's balanced on its 8-cm edge? (*Note:* You can treat the brick as though all its mass is concentrated at its center.)
32. A carbon monoxide molecule can be modeled as a carbon atom and an oxygen atom connected by a spring. If a displacement of the carbon by 1.46 pm from its equilibrium position relative to the oxygen increases the molecule's potential energy by 0.0125 eV, what's the spring constant?
33. A more accurate expression for the force law of the rope in Example 7.3 is $F = -kx + bx^2 - cx^3$, where k and b have the values given in Example 7.3 and $c = 3.1\ \text{N/m}^3$. Find the energy stored in stretching the rope 2.62 m. By what percentage does your result differ from that of Example 7.3?
34. For small stretches, the Achilles tendon can be modeled as an ideal spring. Experiments using a particular tendon showed that it stretched 2.66 mm when a 125-kg mass was hung from it. (a) Find the spring constant of this tendon. (b) How much would it have to stretch to store 50.0 J of energy? **BIO**
35. The force exerted by an unusual spring when it's compressed a distance x from equilibrium is $F = -kx - cx^3$, where $k = 220$ N/m and $c = 3.1\ \text{N/m}^3$. Find the stored energy when it's been compressed 15 cm.
36. The force on a particle is given by $\vec{F} = A\,\hat{\imath}/x^2$, where A is a positive constant. (a) Find the potential-energy difference between two points x_1 and x_2, where $x_1 > x_2$. (b) Show that the potential-energy difference remains finite even when $x_1 \to \infty$.
37. A particle moves along the x-axis under the influence of a force $F = ax^2 + b$, where a and b are constants. Find the potential energy as a function of position, taking $U = 0$ at $x = 0$.
38. As a highway engineer, you're asked to design a runaway truck lane on a mountain road. The lane will head uphill at 30° and should be able to accommodate a 16,000-kg truck with failed brakes entering the lane at 110 km/h. How long should you make the lane? Neglect friction.
39. A spring of constant k, compressed a distance x, is used to launch a mass m up a frictionless slope at angle θ. Find an expression for the maximum distance along the slope that the mass moves after leaving the spring.
40. A child is on a swing whose 3.2-m-long chains make a maximum angle of 50° with the vertical. What's the child's maximum speed?
41. With $x - x_0 = h$ and $a = g$, Equation 2.11 gives the speed of an object thrown downward with initial speed v_0 after it's dropped a distance h: $v = \sqrt{v_0^2 + 2gh}$. Use conservation of mechanical energy to derive the same result.
42. The *nuchal ligament* is a cord-like structure that runs along the back of the neck and supports much of the head's weight in animals like horses and cows. The ligament is extremely stiff for small stretches, but loosens as it stretches further, thus functioning as a biological shock absorber. Figure 7.17 shows the force–distance curve for a particular nuchal ligament; the curve can be modeled approximately by the expression $F(x) = 0.43x - 0.033x^2 + 0.00086x^3$, with F in kN and x in cm. Find the energy stored in the ligament when it's been stretched (a) 7.5 cm and (b) 15 cm. **BIO**

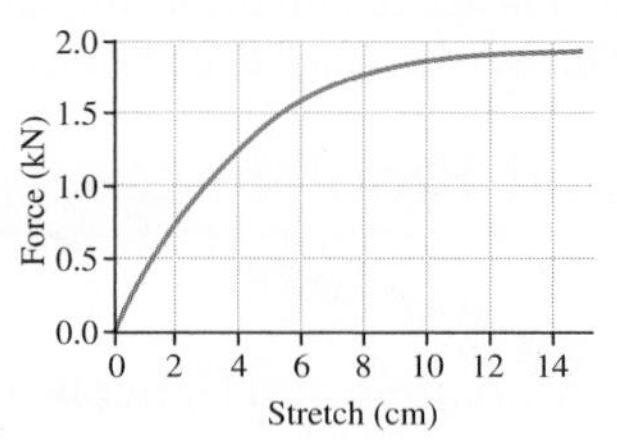

FIGURE 7.17 Problem 42

43. A 200-g block slides back and forth on a frictionless surface between two springs, as shown in Fig. 7.18. The left-hand spring has $k = 130$ N/m and its maximum compression is 16 cm. The right-hand spring has $k = 280$ N/m. Find (a) the maximum compression of the right-hand spring and (b) the speed of the block as it moves between the springs.

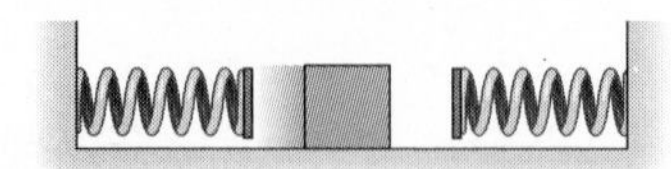

FIGURE 7.18 Problem 43

44. Automotive standards call for bumpers that sustain essentially no damage in a 4-km/h collision with a stationary object. As an automotive engineer, you'd like to improve on that. You've developed a spring-mounted bumper with effective spring

constant 1.3 MN/m. The springs can compress up to 5.0 cm before damage occurs. For a 1400-kg car, what do you claim as the maximum collision speed?

45. **CH** A block slides on the frictionless loop-the-loop track shown in Fig. 7.19. Find the minimum height h at which it can start from rest and still make it around the loop.

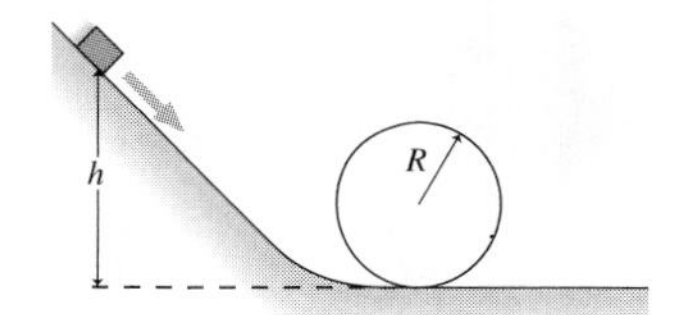

FIGURE 7.19 Problem 45

46. The maximum speed of the pendulum bob in a grandfather clock is 0.55 m/s. If the pendulum makes a maximum angle of 8.0° with the vertical, what's the pendulum's length?
47. A mass m is dropped from height h above the top of a spring of constant k mounted vertically on the floor. Show that the spring's maximum compression is given by $(mg/k)(1 + \sqrt{1 + 2kh/mg})$.
48. A particle with total energy 3.5 J is trapped in a potential well described by $U = 7.0 - 8.0x + 1.7x^2$, where U is in joules and x in meters. Find its turning points.
49. (a) Derive an expression for the potential energy of an object subject to a force $F_x = ax - bx^3$, where $a = 5$ N/m and $b = 2$ N/m^3, taking $U = 0$ at $x = 0$. (b) Graph the potential-energy curve for $x > 0$ and use it to find the turning points for an object whose total energy is -1 J.
50. **CH** In ionic solids such as NaCl (salt), the potential energy of a pair of ions takes the form $U = b/r^n - a/r$, where r is the separation of the ions. For NaCl, a and b have the SI values 4.04×10^{-28} and 5.52×10^{-98}, respectively, and $n = 8.22$. Find the equilibrium separation in NaCl.
51. Repeat Exercise 19 for the case when the coefficient of kinetic friction on both slopes is 0.11, while the level stretches remain frictionless.
52. **ENV** As an energy-efficiency consultant, you're asked to assess a pumped-storage facility. Its reservoir sits 140 m above its generating station and holds 8.5×10^9 kg of water. The power plant generates 330 MW of electric power while draining the reservoir over an 8.0-h period. Its efficiency is the percentage of the stored potential energy that gets converted to electricity. What efficiency do you report?
53. A spring of constant $k = 340$ N/m is used to launch a 1.5-kg block along a horizontal surface whose coefficient of sliding friction is 0.27. If the spring is compressed 18 cm, how far does the block slide?
54. A bug slides back and forth in a bowl 15 cm deep, starting from rest at the top, as shown in Fig. 7.20. The bowl is frictionless except for a 1.4-cm-wide sticky patch on its flat bottom, where the coefficient of friction is 0.89. How many times does the bug cross the sticky region?

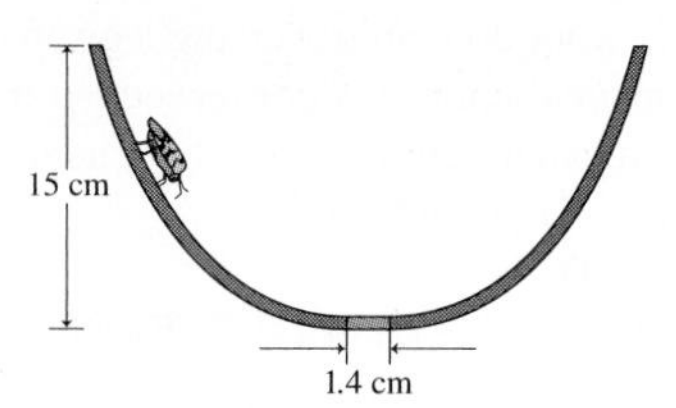

FIGURE 7.20 Problem 54

55. **CH** A 190-g block is launched by compressing a spring of constant $k = 200$ N/m by 15 cm. The spring is mounted horizontally, and the surface directly under it is frictionless. But beyond the equilibrium position of the spring end, the surface has frictional coefficient $\mu = 0.27$. This frictional surface extends 85 cm, followed by a frictionless curved rise, as shown in Fig. 7.21. After it's launched, where does the block finally come to rest? Measure from the left end of the frictional zone.

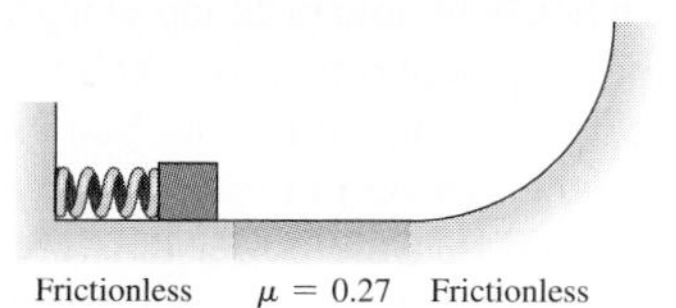

FIGURE 7.21 Problem 55

56. A block slides down a frictionless incline that terminates in a 45° ramp, as shown in Fig. 7.22. Find an expression for the horizontal range x shown in the figure as a function of the heights h_1 and h_2.

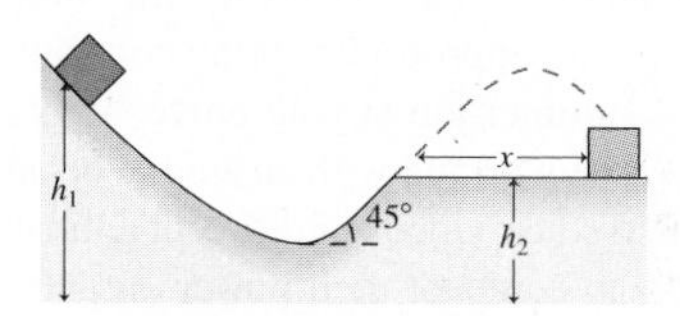

FIGURE 7.22 Problem 56

57. **CH** An 840-kg roller-coaster car is launched from a giant spring with $k = 31$ kN/m into a frictionless loop-the-loop track of radius 6.2 m, as shown in Fig. 7.23. What's the minimum spring compression that will ensure the car stays on the track?

FIGURE 7.23 Problem 57

58. A particle slides back and forth in a frictionless bowl whose height is given by $h = 0.18x^2$, with x and h in meters. Find the x coordinates of its turning points if the particle's maximum speed is 47 cm/s.
59. A child sleds down a frictionless hill whose vertical drop is 7.2 m. At the bottom is a level but rough stretch where the coefficient of kinetic friction is 0.51. How far does she slide across the level stretch?
60. **CH** A bug lands on top of the frictionless, spherical head of a bald man. It begins to slide down his head (Fig. 7.24). Show that the bug leaves the head when it has dropped a vertical distance one-third of the head's radius.

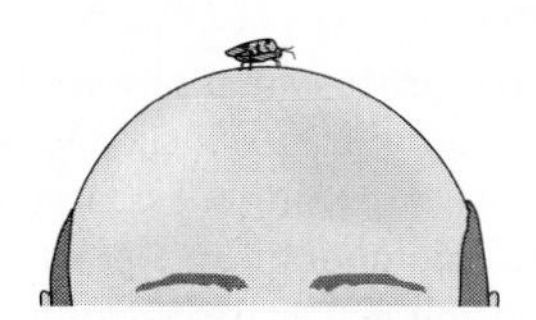

FIGURE 7.24 Problem 60

61. A particle of mass m is subject to a force $\vec{F} = (a\sqrt{x})\,\hat{\imath}$, where a is a constant. The particle is initially at rest at the origin and is given a slight nudge in the positive x-direction. Find an expression for its speed as a function of position x.
62. **CH** A block of weight 4.5 N is launched up a 30° inclined plane 2.0 m long by a spring with $k = 2.0$ kN/m and maximum compression 10 cm. The coefficient of kinetic friction is 0.50. Does the block reach the top of the incline? If so, how much kinetic energy does it have there? If not, how close to the top, along the incline, does it get?
63. Your engineering department is asked to evaluate the performance of a new 370-hp sports car. You know that 27% of the engine's power can be converted to kinetic energy of the 1200-kg car, and that the power delivered is independent of the car's velocity. What do you report for the time it will take to accelerate from rest to 60 mi/h on a level road?
64. Your roommate is writing a science fiction novel and asks your advice about a plot point. Her characters are mining ore on the Moon and launching it toward Earth. Bins with 1500 kg of ore will be launched by a large spring, to be compressed 17 m. It takes a speed of 2.4 km/s to escape the Moon's gravity. What do you tell her is an appropriate spring constant?
65. You have a summer job at your university's zoology department, where you'll be working with an animal behavior expert. She's assigned you to study videos of different animals leaping into the air. Your task is to compare their power outputs as they jump. You'll have the mass m of each animal from data collected in the field. From the videos, you'll be able to measure both the vertical distance d over which the animal accelerates when it pushes off the ground and the maximum height h it reaches. Your task is to find an algebraic expression for power in terms of these parameters.
66. **CH** Biomechanical engineers developing artificial limbs for prosthetic and robotic applications have developed a two-spring design for their replacement Achilles tendon. The first spring has constant k and the second ak, where $a > 1$. When the artificial tendon is stretched from $x = 0$ to $x = x_1$, only the first spring is engaged. For $x > x_1$, a mechanism engages the second spring, giving a configuration like that described in part (a) of Chapter 4's Problem 62. Find an expression for the energy stored in the artificial tendon when it's stretched a distance $2x_1$.
67. **DATA** Blocks with different masses are pushed against a spring one at a time, compressing it different amounts. Each is then launched onto an essentially frictionless horizontal surface that then curves upward, still frictionless (like Fig. 7.21 but without the frictional part). The table below shows the masses, spring compressions, and maximum vertical height each block achieves. Determine a quantity that, when you plot h against it, should yield a straight line. Plot the data, determine a best-fit line, and use its slope to determine the spring constant.

Mass m (g)	50.0	85.2	126	50.0	85.2
Compression x (cm)	2.40	3.17	5.40	4.29	1.83
Height h (cm)	10.3	11.2	19.8	35.2	3.81

Passage Problems

Nuclear fusion is the process that powers the Sun. Fusion occurs when two low-mass atomic nuclei fuse together to make a larger nucleus, in the process releasing substantial energy. This is hard to achieve because atomic nuclei carry positive electric charge, and their electrical repulsion makes it difficult to get them close enough for the short-range nuclear force to bind them into a single nucleus. Figure 7.25 shows the potential-energy curve for fusion of two deuterons (heavy hydrogen nuclei). The energy is measured in million electron volts (MeV), a unit commonly used in nuclear physics, and the separation is in femtometers ($1 \text{ fm} = 10^{-15}$ m).

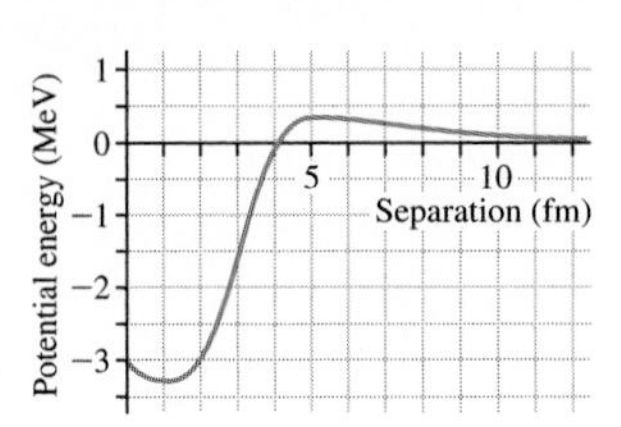

FIGURE 7.25 Potential energy for two deuterons (Passage Problems 68–71)

68. The force between the deuterons is zero at approximately
    a. 3 fm.
    b. 4 fm.
    c. 5 fm.
    d. the force is never zero.
69. In order for initially two widely separated deuterons to get close enough to fuse, their kinetic energy must be about
    a. 0.1 MeV.
    b. 3 MeV.
    c. −3 MeV.
    d. 0.3 MeV.
70. The energy available in fusion is the energy difference between that of widely separated deuterons and the bound deutrons after they've "fallen" into the deep potential well shown in the figure. That energy is about
    a. 0.3 MeV.
    b. 1 MeV.
    c. 3.3 MeV.
    d. 3.6 MeV.
71. When two deuterons are 4 fm apart, the force acting on them
    a. is repulsive.
    b. is attractive.
    c. is zero.
    d. can't be determined from the graph.

Answers to Chapter Questions

Answer to Chapter Opening Question

Potential energy turns into kinetic energy, sound, and internal energy.

Answers to GOT IT? Questions

7.1 (c) On the curved paths, the work is greater for the trunk. The gravitational force is conservative, so the work is independent of path. But the frictional force isn't conservative, and the longer path means more work needs to be done.
7.2 (b) The potential-energy change will be slightly less because at greater heights, the gravitational force is lower and so, therefore, is the work done in traversing a given distance.
7.3 No. Mechanical energy is conserved, so if the ball is released from rest, it cannot climb higher than its initial height.
7.4 (1) (a) only; (2) (a) and (c)
7.5 (a), (c), (e), (f)
7.6 (1) B; (2) E; (3) C; (4) A or D; (5) B or E

8 Gravity

What You Know

- You understand Newton's second law.
- You know how to find the potential energy associated with a conservative force.
- You know that finding potential energy for position-dependent forces requires integration.
- You understand the principle of conservation of mechanical energy.
- You know how to use energy conservation to solve for velocities or positions.
- You know how to describe acceleration in circular motion.

What You're Learning

- You'll see how planetary motion was historically important in the development of Newtonian physics.
- You'll learn Newton's *law of universal gravitation* with its *inverse-square* dependence of gravitational force on distance.
- You'll see how to find potential energy changes over distances so large that gravity varies significantly with position.
- You'll learn to describe circular orbits quantitatively.
- You'll see how space technologies, including communications and GPS satellites, exploit the physics of orbits.
- You'll learn to describe all orbital types in terms of total mechanical energy.
- You'll learn about *escape speed*.
- You'll have a brief introduction to the *gravitational field*.

How You'll Use It

- Whether or not you make practical use of the material you learn in Chapter 8, knowing the physics of our solar system and the broader cosmos will give you a greater appreciation for the universe and your place in it.
- You'll also come away from Chapter 8 with a deeper understanding of technologies like satellite communications, GPS, space-based weather observations, and other Earth-observing systems.
- The concept of *gravitational field* introduced here will be echoed in Part 3, where the *electric field* and the *magnetic field* are introduced.
- In Part 6 you'll see how early models of the atom were based on an electric-force analog of the solar system.

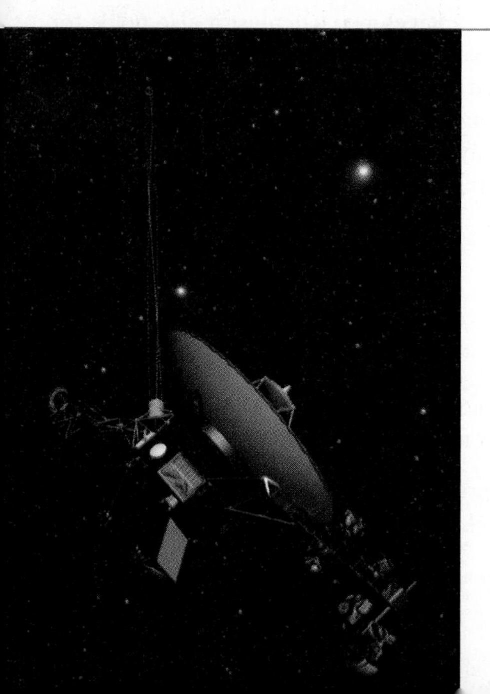

Gravity is the most obvious of nature's fundamental forces. Theories of gravity have brought us new understandings of the nature and evolution of the universe. We've used our knowledge of gravity to explore the solar system and to engineer a host of space-based technologies. In nearly all applications we still use the theory of gravity that Isaac Newton developed in the 1600s. Only in the most extreme astrophysical situations or where—as with Global Positioning System satellites—we need exquisite precision do we use the successor to Newtonian gravitation, namely, Einstein's general theory of relativity.

8.1 Toward a Law of Gravity

Newton's theory of gravity was the culmination of two centuries of scientific revolution that began in 1543 with Polish astronomer Nicolaus Copernicus's radical suggestion that the planets orbit not Earth but the Sun. Fifty years after Copernicus's work was

In 2012, 35 years after launch, Voyager 1 became the first human artifact to leave the solar system. What condition on Voyager's total energy ensures that it will never return to the Sun's vicinity?

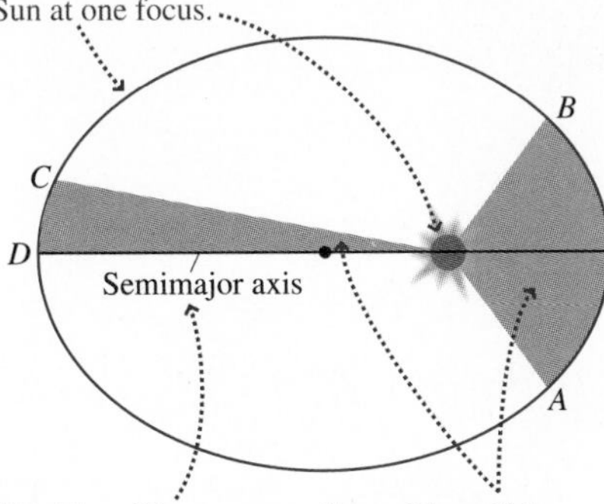

FIGURE 8.1 Kepler's laws.

FIGURE 8.2 Phases of Venus. In an Earth-centered system, Venus would always appear the same size because of its constant distance from Earth.

published, the Danish noble Tycho Brahe began a program of accurate planetary observations. After Tycho's death in 1601, his assistant Johannes Kepler worked to make sense of the observations. Success came when Kepler took a radical step: He gave up the long-standing idea that the planets moved in perfect circles. Kepler summarized his new insights in three laws, described in Fig. 8.1. Kepler based his laws solely on observation and gave no theoretical explanation. So Kepler knew *how* the planets moved, but not *why*.

Shortly after Kepler published his first two laws, Galileo trained his first telescopes on the heavens. Among his discoveries were four moons orbiting Jupiter, sunspots that blemished the supposedly perfect sphere of the Sun, and the phases of Venus (Fig. 8.2). His observations called into question the notion that all celestial objects were perfect and also lent credence to the Copernican view of the Sun as the center of planetary motion.

By Newton's time the intellectual climate was ripe for the culmination of the revolution that had begun with Copernicus. Legend has it that Newton was sitting under an apple tree when an apple struck him on the head, causing him to discover gravity. That story is probably a myth, but if it were true the other half would be that Newton was staring at the Moon when the apple struck. Newton's genius was to recognize that *the motion of the apple and the motion of the Moon were the same, that both were "falling" toward Earth under the influence of the same force*. Newton called this force **gravity**, from the Latin *gravitas*, "heaviness." In one of the most sweeping syntheses in human thought, Newton inferred that everything in the universe, on Earth and in the celestial realm, obeys the same physical laws.

8.2 Universal Gravitation

Newton generalized his new understanding of gravity to suggest that any two particles in the universe exert attractive forces on each other, with magnitude given by

$$F = \frac{Gm_1m_2}{r^2} \quad \text{(universal gravitation)} \tag{8.1}$$

Here m_1 and m_2 are the particle masses, r the distance between them, and G the **constant of universal gravitation**, whose value—which was determined after Newton's time—is $6.67 \times 10^{-11}\ \text{N}\cdot\text{m}^2/\text{kg}^2$. The constant G is truly universal; observation and theory suggest that it has the same value throughout the universe.

The force of gravity acts *between* two particles; that is, m_1 exerts an attractive force on m_2, and m_2 exerts an equal but oppositely directed force on m_1. The two forces therefore obey Newton's third law.

Newton's law of universal gravitation applies strictly only to point particles that have no extent. But, as Newton showed using his newly developed calculus, it also holds for spherically symmetric objects of any size if the distance r is measured from their centers. It also applies approximately to arbitrarily shaped objects provided the distance between them is large compared with their sizes. For example, the gravitational force of Earth on the International Space Station is given accurately by Equation 8.1 because (1) Earth is essentially spherical and (2) the station, though irregular in shape, is vastly smaller than its distance from Earth's center.

EXAMPLE 8.1 The Acceleration of Gravity: On Earth and in Space

Use the law of universal gravitation to find the acceleration of gravity at Earth's surface, at the 380-km altitude of the International Space Station, and on the surface of Mars.

INTERPRET The problem statement tells us this is about universal gravitation, but what's that got to do with the acceleration of gravity? The gravitational force is what causes that acceleration, so we can interpret this problem as being about the force between Earth (or Mars) and some arbitrary mass.

DEVELOP Since the problem involves universal gravitation, Equation 8.1 applies. But we're asked about acceleration, not force. Newton's second law, $F = ma$, relates the two. So our plan is to use Equation 8.1, $F = Gm_1m_2/r^2$, to find the gravitational force on an arbitrary mass and then use Newton's second law to get the acceleration. There's another bit of planning: We need to find the masses of Earth and Mars and their radii. Astrophysical data like these are in Appendix E.

EVALUATE Equation 8.1 gives the force a planet of mass M exerts on an arbitrary mass m a distance r from the planet's center: $F = GMm/r^2$. (Here we set m_1 in Equation 8.1 to the large planetary mass M, and m_2 to the smaller mass m.) But Newton's second law says that this force is equal to the product of mass and acceleration for a body in free fall, so we can write $ma = GMm/r^2$. The mass m cancels, and we're left with the acceleration:

$$a = \frac{GM}{r^2} \tag{8.2}$$

The distance r is measured from the *center* of the object providing the gravitational force, so to find the acceleration at Earth's surface we use R_E, the radius of the Earth, for r. Taking R_E and M_E from Appendix E, we have

$$a = \frac{GM_E}{R_E^2} = \frac{(6.67\times10^{-11}\ \text{N}\cdot\text{m}^2/\text{kg}^2)(5.97\times10^{24}\ \text{kg})}{(6.37\times10^6\ \text{m})^2} = 9.81\ \text{m/s}^2$$

Our result here is the value of g—the acceleration due to gravity at Earth's surface.

At the space station's altitude, we have $r = R_E + 380$ km, so

$$a = \frac{GM_E}{r^2} = \frac{(6.67\times10^{-11}\ \text{N}\cdot\text{m}^2/\text{kg}^2)(5.97\times10^{24}\ \text{kg})}{(6.37\times10^6\ \text{m} + 380\times10^3\ \text{m})^2} = 8.74\ \text{m/s}^2$$

A similar calculation using Appendix E data yields 3.75 m/s^2 for the acceleration of gravity at the surface of Mars.

ASSESS As we've seen, our result for Earth is just what we expect. The acceleration at the space station is lower but still about 90% of the surface value. This confirms Chapter 4's point that weightlessness doesn't mean the absence of gravity. Rather, as Equation 8.2 shows, an object's gravitational acceleration is independent of its mass—so all objects "fall" together. Finally, our answer for Mars is lower than for Earth, as befits its lower mass—although not as much lower as mass alone would imply. That's because Mars is also smaller, so r in the denominator of Equation 8.2 is a smaller number. ■

✓TIP *G and g*

Don't confuse G and g! Both quantities are associated with gravity, but G is a universal constant, while g describes the gravitational acceleration at a particular place—namely, Earth's surface—and its value depends on Earth's size and mass.

The variation of gravitational acceleration with distance from Earth's center provided Newton with a clue that the gravitational force should vary as the inverse square of the distance. Newton knew the Moon's orbital period and distance from Earth; from these he could calculate its orbital speed and thus its acceleration v^2/r. Newton found—as you can in Exercise 12—that the Moon's acceleration is about 1/3600 the gravitational acceleration g at Earth's surface. The Moon is about 60 times farther from Earth's center than is Earth's surface; since $60^2 = 3600$, the decrease in gravitational acceleration with distance from Earth's center is consistent with a gravitational force that varies as $1/r^2$.

PhET: Gravity Force Lab

TACTICS 8.1 Understanding "Inverse Square"

Newton's universal gravitation is the first of several inverse-square force laws you'll encounter, and it's important to understand what this term means. In Equation 8.1 the distance r between the two masses is *squared*, and it occurs in the *denominator*; hence the force depends on the *inverse square* of the distance. Double the distance and the force drops to $1/2^2$, or 1/4 of its original value. Triple the distance and the force drops to $1/3^2$, or 1/9. Although you can always grind through the arithmetic of Equation 8.1, you should use these simple ratio calculations whenever possible. The same considerations apply to gravitational acceleration, since it's proportional to force (Fig. 8.3).

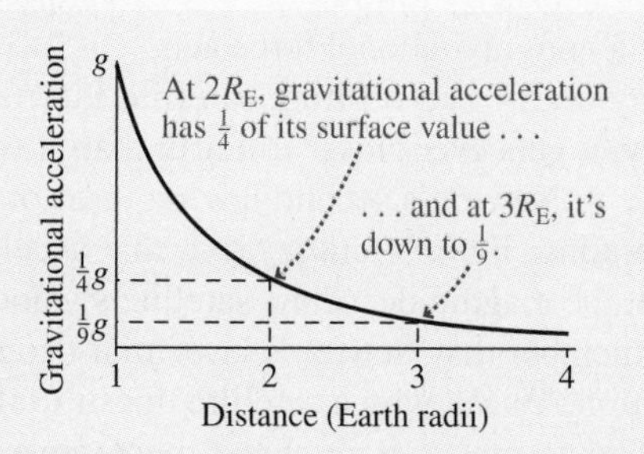

FIGURE 8.3 Meaning of the inverse-square law.

GOT IT? 8.1 Suppose the distance between two objects is cut in half. Is the gravitational force between them (a) quartered, (b) halved, (c) doubled, or (d) quadrupled?

The Cavendish Experiment: Weighing the Earth

Given Earth's mass and radius and the measured value of g, we could use Equation 8.1 to determine the universal constant G. Unfortunately, the only way to determine Earth's mass accurately is to measure its gravitational effect and then use Equation 8.1. But that requires knowing G.

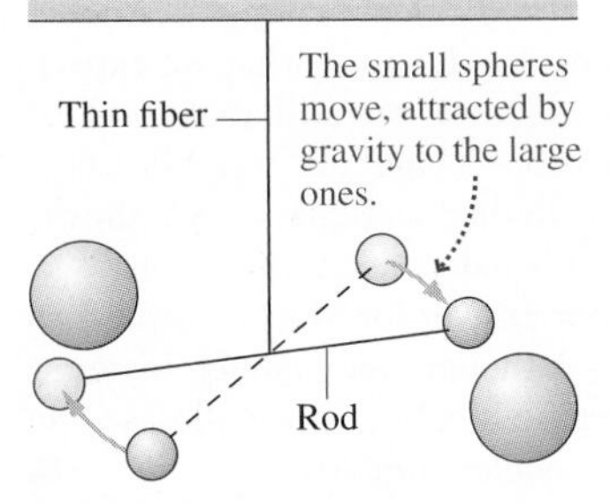

FIGURE 8.4 The Cavendish experiment to determine G.

To determine G, we need to measure the gravitational force of a *known* mass. Given the weak gravitational force of normal-size objects, this is a challenging task. It was accomplished in 1798 through an ingenious experiment by the British physicist Henry Cavendish. Cavendish mounted two 5-cm-diameter lead spheres on the ends of a rod suspended from a thin fiber. He then brought two 30-cm lead spheres nearby (Fig. 8.4). Their gravitational attraction caused a slight movement of the small spheres, twisting the fiber. Knowing the properties of the fiber, Cavendish could determine the force. With the known masses and their separation, he then used Equation 8.1 to calculate G. His result determined Earth's mass; indeed, his published paper was entitled "On Weighing the Earth."

Gravity is the weakest of the fundamental forces, and, as the Cavendish experiment suggests, the gravitational force between everyday objects is negligible. Yet gravity shapes the large-scale structure of matter and indeed the entire universe. Why, if it's so weak? The answer is that gravity, unlike the stronger electric force, is always attractive; there's no "negative mass." So large concentrations of matter produce substantial gravitational effects. Electric charge, in contrast, can be positive or negative, and electric effects in normal-sized objects tend to cancel. We'll explore this distinction further in Chapter 20.

8.3 Orbital Motion

PhET: My Solar System
PhET: Gravity and Orbits

Orbital motion occurs when gravity is the dominant force acting on a body. It's not just planets and spacecraft that are in orbit. An individual astronaut floating outside the space station is orbiting Earth. The Sun itself orbits the center of the galaxy, taking about 200 million years to complete one revolution. If we neglect air resistance, even a baseball is temporarily in orbit. Here we discuss quantitatively the special case of circular orbits; then we describe qualitatively the general case.

Newton's genius was to recognize that the Moon is held in its circular orbit by the same force that pulls an apple to the ground. From there, it was a short step for Newton to realize that human-made objects could be put into orbit. Nearly 300 years before the first artificial satellites, he imagined a projectile launched horizontally from a high mountain (Fig. 8.5). The projectile moves in a curve, as gravity pulls it from the straight-line path it would follow if no force were acting. As its initial speed is increased, the projectile travels farther before striking Earth. Finally, there comes a speed for which the projectile's path bends in a way that exactly follows Earth's curvature. It's then in **circular orbit**, continuing forever unless a nongravitational force acts.

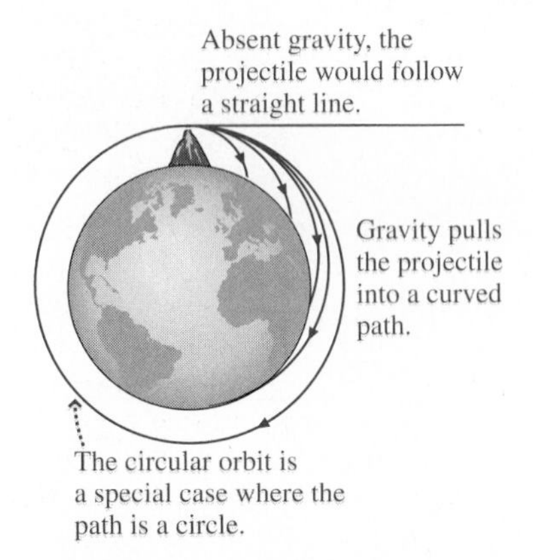

FIGURE 8.5 Newton's "thought experiment" showing that projectile and orbital motions are essentially the same.

Why doesn't an orbiting satellite fall toward Earth? It does! Under the influence of gravity, it gets ever closer to Earth than it would be on a straight-line path. It's behaving exactly as Newton's second law requires of an object under the influence of a force—by accelerating. For a *circular* orbit, that acceleration amounts to a change in the direction, but not the magnitude, of the satellite's velocity.

Remember that Newton's laws aren't so much about *motion* as they are about *changes* in motion. To ask why a satellite doesn't fall to Earth is to adopt the archaic Aristotelian view and assume that an object must move in the direction of the force acting on it. The correct Newtonian question, in contrast, is based on the idea that motion *changes* in response to a force: Why doesn't the satellite move in a straight line? And the answer is simple: because a force is acting. That force—gravity—is exactly analogous to the tension force that keeps a ball on a string whirling in its circular path.

We can analyze circular orbits quantitatively because we know that a force of magnitude mv^2/r is required to keep an object of mass m and speed v in a circular path of radius r. In the case of an orbit, that force is gravity, so we have

$$\frac{GMm}{r^2} = \frac{mv^2}{r}$$

where m is the mass of the orbiting object and M the mass of the object about which it's orbiting. We assume here that $M \gg m$, so the gravitating object can be considered essentially at rest—a reasonable approximation with Earth satellites or planets orbiting the much more massive Sun. Solving for the orbital speed gives

$$v = \sqrt{\frac{GM}{r}} \quad \text{(speed, circular orbit)} \qquad (8.3)$$

Often we're interested in the **orbital period**, or the time to complete one orbit. In one period T, the orbiting object moves the orbital circumference $2\pi r$, so its speed is $v = 2\pi r/T$. Squaring Equation 8.3 then gives

$$\left(\frac{2\pi r}{T}\right)^2 = \frac{GM}{r}$$

or

$$T^2 = \frac{4\pi^2 r^3}{GM} \quad \text{(orbital period, circular orbit)} \qquad (8.4)$$

In deriving Equation 8.4, we've proved Kepler's third law—that the square of the orbital period is proportional to the cube of the semimajor axis—for the special case of a circular orbit, whose semimajor axis is identical to its radius.

Note that orbital speed and period are independent of the orbiting object's mass m—another indication that all objects experience the same gravitational acceleration. Astronauts, for example, have the same orbital parameters as the space station. That's why astronauts seem weightless inside the station and why they don't float away if they step outside.

EXAMPLE 8.2 Orbital Speed and Period: The Space Station

The International Space Station is in a circular orbit at altitude 380 km. What are its orbital speed and period?

INTERPRET This problem involves the speed and period of a circular orbit about Earth.

DEVELOP We can compute the orbit's radius and then use Equation 8.3, $v = \sqrt{GM/r}$, to find the speed and Equation 8.4, $T^2 = 4\pi^2 r^3/GM$, to find the period because the orbit is circular.

EVALUATE As always, the distance is measured from the center of the gravitating body, so r in these equations is Earth's 6.37-Mm radius plus the station's 380-km altitude. So we have

$$v = \sqrt{\frac{GM_E}{r}} = \sqrt{\frac{(6.67\times10^{-11}\ \text{N}\cdot\text{m}^2/\text{kg}^2)(5.97\times10^{24}\ \text{kg})}{6.37\times10^6\ \text{m} + 380\times10^3\ \text{m}}}$$
$$= 7.68\ \text{km/s}$$

or about 17,000 mi/h. We can get the orbital period from the speed and radius, or directly from Equation 8.4, $T = \sqrt{4\pi^2 r^3/GM_E}$. Using the numbers in the calculation for v gives $T = 5.52\times10^3$ s, or about 90 min.

ASSESS Make sense? Both answers have the correct units, and 90 min seems reasonable for the period of an orbit at a small fraction of the Moon's distance from Earth. Astronauts who want a circular orbit 380 km up have no choice but this speed and period. In fact, for any "near-Earth" orbit, with altitude much less than Earth's radius, the orbital period is about 90 min. If there were no air resistance and if you could throw a baseball fast enough, it too would go into orbit, skimming Earth's surface with a roughly 90-min period. ■

Example 8.2 shows that the near-Earth orbital period is about 90 min. The Moon, on the other hand, takes 27 days to complete its nearly circular orbit. So there must be a distance where the orbital period is 24 h—the same as Earth's rotation. A satellite at this distance will remain fixed with respect to Earth's surface provided its orbit is parallel to the equator. TV, weather, and communication satellites are often placed in such a **geostationary orbit**.

EXAMPLE 8.3 Geostationary Orbit: Finding the Altitude

What altitude is required for geostationary orbit?

INTERPRET Here we're given an orbital period—24 h or 86,400 s—and asked to find the corresponding altitude for a circular orbit.

DEVELOP Equation 8.4, $T^2 = 4\pi^2 r^3/GM$, relates the period T and distance r from Earth's center. Our plan is to solve for r and then subtract Earth's radius to find the altitude (distance from the surface).

EVALUATE Solving for r, we get

$$r = \left(\frac{GM_E T^2}{4\pi^2}\right)^{1/3}$$

$$= \left[\frac{(6.67\times10^{-11}\ \mathrm{N{\cdot}m^2/kg^2})(5.97\times10^{24}\ \mathrm{kg})(8.64\times10^4\ \mathrm{s})^2}{4\pi^2}\right]^{1/3}$$

$$= 4.22\times10^7\ \mathrm{m}$$

or 42,200 km from Earth's center. Subtracting Earth's radius then gives an altitude of about 36,000 km, or 22,000 miles.

ASSESS Make sense? This is a lot higher than the 90-min low-Earth orbit, but a lot lower than the Moon's 384,000 km distance. Our answer defines one of the most valuable pieces of "real estate" in space—a place where satellites appear suspended over a fixed spot on Earth. Every TV dish antenna points to such a satellite, positioned 22,000 mi over the equator. A more careful calculation would use Earth's so-called sidereal rotation period, measured with respect to the distant stars rather than the Sun. Because Earth isn't a perfect sphere, geostationary satellites drift slightly, and therefore they need to fire small rockets every few weeks to stay in position. ■

FIGURE 8.6 Orbits of most known comets, like the one shown here, are highly elliptical.

Elliptical Orbits

Using his laws of motion and gravity, Newton was able to prove Kepler's assertion that the planets move in elliptical paths with the Sun at one focus. Circular orbits represent the special case where the two foci of the ellipse coincide, so the distance from the gravitating center remains constant. Most planetary orbits are nearly, but not quite, circular; Earth's distance from the Sun, for exam ple, varies by about 3% throughout the year. But the orbits of comets and other smaller bodies are often highly elliptical (Fig. 8.6). Their orbital speeds vary, as they gain speed "falling" toward the Sun, whip quickly around the Sun at the point of closest approach (**perihelion**), and then "climb" ever more slowly to their most distant point (**aphelion**) before returning.

In Chapter 3, we showed that the trajectory of a projectile is a parabola. But our derivation neglected Earth's curvature and the associated variation in g with altitude. In fact, a projectile is just like any orbiting body. If we neglect air resistance, it too describes an elliptical orbit with Earth's center at one focus. Only for trajectories whose height and range are small compared with Earth's radius are the true elliptical path and the parabola of Chapter 3 generally indistinguishable (Fig. 8.7).

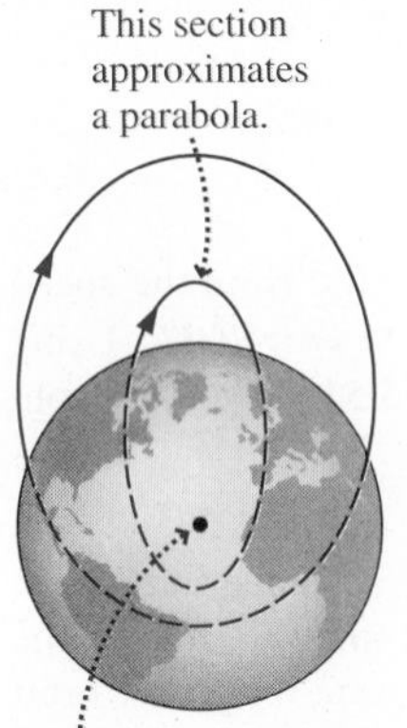

FIGURE 8.7 Projectile trajectories are actually elliptical.

Are missiles and baseballs really in orbit? Yes. But their orbits happen to intersect Earth's surface. At that point, nongravitational forces put an end to orbital motion. If Earth suddenly shrank to the size of a grapefruit (but kept the same mass), a baseball would continue happily in orbit, as the dashed continuation of the smaller orbit in Fig. 8.7 suggests. Newton's ingenious intuition was correct: Barring air resistance, there's truly no difference between the motion of everyday objects near Earth and the motion of celestial objects.

Open Orbits

With elliptical and circular orbits, the motion repeats indefinitely because the orbit is a closed path. But closed orbits aren't the only possibility. Imagine again Newton's thought experiment—only now fire the projectile faster than necessary for a circular orbit (Fig. 8.8). The projectile goes farther from Earth than before, describing an ellipse that's closest to Earth at the launch site. Faster, and the ellipse gets more elongated. But with great enough initial speed, the projectile describes a hyperbolic trajectory that takes it ever farther from Earth. We'll see in the next section how energy determines the type of orbit.

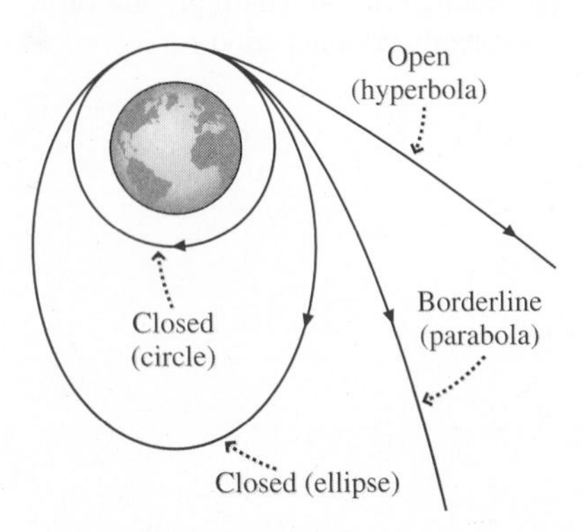

FIGURE 8.8 Closed and open orbits.

GOT IT? 8.2 Suppose the paths in Fig. 8.8 are the paths of four projectiles. Rank each path (circular, elliptical, parabolic, and hyperbolic) according to the initial speed of the corresponding projectile. Assume all are launched from their common point at the top of the figure.

8.4 Gravitational Energy

How much energy does it take to boost a satellite to geostationary altitude? Our simple answer mgh won't do here, since g varies substantially over the distance involved. So, as we found in Chapter 7, we have to integrate to determine the potential energy.

Figure 8.9 shows two points at distances r_1 and r_2 from the center of a gravitating mass M, in this case Earth. Equation 7.2 gives the change in potential energy associated with moving a mass m from r_1 to r_2:

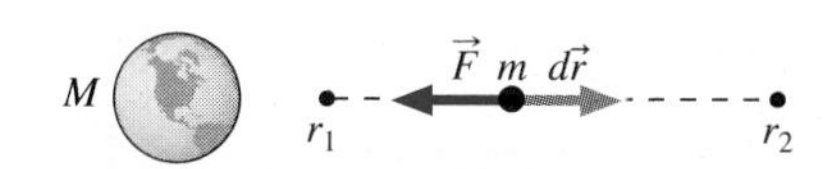

FIGURE 8.9 Finding the potential-energy change requires integration.

$$\Delta U_{12} = -\int_{r_1}^{r_2} \vec{F} \cdot d\vec{r}$$

Here the force points radially inward and has magnitude GMm/r^2, while the path element $d\vec{r}$ points radially outward. Then $\vec{F} \cdot d\vec{r} = -(GMm/r^2)\, dr$, where the minus sign comes from the factor cos 180° in the dot product of oppositely directed vectors. This minus sign cancels the minus sign in the expression above for ΔU_{12}, so here the potential energy difference becomes

$$\Delta U_{12} = \int_{r_1}^{r_2} \frac{GMm}{r^2}\, dr = GMm \int_{r_1}^{r_2} r^{-2}\, dr = GMm \left.\frac{r^{-1}}{-1}\right|_{r_1}^{r_2} = GMm\left(\frac{1}{r_1} - \frac{1}{r_2}\right) \quad (8.5)$$

Does this make sense? Yes: For $r_1 < r_2$, ΔU_{12} is positive, showing that potential energy increases with height—consistent with our simpler result $\Delta U = mgh$ near Earth's surface. Although we derived Equation 8.5 for two points on a radial line, Fig. 8.10 shows that it holds for any two points at distances r_1 and r_2 from the gravitating center.

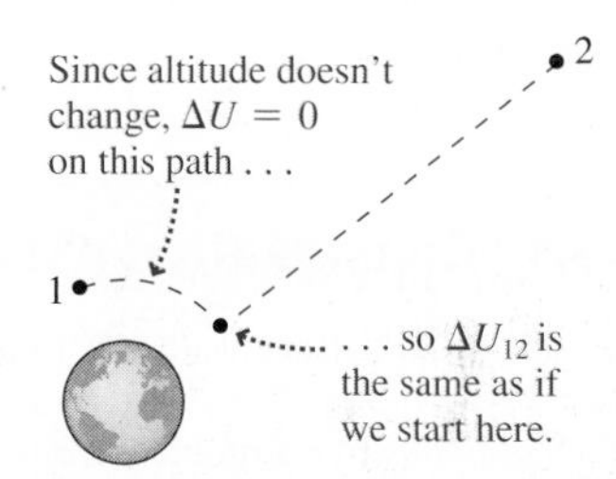

FIGURE 8.10 Gravity is conservative, so we can use any path to evaluate the potential-energy change. Only the radial part of the path contributes to ΔU.

EXAMPLE 8.4 Gravitational Potential Energy: Steps to the Moon

Materials to construct an 11,000-kg lunar observatory are boosted from Earth to geostationary orbit. There they are assembled and then launched to the Moon, 384,400 km from Earth. Compare the work that must be done against Earth's gravity on the two legs of the trip.

INTERPRET This problem asks about work done against gravity, a conservative force.

DEVELOP As we saw in Chapter 7, the work done against a conservative force is equal to the change in potential energy; here that change is given by Equation 8.5. For the first leg, we have $r_1 = R_E$ and then, from Example 8.3, $r_2 = 42{,}200$ km.

EVALUATE Since the quantity GM_Em that appears in Equation 8.5 will be used in both steps, we calculate it first: $GM_Em = 4.383 \times 10^{18}\ \text{N}\cdot\text{m}^2$. Then for the first step, we have

$$\begin{aligned} W = \Delta U_{12} &= GM_Em\left(\frac{1}{r_1} - \frac{1}{r_2}\right) \\ &= (4.383 \times 10^{18}\ \text{N}\cdot\text{m}^2)\left(\frac{1}{6.37 \times 10^6\ \text{m}} - \frac{1}{4.22 \times 10^7\ \text{m}}\right) \\ &= 5.842 \times 10^{11}\ \text{J} \end{aligned}$$

From geostationary orbit to the Moon, a similar calculation gives

$$\begin{aligned} W &= (4.383 \times 10^{18}\ \text{N}\cdot\text{m}^2)\left(\frac{1}{4.22 \times 10^7\ \text{m}} - \frac{1}{3.844 \times 10^8\ \text{m}}\right) \\ &= 9.25 \times 10^{10}\ \text{J} \end{aligned}$$

ASSESS Make sense? Even though the second leg is much longer, the rapid drop-off in the gravitational force means that less work is required than for the shorter boost to geostationary altitude. Our calculations here include only the work done against Earth's gravity; additional energy would be required to attain a circular geostationary orbit. On the other hand, the Moon's gravitational attraction would lower the required energy somewhat. ■

The Zero of Potential Energy

Equation 8.5 has an interesting feature: The potential-energy difference remains finite even when the points are infinitely far apart, as you can see by setting either r_1 or r_2 to infinity.

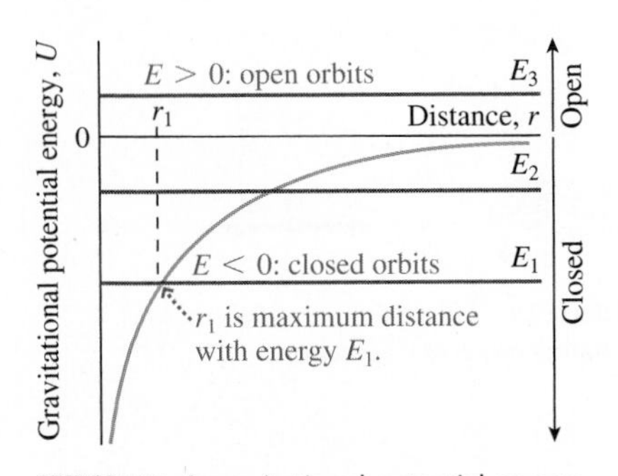

FIGURE 8.11 A gravitational potential-energy curve. Distance is measured from the center of a gravitating object like a star or planet. Closed orbits, which occur when total energy E is less than 0, are elliptical or circular; orbits with $E > 0$ are hyperbolas. The intermediate case $E = 0$ gives parabolic orbits.

Although the gravitational force always acts, it weakens so rapidly that its effect is finite over even infinite distances. This property makes it convenient to set the zero of potential energy when the two gravitating masses M and m are infinitely far apart. Setting $r_1 = \infty$ and dropping the subscript on r_2, we then have an expression for the gravitational potential energy of a system comprising a mass m located a distance r from the center of another mass M.

$$U(r) = -\frac{GMm}{r} \quad \text{(gravitational potential energy)} \tag{8.6}$$

The potential energy is negative because we chose $U = 0$ when $r = \infty$. When the two masses are closer than infinitely far apart, the system has lower—hence negative—potential energy.

Knowing the gravitational potential energy allows us to apply the powerful conservation-of-energy principle. Figure 8.11 shows the potential-energy curve given by Equation 8.6. Superposing three values of total energy E shows that orbits with $E < 0$ have a turning point where they intersect the potential-energy curve, and are therefore closed. Orbits with $E > 0$, in contrast, are open because they never intersect the curve and therefore extend to infinity.

EXAMPLE 8.5 Conservation of Energy: Blast Off!

A rocket is launched vertically upward at 3.1 km/s. How high does it go?

INTERPRET This sounds like a problem from Chapter 2, but here we'll see that the rocket rises high enough that we can't ignore the variation in gravity. So the acceleration isn't constant, and we can't use the constant-acceleration equations of Chapter 2. But the conservation-of-mechanical-energy principle lets us cut through those details, so we can apply the methods of Chapter 7. "How high does it go?" in the problem statement means we're dealing with the initial launch state and a final state where the rocket is momentarily at rest at the top of its trajectory.

DEVELOP Equation 7.6 describes conservation of mechanical energy: $K + U = K_0 + U_0$. Here we're given speed v at the bottom, so $K_0 = \frac{1}{2}mv^2$. We're going to be using Equation 8.6, $U(r) = -GMm/r$, for potential energy, and that's already established the zero of potential energy at infinity. So U_0 isn't zero but is given by Equation 8.6 with r equal to Earth's radius. Finally, at the top, $K = 0$ and U is also given by Equation 8.6, but now we don't know r. Our plan is to solve for that r and from it get the rocket's altitude. Figure 8.12 shows "before" and "after" diagrams with bar graphs, like those we introduced in Chapter 7.

EVALUATE With our values for the kinetic and potential energies, the equation $K + U = K_0 + U_0$ becomes

$$-\frac{GM_Em}{r} = \frac{1}{2}mv_0^2 - \frac{GM_Em}{R_E}$$

where m is the rocket's mass, r is the distance from Earth's center at the peak, and Earth's radius R_E is the distance at launch. Solving for r gives

$$r = \left(\frac{1}{R_E} - \frac{v_0^2}{2GM_E}\right)^{-1}$$

$$= \left(\frac{1}{6.37\times10^6\ \text{m}} - \frac{(3100\ \text{m/s})^2}{2(6.67\times10^{-11}\ \text{N}\cdot\text{m}^2/\text{kg}^2)(5.97\times10^{24}\ \text{kg})}\right)^{-1}$$

$$= 6.90\ \text{Mm}$$

Again, this is the distance from Earth's center; subtracting Earth's radius then gives a peak altitude of 530 km.

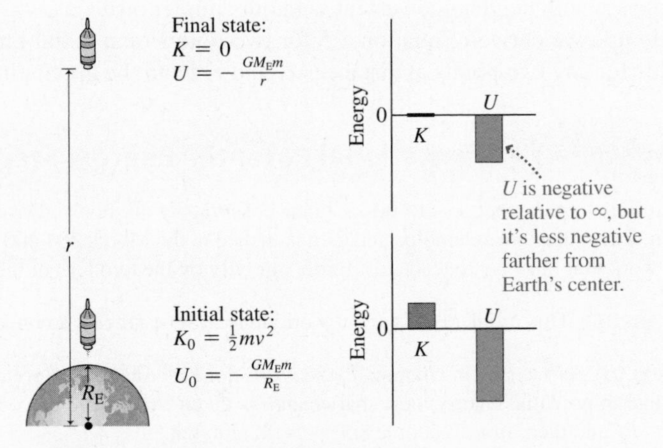

FIGURE 8.12 Diagrams for Example 8.5.

ASSESS Make sense? Yes. Our answer of 530 km is significantly greater than the 490 km you'd get assuming a potential-energy change of $\Delta U = mgh$. That's because the decreasing gravitational force lets the rocket go higher before all its kinetic energy becomes potential energy.

✓TIP All Conservation-of-Energy Problems Are the Same

This problem is essentially the same as throwing a ball straight up and solving for its maximum height using $U = mgh$ for potential energy. The only difference is the more complicated potential-energy function $U = -GMm/r$, used here because the variation in gravity is significant over the rocket's trajectory. Recognize what's common to all similar problems, and you'll begin to see how physics really is based on just a few simple principles.

Escape Speed

What goes up comes down, right? Not always! Figure 8.11 shows that when total energy is zero or greater, an object can escape infinitely far from a gravitating body, never to return. Consider an object of mass m at the surface of a gravitating body of mass M and radius r. The gravitational potential energy is given by Equation 8.6, $U = -GMm/r$. Toss the object upward with speed v, and there's also kinetic energy $\frac{1}{2}mv^2$. The total energy will be zero if

$$0 = K + U = \tfrac{1}{2}mv^2 - \frac{GMm}{r}$$

The speed v here that makes the total energy zero is called the **escape speed** because an object with this speed or greater has enough energy to escape forever from the gravitating body. Solving for v in the preceding equation gives the escape speed:

$$v_{esc} = \sqrt{\frac{2GM}{r}} \quad \text{(escape speed)} \tag{8.7}$$

At Earth's surface, $v_{esc} = 11.2$ km/s. Earth-orbiting spacecraft have lower speeds. Moon-bound astronauts go at just under v_{esc}, so if anything goes wrong (as with *Apollo 13*), they can return to Earth. Planetary spacecraft have speeds greater than v_{esc}. The *Pioneer* and *Voyager* missions to the outer planets gained enough additional energy in their encounters with Jupiter that they now have escape speed relative to the Sun and will coast indefinitely through our galaxy. In 2012, *Voyager 1* became the first human-made object to leave the Sun's realm entirely, as it escaped the "bubble" created by the Sun's magnetic field and entered interstellar space (see Fig. 8.13 and this chapter's opening image).

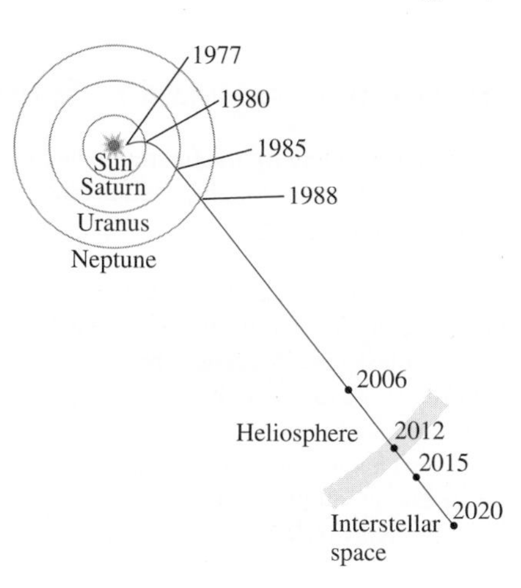

FIGURE 8.13 In 2012 *Voyager 1* crossed into interstellar space. *Voyager* should continue sending data to Earth until about 2020.

Energy in Circular Orbits

In the special case of a circular orbit, kinetic and potential energies are related in a simple way. In Section 8.3, we found that the speed in a circular orbit is given by

$$v^2 = \frac{GM}{r}$$

where r is the distance from a gravitating center of mass M. So the kinetic energy of an object in circular orbit is

$$K = \tfrac{1}{2}mv^2 = \frac{GMm}{2r}$$

while potential energy is given by Equation 8.6:

$$U = -\frac{GMm}{r}$$

Comparing these two expressions shows that $U = -2K$ for a circular orbit. The total energy is therefore

$$E = U + K = -2K + K = -K \tag{8.8a}$$

or, equivalently,

$$E = \tfrac{1}{2}U = -\frac{GMm}{2r} \tag{8.8b}$$

The total energy in these equations is negative, showing that circular orbits are bound orbits. We stress that these results apply only to *circular* orbits; in elliptical orbits, there's a continuous interchange between kinetic and potential energy as the orbiting object moves relative to the gravitating center.

Equation 8.8a shows that *higher* kinetic energy corresponds to *lower* total energy. This surprising result occurs because *higher* orbital speed corresponds to a *lower* orbit, with lower potential energy.

APPLICATION Close Encounters

In 2013, Earth experienced two unusually close asteroid encounters. Although unrelated, they occurred within only 16 hours of each other. The larger of the two asteroids, dubbed 2012 DA_{14}, passed within 35,000 km of Earth—less than one-tenth of the Earth–Moon distance and closer than geostationary satellites. With a mass of some 40 kt (kilotonnes) and speed of 12.7 km/s relative to Earth (29.9 km/s relative to the Sun), this one could have caused major damage had it struck Earth. Since 12.7 km/s is above Earth's escape speed, 2012 DA_{14} could not be orbiting Earth. But, as you can show in Problem 64, its total energy relative to the Sun is negative, putting it in a bound solar orbit. We can expect another close approach of 2012 DA_{14} in the year 2123. Sixteen hours before 2012 DA_{14}'s closest approach in 2013, a 12-kt asteroid entered Earth's atmosphere over Siberia, moving at 19.0 km/s relative to Earth and 35.5 km/s relative to the Sun. It underwent a series of explosive fragmentations at altitudes ranging from 45 to 30 km and then disintegrated into small pieces at 22 km altitude (see photo). Shock waves from these explosions caused significant damage in the Russian city of Chelyabinsk and injured some 1600 people. The Chelyabinsk asteroid was the largest object to enter Earth's atmosphere in over 100 years. As Problems 45 and 64 show, it too was in a bound orbit about the Sun before its demise in Earth's atmosphere.

CONCEPTUAL EXAMPLE 8.1 Space Maneuvers

Astronauts heading for the International Space Station find themselves in the right circular orbit, but well behind the station. How should they maneuver to catch up?

EVALUATE To catch up, the astronauts will have to go faster than the space station. That means increasing their kinetic energy—and, as we've just seen, that corresponds to *lowering* their total energy. So they'll need to drop into a lower orbit.

Figure 8.14 shows the catch-up sequence. The astronauts fire their rocket backward, decreasing their energy and dropping briefly into a lower-energy elliptical orbit. They then fire their rocket to circularize the orbit. Now they're in a lower-energy but faster orbit than the space station. When they're correctly positioned, they fire their rocket to boost themselves into a higher-energy elliptical orbit, then fire again to circularize that orbit in the vicinity of the station.

ASSESS Our solution sounds counterintuitive—as if a car, to speed up, had to apply its brakes. But that's what's needed here, thanks to the interplay between kinetic and potential energy in circular orbits.

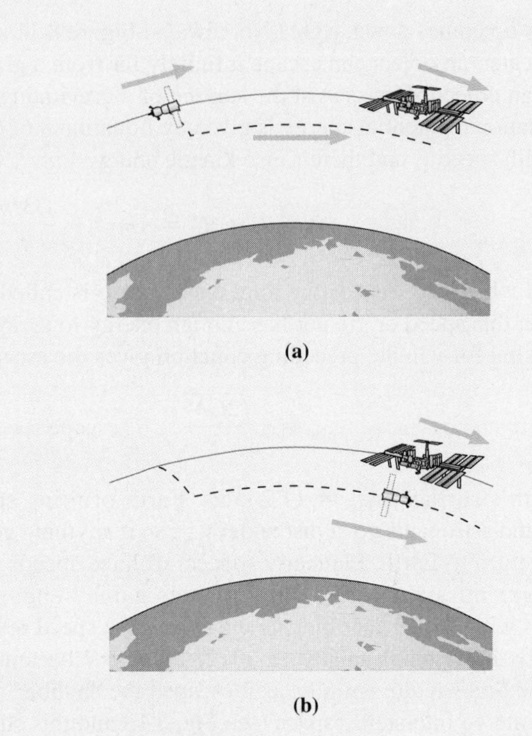

FIGURE 8.14 Playing catch-up with the space station.

MAKING THE CONNECTION Suppose the astronauts reach the space station's 380-km altitude, but find themselves one-fourth of an orbit behind the station. If the maneuver described above drops their spacecraft into a 320-km circular orbit, how many orbits must they make before catching up with the station? Neglect the time involved in transferring between circular orbits.

EVALUATE Applying Equation 8.4 gives periods $T_1 = 92.0$ min for the space station and $T_2 = 90.8$ min for the astronauts in their lower orbit. So with each orbit the astronauts gain 1.2 min on the station. They've got to make up one-fourth of an orbit, or 23 min. That will take $(23\text{ min})/(1.2\text{ min/orbit}) = 19$ orbits, or just over a day.

GOT IT? 8.3 Two identical spacecraft A and B are in circular orbits about Earth, with B at a higher altitude. Which of the following statements are true? (a) B has greater total energy; (b) B is moving faster; (c) B takes longer to complete its orbit; (d) B has greater potential energy; (e) a larger proportion of B's total energy is potential energy

---

8.5 The Gravitational Field

Our description of gravity so far suggests that a massive body like Earth somehow "reaches out" across empty space to pull on objects like falling apples, satellites, or the Moon. This view—called **action-at-a-distance**—has bothered both physicists and philosophers for centuries. How can the Moon, for example, "know" about the presence of the distant Earth?

An alternative view holds that Earth creates a **gravitational field** and that objects respond to the field in their immediate vicinity. The field is described by vectors that give the force per unit mass that would arise at each point if a mass were placed there. Near Earth's surface, for instance, the gravitational field vectors point vertically downward and have magnitude 9.8 N/kg. We can express this field vectorially by writing

$$\vec{g} = -g\hat{j} \quad \text{(gravitational field near Earth's surface)} \qquad (8.9)$$

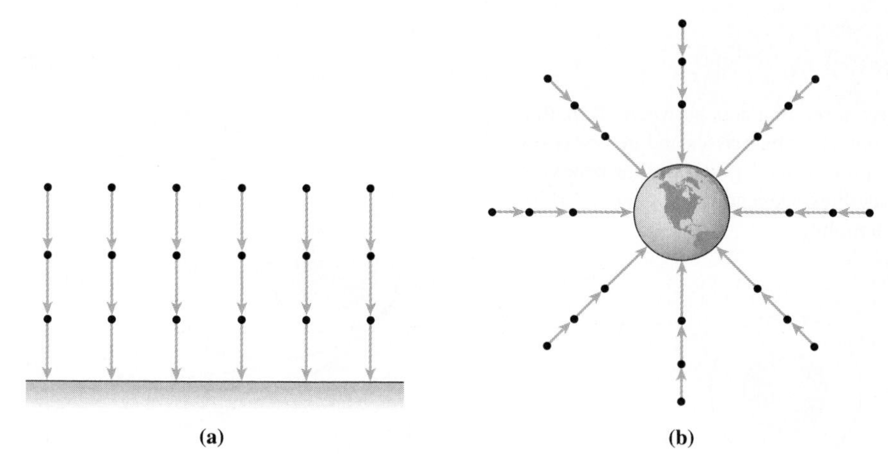

FIGURE 8.15 Gravitational field vectors at points (a) near Earth's surface and (b) on a larger scale.

where we've assumed a coordinate system with the y-axis upward. More generally, the field points toward a spherical gravitating center, and its strength decreases inversely with the square of the distance:

$$\vec{g} = -\frac{GM}{r^2}\hat{r} \qquad \text{(gravitational field of a spherical mass } M\text{)} \qquad (8.10)$$

where $\hat{r}$ is a unit vector that points radially outward. Figure 8.15 shows pictorial representations of Equations 8.9 and 8.10. You can show that the units of gravitational field (N/kg) are equivalent to those of acceleration (m/s^2), so the field is really just a vectorial representation of g, the local acceleration of gravity.

What do we gain by this field description? As long as we deal with situations where nothing changes, the action-at-a-distance and field descriptions are equivalent. But what if, for example, Earth suddenly gains mass? How does the Moon know to adjust its orbit? Under the field view, its orbit doesn't change immediately; instead, it takes a small but nonzero time for the information about the more massive Earth to propagate out to the Moon. The Moon always responds to the gravitational field *in its immediate vicinity*, and it takes a short time for the field itself to change. That description is consistent with Einstein's notion that instantaneous transmission of information is impossible; the action-at-a-distance view is not.

More generally, the field view provides a powerful way of describing interactions in physics. We'll see fields again when we study electricity and magnetism, and you'll find that fields aren't just mathematical or philosophical conveniences but are every bit as real as matter itself.

APPLICATION Tides

If the gravitational field were uniform, all parts of a freely falling object would experience exactly the same acceleration. But gravity does vary, and the result is a force—not from gravity itself but from changes in gravity with position—that tends to stretch or compress an object. Ocean tides result from this **tidal force**, as the nonuniform gravitational forces of Sun and Moon stretch the oceans and create bulges that move across Earth as the planet rotates. The figure shows that the greatest force is on the ocean nearest the Moon, causing one tidal bulge. The solid Earth experiences an intermediate force, pulling it away from the ocean on the far side. The water that's "left behind" forms a second bulge opposite the Moon. The bulges shown are highly exaggerated. Furthermore, shoreline effects and the differing relative positions of the Moon and Sun complicate this simple picture that suggests two equal high tides and two equal low tides a day. Tidal forces also cause internal heating of satellites like Jupiter's moon Io and contribute to the formation of planetary rings.

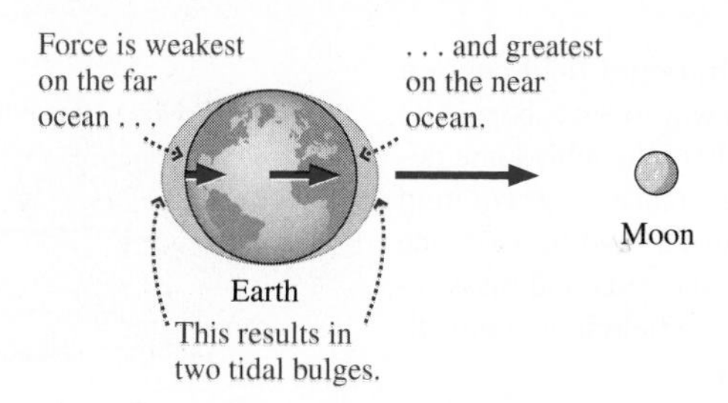

CHAPTER 8 SUMMARY

Big Idea

The big idea here is **universal gravitation**—an attractive force that acts between all matter with a strength that depends directly on the product of two interacting masses and inversely on the square of the distance between them. Gravitation is responsible for the familiar behavior of falling objects and also for the orbits of planets and satellites. Depending on energy, orbits may be closed (elliptical/circular) or open (hyperbolic/parabolic).

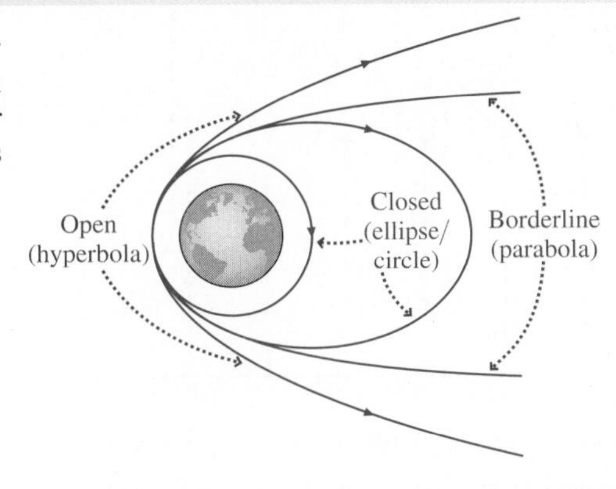

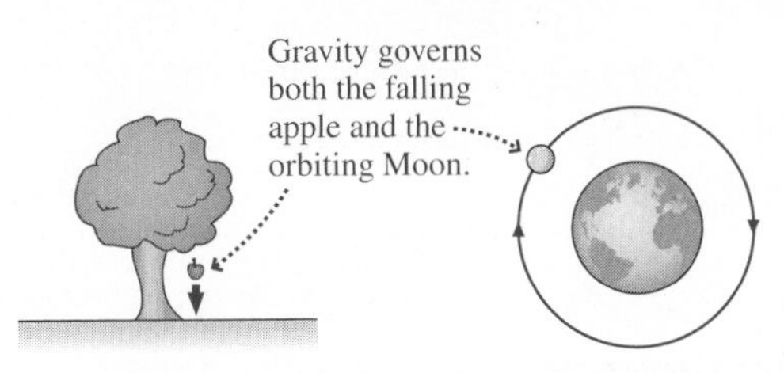

Key Concepts and Equations

Mathematically, Newton's law of universal gravitation describes the attractive force F between two masses m_1 and m_2 located a distance r apart:

$$F = \frac{Gm_1m_2}{r^2} \quad \text{(universal gravitation)}$$

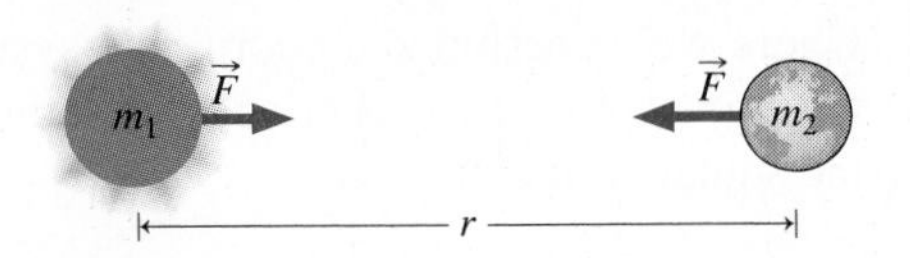

This equation applies to point masses of negligible size and to spherically symmetric masses of any size. It's an excellent approximation for any objects whose size is much smaller than their separation. In all cases, r is measured from the centers of the gravitating objects.

Because the strength of gravity varies with distance, potential-energy changes over large distances aren't just a product of force and distance. Integration shows that the potential-energy change ΔU involved in moving a mass m originally a distance r_1 from the center of a mass M to a distance r_2 is

$$\Delta U = GMm\left(\frac{1}{r_1} - \frac{1}{r_2}\right) \quad \text{(change in potential energy)}$$

With gravity, it's convenient to choose the zero of potential energy at infinity; then

$$U = -\frac{GMm}{r} \quad \text{(potential energy, } U = 0 \text{ at infinity)}$$

for the potential energy of a system comprising a mass m located a distance r from the center of another mass M.

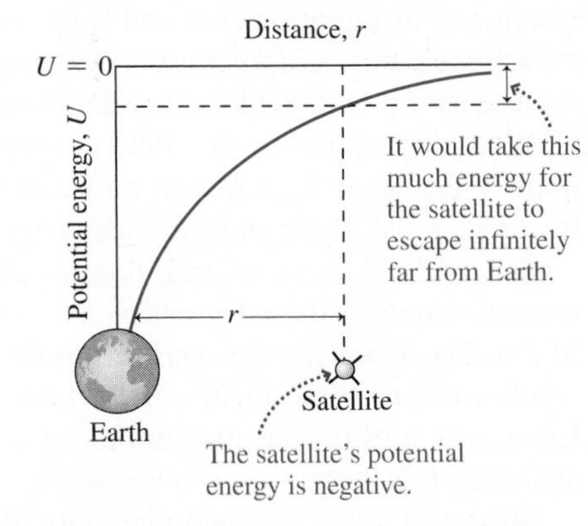

Applications

A total energy—kinetic plus potential—of zero marks the dividing line between closed and open orbits. An object located a distance r from a gravitating mass M must have at least the **escape speed** to achieve an open orbit and escape M's vicinity forever:

$$v_{esc} = \sqrt{\frac{2GM}{r}}$$

The **gravitational field** concept provides a way to describe gravity that avoids the troublesome action-at-a-distance. A gravitating mass creates a field in the space around it, and a second mass responds to the field in its immediate vicinity.

Gravitational field

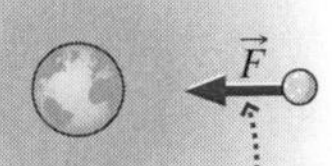

Force arises from field at Moon's location.

Circular orbits are readily analyzed using Newton's laws and concepts from circular motion. A circular orbit of radius r about a mass M has a period given by

$$T^2 = \frac{4\pi^2 r^3}{GM}$$

Kinetic and potential energies are related by $U = -2K$. Total energy is negative, as appropriate for a closed orbit, and the object actually moves faster the lower its total energy.

A special orbit is the **geostationary orbit**, parallel to Earth's equator at an altitude of about 36,000 km. Here the orbital period is 24 h, so a satellite in geostationary orbit appears from Earth's surface to be fixed in the sky. TV, communications, and weather satellites use geostationary orbits.

MP *For homework assigned on MasteringPhysics, go to www.masteringphysics.com*

BIO *Biology and/or medicine-related problems* **DATA** *Data problems* **ENV** *Environmental problems* **CH** *Challenge problems* **COMP** *Computer problems*

For Thought and Discussion

1. What do Newton's apple and the Moon have in common?
2. Explain the difference between G and g.
3. When you stand on Earth, the distance between you and Earth is zero. So why isn't the gravitational force infinite?
4. The force of gravity on an object is proportional to the object's mass, yet all objects fall with the same gravitational acceleration. Why?
5. A friend who knows nothing about physics asks what keeps an orbiting satellite from falling to Earth. Give an answer that will satisfy your friend.
6. Could you put a satellite in an orbit that keeps it stationary over the south pole? Explain.
7. Why are satellites generally launched eastward and from low latitudes? (*Hint:* Think about Earth's rotation.)
8. Given Earth's mass, the Moon's distance and orbital period, and the value of G, could you calculate the Moon's mass? If yes, how? If no, why not?
9. How should a satellite be launched so that its orbit takes it over every point on the (rotating) Earth?
10. Does the gravitational force of the Sun do work on a planet in a circular orbit? In an elliptical orbit? Explain.

Exercises and Problems

Exercises

Section 8.2 Universal Gravitation

11. Space explorers land on a planet with the same mass as Earth, but find they weigh twice as much as they would on Earth. What's the planet's radius?
12. Use data for the Moon's orbit from Appendix E to compute the Moon's acceleration in its circular orbit, and verify that the result is consistent with Newton's law of gravitation.
13. To what fraction of its current radius would Earth have to shrink (with no change in mass) for the gravitational acceleration at its surface to triple?
14. Calculate the gravitational acceleration at the surface of (a) Mercury and (b) Saturn's moon Titan.
15. Two identical lead spheres with their centers 14 cm apart attract each other with a 0.25-μN force. Find their mass.
16. What's the approximate value of the gravitational force between a 67-kg astronaut and a 73,000-kg spacecraft when they're 84 m apart?
17. A sensitive gravimeter is carried to the top of New York's new One World Trade Center, where its reading for the acceleration of gravity is 1.67 mm/s^2 lower than at street level. Find the building's height.

Section 8.3 Orbital Motion

18. At what altitude will a satellite complete a circular orbit of Earth in 2.0 h?
19. Find the speed of a satellite in geostationary orbit.
20. Mars's orbit has a diameter 1.52 times that of Earth's orbit. How long does it take Mars to orbit the Sun?
21. Calculate the orbital period for Jupiter's moon Io, which orbits 4.22×10^5 km from the planet's center.
22. An astronaut hits a golf ball horizontally from the top of a lunar mountain so fast that it goes into circular orbit. What's its orbital period?
23. The Mars Reconnaissance Orbiter circles the red planet with a 112-min period. What's the spacecraft's altitude?

Section 8.4 Gravitational Energy

24. Earth's distance from the Sun varies from 147 Gm at perihelion to 152 Gm at aphelion because its orbit isn't quite circular. Find the change in potential energy as Earth goes from perihelion to aphelion.
25. So-called suborbital missions take scientific instruments into space for brief periods without the expense of getting into orbit; their trajectories are often simple "up and down" vertical paths. How much energy does it take to launch a 230-kg instrument on a vertical trajectory that peaks at 1800 km altitude?
26. A rocket is launched vertically upward from Earth's surface at 5.1 km/s. What's its maximum altitude?
27. What vertical launch speed is necessary to get a rocket to an altitude of 1100 km?
28. Find the energy necessary to put 1 kg, initially at rest on Earth's surface, into geostationary orbit.
29. What's the total mechanical energy associated with Earth's orbital motion?
30. The escape speed from a planet of mass 2.9×10^{24} kg is 7.1 km/s. Find the planet's radius.
31. Determine escape speeds from (a) Jupiter's moon Callisto and (b) a neutron star, with the Sun's mass crammed into a sphere of radius 6.0 km. See Appendix E for relevant data.
32. To what radius would Earth have to shrink, with no change in mass, for escape speed at its surface to be 30 km/s?

Problems

33. The gravitational acceleration at a planet's surface is 22.5 m/s^2. Find the acceleration at an altitude equal to half the planet's radius.
34. **BIO** One of the longest-standing athletic records is Cuban Javier Sotomayor's 2.45-m high jump. How high could Sotomayor jump on (a) Mars and (b) Earth's Moon?
35. You're the navigator on a spaceship studying an unexplored planet. Your ship has just gone into a circular orbit around the planet, and you determine that the gravitational acceleration at your orbital altitude is half what it would be at the surface. What do you report for your altitude, in terms of the planet's radius?
36. If you're standing on the ground 15 m directly below the center of a spherical water tank containing 4×10^6 kg of water, by what fraction is your weight reduced due to the water's gravitational attraction?
37. Given the Moon's orbital radius of 384,400 km and period of 27.3 days, calculate its acceleration in its circular orbit, and compare with the acceleration of gravity at Earth's surface. Show that the Moon's acceleration is lower by the ratio of the square of Earth's radius to the square of the Moon's orbital radius, thus confirming the inverse-square law for the gravitational force.
38. Equation 7.9 relates force to the derivative of potential energy. Use this fact to differentiate Equation 8.6 for gravitational potential energy, and show that you recover Newton's law of gravitation.

39. During the *Apollo* Moon landings, one astronaut remained with the command module in lunar orbit, about 130 km above the surface. For half of each orbit, this astronaut was completely cut off from the rest of humanity as the spacecraft rounded the far side of the Moon. How long did this period last?
40. A white dwarf is a collapsed star with roughly the Sun's mass compressed into the size of Earth. What would be (a) the orbital speed and (b) the orbital period for a spaceship in orbit just above the surface of a white dwarf?
41. Given that our Sun orbits the galaxy with a period of 200 My at 2.6×10^{20} m from the galactic center, estimate the galaxy's mass. Assume (incorrectly) that the galaxy is essentially spherical and that most of its mass lies interior to the Sun's orbit.
42. You're preparing an exhibit for the Golf Hall of Fame, and you realize that the longest golf shot in history was Astronaut Alan Shepard's lunar drive. Shepard, swinging single-handed with a golf club attached to a lunar sample scoop, claimed his ball went "miles and miles." The record for a single-handed golf shot on Earth is 257 m. Could Shepard's ball really have gone "miles and miles"? Assume the ball's initial speed is independent of gravitational acceleration.
43. **CH** Exact solutions for gravitational problems involving more than two bodies are notoriously difficult. One solvable problem involves a configuration of three equal-mass objects spaced in an equilateral triangle. Forces due to their mutual gravitation cause the configuration to rotate. Suppose three identical stars, each of mass M, form a triangle of side L. Find an expression for the period of their orbital motion.
44. Satellites A and B are in circular orbits, with A four times as far from Earth's center as B. How do their orbital periods compare?
45. The asteroid that exploded over Chelyabinsk, Russia, in 2012 (see Application on page 137) was moving at 35.5 km/s relative to the Sun just before it entered Earth's atmosphere. Calculations based on orbital observations show that it was moving at 11.2 km/s at aphelion (its most distant point from the Sun). Find the distance at aphelion, expressed in astronomical units (1 AU is the average distance of Earth from the Sun; see Appendix E).
46. **ENV** We still don't have a permanent solution for the disposal of radioactive waste. As a nuclear waste specialist with the Department of Energy, you're asked to evaluate a proposal to shoot waste canisters into the Sun. You need to report the speed at which a canister, dropped from rest in the vicinity of Earth's orbit, would hit the Sun. What's your answer?
47. In November 2013, Comet ISON reached its perihelion (closest approach to the Sun) at 1.87 Gm from the Sun's center (only 1.17 Gm from the solar surface); at that point ISON was moving at 378 km/s relative to the Sun. Do a calculation to determine whether its orbit was elliptical or hyperbolic. (Most of ISON's cometary nucleus was destroyed in its close encounter with the Sun.)
48. Neglecting air resistance, to what height would you have to fire a rocket for the constant-acceleration equations of Chapter 2 to give a height in error by 1%? Would those equations overestimate or underestimate the height?
49. Show that an object released from rest very far from Earth reaches Earth's surface at essentially escape speed.
50. By what factor must an object's speed in circular orbit be increased to reach escape speed from its orbital altitude?
51. You're in charge of tracking celestial objects that might pose a danger to Earth. Astronomers have discovered a new comet that's moving at 53 km/s as it crosses Earth's orbit. Determine whether the comet will again return to Earth's vicinity.
52. Two meteoroids are 250,000 km from Earth's center and moving at 2.1 km/s. One is headed straight for Earth, while the other is on a path that will come within 8500 km of Earth's center (Fig. 8.16). Find the speed of (a) the first meteoroid when it strikes Earth and (b) the second meteoroid at its closest approach. (c) Will the second meteoroid ever return to Earth's vicinity?

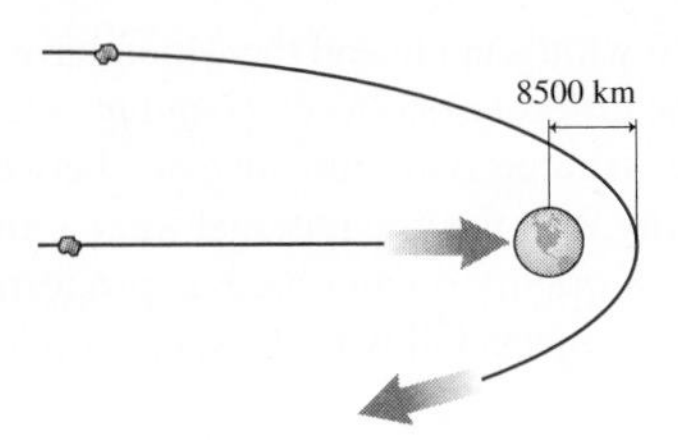

FIGURE 8.16 Problem 52

53. **CH** Neglecting Earth's rotation, show that the energy needed to launch a satellite of mass m into circular orbit at altitude h is $\left(\frac{GM_Em}{R_E}\right)\left(\frac{R_E + 2h}{2(R_E + h)}\right)$.
54. **CH** A projectile is launched vertically upward from a planet of mass M and radius R; its initial speed is $\sqrt{2}$ times the escape speed. Derive an expression for its speed as a function of the distance r from the planet's center.
55. **CH** A spacecraft is in circular orbit 5500 km above Earth's surface. How much will its altitude decrease if it moves to a new circular orbit where (a) its orbital speed is 10% higher or (b) its orbital period is 10% shorter?
56. Two meteoroids are 160,000 km from Earth's center and heading straight toward Earth, one at 10 km/s, the other at 20 km/s. At what speeds will they strike Earth?
57. Two rockets are launched from Earth's surface, one at 12 km/s and the other at 18 km/s. How fast is each moving when it crosses the Moon's orbit?
58. A satellite is in an elliptical orbit at altitudes ranging from 230 to 890 km. At its highest point, it's moving at 7.23 km/s. How fast is it moving at its lowest point?
59. A missile's trajectory takes it to a maximum altitude of 1200 km. If its launch speed is 6.1 km/s, how fast is it moving at the peak of its trajectory?
60. A 720-kg spacecraft has total energy -0.53 TJ and is in circular orbit around the Sun. Find (a) its orbital radius, (b) its kinetic energy, and (c) its speed.
61. Mercury's orbital speed varies from 38.8 km/s at aphelion to 59.0 km/s at perihelion. If the planet is 6.99×10^{10} m from the Sun's center at aphelion, how far is it at perihelion?
62. **CH** Show that the form $\Delta U = mg\,\Delta r$ follows from Equation 8.5 when $r_1 \simeq r_2$. [*Hint:* Write $r_2 = r_1 + \Delta r$ and apply the binomial approximation (Appendix A).]
63. **CH** Two satellites are in geostationary orbit but in diametrically opposite positions (Fig. 8.17). In order to catch up with the other, one satellite descends into a lower circular orbit (see Conceptual Example 8.1 for a description of this maneuver). How far should it descend if it's to catch up in 10 orbits? Neglect rocket firing times and time spent moving between the two circular orbits.

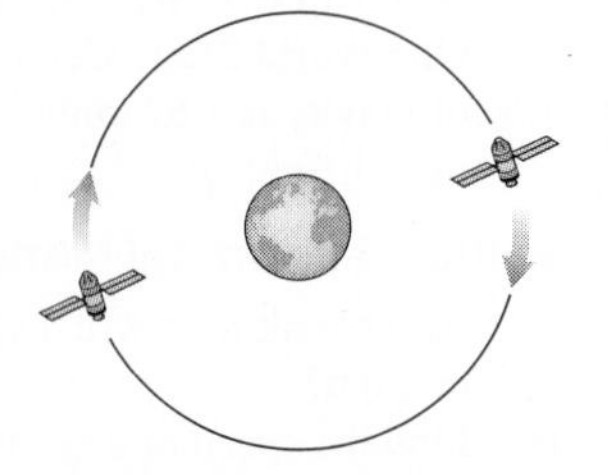

FIGURE 8.17 Problem 63

64. The two asteroids described in the Application on page 137 both set records for being the largest objects of their sizes to come as close to Earth in recent times as they did. Use appropriate data given in the Application to find the total energy for each asteroid—that is, each asteroid's kinetic energy plus potential energy in the asteroid–Sun system. What do your results show about the asteroids' orbits?

65. DATA A spacecraft is orbiting a spherical asteroid when it deploys a probe that falls toward the asteroid's surface. The spacecraft radios to Earth the probe's position and its acceleration; the data are shown in the table below. Determine a quantity that, when you plot a against it, should yield a straight line. Plot the data, determine a best-fit line, and use its slope to determine the asteroid's mass.

Probe position r (km from asteroid's center)	80.0	55.0	40.0	35.0	30.0
Acceleration a (mm/s^2)	0.172	0.353	0.704	0.858	1.18

66. We derived Equation 8.4 on the assumption that the massive gravitating center remains fixed. Now consider two objects with equal mass M orbiting each other, as shown in Fig. 8.18. Show that the orbital period is given by $T^2 = 2\pi^2 d^3/GM$, where d is the distance between the objects.

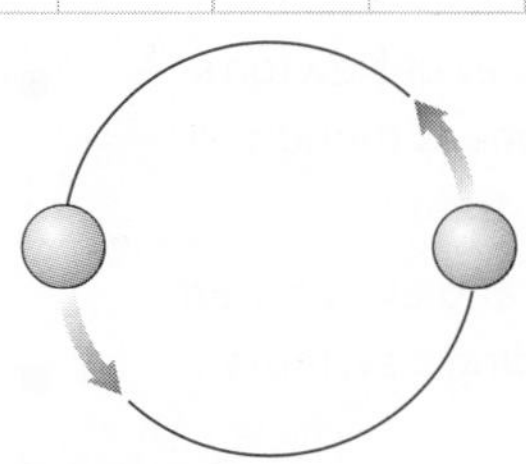

FIGURE 8.18 Problem 66

67. CH Tidal effects in the Earth–Moon system cause the Moon's orbital period to increase at a current rate of about 35 ms per century. Assuming the Moon's orbit is circular, to what rate of change in the Earth–Moon distance does this correspond? (*Hint:* Differentiate Kepler's third law, Equation 8.4, and consult Appendix E.)

68. BIO As a member of the 2040 Olympic committee, you're considering a new sport: asteroid jumping. On Earth, world-class high jumpers routinely clear 2 m. Your job is to make sure athletes jumping from asteroids will return to the asteroid. Make the simplifying assumption that asteroids are spherical, with average density 2500 kg/m^3. For safety, make sure even a jumper capable of 3 m on Earth will return to the surface. What do you report for the minimum asteroid diameter?

69. The Olympic Committee is keeping you busy! You're now asked to consider a proposal for lunar hockey. The record speed for a hockey puck is 178 km/h. Is there any danger that hockey pucks will go into lunar orbit?

70. CH Tidal forces are proportional to the variation in gravity with position. By differentiating Equation 8.1, estimate the ratio of the tidal forces due to the Sun and the Moon. Compare your answer with the ratio of the gravitational forces that the Sun and Moon exert on Earth. Use data from Appendix E.

71. COMP Spacecraft that study the Sun are often placed at the so-called L1 Lagrange point, located sunward of Earth on the Sun–Earth line. L1 is the point where Earth's and Sun's gravity together produce an orbital period of one year, so that a spacecraft at L1 stays fixed relative to Earth as both planet and spacecraft orbit the Sun. This placement ensures an uninterrupted view of the Sun, without being periodically eclipsed by Earth as would occur in Earth orbit. Find L1's location relative to Earth. (*Hint*: This problem calls for numerical methods or solving a higher-order polynomial equation.)

Passage Problems

The Global Positioning System (GPS) uses a "constellation" of some 30 satellites to provide accurate positioning for any point on Earth (Fig. 8.19). GPS receivers time radio signals traveling at the speed of light from three of the satellites to find the receiver's position. Signals from one or more additional satellites provide corrections, eliminating the need for high-accuracy clocks in individual GPS receivers. GPS satellites are in circular orbits at 20,200 km altitude.

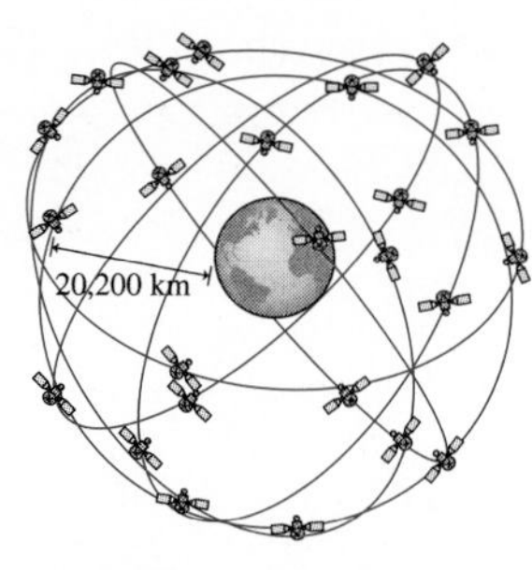

FIGURE 8.19 GPS satellites (Passage Problems 72–75)

72. What's the approximate orbital period of GPS satellites?
    a. 90 min
    b. 8 h
    c. 12 h
    d. 24 h
    e. 1 week
73. What's the approximate speed of GPS satellites?
    a. 9.8 m/s
    b. 500 m/s
    c. 1.7 km/s
    d. 4 km/s
    e. 12 km/s
74. What's the approximate escape speed at GPS orbital distance?
    a. 4 km/s
    b. 5.5 km/s
    c. 6.3 km/s
    d. 9.8 km/s
    e. 11 km/s
75. The current generation of GPS satellites has masses of 844 kg. What's the approximate total energy of such a satellite?
    a. 6 GJ
    b. 3 GJ
    c. −3 GJ
    d. −6 GJ
    e. −8 GJ

Answers to Chapter Questions

Answer to Chapter Opening Question

Voyager's total energy—kinetic energy plus potential energy associated with the Sun's gravitational field—is greater than zero. Put another way, Voyager has escape speed relative to the Sun.

Answers to GOT IT? Questions

8.1 (d) Quadrupled. If the original distance were r, the original force would be proportional to $1/r^2$. At half that distance, the force is proportional to $1/(r/2)^2 = 4/r^2$.

8.2 Hyperbolic > parabolic > elliptical > circular

8.3 (a), (c), and (d). Since B has higher total energy, it must have lower kinetic energy and is therefore moving slower. B is farther from the gravitating body, so its potential energy is higher—still negative, but less so than A's. For circular orbits, the ratio of potential energy to total energy is always the same—namely, $U = 2E$.

Systems of Particles

What You Know

- You know Newton's second law not only in the form $\vec{F} = m\vec{a}$ but also in Newton's original form relating force to the rate of change of momentum: $\vec{F} = d\vec{p}/dt$.
- You understand conservation of mechanical energy: When only conservative forces act, the sum of a system's kinetic and potential energy remains constant.
- You've seen the concept of a system, and you've had practice in defining systems for purposes of applying energy conservation.
- You know that nonconservative forces result in a loss of mechanical energy.
- You recognize that Newton's third law requires that forces between interacting objects come in pairs with equal magnitudes and opposite directions.

What You're Learning

- Here you'll study systems consisting of two or more particles, beginning with a system's *center of mass*—a point where, for the purposes of Newton's laws, a system behaves as though all its mass were concentrated.
- You'll learn to set up and evaluate an integral, here for finding a system's center of mass.
- You'll learn to evaluate the momentum of a system, and you'll see that momentum is conserved in the absence of external forces.
- You'll study collisions between objects, both energy-conserving *elastic* collisions and *inelastic collisions* in which some mechanical energy is lost.
- You'll learn how complicated collision problems can be made easier by working in the *center-of-mass reference frame.*

How You'll Use It

- Much of what you learn here will have analogies in the rotating systems you'll study in Chapters 10 and 11.
- The technique you learn for setting up integrals will be useful in diverse physics contexts, especially in Chapter 10 and again in Chapter 22.
- Center of mass, the momentum of a system, and momentum conservation will remain important concepts even as you move beyond Newtonian physics in Part 6.

Most parts of the dancer's body undergo complex motions during this jump, yet one special point follows the parabolic trajectory of a projectile. What is that point, and why is it special?

So far we've generally treated objects as point particles, ignoring the fact that most are composed of smaller parts. In Chapter 6's introduction of energy, however, we needed also to develop the idea of a system that might comprise more than one object, and in Chapter 7 we found that the concept of potential energy necessarily required us to consider systems of at least two interacting particles. Here we deal explicitly with systems of many particles. These include **rigid bodies**—objects such as baseballs, cars, and planets whose constituent particles are stuck together in fixed orientations—as well as systems like human bodies, exploding fireworks, or flowing rivers, whose parts move relative to one another. In subsequent chapters we'll look at specific instances of many-particle systems, including the rotational motion of rigid bodies (Chapter 10) and the behavior of fluids (Chapter 15).

9.1 Center of Mass

The motion of the dancer in the photo to the left is complex, with each part of his body moving on a different path. But the superimposed curve shows one point following the parabola we expect of a projectile (Section 3.5). This point is the **center of mass**, an average position of all the mass making up the dancer. Since the net force on the dancer as a whole is gravity, the photo, with its parabolic arc, suggests that the center of mass obeys Newton's second law, $\vec{F}_{\text{net}} = M\vec{a}_{\text{cm}}$, where M is the dancer's total mass and $\vec{a}_{\text{cm}}$ is the acceleration of the center of mass. (We'll use the subscript cm for quantities associated with the center of mass.) To find the center of mass, we therefore need to locate a point whose acceleration obeys $\vec{F}_{\text{net}} = M\vec{a}_{\text{cm}}$, with $\vec{F}_{\text{net}}$ the net force on the entire system.

Consider a system of many particles. To find the center of mass, we want an equation like Newton's second law that involves the total mass of the system and the net force on the entire system. If we apply Newton's second law to the ith particle in the system, we have

$$\vec{F}_i = m_i\vec{a}_i = m_i\frac{d^2\vec{r}_i}{dt^2} = \frac{d^2 m_i\vec{r}_i}{dt^2}$$

where $\vec{F}_i$ is the net force on the particle, m_i is its mass, and we've written the acceleration $\vec{a}_i$ as the second derivative of the position $\vec{r}_i$. The total force on the system is the sum of the forces acting on all N particles. We write this sum compactly using the summation symbol $\sum$:

$$\vec{F}_{\text{total}} = \sum_{i=1}^{N}\vec{F}_i = \sum_{i=1}^{N}\frac{d^2 m_i\vec{r}_i}{dt^2}$$

where the sum runs over all particles composing the system, from $i = 1$ to N. But the sum of derivatives is the derivative of the sum, so

$$\vec{F}_{\text{total}} = \frac{d^2\left(\sum m_i\vec{r}_i\right)}{dt^2}$$

We can now put this equation in the form of Newton's second law. Multiplying and dividing the right-hand side by the total mass $M = \sum m_i$, and distributing this constant M through the differentiation, we have

$$\vec{F}_{\text{total}} = M\frac{d^2}{dt^2}\left(\frac{\sum m_i\vec{r}_i}{M}\right) \tag{9.1}$$

Equation 9.1 has a form like Newton's law applied to the total mass if we define

$$\vec{r}_{\text{cm}} = \frac{\sum m_i\vec{r}_i}{M} \quad \text{(center of mass)} \tag{9.2}$$

Then the derivative in Equation 9.1 becomes $d^2\vec{r}_{\text{cm}}/dt^2$, which we recognize as the center-of-mass acceleration, $\vec{a}_{\text{cm}}$. So now Equation 9.1 reads $\vec{F}_{\text{total}} = M\vec{a}_{\text{cm}}$. This is almost Newton's law—but not quite, because the force here is the sum of all the forces acting on all the particles of the system, and we want just the net **external force**—the net force applied from *outside* the system. We can write the force $\vec{F}_{\text{total}}$ as

$$\vec{F}_{\text{total}} = \sum\vec{F}_{\text{ext}} + \sum\vec{F}_{\text{int}}$$

where $\sum\vec{F}_{\text{ext}}$ is the sum of all the external forces and $\sum\vec{F}_{\text{int}}$ the sum of the internal forces. According to Newton's third law, each of the internal forces has an equal but oppositely directed force that itself acts on a particle of the system and is therefore included in the sum $\sum\vec{F}_{\text{int}}$. (Each external force is also part of a third-law pair, but forces paired with the external forces act *outside* the system and therefore aren't included in the sum.) Added vectorially, the internal forces therefore cancel in pairs, so $\sum\vec{F}_{\text{int}} = \vec{0}$, and the force $\vec{F}_{\text{total}}$ in Equation 9.1 is just the net *external* force applied to the system. So the point $\vec{r}_{\text{cm}}$ defined in Equation 9.2 does obey Newton's law, written in the form

$$\vec{F}_{\text{net ext}} = M\vec{a}_{\text{cm}} = M\frac{d^2\vec{r}_{\text{cm}}}{dt^2} \tag{9.3}$$

where $\vec{F}_{\text{net ext}}$ is the net external force applied to the system and M is the total mass.

We've defined the center of mass $\vec{r}_{\text{cm}}$ so we can apply Newton's second law to the entire system rather than to each individual particle. As far as its overall motion is concerned, a complex system acts as though all its mass were concentrated at the center of mass.

Video Tutor Demo | **Balancing a Meter Stick**

Finding the Center of Mass

Equation 9.2 shows that the center-of-mass position is an average of the positions of the individual particles, weighted by their masses. For a one-dimensional system, Equation 9.2 becomes $x_{cm} = \sum m_i x_i / M$; in two and three dimensions, there are similar equations for the center-of-mass coordinates y_{cm} and z_{cm}. Finding the center of mass (CM) is a matter of establishing a coordinate system and then using the components of Equation 9.2.

EXAMPLE 9.1 CM in One Dimension: Weightlifting

Find the center of mass of a barbell consisting of 50-kg and 80-kg weights at the opposite ends of a 1.5-m-long bar of negligible mass.

INTERPRET This is a problem about center of mass. We identify the system as consisting of two "particles"—namely, the two weights.

DEVELOP Figure 9.1 shows the barbell. Here, with just two particles, we have a one-dimensional situation and Equation 9.2, $\vec{r}_{cm} = \sum m_i \vec{r}_i / M$, becomes $x_{cm} = (m_1 x_1 + m_2 x_2)/(m_1 + m_2)$. Before we can apply this equation, however, we need a coordinate system. As always, any coordinate system will do—but a smart choice makes the math easier. Let's take $x = 0$ at the 50-kg mass, so the term $m_1 x_1$ becomes zero.

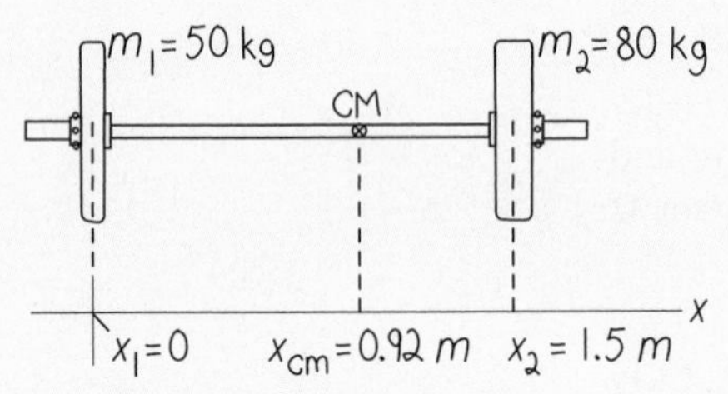

FIGURE 9.1 Our sketch of the barbell.

Our plan is then to find the center-of-mass coordinate x_{cm} using our one-dimensional version of Equation 9.2.

EVALUATE With $x = 0$ at the left end of the barbell, the coordinate of the 80-kg mass is $x_2 = 1.5$ m. So our equation becomes

$$x_{cm} = \frac{m_1 x_1 + m_2 x_2}{m_1 + m_2} = \frac{m_2 x_2}{m_1 + m_2} = \frac{(80\text{ kg})(1.5\text{ m})}{(50\text{ kg} + 80\text{ kg})} = 0.92\text{ m}$$

where the equation simplified because of our choice $x_1 = 0$.

ASSESS As Fig. 9.1 shows, this result makes sense: The center of mass is closer to the heavier weight. If the weights had been equal, the center of mass would have been right in the middle.

✓TIP Choosing the Origin

Choosing the origin at one of the masses here conveniently makes one of the terms in the sum $\sum m_i x_i$ zero. But, as always, the choice of origin is purely for convenience and doesn't influence the actual physical location of the center of mass. Exercise 16 demonstrates this point, repeating Example 9.1 with a different origin.

EXAMPLE 9.2 CM in Two Dimensions: A Space Station

Figure 9.2 shows a space station consisting of three modules arranged in an equilateral triangle, connected by struts of length L and of negligible mass. Two modules have mass m, the other $2m$. Find the center of mass.

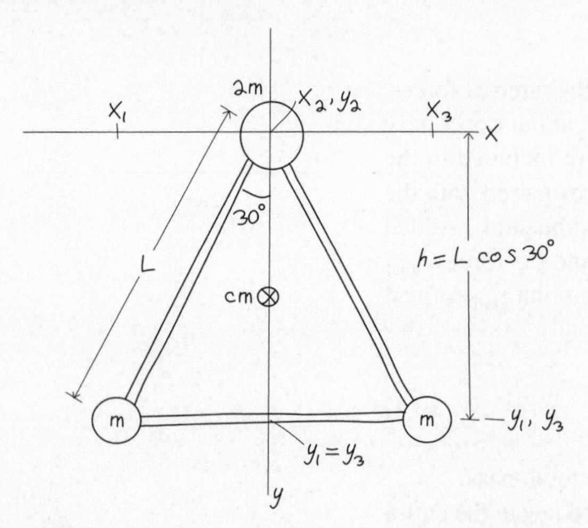

FIGURE 9.2 Our sketch of the space station.

INTERPRET We're after the center of mass of the system consisting of the three modules.

DEVELOP Figure 9.2 is our drawing. We'll use Equation 9.2, $\vec{r}_{cm} = \sum m_i \vec{r}_i / M$, to find the center-of-mass coordinates x_{cm} and y_{cm}. A sensible coordinate system has the origin at the module with mass $2m$ and the y-axis downward, as shown in Fig. 9.2.

EVALUATE Labeling the modules from left to right, we see that $x_1 = -L \sin 30° = -\frac{1}{2}L$, $y_1 = L \cos 30° = L\sqrt{3}/2$; $x_2 = y_2 = 0$; and $x_3 = -x_1 = \frac{1}{2}L$, $y_3 = y_1 = L\sqrt{3}/2$. Writing explicitly the x- and y-components of Equation 9.2 for this case gives

$$x_{cm} = \frac{mx_1 + mx_3}{4m} = \frac{m(x_1 - x_1)}{4m} = 0$$

$$y_{cm} = \frac{my_1 + my_3}{4m} = \frac{2my_1}{4m} = \frac{1}{2}y_1 = \frac{\sqrt{3}}{4}L \simeq 0.43L$$

Although there are three "particles" here, our choice of coordinate system left only two nonzero terms in the numerator, both associated with the same mass m. The more massive module is still in the problem, though; its mass $2m$ contributes to make the total mass M in the denominator equal to $4m$.

ASSESS That $x_{cm} = 0$ is apparent from symmetry (more on this in the following Tip). How about the result for y_{cm}? We have $2m$ at the top of the triangle, and $m + m = 2m$ at the bottom—so shouldn't the center of mass lie midway up the triangle? It does! Expressing the center of mass in terms of the triangle side L obscures this fact. The triangle's height is $h = L\cos 30° = L\sqrt{3}/2$, and our answer for y_{cm} is indeed half this value. We marked the CM on Fig. 9.2.

✓TIP Exploit Symmetries

It's no accident that x_{cm} here lies on the vertical line that bisects the triangle; after all, the triangle is symmetric about that line, so its mass is distributed evenly on either side. Exploit symmetry whenever you can; that can save you a lot of computation throughout physics!

Continuous Distributions of Matter

We've expressed the center of mass as a sum over individual particles. Ultimately, matter *is* composed of individual particles. But it's often convenient to consider that it's continuously distributed; we don't want to deal with 10^{23} atoms to find the center of mass of a macroscopic object! We can think of continuous matter as being composed of individual pieces of mass Δm_i, with position vectors $\vec{r}_i$; we call these pieces **mass elements** (Fig. 9.3). The center of mass of the entire chunk is then given by Equation 9.2: $\vec{r}_{cm} = (\sum \Delta m_i \vec{r}_i)/M$, where $M = \sum \Delta m_i$ is the total mass. In the limit as the mass elements become arbitrarily small, this expression becomes an integral:

$$\vec{r}_{cm} = \lim_{\Delta m_i \to 0} \frac{\sum \Delta m_i \vec{r}_i}{M} = \frac{\int \vec{r}\, dm}{M} \quad \left(\begin{array}{c}\text{center of mass,}\\ \text{continuous matter}\end{array}\right) \tag{9.4}$$

where the integration is over the entire object. Like the sum in Equation 9.2, the integral of the vector $\vec{r}$ stands for three separate integrals for the components of the center-of-mass position.

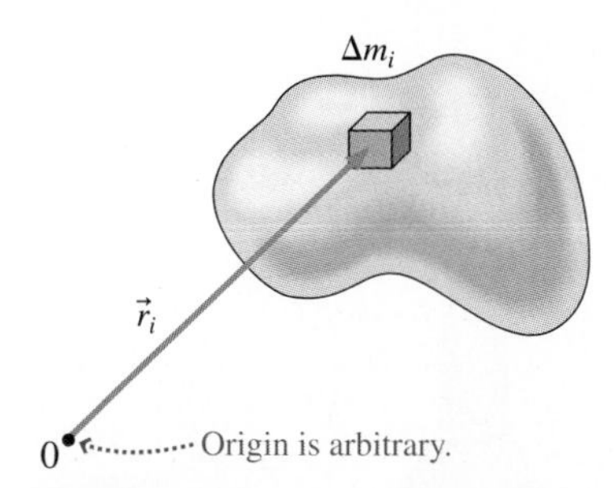

FIGURE 9.3 A chunk of continuous matter, showing one mass element Δm_i and its position vector $\vec{r}_i$.

EXAMPLE 9.3 Continuous Matter: An Aircraft Wing

A supersonic aircraft wing is an isosceles triangle of length L, width w, and negligible thickness. It has mass M, distributed uniformly over the wing. Where's its center of mass?

INTERPRET Here the matter is distributed continuously, so we need to integrate to find the center of mass. We identify an axis of symmetry through the wing, which we designate the x-axis. By symmetry, the center of mass lies along this x-axis, so $y_{cm} = 0$ and we'll need to calculate only x_{cm}.

DEVELOP Figure 9.4 shows the wing. Equation 9.4 applies, and we need only the x-component because the y-component is evident from symmetry. The x-component of Equation 9.4 is $x_{cm} = (\int x\, dm)/M$.

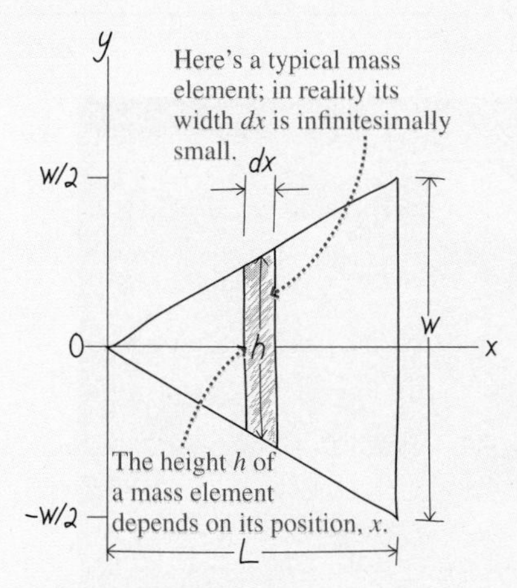

FIGURE 9.4 Our sketch of the supersonic aircraft wing.

Developing a plan for dealing with an integral like this requires some thought; we'll first do the work and then summarize the general steps involved.

Our goal is to find an appropriate mass element dm in terms of the infinitesimal coordinate interval dx. As shown in Fig. 9.4, here it's easiest to use a vertical strip of width dx. Each such strip has a different height h, depending on its position x. If we choose a coordinate system with origin at the wing apex, then, as you can see from the figure, the height grows linearly from 0 at $x = 0$ to w at $x = L$. So $h = (w/L)x$. This strip is infinitesimally narrow, so its sloping edges don't matter and its area is that of a very thin rectangle—namely, $h\,dx = (w/L)x\,dx$. The strip's mass dm is then the same fraction of the total wing mass M as its area is of the total wing area $\frac{1}{2}wL$; that is,

$$\frac{dm}{M} = \frac{(w/L)x\,dx}{\frac{1}{2}wL} = \frac{2x\,dx}{L^2}$$

so $dm = 2Mx\,dx/L^2$.

In the integral we weight each mass element dm by its distance x from the origin, and then sum—that is, integrate—over all mass elements. So, from Equation 9.4, we have

$$x_{cm} = \frac{1}{M}\int x\,dm = \frac{1}{M}\int_0^L x\left(\frac{2Mx}{L^2}\,dx\right) = \frac{2}{L^2}\int_0^L x^2\,dx$$

As always, constants can come outside the integral. We set the limits 0 and L to cover all the mass elements in the wing. Now we're finally ready to find x_{cm}.

(continued)

EVALUATE The hard part is done. All that's left is to evaluate the integral:

$$x_{\mathrm{cm}} = \frac{2}{L^2}\int_0^L x^2\,dx = \frac{2}{L^2}\frac{x^3}{3}\bigg|_0^L = \frac{2L^3}{3L^2} = \frac{2}{3}L$$

ASSESS Make sense? Yes: Our answer puts the center of mass toward the back of the wing where, because of its increasing width, most of the mass lies. In a complicated calculation like this one, it's reassuring to see that the answer is a quantity with the units of length. ■

TACTICS 9.1 Setting Up an Integral

An integral like $\int x\,dm$ can be confusing because you see both x and dm after the integral sign and they don't seem related. But they are, and here's how to proceed:

1. Find a suitable shape for your mass elements, preferably one that exploits any symmetry in the situation. One dimension of the elements should involve an infinitesimal interval in one of the coordinates x, y, or z. In Example 9.3, the mass elements were strips, symmetric about the wing's centerline and with width dx.
2. Find an expression for the infinitesimal area of your mass elements (in a one-dimensional problem it would be the length; in a three-dimensional problem, the volume). In Example 9.3, the infinitesimal area of each mass element was the strip height h multiplied by the width dx.
3. Form ratios that relate the infinitesimal coordinate interval to the physical quantity in the integral—which in Example 9.3 is the mass element dm. Here we formed the ratio of the area of a mass element to the total area, and equated that to the ratio of dm to the total mass M.
4. Solve your ratio statement for the infinitesimal quantity, in this case dm, that appears in your integral. Then you're ready to evaluate the integral.

Sometimes you'll be given a density—mass per volume, per area, or per length—and then in place of steps 3 and 4 you find dm by multiplying the density by the infinitesimal volume, area, or length you identified in step 2.

Although we described this procedure in the context of Example 9.3, it also applies to other integrals you'll encounter in different areas of physics.

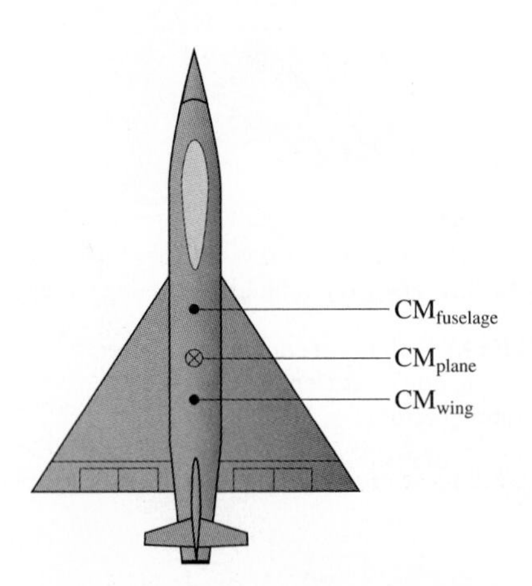

FIGURE 9.5 The center of mass of the airplane is found by treating the wing and fuselage as point particles located at their respective centers of mass.

With more complex objects, it's convenient to find the centers of mass of subparts and then treat those as point particles to find the center of mass of the entire object (Fig. 9.5).

The center of mass need not lie within an object, as Fig. 9.6 shows. High jumpers exploit this fact as they straddle the bar with arms and legs dangling on either side (Fig. 9.7). Although the jumper's entire body clears the bar, his center of mass doesn't need to!

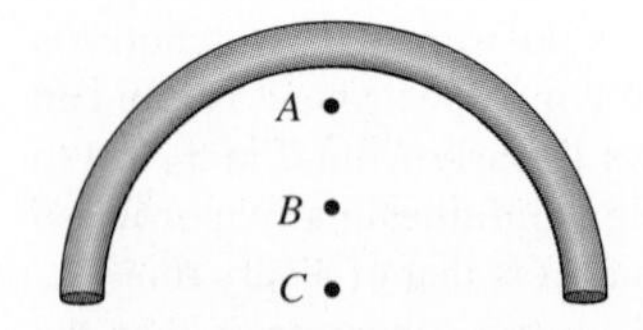

FIGURE 9.6 Got it? The center of mass lies outside the semicircular wire, but which point is it?

GOT IT? 9.1 A thick wire is bent into a semicircle, as shown in Fig. 9.6. Which of the points shown is the center of mass?

FIGURE 9.7 A high jumper clears the bar, but his center of mass doesn't!

Motion of the Center of Mass

We defined the center of mass so its motion obeys Newton's law $\vec{F}_{\text{net ext}} = M\vec{a}_{\text{cm}}$, with $\vec{F}_{\text{net ext}}$ the net external force on the system and M the total mass. When gravity is the only external force, the center of mass follows the trajectory of a point particle. But if the net external force is zero, then the center-of-mass acceleration $\vec{a}_{\text{cm}}$ is also zero, and the center of mass moves with constant velocity. In the special case of a system at rest, the center of mass remains at rest despite any motions of its internal parts.

EXAMPLE 9.4 CM Motion: Circus Train

Jumbo, a 4.8-t elephant, stands near one end of a 15-t railcar at rest on a frictionless horizontal track. (Here t is for tonne, or metric ton, equal to 1000 kg.) Jumbo walks 19 m toward the other end of the car. How far does the car move?

INTERPRET We're asked about the car's motion, but we can interpret this problem as being fundamentally about the center of mass. We identify the relevant system as comprising Jumbo and the car. Because there's no net external force acting on the system, its center of mass can't move.

DEVELOP Figure 9.8*a* shows the initial situation. The symmetric car has its CM at its center (here we care only about the *x*-component). Let's take a coordinate system that's fixed to the ground and that has $x = 0$ at this *initial* location of the car's center. After the car moves, its center will be somewhere else! Equation 9.2 applies—here in the simpler one-dimensional, two-object form we used in Example 9.1: $x_{\text{cm}} = (m_J x_J + m_c x_c)/M$, where we use the subscripts J and c for Jumbo and the car, respectively, and where $M = m_J + m_c$ is the total mass. We have a before/after situation in which the CM position can't change, so we'll write two versions of this expression, before and after Jumbo's walk. We'll then set them equal to state mathematically that the CM itself doesn't move; that is, we'll write $x_{\text{cm i}} = x_{\text{cm f}}$, where the subscripts i and f designate quantities associated with the initial and final states, respectively.

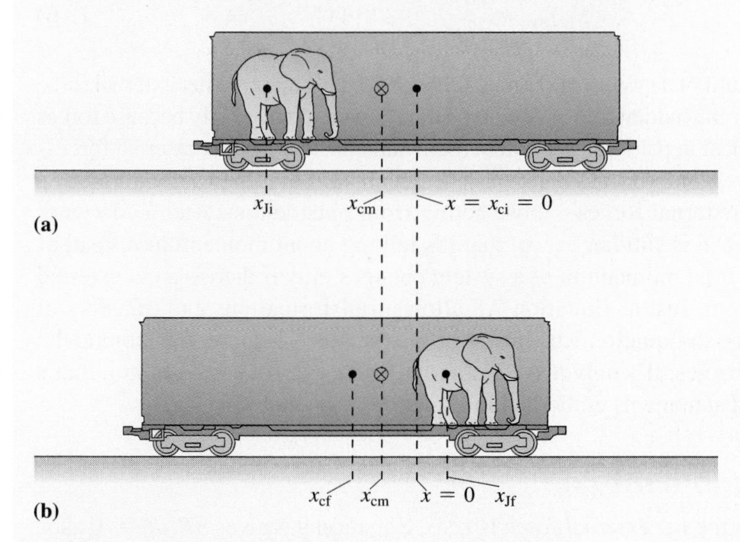

FIGURE 9.8 Jumbo walks, but the center of mass doesn't move.

We chose our coordinate system so that the car's initial position was $x_{ci} = 0$, so our expression for the initial position of the system's center of mass becomes

$$x_{\text{cm i}} = m_J x_{Ji}/M$$

Our expression for the final center-of-mass position, after Jumbo's walk, is $x_{\text{cm f}} = (m_J x_{Jf} + m_c x_{cf})/M$. We don't know either of the final coordinates x_{Jf} or x_{cf} here, but we do know that Jumbo walks 19 m *with respect to the car*. The elephant's final position x_{Jf} is therefore 19 m to the right of x_{Ji}, *adjusted for the car's displacement*. Therefore Jumbo ends up at $x_{Jf} = x_{Ji} + 19\text{ m} + x_{cf}$. You might think we need a minus sign because the car moves to the left. That's true, but the sign of x_{cf} will take care of that. Trust algebra! So our expression for the final center-of-mass position is

$$x_{\text{cm f}} = \frac{m_J x_{Jf} + m_c x_{cf}}{M} = \frac{m_J(x_{Ji} + 19\text{ m} + x_{cf}) + m_c x_{cf}}{M}$$

EVALUATE Finally, we equate our expressions for the initial and final positions of the center of mass. Again, that's because there are no forces external to the elephant–car system acting in the horizontal direction, so the center-of-mass position x_{cm} can't change. Thus we have $x_{\text{cm i}} = x_{\text{cm f}}$, or

$$\frac{m_J x_{Ji}}{M} = \frac{m_J(x_{Ji} + 19\text{ m} + x_{cf}) + m_c x_{cf}}{M}$$

The total mass M cancels, so we're left with the equation $m_J x_{Ji} = m_J(x_{Ji} + 19\text{ m} + x_{cf}) + m_c x_{cf}$. We aren't given x_{Ji}, but the term $m_J x_{Ji}$ is on both sides of this equation, so it cancels, leaving $0 = m(19\text{ m} + x_{cf}) + m_c x_{cf}$. We solve for the unknown x_{cf} to get

$$x_{cf} = -\frac{(19\text{ m})m_J}{(m_J + m_c)} = -\frac{(19\text{ m})(4.8\text{ t})}{(4.8\text{ t} + 15\text{ t})} = -4.6\text{ m}$$

The minus sign here indicates a displacement to the left, as we anticipated (Fig. 9.8*b*). Because the masses appear only in ratios, we didn't need to convert to kilograms.

ASSESS The car's 4.6-m displacement is quite a bit less than Jumbo's (which is 19 m − 4.6 m, or 14.4 m relative to the ground). That makes sense because Jumbo is considerably less massive than the car. ■

9.2 Momentum

In Chapter 4 we defined the linear momentum $\vec{p}$ of a particle as $\vec{p} = m\vec{v}$, and we first wrote Newton's law in the form $\vec{F} = d\vec{p}/dt$. We suggested that this form would play an important role in many-particle systems. We're now ready to explore that role.

The momentum of a system of particles is the vector sum of the individual momenta: $\vec{P} = \sum \vec{p}_i = \sum m_i\vec{v}_i$, where m_i and $\vec{v}_i$ are the masses and velocities of the individual particles. But we'd rather not have to keep track of all the particles in the system. Is there a simpler way to express the total momentum? There is, and it comes from writing the individual velocities as time derivatives of position: $\vec{v} = d\vec{r}/dt$. Then

$$\vec{P} = \sum m_i \frac{d\vec{r}_i}{dt} = \frac{d}{dt}\sum m_i \vec{r}_i$$

where the last step follows because the individual particle masses are constant and because the sum of derivatives is the derivative of the sum. In Section 9.1, we defined the center-of-mass position $\vec{r}_{cm}$ as $\sum m_i\vec{r}_i/M$, where M is the total mass. So the total momentum becomes

$$\vec{P} = \frac{d}{dt}M\vec{r}_{cm}$$

or, assuming the system mass M remains constant,

$$\vec{P} = M\frac{d\vec{r}_{cm}}{dt} = M\vec{v}_{cm} \tag{9.5}$$

where $\vec{v}_{cm} = d\vec{r}_{cm}/dt$ is the center-of-mass velocity. So a system's momentum is given by an expression similar to that of a single particle; it's the product of the system's mass and its velocity—that is, the velocity of its center of mass. If this seems obvious, watch out! We'll see soon that the same is *not* true for the system's total energy.

If we differentiate Equation 9.5 with respect to time, we have

$$\frac{d\vec{P}}{dt} = M\frac{d\vec{v}_{cm}}{dt} = M\vec{a}_{cm}$$

where $\vec{a}_{cm}$ is the center-of-mass acceleration. But we defined the center of mass so its motion obeyed Newton's second law, $\vec{F} = M\vec{a}_{cm}$, with $\vec{F}$ the net external force on the system. So we can write simply

$$\vec{F}_{\text{net ext}} = \frac{d\vec{P}}{dt} \tag{9.6}$$

showing that the momentum of a system of particles changes only if there's a net external force on the system. Remember the hidden role of Newton's third law in all this: Only because forces *internal* to the system cancel in pairs can we ignore them and consider just the external force.

Equation 9.6 might remind you of Equation 7.8, which said that the total energy of a system changes only if external forces—those acting from outside the system—do work on the system. Equation 9.6 is similar, except that it's talking about momentum instead of energy: It states that the total momentum of a system changes only if there's a net external force acting on the system. Just as Equation 7.8 allows transformations and transfers of energy within the system, so Equation 9.6 allows for the transfer of momentum among the system's constituent particles. It's only a system's total energy or total momentum that's constrained by the broad statements embodied in Equations 7.8 and 9.6.

Conservation of Momentum

In the special case when the net external force is zero, Equation 9.6 gives $d\vec{P}/dt = \vec{0}$, so

$$\vec{P} = \text{constant} \quad \text{(conservation of linear momentum)} \tag{9.7}$$

Equation 9.7 describes **conservation of linear momentum**, one of the most fundamental laws of physics:

Conservation of linear momentum: When the net external force on a system is zero, the total momentum $\vec{P}$ of the system—the vector sum of the individual momenta $m\vec{v}$ of its constituent particles—remains constant.

Momentum conservation holds no matter how many particles are involved and no matter how they're moving. It applies to systems ranging from atomic nuclei to pool balls, from colliding cars to galaxies. Although we derived Equation 9.7 from Newton's laws, momentum conservation is even more basic, since it applies to subatomic and nuclear systems where the laws and even the language of Newtonian physics are hopelessly inadequate. The following examples show the range and power of momentum conservation.

GOT IT? 9.2 A 500-g fireworks rocket is moving with velocity $\vec{v} = 60\hat{j}$ m/s at the instant it explodes. If you were to add the momentum vectors of all its fragments just after the explosion, what would be the result?

CONCEPTUAL EXAMPLE 9.1 Conservation of Momentum: Kayaking

Jess (mass 53 kg) and Nick (mass 72 kg) sit in a 26-kg kayak at rest on frictionless water. Jess tosses Nick a 17-kg pack, giving it horizontal speed 3.1 m/s relative to the water. What's the kayak's speed after Nick catches the pack? Why can you answer without doing any calculations?

EVALUATE Figure 9.9 shows the kayak before Jess tosses the pack and again after Nick catches it. The water is frictionless, so there's no net external force on the system comprising Jess, Nick, the kayak, and the pack. Since there's no net external force, the system's momentum is conserved. Everything is initially at rest, so that momentum is zero. Therefore, it's also zero after Nick catches the pack. At that point Jess, Nick, pack, and kayak are all at rest with respect to each other, so the only way the system's momentum can be zero is if they're also all at rest relative to the water. Therefore, the kayak's final speed is zero.

Initially all momenta are zero . . .

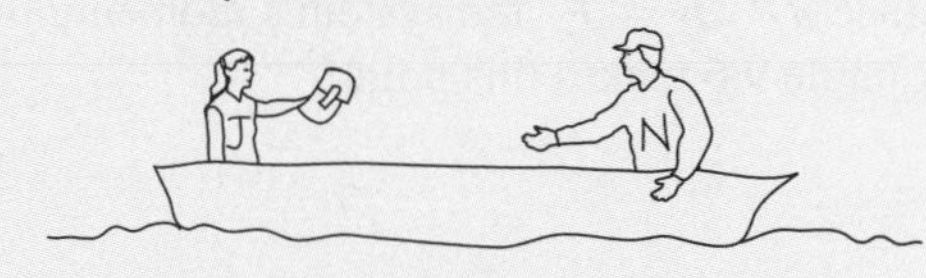

. . . and they're zero again after Nick has caught the pack.

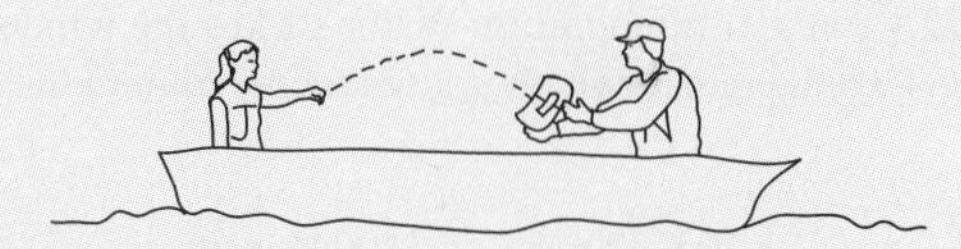

FIGURE 9.9 Our sketch for Conceptual Example 9.1.

ASSESS We didn't need any calculations here because the powerful conservation-of-momentum principle relates the initial and final states, without our having to know what happens in between.

MAKING THE CONNECTION What's the kayak's speed while the pack is in the air?

EVALUATE Momentum conservation still applies, and the system's total momentum is still zero. Now it consists of the pack's momentum $m_p\vec{v}_p$ and the momentum $(m_J + m_N + m_k)\vec{v}_k$ of Jess, Nick, and kayak, with common velocity $\vec{v}_k$ (Fig. 9.10). Sum these momenta, set the sum to zero, and solve, using the given quantities, to get $v_k = -0.35$ m/s. Here we've dropped vector signs; the minus sign then shows that the kayak's velocity is opposite the pack's. Since kayak and passengers are much more massive than the pack, it makes sense that their speed is lower.

FIGURE 9.10 Our sketch for Making the Connection 9.1.

EXAMPLE 9.5 Conservation of Momentum: Radioactive Decay

A lithium-5 nucleus (^{5}Li) is moving at 1.6 Mm/s when it decays into a proton (^{1}H, or p) and an alpha particle (^{4}He, or α). [Superscripts are the total numbers of nucleons and give the approximate masses in unified atomic mass units (u).] The alpha particle is detected moving at 1.4 Mm/s, at 33° to the original velocity of the ^{5}Li nucleus. What are the magnitude and direction of the proton's velocity?

INTERPRET Although the physical situation here is entirely different from the preceding example, we interpret this one, too, as being about momentum conservation. But there are two differences: First, in this case the total momentum isn't zero, and, second, this situation involves two dimensions. The fundamental principle is the same, however: In the absence of external forces, a system's total momentum can't change. Whether a pack gets tossed or a nucleus decays makes no difference.

DEVELOP Figure 9.11 shows what we know: the velocities for the Li and He nuclei. You can probably guess that the proton must emerge with a downward momentum component, but we'll let the math confirm that. We determine that Equation 9.7, $\vec{P}$ = constant, applies, with the constant equal to the ^{5}Li momentum. After the decay, we have two momenta to account for, so Equation 9.7 becomes

$$m_{\text{Li}}\vec{v}_{\text{Li}} = m_p\vec{v}_p + m_\alpha\vec{v}_\alpha$$

$\vec{v}_{\text{Li}}$ $\vec{v}_\alpha$ ϕ 33°

FIGURE 9.11 Our sketch for Example 9.5: what we're given.

(continued)

Let's choose the x-axis along the direction of $\vec{v}_{\text{Li}}$. Then the two components of the momentum conservation equation become

$$x\text{-component:} \quad m_{\text{Li}}v_{\text{Li}} = m_p v_{px} + m_\alpha v_{\alpha x}$$
$$y\text{-component:} \quad 0 = m_p v_{py} + m_\alpha v_{\alpha y}$$

Our plan is to solve these equations for the unknowns v_{px} and v_{py}. From these we can get the magnitude and direction of the proton's velocity.

EVALUATE From Fig. 9.11 it's evident that $v_{\alpha x} = v_\alpha \cos\phi$ and $v_{\alpha y} = v_\alpha \sin\phi$. So we can solve our two equations to get

$$v_{px} = \frac{m_{\text{Li}}v_{\text{Li}} - m_\alpha v_{\alpha x}}{m_p} = \frac{m_{\text{Li}}v_{\text{Li}} - m_\alpha v_\alpha \cos\phi}{m_p}$$
$$= \frac{(5.0\text{ u})(1.6\text{ Mm/s}) - (4.0\text{ u})(1.4\text{ Mm/s})(\cos 33°)}{1.0\text{ u}}$$
$$= 3.30\text{ Mm/s}$$
$$v_{py} = -\frac{m_\alpha v_{\alpha y}}{m_p} = -\frac{m_\alpha v_\alpha \sin\phi}{m_p}$$
$$= \frac{(4.0\text{ u})(1.4\text{ Mm/s})(\sin 33°)}{1.0\text{ u}} = -3.05\text{ Mm/s}$$

We've kept three significant figures in these intermediate results so we can get an accurate two-figure result for our final answer.

Thus the proton's speed is $v_p = \sqrt{v_{px}^2 + v_{py}^2} = 4.5$ Mm/s, and its direction is $\theta = \tan^{-1}(v_{py}/v_{px}) = -43°$. Note that here, as in Example 9.4, the masses appear only in ratios so we don't need to change units.

ASSESS Make sense? That negative θ tells us the proton's velocity is downward, as we anticipated. Figure 9.12 makes our result clear. Here we multiplied the velocities by the masses to get momentum vectors. The two momenta after the decay event have equal but opposite vertical components, reflecting that the total momentum of the system never had a vertical component. And the two horizontal components sum to give the initial momentum of the lithium nucleus. Momentum is indeed conserved.

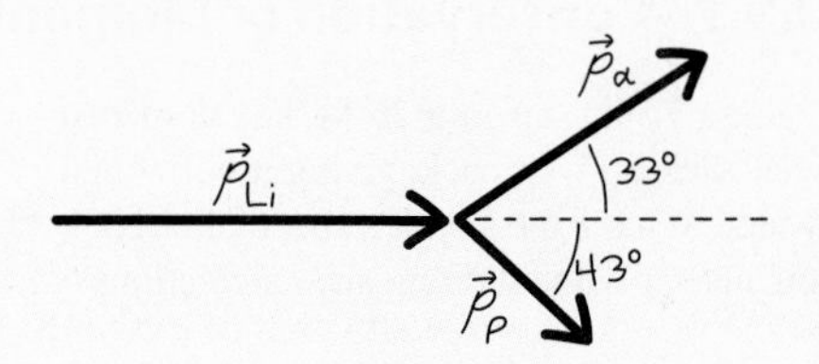

FIGURE 9.12 Our momentum diagram for Example 9.5.

A system's momentum is conserved only if no external forces act. Whether a force is internal or external depends on your choice of what constitutes the system—a choice that, as we noted in Chapter 6, is entirely up to you. In the two preceding examples, it was convenient to choose systems that weren't subject to external forces; then we could apply momentum conservation. Sometimes it's more convenient to deal with systems that do experience external forces; then, since $d\vec{P}/dt = \vec{F}$, the system's momentum changes at a rate equal to the external force. Example 9.6 makes this point.

EXAMPLE 9.6 Changing Momentum: Fighting a Fire

A firefighter directs a stream of water against the window of a burning building, hoping to break the window so water can get to the fire. The hose delivers water at the rate of 45 kg/s, and the water hits the window moving horizontally at 32 m/s. After hitting the window, the water drops vertically. What horizontal force does the water exert on the window?

INTERPRET We're asked about the window, but we're told a lot more about the water. The water stops at the window, so clearly the window exerts a force on the water—and by Newton's third law, that force is equal in magnitude to the force we're after—namely, the force of the water on the window. So we identify the water as our system and recognize that it's subject to an external force from the window.

DEVELOP Newton's law in the form $\vec{F} = d\vec{P}/dt$ applies to the water. So our plan is to find the rate at which the water's momentum changes. By Newton's second law, that's equal to the window's force on the water, and by Newton's third law, that's equal to the water's force on the window.

EVALUATE The water strikes the window at 32 m/s, so each kilogram of water loses 32 kg·m/s of momentum. Water strikes the window at the rate of 45 kg/s, so the rate at which it loses momentum to the window is

$$\frac{dP}{dt} = (45\text{ kg/s})(32\text{ m/s}) = 1400\text{ kg·m/s}^2$$

By Newton's second law, that's equal to the force on the water, and by the third law, that in turn is equal in magnitude to the force on the window. So the window experiences a 1400-N force from the water. Since the window is rigidly attached to the building and Earth, it doesn't experience significant acceleration—until it breaks and the glass fragments accelerate violently.

ASSESS 1400 N is about twice the weight of a typical person, and a fire hose produces quite a blast of water, so this number seems reasonable. Check the units, too: 1 kg·m/s^2 is equal to 1 N, so our answer does have the units of force.

GOT IT? 9.3 Two skaters toss a basketball back and forth on frictionless ice. Which of the following does not change? (a) the momentum of an individual skater; (b) the momentum of the basketball; (c) the momentum of the system consisting of one skater and the basketball; (d) the momentum of the system consisting of both skaters and the basketball

9.3 Kinetic Energy of a System

We've seen how the momentum of a many-particle system is determined entirely by the motion of its center of mass; the detailed behavior of the individual particles doesn't matter. For example, a firecracker sliding on ice has the same total momentum before and after it explodes.

The same, however, is *not* true of a system's kinetic energy. Energetically, that firecracker is very different after it explodes; internal potential energy has become kinetic energy of the fragments. Nevertheless, the center-of-mass concept remains useful in categorizing the kinetic energy associated with a system of particles.

The total kinetic energy of a system is the sum of the kinetic energies of the constituent particles: $K = \sum \frac{1}{2} m_i v_i^2$. But the velocity $\vec{v}_i$ of a particle can be written as the vector sum of the center-of-mass velocity $\vec{v}_{\mathrm{cm}}$ and a velocity $\vec{v}_{i\,\mathrm{rel}}$ of that particle relative to the center of mass: $\vec{v}_i = \vec{v}_{\mathrm{cm}} + \vec{v}_{i\,\mathrm{rel}}$. Then the total kinetic energy of the system is

$$K = \sum \tfrac{1}{2} m_i (\vec{v}_{\mathrm{cm}} + \vec{v}_{i\,\mathrm{rel}}) \cdot (\vec{v}_{\mathrm{cm}} + \vec{v}_{i\,\mathrm{rel}}) = \sum \tfrac{1}{2} m_i v_{\mathrm{cm}}^2 + \sum m_i \vec{v}_{\mathrm{cm}} \cdot \vec{v}_{i\,\mathrm{rel}} + \sum \tfrac{1}{2} m_i v_{i\,\mathrm{rel}}^2 \quad (9.8)$$

Let's examine the three sums making up the total kinetic energy. Since the center-of-mass speed v_{cm} is common to all particles, it can be factored out of the first sum, so $\sum \frac{1}{2} m_i v_{\mathrm{cm}}^2 = \frac{1}{2} v_{\mathrm{cm}}^2 \sum m_i = \frac{1}{2} M v_{\mathrm{cm}}^2$ where M is the total mass. This is the kinetic energy of a particle with mass M moving at speed v_{cm}, so we call it K_{cm}, the **kinetic energy of the center of mass**.

The center-of-mass velocity can also be factored out of the second term in Equation 9.8, giving $\sum m_i \vec{v}_{\mathrm{cm}} \cdot \vec{v}_{i\,\mathrm{rel}} = \vec{v}_{\mathrm{cm}} \cdot \sum m_i \vec{v}_{i\,\mathrm{rel}}$. Because the $\vec{v}_{i\,\mathrm{rel}}$'s are the particle velocities relative to the center of mass, the sum here is the total momentum relative to the center of mass. But that's zero, so the entire second term in Equation 9.8 is zero.

The third term in Equation 9.8, $\sum \frac{1}{2} m_i v_{i\,\mathrm{rel}}^2$, is the sum of the individual kinetic energies measured in a frame of reference moving with the center of mass. We call this term K_{int}, the **internal kinetic energy**.

With the middle term gone, Equation 9.8 shows that the kinetic energy of a system breaks into two terms:

$$K = K_{\mathrm{cm}} + K_{\mathrm{int}} \quad \text{(kinetic energy of a system)} \quad (9.9)$$

The first term, the kinetic energy of the center of mass, depends only on the center-of-mass motion. In our firecracker example, K_{cm} doesn't change when the firecracker explodes. The second term, the internal kinetic energy, depends only on the motions of the individual particles relative to the center of mass. The explosion dramatically increases this internal kinetic energy.

GOT IT? 9.4 Which of the following systems has (1) zero internal kinetic energy and (2) zero center-of-mass kinetic energy? (a) a pair of ice skaters, arms linked, skating together in a straight line; (b) a pair of skaters who start from rest facing each other and then push off so they're moving in opposite directions; (c) a pair of skaters as in (b) but who initially are moving together along the ice before they push off

APPLICATION Rockets

Rockets provide propulsion in the vacuum of space, where there's nothing for a wheel or propeller to push against. If no external forces act, total momentum stays constant. As the rocket's exhaust carries away momentum, the result is an equal but oppositely directed momentum gain for the rocket. The rate of momentum change is the force on the rocket, which engineers call *thrust*. As with the fire hose in Example 9.6, thrust is the product of the exhaust rate dM/dt and exhaust speed v_{ex}: $F = v_{\mathrm{ex}}\, dM/dt$. Because the rocket has to carry the mass it's going to exhaust, the most efficient rockets use high exhaust velocities and therefore need less fuel.

What actually propels the rocket? It's ultimately hot gases inside the rocket engine pushing on the front of the engine chamber. The rocket doesn't "push against" anything outside itself; all the pushing is done *inside* the rocket engine, accelerating the rocket forward. That's why rockets work just fine in the vacuum of space.

The photo shows the 2011 launch of the *Juno* spacecraft, heading for its 2016 rendezvous with Jupiter.

9.4 Collisions

A **collision** is a brief, intense interaction between objects. Examples abound: automobile collisions; collisions of balls on a pool table; the collision of a tennis ball and racket, baseball and bat, or football and foot; an asteroid colliding with a planet; and collisions of high-energy particles that probe the fundamental structure of matter. Less obvious are collisions among galaxies that last a hundred million years, the interaction of a spacecraft with a planet as the craft gains energy for a voyage to the outer solar system, and the repulsive interaction of two protons that approach but never touch. All these collisions meet two criteria. First, they're brief, lasting but a short time in the overall context of the colliding objects' motions. On a pool table, the collision time is short compared with the

Video Tutor Demo | **Water Rocket**

APPLICATION Crash Tests

Automotive engineers perform crash tests to assess the safety of their vehicles. Sensors measure the rapidly varying forces as the test car collides with a fixed barrier. The graph below is a force-versus-time curve from a typical crash test; impulse is the area under the curve. In addition to force sensors on the vehicle, accelerometers in crash-test dummies determine the maximum accelerations of the heads and other body parts to assess potential injuries.

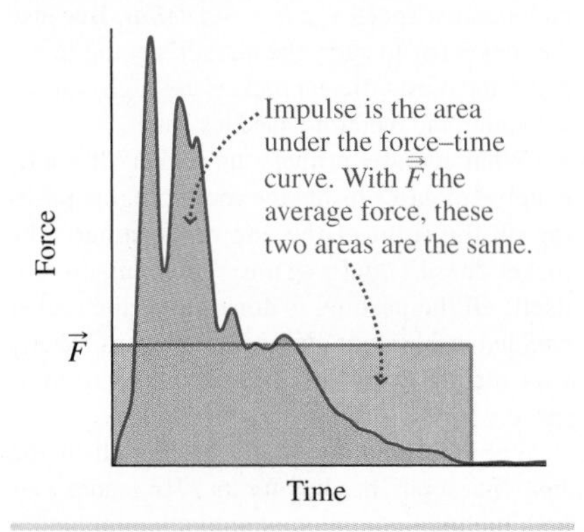

Video Tutor Demo | **Happy/Sad Pendulums**

time it takes for a ball to roll across the table. An automobile collision lasts a fraction of a second. A baseball spends far more time coming from the pitcher than it does interacting with the bat. And even 10^8 years is short compared with the lifetime of a galaxy. Second, collisions are intense: Forces among the interacting objects are far larger than any external forces that may be acting on the system. External forces are therefore negligible during the collision, so the total momentum of the colliding objects remains essentially unchanged.

Impulse

The forces between colliding objects are *internal* to the system comprising those objects, so they can't alter the total momentum of the system. But they dramatically alter the motions of the colliding objects. How much depends on the magnitude of the force and how long it's applied. If $\bar{\vec{F}}$ is the average force acting on one object during a collision that lasts for time Δt, then Newton's second law reads $\bar{\vec{F}} = \Delta\vec{p}/\Delta t$ or

$$\Delta\vec{p} = \bar{\vec{F}}\Delta t \qquad (9.10a)$$

The product of average force and time that appears in this equation is called **impulse**. It's given the symbol $\vec{J}$, and its units are newton-seconds.

An impulse $\vec{J}$ produces the same momentum change regardless of whether it involves a larger force exerted over a shorter time or a smaller force exerted over a longer time. The force in a collision usually isn't constant and can fluctuate wildly. In that case, we find the impulse by integrating the force over time, so the momentum change becomes

$$\Delta\vec{p} = \vec{J} = \int \vec{F}(t)\,dt \qquad \text{(impulse)} \qquad (9.10b)$$

Although we introduced impulse in the context of collisions, it's useful in other situations involving intense forces applied over short times. For example, small rocket engines are characterized by the impulse they impart.

Energy in Collisions

Kinetic energy may or may not be conserved in a collision. If it is, then the collision is **elastic**; if not, it's **inelastic**. An elastic collision requires that the forces between colliding objects be conservative; then kinetic energy is stored briefly as potential energy and released when the collision is over. Interactions at the atomic and nuclear scales are often truly elastic. In the macroscopic realm, nonconservative forces convert kinetic energy into internal energy, heating the colliding objects, or they may permanently deform the objects; either way, nonconservative forces rob the system of mechanical energy. But even many macroscopic collisions are close enough to elastic that we can neglect mechanical energy loss during the collision.

GOT IT? 9.5 Which of the following qualifies as a collision? Of the collisions, which are nearly elastic and which inelastic? (a) A basketball rebounds off the backboard; (b) two magnets approach, their north poles facing; they repel and reverse direction without touching; (c) a basketball flies through the air on a parabolic trajectory; (d) a truck strikes a parked car and the two slide off together, crumpled metal hopelessly intertwined; (e) a snowball splats against a tree, leaving a lump of snow adhering to the bark.

9.5 Totally Inelastic Collisions

In a **totally inelastic collision**, the colliding objects stick together to form a single object. Even then, kinetic energy is usually not all lost. But a totally inelastic collision entails the maximum energy loss consistent with momentum conservation. The motion after a totally inelastic collision is determined entirely by momentum conservation, and that makes totally inelastic collisions easy to analyze.

Consider masses m_1 and m_2 with initial velocities $\vec{v}_1$ and $\vec{v}_2$ that undergo a totally inelastic collision. After colliding, they stick together to form a single object of mass $m_1 + m_2$ and final velocity $\vec{v}_f$. Conservation of momentum states that the initial and final momenta of this system must be the same:

$$m_1\vec{v}_1 + m_2\vec{v}_2 = (m_1 + m_2)\vec{v}_f \quad \text{(totally inelastic collision)} \tag{9.11}$$

Given four of the five quantities m_1, $\vec{v}_1$, m_2, $\vec{v}_2$, and $\vec{v}_f$, we can solve for the fifth.

EXAMPLE 9.7 An Inelastic Collision: Hockey

The hockey captain, a physics major, decides to measure the puck's speed. He loads a small Styrofoam chest with sand, giving a total mass of 6.4 kg. He places it at rest on frictionless ice. The 160-g puck strikes the chest and embeds itself in the Styrofoam. The chest moves off at 1.2 m/s. What was the puck's speed?

INTERPRET This is a totally inelastic collision. We identify the system as consisting of puck and chest. Initially, all the system's momentum is in the puck; after the collision, it's in the combination puck + chest. In this case of a single nonzero velocity before collision and a single velocity after, momentum conservation requires that both motions be in the same direction. Therefore, we have a one-dimensional problem.

DEVELOP Figure 9.13 is a sketch of the situation before and after the collision. With a totally inelastic collision, Equation 9.11—the statement of momentum conservation—tells it all. In our one-dimensional situation, this equation becomes $m_p v_p = (m_p + m_c)v_c$, where the subscripts p and c stand for puck and chest, respectively.

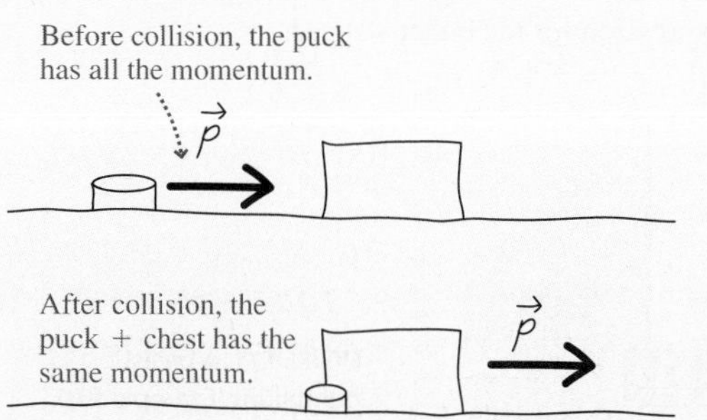

FIGURE 9.13 Our sketch for Example 9.7.

EVALUATE Here we want the initial puck velocity, so we solve for v_p:

$$v_p = \frac{(m_p + m_c)v_c}{m_p} = \frac{(0.16\text{ kg} + 6.4\text{ kg})(1.2\text{ m/s})}{0.16\text{ kg}} = 49\text{ m/s}$$

ASSESS Make sense? Yes: The puck's mass is small, so it needs a much higher speed to carry the same momentum as the much more massive chest. Variations on this technique are often used to determine speeds that would be difficult to measure directly. ■

EXAMPLE 9.8 Conservation of Momentum: Fusion

In a fusion reaction, two deuterium nuclei (^{2}H) join to form helium (^{4}He). Initially, one of the deuterium nuclei is moving at 3.5 Mm/s, the second at 1.8 Mm/s at a 64° angle to the velocity of the first. Find the speed and direction of the helium nucleus.

INTERPRET Although the context is very different, this is another totally inelastic collision. But here both objects are initially moving, and in different directions, so we have a two-dimensional situation. We identify the system as consisting of initially the two deuterium nuclei and finally the single helium nucleus. We're asked for the final velocity of the helium, expressed as magnitude (speed) and direction.

DEVELOP Figure 9.14 shows the situation. Momentum is conserved, so Equation 9.11 applies; solving that equation for $\vec{v}_f$ gives $\vec{v}_f = (m_1\vec{v}_1 + m_2\vec{v}_2)/(m_1 + m_2)$. In two dimensions, this represents two equations for the two components of $\vec{v}_f$. We need a coordinate system, and Fig. 9.14 shows our choice, with the x-axis along the motion of the first deuterium nucleus. We need the components of the initial velocities in order to apply our equation for $\vec{v}_f$.

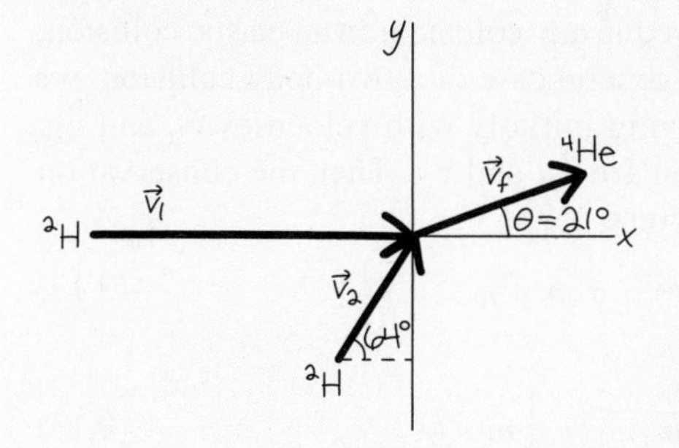

FIGURE 9.14 Our sketch of the velocity vectors for Example 9.8.

EVALUATE With $\vec{v}_1$ in the x-direction, we have $v_{1x} = 3.5$ Mm/s and $v_{1y} = 0$. Figure 9.14 shows that $v_{2x} = (1.8\text{ Mm/s})(\cos 64°) = 0.789$ Mm/s and $v_{2y} = (1.8\text{ Mm/s})(\sin 64°) = 1.62$ Mm/s. So the components of our equation become

$$v_{fx} = \frac{m_1 v_{1x} + m_2 v_{2x}}{m_1 + m_2} = \frac{(2\text{ u})(3.5\text{ Mm/s}) + (2\text{ u})(0.789\text{ Mm/s})}{2\text{ u} + 2\text{ u}} = 2.14\text{ Mm/s}$$

$$v_{fy} = \frac{m_1 v_{1y} + m_2 v_{2y}}{m_1 + m_2} = \frac{0 + (2\text{ u})(1.62\text{ Mm/s})}{2\text{ u} + 2\text{ u}} = 0.809\text{ Mm/s}$$

As in Example 9.5, the superscripts are the nuclear masses in u, and because the mass units cancel, there's no need to convert to kilograms.

From these velocity components we can get the final speed and direction: $v_f = \sqrt{v_{fx}^2 + v_{fy}^2} = 2.3$ Mm/s and $\theta = \tan^{-1}(v_{fy}/v_{fx}) = 21°$. We show this final velocity on the diagram in Fig. 9.14.

(continued)

ASSESS In this example the two incident particles have the same masses, so their velocities are proportional to their momenta. Figure 9.14 shows that the total initial momentum is largely horizontal, with a smaller vertical component, so the 21° angle of the final velocity makes sense. The magnitude of $\vec{v}_f$ also makes sense: Now the total momentum is contained in a single, more massive particle, so we expect a final speed comparable to the initial speeds. ■

EXAMPLE 9.9 The Ballistic Pendulum

The ballistic pendulum measures the speeds of fast-moving objects like bullets. It consists of a wooden block of mass M suspended from vertical strings (Fig. 9.15). A bullet of mass m strikes and embeds itself in the block, and the block swings upward through a vertical distance h. Find an expression for the bullet's speed.

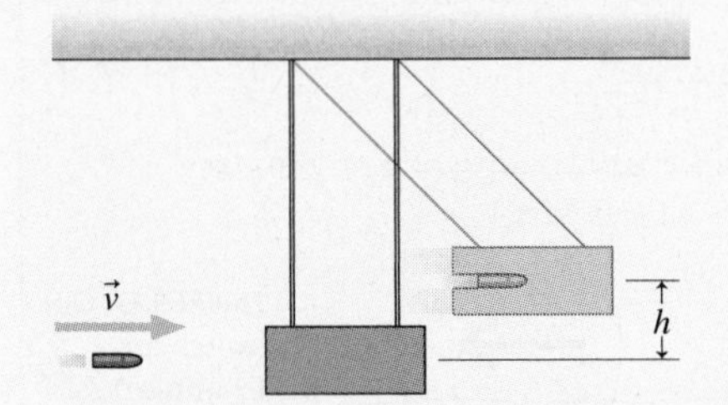

FIGURE 9.15 A ballistic pendulum (Example 9.9).

INTERPRET Interpreting this example is a bit more involved. We actually have two separate events: the bullet striking the block and the subsequent rise of the block. We can interpret the first event as a one-dimensional totally inelastic collision, as in Example 9.7. Momentum is conserved during this event but, because the collision is inelastic, mechanical energy is not. Then the block rises, and now a net external force—from string tension and gravity—acts to change the momentum. But gravity is conservative, and the string tension does no work, so now mechanical energy is conserved.

DEVELOP Figure 9.15 is our drawing. Our plan is to separate the two parts of the problem and then to combine the results to get our final answer. First is the inelastic collision; here momentum is conserved, so Equation 9.11 applies. In one dimension, that reads $mv = (m + M)V$, where v is the initial bullet speed and V is the speed of the block with embedded bullet just after the collision. Solving gives $V = mv/(m + M)$. Now the block swings upward. Momentum isn't conserved, but mechanical energy is. Setting the zero of potential energy in the block's initial position, we have $U_0 = 0$ and—using the situation just after the collision as the initial state—$K_0 = \frac{1}{2}(m + M)V^2$. At the peak of its swing the block is momentarily at rest, so $K = 0$. But it's risen a height h, so the potential energy is $U = (m + M)gh$. Conservation of mechanical energy reads $K_0 + U_0 = K + U$—in this case, $\frac{1}{2}(m + M)V^2 = (m + M)gh$.

EVALUATE Now we've got two equations describing the two parts of the problem. Using our expression for V from momentum conservation in the energy-conservation equation, we get

$$\frac{1}{2}\left(\frac{mv}{m + M}\right)^2 = gh$$

Solving for the bullet speed v then gives our answer:

$$v = \left(\frac{m + M}{m}\right)\sqrt{2gh}$$

ASSESS Make sense? Yes: The smaller the bullet mass m, the higher velocity it must have to carry a given momentum; that's reflected by the factor m alone in the denominator. The higher the rise h, the greater the bullet speed. But the speed scales not as h itself but as $\sqrt{h}$. That's because kinetic energy—which turned into potential energy of the rise—depends on velocity *squared*. ■

GOT IT? 9.6 Which of the following collisions qualify as totally inelastic? (a) Two equal-mass objects approach from opposite directions at different speeds. They collide head-on and stick together; the combined object continues to move; (b) two equal-mass objects approach from opposite directions at the same speed. They collide head-on and stick together; the combined object is then at rest; (c) two equal-mass objects approach from opposite directions at the same speed. They collide head-on and rebound, but with lower speed than before.

9.6 Elastic Collisions

We've seen that momentum is essentially conserved in any collision. In an elastic collision, kinetic energy is conserved as well. In the most general case of a two-body collision, we consider two objects of masses m_1 and m_2, moving initially with velocities $\vec{v}_{1i}$ and $\vec{v}_{2i}$, respectively. Their final velocities after collision are $\vec{v}_{1f}$ and $\vec{v}_{2f}$. Then the conservation statements for momentum and kinetic energy become

$$m_1\vec{v}_{1i} + m_2\vec{v}_{2i} = m_1\vec{v}_{1f} + m_2\vec{v}_{2f} \tag{9.12}$$

and

$$\tfrac{1}{2}m_1v_{1i}^2 + \tfrac{1}{2}m_2v_{2i}^2 = \tfrac{1}{2}m_1v_{1f}^2 + \tfrac{1}{2}m_2v_{2f}^2 \tag{9.13}$$

Given initial velocities, we'd like to predict the outcome of a collision. In the totally inelastic two-dimensional collision, we had enough information to solve the problem. Here, in the two-dimensional elastic case, we have the two components of the momentum conservation equation 9.12 and the single scalar equation for energy conservation 9.13. But we have four unknowns—the magnitudes and directions of both final velocities. With three equations and four unknowns, we don't have enough information to solve the general two-dimensional elastic collision. Later we'll see how other information can help solve such problems. First, though, we look at the special case of one-dimensional elastic collisions.

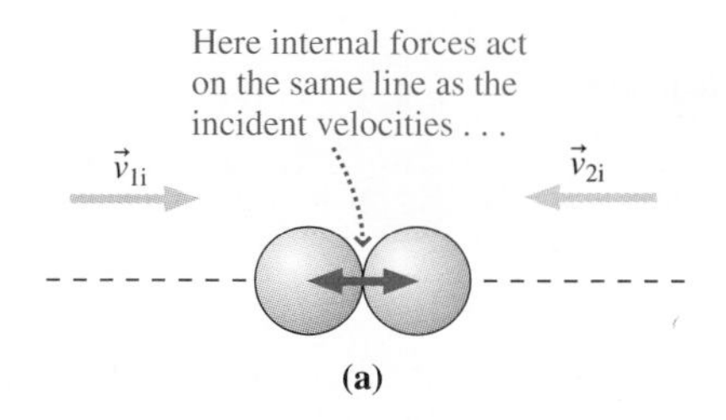

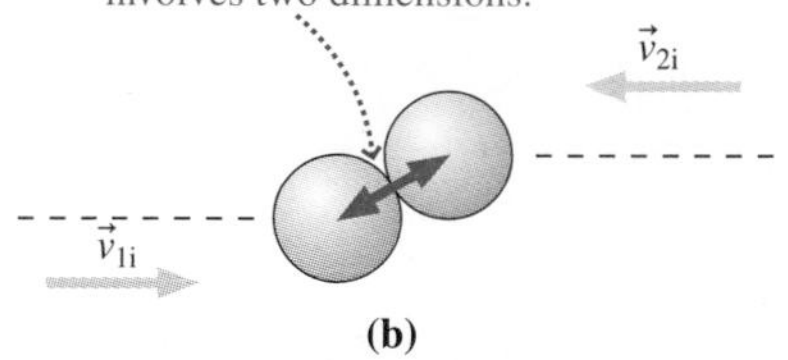

FIGURE 9.16 Only a head-on collision is one-dimensional.

Elastic Collisions in One Dimension

When two objects collide head-on, the internal forces act along the same line as the incident motion, and the objects' subsequent motion must therefore be along that same line (Fig. 9.16*a*). Although such one-dimensional collisions are a special case, they do occur and they provide much insight into the more general case.

In the one-dimensional case, the momentum conservation (Equation 9.12) has only one nontrivial component:

$$m_1v_{1i} + m_2v_{2i} = m_1v_{1f} + m_2v_{2f} \tag{9.12a}$$

where the v's stand for velocity components, rather than magnitudes, and can therefore be positive or negative. If we collect together the terms in Equations 9.12a and 9.13 that are associated with each mass, we have

$$m_1(v_{1i} - v_{1f}) = m_2(v_{2f} - v_{2i}) \tag{9.12b}$$

and

$$m_1(v_{1i}^2 - v_{1f}^2) = m_2(v_{2f}^2 - v_{2i}^2) \tag{9.13a}$$

But $a^2 - b^2 = (a + b)(a - b)$, so Equation 9.13a can be written

$$m_1(v_{1i} - v_{1f})(v_{1i} + v_{1f}) = m_2(v_{2f} - v_{2i})(v_{2f} + v_{2i}) \tag{9.13b}$$

Dividing the left and right sides of Equation 9.13b by the corresponding sides of Equation 9.12b then gives

$$v_{1i} + v_{1f} = v_{2f} + v_{2i}$$

Rearranging shows that

$$v_{1i} - v_{2i} = v_{2f} - v_{1f} \tag{9.14}$$

What does this equation tell us? Both sides describe the relative velocity between the two particles; the equation therefore shows that the relative speed remains unchanged after the collision, although the direction reverses. If the two objects are approaching at a relative speed of 5 m/s, then after collision they'll separate at 5 m/s.

Continuing our search for the final velocities, we solve Equation 9.14 for v_{2f}:

$$v_{2f} = v_{1i} - v_{2i} + v_{1f}$$

and use this result in Equation 9.12a:

$$m_1v_{1i} + m_2v_{2i} = m_1v_{1f} + m_2(v_{1i} - v_{2i} + v_{1f})$$

Solving for v_{1f} then gives

$$v_{1f} = \frac{m_1 - m_2}{m_1 + m_2}v_{1i} + \frac{2m_2}{m_1 + m_2}v_{2i} \tag{9.15a}$$

Problem 71 asks you to show similarly that

$$v_{2f} = \frac{2m_1}{m_1 + m_2}v_{1i} + \frac{m_2 - m_1}{m_1 + m_2}v_{2i} \tag{9.15b}$$

Equations 9.15 are our desired result, expressing the final velocities in terms of the initial velocities alone.

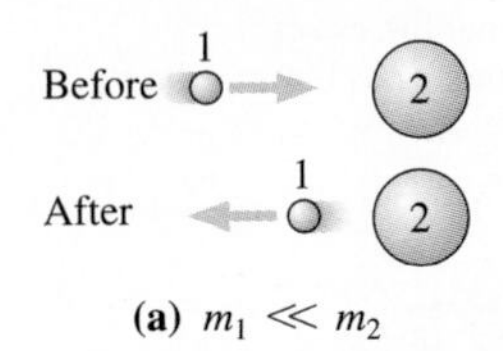

Before ① ②

After ① ②

(b) $m_1 = m_2$

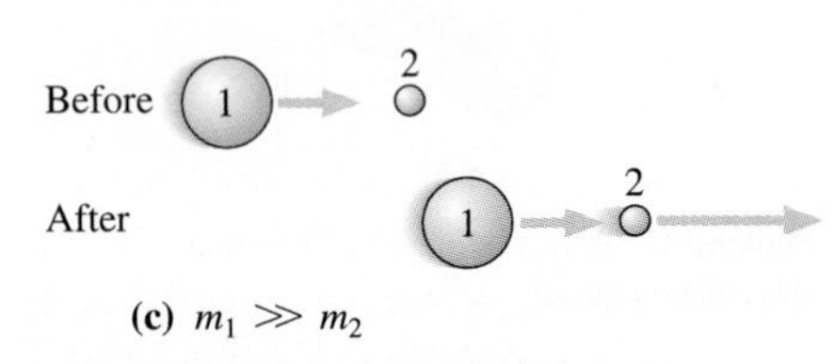

FIGURE 9.17 Special cases of elastic collisions in one dimension.

To see that these results make sense, we suppose that $v_{2i} = 0$. (This really isn't a special case, since we can always work in a reference frame with m_2 initially at rest.) We then consider the three special cases of one-dimensional elastic collisions illustrated in Fig. 9.17.

Case 1: $m_1 \ll m_2$ (Fig. 9.17*a*) Picture a ping-pong ball colliding with a bowling ball, or any object colliding elastically with a perfectly rigid surface. If we set $v_{2i} = 0$ in Equations 9.15, and drop m_1 as being negligible compared with m_2, Equations 9.15 become simply

$$v_{1f} = -v_{1i}$$

and

$$v_{2f} = 0$$

That is, the lighter object rebounds with no change in speed, while the heavier object remains at rest. Does this make sense in light of the conservation laws that Equations 9.15 are supposed to reflect? First consider energy conservation: The kinetic energy of m_2 remains zero and, because m_1's *speed* doesn't change, neither does its kinetic energy $\frac{1}{2}m_1v_1^2$. So kinetic energy is conserved. But what about momentum? The momentum of the lighter object has changed, from $m_1 v_{1i}$ to $-m_1 v_{1i}$. But momentum *is* conserved; the momentum given up by the lighter object is absorbed by the heavier object. In the limit of an arbitrarily large m_2, the heavier object can absorb huge amounts of momentum mv without acquiring significant speed. If we "back off" from the extreme case where m_1 can be neglected altogether, we would find that a lighter object striking a heavier one rebounds with reduced speed and that the heavier object begins moving slowly in the opposite direction.

Case 2: $m_1 = m_2$ (Fig. 9.17*b*) Again with $v_{2i} = 0$, Equations 9.15 now give

$$v_{1f} = 0$$

and

$$v_{2f} = v_{1i}$$

PhET: Collisions (Introduction)

So the first object stops abruptly, transferring all its energy and momentum to the second. A head-on collision between billiard balls is an almost perfect example of this type of collision. For purposes of energy transfer, two equal-mass particles are perfectly "matched." We'll encounter analogous instances of energy transfer "matching" when we discuss wave motion and again in connection with electric circuits.

Case 3: $m_1 \gg m_2$ (Fig. 9.17*c*) Now Equations 9.15 give

$$v_{1f} = v_{1i}$$

and

$$v_{2f} = 2v_{1i}$$

where we've neglected m_2 compared with m_1. So here the more massive object barrels right on with no change in motion, while the lighter one heads off with twice the speed of the massive one. This result is entirely consistent with our earlier claim that the relative speed remains unchanged in a one-dimensional elastic collision. How are momentum and energy conserved in this case? In the extreme limit where we neglect the mass m_2, its energy and momentum are negligible. Essentially all the energy and momentum remain with the more massive object, and both these quantities are essentially unchanged in the collision. In the less extreme case where an object of finite mass strikes a less massive object initially at rest, both objects move off in the initial direction of the incident object, with the lighter one moving faster.

EXAMPLE 9.10 Elastic Collisions: Nuclear Engineering

Nuclear power reactors include a substance called a *moderator*, whose job is to slow the neutrons liberated in nuclear fission, making them more likely to induce additional fission and thus sustain a nuclear chain reaction. A Canadian reactor design uses so-called *heavy water* as its moderator. In heavy water, ordinary hydrogen atoms are replaced by deuterium, the rare form of hydrogen whose nucleus consists of a proton and a neutron. The mass of this *deuteron* is thus about 2 u, compared with a neutron's 1 u. Find the fraction of a neutron's kinetic energy that's transferred to an initially stationary deuteron in a head-on elastic collision.

INTERPRET We have a head-on collision, so we're dealing with a one-dimensional situation. The system of interest consists of the neutron and the deuteron. We're not told much else except the masses of the two particles. That should be enough, though, because we're not asked for the final velocities but rather for a ratio of related quantities—namely, kinetic energies.

DEVELOP Since we have a one-dimensional elastic collision, Equations 9.15 apply. We're asked for the fraction of the neutron's kinetic energy that gets transferred to the deuteron, so we need to express the deuteron's final velocity in terms of the neutron's initial velocity. If we take the neutron to be particle 1, then we want Equation 9.15b. With the deuteron initially at rest, $v_{2i} = 0$ and the equation becomes $v_{2f} = 2m_1 v_{1i}/(m_1 + m_2)$. Our plan is to use this equation to determine the kinetic-energy ratio.

EVALUATE The kinetic energies of the two particles are given by $K_1 = \frac{1}{2}m_1 v_1^2$ and $K_2 = \frac{1}{2}m_2 v_2^2$. Using our equation for v_{2f} gives

$$K_2 = \frac{1}{2}m_2\left(\frac{2m_1 v_1}{m_1 + m_2}\right)^2 = \frac{2m_2 m_1^2 v_1^2}{(m_1 + m_2)^2}$$

We want to compare this with K_1:

$$\frac{K_2}{K_1} = K_2\left(\frac{1}{K_1}\right) = \left(\frac{2m_2 m_1^2 v_1^2}{(m_1 + m_2)^2}\right)\left(\frac{1}{\frac{1}{2}m_1 v_1^2}\right) = \frac{4m_1 m_2}{(m_1 + m_2)^2} \qquad (9.16)$$

In this case $m_1 = 1$ u and $m_2 = 2$ u, so we have $K_2/K_1 = 8/9 \simeq 0.89$. Thus 89% of the incident energy is transferred in a single collision, leaving the neutron with 11% of its initial energy.

ASSESS Let's take a look at Equation 9.16 in the context of our three special cases. We numbered this equation because it's a general result for the fractional energy transfer in any one-dimensional elastic collision. In case 1, $m_1 \ll m_2$, so we neglect m_1 compared with m_2 in the denominator; then our energy ratio is approximately $4m_1/m_2$. This becomes zero in the extreme limit where m_1's mass is negligible—consistent with our case 1 where the massive object didn't move at all. In case 2, $m_1 = m_2$, and Equation 9.16 becomes $4m^2/(2m)^2 = 1$, where m is the mass of both objects. That too agrees with our earlier analysis: The incident object stops and transfers all its energy to the struck object. Finally, in case 3, $m_1 \gg m_2$, so we neglect m_2 in the denominator. Now the energy ratio becomes $4m_2/m_1$. As in case 1, this approaches zero as the mass ratio gets extremely large. So the maximum energy transfer occurs with two equal masses, and tails off toward zero if the mass ratio becomes extreme in either direction.

For the particles in this example, the mass ratio 1:2 is close enough to equality that the energy transfer is nearly 90% efficient. Problem 84 explores further this energy transfer. ■

GOT IT? 9.7 One ball is at rest on a level floor. A second ball collides elastically with the first, and the two move off separately but in the same direction. What can you conclude about the masses of the two balls?

Elastic Collisions in Two Dimensions

Analyzing an elastic collision in two dimensions requires the full vector statement of momentum conservation (Equation 9.12), along with the statement of energy conservation (Equation 9.13). But these equations alone don't provide enough information to solve a problem. In a collision between reasonably simple macroscopic objects, that information may be provided by the so-called **impact parameter**, a measure of how much the collision differs from being head-on (Fig. 9.18). More typically—especially with atomic and nuclear interactions—the necessary information must be supplied by measurements done after the collision. Knowing the direction of motion of one particle after collision, for example, provides enough information to analyze a collision if the masses and initial velocities are also known.

PhET: Collisions (Advanced)

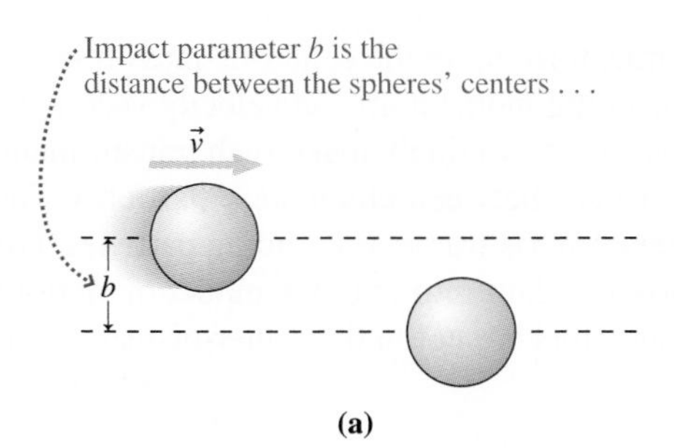

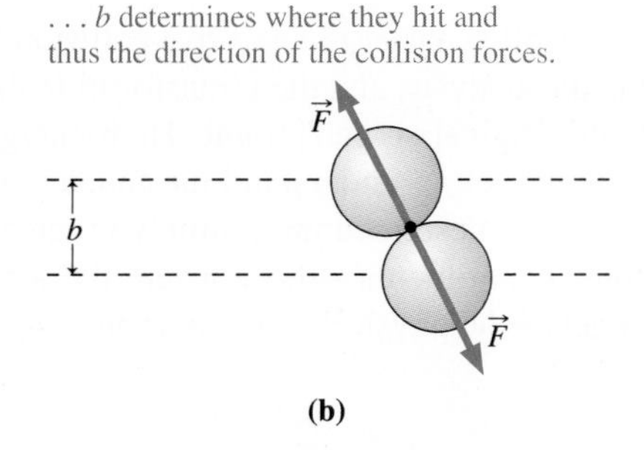

FIGURE 9.18 The impact parameter b determines the directions of the collision forces.

EXAMPLE 9.11 A Two-Dimensional Elastic Collision: Croquet

A croquet ball strikes a stationary ball of equal mass. The collision is elastic, and the incident ball goes off at 30° to its original direction. In what direction does the other ball move?

INTERPRET We've got an elastic collision, so both momentum and kinetic energy are conserved. The system consists of the two croquet balls. We aren't given a lot of information, but since we're asked only for a direction, the magnitudes of the velocities won't matter. Thus we've got what we need to know about the initial velocities, and we've got one other piece of information, so we have enough to solve the problem.

DEVELOP Figure 9.19 shows the situation, in which we're after the unknown angle θ. Since the collision is elastic, Equations 9.12 (momentum conservation) and 9.13 (energy conservation) both apply. The masses are equal, so they cancel from both equations. With $v_{2i} = 0$, we then have $\vec{v}_{1i} = \vec{v}_{1f} + \vec{v}_{2f}$ for momentum conservation and $v_{1i}^2 = v_{1f}^2 + v_{2f}^2$ for energy conservation. The rest will be algebra.

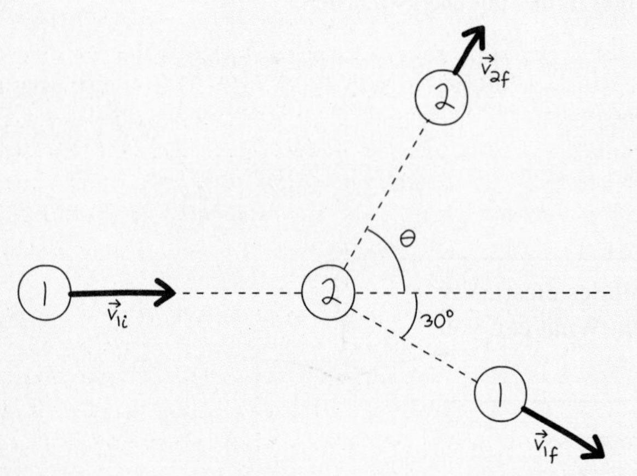

FIGURE 9.19 Our sketch of the collision between croquet balls of equal mass (Example 9.11).

EVALUATE Solving for one unknown in terms of another is going to get messy here, with some velocities squared and some not. Here's a more clever approach: Rather than write the momentum equation in two components, let's take the dot product of each side with itself. That will bring in velocity-squared terms, letting us combine the momentum and energy equations. And the dot product includes an angle—which is what we're asked to find.

The dot product is distributive and commutative, so here's what we get when we dot the momentum equation with itself:

$$\vec{v}_{1i}\cdot\vec{v}_{1i} = (\vec{v}_{1f} + \vec{v}_{2f})\cdot(\vec{v}_{1f} + \vec{v}_{2f})$$
$$= \vec{v}_{1f}\cdot\vec{v}_{1f} + \vec{v}_{2f}\cdot\vec{v}_{2f} + 2\vec{v}_{1f}\cdot\vec{v}_{2f}$$

Recall that the dot product of two vectors is the product of their magnitudes with the cosine of the angle between them: $\vec{A}\cdot\vec{B} = AB\cos\theta$. Since the angle between a vector and itself is zero, the dot product of a vector with itself is the square of its magnitude: $\vec{A}\cdot\vec{A} = A^2\cos(0) = A^2$. So our equation becomes

$$v_{1i}^2 = v_{1f}^2 + v_{2f}^2 + 2v_{1f}v_{2f}\cos(\theta + 30°)$$

where the argument of the cosine follows because, as Fig. 9.19 shows, the angle between $\vec{v}_{1f}$ and $\vec{v}_{2f}$ is $\theta + 30°$. We now subtract the energy equation from this new equation to get

$$2v_{1f}v_{2f}\cos(\theta + 30°) = 0$$

But neither of the final speeds is zero, so this equation requires that $\cos(\theta + 30°) = 0$. Thus $\theta + 30° = 90°$, and our answer follows: $\theta = 60°$.

ASSESS This result seems reasonable, although we don't have a lot to go on because we haven't calculated the final speeds. But it's intriguing that the two balls go off at right angles to each other. Is this a coincidence? No: It happens in any two-dimensional elastic collision between objects of equal mass when one is initially at rest. You can prove this in Problem 72. ■

The Center-of-Mass Frame

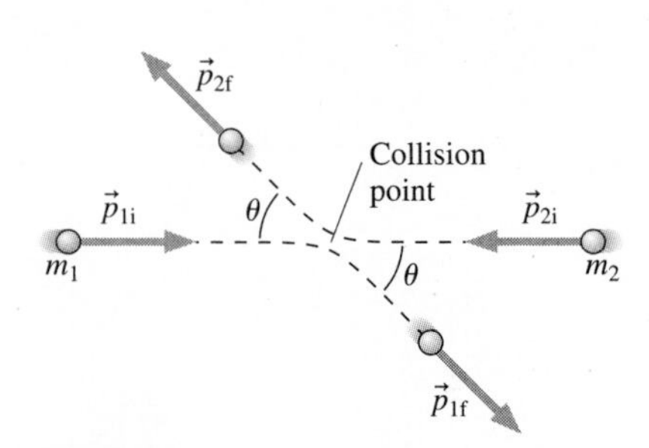

FIGURE 9.20 An elastic collision viewed in the center-of-mass frame, showing that the initial and final momentum vectors form pairs with equal magnitudes and opposite directions.

Two-dimensional collisions take a particularly simple form in a frame of reference moving with the center of mass of the colliding particles, since the total momentum in such a frame must be zero. That remains true after a collision, which involves only *internal* forces that don't affect the center of mass. Therefore, both the initial and final momenta form pairs of oppositely directed vectors of equal magnitude, as shown in Fig. 9.20. In an elastic collision, energy conservation requires further that the incident and final momenta have the same values, so a single number—the angle θ in Fig 9.20—completely describes the collision.

It's often easier to analyze a collision by transforming to the center-of-mass frame, doing the analysis, and then transforming the resulting momentum and velocity vectors back to the original or "lab" frame. High-energy physicists routinely make such transformations as they seek to understand the fundamental forces between elementary particles. Those forces are described most simply in the center-of-mass frame of colliding particles, but in some experiments—those where lighter particles slam into massive nuclei or stationary targets—the physicists and their particle accelerators are not in the center-of-mass frame.

CONCEPTUAL EXAMPLE 9.2 In the CM Frame

Figure 9.21 shows initial and final velocities for a collision between two equal masses as observed in their center-of-mass reference frame. What would a comparable diagram look like in a reference frame where m_2 is initially at rest?

INTERPRET Since the masses are equal, momenta and velocity vectors are proportional. Thus Fig. 9.21 does indeed show a collision in the center-of-mass reference frame. We need to transform the diagram to a frame where m_2 is initially at rest.

EVALUATE To get from Fig. 9.21 to a reference frame where m_2 is initially at rest, we need to add $-\vec{v}_{2i}$ to m_2's initial velocity—and therefore to all other velocities. That makes $\vec{v}_{1i}$ twice as long and adds an equal-length but perpendicular vector to each final velocity, making them both $\sqrt{2}$ times as long as in the CM frame and pointing at 45°. Figure 9.22 is our result.

ASSESS In the ASSESS step of Example 9.11, you learned that a two-dimensional collision between equal masses, with one initially at rest, results in the final velocities being perpendicular. Our result is consistent with that fact, and its symmetry is consistent with the symmetry shown in the center-of-mass frame.

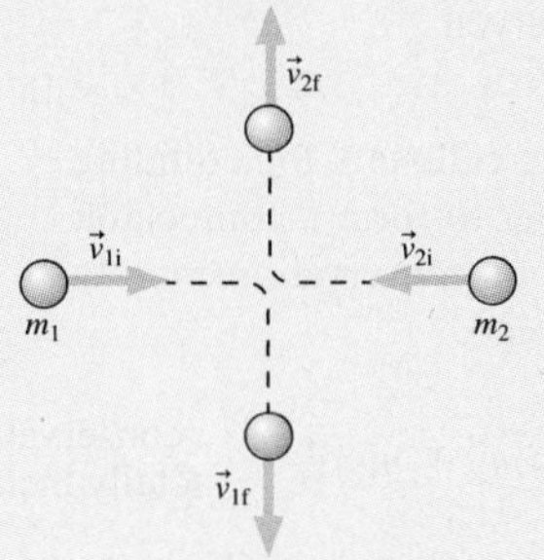

FIGURE 9.21 A two-dimensional collision between equal masses in the CM frame.

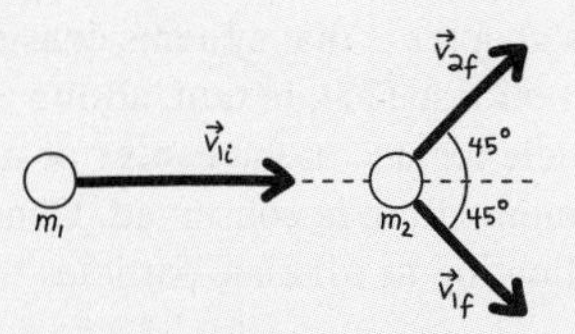

FIGURE 9.22 The same collision in a frame with m_2 initially at rest.

MAKING THE CONNECTION Consider a collision in the center-of-mass frame, as shown in Fig. 9.20, but now with equal-mass objects. If the angle θ shown in Fig. 9.20 is 70°, what are the angles shown in a diagram analogous to Fig. 9.19, in a frame where one of the objects is initially at rest?

EVALUATE Since the objects' masses are equal, in the zero-momentum cm frame they must be approaching each other with equal speeds v. We also know that the two velocities after collision must be equal and opposite in the cm frame; again, that's because the total momentum of the two equal-mass balls is zero in the cm frame. Furthermore, to conserve kinetic energy the speeds in the cm frame must be the same as they were before the collision. So the collision looks like Fig. 9.20, and we can replace the momentum vectors with equal-magnitude velocity vectors since the objects have equal masses. To get to a reference frame where m_2 is initially at rest, we need to add a rightward velocity $\vec{v}$ to all the vectors shown in the cm frame. That will give m_1 an after-collision velocity whose components are $v_{1x} = v\cos\theta + v$ and $v_{1y} = -v\sin\theta$, with the minus sign designating the downward direction in Fig. 9.20. The angle of m_1's velocity, analogous to the 30° angle in Fig. 9.19, is then $\tan^{-1}[-\sin\theta/(1+\cos\theta)]$. Work this out for $\theta = 70°$, and you'll get 35°. In fact, you could show in general that, for equal-mass objects, the angles in the cm frame and in the frame with one object initially at rest are always related by a factor of 2.

CHAPTER 9 SUMMARY

Big Idea

The big idea of this chapter is that systems consisting of many particles exhibit simple behaviors that don't depend on the complexities of their internal structure or motions. That, in turn, allows us to understand those internal details. In particular, a system responds to external forces as though it were a point particle located at the **center of mass**. If the net external force on a system is zero, then the center of mass does not accelerate and the system's total momentum is conserved. Conservation of momentum holds to a very good approximation during the brief, intense encounters called **collisions**, allowing us to relate particles' motions before and after colliding.

Newton's second and third laws are behind these big ideas. The third law, in particular, says that forces *internal* to a system cancel in pairs, and therefore they don't contribute to the net force on the system. That's what allows us to describe a system's overall motion without having to worry about what's going on internally.

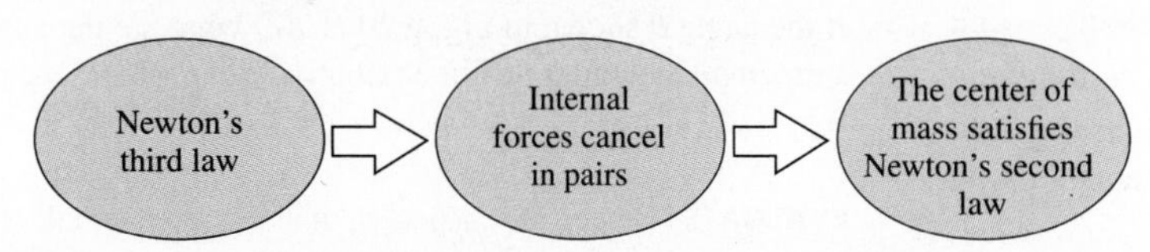

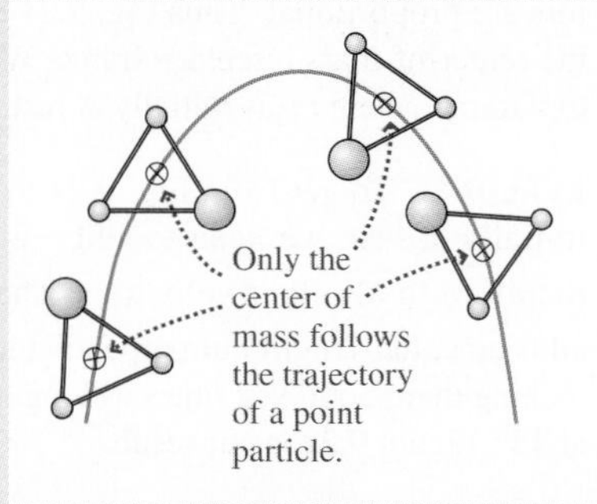

Key Concepts and Equations

The **center of mass** position $\vec{r}_{cm}$ is a weighted average of the positions of a system's constituent particles:

$$\vec{r}_{cm} = \frac{\sum m_i \vec{r}_i}{M} \quad \text{or, with continuous matter,} \quad \vec{r}_{cm} = \frac{\int \vec{r}\, dm}{M}$$

Here M is the system's total mass and the sum or integral is taken over the entire system. The center of mass obeys Newton's second law:

$$\vec{F}_{net\,ext} = M\vec{a}_{cm} = \frac{d\vec{P}}{dt}$$

where $\vec{F}_{net\,ext}$ is the net external force on the system, $\vec{a}_{cm}$ the acceleration of the center of mass, and $\vec{P}$ the system's total momentum.

A **collision** is a brief, intense interaction between particles involving large internal forces. External forces have little effect during a collision, so to a good approximation the total momentum of the interacting particles is conserved.

In a **totally inelastic collision**, the colliding objects stick together to form a composite; in that case momentum conservation entirely determines the outcome:

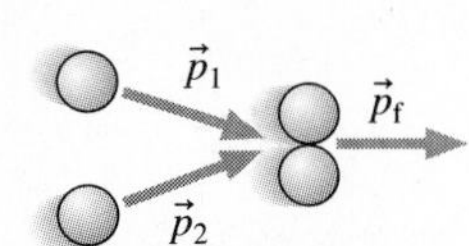

$$m_1\vec{v}_1 + m_2\vec{v}_2 = (m_1 + m_2)\vec{v}_f \quad \text{(conservation of momentum, totally inelastic collision)}$$

An **elastic collision** conserves kinetic energy as well as momentum, and the colliding particles separate after the collision:

$$m_1\vec{v}_{1i} + m_2\vec{v}_{2i} = m_1\vec{v}_{1f} + m_2\vec{v}_{2f} \quad \text{(conservation of momentum, elastic collision)}$$

$$\tfrac{1}{2}m_1v_{1i}^2 + \tfrac{1}{2}m_2v_{2i}^2 = \tfrac{1}{2}m_1v_{1f}^2 + \tfrac{1}{2}m_2v_{2f}^2 \quad \text{(conservation of energy, elastic collision)}$$

In the special case of a one-dimensional elastic collision, knowledge of the mass and initial velocities is sufficient to determine the outcome. To analyze elastic collisions in two dimensions requires an additional piece of information, such as the impact parameter or the direction of one of the particles after the collision.

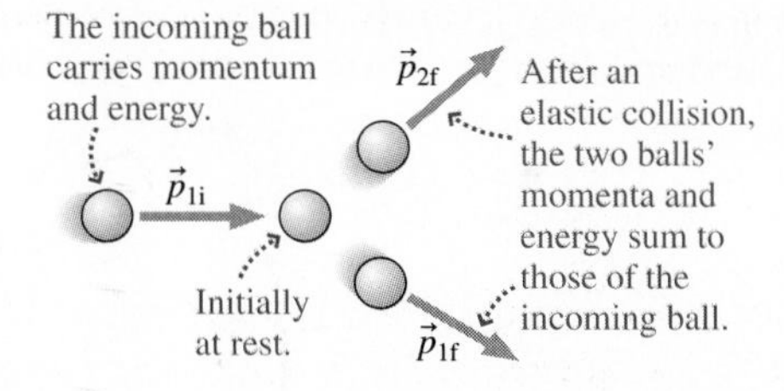

Applications

One-dimensional collisions with one object initially at rest provide insights into the nature of collisions. There are three cases, depending on the relative masses:

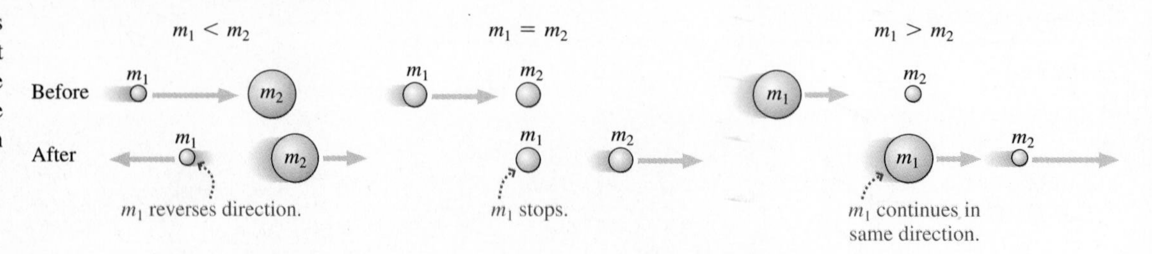

Rockets provide a technological application of momentum conservation. A rocket exhausts matter out the back at high velocity; momentum conservation then requires that the rocket gain momentum in the forward direction. Rocket propulsion requires no interaction with any external material, which is why rockets work in space.

MP® *For homework assigned on MasteringPhysics, go to www.masteringphysics.com*

BIO *Biology and/or medicine-related problems* **DATA** *Data problems* **ENV** *Environmental problems* **CH** *Challenge problems* **COMP** *Computer problems*

For Thought and Discussion

1. Roughly where is your center of mass when you're standing?
2. Explain why a high jumper's center of mass need not clear the bar.
3. The center of mass of a solid sphere is clearly at its center. If the sphere is cut in half and the two halves are stacked as in Fig. 9.23, is the center of mass at the point where they touch? If not, roughly where is it? Explain.

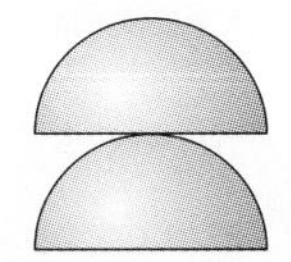

FIGURE 9.23 For Thought and Discussion 3

4. The momentum of a system of pool balls is the same before and after they are hit by the cue ball. Is it still the same after one of the balls strikes the edge of the table? Explain.
5. An hourglass is inverted and placed on a scale. Compare the scale readings (a) before sand begins to hit the bottom; (b) while sand is hitting the bottom; and (c) when all the sand is on the bottom.
6. Why are cars designed so that their front ends crumple during an accident?
7. Give three everyday examples of inelastic collisions.
8. Is it possible to have an inelastic collision in which all the kinetic energy of the colliding objects is lost? If so, give an example. If not, why not?
9. If you want to stop the neutrons in a reactor, why not use massive nuclei like lead?
10. Why don't we need to consider external forces acting on a system as its constituent particles undergo a collision?
11. How is it possible to have a collision between objects that don't ever touch? Give an example of such a collision.
12. A pitched baseball moves no faster than the pitcher's hand. But a batted ball can move much faster than the bat. What's the difference?
13. Two identical satellites are going in opposite directions in the same circular orbit when they collide head-on. Describe their subsequent motion if the collision is (a) elastic or (b) totally inelastic.

Exercises and Problems

Exercises

Section 9.1 Center of Mass

14. A 28-kg child sits at one end of a 3.5-m-long seesaw. Where should her 65-kg father sit so the center of mass will be at the center of the seesaw?
15. Two particles of equal mass m are at the vertices of the base of an equilateral triangle. The triangle's center of mass is midway between the base and the third vertex. What's the mass at the third vertex?
16. Rework Example 9.1 with the origin at the center of the barbell, showing that the physical location of the center of mass doesn't depend on your coordinate system.
17. Three equal masses lie at the corners of an equilateral triangle of side L. Find the center of mass.
18. How far from Earth's center is the center of mass of the Earth–Moon system? (*Hint*: Consult Appendix E.)

Section 9.2 Momentum

19. A popcorn kernel at rest in a hot pan bursts into two pieces, with masses 91 mg and 64 mg. The more massive piece moves horizontally at 47 cm/s. Describe the motion of the second piece.
20. A 60-kg skater, at rest on frictionless ice, tosses a 12-kg snowball with velocity $\vec{v} = 53.0\hat{\imath} + 14.0\hat{\jmath}$ m/s, where the x- and y-axes are in the horizontal plane. Find the skater's subsequent velocity.
21. A plutonium-239 nucleus at rest decays into a uranium-235 nucleus by emitting an alpha particle (^{4}He) with kinetic energy 5.15 MeV. Find the speed of the uranium nucleus.
22. A toboggan of mass 8.6 kg is moving horizontally at 23 km/h. As it passes under a tree, 15 kg of snow drop onto it. Find its subsequent speed.

Section 9.3 Kinetic Energy of a System

23. A 150-g trick baseball is thrown at 60 km/h. It explodes in flight into two pieces, with a 38-g piece continuing straight ahead at 85 km/h. How much energy do the pieces gain in the explosion?
24. An object with kinetic energy K explodes into two pieces, each of which moves with twice the speed of the original object. Find the ratio of the internal kinetic energy to the center-of-mass energy after the explosion.

Section 9.4 Collisions

25. Two 140-kg satellites collide at an altitude where $g = 8.7\ \text{m/s}^2$, and the collision imparts an impulse of 1.8×10^5 N·s to each. If the collision lasts 120 ms, compare the collisional impulse to that imparted by gravity. Your result should show why you can neglect the external force of gravity.
26. **BIO** High-speed photos of a 220-μg flea jumping vertically show that the jump lasts 1.2 ms and involves an average vertical acceleration of $100g$. What (a) average force and (b) impulse does the ground exert on the flea during its jump? (c) What's the change in the flea's momentum during its jump?
27. You're working in mission control for an interplanetary space probe. A trajectory correction calls for a rocket firing that imparts an impulse of 5.64 N·s. If the rocket's average thrust is 135 mN, how long should the rocket fire?

Section 9.5 Totally Inelastic Collisions

28. In a railroad switchyard, a 56-ton freight car is sent at 7.0 mi/h toward a 31-ton car moving in the same direction at 2.6 mi/h. (a) What's the speed of the cars after they couple? (b) What fraction of the initial kinetic energy was lost in the collision?
29. In a totally inelastic collision between two equal masses, with one initially at rest, show that half the initial kinetic energy is lost.
30. A neutron (mass 1.01 u) strikes a deuteron (mass 2.01 u), and they combine to form a tritium nucleus (mass 3.02 u). If the neutron's initial velocity was $23.5\hat{\imath} + 14.4\hat{\jmath}$ Mm/s and if the tritium leaves the reaction with velocity $15.1\hat{\imath} + 22.6\hat{\jmath}$ Mm/s, what was the deuteron's velocity?

31. Two identical trucks have mass 5500 kg when empty, and the maximum permissible load for each is 8000 kg. The first truck, carrying 3800 kg, is at rest. The second truck plows into it at 65 km/h, and the pair moves away at 27 km/h. As an expert witness, you're asked to determine whether the second truck was overloaded. What do you report?

Section 9.6 Elastic Collisions

32. An alpha particle (^{4}He) strikes a stationary gold nucleus (^{197}Au) head-on. What fraction of the alpha's kinetic energy is transferred to the gold? Assume a totally elastic collision.
33. Playing in the street, a child accidentally tosses a ball at 18 m/s toward the front of a car moving toward him at 14 m/s. What's the ball's speed after it rebounds elastically from the car?
34. A block of mass m undergoes a one-dimensional elastic collision with a block of mass M initially at rest. If both blocks have the same speed after colliding, how are their masses related?
35. A proton moving at 6.9 Mm/s collides elastically head-on with a second proton moving in the opposite direction at 11 Mm/s. Find their subsequent velocities.
36. A head-on, elastic collision between two particles with equal initial speed v leaves the more massive particle (m_1) at rest. Find (a) the ratio of the particle masses and (b) the final speed of the less massive particle.

Problems

37. Find the center of mass of a pentagon with five equal sides of length a, but with one triangle missing (Fig. 9.24). (*Hint:* See Example 9.3, and treat the pentagon as a group of triangles.)

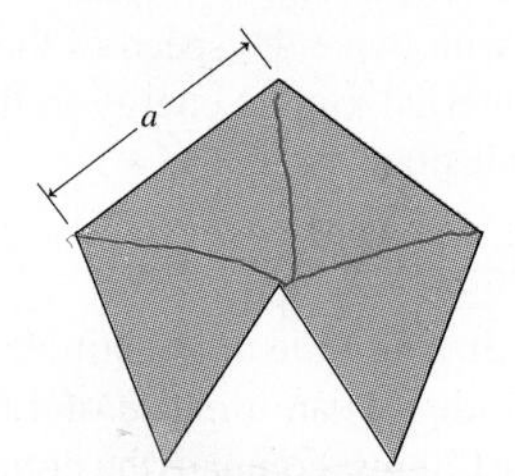

FIGURE 9.24 Problem 37

38. BIO Wildlife biologists fire 20-g rubber bullets to stop a rhinoceros charging at 0.81 m/s. The bullets strike the rhino and drop vertically to the ground. The biologists' gun fires 15 bullets each second, at 73 m/s, and it takes 34 s to stop the rhino. (a) What impulse does each bullet deliver? (b) What's the rhino's mass? Neglect forces between rhino and ground.
39. Consider a system of three equal-mass particles moving in a plane; their positions are given by $a_i\hat{\imath} + b_i\hat{\jmath}$, where a_i and b_i are functions of time with the units of position. Particle 1 has $a_1 = 6t^2 + 5$ and $b_1 = 0$; particle 2 has $a_2 = 4t + 3$ and $b_2 = 4t$; particle 3 has $a_3 = 8t$ and $b_3 = t + 4$. Find the center-of-mass position, velocity, and acceleration of the system as functions of time.
40. You're with 19 other people on a boat at rest in frictionless water. The group's total mass is 1500 kg, and the boat's mass is 12,000 kg. The entire party walks the 6.5-m distance from bow to stern. How far does the boat move?
41. A hemispherical bowl is at rest on a frictionless counter. A mouse drops onto the bowl's rim from a cabinet directly overhead. The mouse climbs down inside the bowl to eat crumbs at the bottom. If the bowl moves along the counter a distance equal to one-tenth of its diameter, how does the mouse's mass compare with the bowl's mass?
42. BIO Physicians perform *needle biopsies* to sample tissue from internal organs. A spring-loaded gun shoots a hollow needle into the tissue; extracting the needle brings out the tissue core. A particular device uses 8.3-mg needles that take 90 ms to stop in the tissue, which exerts a stopping force of 41 mN. (a) Find the impulse imparted by the tissue. (b) How far into the tissue does the needle penetrate?
43. Find the center of mass of the uniform, solid cone of height h, base radius R, and constant density ρ shown in Fig. 9.25. (*Hint*: Integrate over disk-shaped mass elements of thickness dy, as shown in the figure.)

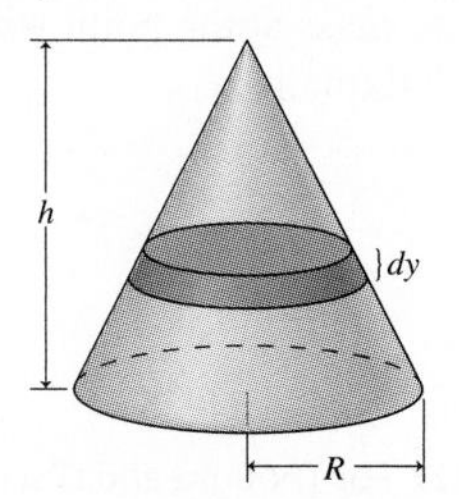

FIGURE 9.25 Problem 43

44. A firecracker, initially at rest, explodes into two fragments. The first, of mass 14 g, moves in the $+x$-direction at 48 m/s. The second moves at 32 m/s. Find the second fragment's mass and the direction of its motion.
45. An 11,000-kg freight car rests against a spring bumper at the end of a railroad track. The spring has constant $k = 0.32$ MN/m. The car is hit by a second car of 9400-kg mass moving at 8.5 m/s, and the two couple together. Find (a) the maximum compression of the spring and (b) the speed of the two cars when they rebound together from the spring.
46. On an icy road, a 1200-kg car moving at 50 km/h strikes a 4400-kg truck moving in the same direction at 35 km/h. The pair is soon hit from behind by a 1500-kg car speeding at 65 km/h, and all three vehicles stick together. Find the speed of the wreckage.
47. A car of mass M is initially at rest on a frictionless surface. A jet of water carrying mass at the rate dm/dt and moving horizontally at speed v_0 strikes the rear window of the car, which is at 45° to the horizontal; the water bounces off at the same speed, relative to the window, with which it hit (see Fig. 9.26). Find expressions for (a) the car's initial acceleration and (b) the maximum speed it reaches. Note: (b) doesn't require any calculation. It might help to sketch a rough plot (no calculations!) of the car's speed versus time.

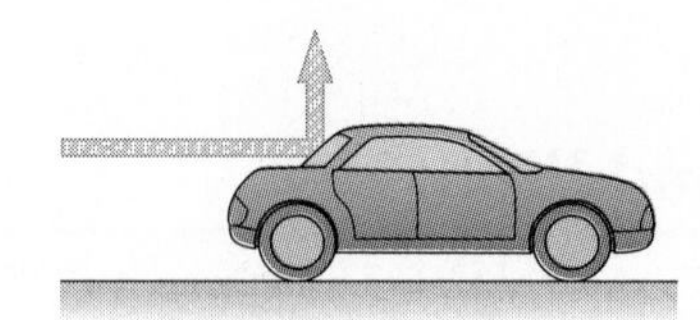

FIGURE 9.26 Problem 47

48. A 1250-kg car is moving with velocity $\vec{v}_1 = 36.2\hat{\imath} + 12.7\hat{\jmath}$ m/s. It skids on a frictionless icy patch and collides with a 448-kg hay wagon with velocity $\vec{v}_2 = 13.8\hat{\imath} + 10.2\hat{\jmath}$ m/s. If the two stay together, what's their velocity?
49. Masses m and $3m$ approach at the same speed v and undergo a head-on elastic collision. Show that mass $3m$ stops, while mass m rebounds at speed $2v$.
50. A ^{238}U nucleus is moving in the x-direction at 5.0×10^5 m/s when it decays into an alpha particle (^{4}He) and a ^{234}Th nucleus. The

alpha moves at 1.4×10^7 m/s at 22° above the x-axis. Find the recoil velocity of the thorium.

51. A cylindrical concrete silo is 4.0 m in diameter and 30 m high. It consists of a 6000-kg concrete base and 38,000-kg cylindrical concrete walls. Locate the center of mass of the silo (a) when it's empty and (b) when it's two-thirds full of silage whose density is 800 kg/m^3. Neglect the thickness of the walls and base.
52. A 42-g firecracker is at rest at the origin when it explodes into three pieces. The first, with mass 12 g, moves along the x-axis at 35 m/s. The second, with mass 21 g, moves along the y-axis at 29 m/s. Find the velocity of the third piece.
53. A 60-kg astronaut floating in space simultaneously tosses away a 14-kg oxygen tank and a 5.8-kg camera. The tank moves in the x-direction at 1.6 m/s, and the astronaut recoils at 0.85 m/s in a direction 200° counterclockwise from the x-axis. Find the camera's velocity.
54. Assuming equal-mass pieces in Exercise 24, find the angles of the two velocities relative to the direction of motion before the explosion.
55. A 62-kg sprinter stands on the left end of a 190-kg cart moving leftward at 7.1 m/s. She runs to the right end and continues horizontally off the cart. What should be her speed relative to the cart so that once she's off the cart, she has no horizontal velocity relative to the ground?
56. You're a production engineer in a cookie factory, where mounds of dough drop vertically onto a conveyor belt at the rate of one 12-g mound every 2 seconds. You're asked to design a mechanism that will keep the conveyor belt moving at a constant 50 cm/s. What average force must the mechanism exert on the belt?
57. Mass m, moving at speed $2v$, approaches mass $4m$, moving at speed v. The two collide elastically head-on. Find expressions for their subsequent speeds.
58. Verify explicitly that kinetic energy is conserved in the collision of the preceding problem.
59. While standing on frictionless ice, you (mass 65.0 kg) toss a 4.50-kg rock with initial speed 12.0 m/s. If the rock is 15.2 m from you when it lands, (a) at what angle did you toss it? (b) How fast are you moving?
60. You're an accident investigator at a scene where a drunk driver in a 1600-kg car has plowed into a 1300-kg parked car with its brake set. You measure skid marks showing that the combined wreckage moved 25 m before stopping, and you determine a frictional coefficient of 0.77. What do you report for the drunk driver's speed just before the collision?
61. **CH** A fireworks rocket is launched vertically upward at 40 m/s. At the peak of its trajectory, it explodes into two equal-mass fragments. One reaches the ground 2.87 s after the explosion. When does the second reach the ground?
62. Two objects moving in opposite directions with the same speed v undergo a totally inelastic collision, and half the initial kinetic energy is lost. Find the ratio of their masses.
63. Explosive bolts separate a 950-kg communications satellite from its 640-kg booster rocket, imparting a 350-N·s impulse. At what relative speed do satellite and booster separate?
64. You're working in quality control for a model rocket manufacturer, testing a class-D rocket whose specifications call for an impulse between 10 and 20 N·s. The rocket's burn time is $\Delta t = 2.8$ s, and its thrust during that time is $F(t) = at(t - \Delta t)$, where $a = -4.6$ N/s^2. Does the rocket meet its specs?
65. You're investigating an accident in which a 1040-kg Toyota Yaris and an 2140-kg Buick Enclave collided at right angles in an intersection. The combined wreckage skidded 12.3 m before stopping. You measure the coefficient of friction between tires and road and find it to be 0.712. Show that at least one car must have exceeded the 55-km/h speed limit at the intersection.
66. A 400-mg popcorn kernel is skittering across a nonstick frying pan at 8.2 cm/s when it pops and breaks into two equal-mass pieces. If one piece ends up at rest, how much energy was released in the popping?
67. **CH** Two identical objects with the same initial speed collide and stick together. If the composite object moves with half the initial speed of either object, what was the angle between the initial velocities?
68. A proton (mass 1 u) moving at 6.90 Mm/s collides elastically head-on with a second particle moving in the opposite direction at 2.80 Mm/s. After the collision, the proton is moving opposite its initial direction at 8.62 Mm/s. Find the mass and final velocity of the second particle.
69. **CH** Two objects, one initially at rest, undergo a one-dimensional elastic collision. If half the kinetic energy of the initially moving object is transferred to the other object, what is the ratio of their masses?
70. Blocks B and C have masses $2m$ and m, respectively, and are at rest on a frictionless surface. Block A, also of mass m, is heading at speed v toward block B as shown in Fig. 9.27. Determine the final velocity of each block after all subsequent collisions are over. Assume all collisions are elastic.

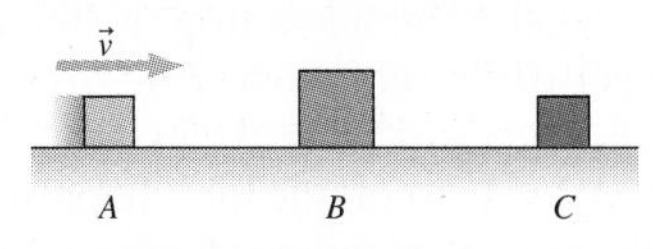

FIGURE 9.27 Problem 70

71. Derive Equation 9.15b.
72. An object collides elastically with an equal-mass object initially at rest. If the collision isn't head-on, show that the final velocity vectors are perpendicular.
73. **CH** A proton (mass 1 u) collides elastically with a stationary deuteron (mass 2 u). If the proton is deflected 37° from its original direction, what fraction of its kinetic energy does it transfer to the deuteron?
74. Two identical billiard balls are initially at rest when they're struck symmetrically by a third identical ball moving with velocity $\vec{v}_0 = v_0\hat{\imath}$ (Fig. 9.28). Find the velocities of all three balls after this elastic collision.

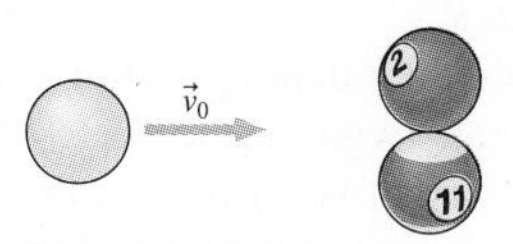

FIGURE 9.28 Problem 74

75. Find an expression for the impulse imparted by a force $F(t) = F_0 \sin(at)$ during the time $t = 0$ to $t = \pi/a$. Here a is a constant with units of s^{-1}.
76. A 32-u oxygen molecule (O_2) moving in the $+x$-direction at 580 m/s collides with an oxygen atom (mass 16 u) moving at 870 m/s at 27° to the x-axis. The particles stick together to form an ozone molecule. Find the ozone's velocity.
77. A 114-g Frisbee is lodged on a tree branch 7.65 m above the ground. To free it, you lob a 240-g dirt clod vertically upward. The dirt leaves your hand at a point 1.23 m above the ground, moving at 17.7 m/s. It sticks to the Frisbee. Find (a) the maximum height reached by the Frisbee-dirt combination and (b) the speed with which the combination hits the ground.

78. You set a small ball of mass m atop a large ball of mass $M \gg m$ and drop the pair from height h. Assuming the balls are perfectly elastic, show that the smaller ball rebounds to height $9h$.

79. A car moving at speed v undergoes a one-dimensional collision with an identical car initially at rest. The collision is neither elastic nor fully inelastic; 5/18 of the initial kinetic energy is lost. Find the velocities of the two cars after the collision.

80. CH A 200-g block is released from rest at a height of 25 cm on a frictionless 30° incline. It slides down the incline and then along a frictionless surface until it collides elastically with an 800-g block at rest 1.4 m from the bottom of the incline (Fig. 9.29). How much later do the two blocks collide again?

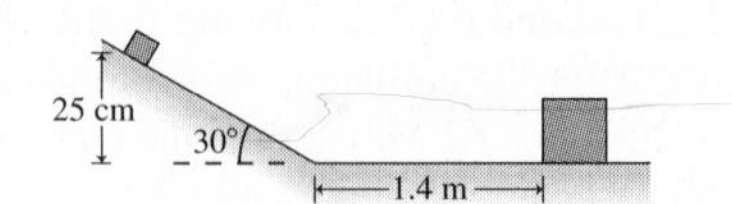

FIGURE 9.29 Problem 80

81. A 14-kg projectile is launched at 380 m/s at a 55° angle to the horizontal. At the peak of its trajectory it collides with a second projectile moving horizontally, in the opposite direction, at 140 m/s. The two stick together and land 9.6 km horizontally downrange from the first projectile's launch point. Find the mass of the second projectile.

82. During a crash test, a car moving at 50 km/h collides with a rigid barrier and comes to a complete stop in 200 ms. The collision force as a function of time is given by $F = at^4 + bt^3 + ct^2 + dt$, where $a = -8.86\ \text{GN/s}^4$, $b = 3.27\ \text{GN/s}^3$, $c = -362\ \text{MN/s}^2$, and $d = 12.5\ \text{MN/s}$. Find (a) the total impulse imparted by the collision, (b) the average collisional force, and (c) the car's mass.

83. COMP Use numerical or graphical techniques to estimate the peak force of the collision in the preceding problem, and determine when it occurs.

84. A block of mass m_1 undergoes a one-dimensional elastic collision with an initially stationary block of mass m_2. Find an expression for the fraction of the initial kinetic energy transferred to the second block, and plot your result for mass ratios m_1/m_2 from 0 to 20.

85. Two objects of unequal mass, one initially at rest, undergo a one-dimensional elastic collision. For a given mass ratio, show that the fraction of the initial energy transferred to the initially stationary object doesn't depend on which object it is.

86. CH In Figure 9.6, the uniform semicircular wire has radius R. How far above the center of the semicircle is its center of mass?

87. CH Find the center of mass of a uniform slice of pizza with radius R and angular width θ.

88. In a ballistic pendulum demonstration gone bad, a 0.52-g pellet, fired horizontally with kinetic energy 3.25 J, passes straight through a 400-g Styrofoam pendulum block. If the pendulum rises a maximum height of 0.50 mm, how much kinetic energy did the pellet have after emerging from the Styrofoam?

89. An 80-kg astronaut has become detached from the safety line connecting her to the International Space Station. She's 200 m from the station, at rest relative to it, and has 4 min of air remaining. To get herself back, she tosses a 10-kg tool kit away from the station at 8.0 m/s. Will she make it back in time?

90. Astronomers detect extrasolar planets by measuring the slight movement of stars around the center of mass of the star–planet system. Considering just the Sun and Jupiter, determine the radius of the circular orbit the Sun makes about the Sun–Jupiter center of mass.

91. CH A thin rod extends from $x = 0$ to $x = L$. It carries a nonuniform mass per unit length $\mu = Mx^a/L^{1+a}$, where M is a constant with units of mass, and a is a non-negative dimensionless constant. Find expressions for (a) the rod's mass and (b) the location of its center of mass. (c) Are your results what you expect when $a = 0$?

92. DATA Model rocket motors are specified by giving the impulse they provide, in N · s, over the entire time the rocket is firing. The table below shows the results of rocket-motor tests with different motors used to launch rockets of different masses. Determine two data-based quantities that, when plotted against each other, should give a straight line and whose slope should allow you to determine g. Plot the data, establish a best-fit line, and determine g. Assume that the maximum height is much greater than the distance over which the rocket motor is firing, so you can neglect the latter. You're also neglecting air resistance—but explain how that affects your experimentally determined value for g.

Impulse, J (N · s)	4.5	7.8	4.5	7.8	11
Rocket mass (g) (including motor)	180	485	234	234	485
Maximum height achieved (m)	22	13	19	51	23

93. CH A block of mass M is moving at speed v_0 on a frictionless surface that ends in a rigid wall, heading toward a stationary block of mass nM, where $n \geq 1$ (Fig. 9.30). Collisions between the two blocks or the left-hand block and the wall are elastic and one-dimensional. (a) Show that the blocks will undergo only one collision with each other if $n \leq 3$. (b) Show that the blocks will undergo two collisions with each other if $n = 4$. (c) How many collisions will the blocks undergo if $n = 10$, and what will be their final speeds?

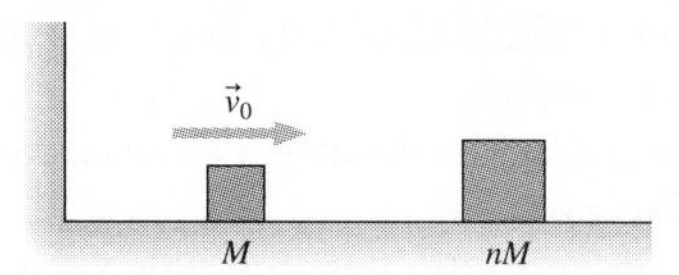

FIGURE 9.30 Problem 93

Passage Problems

You're interested in the intersection of physics and sports, and you recognize that many sporting events involve collisions—bat and baseball, foot and football, hockey stick and puck, basketball and floor. Using strobe photography, you embark on a study of such collisions. Figure 9.31 is your strobe photo of a ball bouncing off the floor. The ball is launched from a point near the top left of the photo and your camera then captures it undergoing three subsequent collisions with the floor.

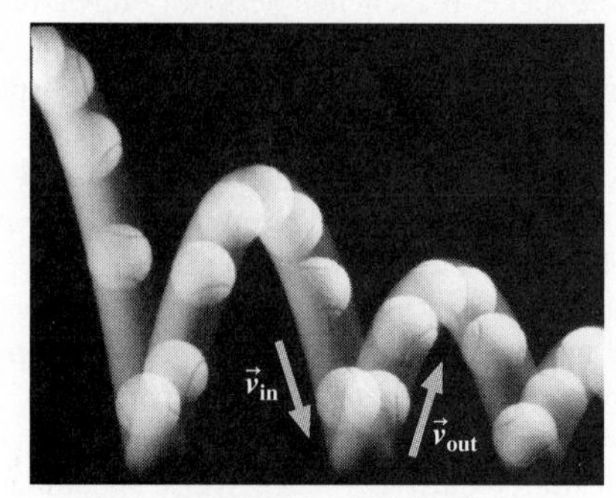

FIGURE 9.31 Passage Problems 94–97

94. The collisions between ball and floor are
a. totally elastic.
b. totally inelastic.
c. neither totally elastic nor totally inelastic.

95. The fraction of the ball's mechanical energy that's lost in the second collision is
    a. about 10%.
    b. a little less than half.
    c. a little more than half.
    d. about 90%.
96. The component of the ball's velocity whose magnitude is most affected by the collisions is
    a. horizontal.
    b. vertical.
    c. Both are affected equally.
97. Compared with the time between bounces, the duration of each collision is
    a. a tiny fraction of the time between bounces.
    b. a significant fraction of the time between bounces.
    c. much longer than the time between bounces.

Answers to Chapter Questions

Answer to Chapter Opening Question

The dancer's center of mass follows the simple path of a projectile because, as Newton's laws show, the dancer's mass acts like it's all concentrated at this point.

Answers to GOT IT? Questions

9.1 The CM is the uppermost point *A*. You can see this by imagining horizontal strips through the loop; the higher the strip the more mass is included, so the CM must lie nearer the top of the loop. The bottommost point would be the CM for a complete circle.

9.2 Momentum is conserved, so the momentum both before and after the explosion is the same: $\vec{P} = m\vec{v} = (0.50\text{ kg})(60\hat{\jmath}\text{ m/s}) = 30\hat{\jmath}\text{ kg}\cdot\text{m/s}$.

9.3 Only (d). The individual skaters experience external forces from the ball, as does the ball from the skaters. A system consisting of the ball and one skater experiences external forces from the other skater. Only the system of all three has no net external force and therefore has conserved momentum.

9.4 (1) (a); (2) (b)

9.5 all but (c) are collisions; (a) and (b) are nearly elastic; (d) and (e) are inelastic

9.6 (a) and (b) are totally inelastic; (c) is inelastic but not totally so

9.7 The ball initially at rest is less massive; otherwise, the incident ball would have reversed direction (or stopped if the masses were equal).

Rotational Motion

What You Know

- You know how to describe motion in one dimension using the concepts of position, velocity, and acceleration.
- You know how to relate position, velocity, and acceleration in the special case of one-dimensional motion with constant acceleration.
- You understand Newton's second law and how to apply it to one-dimensional motion.
- You can describe circular motion and the associated centripetal acceleration.

What You're Learning

- Here you'll learn the rotational analogs of quantities in one-dimensional motion: angular position, angular velocity, and angular acceleration.
- You'll be able to solve rotational-motion problems analogously to problems involving one-dimensional motion in Chapter 2.
- You'll learn about *torque*, the rotational analog of force, and *rotational inertia*, the analog of mass.
- You'll learn the rotational analog of Newton's second law.
- You'll learn how to calculate rotational inertias by integration.
- You'll learn to couple rotational and linear motion, including the special case of rolling motion.
- You'll learn how to calculate rotational kinetic energy.

How You'll Use It

- Concepts of rotational motion that you learn here set the groundwork for the vector treatment of rotation in Chapter 11.
- Angular velocity and rotational inertia will prove useful in the important quantity called *angular momentum*.
- Appreciating the analogies between linear and angular motion will help you to understand the seemingly counterintuitive behaviors that result from the vector nature of rotational motion, from the precession of Earth's axis to the stability of a bicycle.

For a given blade mass, how should you engineer a wind turbine's blades so it's easiest for the wind to get the turbine rotating?

You're sitting on a rotating planet. The wheels of your car rotate. Your favorite movie comes from a rotating DVD. A circular saw rotates to rip its way through a board. A dancer pirouettes, and a satellite spins about its axis. Even molecules rotate. Rotational motion is commonplace throughout the physical universe.

In principle, we could treat rotational motion by analyzing the motion of each particle in a rotating object. But that would be a hopeless task for all but the simplest objects. Instead, we'll describe rotational motion by analogy with linear motion as governed by Newton's laws.

This chapter parallels our study of one-dimensional motion in Chapters 2 and 4. In the next chapter we introduce a full vector description to treat multidimensional rotational motion.

10.1 Angular Velocity and Acceleration

You slip a DVD into a player, and it starts spinning. You could describe its motion by giving the speed and direction of each point on the disc. But it's much easier just to say that the disc is rotating at 800 revolutions per minute (rpm). As long as the disc is a **rigid body**—one whose parts remain in fixed positions relative to one another—then that single statement suffices to describe the motion of the entire disc.

Angular Velocity

The rate at which a body rotates is its **angular velocity**—the rate at which the angular position of any point on the body changes. With our 800-rpm DVD, the unit of angle was one full revolution (360°, or 2π radians), and the unit of time was the minute. But we could equally well express angular velocity in revolutions per second (rev/s), degrees per second (°/s), or radians per second (rad/s or simply s^{-1} since radians are dimensionless). Because of the mathematically simple nature of radian measure, we often use radians in calculations involving rotational motion (Fig. 10.1).

We use the Greek symbol ω (omega) for angular velocity and define **average angular velocity** $\overline{\omega}$ as

$$\overline{\omega} = \frac{\Delta\theta}{\Delta t} \qquad \text{(average angular velocity)} \qquad (10.1)$$

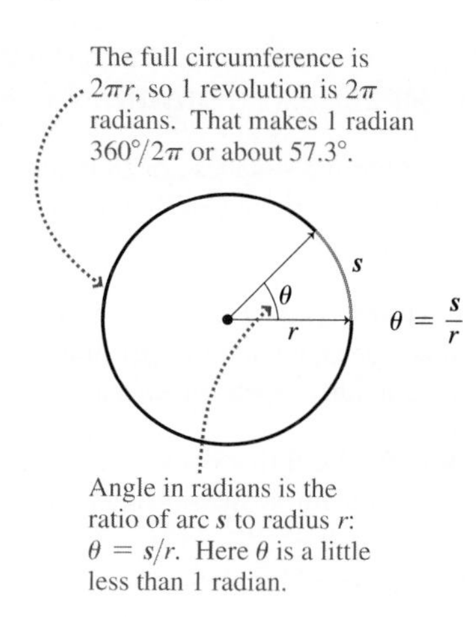

FIGURE 10.1 Radian measure of angles.

where $\Delta\theta$ is the **angular displacement**—that is, the change in angular position—occurring in the time Δt (Fig. 10.2). When angular velocity is changing, we define **instantaneous angular velocity** as the limit over arbitrarily short time intervals:

$$\omega = \lim_{\Delta t \to 0} \frac{\Delta\theta}{\Delta t} = \frac{d\theta}{dt} \qquad \text{(instantaneous angular velocity)} \qquad (10.2)$$

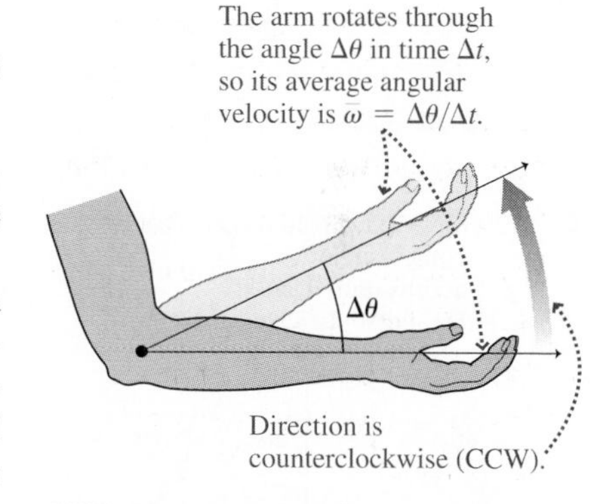

FIGURE 10.2 Average angular velocity.

These definitions are analogous to those of average and instantaneous linear velocity introduced in Chapter 2. Just as we use the term *speed* for the magnitude of velocity, so we define **angular speed** as the magnitude of the angular velocity.

Velocity is a vector quantity, with magnitude and direction. Is angular velocity also a vector? Yes, but we'll wait until the next chapter for the full vector description of rotational motion. In this chapter, it's sufficient to know whether an object's rotation is clockwise (CW) or counterclockwise (CCW) about a fixed axis—as suggested by the curved arrow in Fig. 10.2. This restriction to a fixed axis is analogous to Chapter 2's restriction to one-dimensional motion.

Angular and Linear Speed

PhET: Ladybug Revolution

Individual points on a rotating object undergo circular motion. Each point has an instantaneous linear velocity $\vec{v}$ whose magnitude is the linear speed v. We now relate this linear speed v to the angular speed ω. The definition of angular measure in radians (Fig. 10.1) is $\theta = s/r$. Differentiating this expression with respect to time, we have

$$\frac{d\theta}{dt} = \frac{1}{r}\frac{ds}{dt}$$

because the radius r is constant. The left-hand side of this equation is the angular velocity ω, as defined in Equation 10.2. Because s is the arc length—the actual distance traversed by a point on the rotating object—the term ds/dt is just the linear speed v, so $\omega = v/r$, or

$$v = \omega r \qquad (10.3)$$

Thus the linear speed of any point on a rotating object is proportional both to the angular speed of the object and to the distance from that point to the axis of rotation (Fig. 10.3).

Linear speed is proportional to distance from the rotation axis.

ω r $v = \omega r$ $\vec{v}$ $\vec{v}$

The point on the rim has the same angular speed ω but a higher linear speed v than the inner point.

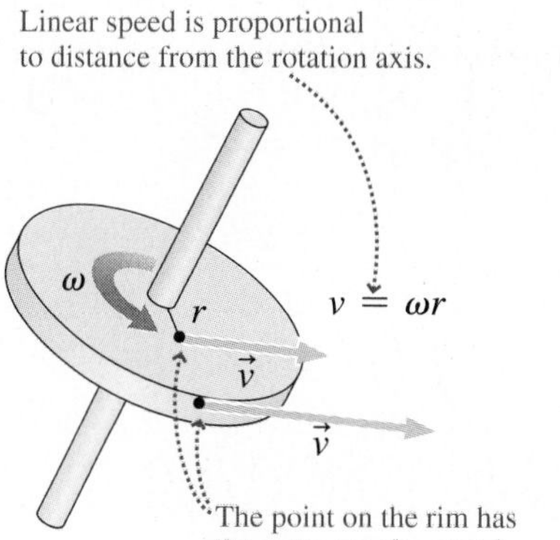

FIGURE 10.3 Linear and rotational speeds.

✓TIP Radian Measure

Equation 10.3 was derived using the definition of angle *in radians* and therefore holds for only angular speed measured in radians per unit time. If you're given other angular measures—degrees or revolutions, for example—you should convert to radians before using Equation 10.3.

EXAMPLE 10.1 Angular Speed: A Wind Turbine

A wind turbine's blades are 28 m long and rotate at 21 rpm. Find the angular speed of the blades in radians per second, and determine the linear speed at the tip of a blade.

INTERPRET This problem is about converting between two units of angular speed, revolutions per minute and radians per second, as well as finding linear speed given angular speed and radius.

DEVELOP We'll first convert the units to radians per second and then calculate the linear speed using Equation 10.3, $v = \omega r$.

EVALUATE One revolution is 2π rad, and 1 min is 60 s, so we have

$$\omega = 21 \text{ rpm} = \frac{(21 \text{ rev/min})(2\pi \text{ rad/rev})}{60 \text{ s/min}} = 2.2 \text{ rad/s}$$

The speed at the tip of a 28-m-long blade then follows from Equation 10.3: $v = \omega r = (2.2 \text{ rad/s})(28 \text{ m}) = 62 \text{ m/s}$.

ASSESS With ω in radians per second, multiplying by length in meters gives correct velocity units of meters per second because radians are dimensionless. ■

Angular Acceleration

If the angular velocity of a rotating object changes, then the object undergoes **angular acceleration** α, defined analogously to linear acceleration:

$$\alpha = \lim_{\Delta t \to 0} \frac{\Delta \omega}{\Delta t} = \frac{d\omega}{dt} \qquad \text{(angular acceleration)} \qquad (10.4)$$

Taking the limit gives the instantaneous angular acceleration; if we don't take the limit, then we have an average over the time interval Δt. The SI units of angular acceleration are rad/s^2, although we sometimes use other units such as rpm/s or rev/s^2.

Angular acceleration has the same direction as angular velocity—CW or CCW—if the angular speed is increasing, and the opposite direction if it's decreasing. These situations are analogous to a car that's speeding up (acceleration and velocity in the same direction) or braking (acceleration opposite velocity).

When a rotating object undergoes angular acceleration, points on the object speed up or slow down. Therefore, they have **tangential acceleration** dv/dt directed parallel or antiparallel to their linear velocity (Fig. 10.4). We introduced this idea of tangential acceleration back in Chapter 3; here we can recast it in terms of the angular acceleration:

$$a_t = \frac{dv}{dt} = r\frac{d\omega}{dt} = r\alpha \qquad \text{(tangential acceleration)} \qquad (10.5)$$

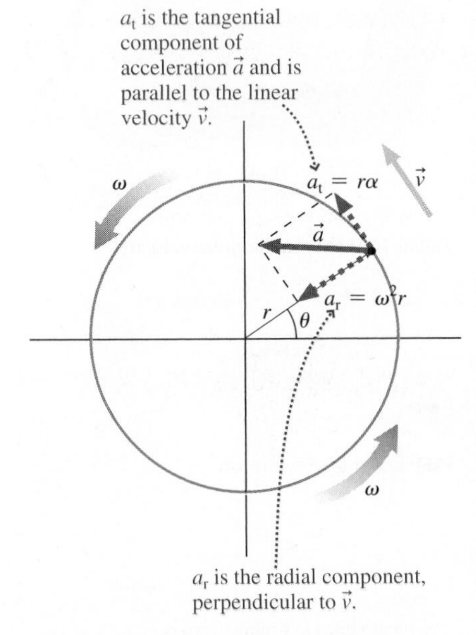

FIGURE 10.4 Radial and tangential acceleration.

Whether or not there's angular acceleration, points on a rotating object also have **radial acceleration** because they're in circular motion. Radial acceleration is given, as usual, by $a_r = v^2/r$; using $v = \omega r$ from Equation 10.3, we can recast this equation in angular terms as $a_r = \omega^2 r$.

Because angular velocity and acceleration are defined analogously to linear velocity and acceleration, all the relations among linear position, velocity, and acceleration automatically apply among angular position, angular velocity, and angular acceleration. If angular acceleration is constant, then all our constant-acceleration formulas of Chapter 2 apply when we make the substitutions θ for x, ω for v, and α for a. Table 10.1 summarizes

Table 10.1 Angular and Linear Position, Velocity, and Acceleration

Linear Quantity		Angular Quantity	
Position x		Angular position θ	
Velocity $v = \frac{dx}{dt}$		Angular velocity $\omega = \frac{d\theta}{dt}$	
Acceleration $a = \frac{dv}{dt} = \frac{d^2x}{dt^2}$		Angular acceleration $\alpha = \frac{d\omega}{dt} = \frac{d^2\theta}{dt^2}$	
Equations for Constant Linear Acceleration		**Equations for Constant Angular Acceleration**	
$\bar{v} = \frac{1}{2}(v_0 + v)$	(2.8)	$\bar{\omega} = \frac{1}{2}(\omega_0 + \omega)$	(10.6)
$v = v_0 + at$	(2.7)	$\omega = \omega_0 + \alpha t$	(10.7)
$x = x_0 + v_0 t + \frac{1}{2}at^2$	(2.10)	$\theta = \theta_0 + \omega_0 t + \frac{1}{2}\alpha t^2$	(10.8)
$v^2 = v_0^2 + 2a(x - x_0)$	(2.11)	$\omega^2 = \omega_0^2 + 2\alpha(\theta - \theta_0)$	(10.9)

this direct analogy between linear and rotational quantities. With Table 10.1, problems involving rotational motion are analogous to the one-dimensional linear problems you solved in Chapter 2.

EXAMPLE 10.2 Linear Analogies: Spin-down

When the wind dies, the turbine of Example 10.1 spins down with constant angular acceleration of magnitude 0.12 rad/s^2. How many revolutions does the turbine make before coming to a stop?

INTERPRET The key to problems involving rotational motion is to identify the analogous situation for linear motion. This problem is analogous to asking how far a braking car travels before coming to a stop. We identify the number of rotations—the angular displacement—as the analog of the car's linear displacement. The given angular acceleration is analogous to the car's braking acceleration. The initial angular speed (2.2 rad/s, from Example 10.1) is analogous to the car's initial speed. And in both cases the final state we're interested in has zero speed—whether linear or angular.

DEVELOP Our plan is to develop the analogy further so we can find the angular displacement. The easiest way to solve the linear problem would be to use Equation 2.11, $v^2 = v_0^2 + 2a(x - x_0)$, with $v = 0$, v_0 the initial velocity, a the car's acceleration, and $\Delta x = x - x_0$ the distance we're solving for. In Table 10.1, Equation 10.9 is the analogous equation for rotational motion: $\omega^2 = \omega_0^2 + 2\alpha\Delta\theta$, where we've written $\theta - \theta_0 = \Delta\theta$ for the rotational displacement during the spin-down.

EVALUATE We solve for $\Delta\theta$:

$$\Delta\theta = \frac{\omega^2 - \omega_0^2}{2\alpha} = \frac{0 - (2.2\ \text{rad/s})^2}{(2)(-0.12\ \text{rad/s}^2)} = 20\ \text{rad} = 3.2\ \text{revolutions}$$

where the last conversion follows because 1 revolution is 2π radians.

ASSESS The turbine blades are turning rather slowly—less than 1 revolution every second—so it's not surprising that a small angular acceleration can bring them to a halt in a short angular "distance." Note, too, how the units work out. Also, by taking ω as positive, we needed to treat α as negative because the angular acceleration is opposite the angular velocity when the rotation rate is slowing—just as the braking car's linear acceleration is opposite its velocity. ■

GOT IT? 10.1 A wheel undergoes constant angular acceleration, starting from rest. Which graph describes correctly the time dependence of both the transverse and radial accelerations of a point on the wheel's rim? Explain.

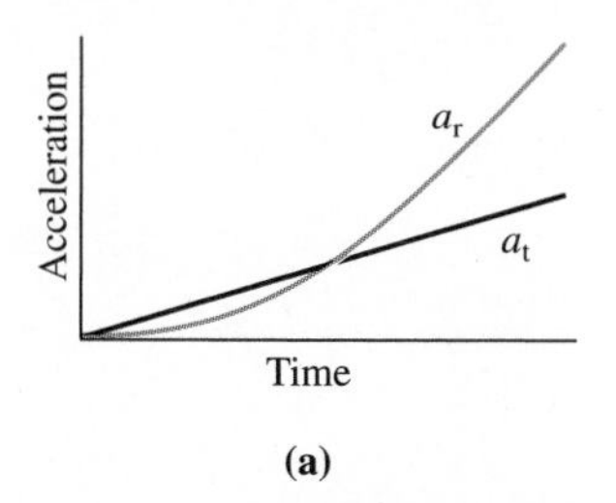

(a)

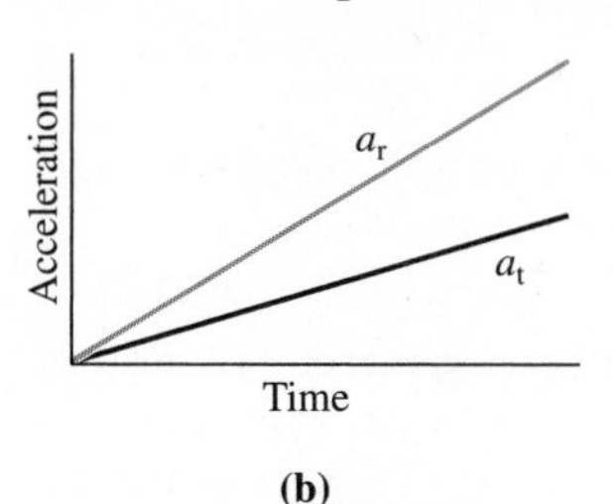

(b)

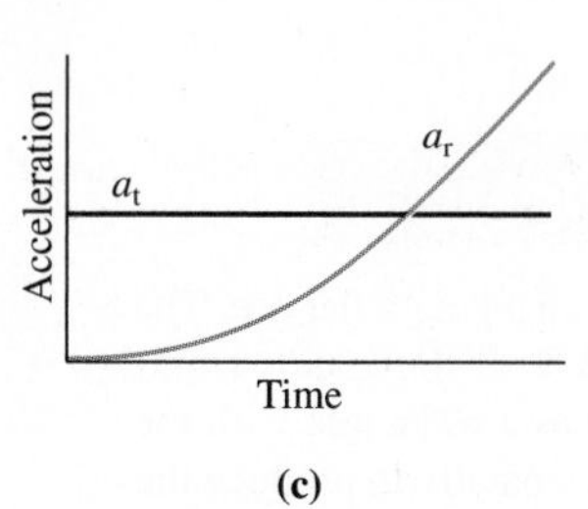

(c)

10.2 Torque

Newton's second law, $\vec{F} = m\vec{a}$, proved very powerful in our study of motion. Ultimately Newton's law governs all motion, but its application to every particle in a rotating object would be terribly cumbersome. Can we instead formulate an analogous law that deals with rotational quantities?

To develop such a law, we need rotational analogs of force, mass, and acceleration. Angular acceleration α is the analog of linear acceleration; in the next two sections we develop analogs for force and mass.

How can a small child balance her father on a seesaw? By sitting far from the seesaw's rotation axis; that way, her smaller weight at a greater distance from the pivot is as effective as her father's greater weight closer to the pivot. In general, the effectiveness of a force in bringing about changes in rotational motion—a quantity called **torque**—depends not only on the magnitude of the force but also on how far from the rotation axis it's applied (Fig. 10.5). The effectiveness of the force also depends on the *direction* in which it's applied, as Fig. 10.6 suggests. Based on these considerations, we define torque as the

The same force is applied at different points on the wrench.

Closest to O, τ is smallest.

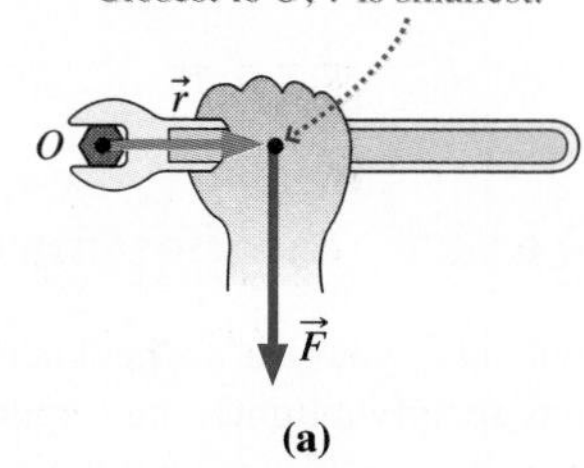

(a)

Farther away, τ becomes larger.

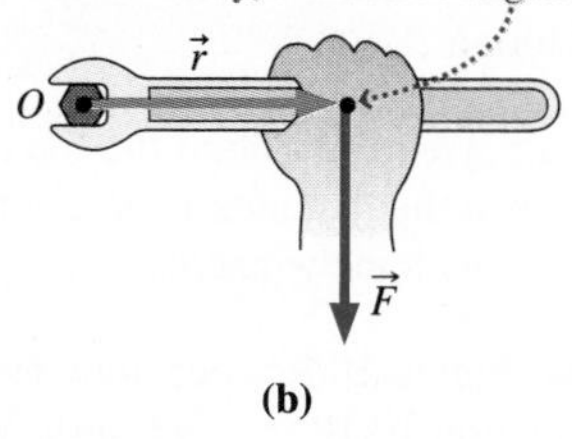

(b)

Farthest away, τ becomes greatest.

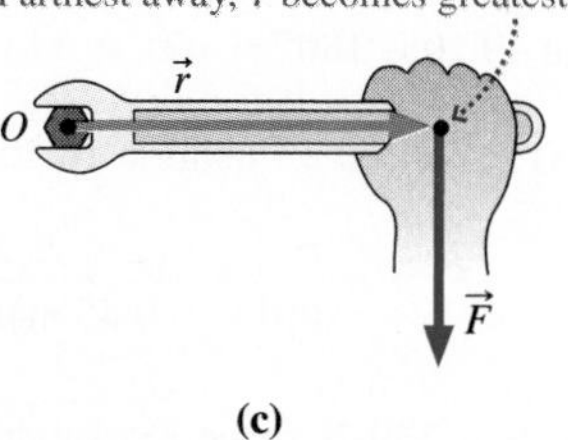

(c)

FIGURE 10.5 Torque increases with the distance r from the rotation axis O to the point where force is applied.

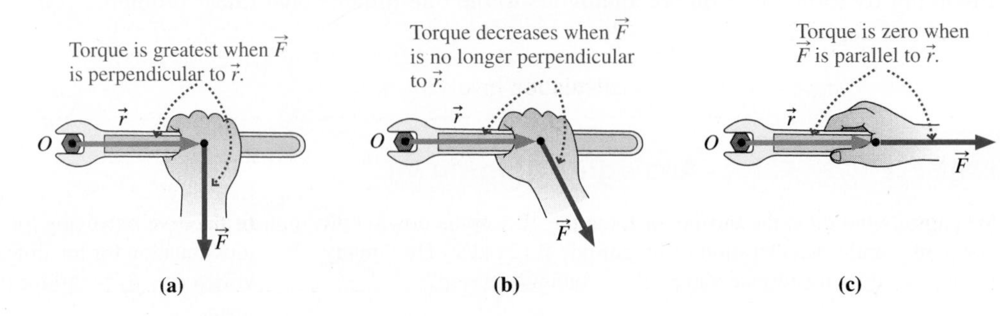

FIGURE 10.6 Torque is greatest with $\vec{F}$ and $\vec{r}$ at right angles, and diminishes to zero as they become colinear.

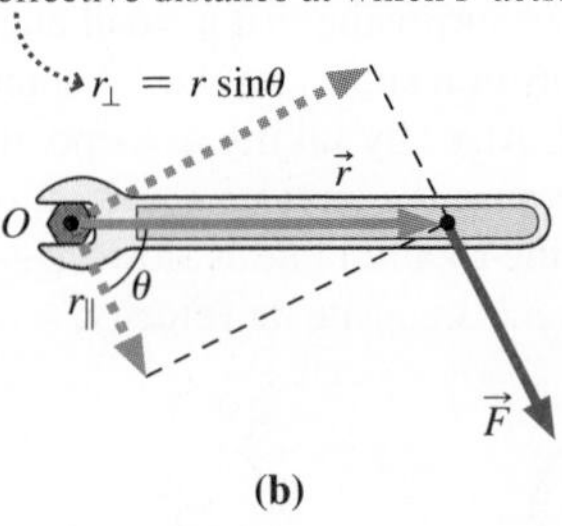

FIGURE 10.7 Two ways of thinking about torque. (a) $\tau = rF_\perp$; (b) $\tau = r_\perp F$. Both give $\tau = rF \sin\theta$.

product of the distance r from the rotation axis and the component of force perpendicular to that axis. Torque is given the symbol τ (Greek tau, pronounced to rhyme with "how"). Then we can write

$$\tau = rF \sin\theta \quad (10.10)$$

where θ is the angle between the force vector and the vector $\vec{r}$ from the rotation axis to the force application point. Figure 10.7 shows two interpretations of Equation 10.10. Figure 10.7*b* also defines the so-called **lever arm**.

Torque, which you can think of as a "twisting force," plays the role of force in the rotational analog of Newton's second law. Equation 10.10 shows that torque is measured in newton-meters. Although this is the same unit as energy, torque is a different physical quantity, so we reserve the term *joule* ($=1$ N·m) for energy.

Does torque have direction? Yes, and we'll extend our notion of torque to provide a vector description in the next chapter. For now we'll specify the direction as either clockwise or counterclockwise.

EXAMPLE 10.3 Torque: Changing a Tire

You're tightening your car's wheel nuts after changing a flat tire. The instructions specify a tightening torque of 95 N·m so the nuts won't come loose. If your 45-cm-long wrench makes a 67° angle with the horizontal, with what force must you pull horizontally to produce the required torque?

INTERPRET We need to find the force required to produce a specific torque, given the distance from the rotation axis and the angle the force makes with the wrench.

DEVELOP Figure 10.8 is our drawing, and we'll calculate the torque using Equation 10.10, $\tau = rF \sin\theta$. With the force applied horizontally, comparison of Figs. 10.7*a* and 10.8 shows that the angle θ in Equation 10.10 is $180° - 67° = 113°$.

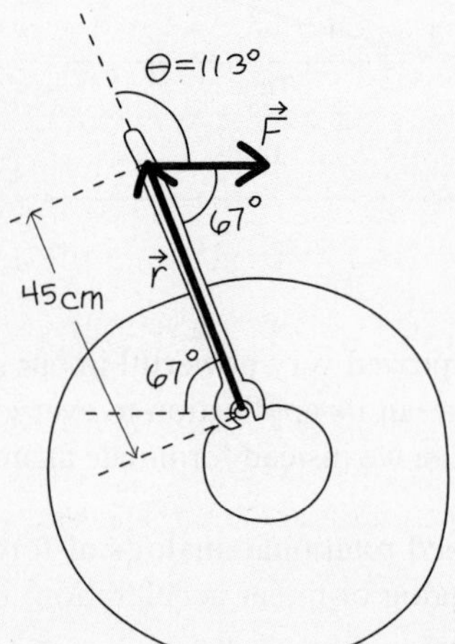

FIGURE 10.8 Our sketch of the wrench and wheel nut.

EVALUATE We solve Equation 10.10 for the force F:

$$F = \frac{\tau}{r \sin\theta} = \frac{95\ \text{N}\cdot\text{m}}{(0.45\ \text{m})(\sin 113°)} = 230\ \text{N}$$

ASSESS Is a 230-N force reasonable? Yes: It's roughly the force needed to lift a 23-kg (~50-lb) suitcase. Tightening torques, as in this example, are often specified for nuts and bolts in critical applications. Mechanics use specially designed "torque wrenches" that provide a direct indication of the applied torque. ■

✓TIP Specify the Axis

Torque depends on where the force is applied *relative to some rotation axis*. The same physical force results in different torques about different axes. Be sure the rotation axis is specified before you make a calculation involving torque.

PhET: Torque (Torque)

GOT IT? 10.2 The forces in Figs. 10.5 and 10.6 all have the same magnitude. (1) Which of Figs. 10.5*a*, 10.5*b*, and 10.6*b* has the greatest torque? (2) Which of these has the least torque?

10.3 Rotational Inertia and the Analog of Newton's Law

Torque and angular acceleration are the rotational analogs of force and linear acceleration. To develop a rotational analog of Newton's law, we still need the rotational analog of mass.

The mass m in Newton's law is a measure of a body's inertia—of its resistance to changes in motion. So we want a quantity that describes resistance to changes in rotational motion. Figure 10.9 shows that it's easier to set an object rotating when its mass is concentrated near the rotation axis. Therefore, our rotational analog of inertia must depend not only on mass itself but also on the distribution of mass relative to the rotation axis.

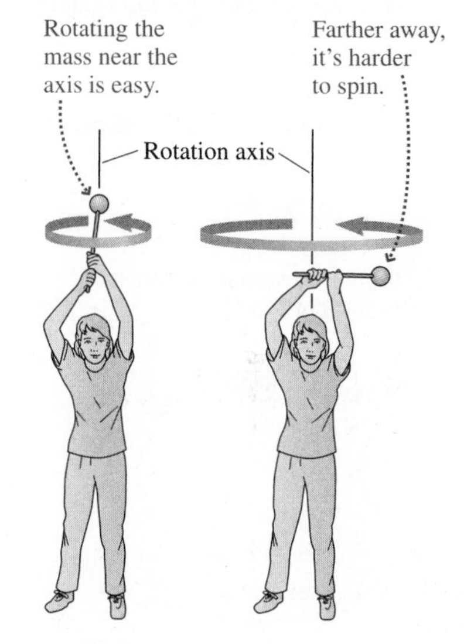

FIGURE 10.9 It's easier to set an object rotating if the mass is concentrated near the axis.

Suppose the object in Fig. 10.9 consists of an essentially massless rod of length R with a ball of mass m on the end. We allow the object to rotate about an axis through the free end of the rod and apply a force $\vec{F}$ to the ball, always at right angles to the rod (Fig. 10.10). The ball undergoes a tangential acceleration given by Newton's law: $F = ma_t$. (There's also a tension force in the rod, but because it acts along the rod, it doesn't contribute to the torque or angular acceleration.) We can use Equation 10.5 to express the tangential acceleration in terms of the angular acceleration α and the distance R from the rotation axis: $F = ma_t = m\alpha R$. We can also express the force F in terms of its associated torque. Since the force is perpendicular to the rod, Equation 10.10 gives $\tau = RF$. Using our expression for F, we have

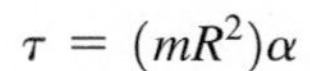

$$\tau = (mR^2)\alpha$$

Here we have Newton's law, $F = ma$, written in terms of rotational quantities. The torque—analogous to force—is the product of the angular acceleration and the quantity mR^2, which must therefore be the rotational analog of mass. We call this quantity the **rotational inertia** or **moment of inertia** and give it the symbol I. Rotational inertia is measured in $\text{kg}\cdot\text{m}^2$ and accounts for both an object's mass and the distribution of that mass. Like torque, the value of the rotational inertia depends on the location of the rotation axis. Given the rotational inertia I, our rotational analog of Newton's law becomes

$$\tau = I\alpha \qquad \text{(rotational analog of Newton's second law)} \qquad (10.11)$$

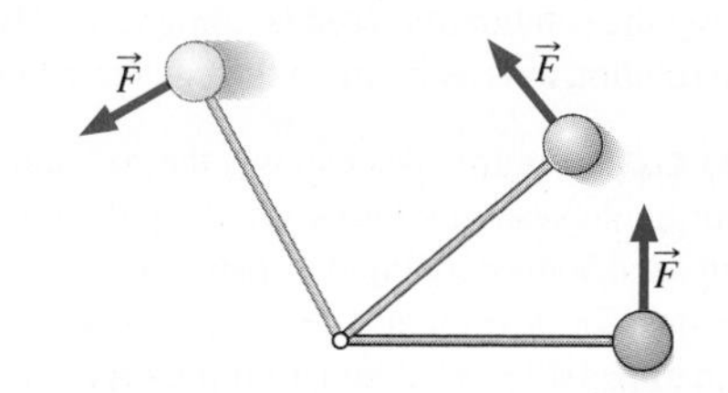

FIGURE 10.10 A force applied perpendicular to the rod results in angular acceleration.

Although we derived Equation 10.11 for a single, localized mass, it applies to extended objects if we interpret τ as the net torque on the object and I as the sum of the rotational inertias of the individual mass elements making up the object.

Calculating the Rotational Inertia

When an object consists of a number of discrete mass points, its rotational inertia about an axis is the sum of the rotational inertias of the individual mass points:

PhET: Torque (Moment of Inertia)

$$I = \sum m_i r_i^{\,2} \qquad \text{(rotational inertia)} \qquad (10.12)$$

Here m_i is the mass of the ith mass point, and r_i is its distance from the rotation axis.

EXAMPLE 10.4 Rotational Inertia: A Sum

A dumbbell-shaped object consists of two equal masses $m = 0.64$ kg on the ends of a massless rod of length $L = 85$ cm. Calculate its rotational inertia about an axis one-fourth of the way from one end of the rod and perpendicular to it.

INTERPRET Here we have two discrete masses, so this problem is asking us to calculate the rotational inertia by summing over the individual masses.

DEVELOP Figure 10.11 is our sketch. We'll use Equation 10.12, $I = \sum m_i r_i^2$, to sum the two individual rotational inertias.

EVALUATE $I = \sum m_i r_i^2 = m(\frac{1}{4}L)^2 + m(\frac{3}{4}L)^2 = \frac{5}{8}mL^2$

$= \frac{5}{8}(0.64\text{ kg})(0.85\text{ m})^2 = 0.29\text{ kg}\cdot\text{m}^2$

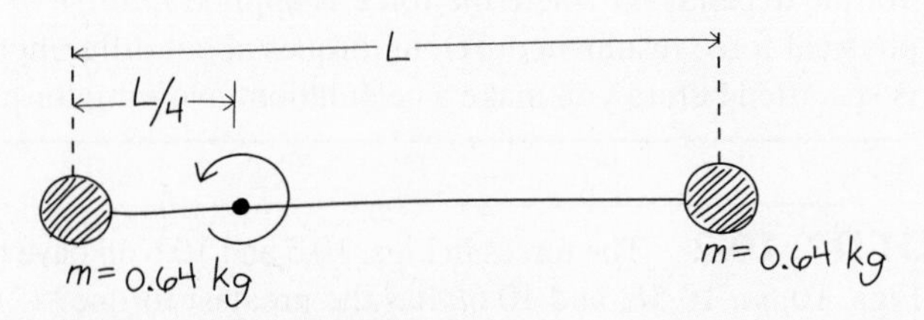

FIGURE 10.11 Our sketch for Example 10.4, showing rotation about an axis perpendicular to the page.

ASSESS Make sense? Even though there are two masses, our answer is less than the rotational inertia mL^2 of a single mass rotated about a rod of length L. That's because distance from the rotation axis is *squared*, so it contributes more in determining rotational inertia than does mass. ■

GOT IT? 10.3 Would the rotational inertia of the two-mass dumbbell in Example 10.4 (a) increase, (b) decrease, or (c) stay the same (1) if the rotation axis were at the center of the rod? (2) If it were at one end?

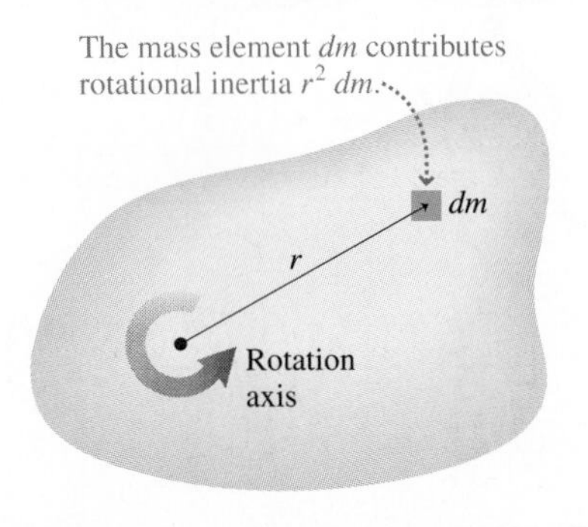

FIGURE 10.12 Rotational inertia can be found by integrating the rotational inertias r^2 *dm* of the mass elements making up an object.

With continuous distributions of matter, we consider a large number of very small mass elements *dm* throughout the object, and sum the individual rotational inertias $r^2\,dm$ over the entire object (Fig. 10.12). In the limit of an arbitrarily large number of infinitesimally small mass elements, that sum becomes an integral:

$$I = \int r^2\,dm \quad \left(\begin{array}{c}\text{rotational inertia,}\\ \text{continuous matter}\end{array}\right) \tag{10.13}$$

where the limits of integration cover the entire object.

EXAMPLE 10.5 Rotational Inertia by Integration: A Rod

Find the rotational inertia of a uniform, narrow rod of mass M and length L about an axis through its center and perpendicular to the rod.

INTERPRET The rod is a continuous distribution of matter, so calculating the rotational inertia is going to involve integration. We identify the rotation axis as being in the center of the rod.

DEVELOP Figure 10.13 shows the rod and rotation axis; we added a coordinate system with x-axis along the rod and the origin at the rotation axis. With a continuous distribution, Equation 10.13, $I = \int r^2\,dm$, applies. To develop a solution plan, we need to set up the integral in Equation 10.13. That equation may seem confusing because the integral contains both the geometric variable r and the mass element dm. How are they related? At this point you might want to review Tactics 9.1 (page 148); we'll follow its steps here. (1) We're first supposed to find a suitable mass element; here, with a one-dimensional rod, that can be a short section of the rod. We marked a typical mass element in Fig. 10.13. (2) This step is straightforward in this one-dimensional case; the length of the mass element is dx, signifying an infinitesimally short piece of the rod. (3) Now we form ratios to relate dx and the mass element dm. The total mass of the rod is M, and its total length is L. With the mass distributed uniformly, that means dx is the same fraction of L that dm is of M, or $dx/L = dm/M$. (4) We solve for the mass element: $dm = (M/L)\,dx$.

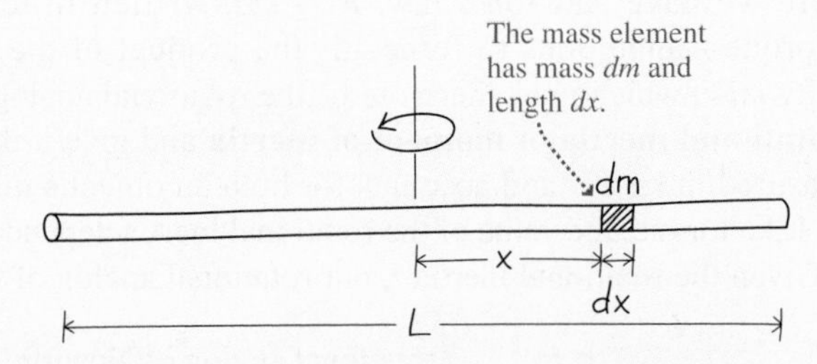

FIGURE 10.13 Our sketch of the uniform rod of Example 10.5.

We're almost done. But the integral in Equation 10.13 contains r, and we've related dm and dx. No problem: On the one-dimensional rod, distances from the rotation axis are just the coordinates x. So r becomes x in our integral, and we have

$$I = \int r^2\,dm = \int_{-L/2}^{L/2} x^2 \frac{M}{L}\,dx$$

We chose the limits to include the entire rod; with the origin at the center, it runs from $-L/2$ to $L/2$.

EVALUATE The constants M and L come outside the integral, so we have

$$I = \int_{-L/2}^{L/2} x^2 \frac{M}{L}\,dx = \frac{M}{L}\int_{-L/2}^{L/2} x^2\,dx = \frac{M}{L}\frac{x^3}{3}\bigg|_{-L/2}^{L/2} = \tfrac{1}{12}ML^2 \quad (10.14)$$

ASSESS Make sense? In Example 10.4 we found $I = \frac{5}{8}mL^2$ for a rod with two masses m on the ends. If you thought about GOT IT? 10.3, you probably realized that the rotational inertia would be $\frac{1}{2}mL^2$ for rotation about the rod's center. The total mass for that one was $M = 2m$, so in terms of total mass the rotational inertia about the center would be $I = \frac{1}{4}ML^2$—a lot larger than what we've found for the continuous rod. That's because much of the continuous rod's mass is close to the rotation axis, so it contributes less to the rotational inertia. ■

EXAMPLE 10.6 Rotational Inertia by Integration: A Ring

Find the rotational inertia of a thin ring of radius R and mass M about the ring's axis.

INTERPRET This example is similar to Example 10.5, but the geometry has changed from a rod to a ring.

DEVELOP Figure 10.14 shows the ring with a mass element dm. All the mass elements in the ring are the same distance R from the rotation axis, so r in Equation 10.13 is the constant R, and the equation becomes

$$I = \int R^2\,dm = R^2 \int dm$$

where the integration is over the ring.

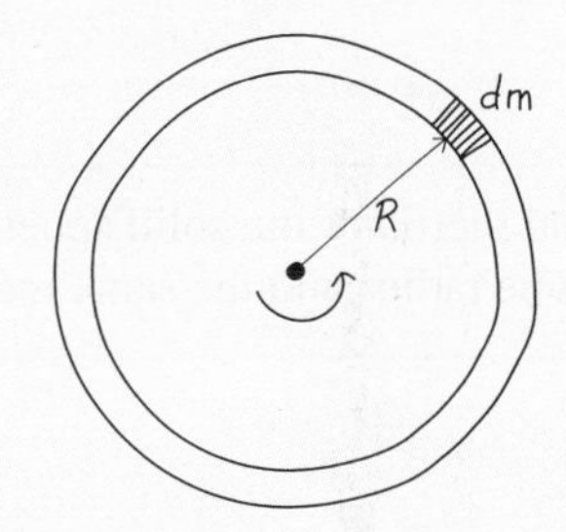

FIGURE 10.14 Our sketch of a thin ring, showing one mass element dm.

EVALUATE Because the sum of the mass elements over the ring is the total mass M, we find

$$I = MR^2 \quad \text{(thin ring)} \quad (10.15)$$

ASSESS The rotational inertia of the ring is the same as if all the mass were concentrated in one place a distance R from the rotation axis; the angular distribution of the mass about the axis doesn't matter. Notice, too, that it doesn't matter whether the ring is narrow like a loop of wire or long like a section of hollow pipe, as long as it's thin enough that all of it is essentially equidistant from the rotation axis (Fig. 10.15).

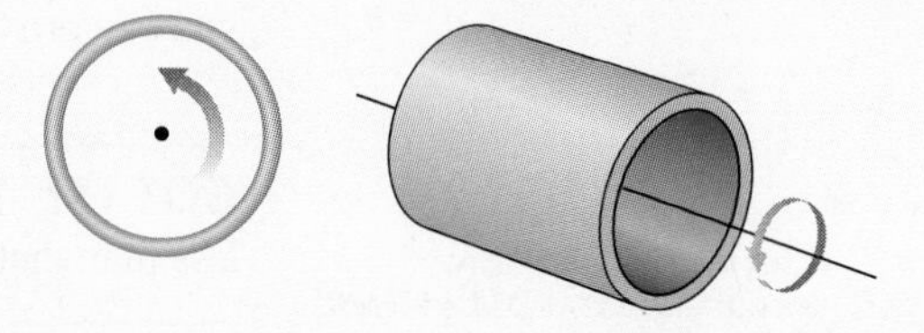

FIGURE 10.15 The rotational inertia is MR^2 for any thin ring, whether it's narrow like a wire loop or long like a pipe. ■

EXAMPLE 10.7 Rotational Inertia by Integration: A Disk

A disk of radius R and mass M has uniform density. Find the rotational inertia of the disk about an axis through its center and perpendicular to the disk.

INTERPRET Again we need to find the rotational inertia for a piece of continuous matter, this time a disk.

DEVELOP Because the disk is continuous, we need to integrate using Equation 10.13, $I = \int r^2\,dm$. We'll condense the strategy we applied in Example 10.5. The result of Example 10.6 suggests dividing the disk into rings, as shown in Fig. 10.16*a*. Equation 10.15, with $M \rightarrow dm$, shows that a ring of radius r and mass dm contributes $r^2\,dm$ to the rotational inertia of the disk. Then the total inertia will be $I = \int_0^R r^2\,dm$, where we chose the limits to pick up contributions from all the mass elements on the disk. Again we need to relate r and dm. Think of "unwinding" the ring, as shown in Fig. 10.16*b*; it becomes essentially a rectangle whose area dA is its circumference multiplied by its width: $dA = 2\pi r\,dr$. Next, we form ratios.

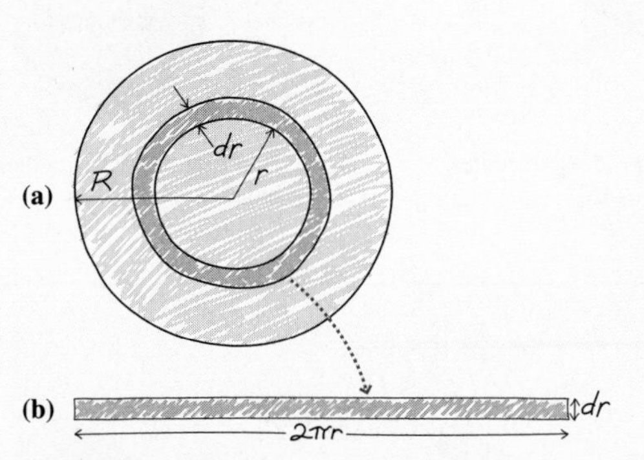

FIGURE 10.16 A disk may be divided into ring-shaped mass elements of mass dm, radius r, and width dr.

The ring area dA is to the total disk area πR^2 as the ring mass dm is to the total mass M: $2\pi r\,dr/\pi R^2 = dm/M$. Solving for dm gives $dm = (2Mr/R^2)\,dr$.

(continued)

EVALUATE We now evaluate the integral:

$$I = \int_0^R r^2\,dm = \int_0^R r^2\left(\frac{2Mr}{R^2}\right)dr$$

$$= \frac{2M}{R^2}\int_0^R r^3\,dr = \frac{2M}{R^2}\frac{r^4}{4}\bigg|_0^R = \tfrac{1}{2}MR^2 \qquad \text{(disk)} \quad (10.16)$$

ASSESS Again, this result makes sense. In the disk, some of the mass is closer to the rotation axis, so the rotational inertia should be less than the value MR^2 for the ring.

✓TIP Constants and Variables

Note the different roles of R and r here. R represents a fixed quantity—the actual radius of the disk—and it's a constant that can go outside the integral. In contrast, r is the *variable of integration*, and it changes as we range from the disk's center to its edge, adding up all the infinitesimal mass elements. Because r is a variable over the region of integration, we can't take it outside the integral.

■

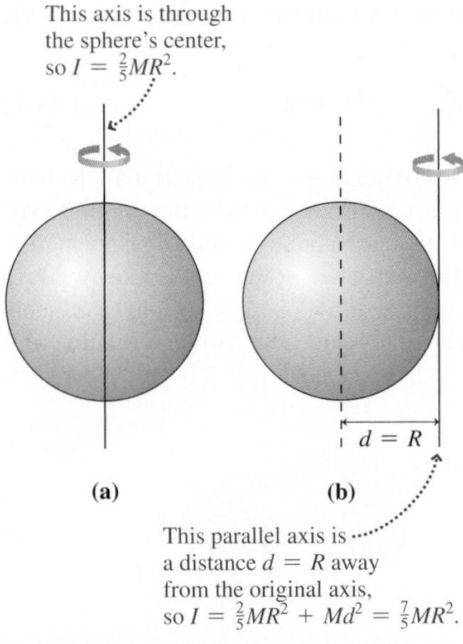

FIGURE 10.17 Meaning of the parallel-axis theorem.

The rotational inertias of other shapes about various axes are found by integration as in these examples. Table 10.2 lists results for some common shapes. Note that more than one rotational inertia is listed for some shapes, since the rotational inertia depends on the rotation axis.

If we know the rotational inertia I_{cm} about an axis through the center of mass of a body, a useful relation called the **parallel-axis theorem** allows us to calculate the rotational inertia I through any parallel axis. The parallel-axis theorem states that

$$I = I_{\text{cm}} + Md^2 \tag{10.17}$$

where d is the distance from the center-of-mass axis to the parallel axis and M is the total mass of the object. Figure 10.17 shows the meaning of the parallel-axis theorem, which you can prove in Problem 78.

GOT IT? 10.4 Explain why the rotational inertia of the solid sphere in Table 10.2 is less than that of the spherical shell with the same radius and the same mass.

Table 10.2 Rotational Inertias

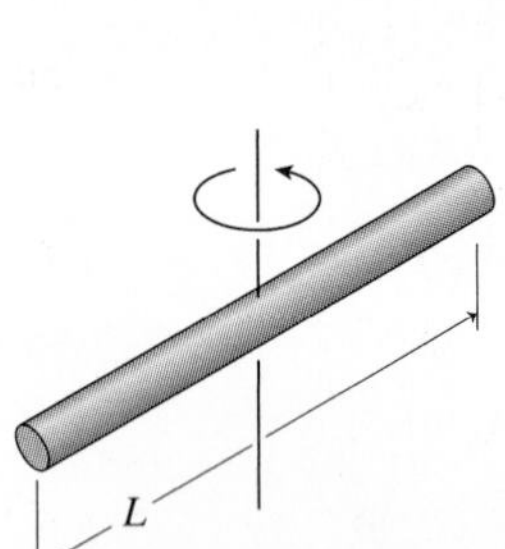

Thin rod about center
$I = \frac{1}{12}ML^2$

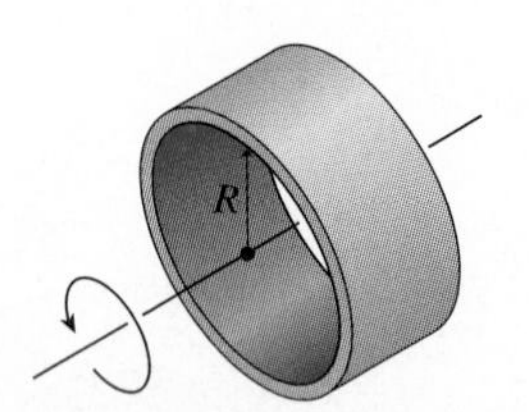

Thin ring or hollow cylinder about its axis
$I = MR^2$

Solid sphere about diameter
$I = \frac{2}{5}MR^2$

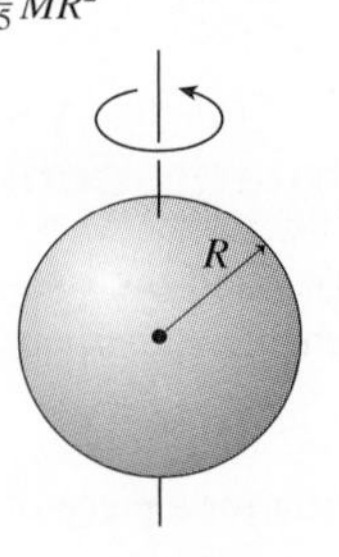

Flat plate about perpendicular axis
$I = \frac{1}{12}M(a^2 + b^2)$

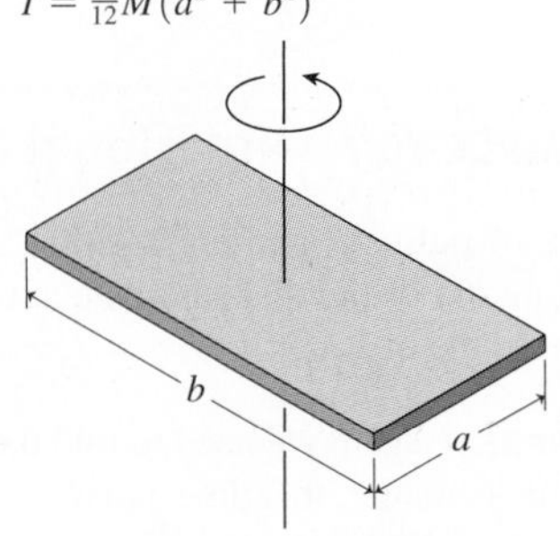

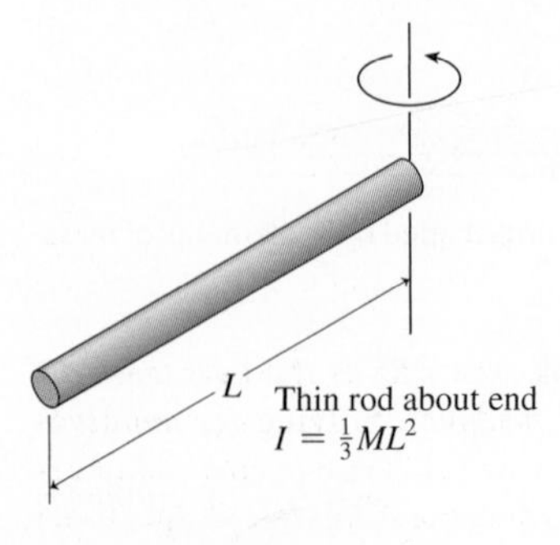

Thin rod about end
$I = \frac{1}{3}ML^2$

Disk or solid cylinder about its axis
$I = \frac{1}{2}MR^2$

Hollow spherical shell about diameter
$I = \frac{2}{3}MR^2$

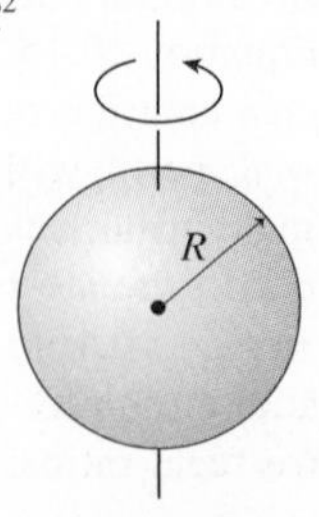

Flat plate about central axis
$I = \frac{1}{12}Ma^2$

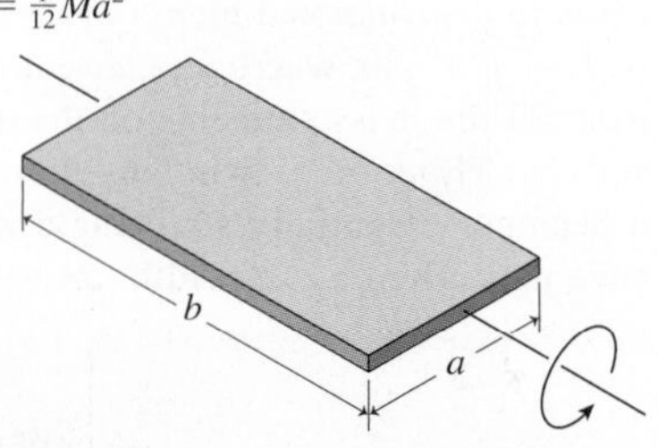

Rotational Dynamics

Knowing a body's rotational inertia, we can use the rotational analog of Newton's second law (Equation 10.11) to determine its behavior, just as we used Newton's law itself to analyze linear motion. Like the force in Newton's law, the torque in Equation 10.11 is the *net* external torque—the sum of all external torques acting on the body.

EXAMPLE 10.8 Rotational Dynamics: De-Spinning a Satellite

A cylindrical satellite is 1.4 m in diameter, with its 940-kg mass distributed uniformly. The satellite is spinning at 10 rpm but must be stopped so that astronauts can make repairs. Two small gas jets, each with 20-N thrust, are mounted on opposite sides of the satellite and fire tangent to the satellite's rim. How long must the jets be fired in order to stop the satellite's rotation?

INTERPRET This is ultimately a problem about angular acceleration, but we're given the forces the jets exert. So it becomes a problem about calculating torque and then acceleration—that is, a problem in rotational dynamics using the rotational analog of Newton's law.

DEVELOP Figure 10.18 shows the situation. We're asked about the time, which we can get from the angular acceleration and initial angular speed. We can find the acceleration using the rotational analog of Newton's law, Equation 10.11, if we know both torque and rotational inertia. So here's our plan: (1) Find the satellite's rotational inertia from Table 10.2, treating it as a solid cylinder. (2) Find the torque due to the jets using Equation 10.10, $\tau = rF\sin\theta$. (3) Use the rotational analog of Newton's law—Equation 10.11, $\tau = I\alpha$—to find the angular acceleration. (4) Use the change in angular speed to get the time.

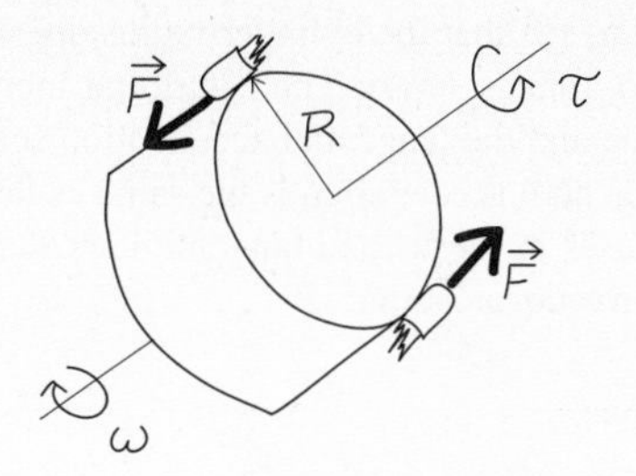

FIGURE 10.18 Torque from the jets stops the satellite's rotation.

EVALUATE Following our plan, (1) the rotational inertia from Table 10.2 is $I = \frac{1}{2}MR^2$. (2) With the jets tangent to the satellite, $\sin\theta$ in Equation 10.10 is 1, so each jet contributes a torque of magnitude RF, where R is the satellite radius and F the jet thrust force. With two jets, the net torque then has magnitude $\tau = 2RF$. (3) Equation 10.11 gives $\alpha = \tau/I = (2RF)/(\frac{1}{2}MR^2) = 4F/MR$. (4) We want this torque to drop the angular speed from $\omega_0 = 10$ rpm to zero, so the magnitude of the speed change is

$$\Delta\omega = 10\text{ rev/min} = (10\text{ rev/min})(2\pi\text{ rad/rev})/(60\text{ s/min}) = 1.05\text{ rad/s}$$

Since angular acceleration is $\alpha = \Delta\omega/\Delta t$, our final answer is

$$\Delta t = \frac{\Delta\omega}{\alpha} = \frac{MR\,\Delta\omega}{4F} = \frac{(940\text{ kg})(0.70\text{ m})(1.05\text{ rad/s})}{(4)(20\text{ N})} = 8.6\text{ s}$$

ASSESS Make sense? Yes: The thrust F appears in the denominator, showing that a larger force and hence torque will bring the satellite more rapidly to a halt. Larger M and R contribute to a larger rotational inertia, thus lengthening the stopping time—although a larger R also means a larger torque, an effect that reduces the R dependence from the R^2 that appears in the expression for rotational inertia. ■

A single problem can involve both rotational and linear motion with more than one object. The strategy for dealing with such problems is similar to the multiple-object strategy we developed in Chapter 5, where we identified the objects whose motions we were interested in, drew a free-body diagram for each, and then applied Newton's law separately to each object. We used the physical connections among the objects to relate quantities appearing in the separate Newton's law equations. Here we do the same thing, except that when an object is rotating, we use Equation 10.11, the rotational analog of Newton's law. Often the physical connection will entail relations between the force on an object in linear motion and the torque on a rotating object, as well as between the objects' linear and rotational accelerations.

EXAMPLE 10.9 Rotational and Linear Dynamics: Into the Well

A solid cylinder of mass M and radius R is mounted on a frictionless horizontal axle over a well, as shown in Fig. 10.19. A rope of negligible mass is wrapped around the cylinder and supports a bucket of mass m. Find an expression for the bucket's acceleration as it falls down the well shaft.

INTERPRET If it weren't connected to the cylinder, the bucket would fall with acceleration g. But the rope exerts an upward tension force $\vec{T}$ on the bucket, reducing its acceleration and at the same time exerting a torque on the cylinder. So we have a problem involving both rotational and linear dynamics. We identify the bucket and the cylinder as the

(continued)

FIGURE 10.19 Example 10.9.

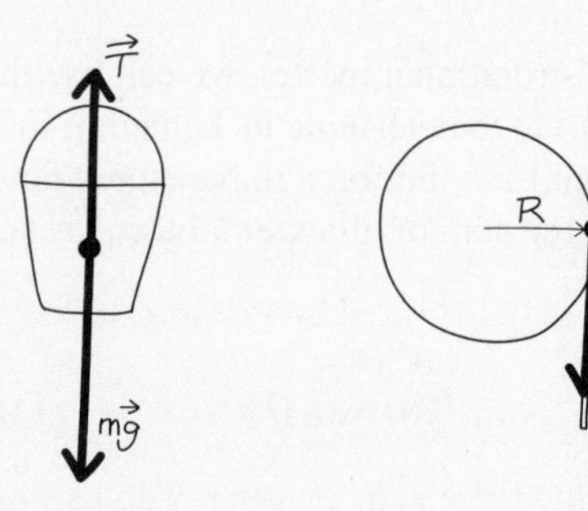

FIGURE 10.20 Our free-body diagrams for the bucket and cylinder.

objects of interest; the bucket is in linear motion while the cylinder rotates. The connection between them is the rope.

DEVELOP Figure 10.20 shows free-body diagrams for the two objects; note that both involve the rope tension, $\vec{T}$. We chose the downward direction as positive in the bucket diagram and the clockwise direction as positive in the cylinder diagram. Now we're ready to write Newton's second law and its analog—Equation 10.11, $\tau = I\alpha$—for the two objects. Our plan is to formulate both equations and solve using the connection between them—physically the rope and mathematically the magnitude of the rope tension. We have to express the torque on the cylinder in terms of the tension force, using Equation 10.10, $\tau = rF\sin\theta$. We also need to relate the cylinder's angular acceleration to the bucket's linear acceleration, using Equation 10.5, $a_t = r\alpha$.

EVALUATE With the downward direction positive, Newton's second law for the bucket reads $F_{net} = mg - T = ma$. For the cylinder we have the rotational analog of Newton's second law: $\tau = I\alpha$. But here the torque is due to the rope tension, which exerts a force T at right angles to a line from the rotation axis and so produces torque RT. Then the Newton's law analog becomes $RT = I\alpha$. As the rope unwinds, the tangential acceleration of the cylinder's edge must be equal to the bucket's linear acceleration; thus, using Equation 10.5, we have $\alpha = a/R$, and the cylinder equation becomes $RT = Ia/R$ or $T = Ia/R^2$. But the cylinder's rotational inertia, from Table 10.2, is $I = \frac{1}{2}MR^2$, so $T = \frac{1}{2}Ma$. Using this result in the bucket equation gives $ma = mg - T = mg - \frac{1}{2}Ma$; solving for a, we then have

$$a = \frac{mg}{m + \frac{1}{2}M}$$

ASSESS Make sense? If $M = 0$, there would be no rotational inertia and we would have $a = g$. With no torque needed to accelerate the cylinder, there would be no rope tension and the bucket would fall freely with acceleration g. But as the cylinder's mass M increases, the bucket's deceleration drops as greater torque and thus rope tension are needed to give the cylinder its rotational acceleration. You may be surprised to see that the cylinder radius doesn't appear in our answer. That, too, makes sense: The rotational inertia scales as R^2, but both the torque and the tangential acceleration scale with R. Since the cylinder's tangential acceleration is the same as the bucket's acceleration, the increases in torque and tangential acceleration cancel the effect of a greater rotational inertia. ■

GOT IT? 10.5 The figure shows two identical masses m connected by a string that passes over a frictionless pulley whose mass M is *not* negligible. One mass rests on a frictionless table; the other hangs vertically, as shown. Is the magnitude of the tension force in the vertical section of the string (a) greater than, (b) equal to, or (c) less than that in the horizontal section? Explain.

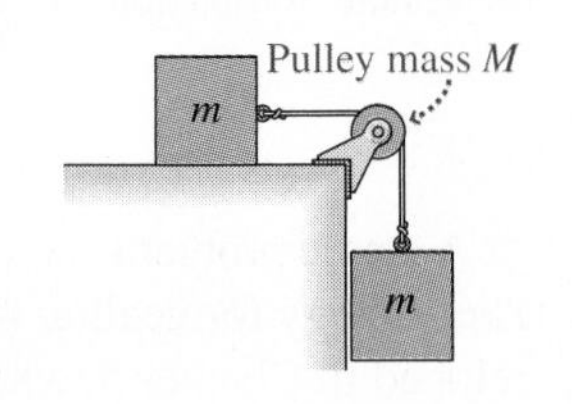

10.4 Rotational Energy

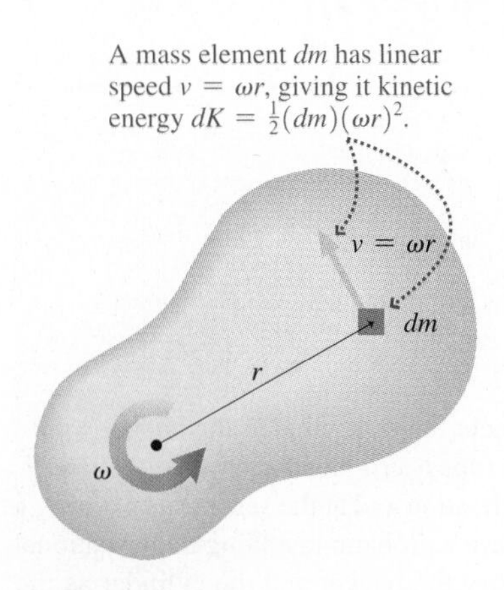

FIGURE 10.21 Kinetic energy of a mass element.

A rotating object has kinetic energy because all its parts are in motion. We define an object's **rotational kinetic energy** as the sum of the kinetic energies of all its individual mass elements, taken with respect to the rotation axis. Figure 10.21 shows that an individual mass element dm a distance r from the rotation axis has kinetic energy given by $dK = \frac{1}{2}(dm)(v^2) = \frac{1}{2}(dm)(\omega r)^2$. The rotational kinetic energy is given by summing—that is, integrating—over the entire object:

$$K_{rot} = \int dK = \int \tfrac{1}{2}(dm)(\omega r)^2 = \tfrac{1}{2}\omega^2 \int r^2\,dm$$

where we've taken ω^2 outside the integral because it's the same for every mass element in the rigid, rotating object. The remaining integral is just the rotational inertia I, so we have

$$K_{rot} = \tfrac{1}{2}I\omega^2 \quad \text{(rotational kinetic energy)} \qquad (10.18)$$

This formula makes sense in light of our analogies between linear and rotational motion: Since I and ω are the rotational analogs of mass and speed, Equation 10.18 is the rotational equivalent of $K = \frac{1}{2}mv^2$.

EXAMPLE 10.10 Rotational Energy: Flywheel Storage

A flywheel has a 135-kg solid cylindrical rotor with radius 30 cm and spins at 31,000 rpm. How much energy does it store?

INTERPRET We're being asked about kinetic energy stored in a rotating cylinder.

DEVELOP Equation 10.18, $K_{rot} = \frac{1}{2}I\omega^2$, gives the rotational energy. To use it, we need the rotational inertia from Table 10.2, and we need to convert the rotation rate in revolutions per minute to angular speed ω in radians per second.

EVALUATE Table 10.2 gives the rotational inertia, $I = \frac{1}{2}MR^2 = (\frac{1}{2})(135\ \text{kg})(0.30\ \text{m})^2 = 6.1\ \text{kg}\cdot\text{m}^2$, and 31,000 rpm is equivalent to $(31{,}000\ \text{rev/min})(2\pi\ \text{rad/rev})/(60\ \text{s/min}) = 3246\ \text{rad/s}$. Then Equation 10.18 gives

$$K_{rot} = \tfrac{1}{2}I\omega^2 = (\tfrac{1}{2})(6.1\ \text{kg}\cdot\text{m}^2)(3246\ \text{rad/s})^2 = 32\ \text{MJ}$$

ASSESS 32 MJ is roughly the energy contained in a liter of gasoline. The advantages of the flywheel over a fuel or a chemical battery are more concentrated energy storage and greater efficiency at getting energy into and out of storage; see the Application below. Can you see why the solid disk of this example isn't the most efficient flywheel design? You can explore this question further in Question 11 and Problem 77. ■

✓TIP When to Use Radians

We derived Equation 10.18, $K = \frac{1}{2}I\omega^2$, using Equation 10.3, $v = \omega r$. Since that equation works only with radian measure, the same is true of Equation 10.18.

Energy and Work in Rotational Motion

In Section 6.3 we proved the work–kinetic energy theorem, which states that the change in an object's linear kinetic energy is equal to the net work done on the object. There the work was the product (or the integral, for a changing force) of the net force and the distance the object moves. Not surprisingly, there's an analogous relation for rotational motion: The change in an object's rotational kinetic energy is equal to the net work done on the object. Now the work is, analogously, the product (or the integral, when torque varies with angle) of the torque and the angular displacement:

$$W = \int_{\theta_i}^{\theta_f} \tau\, d\theta = \Delta K_{rot} = \tfrac{1}{2}I\omega_f^2 - \tfrac{1}{2}I\omega_i^2 \quad \left(\begin{array}{c}\text{work–kinetic energy theorem,}\\ \text{rotational motion}\end{array}\right) \quad (10.19)$$

Here the subscripts refer to the initial and final states.

APPLICATION Flywheel Energy Storage

Flywheels provide an attractive alternative to batteries in applications requiring short bursts of power. Examples include acceleration and hill climbing in hybrid vehicles, industrial lifting equipment and amusement park rides, power management on the electric grid, and uninterruptible power supplies. Flywheel-based hybrid vehicles would achieve high efficiency by storing mechanical energy in the flywheel during braking rather than dissipating it as heat in conventional brakes or even storing it in a chemical battery as in today's hybrids.

Equation 10.18 shows that the stored energy can be substantial, provided the flywheel has significant rotational inertia and angular speed—the latter being especially important because the energy scales as the *square* of the angular speed. Modern flywheels can supply tens of kilowatts of power for as long as a minute; unlike batteries, their output isn't reduced in cold weather. They achieve rotation rates of 30,000 rpm and more using advanced carbon composite materials that can withstand the forces needed to maintain the radial acceleration of magnitude $\omega^2 r$. Advanced flywheels spin in vacuum, using magnetic bearings to minimize friction. Some even use superconducting materials, which eliminate electrical losses that we'll examine in Chapter 26. The photo shows a high-speed flywheel used in a prototype hybrid bus operating in Austin, Texas. The flywheel helps the bus achieve 30% fuel savings.

EXAMPLE 10.11 Work and Rotational Energy: Balancing a Tire

An automobile wheel with tire has rotational inertia $2.7\ \text{kg}\cdot\text{m}^2$. What constant torque does a tire-balancing machine need to apply in order to spin this tire up from rest to 700 rpm in 25 revolutions?

INTERPRET The wheel's rotational kinetic energy changes as it spins up, so the machine must be doing work by applying a torque. Therefore, the concept behind this problem is the work–kinetic energy theorem for rotational motion.

DEVELOP The work–kinetic energy theorem of Equation 10.19 relates the work to the change in rotational kinetic energy:

$$W = \int_{\theta_i}^{\theta_f} \tau\, d\theta = \Delta K_{\text{rot}} = \tfrac{1}{2}I\omega_f^2 - \tfrac{1}{2}I\omega_i^2.$$

We're given the initial and final angular velocities, although we have to convert them to radians per second. With constant torque, the integral in Equation 10.19 becomes the product $\tau\,\Delta\theta$, so we can solve for the torque.

EVALUATE The initial angular speed ω_i is zero, and the final speed $\omega_f = (700\ \text{rev/min})(2\pi\ \text{rad/rev})/(60\ \text{s/min}) = 73.3\ \text{rad/s}$. The angular displacement $\Delta\theta$ is $(25\ \text{rev})(2\pi\ \text{rad/rev}) = 157\ \text{rad}$. Then Equation 10.19 becomes $W = \tau\,\Delta\theta = \tfrac{1}{2}I\omega_f^2$, which gives

$$\tau = \frac{\tfrac{1}{2}I\omega_f^2}{\Delta\theta} = \frac{(\tfrac{1}{2})(2.7\ \text{kg}\cdot\text{m}^2)(73.3\ \text{rad/s})^2}{157\ \text{rad}} = 46\ \text{N}\cdot\text{m}$$

ASSESS If this torque results from a force applied at the rim of a typical 40-cm-radius tire, then the magnitude of the force would be just over 100 N, about the weight of a 10-kg mass and thus a reasonable value. ■

GOT IT? 10.6 A wheel is rotating at 100 rpm. To spin it up to 200 rpm will take (a) less; (b) more; (c) the same work as it took to get it from rest to 100 rpm.

10.5 Rolling Motion

A rolling object exhibits both rotational motion and translational motion—the motion of the whole object from place to place. How much kinetic energy is associated with each?

In Section 9.3, we found that the kinetic energy of a composite object comprises two terms: the kinetic energy of the center of mass and the internal kinetic energy relative to the center of mass: $K = K_{\text{cm}} + K_{\text{internal}}$. A wheel of mass M moving with speed v has center-of-mass kinetic energy $K_{\text{cm}} = \tfrac{1}{2}Mv^2$. In the center-of-mass frame, the wheel is simply rotating with angular speed ω about the center of mass, so its internal kinetic energy is $K_{\text{internal}} = \tfrac{1}{2}I_{\text{cm}}\omega^2$, where the rotational inertia is taken about the center of mass. We now sum K_{cm} and K_{internal} to get the total kinetic energy:

$$K_{\text{total}} = \tfrac{1}{2}Mv^2 + \tfrac{1}{2}I_{\text{cm}}\omega^2 \tag{10.20}$$

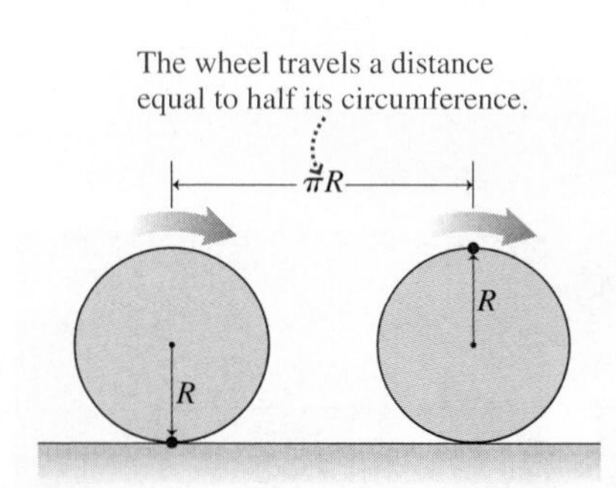

FIGURE 10.22 A rolling wheel turns through half a revolution.

When a wheel is *rolling*—moving without slipping against the ground—its translational speed v and angular speed ω about its center of mass are related. Imagine a wheel that rolls half a revolution and therefore moves horizontally half its circumference (Fig. 10.22). Then the wheel's angular speed is the angular displacement $\Delta\theta$, here half a revolution, or π radians, divided by the time Δt: $\omega = \pi/\Delta t$. Its translational speed is the actual distance the wheel travels divided by the same time interval. But we've just argued that the wheel travels half a circumference, or πR, where R is its radius. So its translational speed is $v = \pi R/\Delta t$. Comparing our expressions for v and ω, we see that

$$v = \omega R \qquad \text{(rolling motion)} \tag{10.21}$$

Equation 10.21 looks deceptively like Equation 10.3. But it says more. In Equation 10.3, $v = \omega r$, v is the linear speed of a point a distance r from the center of a rotating object. In Equation 10.21, v is the translational speed of the whole object and R is its radius. The two equations look similar because, as our argument leading to Equation 10.21 shows, an object that rolls without slipping moves with respect to the ground at the same rate that a point on its rim moves in the center-of-mass frame.

Our description of rolling motion leads to a point you may at first find absurd: In a rolling wheel, the point in contact with the ground is, instantaneously, at rest! Figure 10.23 shows how this surprising situation comes about.

Why would an object roll without slipping? The answer is friction. On an icy slope, a wheel just slides down without rolling. Normally, though, the force of static friction keeps

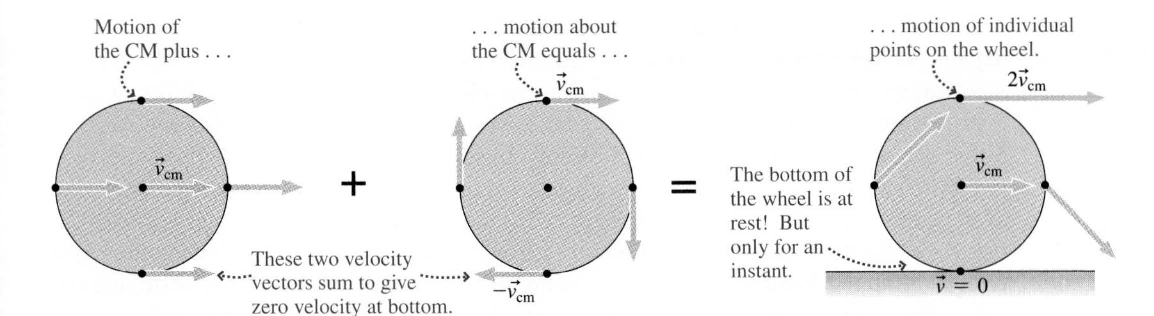

FIGURE 10.23 Motion of a rolling wheel, decomposed into translation of the entire wheel plus rotation about the center of mass.

it from sliding. Instead, it rolls (Fig. 10.24). Because the contact point is at rest, the frictional force does no work and therefore mechanical energy is conserved. This lets us use the conservation-of-energy principle to analyze rolling objects.

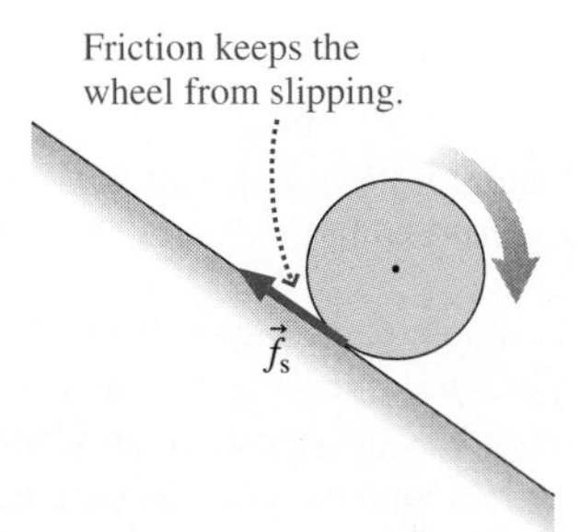

FIGURE 10.24 Rolling down a slope.

GOT IT? 10.7 The wheels of trains, subway cars, and other rail vehicles include a flange that extends beyond the part of the wheel that rolls on top of the rail, as shown. The flanges keep the train from running off the rails. Consider the bottom-most point on the flange: Is it (a) moving in the direction of the train's motion; (b) instantaneously at rest; or (c) moving backward, opposite the train's motion?

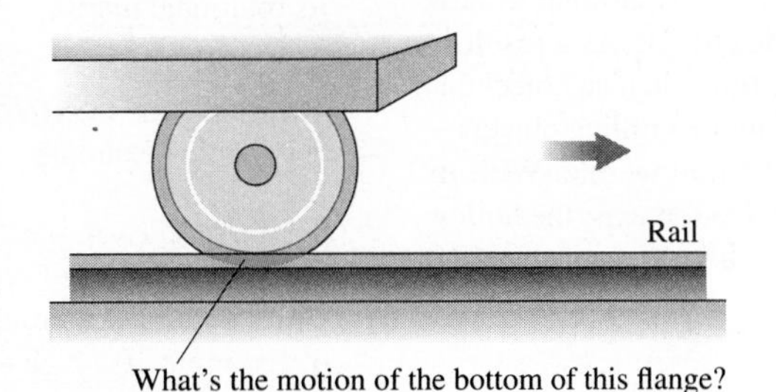

EXAMPLE 10.12 Energy Conservation: Rolling Downhill

A solid ball of mass M and radius R starts from rest and rolls down a hill. Its center of mass drops a total distance h. Find the ball's speed at the bottom of the hill.

INTERPRET This is similar to conservation-of-energy problems from Chapter 7, but now we identify two types of kinetic energy: translational and rotational. The ball starts on the slope with some gravitational potential energy, which ends up as kinetic energy at the bottom. The frictional force that keeps the ball from slipping does no work, so we can apply conservation of mechanical energy.

DEVELOP Figure 10.25 shows the situation, including bar graphs showing the distribution of energy in the ball's initial and final states.

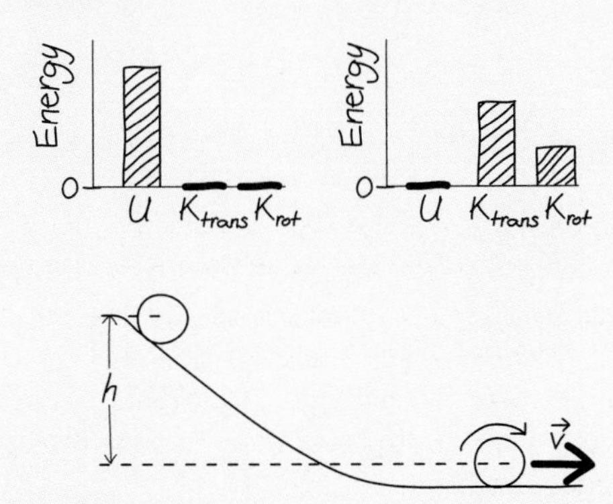

FIGURE 10.25 How fast is the ball moving at the bottom of the hill?

We've determined that conservation of mechanical energy holds, so $K_0 + U_0 = K + U$. Here $K_0 = 0$ and, if we take the zero of potential energy at the bottom, then $U_0 = Mgh$ and $U = 0$. Finally, K consists of both translational and rotational kinetic energy as expressed in Equation 10.20, $K_{\text{total}} = \frac{1}{2}Mv^2 + \frac{1}{2}I\omega^2$. Our plan is to use this expression in the conservation-of-energy statement and solve for v. It looks like there's an extra variable, ω, that we don't know. But the ball isn't slipping, so Equation 10.21 holds and gives $\omega = v/R$. Then conservation of energy becomes

$$Mgh = \tfrac{1}{2}Mv^2 + \tfrac{1}{2}I\omega^2 = \tfrac{1}{2}Mv^2 + \tfrac{1}{2}(\tfrac{2}{5}MR^2)\left(\frac{v}{R}\right)^2 = \tfrac{7}{10}Mv^2$$

where we found the rotational inertia of a solid sphere, $\frac{2}{5}MR^2$, from Table 10.2.

EVALUATE Solving for v gives our answer:

$$v = \sqrt{\frac{10}{7}gh}$$

ASSESS This result is less than the speed $v = \sqrt{2gh}$ for an object that slides down a frictionless incline. Make sense? Yes: Some of the energy the rolling object gains goes into rotation, leaving less for translational motion. As often happens with gravitational problems, mass doesn't matter. Neither does radius: That factor $\frac{7}{10}$ results from the distribution of mass that gives the sphere its particular rotational inertia and would be the same for all spheres regardless of radius or mass. ■

Video Tutor Demo | Canned Food Race

Example 10.12 shows that the final speed of an object that rolls down an incline depends on the details of its mass distribution. That means objects that look superficially identical will reach the bottom of an incline at different times, if they have different mass distributions. Conceptual Example 10.1 helps you think further about this point. Another difference that can affect the speed of rolling objects is whether they roll as rigid bodies or not. When a can of liquid rolls down a ramp, for instance, the liquid need not spin as fast as the can itself (or it may not even spin at all), and therefore less energy goes into rotation—leaving more for translational motion. You can see an example by viewing the video tutorial "Canned Food Race" accessed from the QR code at the left. After watching the video, can you see how you might distinguish a hard-boiled egg from a raw one?

CONCEPTUAL EXAMPLE 10.1 A Rolling Race

A solid ball and a hollow ball roll without slipping down a ramp. Which reaches the bottom first?

EVALUATE Example 10.12 shows that when a ball rolls down a slope, some of its potential energy gets converted into rotational kinetic energy—leaving less for translational kinetic energy. As a result, it moves more slowly, and therefore takes more time, than an object that slides without rolling. Here we want to compare two rolling objects—the solid ball treated in Example 10.12 and a hollow one. With its mass concentrated at its surface, far from the rotation axis, the hollow ball has greater rotational inertia. Thus more of its energy goes into rotation, meaning its translational speed is lower, so it reaches the bottom later.

ASSESS Make sense? Yes: Energy is conserved for both balls, but for the hollow ball more of that energy is in rotation and less in translation. As Example 10.12 shows, neither the mass nor the radius of a ball affects its speed; all that matters is its mass distribution and hence its rotational inertia.

MAKING THE CONNECTION Compare the final speeds of the two balls in this example.

EVALUATE Example 10.12 gives $\sqrt{10gh/7}$ for the speed of the solid ball after it's rolled down a vertical drop h. Substituting the hollow ball's rotational inertia, $I = \frac{2}{3}MR^2$ from Table 10.2, in the calculation of Example 10.12 gives $v = \sqrt{6gh/5}$. So the solid ball is faster by a factor $\sqrt{10/7}/\sqrt{6/5} \simeq 1.1$.

CHAPTER 10 SUMMARY

Big Idea

The big idea of this chapter is rotational motion, quantified as the rate of change of angular position of any point on a rotating object. All the quantities used to describe linear motion have analogs in rotational motion. The analogs of force, mass, and acceleration are, respectively, torque, rotational inertia, and angular acceleration—and together they obey the rotational analog of Newton's second law.

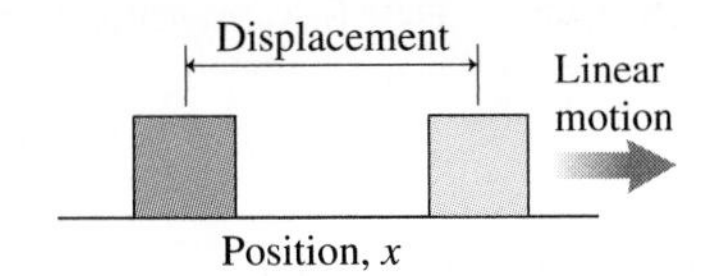

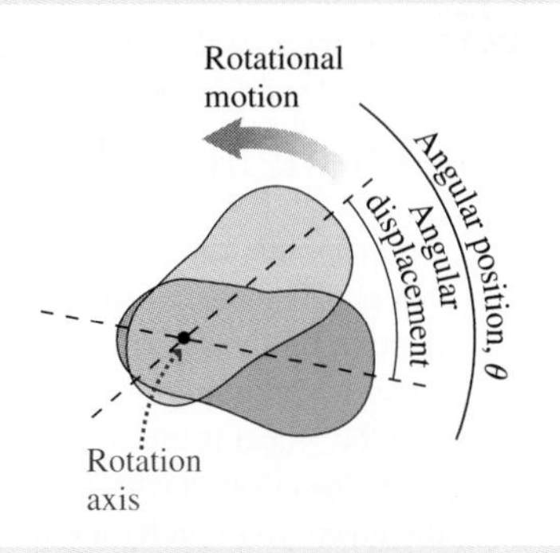

Key Concepts and Equations

The defining relations for rotational quantities are analogous to those for linear quantities, as is the statement of Newton's second law for rotational motion. Key concepts include angular velocity and acceleration, torque, and rotational inertia.

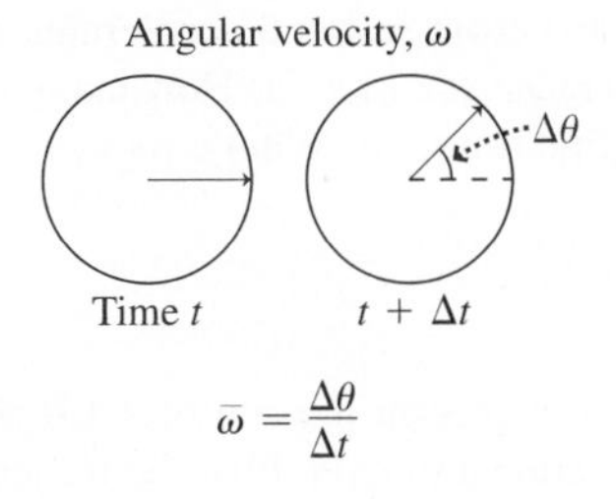

$$\bar{\omega} = \frac{\Delta\theta}{\Delta t}$$

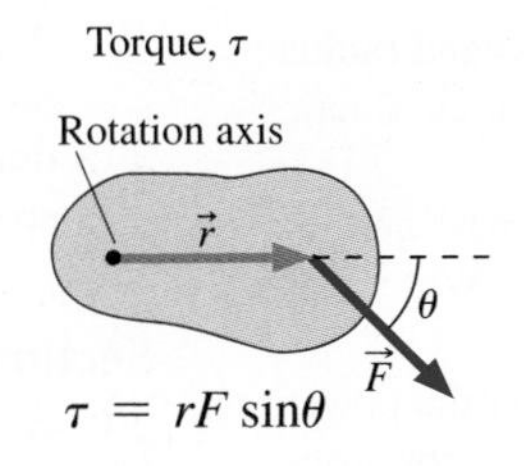

$$\tau = rF\sin\theta$$

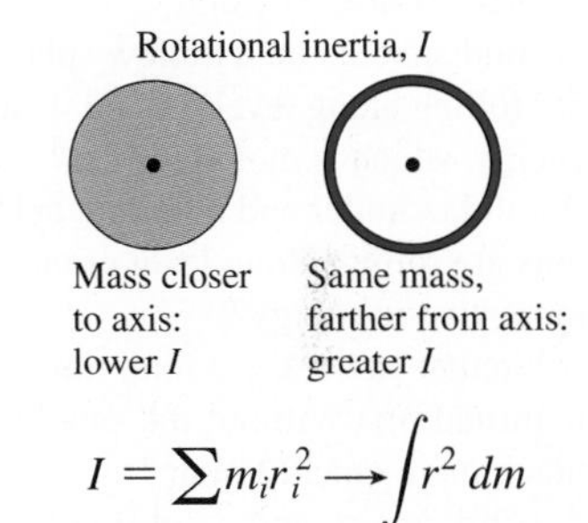

$$I = \sum m_i r_i^2 \longrightarrow \int r^2\,dm$$

Discrete masses — Continuous matter

This table summarizes the analogies between linear and rotational quantities, along with quantitative relations that link rotational and linear quantities. Many of these relations require that angles be measured in radians, and most require explicit specification of a rotation axis.

Linear Quantity or Equation	Angular Quantity or Equation	Relation Between Linear and Angular Quantities
Position x	Angular position θ	
Speed $v = dx/dt$	Angular speed $\omega = d\theta/dt$	$v = \omega r$
Acceleration a	Angular acceleration α	$a_t = \alpha r$
Mass m	Rotational inertia I	$I = \int r^2\,dm$
Force F	Torque τ	$\tau = rF\sin\theta$
Kinetic energy $K_{trans} = \frac{1}{2}mv^2$	Kinetic energy $K_{rot} = \frac{1}{2}I\omega^2$	
Newton's second law (constant mass or rotational inertia):		
$F = ma$	$\tau = I\alpha$	

Applications

Constant angular acceleration: When angular acceleration is constant, equations analogous to those of Chapter 2 apply.

Equations for Constant Linear Acceleration		Equations for Constant Angular Acceleration	
$\bar{v} = \frac{1}{2}(v_0 + v)$	(2.8)	$\bar{\omega} = \frac{1}{2}(\omega_0 + \omega)$	(10.6)
$v = v_0 + at$	(2.7)	$\omega = \omega_0 + \alpha t$	(10.7)
$x = x_0 + v_0 t + \frac{1}{2}at^2$	(2.10)	$\theta = \theta_0 + \omega_0 t + \frac{1}{2}\alpha t^2$	(10.8)
$v^2 = v_0^2 + 2a(x - x_0)$	(2.11)	$\omega^2 = \omega_0^2 + 2\alpha(\theta - \theta_0)$	(10.9)

Rolling motion: When an object of radius R rolls without slipping, the point in contact with the ground is instantaneously at rest. In this case the object's translational and rotational speeds are related by $v = \omega R$. The object's kinetic energy is shared among translational kinetic energy $\frac{1}{2}Mv^2$ and rotational kinetic energy $\frac{1}{2}I\omega^2$, with the division between these forms dependent on the rotational inertia.

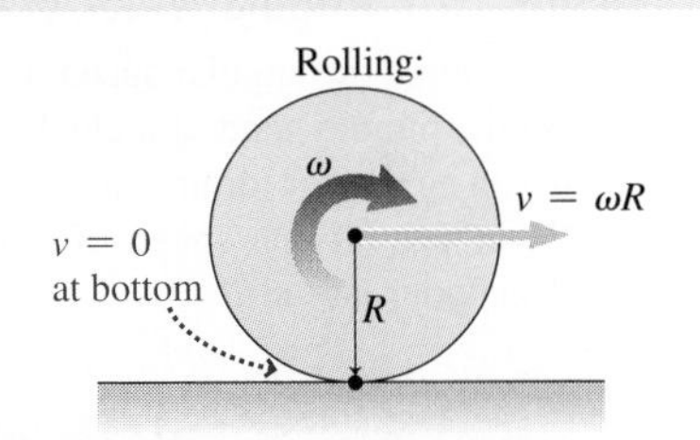

MP For homework assigned on MasteringPhysics, go to www.masteringphysics.com

BIO *Biology and/or medicine-related problems* **DATA** *Data problems* **ENV** *Environmental problems* **CH** *Challenge problems* **COMP** *Computer problems*

For Thought and Discussion

1. Do all points on a rigid, rotating object have the same angular velocity? Linear speed? Radial acceleration?
2. A point on the rim of a rotating wheel has nonzero acceleration, since it's moving in a circular path. Does it necessarily follow that the wheel is undergoing angular acceleration?
3. Why doesn't it make sense to talk about a body's rotational inertia unless you specify a rotation axis?
4. Two forces act on an object, but the net force is zero. Must the net torque be zero? If so, why? If not, give a counterexample.
5. Is it possible to apply a counterclockwise torque to an object that's rotating clockwise? If so, how will the object's motion change? If not, why not?
6. A solid sphere and a hollow sphere of the same mass and radius are rolling along level ground. If they have the same total kinetic energy, which is moving faster?
7. A solid cylinder and a hollow cylinder of the same mass and radius are rolling along level ground at the same speed. Which has more kinetic energy?
8. A circular saw takes a long time to stop rotating after the power is turned off. Without the saw blade mounted, the motor stops much more quickly. Why?
9. A solid sphere and a solid cube have the same mass, and the side of the cube is equal to the diameter of the sphere. The cube's rotation axis is perpendicular to two of its faces. Which has greater rotational inertia about an axis through the center of mass?
10. BIO The lower part of a horse's leg contains essentially no muscle. How does this help the horse to run fast? Explain in terms of rotational inertia.
11. Given a fixed amount of a material, what shape should you make a flywheel so it will store the most energy at a given angular speed?
12. A ball starts from rest and rolls without slipping down a slope, then starts up a frictionless slope (Fig. 10.26). Compare its maximum height on the frictionless slope with its starting height on the first slope.

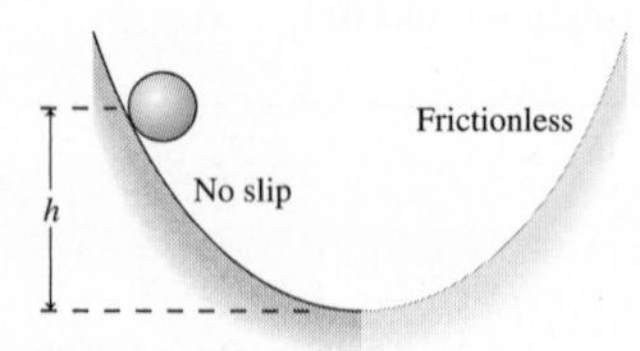

FIGURE 10.26 For Thought and Discussion 12, Problem 64

Exercises and Problems

Exercises

Section 10.1 Angular Velocity and Acceleration

13. Determine the angular speed, in rad/s, of (a) Earth about its axis; (b) the minute hand of a clock; (c) the hour hand of a clock; and (d) an eggbeater turning at 300 rpm.
14. What's the linear speed of a point (a) on Earth's equator and (b) at your latitude?
15. Express each of the following in radians per second: (a) 720 rpm; (b) 50°/h; (c) 1000 rev/s; (d) 1 rev/year (Earth's angular speed in its orbit).
16. A 25-cm-diameter circular saw blade spins at 3500 rpm. How fast would you have to push a straight hand saw to have the teeth move through the wood at the same rate as the circular saw teeth?
17. A compact disc's rotation varies from about 200 rpm to 500 rpm. If the disc plays for 74 min, what's its average angular acceleration in (a) rpm/s and (b) rad/s^2?
18. During startup, a power plant's turbine accelerates from rest at 0.52 rad/s^2. (a) How long does it take to reach its 3600-rpm operating speed? (b) How many revolutions does it make during this time?
19. A merry-go-round starts from rest and accelerates with angular acceleration 0.010 rad/s^2 for 14 s. (a) How many revolutions does it make during this time? (b) What's its average angular speed?

Section 10.2 Torque

20. A 320-N frictional force acts on the rim of a 1.0-m-diameter wheel to oppose its rotational motion. Find the torque about the wheel's central axis.
21. A 110-N·m torque is needed to start a revolving door rotating. If a child can push with a maximum force of 90 N, how far from the door's rotation axis must she apply this force?
22. A car tune-up manual calls for tightening the spark plugs to a torque of 35.0 N·m. To achieve this torque, with what force must you pull on the end of a 24.0-cm-long wrench if you pull (a) at a right angle to the wrench shaft and (b) at 110° to the wrench shaft?
23. A 55-g mouse runs out to the end of the 17-cm-long minute hand of a grandfather clock when the clock reads 10 past the hour. What torque does the mouse's weight exert about the rotation axis of the clock hand?
24. You have your bicycle upside down for repairs. The front wheel is free to rotate and is perfectly balanced except for the 25-g valve stem. If the valve stem is 32 cm from the rotation axis and at 24° below the horizontal, what's the resulting torque about the wheel's axis?

Section 10.3 Rotational Inertia and the Analog of Newton's Law

25. Four equal masses m are located at the corners of a square of side L, connected by essentially massless rods. Find the rotational inertia of this system about an axis (a) that coincides with one side and (b) that bisects two opposite sides.
26. The shaft connecting a power plant's turbine and electric generator is a solid cylinder of mass 6.8 Mg and diameter 85 cm. Find its rotational inertia.
27. The chamber of a rock-tumbling machine is a hollow cylinder with mass 120 g and radius 8.5 cm. The chamber is closed by end caps in the form of uniform circular disks, each of mass 33 g. Find (a) the rotational inertia of the chamber about its central axis and (b) the torque needed to give the chamber an angular acceleration of 3.3 rad/s^2.

28. A wheel's diameter is 92 cm, and its rotational inertia is $7.8\ \text{kg}\cdot\text{m}^2$. (a) What's the minimum mass it could have? (b) How could it have more mass?
29. Three equal masses m are located at the vertices of an equilateral triangle of side L, connected by rods of negligible mass. Find expressions for the rotational inertia of this object (a) about an axis through the center of the triangle and perpendicular to its plane and (b) about an axis that passes through one vertex and the midpoint of the opposite side.
30. (a) Estimate Earth's rotational inertia, assuming it to be a uniform solid sphere. (b) What torque applied to Earth would cause the length of a day to change by 1 second every century?
31. A neutron star is an extremely dense, rapidly spinning object that results from the collapse of a massive star at the end of its life. A neutron star with 2.3 times the Sun's mass has an essentially uniform density of $4.8\times10^{17}\ \text{kg/m}^3$. (a) What's its rotational inertia? (b) The neutron star's spin rate slowly decreases as a result of torque associated with magnetic forces. If the spin-down rate is $5.6\times10^{-5}\ \text{rad/s}^2$, what's the magnitude of the magnetic torque?
32. A 108-g Frisbee is 24 cm in diameter and has half its mass spread uniformly in the disk and the other half concentrated in the rim. (a) What's the Frisbee's rotational inertia? (b) With a quarter-turn flick of the wrist, a student sets the Frisbee rotating at 550 rpm. What's the magnitude of the torque, assumed constant, that the student applied?
33. At the MIT Magnet Laboratory, energy is stored in huge solid flywheels of mass 7.7×10^4 kg and radius 2.4 m. The flywheels ride on shafts 41 cm in diameter. If a frictional force of 34 kN acts tangentially on the shaft, how long will it take the flywheel to come to a stop from its usual 360-rpm rotation rate?

Section 10.4 Rotational Energy

34. A 25-cm-diameter circular saw blade has mass 0.85 kg, distributed uniformly in a disk. (a) What's its rotational kinetic energy at 3500 rpm? (b) What average power must be applied to bring the blade from rest to 3500 rpm in 3.2 s?
35. Humankind uses energy at the rate of about 16 TW. If we found a way to extract this energy from Earth's rotation, how long would it take before the length of the day increased by 1 second?
36. A 150-g baseball is pitched at 33 m/s spinning at 42 rad/s. You can treat the baseball as a uniform solid sphere of radius 3.7 cm. What fraction of its kinetic energy is rotational?
37. (a) Find the energy stored in the flywheel of Exercise 33 when it's rotating at 360 rpm. (b) The wheel is attached to an electric generator and the rotation rate drops from 360 rpm to 300 rpm in 3.0 s. What's the average power output?

Section 10.5 Rolling Motion

38. A solid 2.4-kg sphere is rolling at 5.0 m/s. Find (a) its translational kinetic energy and (b) its rotational kinetic energy.
39. What fraction of a solid disk's kinetic energy is rotational if it's rolling without slipping?
40. A rolling ball has total kinetic energy 100 J, 40 J of which is rotational energy. Is the ball solid or hollow?

Problems

41. A wheel turns through 2.0 revolutions while accelerating from rest at 18 rpm/s. (a) What's its final angular speed? (b) How long does it take?
42. You're an engineer designing kitchen appliances, and you're working on a two-speed food blender, with 3600 rpm and 1800 rpm settings. Specs call for the blender to make no more than 60 revolutions while it's switching from high to low speed. If it takes 1.4 s to make the transition, does it meet its specs?
43. **BIO** An eagle with 2.1-m wingspan flaps its wings 20 times per minute, each stroke extending from 45° above the horizontal to 45° below. Downward and upward strokes take the same time. On a given downstroke, what's (a) the average angular velocity of the wing and (b) the average tangential velocity of the wingtip?
44. A compact disc (CD) player varies the rotation rate of the disc in order to keep the part of the disc from which information is being read moving at a constant linear speed of 1.30 m/s. Compare the rotation rates of a 12.0-cm-diameter CD when information is being read (a) from its outer edge and (b) from a point 3.75 cm from the center. Give your answers in rad/s and rpm.
45. You rev your car's engine and watch the tachometer climb steadily from 1200 rpm to 5500 rpm in 2.7 s. What are (a) the engine's angular acceleration and (b) the tangential acceleration of a point on the edge of the engine's 3.5-cm-diameter crankshaft? (c) How many revolutions does the engine make during this time?
46. A circular saw spins at 5800 rpm, and its electronic brake is supposed to stop it in less than 2 s. As a quality-control specialist, you're testing saws with a device that counts the number of blade revolutions. A particular saw turns 75 revolutions while stopping. Does it meet its specs?
47. **BIO** Full-circle rotation is common in mechanical systems, but less evident in biology. Yet many single-celled organisms are propelled by spinning, tail-like *flagella*. The flagellum of the bacterium *E. coli* spins at some 600 rad/s, propelling the bacterium at speeds around 25 μm/s. How many revolutions does *E. coli*'s flagellum make as the bacterium crosses a microscope's field of view, which is 150-μm wide.
48. A pulley 12 cm in diameter is free to rotate about a horizontal axle. A 220-g mass and a 470-g mass are tied to either end of a massless string, and the string is hung over the pulley. Assuming the string doesn't slip, what torque must be applied to keep the pulley from rotating?
49. A square frame is made from four thin rods, each of length L and mass m. Calculate its rotational inertia about the three axes shown in Fig. 10.27.

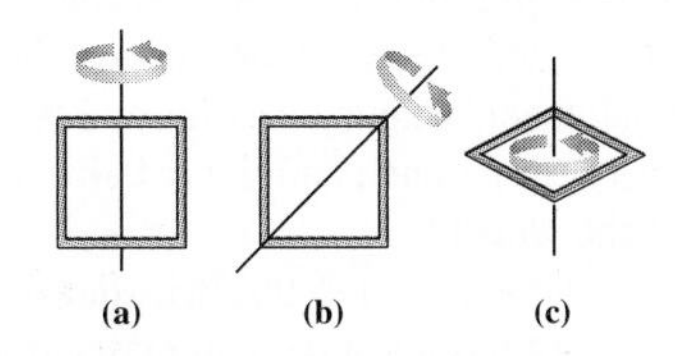

FIGURE 10.27 Problem 49

50. A thick ring has inner radius $\frac{1}{2}R$, outer radius R, and mass M. Find an expression for its rotational inertia. (*Hint*: Consult Example 10.7.)
51. A uniform rectangular flat plate has mass M and dimensions a by b. Use the parallel-axis theorem in conjunction with Table 10.2 to show that its rotational inertia about the side of length b is $\frac{1}{3}Ma^2$.

52. Each propeller on a King Air twin-engine airplane consists of three blades, each of mass 10 kg and length 125 cm. The blades may be treated approximately as uniform, thin rods. (a) What's the propeller's rotational inertia? (b) If the plane's engine develops a torque of 2.7 kN·m, how long will it take to spin up the propeller from 1400 rpm to 1900 rpm?
53. **BIO** The cellular motor driving the flagellum in *E. coli* (see Problem 47) exerts a typical torque of 400 pN·nm on the flagellum. If this torque results from a force applied tangentially to the outside of the 12-nm-radius flagellum, what's the magnitude of that force?
54. Verify by direct integration Table 10.2's entry for the rotational inertia of a flat plate about a central axis. (*Hint*: Divide the plate into strips parallel to the axis.)
55. You're an astronaut in the first crew of a new space station. The station is shaped like a wheel 22 m in diameter, with essentially all its 5×10^5-kg mass at the rim. When the crew arrives, it will be set rotating at a rate that requires an object at the rim to have radial acceleration g, thereby simulating Earth's surface gravity. This will be accomplished using two small rockets, each with 100-N thrust, mounted on the station's rim. Your job is to determine how long to fire the rockets and the number of revolutions the station will make during the firing.
56. A skater's body has rotational inertia 4.2 kg·m^2 with his fists held to his chest and 5.7 kg·m^2 with his arms outstretched. He's twirling at 3.1 rev/s while holding 2.5-kg weights in each outstretched hand; the weights are 76 cm from his rotation axis. If he pulls his hands to his chest, so the weights are essentially at his rotation axis, how fast will he be rotating?
57. **CH** A 2.4-kg block rests on a slope and is attached by a string of negligible mass to a solid drum of mass 0.85 kg and radius 5.0 cm, as shown in Fig. 10.28. When released, the block accelerates down the slope at 1.6 m/s^2. Find the coefficient of friction between block and slope.

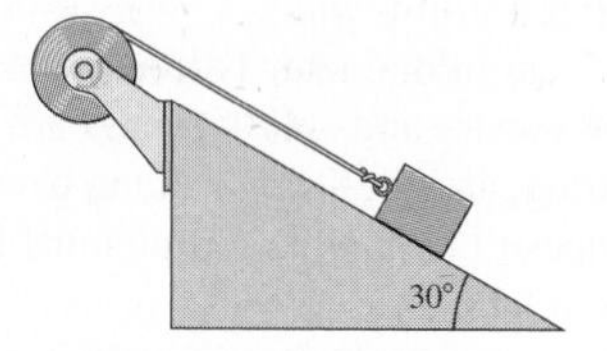

FIGURE 10.28 Problem 57

58. You've got your bicycle upside down for repairs, with its 66-cm-diameter wheel spinning freely at 230 rpm. The wheel's mass is 1.9 kg, concentrated mostly at the rim. You hold a wrench against the tire for 3.1 s, applying a 2.7-N normal force. If the coefficient of friction between wrench and tire is 0.46, what's the final angular speed of the wheel?
59. A potter's wheel is a stone disk 90 cm in diameter with mass 120 kg. If the potter's foot pushes at the outer edge of the initially stationary wheel with a 75-N force for one-eighth of a revolution, what will be the final speed?
60. A ship's anchor weighs 5.0 kN. Its cable passes over a roller of negligible mass and is wound around a hollow cylindrical drum of mass 380 kg and radius 1.1 m, mounted on a frictionless axle. The anchor is released and drops 16 m to the water. Use energy considerations to determine the drum's rotation rate when the anchor hits the water. Neglect the cable's mass.
61. Starting from rest, a hollow ball rolls down a ramp inclined at angle θ to the horizontal. Find an expression for its speed after it's gone a distance d along the incline.
62. A hollow ball rolls along a horizontal surface at 3.7 m/s when it encounters an upward incline. If it rolls without slipping up the incline, what maximum height will it reach?
63. As an automotive engineer, you're charged with improving the fuel economy of your company's vehicles. You realize that the rotational kinetic energy of a car's wheels is a significant factor in fuel consumption, and you set out to lower it. For a typical car, the wheels' rotational energy is 40% of their translational kinetic energy. You propose a redesigned wheel with the same radius but 10% lower rotational inertia and 20% less mass. What do you report for the decrease in the wheel's total kinetic energy at a given speed?
64. A solid ball of mass M and radius R starts at rest at height h above the bottom of the path in Fig. 10.26. It rolls without slipping down the left side. The right side of the path, starting at the bottom, is frictionless. To what height does the ball rise on the right?
65. **CH** A disk of radius R has an initial mass M. Then a hole of radius $R/4$ is drilled, with its edge at the disk center (Fig. 10.29). Find the new rotational inertia about the central axis. (*Hint*: Find the rotational inertia of the missing piece, and subtract it from that of the whole disk. You'll find the parallel-axis theorem helpful.)

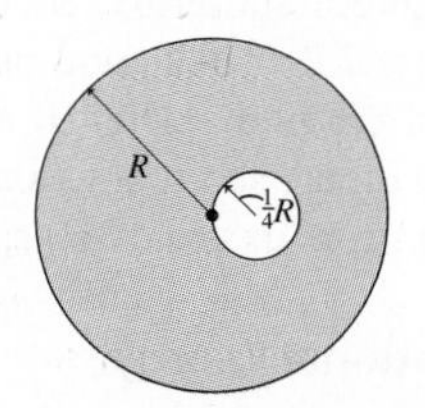

FIGURE 10.29 Problems 65 and 70

66. A 50-kg mass is tied to a massless rope wrapped around a solid cylindrical drum, mounted on a frictionless horizontal axle. When the mass is released, it falls with acceleration $a = 3.7$ m/s^2. Find (a) the rope tension and (b) the drum's mass.
67. Each wheel of a 320-kg motorcycle is 52 cm in diameter and has rotational inertia 2.1 kg·m^2. The cycle and its 75-kg rider are coasting at 85 km/h on a flat road when they encounter a hill. If the cycle rolls up the hill with no applied power and no significant internal friction, what vertical height will it reach?
68. **CH** A solid marble starts from rest and rolls without slipping on the loop-the-loop track in Fig. 10.30. Find the minimum starting height from which the marble will remain on the track through the loop. Assume the marble's radius is small compared with R.

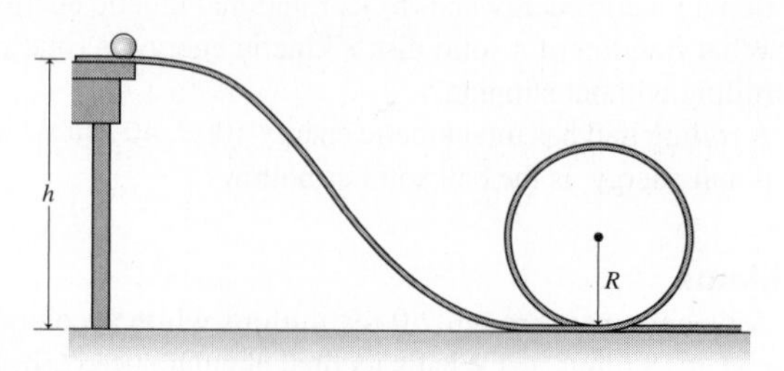

FIGURE 10.30 Problem 68

69. A disk of radius R and thickness w has a mass density that in-
CH creases from the center outward, given by $\rho = \rho_0 r/R$, where r is the distance from the disk axis. Calculate (a) the disk's total mass M and (b) its rotational inertia about its axis in terms of M and R. Compare with the results for a solid disk of uniform density and for a ring.

70. The disk in Fig. 10.29 is rotating freely about a frictionless hori-
CH zontal axle. Since the disk is unbalanced, its angular speed varies as it rotates. If the maximum angular speed is ω_{max}, find an expression for the minimum speed. (*Hint*: How does potential energy change as the wheel rotates?)

71. You're asked to check the specifications for a wind turbine. The turbine produces a peak electric power of 1.50 MW while turning at its normal operating speed of 17.0 rpm. The rotational inertia of its rotating structure—three blades, shaft, gears, and electric generator—is 2.65×10^7 kg·m^2. Under peak conditions, the wind exerts a torque of 896 kN·m on the turbine blades. Starting from rest, the turbine is supposed to take less than 1 min to spin up to its 17-rpm operating speed. The generator is supposed to be 96% efficient at converting the mechanical energy imparted by the wind into electrical energy. During spin-up, the electric generator isn't producing power, and the only torque is due to the wind. Once the turbine reaches operating speed, the generator connects to the electric grid and produces a torque that cancels the wind's torque, so the turbine turns with constant angular speed. Does the turbine meet its specifications?

72. In bicycling, each foot pushes on the pedal for half a rotation of
CH the pedal shaft; that foot then rests and the other foot takes over. During each half-cycle, the torque resulting from the force of the active foot is given approximately by $\tau = \tau_0 \sin \omega t$, where τ_0 is the maximum torque and ω is the angular speed of the pedal shaft (in s^{-1}, as usual). A particular cyclist is turning the pedal shaft at 70.0 rpm, and at the same time τ_0 is measured at 38.5 N·m. Find (a) the energy supplied by the cyclist in one turn of the pedal shaft and (b) the cyclist's average power output.

73. Calculate the rotational inertia of a solid, uniform right circular
CH cone of mass M, height h, and base radius R about its axis.

74. A thick ring of mass M has inner radius R_1 and outer radius R_2.
CH Show that its rotational inertia is given by $\frac{1}{2}M(R_1^2 + R_2^2)$.

75. A thin rod of length L and mass M is free to pivot about one end.
CH If it makes an angle θ with the horizontal, find the torque due to gravity about the pivot. (*Hint*: Integrate the torques on the mass elements composing the rod.)

76. The local historical society has asked your assistance in writing the interpretive material for a display featuring an old steam locomotive. You have information on the torque on a flywheel but need to know the force applied by means of an attached horizontal rod. The rod joins the wheel with a flexible connection 95 cm from the wheel's axis. The maximum torque the rod produces on the flywheel is 10.1 kN·m. What force does the rod apply?

77. You're skeptical about a new hybrid car that stores energy in a flywheel. The manufacturer claims the flywheel stores 12 MJ of energy and can supply 40 kW of power for 5 minutes. You dig deeper and find that the flywheel is a 39-cm-diameter ring with mass 48 kg that rotates at 30,000 rpm. Are the specs correct?

78. Figure 10.31 shows an object of mass M with one axis through
CH its center of mass and a parallel axis through an arbitrary point A. Both axes are perpendicular to the page. The figure shows an arbitrary mass element dm and vectors connecting the center of mass, the point A, and dm. (a) Use the law of cosines (Appendix A) to show that $r^2 = r_{cm}^2 + h^2 - 2\vec{h} \cdot \vec{r}_{cm}$. (b) Use this result in $I = \int r^2\, dm$ to calculate the object's rotational inertia about the axis through A. Each term in your expression for r^2 leads to a separate integral. Identify one as the rotational inertia about the CM, another as the quantity Mh^2, and argue that the third is zero. Your result is a statement of the parallel-axis theorem (Equation 10.17).

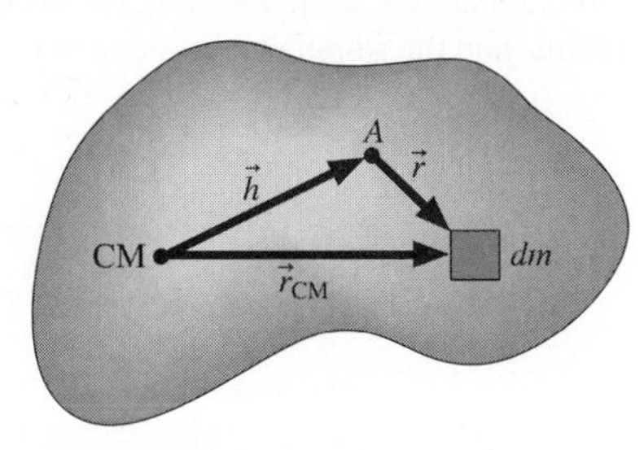

FIGURE 10.31 Problem 78

79. Figure 10.32 shows an apparatus used to measure rotational
DATA CH inertias of various objects, in this case spheres of varying masses M and radii R. The spheres are made of different materials, and some are hollow while others are solid. To perform the experiment, a sphere is mounted to a vertical axle held in a frame with essentially frictionless bearings. A spool of radius $b = 2.50$ cm is also mounted to the axle, and a string is wrapped around the spool. The string runs horizontally over an essentially frictionless pulley and is tied to a mass $m = 77.8$ g. As the mass falls, the string imparts a torque to the spool/axle/disk combination, resulting in angular acceleration. The mass of the string is negligible, but the combination of axle and spool has non-negligible rotational inertia I_0 whose value isn't known in advance. In each experimental run, the mass m is suspended a height $h = 1.00$ m above the floor and the rotating system is initially at rest. The mass is released, and experimenters measure the time to reach the floor. Results are given in the tables below. Determine an appropriate function of the time t which, when plotted against other quantities including M and R, should yield two straight lines—one for the hollow spheres and one for the solid ones. Plot your data, establish best-fit lines, and use the resulting slopes to verify the numerical factors 2/5 and 2/3 in the expressions for the rotational inertias of spheres given in Table 10.2. You should also find a value for the rotational inertia of the axle and drum together.

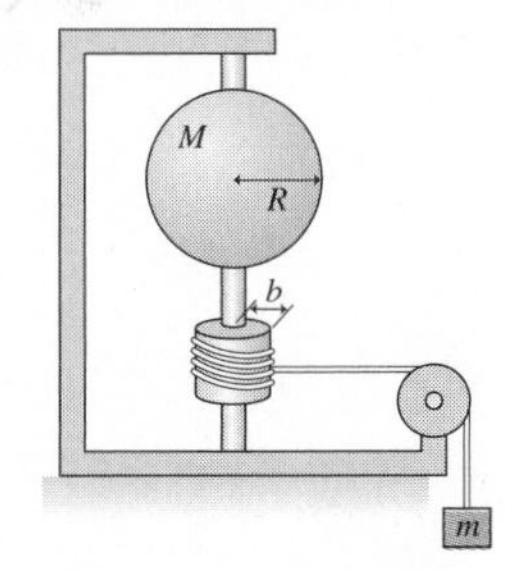

FIGURE 10.32 Problem 79

Sphere mass M (g)	783	432	286	677	347
Sphere radius R (cm)	6.25	3.86	9.34	9.42	9.12
Fall time t (s)	2.36	1.22	2.72	3.24	2.91

Sphere mass M (g)	947	189	821	544	417
Sphere radius R (cm)	6.71	5.45	6.55	4.67	9.98
Fall time t (s)	2.75	1.41	2.51	1.93	3.47

Passage Problems

Centrifuges are widely used in biology and medicine to separate cells and other particles from liquid suspensions. Figure 10.33 shows top and side views of two centrifuge designs. In both designs, the round holes are for tubes holding samples to be separated; the side views show two tubes in place. The total mass and radius of the rotating structure are the same for both, the sample-hole tubes are at the same radius, and the sample tubes are identical.

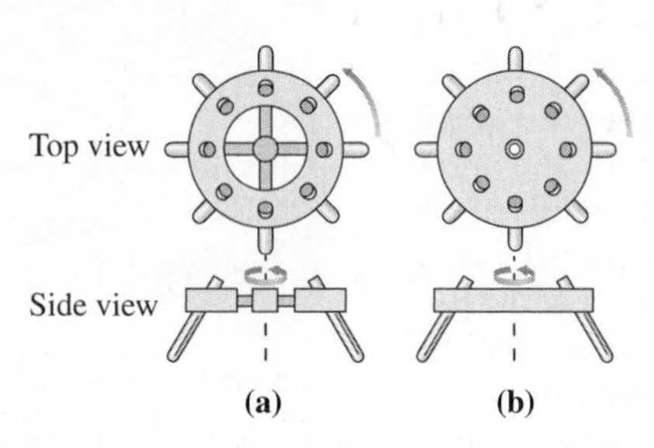

FIGURE 10.33 Two centrifuge designs, shown from the top and the side (Passage Problems 80–84).

80. Which design has greater rotational inertia?
    a. design A
    b. design B
    c. Both have the same rotational inertia.
81. If both centrifuges are made thicker in the vertical direction, without changing their masses or mass distribution, their rotational inertias will
    a. remain the same.
    b. increase.
    c. decrease.
82. If the sample tubes are made longer, the rotational inertia of the centrifuges with sample tubes inserted will
    a. remain the same.
    b. increase.
    c. decrease.
83. While the centrifuges are spinning, the net force on samples in the tubes is
    a. outward.
    b. inward.
    c. zero.
84. If a centrifuge's radius and mass are both doubled without otherwise changing the design, its rotational inertia will
    a. double.
    b. quadruple.
    c. increase by a factor of 8.
    d. increase by a factor of 16.

Answers to Chapter Questions

Answer to Chapter Opening Question

The blade mass should be concentrated toward the rotation axis, thus lowering the turbine's rotational inertia—the rotational analog of mass.

Answers to GOT IT? Questions

10.1 (c) The linear speed v increases linearly with time, and the radial acceleration increases as v^2. Tangential acceleration is constant because it's proportional to angular acceleration, which we're told is constant.

10.2 (1) 10.5*b*; (2) 10.5*a*.

10.3 (1) (b) rotational inertia with axis at the center, $(mL^2/2)$; (2) (a) rotational inertia with the axis at the end, (mL^2)

10.4 The mass of the shell is farther from the rotation axis.

10.5 (a) There must be a net torque acting to increase the pulley's clockwise angular velocity. The difference in the two tension forces provides that torque.

10.6 (b) because the wheel's rotational kinetic energy, and hence the work required, increases as the square of its rotational speed.

10.7 (c)

Rotational Vectors and Angular Momentum

What You Know

- You understand the basic concepts of rotational motion, including the rotational analog of Newton's second law.
- You're proficient in handling vectors.

What You're Learning

- Here you'll learn how to describe rotational quantities using vectors.
- You'll learn about the *cross product*, a way of multiplying two vectors that yields a third vector.
- You'll learn about *angular momentum*, the rotational analog of linear momentum, and the conditions under which it's conserved.
- You'll come to understand some counterintuitive results of angular momentum conservation, including the phenomenon of *precession*.

How You'll Use It

- The vector description of rotational motion, coupled with angular momentum conservation, will help you understand a host of phenomena from magnetic resonance imaging (MRI), to the climate implications of Earth's precession, to the physics of bicycling.

arth isn't quite round. How does this affect :s rotation axis, and what's this got to do with :e ages? (The deviation from roundness is ›xaggerated in this photo.)

Summer, fall, winter, spring: the cycle of the seasons is ultimately determined by the vector direction of Earth's angular velocity. The changing angular velocity of protons in living tissue produces MRI images that give physicians a noninvasive look inside the human body. Rising and rotating, moist, heated air forms itself into the ominous funnel of a tornado. You ride your bicycle, the rotating wheels helping stabilize what seems a precarious balance. These examples all involve rotational motion in which not only the magnitude but also the *direction* matters. They're best understood in terms of the rotational analog of Newton's law, which we introduce here in full vector form involving a rotational analog of momentum. The transition from Chapter 10 to Chapter 11 is analogous to the leap from Chapter 2's one-dimensional description of motion to the full vector description in Chapter 3. Here, as there, we'll find a new richness of phenomena involving motion.

11.1 Angular Velocity and Acceleration Vectors

So far we've ascribed direction to rotational motion using the terms "clockwise" and "counterclockwise." But that's not enough: To describe rotational motion fully we need to specify the direction of the rotation axis. We therefore define **angular velocity** $\vec{\omega}$ as a vector whose magnitude is the angular speed ω and whose direction is parallel to the rotation axis. There's an ambiguity in this definition, since there are two possible directions parallel to the axis. We resolve the ambiguity with the **right-hand rule**: If you curl the fingers of your right hand to follow the rotation, then your right thumb points in the direction of the angular velocity (Fig. 11.1). This refinement means that $\vec{\omega}$ not only gives the angular speed and the direction of the rotation axis but also distinguishes what we would have described previously as clockwise or counterclockwise rotation.

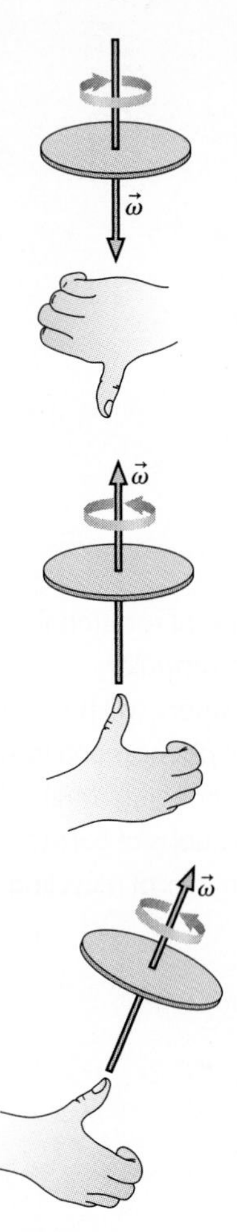

FIGURE 11.1 The right-hand rule gives the direction of the angular velocity vector.

By analogy with the linear acceleration vector, we define angular acceleration as the rate of change of the angular velocity vector:

$$\vec{\alpha} = \lim_{\Delta t \to 0} \frac{\Delta \vec{\omega}}{\Delta t} = \frac{d\vec{\omega}}{dt} \qquad \text{(angular acceleration vector)} \qquad (11.1)$$

where, as with Equation 10.4, we get the average angular acceleration if we don't take the limit.

Equation 11.1 says that angular acceleration points in the direction of the *change* in angular velocity. If that change is only in magnitude, then $\vec{\omega}$ simply grows or shrinks, and $\vec{\alpha}$ is parallel or antiparallel to the rotation axis (Fig. 11.2*a*, *b*). But a change in *direction* is also a change in angular velocity. When the angular velocity $\vec{\omega}$ changes only in direction, then the angular acceleration vector is perpendicular to $\vec{\omega}$ (Fig. 11.2*c*). More generally, both the magnitude and direction of $\vec{\omega}$ may change; then $\vec{\alpha}$ is neither parallel nor perpendicular to $\vec{\omega}$. These cases are exactly analogous to the situations we treated in Chapter 3, where acceleration parallel to velocity changes only the speed, while acceleration perpendicular to velocity changes only the direction of motion.

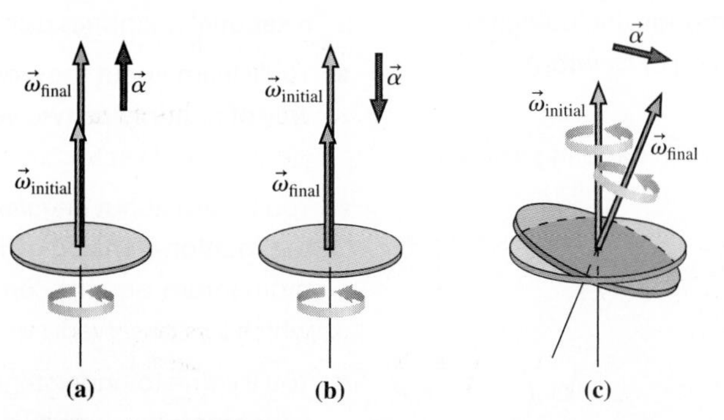

FIGURE 11.2 Angular acceleration can (a) increase or (b) decrease the magnitude of the angular velocity, or (c) change its direction.

GOT IT? 11.1 You're standing on the sidewalk watching a car go by on the adjacent road, moving from left to right. The direction of the angular velocities of the car's wheels is (a) toward the sidewalk; (b) in the direction of the car's forward motion; (c) toward the back of the car; (d) vertically upward; (e) away from the sidewalk; (f) different for each of the four wheels.

11.2 Torque and the Vector Cross Product

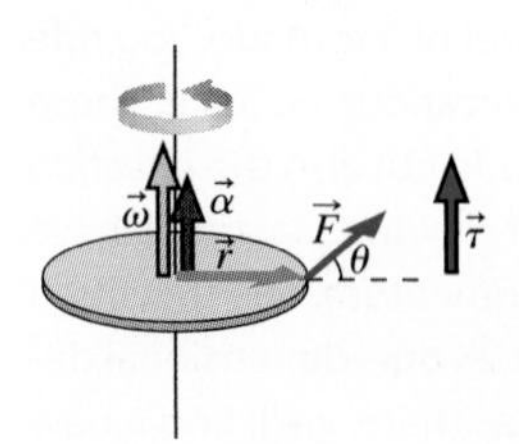

FIGURE 11.3 The torque vector is perpendicular to $\vec{r}$ and $\vec{F}$ and in the same direction as the angular acceleration. Here $\vec{F}$ lies in the plane of the disk.

Figure 11.3 shows a wheel, initially stationary, with a force applied at its rim. The torque associated with this force sets the wheel rotating in the direction shown; applying the right-hand rule, we see that angular velocity vector $\vec{\omega}$ points upward. Since the angular speed is increasing, the angular acceleration $\vec{\alpha}$ also points upward. So that our rotational analog of Newton's law—angular acceleration proportional to torque—will hold for directions as well as magnitudes, we'd like the torque to have an upward direction, too.

We already know the magnitude of the torque: From Equation 10.10, it's $\tau = rF\sin\theta$, where θ is the angle between the vectors $\vec{r}$ and $\vec{F}$ in Fig. 11.3. We define the direction of the torque as being perpendicular to both $\vec{r}$ and $\vec{F}$, as given by the right-hand rule shown in Fig. 11.4. You can verify that this rule gives an upward direction for the torque in Fig. 11.3.

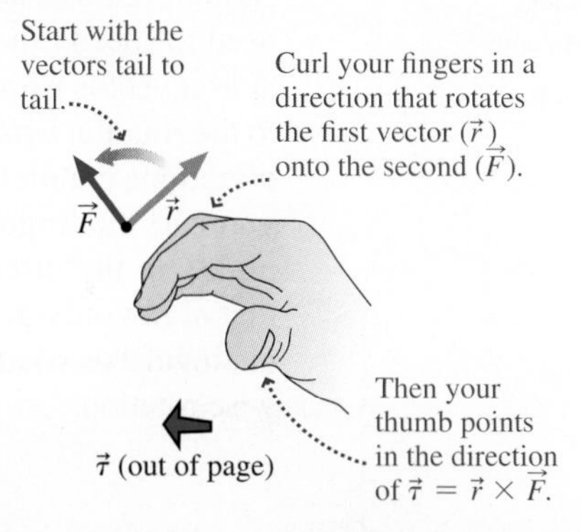

FIGURE 11.4 The right-hand rule for the direction of torque.

The Cross Product

The magnitude of the torque, $\tau = rF\sin\theta$, is determined by the magnitudes of the vectors $\vec{r}$ and $\vec{F}$ and the angle between them; the direction of the torque is determined by the vectors $\vec{r}$ and $\vec{F}$ through the right-hand rule. This operation—forming from two vectors $\vec{A}$ and $\vec{B}$ a third vector $\vec{C}$ of magnitude $C = AB\sin\theta$ and direction given by the right-hand rule—occurs frequently in physics and is called the **cross product**:

The cross product $\vec{C}$ of two vectors $\vec{A}$ and $\vec{B}$ is written

$$\vec{C} = \vec{A} \times \vec{B}$$

and is a vector with magnitude $AB\sin\theta$, where θ is the angle between $\vec{A}$ and $\vec{B}$, and where the direction of $\vec{C}$ is given by the right-hand rule of Fig. 11.4.

Torque is an instance of the cross product, and we can write the torque vector simply as

$$\vec{\tau} = \vec{r} \times \vec{F} \qquad \text{(torque vector)} \qquad (11.2)$$

Both direction and magnitude are described succinctly in this equation.

TACTICS 11.1 Multiplying Vectors

The cross product $\vec{A} \times \vec{B}$ is the second way of multiplying vectors that you've encountered. The first was the scalar product $\vec{A} \cdot \vec{B} = AB\cos\theta$ introduced in Chapter 6 and also called the dot product. Both depend on the product of the vector magnitudes and on the angle between them. But where the dot product depends on the *cosine* of the angle and is therefore maximum when the two vectors are parallel, the cross product depends on the *sine* and is therefore maximum for perpendicular vectors. There's another crucial distinction between dot product and cross product: The dot product is a *scalar*—a single number, with no direction—while the cross product is a *vector*. That's why $AB\cos\theta$ completely specifies the dot product, but $AB\sin\theta$ gives only the magnitude of the cross product; it's also necessary to specify the direction via the right-hand rule.

The cross product obeys the usual distributive rule: $\vec{A} \times (\vec{B} + \vec{C}) = \vec{A} \times \vec{B} + \vec{A} \times \vec{C}$, but it's *not* commutative; in fact, as you can see by rotating $\vec{F}$ onto $\vec{r}$ instead of $\vec{r}$ onto $\vec{F}$ in Fig. 11.4, $\vec{B} \times \vec{A} = -\vec{A} \times \vec{B}$.

With the vectors $\vec{A}$ and $\vec{B}$ in component form, we developed Equation 6.4 to express the dot product in terms of components, as you can show in Problem 49:

$$\vec{A} \times \vec{B} = (A_yB_z - A_zB_y)\hat{\imath} + (A_zB_x - A_xB_z)\hat{\jmath} + (A_xB_y - A_yB_x)\hat{k}$$

This expression is more complicated than Equation 6.4 for the dot product because the cross product is a vector, and also because that vector is perpendicular to both $\vec{A}$ and $\vec{B}$.

GOT IT? 11.2 The figure shows four pairs of force and radius vectors and eight torque vectors. Which numbered torque vector goes with each pair of force–radius vectors? Consider only direction, not magnitude.

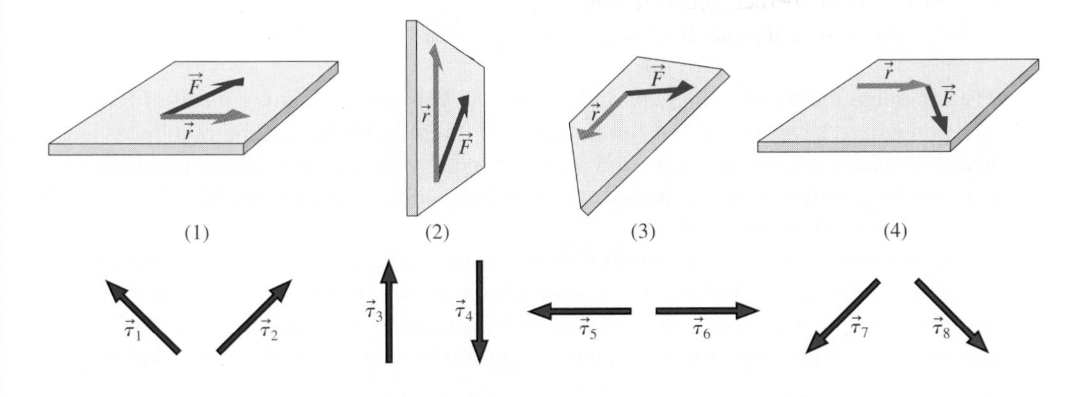

11.3 Angular Momentum

We first used Newton's law in the form $\vec{F} = m\vec{a}$, but later found the momentum form $\vec{F} = d\vec{p}/dt$ especially powerful. The same is true in rotational motion: To explore fully some surprising aspects of rotational dynamics, we need to define angular momentum and develop a relation between its rate of change and the applied torque. Once we've done that, we'll be able to answer questions like why a gyroscope doesn't fall over and how spinning protons yield MRI images of your body's innards.

Like other rotational quantities, angular momentum is always specified with respect to a given point or axis. We begin with the **angular momentum** $\vec{L}$ of a single particle:

If a particle with linear momentum $\vec{p}$ is at position $\vec{r}$ with respect to some point, then its angular momentum $\vec{L}$ about that point is defined as

$$\vec{L} = \vec{r} \times \vec{p} \quad \text{(angular momentum)} \qquad (11.3)$$

EXAMPLE 11.1 Calculating Angular Momentum: A Single Particle

A particle of mass m moves counterclockwise at speed v around a circle of radius r in the x–y plane. Find its angular momentum about the center of the circle, and express the answer in terms of its angular velocity.

INTERPRET We're given the motion of a particle—namely, uniform motion in a circle—and asked to find the corresponding angular momentum and its relation to angular velocity.

DEVELOP Figure 11.5 is our sketch, showing the particle in its circular path. We added an xyz coordinate system with the circular path in the x–y plane. Equation 11.3, $\vec{L} = \vec{r} \times \vec{p}$, gives the angular momentum in terms of the position vector $\vec{r}$ and the linear momentum $\vec{p}$. We know that linear momentum is the product $m\vec{v}$, so we have everything we need to apply Equation 11.3. We'll then express our result in terms of angular velocity using $v = \omega r$.

EVALUATE Figure 11.5 shows that the linear momentum $m\vec{v}$ is perpendicular to $\vec{r}$, so $\sin\theta = 1$ in the cross product, and the magnitude of the angular momentum becomes $L = mvr$. Applying the right-hand rule shows that $\vec{L}$ points in the z-direction, so we can write $\vec{L} = mvr\hat{k}$. But $v = \omega r$, and the right-hand rule shows that $\vec{\omega}$, too, points in the z-direction. So we can write

$$\vec{L} = mvr\hat{k} = mr^2\omega\hat{k} = mr^2\vec{\omega}$$

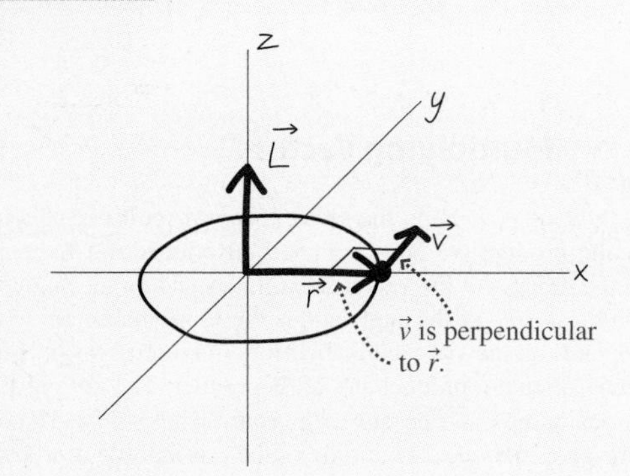

FIGURE 11.5 Finding the angular momentum $\vec{L}$ of a particle moving in a circle.

ASSESS Make sense? The faster the particle is going, the more linear momentum it has. But angular momentum depends on linear momentum and distance from the rotation axis, so at a given angular speed, the angular momentum scales as the *square* of the radius. ■

Angular momentum is the rotational analog of linear momentum $\vec{p} = m\vec{v}$. Since rotational inertia I is the analog of mass m, and angular velocity $\vec{\omega}$ is the analog of linear velocity $\vec{v}$, you might expect that we could write

$$\vec{L} = I\vec{\omega} \qquad (11.4)$$

The rotational inertia of a single particle is mr^2, so you can see that the result of Example 11.1 can indeed be written $\vec{L} = I\vec{\omega}$. Equation 11.4 also holds for symmetric objects like a wheel or sphere rotating about a fixed axis. But in more complicated cases, Equation 11.4 may not hold; surprisingly, $\vec{L}$ and $\vec{\omega}$ can even have different directions. We'll leave such cases for more advanced courses.

We emphasize again that angular momentum isn't absolute, but—as with other rotational quantities—it depends on your choice of rotation axis. If that arbitrariness bothers you, note that there's an analogous arbitrariness to linear momentum. If an object has velocity $\vec{v}$ with respect to you, then it's got linear momentum $\vec{p} = m\vec{v}$—but only as measured by you or others at rest with respect to you. Jump into another reference frame, where the object is moving with some other velocity $\vec{v}'$, and now its momentum has the different value $m\vec{v}'$—which might even be zero if you're at rest with respect to the object. No problem; you just have to know what reference frame you're working in. Analogously, with angular momentum, you have to know what rotation axis or point you're considering as you calculate $\vec{L}$.

Torque and Angular Momentum

We're now ready to develop the full vector analog of Newton's law in the form $\vec{F} = d\vec{P}/dt$. Recall that $\vec{F}$ here is the *net* external force on a system, and $\vec{P}$ is the system's momentum—the vector sum of the momenta of its constituent particles. Can we write, by analogy, $\vec{\tau} = d\vec{L}/dt$? To see that we can, we write the angular momentum of a system as the sum of the angular momenta of its constituent particles:

$$\vec{L} = \sum \vec{L}_i = \sum (\vec{r}_i \times \vec{p}_i)$$

where the subscript i refers to the ith particle. Differentiating gives

$$\frac{d\vec{L}}{dt} = \sum \left(\vec{r}_i \times \frac{d\vec{p}_i}{dt} + \frac{d\vec{r}_i}{dt} \times \vec{p}_i \right)$$

where we've applied the product rule for differentiation, being careful to preserve the order of the cross product since it's not commutative. But $d\vec{r}_i/dt$ is the velocity of the ith particle, so the second term in the sum is the cross product of velocity $\vec{v}$ and momentum $\vec{p} = m\vec{v}$. Since these two vectors are parallel, their cross product is zero, and we're left with only the first term in the sum:

$$\frac{d\vec{L}}{dt} = \sum \left(\vec{r}_i \times \frac{d\vec{p}_i}{dt} \right) = \sum (\vec{r}_i \times \vec{F}_i)$$

where we've used Newton's law to write $d\vec{p}_i/dt = \vec{F}_i$. But $\vec{r}_i \times \vec{F}_i$ is the torque $\vec{\tau}_i$ on the ith particle, so

$$\frac{d\vec{L}}{dt} = \sum \vec{\tau}_i$$

The sum here includes both external and internal torques—the latter due to interactions among the particles of the system. Newton's third law assures us that internal *forces* cancel in pairs, but what about *torques*? They'll cancel, too, provided the internal forces act along lines joining pairs of particles. This condition is stronger than Newton's third law alone, and it usually but not always holds. When it does, the sum of torques reduces to the net *external* torque, and we have

$$\frac{d\vec{L}}{dt} = \vec{\tau} \quad \left(\begin{array}{c}\text{rotational analog,}\\ \text{Newton's second law}\end{array}\right) \tag{11.5}$$

where $\vec{\tau}$ is the net external torque. Thus our analogy between linear and rotational motion holds for momentum as well as for the other quantities we've discussed.

GOT IT? 11.3 The figure shows three particles with the same mass m, all moving with the same constant speed v. Particle (1) moves in a circle of radius R about the point P, particle (2) in a straight line whose closest approach to point P is the same as the circle's radius R, and particle (3) in a straight line that passes through P. Which of these statements correctly describes the magnitudes of the particles' angular momenta?

(a) $L_1 = L_2 = L_3 \neq 0$; (c) $L_1 > L_2 > L_3 = 0$;
(b) $L_1 > 0, L_2 = L_3 = 0$; (d) $L_2 = L_1 \neq 0, L_3 = 0$

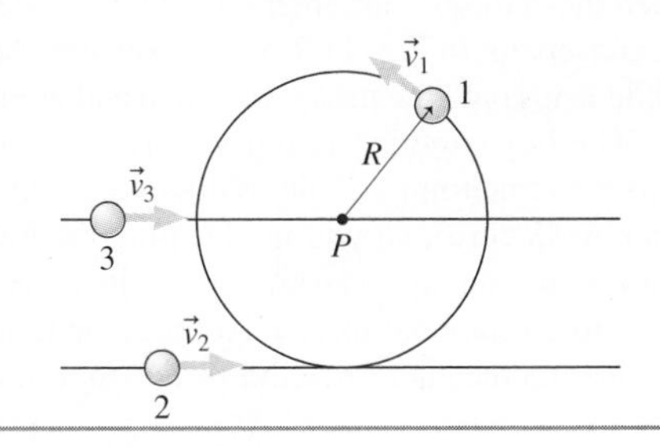

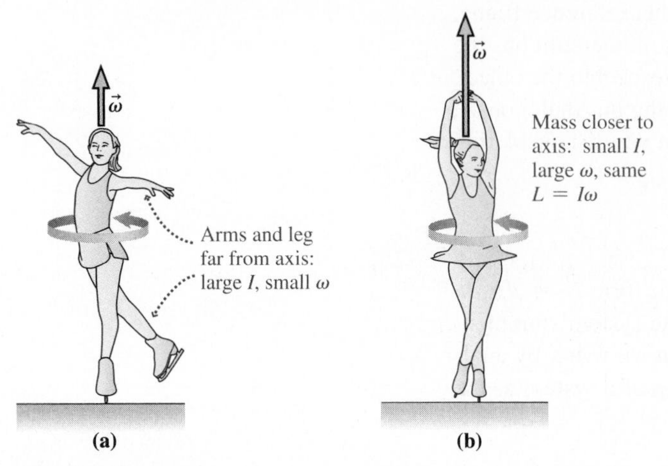

FIGURE 11.6 As the skater's rotational inertia decreases, her angular speed increases to conserve angular momentum.

11.4 Conservation of Angular Momentum

When there's no external torque on a system, Equation 11.5 tells us that angular momentum is constant. This statement—that the angular momentum of an isolated system cannot change—is of fundamental importance in physics, and applies to systems ranging from subatomic particles to galaxies. Because a composite system can change its configuration—and hence its rotational inertia I—conservation of angular momentum requires that angular speed increase if I decreases, and vice versa. The classic example is a figure skater who starts spinning relatively slowly with arms and leg extended and then pulls in her limbs to spin rapidly (Fig. 11.6). A more dramatic example is the collapse of a star at the end of its lifetime, explored in the next example.

EXAMPLE 11.2 Conservation of Angular Momentum: Pulsars

A star rotates once every 45 days. At the end of its life, it undergoes a supernova explosion, hurling much of its mass into the interstellar medium. But the inner core of the star, whose radius is initially 20 Mm, collapses into a neutron star only 6 km in radius. As it rotates, the neutron star emits regular pulses of radio waves, making it a *pulsar*. Calculate the rotation rate, which is the same as the pulse rate that radio astronomers detect. Consider the core to be a uniform sphere, and assume that no external torques act during the collapse.

INTERPRET Here we're given the radius and rotation rate of the stellar core before collapse and asked for the rotation rate afterward. That kind of "before and after" question often calls for the application of a conservation law. In this case there's no external torque, so it's angular momentum that's conserved.

DEVELOP The magnitude of the angular momentum is $I\omega$, so our plan is to write this expression before and after collapse, and then equate the two to find the new rotation rate: $I_1\omega_1 = I_2\omega_2$. We need to use Table 10.2's expression for the rotational inertia of a solid sphere: $I = \frac{2}{5}MR^2$.

EVALUATE Given I, our statement of angular momentum conservation becomes $\frac{2}{5}MR_1^2\omega_1 = \frac{2}{5}MR_2^2\omega_2$, or

$$\omega_2 = \omega_1\left(\frac{R_1}{R_2}\right)^2 = \left(\frac{1\text{ rev}}{45\text{ day}}\right)\left(\frac{2\times10^7\text{ m}}{6\times10^3\text{ m}}\right)^2 = 2.5\times10^5\text{ rev/day}$$

ASSESS Our answer is huge, about 3 revolutions per second. But that makes sense. This neutron star is a fantastic thing—an object with more mass than the entire Sun, crammed into a diameter of about 8 miles. It's because of that dramatic reduction in radius—and thus in rotational inertia—that the pulsar's rotation rate is so high. Note that in a case like this, where ω appears on both sides of the equation, it isn't necessary to convert to radian measure. ■

CONCEPTUAL EXAMPLE 11.1 On the Playground

A merry-go-round is rotating freely when a boy runs radially inward, straight toward the merry-go-round's center, and leaps on. Later, a girl runs tangent to the merry-go-round's edge, in the same direction the edge is moving, and also leaps on. Does the merry-go-round's angular speed increase, decrease, or stay the same in each case?

EVALUATE Because the merry-go-round is rotating freely, the only torques are those exerted by the children as they leap on. If we consider a system consisting of the merry-go-round and both children, then those torques are internal, and the system's angular momentum is conserved. In Fig. 11.7 we've sketched the situation, before either child leaps onto the merry-go-round and after both are on board.

The boy, running radially, carries no angular momentum (his linear momentum and the radius vector are in the same direction, making $\vec{L}$ zero), so you might think he doesn't change the merry-go-round's angular speed. Yet he does, because he adds mass and therefore rotational inertia. At the same time, he doesn't change the angular momentum, so with I increased, ω must therefore drop.

Running in the same direction as the merry-go-round's tangential velocity, the girl adds angular momentum to the system—an addition that would tend to increase the angular speed. But she also adds mass, and thus increases the rotational inertia—which, as in the boy's case, tends to decrease angular speed. So which wins out? That depends on her speed. Without knowing that, we can't tell whether the merry-go-round speeds up or slows down.

ASSESS The angular momentum the girl adds is the product of her linear momentum mv and the merry-go-round's radius R, while she increases the rotational inertia by mR^2. With small m and large v, she could add a lot of angular momentum without increasing the rotational inertia significantly. That would increase the merry-go-round's rotation rate. But with a large m and small v—giving the same additional angular momentum—the increase in rotational inertia would more than offset the angular momentum added, and the merry-go-round would slow down. We can't answer the question about the merry-go-round's angular speed without knowing the

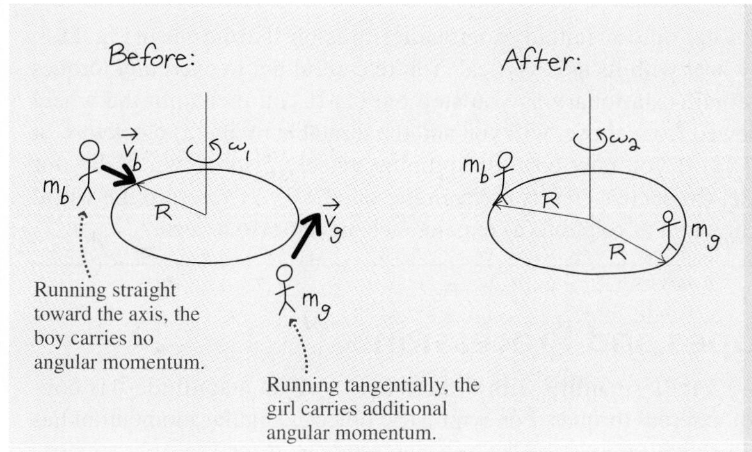

FIGURE 11.7 Our diagrams for Conceptual Example 11.1.

numbers. "Making the Connection," below, solves this example for a particular set of values, and you can explore a similar situation more generally in Problem 53.

MAKING THE CONNECTION Take the merry-go-round's radius to be $R = 1.3$ m, its rotational inertia $I = 240\ \text{kg}\cdot\text{m}^2$, and its initial angular speed $\omega_{\text{initial}} = 11$ rpm. The boy's and girl's masses are, respectively, 28 kg and 32 kg, and they run, respectively, at 2.5 m/s and 3.7 m/s. Find the merry-go-round's angular speed ω_{final} after both children are on board.

EVALUATE Following the conceptual example, take the system to include the merry-go-round and the two children. Before the children leap on, both the merry-go-round itself and the girl carry angular momentum; afterward, with children and merry-go-round rotating with a common angular speed, they all do. Thus conservation of angular momentum reads

$$I\omega_{\text{initial}} + m_g v_g R = I\omega_{\text{final}} + m_b R^2 \omega_{\text{final}} + m_g R^2 \omega_{\text{final}}$$

Solving with the given numbers yields $\omega_{\text{final}} = 12$ rpm. That's not much change, so the girl's effect must have been a speed increase, but only a little more than enough to overcome the boy's slowing effect. Note that the boy's speed didn't matter, since it didn't contribute to angular momentum or rotational inertia. And be careful with units: You've got to express all angular momenta in the same units. That means converting angular speeds to radians per second or expressing the girl's angular momentum $m_g v_g R$ in unusual units, $\text{kg}\cdot\text{m}^2\cdot\text{rpm}$.

✓TIP Angular Momentum in Straight-line Motion

You don't have to be rotating to have angular momentum. The girl in Conceptual Example 11.1 was running in a straight line, yet she had nonzero angular momentum with respect to the merry-go-round's rotation axis. Problem 36 explores this point further.

In a popular demonstration, a student stands on a stationary turntable holding a wheel rotating about a vertical axis. The student flips the wheel upside down, and the turntable starts rotating. Figure 11.8 shows how angular momentum conservation explains this behavior. Once again, though, mechanical energy isn't conserved. In this case the student does work, exerting forces that result in torques on her body and the turntable. The end result is a greater rotational kinetic energy than was initially present.

Video Tutor Demo | **Spinning Person Drops Weights**

Video Tutor Demo | **Off-Center Collision**

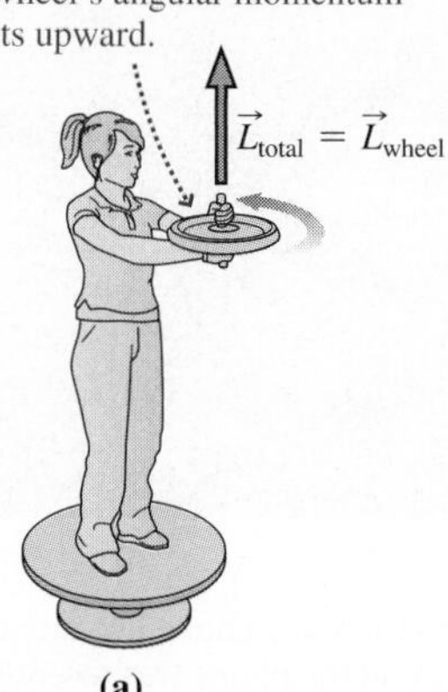

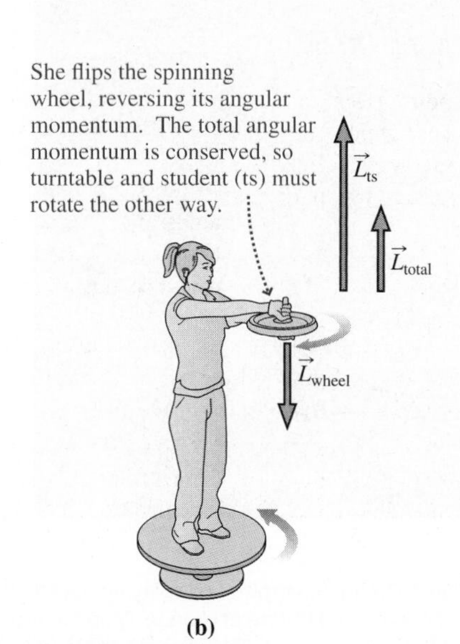

FIGURE 11.8 A demonstration of angular momentum conservation.

GOT IT? 11.4 You step onto an initially nonrotating turntable like the one in Fig. 11.8, holding a nonrotating wheel with its axis vertical. You're careful not to exert any torques so that the turntable remains stationary as you step on. (1) If you then spin the wheel counterclockwise as viewed from above, will you and the turntable rotate (a) clockwise or (b) counterclockwise? (2) If you now turn the spinning wheel upside down, will your rotation rate (a) increase, (b) decrease, or (c) remain the same? (3) As you turn the wheel upside down, will the direction of rotation (a) remain unchanged or (b) reverse?

11.5 Gyroscopes and Precession

MP

PhET: Simplified MRI

Angular momentum—a vector quantity with direction as well as magnitude—is conserved in the absence of external torques. For symmetric objects, angular momentum has the same direction as the rotation axis, so the axis can't change direction unless an external torque acts. This is the principle behind the gyroscope—a spinning object whose rotation axis remains fixed in space. The faster a gyroscope spins, the larger its angular momentum and thus the harder it is to change its orientation. Gyroscopes are widely used for navigation, where their direction-holding capability provides an alternative to the magnetic compass. More sophisticated gyroscope systems guide missiles and submarines and stabilize ships in heavy seas. Space telescopes start and stop gyroscopic wheels oriented along three perpendicular axes; to conserve angular momentum, the entire telescope reorients itself to point toward a desired astronomical object. This approach avoids rocket exhaust that would foul the telescope's superb viewing and ensures that there's no fuel to run out. Instead, solar-generated electricity operates the wheels' drive motors.

If you have a modern smartphone, it, too, contains gyroscopes. They're used to determine the phone's orientation in space; among other uses, they tell the phone how to orient its display. You can even get applications that access data from these gyroscopes directly (Fig. 11.9*a*). Smartphone gyroscopes are microelectromechanical systems (MEMS) devices, and they're based on vibrating rather than rotating structures (Fig. 11.9*b*). Similar MEMS gyroscopes are used in computer mice and video game consoles, and MEMS gyros stabilize the Segway Human Transporter.

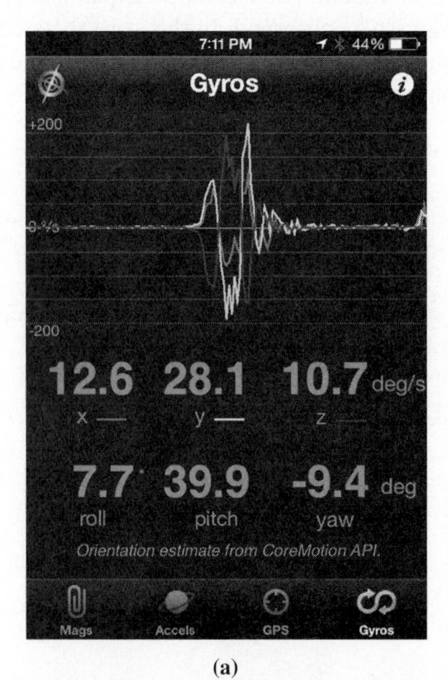

(a)

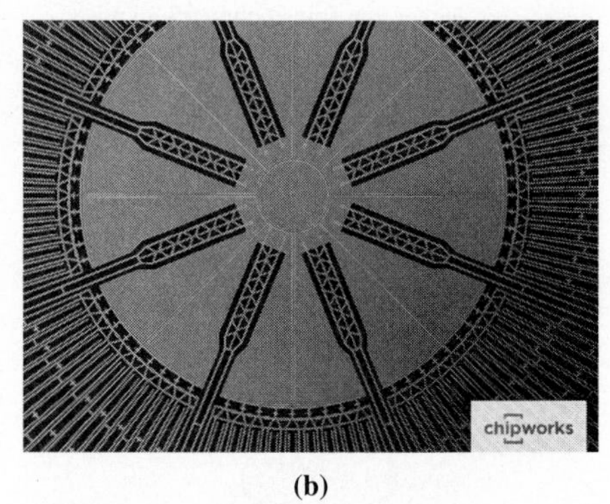

(b)

FIGURE 11.9 (a) Smartphone displaying data from its internal gyroscopes, indicating the phone's orientation and its rate of change. Graph at top shows that the phone was recently reoriented. (b) Micro photo of a MEMS gyro like those used in smartphones. The entire structure is only about 0.5 mm across.

Precession

If an object does experience a net external torque, then, according to the rotational analog of Newton's law (Equation 11.5, $d\vec{L}/dt = \vec{\tau}$), its angular momentum must change. For rapidly rotating objects, the result is the surprising phenomenon of **precession**—a continual change in the direction of the rotation axis, which traces out a circle. You may have seen a toy gyroscope or top precess instead of simply falling over as you might expect.

Figure 11.10 shows why procession occurs. Here a spinning gyroscope is tilted, so there's a gravitational torque acting on it. Yet it doesn't fall over. Why not? Apply the right-hand rule to the vector $\vec{r}$ and the gravitational force vector $\vec{F}_g$ shown in the figure, and you'll see that the torque $\vec{\tau}$ points into the page. So, by $\vec{\tau} = d\vec{L}/dt$, that must also be the direction of the *change* in the angular momentum $\vec{L}$. And that's just what's happening: The *change* $\Delta\vec{L}$ in the angular momentum vector is indeed into the page. So the axis of the gyroscope—which coincides with the angular momentum vector—moves into the page. Repeat this argument, and you'll see that the change $\Delta\vec{L}$ is always perpendicular to $\vec{L}$; as a result, the angular momentum vector describes a circular path, continually changing in direction but not magnitude.

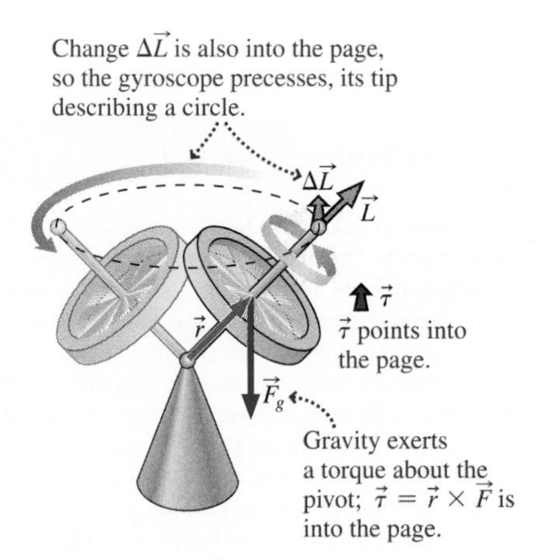

FIGURE 11.10 Why doesn't the spinning gyroscope fall over?

So is there something special about a *rotating* gyroscope? Wouldn't a nonrotating gyroscope also obey the rotational analog of Newton's law? It would, and you can see that by applying the argument of the previous paragraph, now assuming that the gyroscope in Fig. 11.10 isn't rotating. The gravitational force and torque are still the same, with the torque into the page. The rotational analog of Newton's second law still holds, so the change $\Delta\vec{L}$ in angular momentum is still into the page. But here's the difference: In this case the initial angular momentum is zero, so the gyroscope needs to acquire an angular momentum that points into the page. It does that by falling over, rotating about its pivot as it does so. Apply the right-hand rule to the gyroscope as it falls, and the result is an angular momentum pointing into the page. Again, the rotational analog of Newton's law is satisfied. If you're bothered that the gyroscope doesn't rotate about its shaft as before, note that there's nothing in the rotational analog of Newton's law that says how or about what axis something has to rotate. Its falling over is a perfectly good rotational motion—although it will end when the gyroscope hits the floor and nongravitational torques begin to act.

The difference between the rotating and nonrotating gyroscope is like the difference between a satellite in circular orbit and a ball that's simply dropped from rest. Newton's law, $\vec{F} = d\vec{p}/dt$, governs both cases, and says that the *change* in linear momentum is in the direction of the gravitational force. The satellite already has momentum, and since it's going at the right speed for a circular orbit, this change amounts to a change in direction only. The ball has no initial momentum, so it acquires a momentum in the direction of the force—namely, downward. Substitute "rotating gyroscope" for "satellite," "nonrotating gyroscope" for "ball," "angular momentum" for "linear momentum," and "torque" for "force," and you've got the analogous explanations for the two gyroscope situations.

What determines the rate of precession? You can explore that question qualitatively in Question 14, and quantitatively in Problem 61.

Precession on the atomic scale helps explain the medical imaging technique MRI (*m*agnetic *r*esonance *i*maging). Protons in the body's abundant hydrogen precess because of torque resulting from a strong magnetic field. The MRI imager detects signals emitted at the precession frequency. By spatially varying the magnetic field, the device localizes the precessing protons and thus constructs high-resolution images of the body's interior.

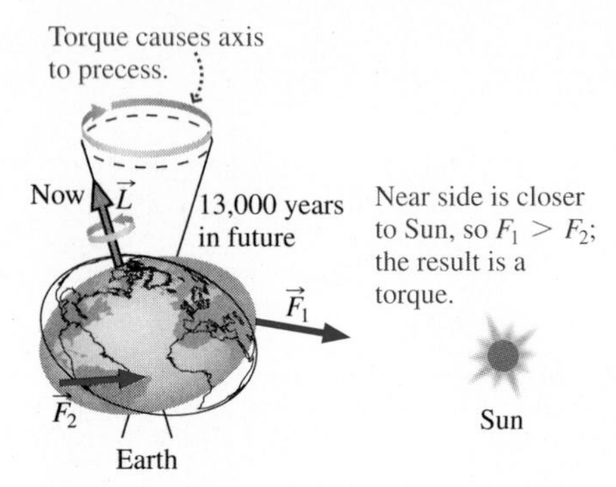

FIGURE 11.11 Earth's precession. The equatorial bulge is highly exaggerated.

On a much larger scale, Earth itself precesses. Because of its rotation, the planet bulges slightly at the equator. Solar gravity exerts a torque on the equatorial bulge, causing Earth's rotation axis to precess with a period of about 26,000 years (Fig. 11.11). The axis now points toward Polaris, which for that reason we call the North Star, but it won't always do so. This precession, in connection with deviations in Earth's orbit from a perfect circle, results in subtle climatic changes that are believed to be partly responsible for the onset of ice ages.

GOT IT? 11.5 You push horizontally at right angles to the shaft of a spinning gyroscope, as shown in the figure. Does the shaft move (a) upward, (b) downward, (c) in the direction of your push, or (d) opposite the direction of your push?

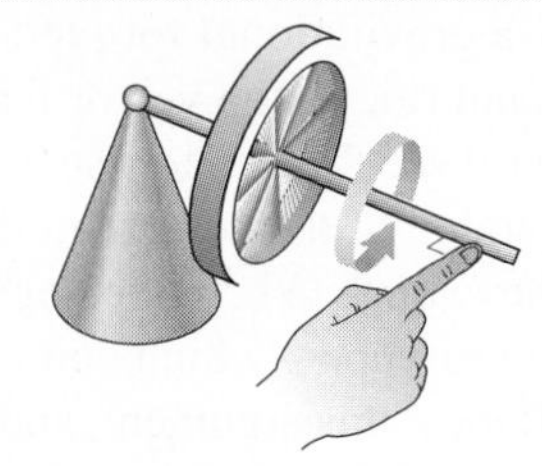

APPLICATION Bicycling

The rotational analog of Newton's second law helps explain why bicycles don't tip over. The photo shows why. If the bicycle is perfectly vertical, the gravitational force exerts no torque. But if it tips to the rider's left, as in the photo, then there's a torque $\vec{\tau} = \vec{r} \times \vec{F}_g$ toward the rear. A stationary bicycle, with no angular momentum, would respond by tipping further left, rotating about a front-to-back axis and gaining angular momentum toward the rear. That's just as Newton requires: a change in angular momentum in the direction of the torque. But a moving bicycle already has angular momentum $\vec{L}$ of its rotating wheels; as the photo shows, that angular momentum points generally to the rider's left. A rearward change in angular momentum then requires just a slight turn of the front wheel to the left. The rider subconsciously makes that turn, at once satisfying Newton and helping to keep the bicycle stable.

The physics of cycling is a complicated subject, and the role of angular momentum described here is only one of several effects that contribute to bicycle stability.

CHAPTER 11 SUMMARY

Big Idea

The big idea of this chapter is that rotational quantities can be described as vectors, with the vector direction at right angles to the plane in which the action—motion, acceleration, or effects associated with torque—is occurring. The direction is given by the right-hand rule. A new concept, angular momentum, is the rotational analog of linear momentum. The rotational analog of Newton's law equates the net torque on a system with the rate of change of its angular momentum. In the absence of a net torque, angular momentum is conserved.

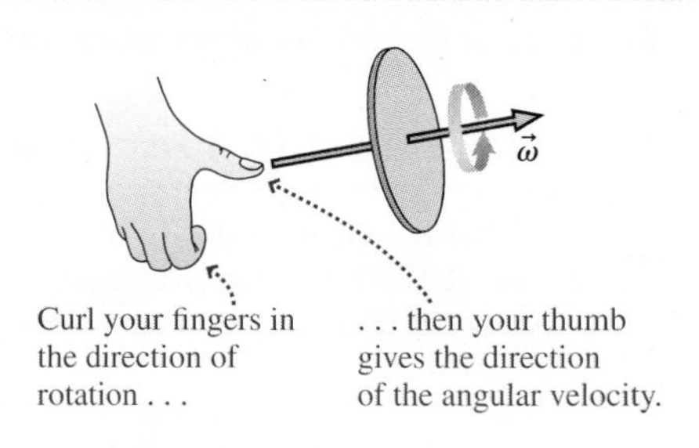

Key Concepts and Equations

The **vector cross product** is a way of multiplying two vectors $\vec{A}$ and $\vec{B}$ to produce a third vector $\vec{C}$ of magnitude $C = AB\sin\theta$ and direction at right angles to the other two, as given by the right-hand rule. It's written as

$$\vec{C} = \vec{A} \times \vec{B}$$

Torque is a vector defined as the cross product of the radius vector $\vec{r}$ from a given axis to the point where a force $\vec{F}$ is applied:

$$\vec{\tau} = \vec{r} \times \vec{F}$$

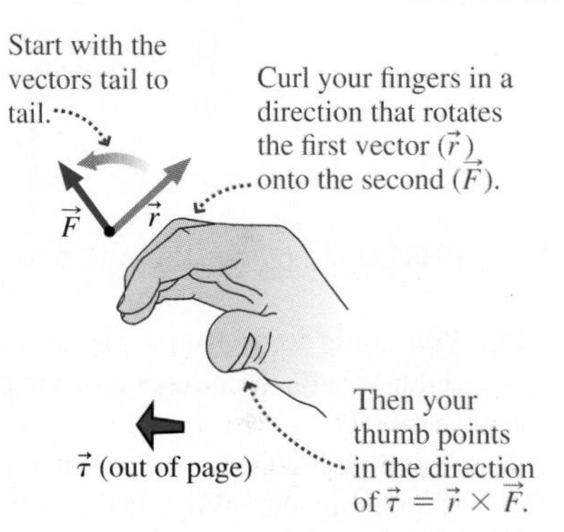

Angular momentum $\vec{L}$ is the rotational analog of linear momentum $\vec{p}$. It's always defined with respect to a particular axis. For a point particle at position $\vec{r}$ with respect to the axis, moving with linear momentum $\vec{p} = m\vec{v}$, the angular momentum is defined as

$$\vec{L} = \vec{r} \times \vec{p}$$

For a symmetric object with rotational inertia I rotating with angular velocity $\vec{\omega}$, angular momentum becomes $\vec{L} = I\vec{\omega}$.
In terms of angular momentum, the rotational analog of Newton's law states that the rate of change of angular momentum is equal to the net external torque:

$$\frac{d\vec{L}}{dt} = \vec{\tau}_{\text{net}}$$

If the external torque on a system is zero, then its angular momentum cannot change.

Applications

Conservation of angular momentum explains the action of gyroscopes—spinning objects whose rotation axis remains fixed in the absence of a net external torque. If an external torque is applied, the rotation axis undergoes a circular motion known as **precession**. Precession occurs in systems ranging from subatomic particles to tops and gyroscopes and on to planets.

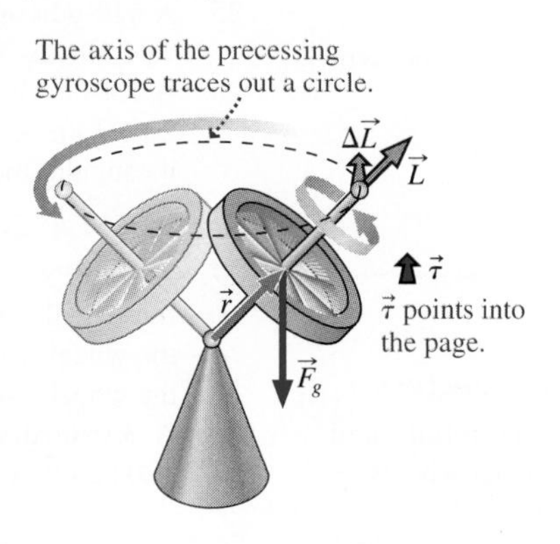

MP For homework assigned on MasteringPhysics, go to www.masteringphysics.com

BIO *Biology and/or medicine-related problems* **DATA** *Data problems* **ENV** *Environmental problems* **CH** *Challenge problems* **COMP** *Computer problems*

For Thought and Discussion

1. Does Earth's angular velocity vector point north or south?
2. Figure 11.12 shows four forces acting on a body. In what directions are the associated torques about point O? About point P?

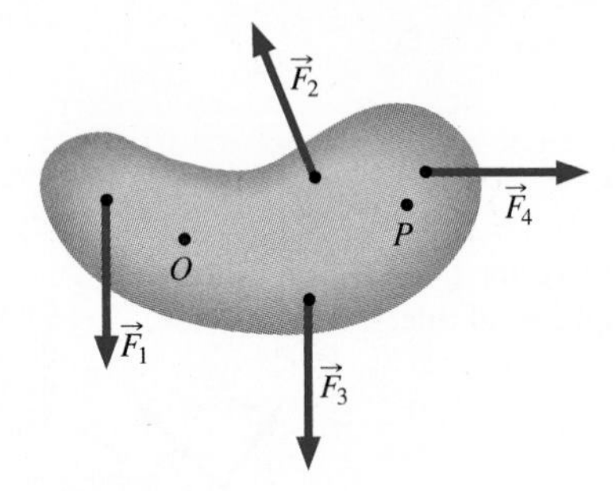

FIGURE 11.12 For Thought and Discussion 2

3. You stand with your right arm extended horizontally to the right. What's the direction of the gravitational torque about your shoulder?
4. Although it contains no parentheses, the expression $\vec{A} \times \vec{B} \cdot \vec{C}$ is unambiguous. Why? Is the expression a vector or a scalar?
5. What's the angle between two vectors if their dot product is equal to the magnitude of their cross product?
6. Why does a tetherball move faster as it winds up its pole?
7. Why do helicopters have two rotors?
8. A group of polar bears is standing around the edge of a slowly rotating ice floe. If the bears all walk to the center, what happens to the rotation rate?
9. Tornadoes in the northern hemisphere rotate counterclockwise as viewed from above. A far-fetched idea suggests that driving on the right side of the road may increase the frequency of tornadoes. Does this idea have *any* merit? Explain in terms of the angular momentum imparted to the air as two cars pass.
10. Does a particle moving at constant speed in a straight line have angular momentum about a point on the line? About a point not on the line? In either case, is its angular momentum constant?
11. When you turn on a high-speed power tool such as a router, the tool tends to twist in your hands. Why?
12. Why is it easier to balance a basketball on your finger if it's spinning?
13. A bug, initially at rest on a stationary, frictionless turntable, walks halfway around the turntable's circumference. Describe the motion of the turntable while the bug is walking and after the bug has stopped.
14. If you increase the rotation rate of a precessing gyroscope, will the precession rate increase or decrease?

Exercises and Problems

Exercises

Section 11.1 Angular Velocity and Acceleration Vectors

15. A car is headed north at 70 km/h. Give the magnitude and direction of the angular velocity of its 62-cm-diameter wheels.
16. If the car of Exercise 15 makes a 90° left turn lasting 25 s, determine the average angular acceleration of the wheels.
17. A wheel is spinning at 45 rpm with its axis vertical. After 15 s, it's spinning at 60 rpm with its axis horizontal. Find (a) the magnitude of its average angular acceleration and (b) the angle the average angular acceleration vector makes with the horizontal.
18. A wheel is spinning about a horizontal axis with angular speed 140 rad/s and with its angular velocity pointing east. Find the magnitude and direction of its angular velocity after an angular acceleration of 35 rad/s^2, pointing 68° west of north, is applied for 5.0 s.

Section 11.2 Torque and the Vector Cross Product

19. A 12-N force is applied at the point $x = 3$ m, $y = 1$ m. Find the torque about the origin if the force points in (a) the x-direction, (b) the y-direction, and (c) the z-direction.
20. A force $\vec{F} = 1.3\hat{\imath} + 2.7\hat{\jmath}$ N is applied at the point $x = 3.0$ m, $y = 0$ m. Find the torque about (a) the origin and (b) the point $x = -1.3$ m, $y = 2.4$ m.
21. **BIO** When you hold your arm outstretched, it's supported primarily by the deltoid muscle. Figure 11.13 shows a case in which the deltoid exerts a 67-N force at 15° to the horizontal. If the force-application point is 18 cm horizontally from the shoulder joint, what torque does the deltoid exert about the shoulder?

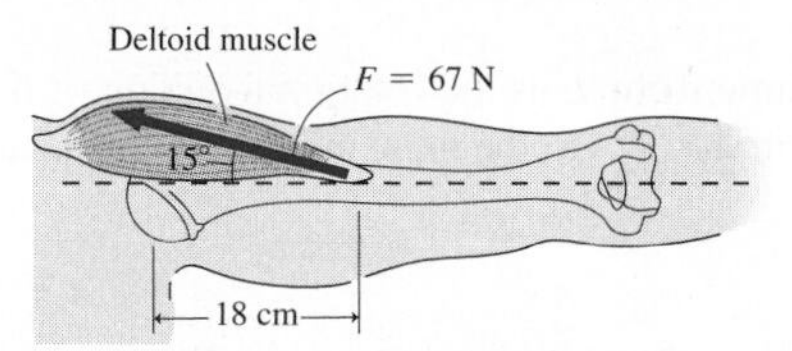

FIGURE 11.13 Exercise 21

Section 11.3 Angular Momentum

22. Express the units of angular momentum (a) using only the fundamental units kilogram, meter, and second; (b) in a form involving newtons; (c) in a form involving joules.
23. In the Olympic hammer throw, a contestant whirls a 7.3-kg steel ball on the end of a 1.2-m cable. If the contestant's arms reach an additional 90 cm from his rotation axis and if the ball's speed just prior to release is 27 m/s, what's the magnitude of the ball's angular momentum?
24. A gymnast of rotational inertia 62 kg·m^2 is tumbling head over heels with angular momentum 470 kg·m^2/s. What's her angular speed?
25. A 640-g hoop 90 cm in diameter is rotating at 170 rpm about its central axis. What's its angular momentum?
26. A 7.4-cm-diameter baseball has mass 145 g and is spinning at 2000 rpm. Treating the baseball as a uniform solid sphere, what's its angular momentum?

Section 11.4 Conservation of Angular Momentum

27. A potter's wheel with rotational inertia 6.40 kg·m^2 is spinning freely at 19.0 rpm. The potter drops a 2.70-kg lump of clay onto the wheel, where it sticks 46.0 cm from the rotation axis. What's the wheel's subsequent angular speed?
28. A 3.0-m-diameter merry-go-round with rotational inertia 120 kg·m^2 is spinning freely at 0.50 rev/s. Four 25-kg children

sit suddenly on the edge of the merry-go-round. (a) Find the new angular speed, and (b) determine the total energy lost to friction between children and merry-go-round.

29. A uniform, spherical cloud of interstellar gas has mass 2.0×10^{30} kg, has radius 1.0×10^{13} m, and is rotating with period 1.4×10^{6} years. The cloud collapses to form a star 7.0×10^{8} m in radius. Find the star's rotation period.

30. A skater has rotational inertia 4.2 kg·m^2 with his fists held to his chest and 5.7 kg·m^2 with his arms outstretched. The skater is spinning at 3.0 rev/s while holding a 2.5-kg weight in each outstretched hand; the weights are 76 cm from his rotation axis. If he pulls his hands in to his chest, so they're essentially on his rotation axis, how fast will he be spinning?

Problems

31. You slip a wrench over a bolt. Taking the origin at the bolt, the other end of the wrench is at $x = 18$ cm, $y = 5.5$ cm. You apply a force $\vec{F} = 88\hat{\imath} - 23\hat{\jmath}$ N to the end of the wrench. What's the torque on the bolt?

32. Vector $\vec{A}$ points 30° counterclockwise from the x-axis. Vector $\vec{B}$ has twice the magnitude of $\vec{A}$. Their product $\vec{A} \times \vec{B}$ has magnitude A^2 and points in the negative z-direction. What's the direction of vector $\vec{B}$?

33. **BIO** A baseball player extends his arm straight up to catch a 145-g baseball moving horizontally at 42 m/s. It's 63 cm from the player's shoulder joint to the point the ball strikes his hand, and his arm remains stiff while it rotates about the shoulder during the catch. The player's hand recoils 5.00 cm horizontally while he stops the ball. What average torque does the player's arm exert on the ball?

34. Show that $\vec{A} \cdot (\vec{A} \times \vec{B}) = 0$ for any vectors $\vec{A}$ and $\vec{B}$.

35. A weightlifter's barbell consists of two 25-kg masses on the ends of a 15-kg rod 1.6 m long. The weightlifter holds the rod at its center and spins it at 10 rpm about an axis perpendicular to the rod. What's the magnitude of the barbell's angular momentum?

36. A particle of mass m moves in a straight line at constant speed v. Show that its angular momentum about a point located a perpendicular distance b from its line of motion is mvb regardless of where the particle is on the line.

37. Two identical 1800-kg cars are traveling in opposite directions at 83 km/h. Each car's center of mass is 3.2 m from the center of the highway (Fig. 11.14). What are the magnitude and direction of the angular momentum of the system consisting of the two cars, about a point on the centerline of the highway?

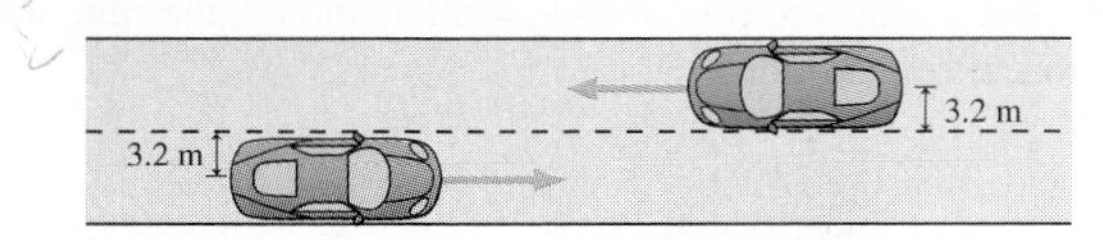

FIGURE 11.14 Problem 37

38. The dot product of two vectors is half the magnitude of their cross product. What's the angle between the two vectors?

39. **BIO** Biomechanical engineers have developed micromechanical devices for measuring blood flow as an alternative to dye injection following angioplasty to remove arterial plaque. One experimental device consists of a 300-μm-diameter, 2.0-μm-thick silicon rotor inserted into blood vessels. Moving blood spins the rotor, whose rotation rate provides a measure of blood flow. This device exhibited an 800-rpm rotation rate in tests with water flows at several m/s. Treating the rotor as a disk, what was its angular momentum at 800 rpm? (*Hint*: You'll need to find the density of silicon.)

40. Figure 11.15 shows the dimensions of a 880-g wooden baseball bat whose rotational inertia about its center of mass is 0.048 kg·m^2. If the bat is swung so its far end moves at 50 m/s, find (a) its angular momentum about the pivot P and (b) the constant torque applied about P to achieve this angular momentum in 0.25 s. (*Hint:* Remember the parallel-axis theorem.)

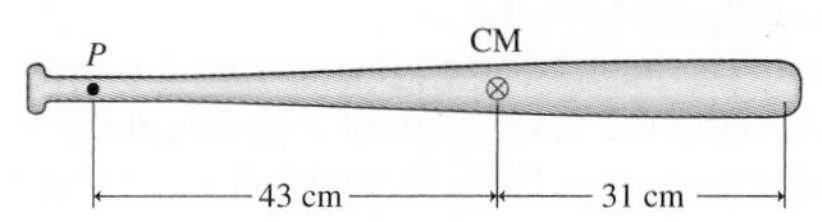

FIGURE 11.15 Problem 40

41. As an automotive engineer, you're charged with redesigning a car's wheels with the goal of decreasing each wheel's angular momentum by 30% for a given linear speed of the car. Other design considerations require that the wheel diameter go from 38 cm to 35 cm. If the old wheel had rotational inertia 0.32 kg·m^2, what do you specify for the new rotational inertia?

42. A turntable of radius 25 cm and rotational inertia 0.0154 kg·m^2 is spinning freely at 22.0 rpm about its central axis, with a 19.5-g mouse on its outer edge. The mouse walks from the edge to the center. Find (a) the new rotation speed and (b) the work done by the mouse.

43. **CH** A 17-kg dog is standing on the edge of a stationary, frictionless turntable of rotational inertia 95 kg·m^2 and radius 1.81 m. The dog walks once around the turntable. What fraction of a full circle does the dog's motion make with respect to the ground?

44. A physics student is standing on an initially motionless, frictionless turntable with rotational inertia 0.31 kg·m^2. She's holding a wheel with rotational inertia 0.22 kg·m^2 spinning at 130 rpm about a vertical axis, as in Fig. 11.8. When she turns the wheel upside down, student and turntable begin rotating at 70 rpm. (a) Find the student's mass, considering her to be a 30-cm-diameter cylinder. (b) Neglecting the distance between the axes of the turntable and wheel, determine the work she did in turning the wheel upside down.

45. You're choreographing your school's annual ice show. You call for eight 60-kg skaters to join hands and skate side by side in a line extending 12 m. The skater at one end is to stop abruptly, so the line will rotate rigidly about that skater. For safety, you don't want the fastest skater to be moving at more than 8.0 m/s, and you don't want the force on that skater's hand to exceed 300 N. What do you determine is the greatest speed the skaters can have before they execute their rotational maneuver?

46. Find the angle between two vectors whose dot product is twice the magnitude of their cross product.

47. A circular bird feeder 19 cm in radius has rotational inertia 0.12 kg·m^2. It's suspended by a thin wire and is spinning slowly at 5.6 rpm. A 140-g bird lands on the feeder's rim, coming in tangent to the rim at 1.1 m/s in a direction opposite the feeder's rotation. What's the rotation rate after the bird lands?

48. A force $\vec{F}$ applied at the point $x = 2.0$ m, $y = 0$ m produces a torque $4.6\hat{k}$ N·m about the origin. If the x-component of $\vec{F}$ is 3.1 N, what angle does it make with the x-axis?

49. Show that the cross product of two vectors $\vec{A} = A_x\hat{\imath} + A_y\hat{\jmath} + A_z\hat{k}$ and $\vec{B} = B_x\hat{\imath} + B_y\hat{\jmath} + B_z\hat{k}$ is given by $\vec{A} \times \vec{B} = (A_yB_z - A_zB_y)\hat{\imath} + (A_zB_x - A_xB_z)\hat{\jmath} + (A_xB_y - A_yB_x)\hat{k}$. (*Hint*: You'll need to work out cross products of all possible pairs of the unit vectors $\hat{\imath}$, $\hat{\jmath}$, and $\hat{k}$—including with themselves.)

50. If you're familiar with determinants, show that the cross product can be written as a determinant:

$$\vec{A} \times \vec{B} = \begin{vmatrix} \hat{\imath} & \hat{\jmath} & \hat{k} \\ A_x & A_y & A_z \\ B_x & B_y & B_z \end{vmatrix}$$

(*Hint*: See the preceding problem.)

51. Jumbo is back! Jumbo is the 4.8-Mg elephant from Example 9.4. This time he's standing at the outer edge of a 15-Mg turntable of radius 8.5 m, rotating with angular velocity $0.15\ \text{s}^{-1}$ on frictionless bearings. Jumbo then walks to the center of the turntable. Treating Jumbo as a point mass and the turntable as a solid disk, find (a) the angular velocity of the turntable once Jumbo reaches the center and (b) the work Jumbo does in walking to the center.

52. An anemometer for measuring wind speeds consists of four small cups, each with mass 124 g, mounted a pair of 32.6-cm-long rods with mass 75.7 g each, as shown in Fig. 11.16. Find the angular momentum of the anemometer when it's spinning at 12.4 rev/s. You can treat the cups as point masses.

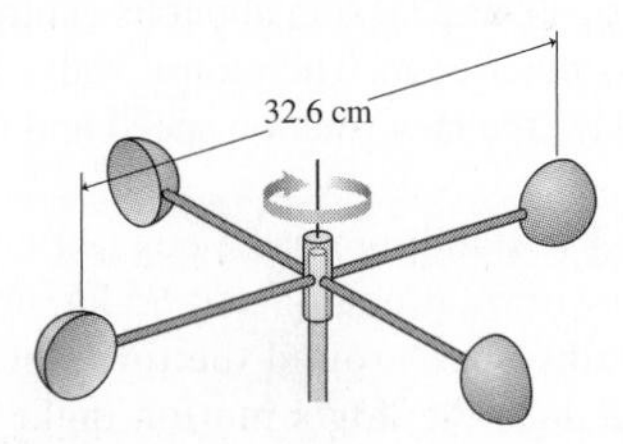

FIGURE 11.16 Problem 52

53. A turntable has rotational inertia I and is rotating with angular speed ω about a frictionless vertical axis. A wad of clay with mass m is tossed onto the turntable and sticks a distance d from the rotation axis. The clay hits horizontally with its velocity $\vec{v}$ at right angles to the turntable's radius, and in the same direction as the turntable's rotation (Fig. 11.17). Find expressions for v that will result in (a) the turntable's angular speed dropping to half its initial value, (b) no change in the turntable's angular speed, and (c) the angular speed doubling.

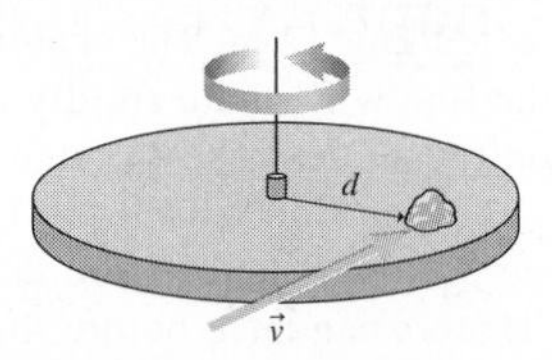

FIGURE 11.17 Problem 53.

54. **CH** A uniform, solid, spherical asteroid with mass 1.2×10^{13} kg and radius 1.0 km is rotating with period 4.3 h. A meteoroid moving in the asteroid's equatorial plane crashes into the equator at 8.4 km/s. It hits at a 58° angle to the vertical and embeds itself at the surface. After the impact the asteroid's rotation period is 3.9 h. Find the meteoroid's mass.

55. About 99.9% of the solar system's total mass lies in the Sun. Using data from Appendix E, estimate what fraction of the solar system's angular momentum about its center is associated with the Sun. Where is most of the rest of the angular momentum?

56. **CH** You're a civil engineer for an advanced civilization on a solid spherical planet of uniform density. Running out of room for the expanding population, the government asks you to redesign your planet to give it more surface area. You recommend reshaping the planet, without adding any material or angular momentum, into a hollow shell whose thickness is one-fifth its outer radius. How much will your design increase the surface area, and how will it change the length of the day?

57. **CH** In Fig. 11.18, the lower disk, of mass 440 g and radius 3.5 cm, is rotating at 180 rpm on a frictionless shaft of negligible radius. The upper disk, of mass 270 g and radius 2.3 cm, is initially not rotating. It drops freely down onto the lower disk, and frictional forces bring the two disks to a common rotational speed. Find (a) that common speed and (b) the fraction of the initial kinetic energy lost to friction.

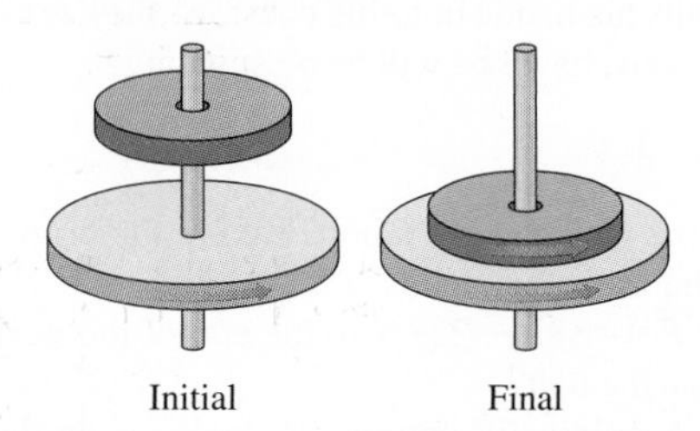

FIGURE 11.18 Problem 57

58. **CH** A massless spring with constant k is mounted on a turntable of rotational inertia I, as shown in Fig. 11.19. The turntable is on a frictionless vertical axle, though initially it's not rotating. The spring is compressed a distance Δx from its equilibrium, with a mass m placed against it. When the spring is released, the mass moves at right angles to a line through the turntable's center, at a distance b from the center, and slides without friction across the table and off the edge. Find expressions for (a) the linear speed of the mass and (b) the rotational speed of the turntable. (*Hint:* What's conserved?)

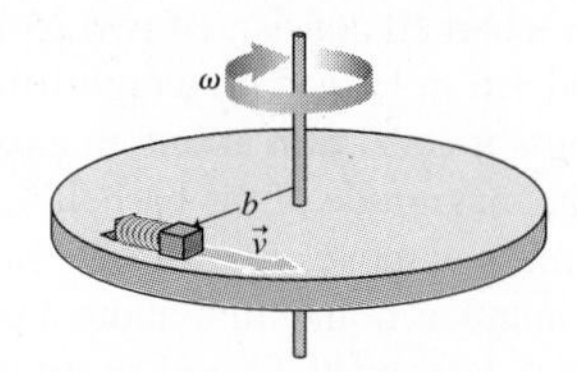

FIGURE 11.19 Problem 58

59. **CH** A solid ball of mass M and radius R is spinning with angular velocity ω_0 about a horizontal axis. It drops vertically onto a surface where the coefficient of kinetic friction with the ball is μ_k (Fig. 11.20). Find expressions for (a) the final angular velocity once it's achieved pure rolling motion and (b) the time it takes to achieve this motion.

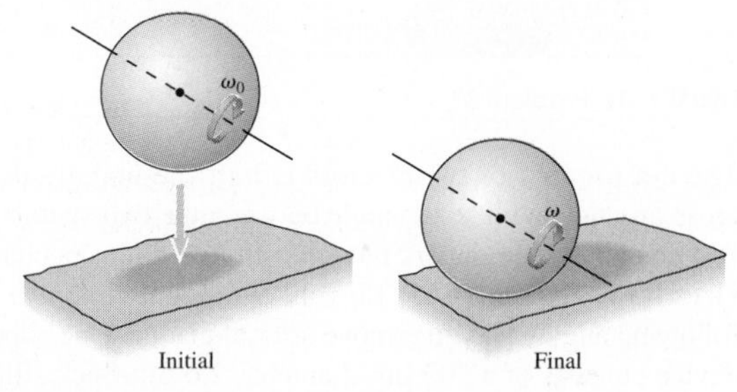

FIGURE 11.20 Problem 59

60. A time-dependent torque given by $\tau = a + b \sin ct$ is applied to an object that's initially stationary but is free to rotate. Here a, b, and c are constants. Find an expression for the object's angular

momentum as a function of time, assuming the torque is first applied at $t = 0$.

61. Consider a rapidly spinning gyroscope whose axis is precessing uniformly in a horizontal circle of radius r, as shown in Fig. 11.10. Apply $\vec{\tau} = d\vec{L}/dt$ to show that the angular speed of precession about the vertical axis through the center of the circle is mgr/L.
62. When a star like our Sun exhausts its fuel, thermonuclear reactions in its core cease, and it collapses to become a *white dwarf*. Often the star will blow off its outer layers and lose some mass before it collapses. Suppose a star with the Sun's mass and radius is rotating with period 25 days and then it collapses to a white dwarf with 60% of the Sun's mass and a rotation period of 131 s. What's the radius of the white dwarf? Compare your answer with the radii of Sun and Earth.
63. DATA Pulsars—the rapidly rotating neutron stars described in Example 11.2—have magnetic fields that interact with charged particles in the surrounding interstellar medium. The result is torque that causes the pulsar's spin rate and therefore its angular momentum to decrease very slowly. The table below gives values for the rotation period of a given pulsar as it's been observed at the same date every 5 years for two decades. The pulsar's rotational inertia is known to be 1.12×10^{38} kg · m^2. Make a plot of the pulsar's angular momentum over time, and use the associated best-fit line, along with the rotational analog of Newton's law, to find the torque acting on the pulsar.

Year of observation	1995	2000	2005	2010	2015
Angular momentum (10^{37} kg · m^2/s)	7.844	7.831	7.816	7.799	7.787

64. CH A system has total angular momentum $\vec{L}$ about an axis O. Show that the system's angular momentum about a parallel axis O' is given by $\vec{L}' = L - \vec{h} \times \vec{p}$, where $\vec{p}$ is the system's linear momentum and $\vec{h}$ is a vector from O to O' (see Fig. 11.21, which also shows vectors $\vec{r}_i$ and $\vec{r}_i'$ from each axis to the system's ith mass element m_i).

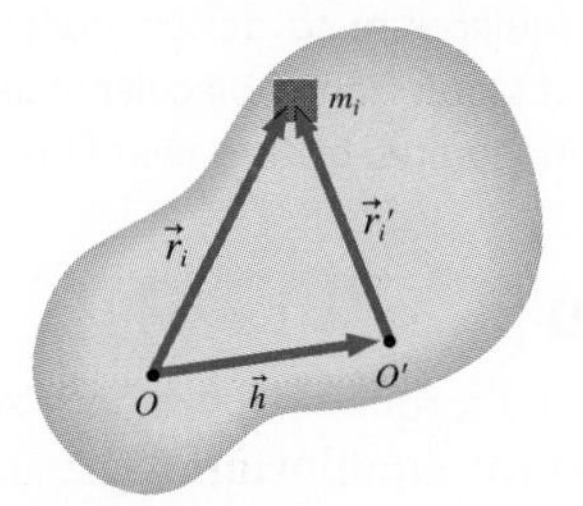

FIGURE 11.21 Problem 64

Passage Problems

Figure 11.22 shows a demonstration gyroscope, consisting of a solid disk mounted on a shaft. The disk spins about the shaft on essentially frictionless bearings. The shaft is mounted on a stand so it's free to pivot both horizontally and vertically. A weight at the far end of the shaft balances the disk, so in the configuration shown there's no torque on the system. An arrowhead mounted on the disk end of the shaft indicates the direction of the disk's angular velocity.

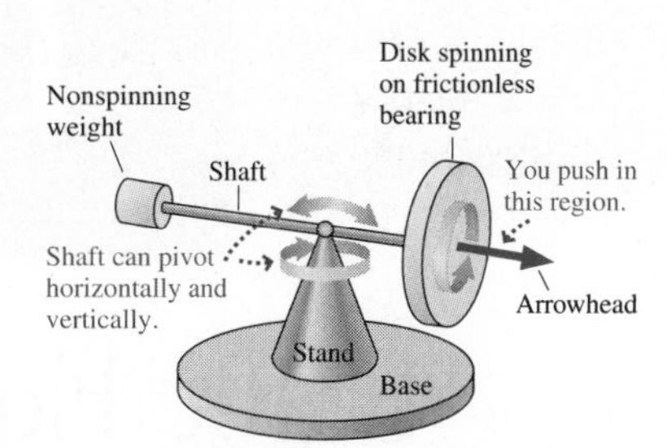

FIGURE 11.22 A gyroscope (Passage Problems 65–68)

65. If you push on the shaft between the arrowhead and the disk, pushing horizontally away from you (i.e., into the page in Fig. 11.22), the arrowhead end of the shaft will move
    a. away from you (i.e., into the page).
    b. toward you (i.e., out of the page).
    c. downward.
    d. upward.
66. If you push on the shaft between the arrowhead and the disk, pushing directly upward on the bottom of the shaft, the arrowhead end of the shaft will move
    a. away from you (i.e., into the page).
    b. toward you (i.e., out of the page).
    c. downward.
    d. upward.
67. If an additional weight is hung on the left end of the shaft, the arrowhead will
    a. pivot upward until the weighted end of the shaft hits the base.
    b. pivot downward until the arrowhead hits the base.
    c. precess counterclockwise when viewed from above.
    d. precess clockwise when viewed from above.
68. If the system is precessing, and only the disk's rotation rate is increased, the precession rate will
    a. decrease.
    b. increase.
    c. stay the same.
    d. become zero.

Answers to Chapter Questions

Answer to Chapter Opening Question

The rotation axis precesses—changes orientation—over a 26,000-year cycle. This alters the relation between sunlight intensity and seasons, triggering ice ages.

Answers to GOT IT? Questions

11.1 (e)
11.2 (1) $\vec{\tau}_3$; (2) $\vec{\tau}_5$; (3) $\vec{\tau}_1$; (4) $\vec{\tau}_4$
11.3 (d)
11.4 (1) (a) to keep the total angular momentum at 0; (2) (c) so L_{total} remains 0; (3) (b)
11.5 (a)

Static Equilibrium

What You Know

- You understand Newton's second law and its rotational analog.
- You can solve problems involving multiple force vectors in two dimensions.
- You're familiar with the center-of-mass concept.
- You know how to calculate torque as a cross product.

What You're Learning

- Here you'll learn to describe situations involving *static equilibrium*, in which an object remains at rest because there's zero net force and zero net torque acting on it.
- You'll learn about *center of gravity*—which, for everyday-sized objects, is the same point as the center of mass.
- You'll learn to distinguish *stable equilibria* from *unstable equilibria*, and to identify the in-between case of *metastable equilibria*.

How You'll Use It

- Static equilibrium is especially important for architects and engineers concerned about the stability of structures they design.
- Static equilibrium is also important in physiology because conditions for static equilibrium often determine the forces acting within the body.
- Equilibrium situations occur throughout the universe, and whether they're stable or unstable is a crucial determinant of a system's behavior.

The Alamillo Bridge in Seville, Spain, is the work of architect Santiago Calatrava. What conditions must be met to ensure the stability of this dramatic structure?

Architect Santiago Calatrava envisioned the boldly improbable bridge shown here. But it took engineers to make sure that the bridge would be stable in the face of what looks like an obvious tendency to topple to the left. The key to the engineers' success is **static equilibrium**—the condition in which a structure or system experiences neither a net force nor a net torque. Engineers use the principles of static equilibrium to design buildings, bridges, and aircraft. Scientists apply equilibrium principles at scales from molecular to astrophysical. Here we explore the conditions for static equilibrium required by the laws of physics.

12.1 Conditions for Equilibrium

A body is in **equilibrium** when the net external force and torque on it are both zero. In the special case when the body is also at rest, it's in **static equilibrium**. Systems in static equilibrium include not only engineered structures but also trees, molecules, and even your bones and muscles when you're at rest.

We can write the conditions for static equilibrium mathematically by setting the sums of all the external forces and torques both to zero:

$$\sum \vec{F}_i = \vec{0} \tag{12.1}$$

and

$$\sum \vec{\tau}_i = \sum (\vec{r}_i \times \vec{F}_i) = \vec{0} \tag{12.2}$$

Here the subscripts i label the forces $\vec{F}$ acting on an object, the positions $\vec{r}$ of the force-application points, and the associated torques $\vec{\tau}$.

In Chapters 10 and 11, we noted that torque depends on the choice of a rotation axis. Actually, the issue is not so much an axis but a single point—the point of origin of the vectors $\vec{r}$ that enter the expression $\vec{\tau} = \vec{r} \times \vec{F}$. In this chapter, where we have objects in equilibrium so they aren't rotating, we'll talk of this "pivot point" rather than a rotation axis. So the torque $\vec{\tau} = \vec{r} \times \vec{F}$ depends on the choice of pivot point. Then there seems to be an ambiguity in Equation 12.2, since we haven't specified a pivot point.

For an object to be in static equilibrium it can't rotate about *any* point, so Equation 12.2 must hold no matter what point we choose. Must we then check every possible point? Fortunately, no. If the first equilibrium condition holds—that is, if the net force on an object is zero—and if the net torque about *some* point is zero, then the net torque about *any other* point is also zero. Problem 51 leads you through the proof of this statement.

In solving equilibrium problems, we're thus free to choose any convenient point about which to evaluate the torques. An appropriate choice is often the application point of one of the forces; then $\vec{r} = \vec{0}$ for that force, and the associated torque $\vec{r} \times \vec{F}$ is zero. This leaves Equation 12.2 with one term fewer than it would otherwise have.

EXAMPLE 12.1 Choosing the Pivot: A Drawbridge

The raised span of the drawbridge shown in Fig. 12.1*a* has its 11,000-kg mass distributed uniformly over its 14-m length. Find the magnitude of the tension in the supporting cable.

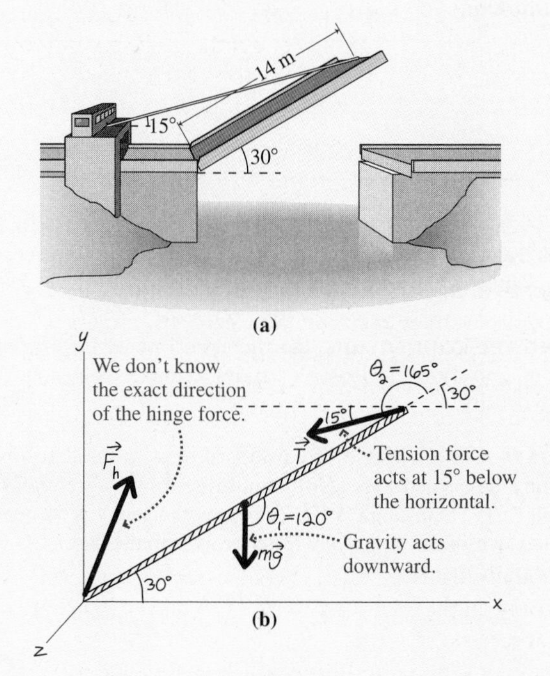

FIGURE 12.1 (a) A drawbridge. (b) Our sketch showing forces supporting the bridge.

INTERPRET Because the drawbridge is at rest, it's in static equilibrium.

DEVELOP Here we'll demonstrate how a sensible choice of the pivot point can make solving static-equilibrium problems easier. Figure 12.1*b* is a simplified diagram of the bridge, showing the three forces acting on it. These forces must satisfy both Equations 12.1 and 12.2, but we aren't asked about the hinge force $\vec{F}_h$, so it makes sense to choose the pivot at the hinge. We can then focus on Equation 12.2, $\sum \vec{\tau}_i = \vec{0}$, in which the only torques are due to gravity and tension. Gravity acts at the center of mass, half the bridge length L from the pivot (we'll prove this shortly). Therefore, it exerts a torque $\tau_g = -(L/2)mg \sin\theta_1$, where θ_1 is the angle between the gravitational force and a vector from the pivot. This torque is into the page, or in the negative z-direction—hence the negative sign. Similarly, the tension force, applied at the full length L, exerts a torque $\tau_T = LT\sin\theta_2$. Equation 12.2 then becomes

$$-\frac{L}{2}mg\sin\theta_1 + LT\sin\theta_2 = 0$$

EVALUATE We solve for the tension T:

$$T = \frac{mg\sin\theta_1}{2\sin\theta_2} = \frac{(11{,}000\text{ kg})(9.8\text{ m/s}^2)(\sin 120°)}{(2)(\sin 165°)} = 180\text{ kN}$$

ASSESS This tension force is considerably larger than the approximately 110-kN weight of the bridge because the tension acts at a small angle to produce a torque that balances the torque due to gravity.

One point of this example is that a wise choice of the pivot point can eliminate a lot of work—in this case, allowing us to solve the problem using only Equation 12.2. If we had chosen a different pivot, then the force F_h would have appeared in the torque equation, and we would have had to eliminate it using the force equation, Equation 12.1 (see Exercise 13). ■

GOT IT? 12.1 The figure shows three pairs of forces acting on an object. Which pair, acting as the *only* forces on the object, results in static equilibrium? Explain why the others don't.

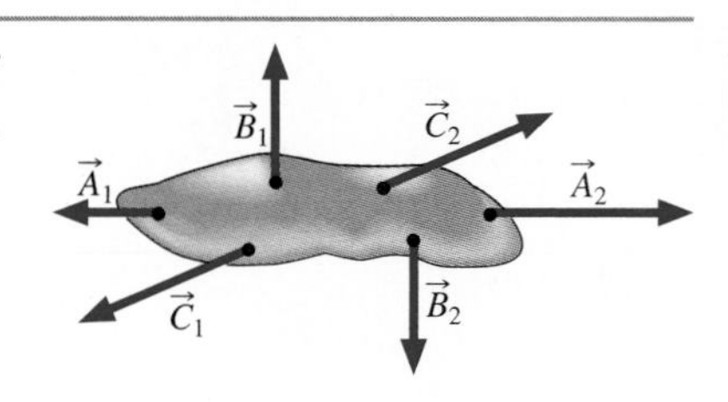

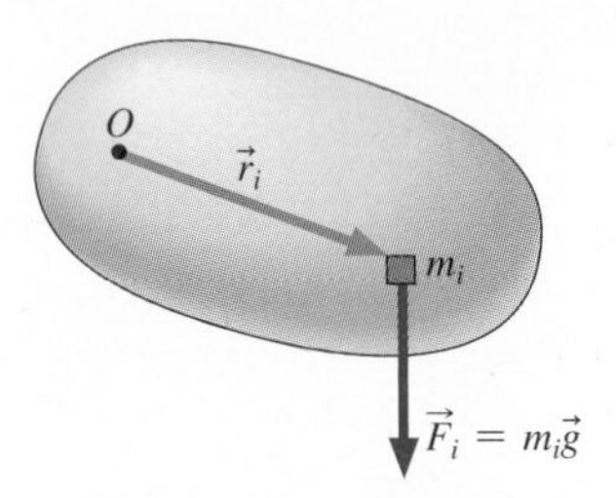

FIGURE 12.2 The gravitational force $\vec{F}_i$ on the mass element m_i produces a torque about point O.

12.2 Center of Gravity

In Fig. 12.1*b* we drew the gravitational force acting at the center of mass of the bridge. That seems sensible, but is it correct? After all, gravity acts on all parts of an object. How do we know that the resulting torque is equivalent to the torque due to a single force acting at the center of mass? To see that it is, consider the gravitational forces on all parts of an object of mass M. The vector sum of those forces is $M\vec{g}$, but what about the torques? Figure 12.2 shows the ingredients we need to calculate the torque $\vec{\tau} = \vec{r} \times \vec{F}$ associated with one mass element; summing gives the total torque:

$$\vec{\tau} = \sum \vec{r}_i \times \vec{F}_i = \sum \vec{r}_i \times m_i\vec{g} = \left(\sum m_i\vec{r}_i\right) \times \vec{g}$$

We can rewrite this equation by multiplying the right-hand side by M/M, with M the total mass:

$$\vec{\tau} = \left(\frac{\sum m_i\vec{r}_i}{M}\right) \times M\vec{g}$$

The term in parentheses is the position of the center of mass (Section 9.1), and the right-hand term is the total weight. Therefore, the net torque on the body due to gravity is that of the gravitational force $M\vec{g}$ acting at the center of mass. In general, the point at which the gravitational force seems to act is called the **center of gravity**. We've just proven an important point: **The center of gravity coincides with the center of mass when the gravitational field is uniform.**

CONCEPTUAL EXAMPLE 12.1 Finding the Center of Gravity

Explain how you can find an object's center of gravity by suspending it from a string.

EVALUATE Suspend an object from a string and it will quickly come to equilibrium, as shown in Figs. 12.3*a, b*. In equilibrium there's no torque on the object and so, as Fig. 12.3*b* shows, its center of gravity (CG) must be directly below the suspension point. So far all we know is that the CG lies on a vertical line extending from the suspension point. But two intersecting lines determine a point, so all we have to do is suspend the object from a *different* point. In its new equilibrium, the CG again lies on a vertical line from the suspension point. Where the two lines meet is the center of gravity (Fig. 12.3*c*).

ASSESS Here's a quick, easy, and practical way to find the center of gravity—at least for two-dimensional objects.

MAKING THE CONNECTION Do the experiment! Determine the center of gravity of an isosceles triangle made from material of uniform density.

EVALUATE Cut a triangle of cardboard or wood and follow the procedure described here. You should get good agreement with Example 9.3: The triangle's CG (which is the same as its center of mass) lies two-thirds of the way from the apex to the base.

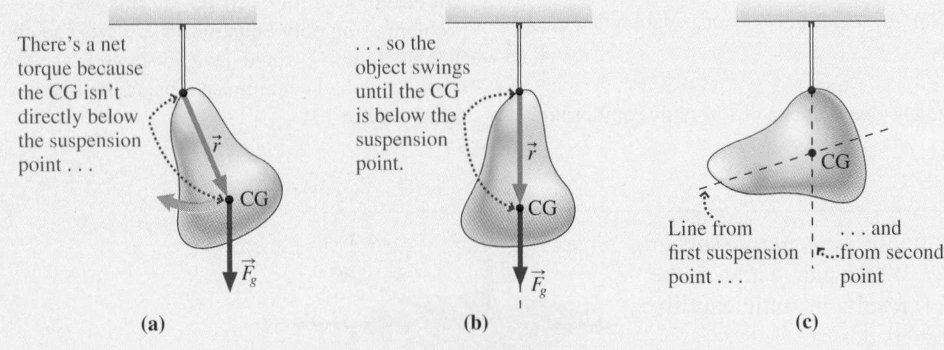

FIGURE 12.3 Finding the center of gravity.

GOT IT? 12.2 The dancer in the figure is balanced; that is, she's in static equilibrium. Which of the three lettered points could be her center of gravity?

12.3 Examples of Static Equilibrium

It's frequently the case that all the forces acting on a system lie in a plane, so Equation 12.1—the statement that there's no net force in static equilibrium—becomes two equations for the two force components in that plane. And with all the forces in a plane, the torques are all at right angles to that plane, so Equation 12.2—the statement that there's no net torque—becomes a single equation. We'll restrict ourselves to such cases in which the conditions for static equilibrium reduce to three scalar equations. Sometimes, as in Example 12.1, the torque equation alone will give what we're looking for, but often that's not the case.

Solving static-equilibrium problems is much like solving Newton's law problems; after all, the equations for static equilibrium are Newton's law and its rotational analog, both with acceleration set to zero. Here we adapt our Newton's law strategy from Chapter 4 to problems of static equilibrium. The examples that follow illustrate the use of this strategy.

Video Tutor Demo | **Walking the Plank**

PROBLEM-SOLVING STRATEGY 12.1 Static-Equilibrium Problems

INTERPRET Interpret the problem to be sure it's about static equilibrium, and identify the object that you want to keep in equilibrium. Next, identify all the forces acting on the object.

DEVELOP Draw a diagram showing the forces acting on your object. Since you've got torques to calculate, it's important to show *where* each force is applied. So don't represent your object as a single dot but show it semirealistically with the force-application points. This is a static-equilibrium problem, so Equations 12.1, $\sum \vec{F}_i = \vec{0}$, and 12.2, $\sum \vec{\tau}_i = \vec{0}$, apply. Develop your solution by choosing a coordinate system that will help resolve the force vectors into components *and* choose its origin at an appropriate pivot point—usually the application point of one of the forces. In some problems the unknown is itself a force; in that case, draw a force vector that you think is appropriate and let the algebra take care of the signs and angles.

EVALUATE At this point the physics is done, and you're ready to evaluate your answer. Begin by writing the two components of Equation 12.1 in your coordinate system. Then evaluate the torques about your chosen origin, and write Equation 12.2 as a single scalar equation showing that the torques sum to zero. Now you've got three equations, and you're ready to solve. Since there are three equations, there will be three unknowns even if you're asked for only one final answer. You can use the equations to eliminate the unknowns you don't want.

ASSESS Assess your solution to see whether it makes sense. Are the numbers reasonable? Do the directions of forces and torques make sense in the context of static equilibrium? What happens in special cases—for example, when a force or mass goes to zero or gets very large, or for special values of angles among the various vectors?

EXAMPLE 12.2 Static Equilibrium: Ladder Safety

A ladder of mass m and length L is leaning against a wall, as shown in Fig. 12.4a (next page). The wall is frictionless, and the coefficient of static friction between ladder and ground is μ. Find an expression for the minimum angle ϕ at which the ladder can lean without slipping.

INTERPRET This problem is about static equilibrium, and the ladder is the object we want to keep in equilibrium. We identify four forces acting on the ladder: gravity, normal forces from the floor and wall, and static friction from the ground.

(continued)

DEVELOP Figure 12.4*b* shows the four forces and the unknown angle ϕ. We'll get the minimum angle when static friction is greatest: $f_s = \mu n_1$. Since we're dealing with static equilibrium, Equations 12.1 and 12.2 apply. In a horizontal/vertical coordinate system, Equation 12.1 has the two components:

$$\text{Force, } x: \quad \mu n_1 - n_2 = 0$$
$$\text{Force, } y: \quad n_1 - mg = 0$$

Now for the torques: If we choose the bottom of the ladder as the pivot, we eliminate two forces from the torque equation. That

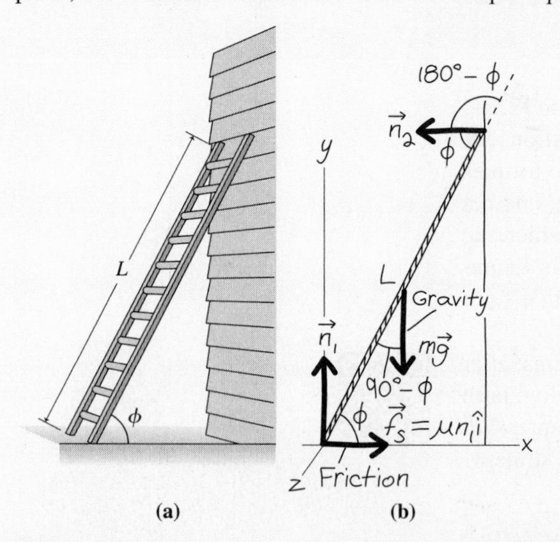

FIGURE 12.4 (a) At what angle will the ladder slip? (b) Our sketch.

leaves only the gravitational torque and the torque due to the wall's normal force; both involve the unknown angle ϕ. The gravitational torque is into the page, or the negative z-direction, so it's given by $\tau_g = -(L/2)mg\sin(90° - \phi) = -(L/2)mg\cos\phi$. The torque due to the wall is out of the page: $\tau_w = Ln_2\sin(180° - \phi) = Ln_2\sin\phi$. We used two trig identities here: $\sin(90° - \phi) = \cos\phi$ and $\sin(180° - \phi) = \sin\phi$. Then Equation 12.2 becomes

$$\text{Torque:} \quad Ln_2\sin\phi - \frac{L}{2}mg\cos\phi = 0$$

EVALUATE We have three unknowns: n_1, n_2, and ϕ. The y-component of the force equation gives $n_1 = mg$, showing that the ground supports the ladder's weight. Using this result in the x-component of the force equation gives $n_2 = \mu mg$. Then the torque equation becomes $\mu mgL\sin\phi - (L/2)mg\cos\phi = 0$. The term mgL cancels, giving $\mu\sin\phi - \frac{1}{2}\cos\phi = 0$. We solve for the unknown angle ϕ by forming its tangent:

$$\tan\phi = \frac{\sin\phi}{\cos\phi} = \frac{1}{2\mu}$$

ASSESS Make sense? The larger the frictional coefficient, the more horizontal force holding the ladder in place, and the smaller the angle at which it can safely lean. On the other hand, a very small frictional coefficient makes for a very large tangent—meaning the angle approaches 90°. With no friction, you could stand the ladder only if it were strictly vertical. A word of caution: We worked this example with no one on the ladder. With the extra weight of a person, especially near the top, the minimum safe angle will be a lot larger. Problem 29 explores this situation. ■

EXAMPLE 12.3 Static Equilibrium: In the Body

Figure 12.5*a* shows a human arm holding a pumpkin, with masses and distances marked. Find the magnitudes of the biceps tension and the contact force at the elbow joint.

INTERPRET This problem is about static equilibrium, with the arm/pumpkin being the object in equilibrium. We identify four forces: the weights of the arm and the pumpkin, the biceps tension, and the contact force at the elbow.

DEVELOP Figure 12.5*b* shows the four forces, including the elbow contact force $\vec{F}_c$, whose exact direction we don't know. We can read the horizontal and vertical components of Equation 12.1, the force balance equation, from the diagram:

$$\text{Force, } x: \quad F_{cx} - T\cos\theta = 0$$
$$\text{Force, } y: \quad T\sin\theta - F_{cy} - mg - Mg = 0$$

Choosing the elbow as the pivot eliminates the contact force from the torque equation, giving

$$\text{Torque:} \quad x_1T\sin\theta - x_2mg - x_3Mg = 0$$

where the x values are the coordinates of the three force-application points.

EVALUATE We begin by solving the torque equation for the biceps tension:

$$T = \frac{(x_2m + x_3M)g}{x_1\sin\theta} = 500\text{ N}$$

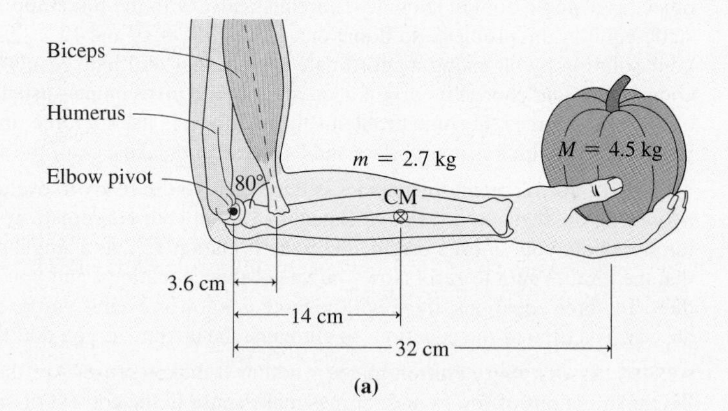

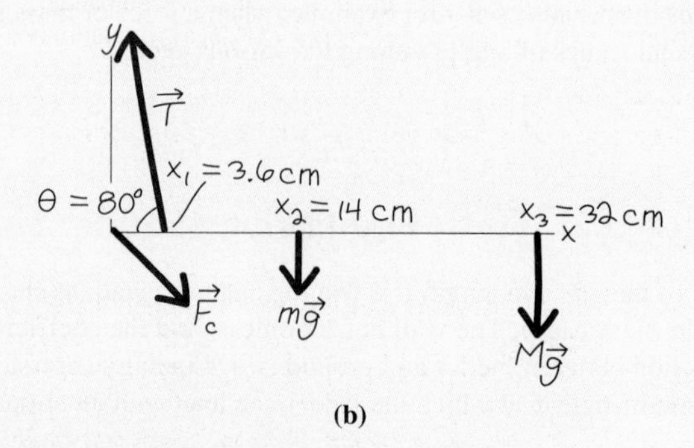

FIGURE 12.5 (a) Holding a pumpkin. (b) Our sketch.

where we used the values in Fig. 12.5 to evaluate the numerical answer. The force equations then give the components of the elbow contact force:

$$F_{cx} = T\cos\theta = 87\text{ N} \quad \text{and} \quad F_{cy} = T\sin\theta - (m + M)g = 420\text{ N}$$

The magnitude of the contact force at the elbow then becomes $F_c = \sqrt{87^2 + 420^2}\text{ N} = 430\text{ N}$.

ASSESS These answers may seem huge—both the biceps tension and the elbow contact force are roughly 10 times the weight of the pumpkin, on the order of 100 pounds. But that's because the biceps muscle is attached so close to the elbow; given this small lever arm, it takes a large force to balance the torque from the weight of pumpkin and arm. This example shows that the human body routinely experiences forces substantially greater than the weights of objects it's lifting. ■

GOT IT? 12.3 The figure shows a person in static equilibrium leaning against a wall. Which of the following must be true? (a) There must be a frictional force at the wall but not necessarily at the floor. (b) There must be a frictional force at the floor but not necessarily at the wall. (c) There must be frictional forces at both floor and wall.

12.4 Stability

If a body is disturbed from equilibrium, it generally experiences nonzero torques or forces that cause it to accelerate. Figure 12.6 shows two very different possibilities for the subsequent motion of two cones initially in equilibrium. Tip the cone on the left slightly, and a torque develops that brings it quickly back to equilibrium. Tip the cone on the right, and over it goes. The torque arising from even a slight displacement swings the cone permanently away from its original equilibrium. The former situation is an example of **stable equilibrium**, the latter of **unstable equilibrium**. Nearly all the equilibria we encounter in nature are stable, since a body in unstable equilibrium won't remain so. The slightest disturbance will set it in motion, bringing it to a very different equilibrium state.

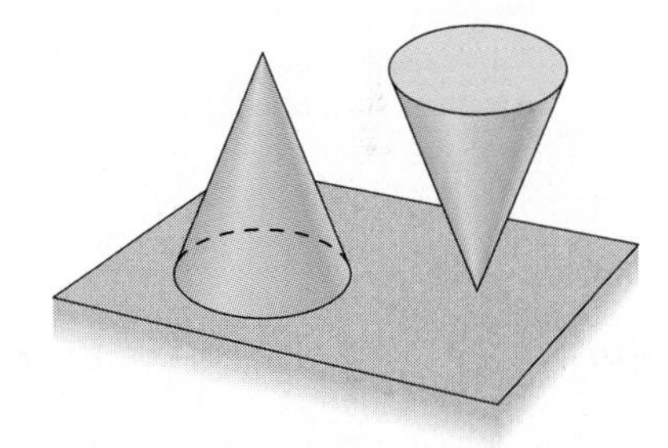

FIGURE 12.6 Stable (*left*) and unstable (*right*) equilibria.

APPLICATION Vehicle Stability Control

When a car or other vehicle rounds a curve, the force of static friction between road and wheels provides the centripetal acceleration that keeps the car in its circular path. These frictional forces act at the road, and so they exert a torque that tends to rotate the vehicle about its center of gravity (see drawing). The effect is to increase the normal force on the wheels at the outside of the turn and decrease it on the wheels at the inside of the turn. In extreme cases, the inside wheels may leave the road—a condition that can rapidly worsen and lead to a rollover.

Consider the case of a vehicle whose inside wheels are just about to leave the road, so there's neither a normal force nor a frictional force on the wheels at the inside of the turn. Applying Newton's second law to the remaining forces (see the drawing) gives $f = mv^2/r$ in the horizontal direction and $n = mg$ in the vertical direction. Meanwhile, the torques associated with these two forces are fh and $nt/2$, where h is the height of the center of gravity above the road and t is the width between the wheels. The drawing shows that these torques are in opposite directions; setting the net torque to zero and substituting for the two forces then gives the **rollover condition**:

$$\frac{v^2}{rg} = \frac{t}{2h}$$

The term on the right depends only on the geometry of the vehicle (including how it's loaded with passengers and cargo), and is called the *static stability factor* (SSF). The equation shows that if v^2/rg exceeds the SSF, the vehicle's inner tires will leave the road, setting the stage for a rollover. The equation also shows that the wider the tire spacing t, the higher the SSF and the more stable the vehicle. But the higher the center of gravity, as given by h, the lower the SSF and the less stable is the vehicle. That's why SUVs and vans have had high rates of rollover accidents—among the most serious of single-vehicle accidents.

Today's cars and SUVs increasingly include electronic stability control systems (ECS), which monitor speed, tilt angle, and steering wheel position and apply brakes to individual wheels so as to prevent rollover; ECS may also throttle back the engine as needed. Studies show that ECS can reduce SUV accidents by two-thirds and fatal rollovers by as much as 80%. Extensive use of ECS in recent SUVs has actually made late-model SUVs less likely to experience rollover than non-ECS cars.

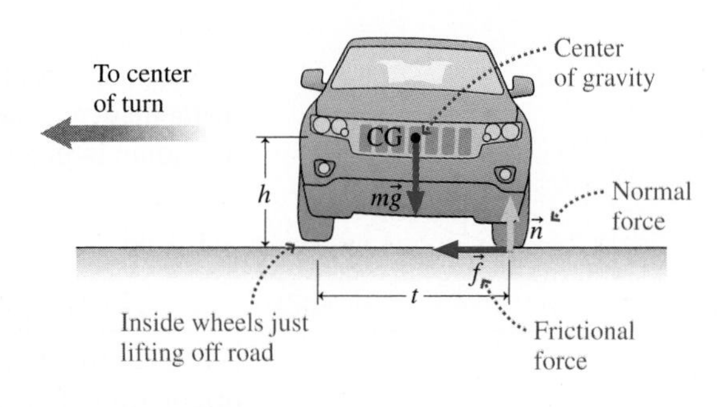

(continued)

Our simple analysis doesn't take into account factors like the vehicle's suspension and the deformation of its tires—both of which can exacerbate rollover danger by allowing the vehicle to tilt even before its tires leave the road.

But wait! A vehicle rounding a curve is hardly in static equilibrium; after all, it's both moving and, more importantly, accelerating. But our analysis nevertheless applies, provided we recognize that the nonzero net force means we can no longer conclude that zero torque about one point implies zero torque about all other points. In this case, though, rotation tends to begin about the center of gravity, so our analysis involving that point is what's relevant here.

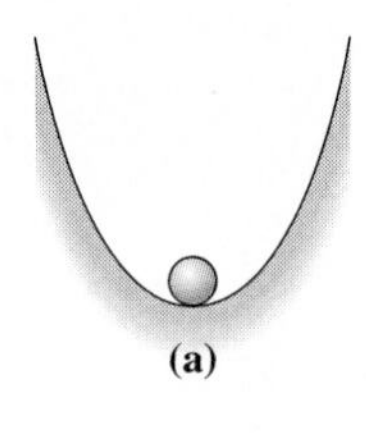

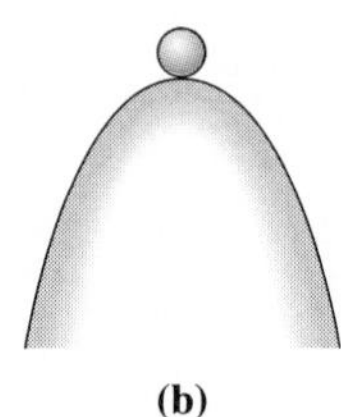

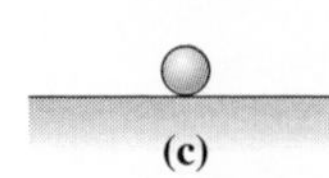

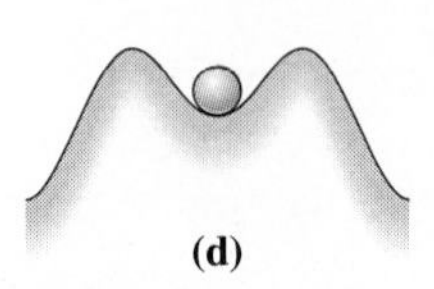

FIGURE 12.7 (a) Stable, (b) unstable, (c) neutrally stable, and (d) metastable equilibria.

Figure 12.7 shows a ball in four different equilibrium situations. Situation (a) is stable and (b) is unstable. Situation (c) is neither stable nor unstable; it's called **neutrally stable**. But what about (d)? For small disturbances, the ball will return to its original state, so the equilibrium is stable. But for larger disturbances—large enough to push the ball over the highest points on the hill—it's unstable. Such an equilibrium is **conditionally stable** or **metastable**.

A system disturbed from stable equilibrium can take a while to return to equilibrium. In Fig. 12.7*a*, for example, displacing the ball results in its rolling back and forth. Eventually friction dissipates its energy, and it comes to rest at equilibrium. Back-and-forth motion is common to many systems—from nuclei and atoms to skyscrapers and bridges—that are displaced from stable equilibrium. Such motion is the topic of the next chapter.

Stability is closely associated with potential energy. Because gravitational potential energy is directly proportional to height, the shapes of the hills and valleys in Fig. 12.7 are in fact potential-energy curves. In all cases of equilibrium, the ball is at a minimum or maximum of the potential-energy curve—at a place where the force (i.e., the derivative of potential energy with respect to position) is zero. For the stable and metastable equilibria, the potential energy at equilibrium is a local minimum. A deviation from equilibrium requires that work be done against the force that tends to restore the ball to equilibrium. The unstable equilibrium, in contrast, occurs at a maximum in potential energy. Here, a deviation from equilibrium results in lower potential energy and in a force that accelerates the ball farther from equilibrium. For the neutrally stable equilibrium, there's no change in potential energy as the ball moves; consequently it experiences no force. Figure 12.8 gives another example of equilibria in the context of potential energy.

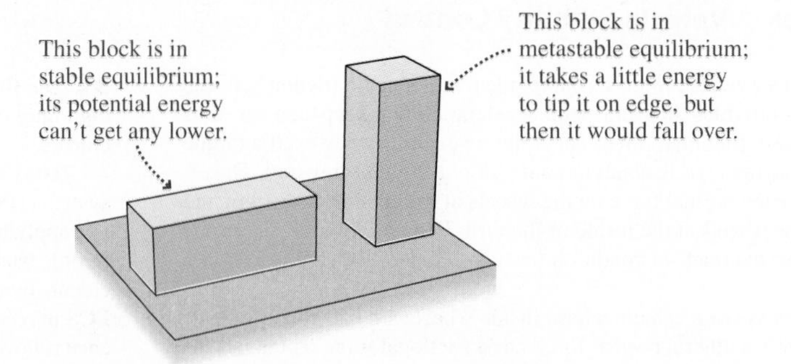

FIGURE 12.8 Identical blocks in stable and metastable equilibria.

We can sum up our understanding of equilibrium and potential energy in two simple mathematical statements. First, the force must be zero; that requires a local maximum or minimum in potential energy:

$$\frac{dU}{dx} = 0 \qquad \text{(equilibrium condition)} \qquad (12.3)$$

where U is the potential energy of a system and x is a variable describing the system's configuration. For the simple systems we've been considering, x measures the position or orientation of an object, but for more complicated systems, it could be another quantity such as the system's volume or even its composition. For a stable equilibrium, we require a local minimum, so the potential-energy curve is concave upward. (See Tactics 12.1 to review the relevant calculus.) Mathematically,

$$\frac{d^2U}{dx^2} > 0 \qquad \text{(stable equilibrium)} \tag{12.4}$$

This condition applies to metastable equilibria as well because they're *locally* stable. In contrast, unstable equilibrium occurs where the potential energy has a local maximum, or

$$\frac{d^2U}{dx^2} < 0 \qquad \text{(unstable equilibrium)} \tag{12.5}$$

The intermediate case $d^2U/dx^2 = 0$ corresponds to neutral stability.

TACTICS 12.1 Finding Maxima and Minima

1. Begin by sketching a plot of the function, which will give a visual check for your numerical answers.
2. Next, take the function's first derivative and set it to zero. As Fig. 12.7 suggests, a hill (maximum) or valley (minimum) is level right at its top or bottom. So by setting the first derivative to zero, you're requiring that its slope be zero and therefore requiring the function to be at a maximum or minimum.
3. Find the sign of the function's second derivative at the points where you found the first derivative is zero. Your sketch should show this; where the curve is concave upward, as in Figs. 12.7*a* and *d*, the second derivative is positive and the point is a minimum. Where it's concave downward, as in Fig. 12.7*b*, d^2U/dt^2 is negative and you've got a maximum. If it wasn't obvious how to sketch the function, you can use calculus to determine the second derivative and then find its sign at the equilibrium points.
4. Check that the values you found for maxima and minima agree with your plot of the function.

EXAMPLE 12.4 Stability Analysis: Semiconductor Engineering

Physicists develop a new semiconductor device in which the potential energy associated with an electron's being at position x is given by $U(x) = ax^2 - bx^4$, where x is in nm, U is the potential energy in aJ (10^{-18} J), and constants a and b are 8 aJ/nm^2 and 1 aJ/nm^4, respectively. Find the equilibrium positions for the electron, and describe their stability.

INTERPRET This problem is about stability in the context of a given potential-energy function. We're interested in the electron, and we're asked to find the values of x where it's in equilibrium and then examine their stability.

DEVELOP The potential-energy curve gives us insight into this problem, so we've drawn it by plotting the function $U(x)$ in Fig. 12.9. Equation 12.3, $dU/dx = 0$, determines the equilibria, while Equations 12.4, $d^2U/dx^2 > 0$, and 12.5, $d^2U/dx^2 < 0$, determine the stability. Our plan is first to find the equilibrium positions using Equation 12.3 and then to examine their stability.

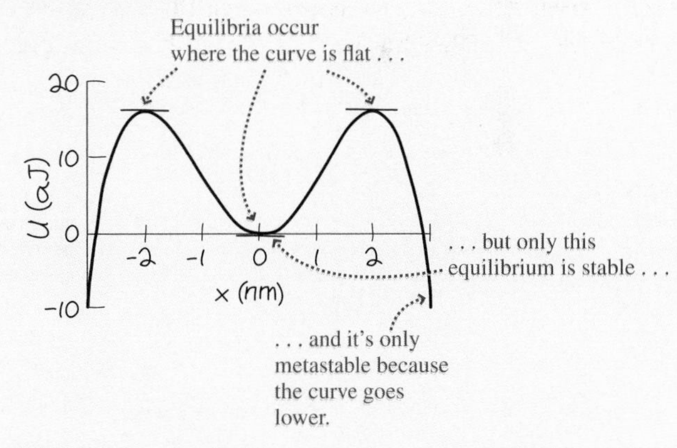

FIGURE 12.9 Our sketch of the potential-energy curve for Example 12.4.

EVALUATE Equation 12.3 states that equilibria occur where the potential energy has a maximum or minimum—that is, where its derivative is zero. Taking the derivative of U and setting it to zero gives

$$0 = \frac{dU}{dx} = 2ax - 4bx^3 = 2x(a - 2bx^2)$$

This equation has solutions when $x = 0$ and when $a = 2bx^2$ or $x = \pm\sqrt{a/2b} = \pm 2$ nm. We could take second derivatives to evaluate the stability, but the situation is evident from our plot: $x = 0$ lies at a local minimum of the potential-energy curve, so this equilibrium is metastable. The other two equilibria, at maxima of U, are unstable.

ASSESS Do our numerical answers make sense? Yes: The potential-energy curve has zero slope at the points $x = -2$ nm, $x = 0$, and $x = 2$ nm, so we've found all the equilibria. Note that the equilibrium at $x = 0$ is only metastable; given enough energy, an electron disturbed from this position could make it all the way over the peaks and never return to $x = 0$. ■

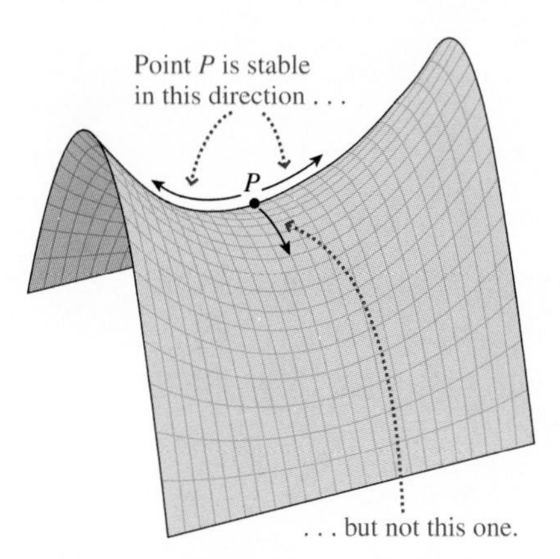

FIGURE 12.10 Equilibrium on a saddle-shaped potential-energy curve.

Stability considerations apply to the overall arrangements of matter. A mixture of hydrogen and oxygen, for example, is in metastable equilibrium at room temperature. Lighting a match puts some atoms over the maxima in their potential-energy curves, at which point they rearrange into a state of lower potential energy—the state we call H_2O. Similarly, a uranium nucleus is at a local minimum of its potential-energy curve, and a little excess energy can result in its splitting into two smaller nuclei whose total potential energy is much lower. That transition from a less stable to a more stable equilibrium describes the basic physics of nuclear fission.

Potential-energy curves for complex structures like molecules or skyscrapers can't be described fully with one-dimensional graphs. If potential energy varies in different ways when the structure is altered in different directions, then in order to determine stability we need to consider all possible ways potential energy might vary. For example, a snowball sitting on a mountain pass—or any other system with a saddle-shaped potential-energy curve—is stable against displacements in one direction but not another (Fig. 12.10). Stability analysis of complex physical systems, ranging from nuclei and molecules to bridges and buildings and machinery, and on to stars and galaxies, is an important part of contemporary work in engineering and science.

GOT IT? 12.4 Which of the labeled points in the figure are stable, metastable, unstable, or neutrally stable equilibria?

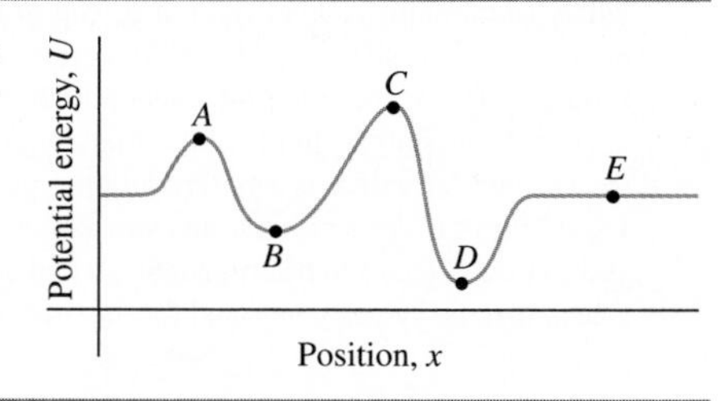

CHAPTER 12 SUMMARY

Big Idea

The big idea here is **static equilibrium**—the state in which a system at rest remains at rest because there's no net force to accelerate it and no net torque to start it rotating. An equilibrium is stable if a disturbance of the system results in its returning to the original equilibrium state.

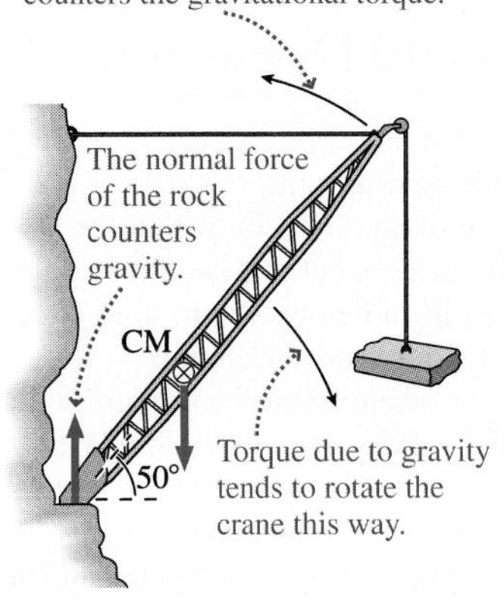

Key Concepts and Equations

Static equilibrium requires that there be no net force and no net torque on a system; mathematically:

$$\sum \vec{F}_i = \vec{0}$$

and

$$\sum \vec{\tau}_i = \sum (\vec{r}_i \times \vec{F}_i) = \vec{0}$$

where the sums include all the forces applied to the system. Solving an equilibrium problem involves identifying all the forces $\vec{F}_i$ acting on the system, choosing an appropriate origin about which to evaluate the torques, and requiring that forces and torques sum to zero.

Equilibria occur where a system's potential energy $U(x)$ has a maximum or a minimum:

$$\frac{dU}{dx} = 0 \quad \text{(equilibrium condition)}$$

$$\frac{d^2U}{dx^2} > 0 \quad \text{(stable equilibrium)}$$

$$\frac{d^2U}{dx^2} < 0 \quad \text{(unstable equilibrium)}$$

Stable equilibria occur at minima of U and unstable equilibria at maxima.

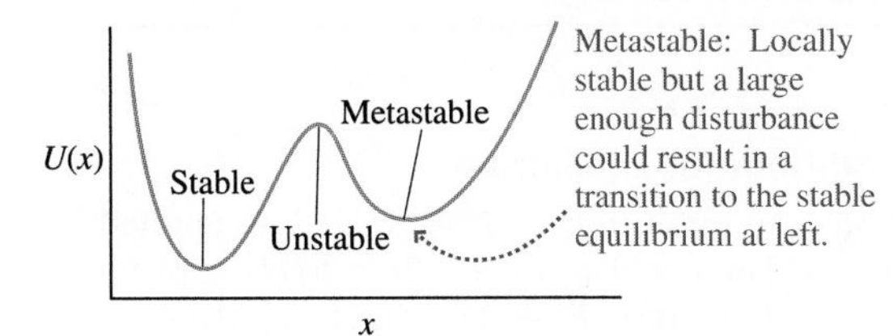

Applications

The **center of gravity** of a system is the point where the force of gravity appears to act. When the gravitational field is uniform over the system, the center of gravity coincides with the center of mass. This provides a handy way to locate the center of mass.

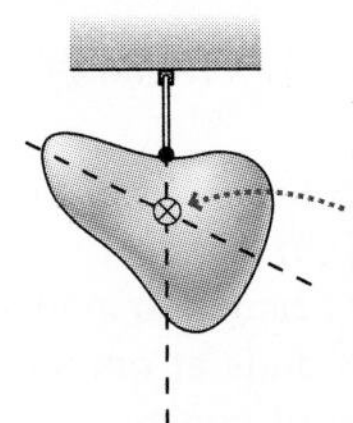

Four different types of equilibrium are **stable**, **unstable**, **neutrally stable**, and **metastable**.

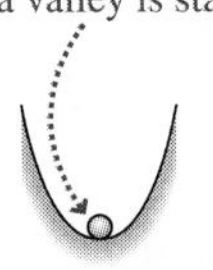

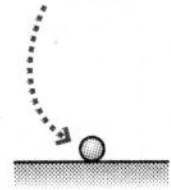

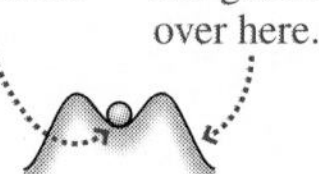

BIO Biology and/or medicine-related problems **DATA** Data problems **ENV** Environmental problems **CH** Challenge problems **COMP** Computer problems

For Thought and Discussion

1. Give an example of an object on which the net force is zero, but that isn't in static equilibrium.
2. Give an example of an object on which the net torque about the center of gravity is zero, but that isn't in static equilibrium.
3. The best way to lift a heavy weight is to squat with your back vertical, rather than to lean over. Why?
4. Pregnant women often assume a posture with their shoulders held far back from their normal position. Why?
5. When you carry a bucket of water with one hand, you instinctively extend your opposite arm. Why?
6. Is a ladder more likely to slip when you stand near the top or the bottom? Explain.
7. How does a heavy keel help keep a boat from tipping over?
8. Does choosing a pivot point in an equilibrium problem mean that something is necessarily going to rotate about that point?
9. If you take the pivot point at the application point of one force in a static-equilibrium problem, that force doesn't enter the torque equation. Does that make the force irrelevant to the problem? Explain.
10. A short dog and a tall person are standing on a slope. If the slope angle increases, which will fall over first? Why?
11. A stiltwalker is standing motionless on one stilt. What can you say about the location of the stiltwalker's center of mass?

Exercises and Problems

Exercises

Section 12.1 Conditions for Equilibrium

12. A body is subject to three forces: $\vec{F}_1 = 1\hat{\imath} + 2\hat{\jmath}$ N, applied at the point $x = 2$ m, $y = 0$ m; $\vec{F}_2 = -2\hat{\imath} - 5\hat{\jmath}$ N, applied at $x = -1$ m, $y = 1$ m; and $\vec{F}_3 = 1\hat{\imath} + 3\hat{\jmath}$ N, applied at $x = -2$ m, $y = 5$ m. Show that (a) the net force and (b) the net torque about the origin are both zero.
13. To demonstrate that the choice of pivot point doesn't matter, show that the torques in Exercise 12 sum to zero when evaluated about the points (3 m, 2 m) and (−7 m, 1 m).
14. In Fig. 12.11 the forces shown all have the same magnitude F. For each case shown, is it possible to place a third force so as to meet both conditions for static equilibrium? If so, specify the force and a suitable application point. If not, why not?

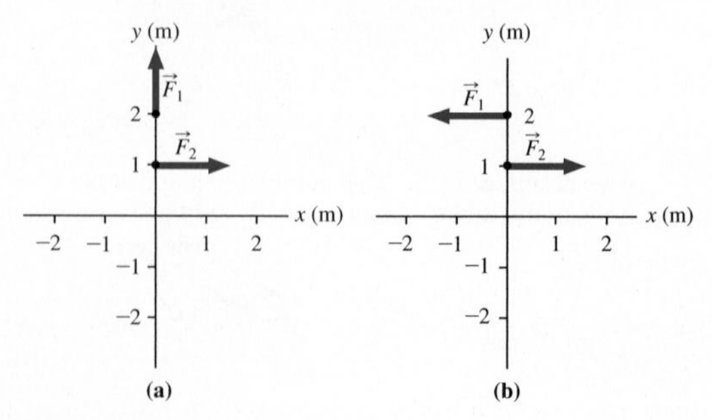

FIGURE 12.11 Exercise 14

Section 12.2 Center of Gravity

15. Figure 12.12*a* shows a thin, uniform square plate of mass m and side L. The plate is in a vertical plane. Find the magnitude of the gravitational torque on the plate about each of the three points shown.

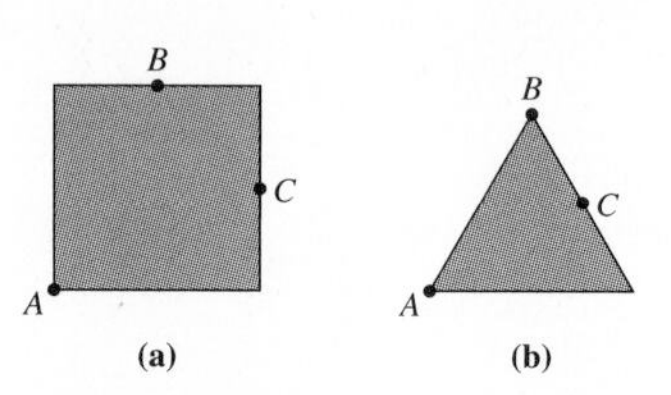

FIGURE 12.12 Exercises 15 and 16

16. Repeat the preceding problem for the equilateral triangle in Fig. 12.12*b*, which has side L.
17. A 23-m-long log of irregular cross section lies horizontally, supported by a wall at one end and a cable attached 4.0 m from the other end, as shown in Fig. 12.13. The log weighs 7.5 kN and the tension in the cable is 6.2 kN. Find the log's center of gravity.

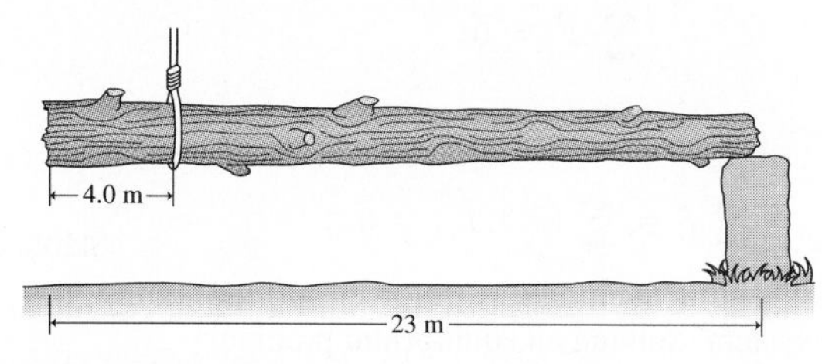

FIGURE 12.13 Exercise 17

Section 12.3 Examples of Static Equilibrium

18. A 60-kg uniform board 2.4 m long is supported by a pivot 80 cm from the left end and by a scale at the right end (Fig. 12.14). How far from the left end should a 40-kg child sit for the scale to read zero?

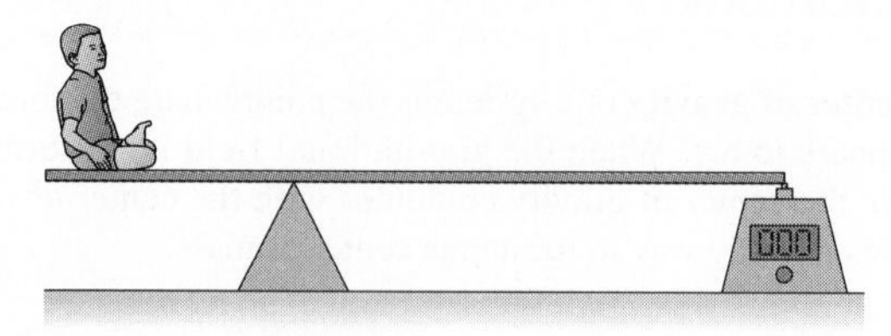

FIGURE 12.14 Exercises 18 and 19

19. Where should the child in Fig. 12.14 sit if the scale is to read (a) 100 N and (b) 300 N?
20. A 4.2-m-long beam is supported by a cable at its center. A 65-kg steelworker stands at one end of the beam. Where should a 190-kg bucket of concrete be suspended for the beam to be in static equilibrium?
21. Figure 12.15 shows how a scale with a capacity of only 250 N can be used to weigh a heavier person. The 3.4-kg board is 3.0 m long and has uniform density. It's free to pivot about the end farthest from the scale. Assume that the beam remains essentially

horizontal. What's the weight of a person standing 1.2 m from the pivot end if the scale reads 210 N?

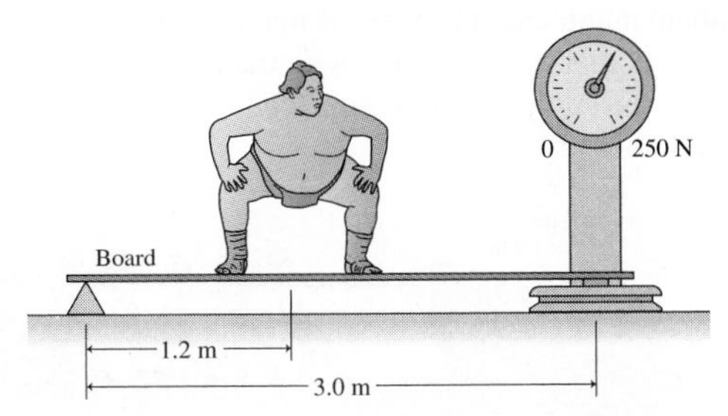

FIGURE 12.15 Exercise 21

Section 12.4 Stability

22. A portion of a roller-coaster track is described by the equation $h = 0.94x - 0.010x^2$, where h and x are the height and horizontal position in meters. (a) Find a point where the roller-coaster car could be in static equilibrium on this track. (b) Is this equilibrium stable or unstable?
23. The potential energy associated with a particle at position x is given by $U = 2x^3 - 2x^2 - 7x + 10$, with x in meters and U in joules. Find the positions of any stable and unstable equilibria.

Problems

24. You're a highway safety engineer, and you're asked to specify bolt sizes so the traffic signal in Fig. 12.16 won't fall over. The figure indicates the masses and positions of the structure's various parts. The structure is mounted with two bolts, located symmetrically about the vertical member's centerline, as shown. What tension force must the left-hand bolt be capable of withstanding?

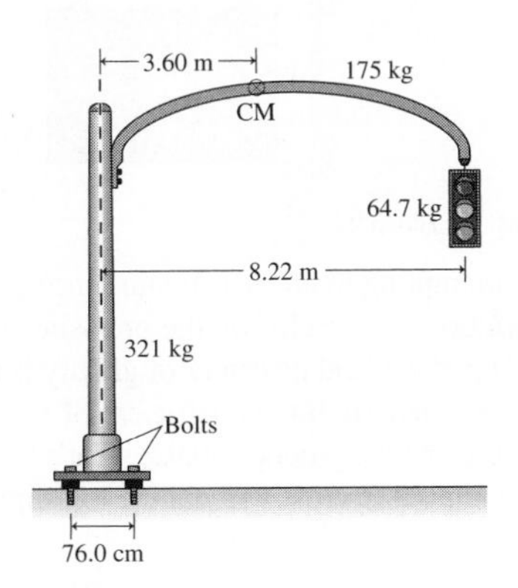

FIGURE 12.16 Problem 24

25. **BIO** Figure 12.17*a* shows an outstretched arm with mass 4.2 kg. The arm is 56 cm long, and its center of gravity is 21 cm from the shoulder. The hand at the end of the arm holds a 6.0-kg mass. (a) Find the torque about the shoulder due to the weight of the arm and the 6.0-kg mass. (b) If the arm is held in equilibrium by the deltoid muscle, whose force on the arm acts below the horizontal at a point 18 cm from the shoulder joint (Fig. 12.17*b*), what's the force exerted by the muscle?

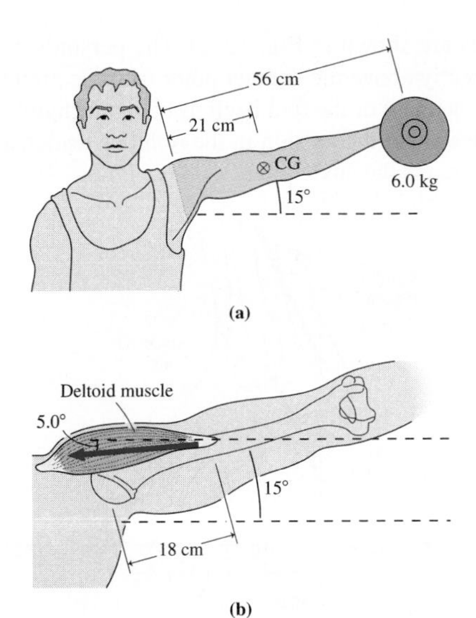

FIGURE 12.17 Problem 25

26. A uniform sphere of radius R is supported by a rope attached to a vertical wall, as shown in Fig. 12.18. The rope joins the sphere at a point where a continuation of the rope would intersect a horizontal line through the sphere's center a distance $\frac{1}{2}R$ beyond the center, as shown. What's the smallest possible value for the coefficient of friction between wall and sphere?

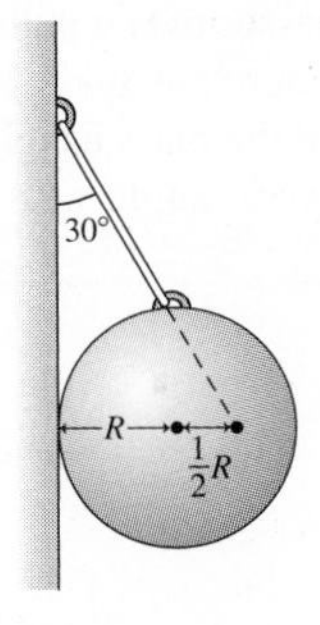

FIGURE 12.18 Problem 26

27. **CH** You work for a garden equipment company, and you're designing a new garden cart. Specifications to be listed include the horizontal force that must be applied to push the fully loaded cart (mass 55 kg, 60-cm-diameter wheels) up an abrupt 8.0-cm step, as shown in Fig. 12.19. What do you specify for the force?

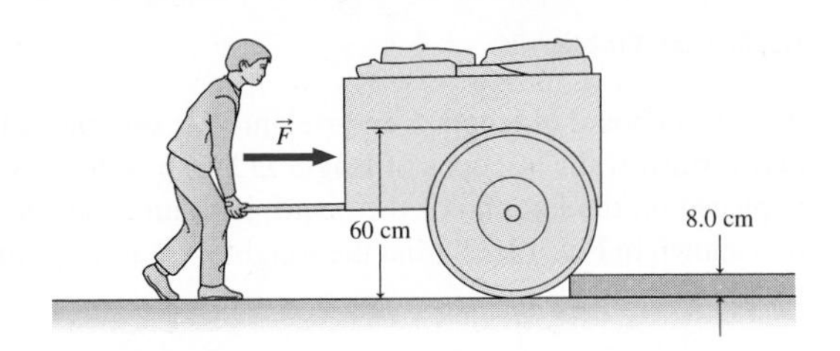

FIGURE 12.19 Problem 27

28. **BIO** Figure 12.20 shows the foot and lower leg of a person standing on the ball of one foot. Three forces act to maintain this equilibrium: the tension force $\vec{T}$ in the Achilles tendon, the contact force $\vec{F}_c$ at the ankle joint, and the normal force $\vec{n}$ that supports the person's 697-N weight. The application points for these

forces are shown in Fig. 12.20. The person's center of gravity is directly above the contact point with the ground, and you can treat the mass of the foot itself as being negligible. Find the magnitudes of (a) the tension in the Achilles tendon and (b) the contact force at the ankle joint.

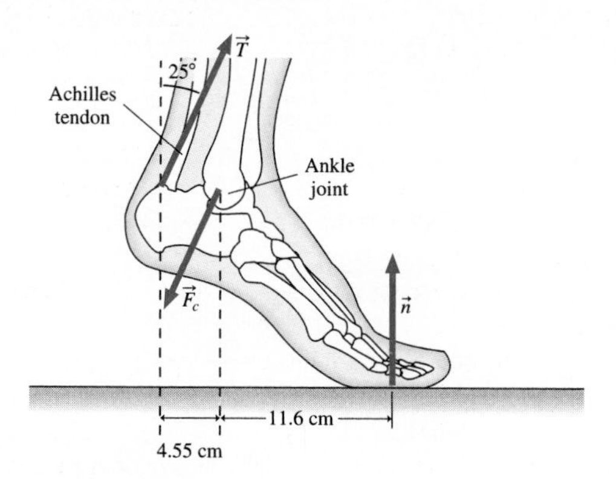

FIGURE 12.20 Problem 28

29. A uniform 5.0-kg ladder is leaning against a frictionless vertical wall, with which it makes a 15° angle. The coefficient of friction between ladder and ground is 0.26. Can a 65-kg person climb to the top of the ladder without it slipping? If not, how high can that person climb? If so, how massive a person would make the ladder slip?
30. The boom in the crane of Fig. 12.21 is free to pivot about point P and is supported by the cable attached halfway along its 18-m length. The cable passes over a pulley and is anchored at the back of the crane. The boom has mass 1700 kg distributed uniformly along its length, and the mass hanging from the boom is 2200 kg. The boom makes a 50° angle with the horizontal. Find the tension in the cable.

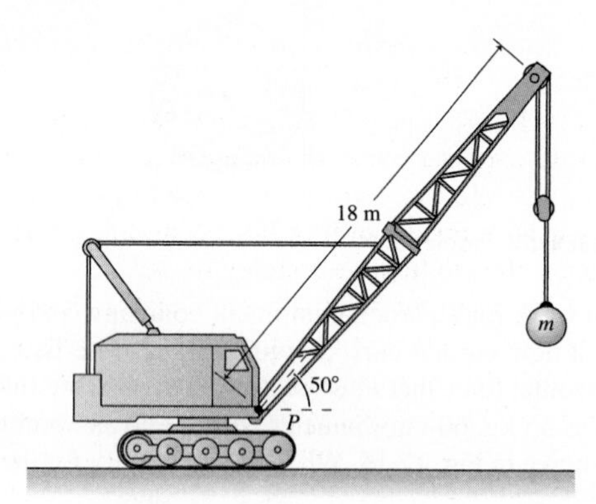

FIGURE 12.21 Problem 30

31. CH A uniform board of length L and weight W is suspended between two vertical walls by ropes of length $L/2$ each. When a weight w is placed on the left end of the board, it assumes the configuration shown in Fig. 12.22. Find the weight w in terms of the board weight W.

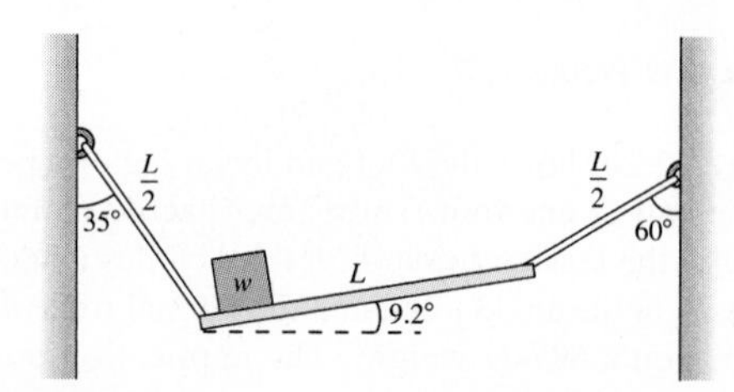

FIGURE 12.22 Problem 31

32. Figure 12.23 shows a 1250-kg car that has slipped over an embankment. People are trying to hold the car in place by pulling on a horizontal rope. The car's bottom is pivoted on the edge of the embankment, and its center of mass lies farther back, as shown. If the car makes a 34° angle with the horizontal, what force must the people apply to hold it in place?

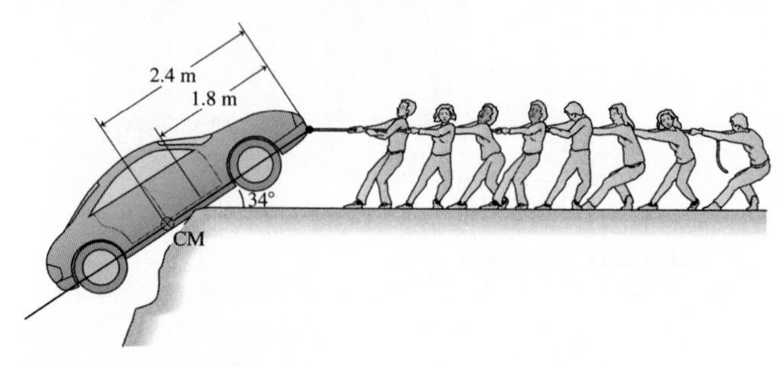

FIGURE 12.23 Problem 32

33. CH Repeat Example 12.2, now assuming that the coefficient of friction at the ground is μ_1 and at the wall is μ_2. Show that the minimum angle at which the ladder won't slip is now given by $\phi = \tan^{-1}[(1 - \mu_1\mu_2)/2\mu_1]$.
34. You are headwaiter at a new restaurant, and your boss asks you to hang a sign for her. You're to hang the sign, whose mass is 66 kg, in the configuration shown in Fig. 12.24. A uniform horizontal rod of mass 8.2 kg and length 2.3 m holds the sign. At one end the rod is attached to the wall by a pivot; at the other end it's supported by a cable that can withstand a maximum tension of 800 N. You're to determine the minimum height h above the pivot for anchoring the cable to the wall.

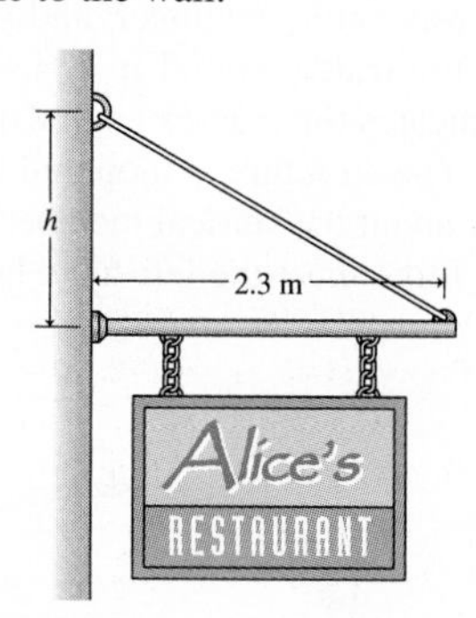

FIGURE 12.24 Problem 34

35. Climbers attempting to cross a stream place a 340-kg log against a vertical, frictionless ice cliff on the opposite side (Fig. 12.25). The log slopes up at 27° and its center of gravity is one-third of the way along its 6.3-m length. If the coefficient of friction between the left end of the log and the ground is 0.92, what's the maximum mass for a climber and pack to cross without the log slipping?

FIGURE 12.25 Problem 35

36. A crane in a marble quarry is mounted on the quarry's rock walls and is supporting a 2500-kg marble slab as shown in Fig. 12.26. The center of mass of the 830-kg boom is located one-third of the way from the pivot end of its 15-m length, as shown. Find the tension in the horizontal cable that supports the boom.

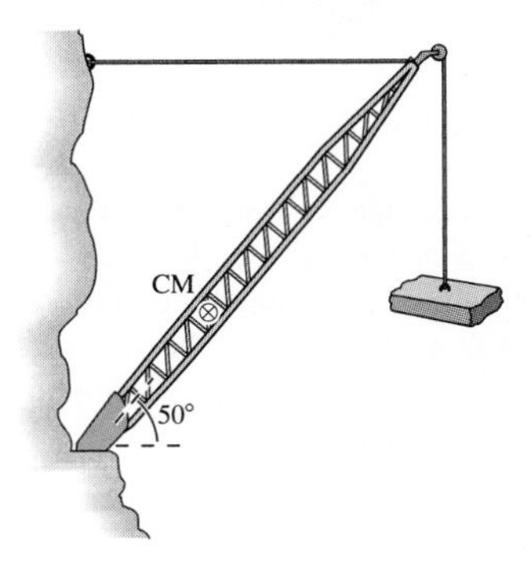

FIGURE 12.26 Problem 36

37. A rectangular block measures $w \times w \times L$, where L is the longer dimension. It's on a horizontal surface, resting on its long side. Use geometrical arguments to find an expression for the angle through you would have to tilt it in order to put it in an unstable equilibrium, resting on a short edge.

38. The potential energy as a function of position for a particle is given by

$$U(x) = U_0\left(\frac{x^3}{x_0^3} + a\frac{x^2}{x_0^2} + 4\frac{x}{x_0}\right)$$

where x_0 and a are constants. For what values of a will there be two static equilibria? Comment on the stability of these equilibria.

39. **CH** A rectangular block of mass m measures $w \times w \times L$, where L is the longer dimension. It's on a horizontal surface, resting on its long side, as in the left-hand block in Fig. 12.8. (a) Taking the zero of potential energy when the block is lying on its long side, find an expression for its potential energy as a function of the angle θ that the long dimension of the block makes with the horizontal, starting with $\theta = 0$ in the left-hand configuration of Fig. 12.8 and continuing through the upright position shown at the right ($\theta = 90°$). (b) Use calculus to find the angle θ where your function has a maximum, and check that it agrees with the answer to Problem 37. (c) Use calculus to show that this is a point of unstable equilibrium.

40. A 160-kg highway sign of uniform density is 2.3 m wide and 1.4 m high. At one side it's secured to a pole with a single bolt, mounted a distance d from the top of the sign. The only other place where the sign contacts the pole is at its bottom corner. If the bolt can sustain a horizontal tension of 2.1 kN, what's the maximum permissible value for the distance d?

41. A 5.0-m-long ladder has mass 9.5 kg and is leaning against a frictionless wall, making a 66° angle with the horizontal. If the coefficient of friction between ladder and ground is 0.42, what's the mass of the heaviest person who can safely ascend to the top of the ladder? (The center of mass of the ladder is at its center.)

42. To what vertical height on the ladder in Problem 41 could a 95-kg person reach before the ladder starts to slip?

43. A uniform, solid cube of mass m and side s is in stable equilibrium when sitting on a level tabletop. How much energy is required to bring it to an unstable equilibrium where it's resting on its corner?

44. **CH** An isosceles triangular block of mass m and height h is in stable equilibrium, resting on its base on a horizontal surface. How much energy does it take to bring it to unstable equilibrium, resting on its apex?

45. **CH** You're investigating ladder safety for the Consumer Product Safety Commission. Your test case is a uniform ladder of mass m leaning against a frictionless vertical wall with which it makes an angle θ. The coefficient of static friction at the floor is μ. Your job is to find an expression for the maximum mass of a person who can climb to the top of the ladder without its slipping. With that result, you're to show that *anyone* can climb to the top if $\mu \geq \tan\theta$ but that *no one* can if $\mu < \frac{1}{2}\tan\theta$.

46. **CH** A 2.0-m-long rod has density λ in kilograms per meter of length described by $\lambda = a + bx$, where $a = 1.0$ kg/m, $b = 1.0$ kg/m^2, and x is the distance from the left end of the rod. The rod rests horizontally with each end supported by a scale. What do the two scales read?

47. What horizontal force applied at its highest point is necessary to keep a wheel of mass M from rolling down a slope inclined at angle θ to the horizontal?

48. A rectangular block twice as high as it is wide is resting on a board. The coefficient of static friction between board and incline is 0.63. If the board's inclination angle θ (shown in Fig. 12.27) is gradually increased, will the block first tip over or first begin sliding?

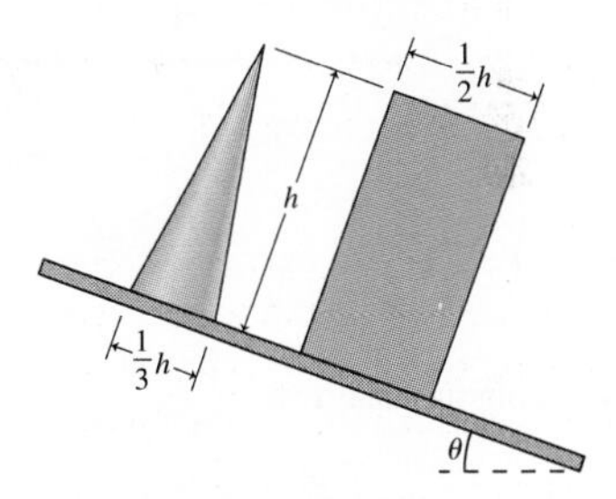

FIGURE 12.27 Problems 48, 49, and 50

49. What condition on the coefficient of friction in Problem 48 will cause the block to slide before it tips?

50. **CH** A uniform solid cone of height h and base diameter $\frac{1}{3}h$ sits on the board of Fig. 12.27. The coefficient of static friction between the cone and incline is 0.63. As the slope of the board is increased, will the cone first tip over or first begin sliding? (*Hint:* Start with an integration to find the center of mass.)

51. Prove the statement in Section 12.1 that the choice of pivot point doesn't matter when applying conditions for static equilibrium. Figure 12.28 shows an object on which the net force is assumed to be zero. The net torque about the point O is also zero. Show that the net torque about any other point P is also zero. To do so, write the net torque about P as $\vec{\tau}_P = \sum \vec{r}_{Pi} \times \vec{F}_i$, where the vectors $\vec{r}_P$ go from P to the force-application points, and the index i labels the different forces. In Fig. 12.28, note that $\vec{r}_{Pi} = \vec{r}_{Oi} + \vec{R}$, where $\vec{R}$ is a vector from P to O. Use this result in your expression for $\vec{\tau}_P$ and apply the distributive law to get two separate sums. Use the assumptions that $\vec{F}_{\text{net}} = \vec{0}$ and $\vec{\tau}_O = \vec{0}$ to argue that both terms are zero. This completes the proof.

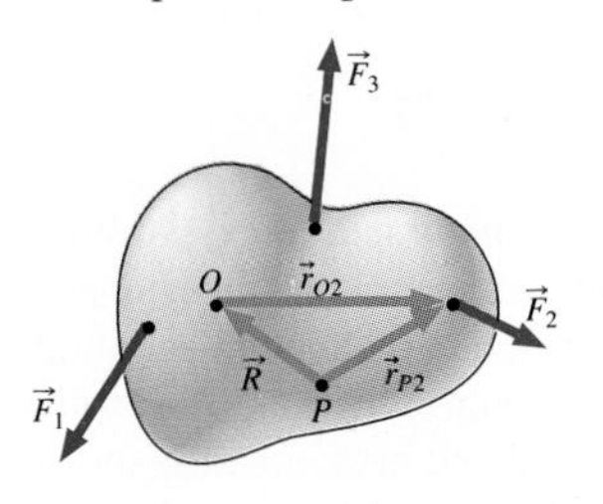

FIGURE 12.28 Problem 51

52. Three identical books of length L are stacked over the edge of a table as shown in Fig. 12.29. The top book overhangs the middle one by $L/2$, so it just barely avoids falling. The middle book overhangs the bottom one by $L/4$. How much of the bottom book can overhang the edge of the table without the books falling?

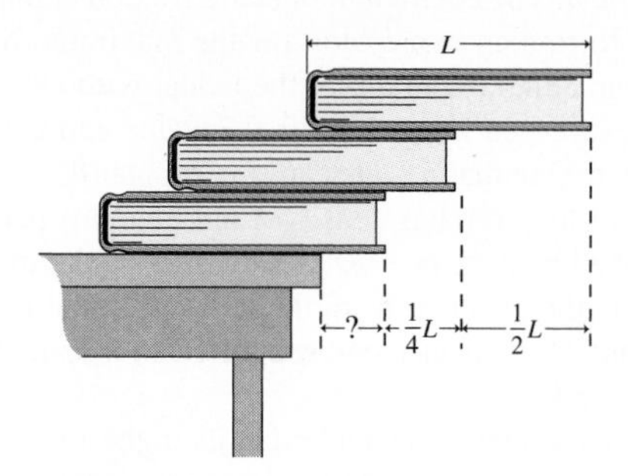

FIGURE 12.29 Problem 52

53. A uniform pole of mass M is at rest on an incline of angle θ secured by a horizontal rope as shown in Fig. 12.30. Find the minimum frictional coefficient that will keep the pole from slipping.

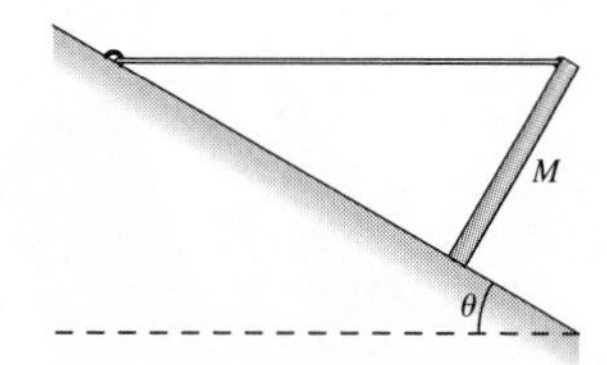

FIGURE 12.30 Problems 53 and 54

54. **CH** For what angle does the situation in Problem 53 require the greatest coefficient of friction?

55. Figure 12.31 shows a popular system for mounting bookshelves. An aluminum bracket is mounted on a vertical aluminum support by small tabs inserted into vertical slots. Contact between the bracket and support occurs only at the upper tab and at the bottom of the bracket, 4.5 cm below the upper tab. If each bracket in the shelf system supports 32 kg of books, with the center of gravity 12 cm out from the vertical support, what is the horizontal component of the force exerted on the upper bracket tab?

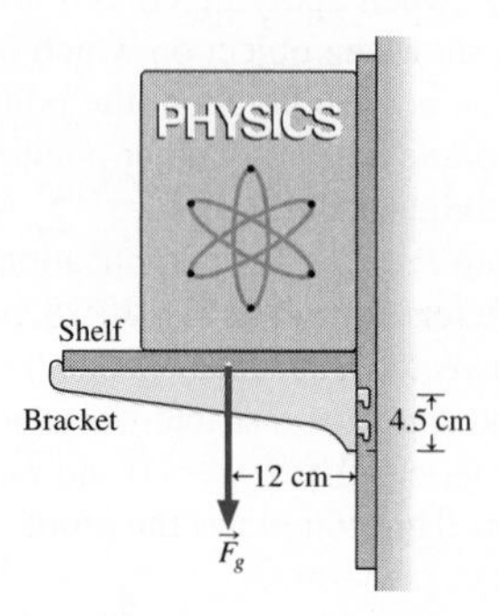

FIGURE 12.31 Problem 55

56. **BIO** The *nuchal ligament* is a thick, cordlike structure that supports the head and neck in animals like horses. Figure 12.32 shows the nuchal ligament and its attachment points on a horse's skeleton, along with an approximation to the spine as a rigid rod. Centers of mass of head and neck are also shown. If the masses of head and neck are 29 kg and 68 kg, respectively, what's the tension in the nuchal ligament? (*Note*: Your answer will be an overestimate because muscles also provide support.)

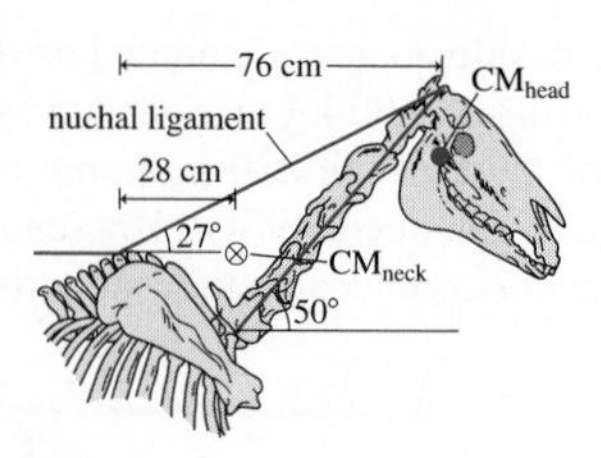

FIGURE 12.32 Problem 56

57. A 4.2-kg plant hangs from the bracket shown in Fig. 12.33. The bracket's mass is 0.85 kg, and its center of mass lies 9.0 cm from the wall. A single screw holds the bracket to the wall, as shown. Find the horizontal tension in the screw. (*Hint:* Imagine that the bracket is slightly loose and pivoting about its bottom end. Assume the wall is frictionless.)

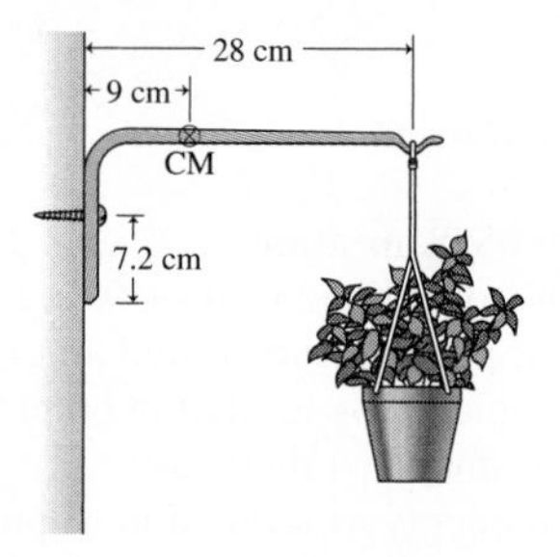

FIGURE 12.33 Problem 57

58. The wheel in Fig. 12.34 has mass M and is weighted with an additional mass m as shown. The coefficient of friction is sufficient to keep the wheel from sliding; however, it might still roll. Show that it won't roll only if $m > \dfrac{M \sin\theta}{1 - \sin\theta}$.

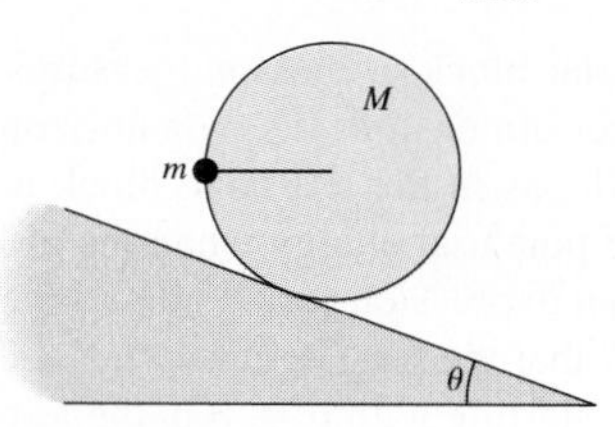

FIGURE 12.34 Problem 58

59. **CH** An interstellar spacecraft from an advanced civilization is hovering above Earth, as shown in Fig. 12.35. The ship consists of two pods of mass m separated by a rigid shaft of negligible mass and one Earth radius (R_E) long. Find (a) the magnitude and direction of the net gravitational force on the ship and (b) the net torque about the center of mass. (c) Show that the ship's center of gravity is displaced approximately $0.083R_E$ from its center of mass.

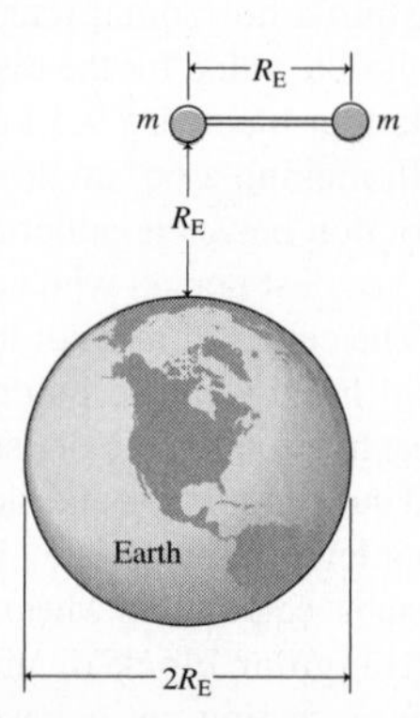

FIGURE 12.35 Problem 59

60. You're called to testify in a product liability lawsuit. An infant sitting in the portable seat shown in Fig. 12.36 was injured when it fell to the floor. The manufacturer claims the child was too heavy for the seat; the parents claim the seat was defective. Tests show that the seat can safely hold a child if the force $\vec{F}$ of the table on the chair doesn't exceed 229 N. The seat's mass is 2.0 kg, the injured child's is 10 kg, and the center of mass of child and seat was 16 cm from the table edge. In whose favor should the jury rule?

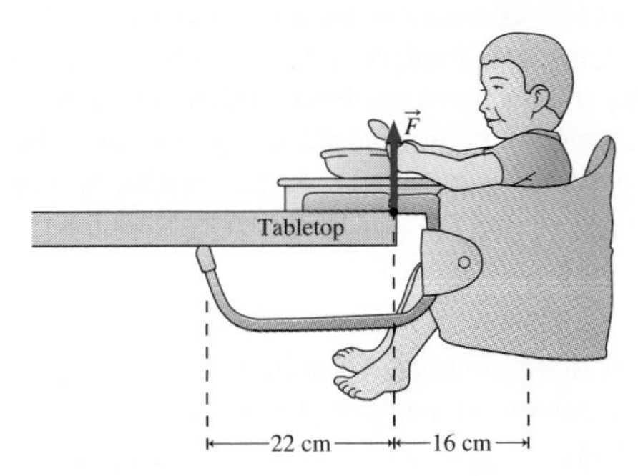

FIGURE 12.36 Problem 60

61. You're designing a vacation cabin at a ski resort. The cabin has a cathedral ceiling as shown in Fig. 12.37, and you estimate that each roof rafter needs to support up to 170 kg of snow and building materials. The horizontal tie beam near the roof peak can withstand a 7.5-kN force. You can neglect any horizontal force from the vertical walls, and treat contact forces as concentrated at the roof peak and the outside edge of the rafter/wall junctions. Will the tie beam hold? Will it be in tension or compression?

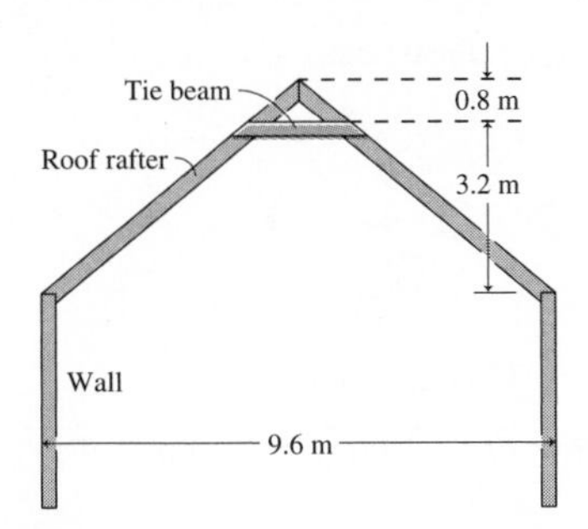

FIGURE 12.37 Problem 61

62. You'll need to study the Application on page 209 to do this problem. An SUV without ECS has SSF = 1.12 with its two passengers on board. (a) Can it successfully negotiate an 85-m-radius turn on a flat road, going at the speed limit of 100 km/h? (b) With its passengers, the SUV's total mass is 1940 kg, and the left-to-right spacing between its tires is 1.71 m. If a 315-kg load of cargo is secured to the roof, with its center of gravity 2.1 m above the road, what's the maximum safe speed on the same road?

63. **DATA** Engineers designing a new semiconductor device measure the potential energy that results when they move an electron to different positions within their device. The device is one-dimensional, so the positions all lie along a line. The table below gives the resulting data. Plot these data and from your plot, determine the approximate positions of any equilibria and whether such equilibria are stable or unstable.

Position x (nm)	0	3.26	5.85	6.41	7.12	9.37	10.5	12.2	14.0	14.5	15.3	17.2
Potential energy U (aJ)	1.5	0.65	0.30	0.47	0.85	2.7	3.3	2.1	−0.47	−0.86	−0.72	3.2

Passage Problems

You've been hired by your state's environmental agency to monitor carbon dioxide levels just above rivers, with the goal of understanding whether river water acts as a source or sink of CO_2. You've constructed the apparatus shown in Fig. 12.38, consisting of a boom mounted on a pivot, a vertical support, and a rope with pulley for raising and lowering the boom so its end can extend different distances over the river. In addition, there's a separate rope and pulley for dropping the sampling apparatus so it's just above the river.

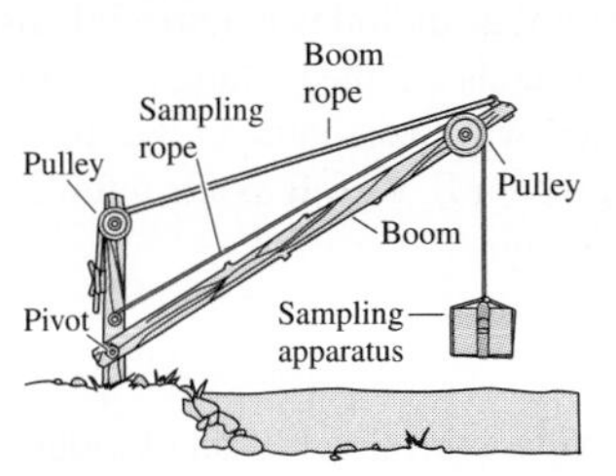

FIGURE 12.38 Passage Problems 64–67

64. When the boom rope is horizontal, it can't exert any vertical force. Therefore,
    a. it's impossible to hold the boom with the boom rope horizontal.
    b. the boom rope tension becomes infinite.
    c. the pivot supplies the necessary vertical force.
    d. the boom rope exerts no torque.
65. The tension in the boom rope will be greatest when
    a. the boom is horizontal.
    b. the boom rope is horizontal.
    c. the boom is vertical.
    d. in some orientation other than (a), (b), or (c).
66. If you secure the boom at a fixed angle and lower the sampling apparatus at constant speed, the boom rope tension will
    a. increase.
    b. decrease.
    c. remain the same.
    d. increase only if the sampling apparatus is more massive than the boom.
67. If you pull the boom rope with constant speed, the angle the boom makes with the horizontal will
    a. increase at a constant rate.
    b. increase at an increasing rate.
    c. increase at a decreasing rate.
    d. decrease.

Answers to Chapter Questions

Answer to Chapter Opening Question

Both the net force and the net torque on all parts of the bridge must be zero.

Answers to GOT IT? Questions

12.1 Pair C; pair A produces nonzero net force, while pair B produces nonzero net torque

12.2 B; It's located directly over the point of contact with the floor, ensuring there's no gravitational torque.

12.3 (b) A frictional force at the floor is necessary to balance the normal force from the wall.

12.4 D: stable; B: metastable; A and C: unstable; E: neutrally stable

PART ONE SUMMARY

Mechanics

The big idea of Part One is Newton's realization that forces—pushes and pulls—don't cause motion but instead cause *changes* in motion. Newton's second law quantifies this idea. With momentum $\vec{p} = m\vec{v}$ as Newton's measure of "quantity of motion," the second law equates the net force on an object to the rate of change of its momentum: $\vec{F} = d\vec{p}/dt$ or, for constant mass, $\vec{F} = m\vec{a}$. The second law encompasses the first law, also called the law of inertia: In the absence of a net force, an object continues in uniform motion, unchanging in speed or direction—a state that includes the special case of being at rest. Newton's third law rounds out the picture, providing a fully consistent description of motion with its statement that forces come in pairs: If object A exerts a force on B, then B exerts a force of equal magnitude but opposite direction on A.

From the concept of force and Newton's laws follow the essential ideas of energy and work, including kinetic and potential energy and the conservation of mechanical energy in the absence of nonconservative forces like friction. One important force is gravity, which Newton described through his law of universal gravitation and applied to explain the motions of the planets. Application of Newton's laws to systems comprising multiple objects gives us the concept of center of mass and lets us describe the interactions of colliding objects. Finally, Newton's laws explain circular and rotational motion, the latter through the analogy between force and torque. That, in turn, gives us the tools needed to determine static equilibrium—the state in which an object at rest remains at rest, subject neither to a net force nor to a net torque.

Newton's laws provide a full description of motion.
Newton's first law: Force causes a change in motion.
Newton's second law: $\vec{F} = d\vec{p}/dt$ or, for constant mass, $\vec{F} = m\vec{a}$
Newton's third law: $\vec{F}_{AB} = -\vec{F}_{BA}$

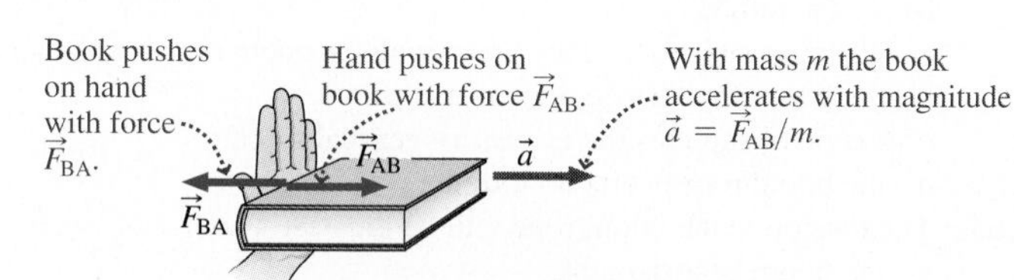

Energy and **work** are related concepts; work is a mechanical means of transferring energy.

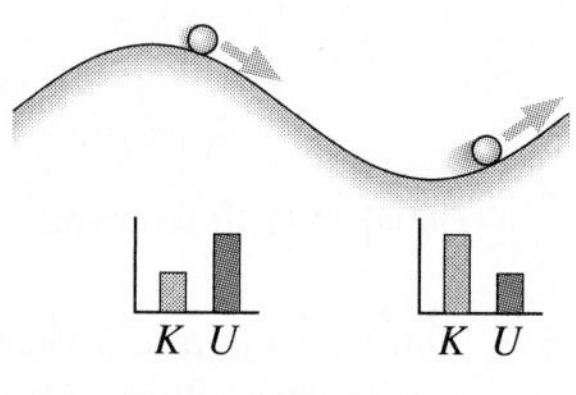

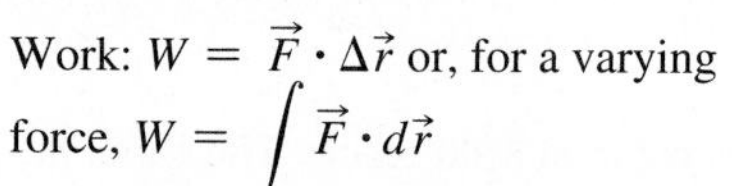

Work: $W = \vec{F} \cdot \Delta\vec{r}$ or, for a varying force, $W = \int \vec{F} \cdot d\vec{r}$

Work–kinetic energy theorem: $\Delta K = W$ with kinetic energy $K = \frac{1}{2}mv^2$

For conservative forces, energy that gets transferred by doing work is stored as potential energy U. Then $K + U =$ constant.

Universal gravitation describes the attractive force between all matter in the universe.

$$F = \frac{Gm_1m_2}{r^2}$$

r

m_1 $\vec{F}_{21}$ $\vec{F}_{12}$ m_2

Momentum is conserved in a system that's not subject to external forces.

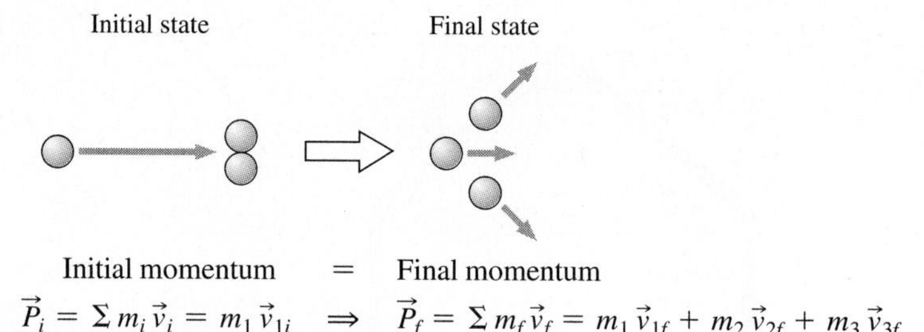

Initial momentum = Final momentum

$$\vec{P}_i = \Sigma m_i\vec{v}_i = m_1\vec{v}_{1i} \quad \Rightarrow \quad \vec{P}_f = \Sigma m_f\vec{v}_f = m_1\vec{v}_{1f} + m_2\vec{v}_{2f} + m_3\vec{v}_{3f}$$

Rotational motion is described by quantities analogous to those of linear motion.

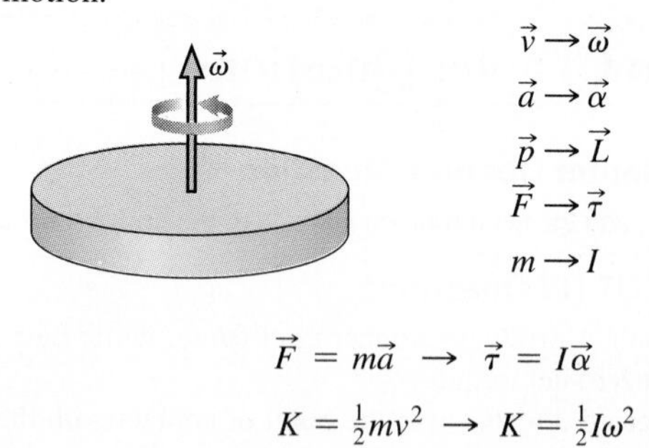

$\vec{v} \rightarrow \vec{\omega}$
$\vec{a} \rightarrow \vec{\alpha}$
$\vec{p} \rightarrow \vec{L}$
$\vec{F} \rightarrow \vec{\tau}$
$m \rightarrow I$

$$\vec{F} = m\vec{a} \rightarrow \vec{\tau} = I\vec{\alpha}$$

$$K = \tfrac{1}{2}mv^2 \rightarrow K = \tfrac{1}{2}I\omega^2$$

A system is in **static equilibrium** when the net force and the net torque on the system are both zero:

$\vec{F}_{net} = \vec{0}$

and

$\vec{\tau}_{net} = \vec{0}$

Stable equilibrium

Unstable equilibrium

Part One Challenge Problem

A solid ball of radius R is set spinning with angular speed ω about a horizontal axis. The ball is then lowered vertically with negligible speed until it just touches a horizontal surface and is released (see figure). If the coefficient of kinetic friction between the ball and the surface is μ, find (a) the linear speed of the ball once it achieves pure rolling motion, (b) the distance it travels before it achieves this motion, and (c) the fraction of the ball's initial rotational kinetic energy that's been lost to friction.

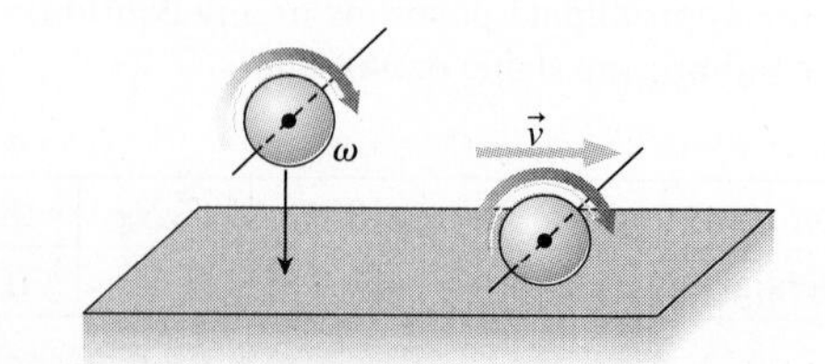

PART TWO

OVERVIEW

Oscillations, Waves, and Fluids

High-speed photo shows complex fluid behavior and spreading circular waves on water.

A tsunami crashes on shore, dissipating energy that has traveled across thousands of kilometers of open ocean. Near the epicenter of the earthquake that spawned the tsunami, a skyscraper sways in response but suffers no damage thanks to a carefully engineered system that counters quake-induced vibrations. An electric guitar sounds loud during a rock concert, the sound waves following the vibrations of the guitar strings. Inside your watch, a tiny quartz crystal vibrates 32,768 times each second to keep near-perfect time. A radar-equipped police officer waits around the next turn in the highway ready to ticket your speeding car, while astrophysicists use the same principle to measure the expansion of the universe. A rafting party enters a narrow gorge, getting a wild ride as the river's speed increases. A plane cruises far overhead, supported by the force of air on its wings. All these examples involve the collective motion of many particles. In the next three chapters, we first explore the repetitive motion called oscillation and then show how oscillations in many-particle systems lead to wave motion. Finally, we apply the laws of motion to reveal the fascinating and sometimes surprising behavior of fluids like air and water.

13

Oscillatory Motion

What You Know

- You understand Newton's second law.
- You're familiar with ideal springs.
- You can describe circular and rotational motion in terms of angular velocity, revolutions per second, or by giving the period of the motion.
- In calculus, you've learned to differentiate the sine and cosine functions.

What You're Learning

- Here you'll see how displacing a system from stable equilibrium usually results in oscillatory motion.
- You'll learn about the special case of *simple harmonic motion*, in which the force or torque tending to restore equilibrium is directly proportional to the displacement.
- You'll see why simple harmonic motion is ubiquitous in systems ranging from molecules to engineered structures and on to stars.
- You'll explore the relationship between simple harmonic motion and circular motion.
- You'll learn about *resonance*.

How You'll Use It

- In Chapter 14 you'll see how oscillations, coupled between adjacent parts of a system to another, lead to waves.
- In Part 4 you'll see electrical analogs of the oscillating systems described here.
- In Part 6 you'll learn how quantum physics describes systems undergoing simple harmonic motion.

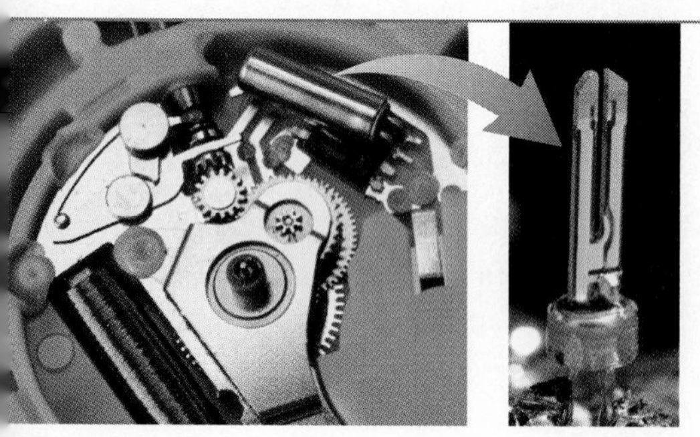

A tiny quartz tuning fork sets the timekeeping of a quartz watch. It oscillates at 32,768 Hz. What does this mean, and why this number?

Displace a system from stable equilibrium, and forces or torques tend to restore that equilibrium. But, like the ball in Fig. 13.1, the system often overshoots its equilibrium and goes into **oscillatory motion** back and forth about equilibrium. In the absence of friction, this oscillation would continue forever; in reality, the system eventually settles into equilibrium.

Oscillatory motion occurs throughout the physical world. A uranium nucleus oscillates before it fissions. Water molecules oscillate to heat the food in a microwave oven. Carbon dioxide molecules in the atmosphere oscillate, absorbing energy and thus contributing to global warming. A watch—whether an old-fashioned mechanical one or a modern quartz timepiece—is a carefully engineered oscillating system. Buildings and bridges undergo oscillatory motion, sometimes with disastrous results. Even stars oscillate. And waves—from sound to ocean waves to seismic waves in the solid Earth—ultimately involve oscillatory motion.

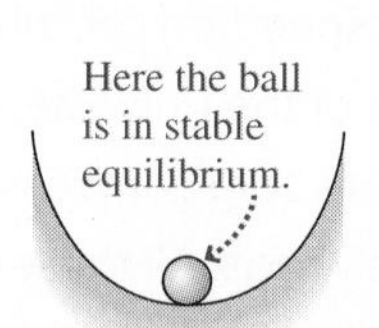

FIGURE 13.1 Disturbing a system results in oscillatory motion.

Oscillatory motion is universal because systems in stable equilibrium naturally tend to return toward equilibrium when they're displaced. And it's not just the qualitative phenomenon of oscillation that's universal: Remarkably, the mathematical description of oscillatory motion is the same for systems ranging from atoms and molecules to cars and bridges and on to stars and galaxies.

13.1 Describing Oscillatory Motion

Figure 13.2 shows two quantities that characterize oscillatory motion: **Amplitude** is the maximum displacement from equilibrium, and **period** is the time it takes for the motion to repeat itself. Another way to express the time aspect is **frequency**, or number of oscillation cycles per unit time. Frequency f and period T are complementary ways of conveying the same information, and mathematically they're inverses:

$$f = \frac{1}{T} \quad (13.1)$$

The unit of frequency is the **hertz** (Hz), named after the German Heinrich Hertz (1857–1894), who was the first to produce and detect radio waves. One hertz is equal to one oscillation cycle per second.

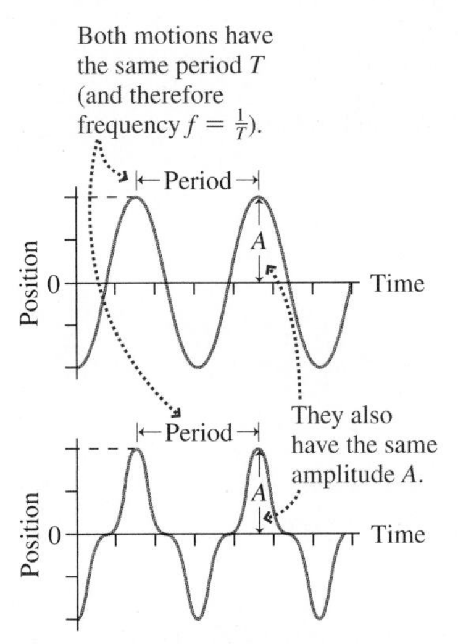

FIGURE 13.2 Position–time graphs for two oscillatory motions with the same amplitude A and period T (and therefore frequency).

EXAMPLE 13.1 Amplitude, Period, Frequency: An Oscillatory Distraction

Tired of homework, a student holds one end of a flexible plastic ruler against a desk and idly strikes the other end, setting it into oscillation (Fig. 13.3). The student notes that 28 complete cycles occur in 10 s and that the end of the ruler moves a total distance of 8.0 cm. What are the amplitude, period, and frequency of this oscillatory motion?

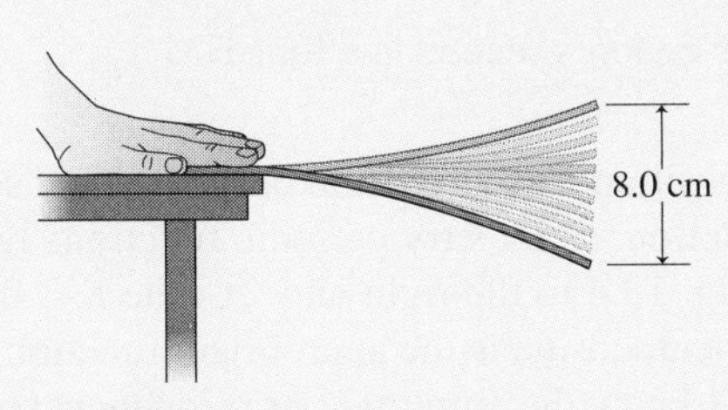

FIGURE 13.3 A ruler undergoing oscillatory motion.

INTERPRET We've got a case of oscillatory motion, and we're asked to describe it quantitatively in terms of amplitude, period, and frequency.

DEVELOP We can work from the definitions of these quantities: Amplitude is the maximum displacement from equilibrium, period is the time to complete a full oscillation, and frequency is the inverse of the period (Equation 13.1).

EVALUATE The ruler moves a total of 8.0 cm from one extreme to the other. Since the motion takes it to both sides of its equilibrium position, the amplitude is 4.0 cm. With 28 cycles in 10 s, the time per cycle, or the period, is

$$T = \frac{10\text{ s}}{28} = 0.36\text{ s}$$

The frequency is the inverse of the period: $f = 1/T = 1/0.36\text{ s} = 2.8\text{ Hz}$. We can also get this directly: 28 cycles/10 s $= 2.8$ Hz.

ASSESS Make sense? With a period that's less than 1 s, the frequency must be more than 1 cycle per second or 1 Hz. Our definition of amplitude as the maximum displacement from equilibrium led to our 4.0-cm amplitude; the full 8.0 cm between extreme positions is called the **peak-to-peak amplitude**. ■

Amplitude and frequency don't provide all the details of oscillatory motion, since two quite different motions can have the same frequency and amplitude (Fig. 13.2). The differences reflect the restoring forces that return systems to equilibrium. Remarkably, though, restoring forces in many physical systems have the same mathematical form—a form we encountered before, when we introduced the force of an ideal spring in Chapter 4.

GOT IT? 13.1 A typical human heart rate is about 65 beats per minute. The corresponding period and frequency are (a) period just over 1 s and frequency just under 1 Hz; (b) period just under 1 s and frequency just under 1 Hz; (c) period just under 1 s and frequency just over 1 Hz; or (d) period just over 1 minute and frequency of 70 Hz.

PhET: Masses and Springs

13.2 Simple Harmonic Motion

In many systems, the restoring force that develops when the system is displaced from equilibrium increases approximately in direct proportion to the displacement—meaning that if you displace the system twice as far from equilibrium, the force tending to restore equilibrium becomes twice as great. In the rest of this chapter, we therefore consider the case of a restoring force directly proportional to displacement. This is an approximation for most real systems, but often a very good approximation, especially for small displacements from equilibrium.

The type of motion that results from a restoring force proportional to displacement is called **simple harmonic motion** (SHM). Mathematically, we describe such a force by writing

$$F = -kx \qquad \text{(restoring force in SHM)} \tag{13.2}$$

where F is the force, x is the displacement, and k is a constant of proportionality between them. The minus sign in Equation 13.2 indicates a *restoring* force: If the object is displaced in one direction, the force is in the *opposite* direction, so it tends to restore the equilibrium.

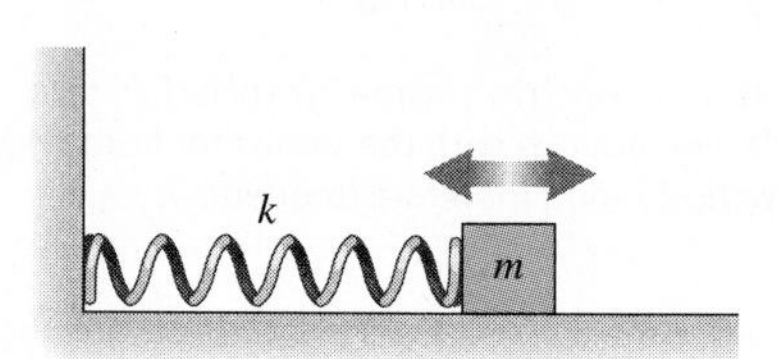

FIGURE 13.4 A mass attached to a spring undergoes simple harmonic motion.

You've seen Equation 13.2 before: It's the force exerted by an ideal spring of spring constant k. So a system consisting of a mass attached to a spring undergoes simple harmonic motion (Fig. 13.4). Many other systems—including atoms and molecules—can be modeled as miniature mass–spring systems.

How does a body in simple harmonic motion actually move? We can find out by applying Newton's second law, $F = ma$, to the mass–spring system of Fig. 13.4. Here the force on the mass m is $-kx$, so Newton's law becomes $-kx = ma$, where we take the x-axis along the direction of motion, with $x = 0$ at the equilibrium position. Now, the acceleration a is the second derivative of position, so we can write our Newton's law equation as

$$m\frac{d^2x}{dt^2} = -kx \qquad \text{(Newton's second law for SHM)} \tag{13.3}$$

The solution to this equation is the position x as a function of time. What sort of function might it be? We expect periodic motion, so let's try periodic functions like sine and cosine. Suppose we pull the mass in Fig. 13.4 to the right and, at time $t = 0$, release it. Since it starts with a nonzero displacement, cosine is the appropriate function [recall that $\cos(0) = 1$, and $\sin(0) = 0$]. We don't know the amplitude or frequency, so we'll try a form that has two unknown constants:

$$x(t) = A\cos\omega t \tag{13.4}$$

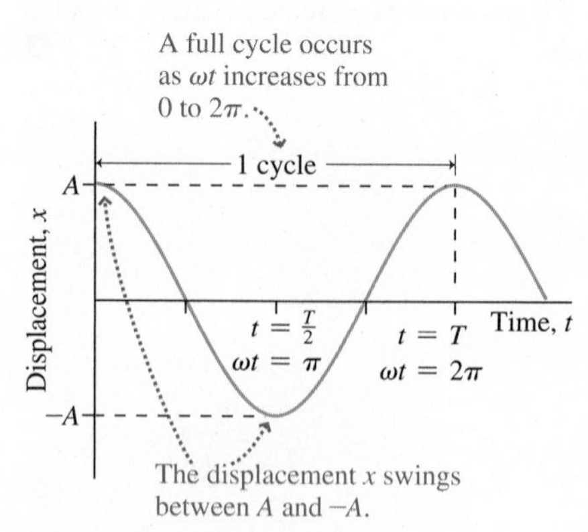

FIGURE 13.5 The function $A\cos\omega t$.

Because the cosine function itself varies between $+1$ and -1, A in Equation 13.4 is the amplitude—the greatest displacement from equilibrium (Fig. 13.5). What about ω? The cosine function undergoes a full cycle as its argument increases by 2π radians, or 360°, as shown in Fig. 13.5. In Equation 13.4, the argument of the cosine is ωt. Since the time for a full cycle is the period T, the argument ωt must go from 0 to 2π as the time t goes from 0 to T. So we have $\omega T = 2\pi$, or

$$T = \frac{2\pi}{\omega} \tag{13.5}$$

The frequency of the motion is then

$$f = \frac{1}{T} = \frac{\omega}{2\pi} \tag{13.6}$$

Equation 13.6 shows that ω is a measure of the frequency, although it differs from the frequency f by the factor 2π. The quantity ω is called the **angular frequency**, and its units are radians per second or, since radians are dimensionless, simply inverse seconds (s^{-1}).

✓**TIP** Why Radians?

Here, as in Chapter 10, we use the angular quantity ω because it provides the simplest mathematical description of the motion. In fact, the relationship between angular frequency and frequency in hertz is the same as Chapter 10's relationship between angular speed in radians per second and in revolutions per second. We'll explore this similarity further in Section 13.4.

Writing the displacement x in the form 13.4 doesn't guarantee that we have a solution; we still need to see whether this form satisfies Equation 13.3. With $x(t)$ given by Equation 13.4, its first derivative is

$$\frac{dx}{dt} = \frac{d}{dt}(A\cos\omega t) = -A\omega\sin\omega t$$

where we've used the chain rule for differentiation (see Appendix A). Then the second derivative is

$$\frac{d^2x}{dt^2} = \frac{d}{dt}\left(\frac{dx}{dt}\right) = \frac{d}{dt}(-A\omega\sin\omega t) = -A\omega^2\cos\omega t$$

We can now try out our assumed solution for x (Equation 13.4) and its second derivative in Equation 13.3. Substituting $x(t)$ and d^2x/dt^2 in the appropriate places gives

$$m(-A\omega^2\cos\omega t) \stackrel{?}{=} -k(A\cos\omega t)$$

where the ? indicates that we're still trying to find out whether this is indeed an equality. If it is, the equality must hold *for all values of time t*. Why? Because Newton's law holds at all times, and we derived our questionable equality from Newton's law. Fortunately, the time-dependent term $\cos\omega t$ appears on both sides of the equation, so we can cancel it. Also, the amplitude A and the minus sign cancel from the equation, leaving only $m\omega^2 = k$, or

$$\omega = \sqrt{\frac{k}{m}} \quad \text{(angular frequency, simple harmonic motion)} \qquad (13.7a)$$

Thus, Equation 13.4 *is* a solution of Equation 13.3, *provided* the angular frequency ω is given by Equation 13.7a.

Frequency and Period in Simple Harmonic Motion

We can recast Equation 13.7a in terms of the more familiar frequency f and period T using Equation 13.6, $f = \omega/2\pi$. This gives

$$f = \frac{\omega}{2\pi} = \frac{1}{2\pi}\sqrt{\frac{k}{m}} \quad \text{and} \quad T = \frac{1}{f} = 2\pi\sqrt{\frac{m}{k}} \qquad (13.7b, c)$$

Do these relationships make sense? If we increase the mass m, it becomes harder to accelerate and we expect slower oscillations. This is reflected in Equations 13.7a and b, where m appears in the denominator. Increasing k, on the other hand, makes the spring stiffer and therefore results in greater force. That increases the oscillation frequency—as shown by the presence of k in the numerators of Equations 13.7a and b.

Physical systems display a wide range of m and k values and a correspondingly large range of oscillation frequencies. A molecule, with its small mass and its "springiness" provided by electric forces, may oscillate at 10^{14} Hz or more. A massive skyscraper, in contrast, typically oscillates at about 0.1 Hz.

Amplitude in Simple Harmonic Motion

The amplitude A canceled from our equations, so our analysis works for *any* value of A. This means that the oscillation frequency doesn't depend on amplitude. Frequency that's independent of amplitude is an essential feature of simple harmonic motion and arises because the restoring force is *directly proportional* to the displacement. When the restoring force does not have the simple form $F = -kx$, then frequency *does* depend on amplitude and the analysis of oscillatory motion becomes much more complicated. In many systems the relation $F = -kx$ breaks down if the displacement x gets too big; for this reason, simple harmonic motion usually occurs only for small oscillation amplitudes.

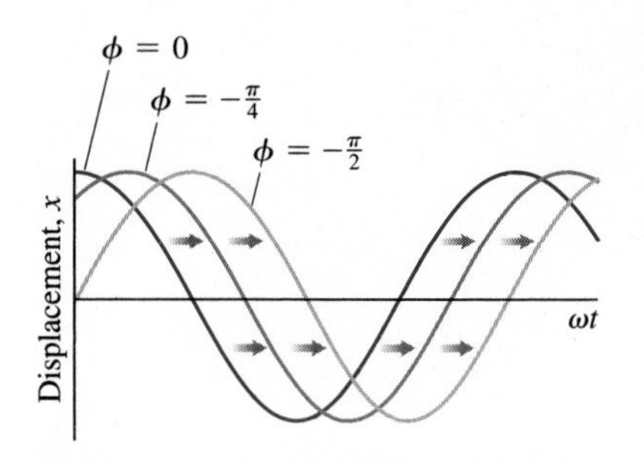

FIGURE 13.6 A negative phase constant shifts the curve to the right.

Phase

Equation 13.4 isn't the only solution to Equation 13.3; you can readily show that $x = A\sin\omega t$ works just as well. We chose the cosine because we took time $t = 0$ at the point of maximum displacement. Had we set $t = 0$ as the mass passed through its equilibrium point, sine would have been the appropriate function. More generally we can take the zero of time at some arbitrary point in the oscillation cycle. Then, as Fig. 13.6 shows, we can represent the motion by the form

$$x(t) = A\cos(\omega t + \phi) \quad \text{(simple harmonic motion)} \tag{13.8}$$

where the **phase constant** ϕ has the effect of shifting the cosine curve to the left (for $\phi > 0$) or right ($\phi < 0$) but doesn't affect the frequency or amplitude.

Velocity and Acceleration in Simple Harmonic Motion

Equation 13.4 (or, more generally, Equation 13.8) gives the position of an object in simple harmonic motion as a function of time, so its first derivative must be the object's velocity:

$$v(t) = \frac{dx}{dt} = \frac{d}{dt}(A\cos\omega t) = -\omega A\sin\omega t \tag{13.9}$$

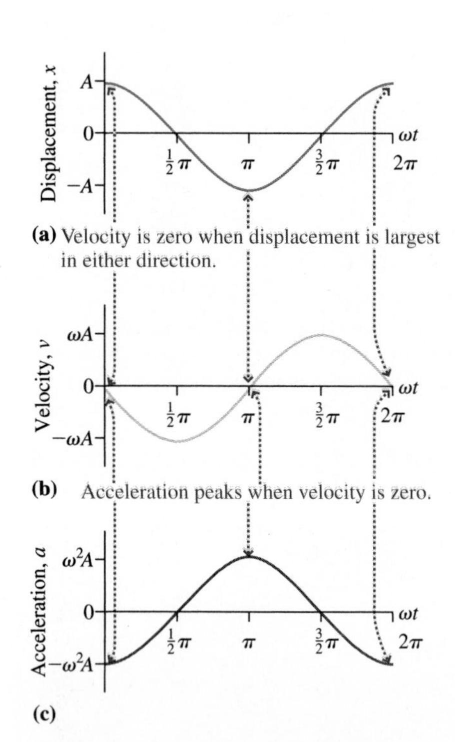

FIGURE 13.7 Displacement, velocity, and acceleration in simple harmonic motion.

Because the maximum value of the sine function is 1, this expression shows that the maximum velocity is ωA. This makes sense because a higher-frequency oscillation requires that the object traverse the distance A in a shorter time—so it must move faster. Equation 13.9 shows that the velocity $v(t)$ is a sine function when the displacement $x(t)$ is a cosine. Thus velocity is a maximum when displacement is zero, and vice versa; mathematically, we express this by saying that displacement and velocity differ in phase by $\frac{\pi}{2}$ radians or 90°. Does this make sense? Sure, because at the extremes of its motion, the object is instantaneously at rest as it reverses direction: maximum displacement, zero speed. And when it passes through its equilibrium position, the object is going fastest. Figures 13.7*a* and *b* show graphically the relationship between displacement and velocity in simple harmonic motion.

Just as velocity is the derivative of position, so acceleration is the derivative of velocity, or the second derivative of position:

$$a(t) = \frac{dv}{dt} = \frac{d}{dt}(-\omega A\sin\omega t) = -\omega^2 A\cos\omega t \tag{13.10}$$

Thus the maximum acceleration is $\omega^2 A$. Since acceleration is a cosine function if velocity is a sine, each reaches its maximum value when the other is zero (Fig. 13.7*b*, *c*).

GOT IT? 13.2 Two identical mass–spring systems are displaced different amounts from equilibrium and then released at different times. Of the amplitudes, frequencies, periods, and phase constants of the subsequent motions, which are the same for both systems and which are different?

EXAMPLE 13.2 Simple Harmonic Motion: A Tuned Mass Damper

The tuned mass damper in New York's Citicorp Tower (see Application on this page) consists of a 373-Mg concrete block that completes one oscillation in 6.80 s. The oscillation amplitude in a high wind is 110 cm. Determine the spring constant and the maximum speed and acceleration of the block.

INTERPRET This is a problem involving simple harmonic motion, with the concrete block and spring making up the oscillating system. We're given the period, mass, and amplitude.

DEVELOP Equation 13.7c, $T = 2\pi\sqrt{m/k}$, will give the spring constant. Equations 13.9 and 13.10 show that the maximum speed and acceleration are $v_{max} = \omega A$ and $a_{max} = \omega^2 A$, and we can get the angular frequency ω from the period using Equation 13.5: $\omega = 2\pi/T$.

EVALUATE First we solve Equation 13.7c for the spring constant:

$$k = \frac{4\pi^2 m}{T^2} = \frac{(4\pi^2)(3.73\times 10^5\ \text{kg})}{(6.80\ \text{s})^2} = 3.18\times 10^5\ \text{N/m}$$

The angular frequency is $\omega = 2\pi/T = 0.924\ \text{s}^{-1}$. Then we have $v_{max} = \omega A = (0.924\ \text{s}^{-1})(1.10\ \text{m}) = 1.02\ \text{m/s}$ and $a_{max} = \omega^2 A = 0.939\ \text{m/s}^2$.

ASSESS The large spring constant and relatively low velocity and acceleration make sense given the huge mass involved. Note that we had to convert the mass, given as 373 Mg (373×10^6 g), to kilograms before evaluating. ■

13.3 Applications of Simple Harmonic Motion

Simple harmonic motion occurs in any system where the tendency to return to equilibrium increases in direct proportion to the displacement from equilibrium. Analysis of such systems is like that of the mass–spring system we just considered but may involve different physical quantities.

The Vertical Mass–Spring System

A mass hanging vertically from a spring is subject to gravity as well as the spring force (Fig. 13.8). In equilibrium the spring stretches enough for its force to balance gravity: $mg - kx_1 = 0$, where x_1 is the new equilibrium position. Stretching the spring an additional amount Δx increases the spring force by $k\,\Delta x$, and this increased force tends to restore the equilibrium. So once again we have a restoring force that's directly proportional to displacement. And here, with the same spring constant k and mass m, our previous analysis still applies and we get simple harmonic motion with frequency $\omega = \sqrt{k/m}$. Thus gravity changes only the equilibrium position and doesn't affect the frequency.

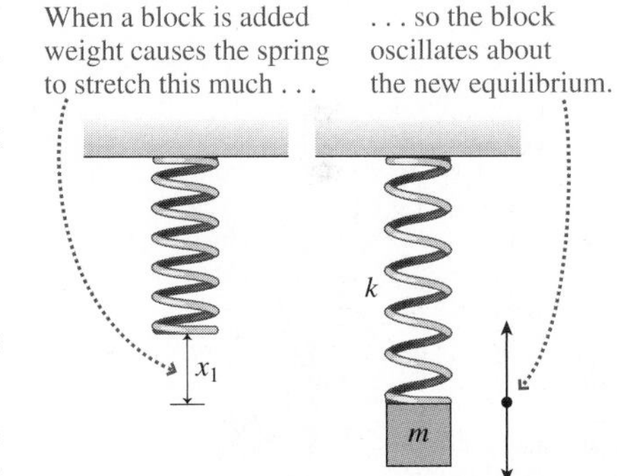

FIGURE 13.8 A vertical mass–spring system oscillates about a new equilibrium position x_1, with the same frequency $\omega = \sqrt{k/m}$.

APPLICATION Swaying Skyscrapers

Skyscrapers are tall, thin, flexible structures. High winds and earthquakes can set them oscillating, much like the ruler of Example 13.1. Wind-driven oscillations are uncomfortable to occupants of a building's upper floors, and earthquake-induced oscillations can be downright destructive.

Modern skyscrapers use so-called tuned mass dampers to counteract building oscillations. These devices are essentially large mass–spring systems mounted high in the building. They're engineered to oscillate with the same frequency as the building (hence the term "tuned") but 180° out of phase, thus reducing the amplitude of the building's own oscillation. The result is increased comfort for the building's occupants and improved safety for buildings in earthquake-prone regions. Tuned mass dampers also find applications in tall smokestacks, airport control towers, power-plant cooling towers, bridges, ski lifts, and even the Grand Canyon skywalk. By suppressing vibrations, tuned mass dampers enable architects and engineers to design structures that don't need as much intrinsic stiffness, so they can be lighter and less expensive. The photos show the world's largest tuned mass damper and the building that houses it, Taiwan's Taipei 101 skyscraper. The damper helps the building survive earthquakes and typhoons. Example 13.2 explores another tuned mass damper.

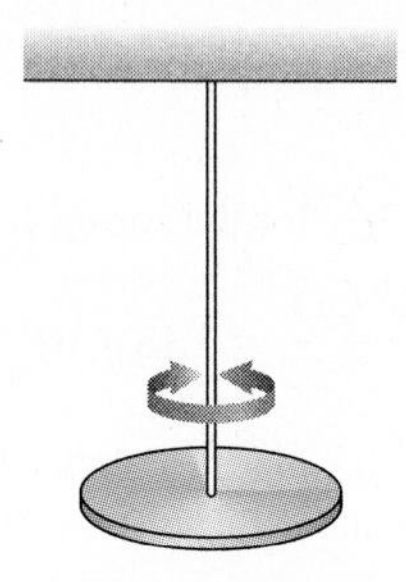

FIGURE 13.9 A torsional oscillator.

The Torsional Oscillator

Figure 13.9 shows a disk suspended from a wire. Rotate the disk slightly, and a torque develops in the wire. Let go, and the disk oscillates by rotating back and forth. This is a **torsional oscillator**, and it's best described using the language of rotational motion. The **angular displacement** θ, **restoring torque** τ, and **torsional constant** κ relate the torque and displacement: $\tau = -\kappa\theta$, where again the minus sign indicates that the torque is opposite the displacement, tending to restore the system to equilibrium. The rotational analog of Newton's law, $\tau = I\alpha$, describes the system's behavior; here the rotational inertia I plays the role of mass. But the angular acceleration α is the second derivative of the angular position, so Newton's law becomes

$$I\frac{d^2\theta}{dt^2} = -\kappa\theta \tag{13.11}$$

This is identical to Equation 13.3 for the linear oscillator, with I replacing m, θ replacing x, and κ replacing k. So we can immediately write $\theta(t) = A\cos\omega t$ for the angular displacement and, in analogy with Equation 13.7a,

$$\omega = \sqrt{\frac{\kappa}{I}} \tag{13.12}$$

for the angular frequency. Note that the units of κ are N·m/rad.

Torsional oscillators constitute the timekeeping mechanism in mechanical watches, and they can provide accurate measures of rotational inertia.

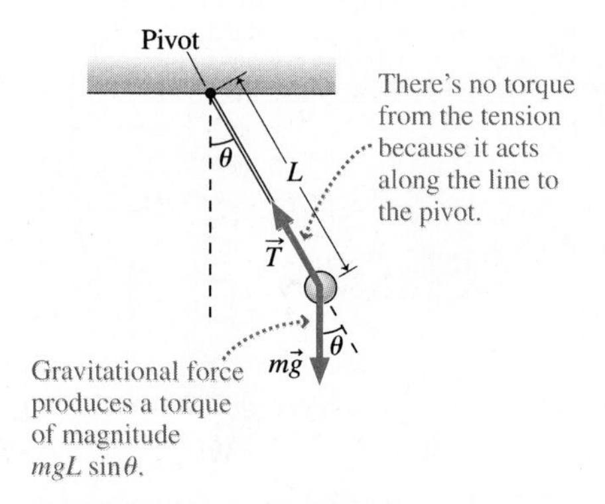

FIGURE 13.10 Forces on a pendulum.

The Pendulum

A **simple pendulum** consists of a point mass suspended from a massless string. Real systems approximate this ideal when a suspended object's size is negligible compared with the suspension length and its mass is much greater than that of the suspension. The pendulum in a grandfather clock is essentially a simple pendulum. Figure 13.10 shows a pendulum of mass m and length L displaced slightly from equilibrium. The gravitational force exerts a torque given by $\tau = -mgL\sin\theta$, where the minus sign indicates that the torque tends to rotate the pendulum back toward equilibrium. The rotational analog of Newton's law, $\tau = I\alpha$, then becomes

$$I\frac{d^2\theta}{dt^2} = -mgL\sin\theta$$

where we've written the angular acceleration as the second derivative of the angular displacement. This looks like Equation 13.11 for the torsional oscillator—but not quite, since the torque involves $\sin\theta$ rather than θ itself. Thus the restoring torque is not *directly* proportional to the angular displacement, and the motion is therefore *not* simple harmonic.

If, however, the amplitude of the motion is small, then it *approximates* simple harmonic motion. Figure 13.11 shows that for small angles, $\sin\theta$ and θ are essentially equal. For a small-amplitude pendulum we can therefore replace $\sin\theta$ with θ to get

$$I\frac{d^2\theta}{dt^2} = -mgL\theta$$

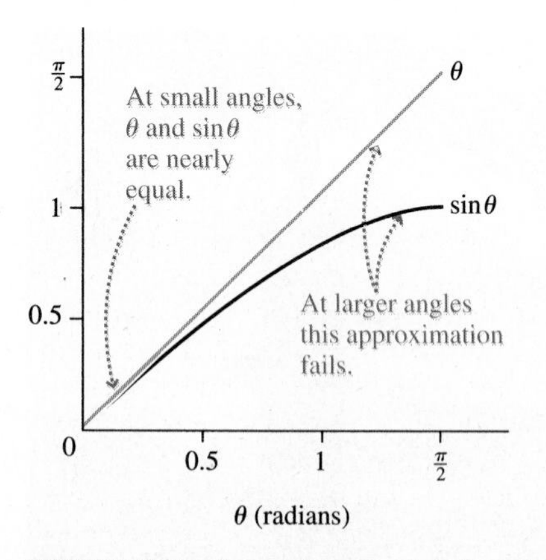

FIGURE 13.11 For θ much less than 1 radian, $\sin\theta$ and θ are nearly equal.

This is essentially Equation 13.11, with mgL playing the role of κ. So the small-amplitude pendulum undergoes simple harmonic motion, with its angular frequency given by Equation 13.12 with $\kappa = mgL$:

$$\omega = \sqrt{\frac{mgL}{I}} \tag{13.13}$$

For a *simple* pendulum, the rotational inertia I is that of a point mass m a distance L from the rotation axis, or $I = mL^2$, as we found in Chapter 10. Then we have

$$\omega = \sqrt{\frac{mgL}{mL^2}} = \sqrt{\frac{g}{L}} \quad \text{(simple pendulum)} \tag{13.14}$$

or, from Equation 13.5,

$$T = \frac{2\pi}{\omega} = 2\pi\sqrt{\frac{L}{g}} \quad \text{(simple pendulum)} \qquad (13.15)$$

MP

PhET: Pendulum Lab

These equations show that the frequency and period of a simple pendulum are independent of its mass, depending only on length and gravitational acceleration.

EXAMPLE 13.3 A Pendulum: Rescuing Tarzan

Tarzan stands on a branch as a leopard threatens. Fortunately, Jane is on a nearby branch of the same height, holding a 25-m-long vine attached directly above the point midway between her and Tarzan. She grasps the vine and steps off with negligible velocity. How soon does she reach Tarzan?

INTERPRET This is a problem about a pendulum, which we identify as consisting of Jane and the vine. The period of the pendulum is the time for a full swing back and forth, so the answer we're after—the time to reach Tarzan—is half the period.

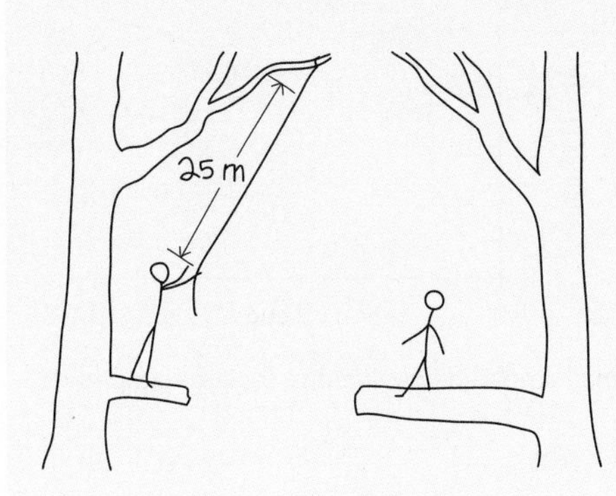

FIGURE 13.12 Our sketch for Example 13.3. Vine length is not to scale.

DEVELOP We sketched the situation in Fig. 13.12. Equation 13.15, $T = 2\pi\sqrt{L/g}$, determines the period, so we can use this equation to find the half-period.

EVALUATE Equation 13.15 gives

$$\frac{1}{2}T = \left(\frac{1}{2}\right)(2\pi)\sqrt{\frac{L}{g}} = (\pi)\sqrt{\frac{25\text{ m}}{9.8\text{ m/s}^2}} = 5.0\text{ s}$$

ASSESS This seems a reasonable answer for a problem involving human-scale objects and many meters of vine. One caution: Jane's rescue will be successful only if the vine is strong enough—not only to support her weight but also to provide the acceleration that keeps her moving in a circular arc. You can explore that issue in Problem 56. ■

GOT IT? 13.3 What happens to the period of a pendulum if (1) its mass is doubled; (2) it's moved to a planet whose gravitational acceleration is one-fourth that of Earth; and (3) its length is quadrupled?

CONCEPTUAL EXAMPLE 13.1 The Nonlinear Pendulum

A pendulum consists of a weight on the end of a rigid rod of negligible mass, hanging vertically from a frictionless pivot at the opposite end of the rod. For small-amplitude disturbances from equilibrium, the system constitutes a simple pendulum. But for larger disturbances it becomes a *nonlinear pendulum*, so named because the restoring torque is no longer proportional to the displacement. Quantitative analysis of a nonlinear pendulum is difficult, but you can still understand it conceptually.

(a) As the pendulum's amplitude increases, how will its period change?
(b) If you start the pendulum by striking it when it's hanging vertically, will it undergo oscillatory motion no matter how hard it's hit?

EVALUATE (a) Before we made the small-amplitude approximation, we showed that a pendulum's restoring torque is, in general, proportional to $\sin\theta$. But Fig. 13.11 shows that $\sin\theta$ doesn't increase as fast as θ itself. So for large-amplitude swings, the restoring torque is *less* than it would be in the small-amplitude approximation. This suggests the pendulum should return more slowly toward equilibrium—and thus its period should increase.

(b) When you strike the pendulum, you give it kinetic energy. If that energy is insufficient to invert it completely, then the pendulum will swing to one side, eventually stop, and return, undergoing back-and-forth oscillatory motion. But hit it hard enough, and it will go "over the top," reaching its fully inverted position with kinetic energy

(continued)

to spare. Round and round it goes, executing motion that's periodic and circular, but not oscillatory. This circular motion isn't uniform, because it moves more slowly at the top and faster at the bottom.

ASSESS Make sense? Yes: Consider a pendulum with just a little less energy than it takes to go "over the top." It will move very slowly near the top of its trajectory, so its period will be quite long. And its angular-position-versus-time curve will be flatter than the sine curve of a simple pendulum. Give it just a little more energy, and it goes into circular motion. Figure 13.13 illustrates all three situations. You can explore the nonlinear pendulum computationally in Problem 86.

MAKING THE CONNECTION If the pendulum has length L, what's the minimum speed that will get it "over the top," into periodic nonuniform circular motion?

EVALUATE Potential energy at the top is $U = mg(2L)$, so kinetic energy $K = \frac{1}{2}mv^2$ has to be at least this large. That gives $v > 2\sqrt{gL}$.

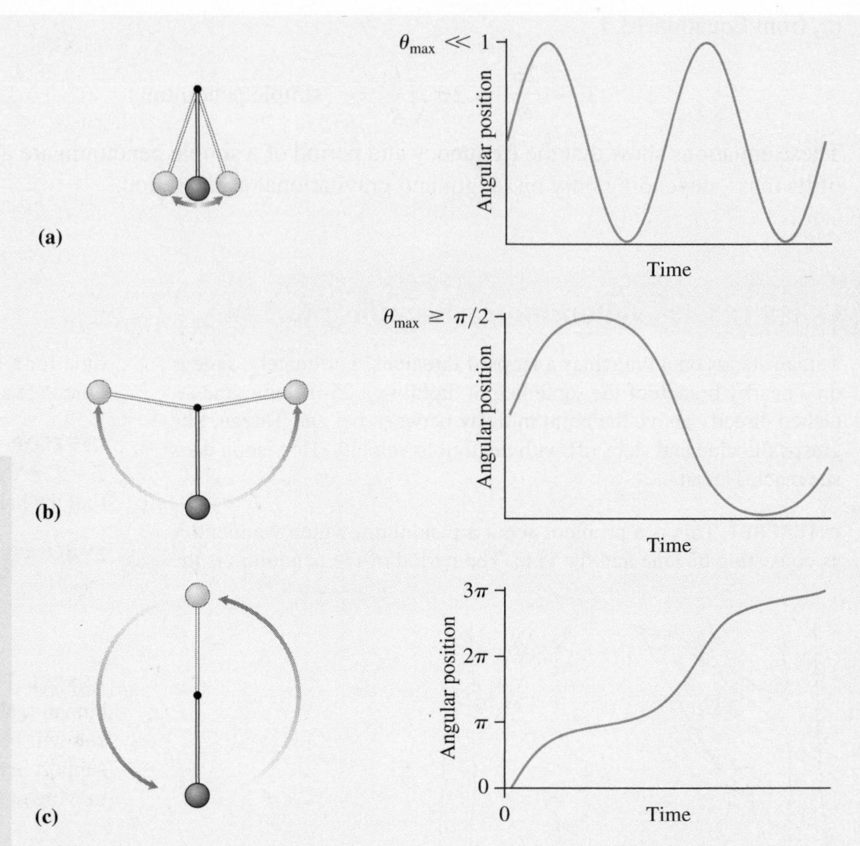

FIGURE 13.13 Conceptual Example 13.1: (a) Small-amplitude oscillations; (b) large-amplitude oscillations; (c) circular motion.

The Physical Pendulum

A **physical pendulum** is an object of arbitrary shape that's free to swing (Fig. 13.14). It differs from a simple pendulum in that mass may be distributed over its entire length. Physical pendulums are everywhere: Examples include the legs of humans and other animals (see Example 13.4), a skier on a chair lift, a boxer's punching bag, a frying pan hanging from a rack, and a crane lifting any object of significant extent. In our analysis of the simple pendulum, we used the fact that mass was concentrated at the bottom only in the final step, when we wrote mL^2 for the rotational inertia. Our analysis before that step therefore applies to the physical pendulum as well.

In particular, a physical pendulum displaced slightly from equilibrium undergoes simple harmonic motion with frequency given by Equation 13.13. But how are we to interpret the length L in that equation? Because gravity—which provides the restoring torque for *any* pendulum—acts on an object's center of gravity, L must be the distance from the pivot to the center of gravity, as marked in Fig. 13.14.

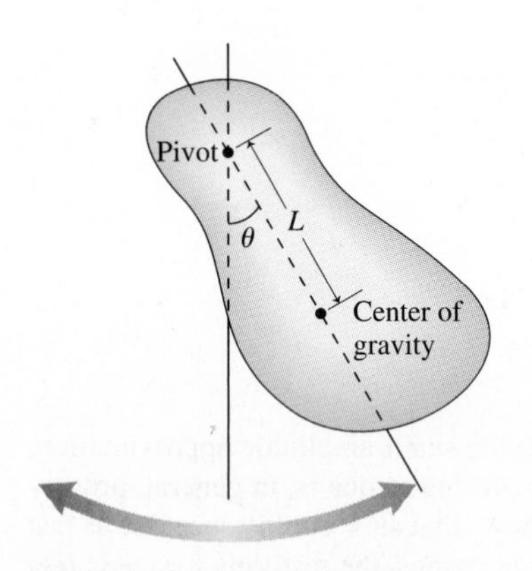

FIGURE 13.14 A physical pendulum.

EXAMPLE 13.4 A Physical Pendulum: Walking

When walking, the leg not in contact with the ground swings forward, acting like a physical pendulum. Approximating the leg as a uniform rod, find the period of this pendulum motion for a leg of length 90 cm.

INTERPRET This problem is about a physical pendulum, here identified as a uniform rod approximating the leg.

DEVELOP Figure 13.15 is our drawing, showing the leg as a rod pivoting at the hip. The center of mass of a uniform rod is at its center, so the effective length L is half the leg's length, or 45 cm. Equation 13.13, $\omega = \sqrt{mgL/I}$, determines the angular frequency, from which we can get the period using Equation 13.5, $T = 2\pi/\omega$. We also need the rotational inertia; from Table 10.2, that's $I = \frac{1}{3}M(2L)^2$, where

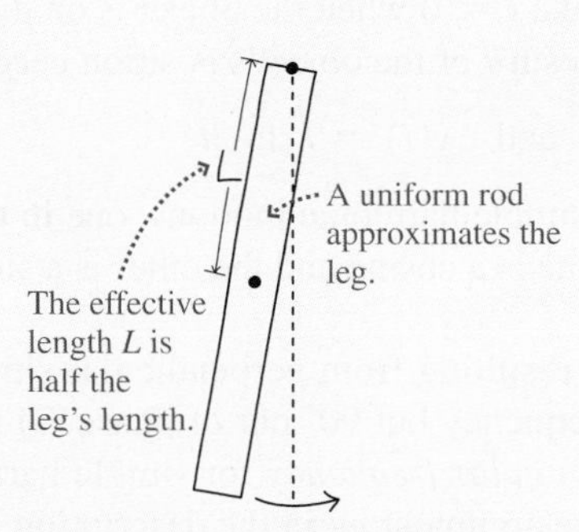

FIGURE 13.15 A human leg treated as a pendulum.

we use $2L$ because Table 10.2's expression involves the *full* length of the rod.

EVALUATE Putting this all together, we evaluate to get the answer:

$$T = \frac{2\pi}{\omega} = 2\pi\sqrt{\frac{I}{mgL}} = 2\pi\sqrt{\frac{\frac{1}{3}m(2L)^2}{mgL}} = 2\pi\sqrt{\frac{4L}{3g}}$$

Using $L = 0.45$ m gives $T = 1.6$ s.

ASSESS The leg swings forward to complete a full stride in half a period, or 0.8 s. This seems a reasonable value for the pace in walking. ■

13.4 Circular Motion and Harmonic Motion

Look down on the solar system, and you see Earth in circular motion about the Sun (Fig. 13.16*a*). But look in from the plane of Earth's orbit, and Earth appears to be moving back and forth (Fig. 13.16*b*). Figure 13.17 shows that this apparent back-and-forth motion is a single component of the actual circular motion, and that this component describes a sinusoidal function of time. Specifically, the position vector $\vec{r}$ for Earth or any other object in circular motion makes an angle that increases linearly with time: $\theta = \omega t$, where we

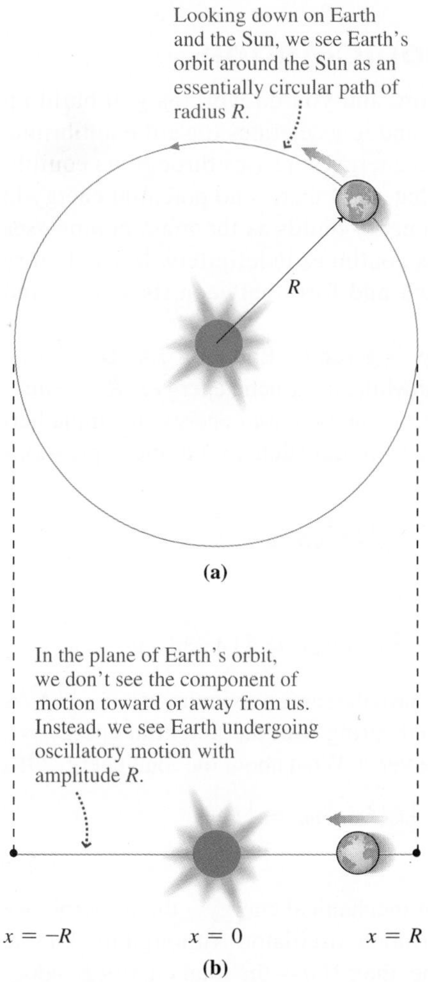

Looking down on Earth and the Sun, we see Earth's orbit around the Sun as an essentially circular path of radius R.

R

(a)

In the plane of Earth's orbit, we don't see the component of motion toward or away from us. Instead, we see Earth undergoing oscillatory motion with amplitude R.

$x = -R$ $x = 0$ $x = R$

(b)

FIGURE 13.16 Two views of Earth's orbital motion.

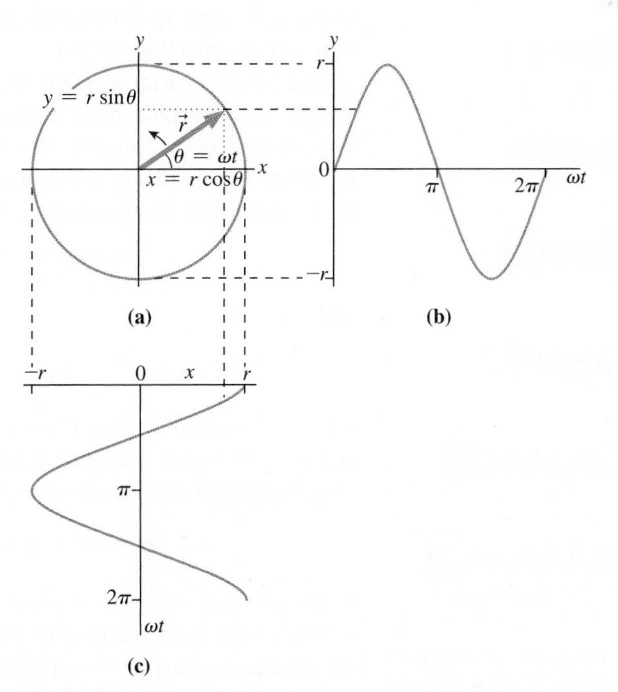

FIGURE 13.17 As the position vector $\vec{r}$ traces out a circle, its x- and y-components are sinusoidal functions of time.

measure θ with respect to the x-axis and take $t = 0$ when the object is on the x-axis. Then the two components $x = r\cos\theta$ and $y = r\sin\theta$ of the object's position become

$$x(t) = r\cos\omega t \quad \text{and} \quad y(t) = r\sin\omega t$$

These are the equations for two different simple harmonic motions, one in the x-direction and the other in the y-direction. Because one is a cosine and the other is a sine, they're out of phase by $\frac{\pi}{2}$ or 90°.

So we can think of circular motion as resulting from perpendicular simple harmonic motions, with the same amplitude and frequency but 90° out of phase. This should help you to understand why we use the term *angular frequency* for simple harmonic motion even though there's no angle involved. The argument ωt in the description of simple harmonic motion is the same as the physical angle θ in the corresponding circular motion. The time for one cycle of simple harmonic motion is the same as the time for one revolution in the circular motion, so the values of T and therefore ω are exactly the same.

You can verify that two mutually perpendicular simple harmonic motions of the same amplitude and frequency sum vectorially to give circular motion (see Problem 53). If the amplitudes or frequencies aren't the same, then interesting complex motions occur, as shown in Fig. 13.18.

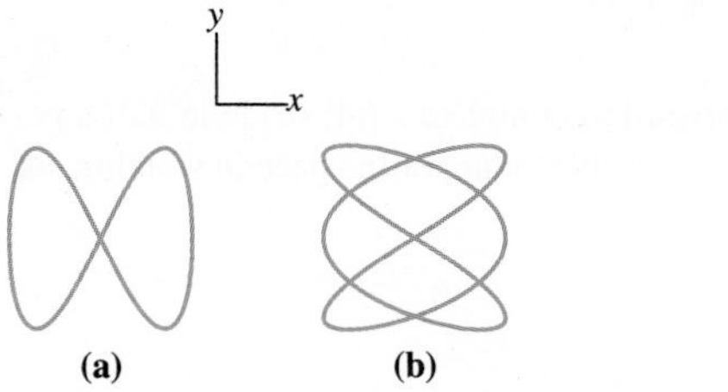

(a) (b)

FIGURE 13.18 Complex paths resulting from different frequencies in different directions. Can you determine the frequency ratios?

GOT IT? 13.4 Figure 13.18 shows the paths traced in the horizontal plane by two pendulums swinging with different frequencies in two perpendicular directions. What's the ratio of x-direction frequency to y-direction frequency for (1) path (a) and (2) path (b)?

13.5 Energy in Simple Harmonic Motion

Displace a mass–spring system from equilibrium, and you do work as you build up potential energy in the spring. Release the mass, and it accelerates toward equilibrium, gaining kinetic energy at the expense of potential energy. It passes through its equilibrium position with maximum kinetic energy; at that point there's no potential energy in the system. The mass then slows and potential energy builds as the mass compresses the spring. If there's no energy loss, this process continues indefinitely. In oscillatory motion, energy is continuously transferred back and forth between its kinetic and potential forms (Fig. 13.19).

For a mass–spring system, the potential energy is given by Equation 7.4: $U = \frac{1}{2}kx^2$, where x is the displacement from equilibrium. Meanwhile, the kinetic energy is $K = \frac{1}{2}mv^2$. We can illustrate explicitly the interchange of kinetic and potential energy in simple harmonic motion by using x from Equation 13.4 and v from Equation 13.9 in the expressions for potential and kinetic energy. Then we have

$$U = \tfrac{1}{2}kx^2 = \tfrac{1}{2}k(A\cos\omega t)^2 = \tfrac{1}{2}kA^2\cos^2\omega t$$

and

$$K = \tfrac{1}{2}mv^2 = \tfrac{1}{2}m(-\omega A\sin\omega t)^2 = \tfrac{1}{2}m\omega^2A^2\sin^2\omega t = \tfrac{1}{2}kA^2\sin^2\omega t$$

where we used $\omega^2 = k/m$. Both energy expressions have the same maximum value—$\frac{1}{2}kA^2$—equal to the initial potential energy of the stretched spring. But the potential energy is a maximum when the kinetic energy is zero, and vice versa. What about the total energy? It's

$$E = U + K = \tfrac{1}{2}kA^2\cos^2\omega t + \tfrac{1}{2}kA^2\sin^2\omega t = \tfrac{1}{2}kA^2$$

where we used $\sin^2\omega t + \cos^2\omega t = 1$.

Our result is a statement of the conservation of mechanical energy—the principle we introduced in Chapter 7—applied to a simple harmonic oscillator. Although the kinetic and potential energies K and U both vary with time, their sum—the total energy E—does not (Fig. 13.20).

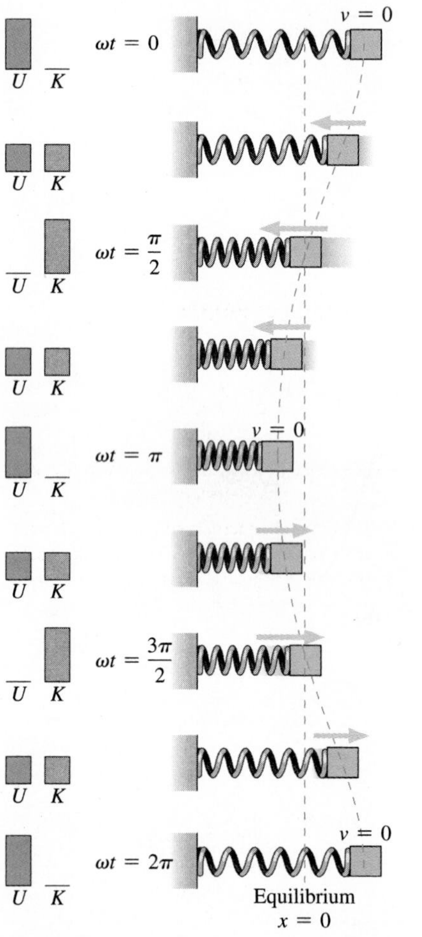

$\omega t = 0$, $v = 0$; $\omega t = \frac{\pi}{2}$; $\omega t = \pi$, $v = 0$; $\omega t = \frac{3\pi}{2}$; $\omega t = 2\pi$, $v = 0$; U K; Equilibrium $x = 0$

FIGURE 13.19 Kinetic and potential energy in simple harmonic motion. Dashed curve is the position of the mass; straight dashed line marks the equilibrium position $x = 0$.

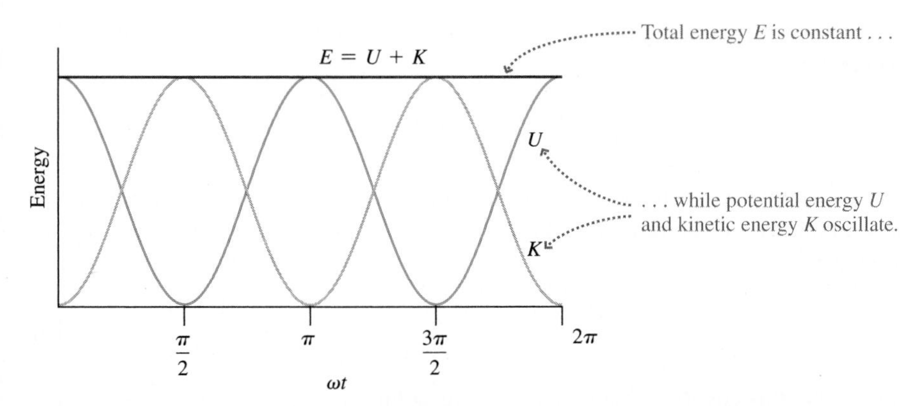

FIGURE 13.20 Energy of a simple harmonic oscillator.

EXAMPLE 13.5 Energy in Simple Harmonic Motion

A mass–spring system undergoes simple harmonic motion with angular frequency ω and amplitude A. Find its speed at the point where the kinetic and potential energies are equal.

INTERPRET This example involves the concept of energy conservation in simple harmonic motion. We're asked to find a speed, which is related to kinetic energy.

DEVELOP When the kinetic energy equals the potential energy, each must be half the total energy. What is that total? The speed is at its maximum, $v_{\max} = \omega A$ from Equation 13.9, when the energy is all kinetic. Thus the total energy is $E = \frac{1}{2}mv_{\max}^2 = \frac{1}{2}m\omega^2A^2$. The speed v we're after occurs when the kinetic energy has half this value, or $K = \frac{1}{2}mv^2 = \frac{1}{2}(\frac{1}{2}m\omega^2A^2) = \frac{1}{4}m\omega^2A^2$.

EVALUATE Solving for v gives our answer:

$$v = \frac{\omega A}{\sqrt{2}}$$

ASSESS Make sense? Yes: The speed at this point must be less than the maximum speed, since half the energy is tied up as potential energy in the spring. And because kinetic energy depends on the *square* of the speed, it's lower not by a factor of 2 but by $\sqrt{2}$.

Potential-Energy Curves and Simple Harmonic Motion

We arrived at the expression $U = \frac{1}{2}kx^2$ for the potential energy of a spring by integrating the spring force, $-kx$, over distance. Since every simple harmonic oscillator has a restoring force or torque proportional to displacement, integration always results in a potential energy proportional to the *square* of the displacement—that is, in a parabolic potential-energy curve. Conversely, any system with a parabolic potential-energy curve exhibits simple harmonic motion. The simplest mathematical approximation to a smooth curve near a minimum is a parabola, and for that reason potential-energy curves for complex systems often approximate parabolas near their stable equilibrium points (Fig. 13.21). Small disturbances from these equilibria therefore result in simple harmonic motion, and that's why simple harmonic motion is so common throughout the physical world.

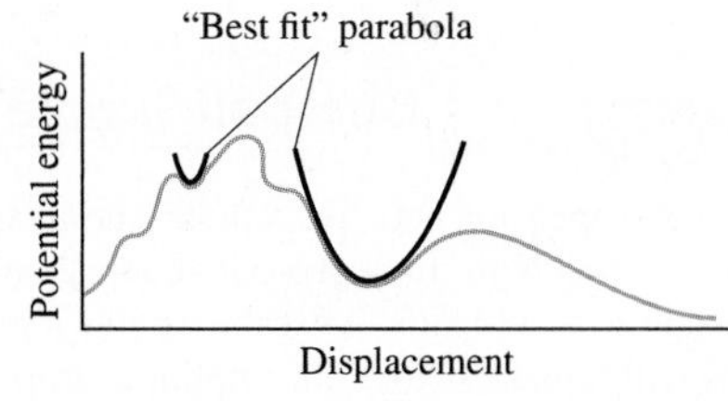

FIGURE 13.21 Near their minima, potential-energy curves often approximate parabolas. This results in simple harmonic motion.

GOT IT? 13.5 Two different mass–spring systems are oscillating with the same amplitude and frequency. If one has twice as much total energy as the other, how do (1) their masses and (2) their spring constants compare? (3) What about their maximum speeds?

13.6 Damped Harmonic Motion

In real oscillating systems, forces such as friction or air resistance normally dissipate the oscillation energy. This energy loss causes the oscillation amplitude to decrease, and the motion is said to be **damped**.

If dissipation is sufficiently weak that only a small fraction of the system's energy is lost in each oscillation cycle, then we expect that the system should behave essentially as in the undamped case, except for a gradual decrease in amplitude (Fig. 13.22).

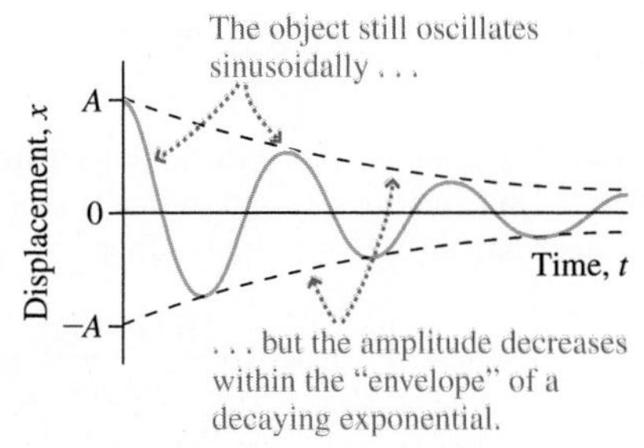

FIGURE 13.22 Weakly damped motion.

In many systems the damping force is approximately proportional to the velocity and in the opposite direction:

$$F_d = -bv = -b\frac{dx}{dt}$$

where b is a constant giving the strength of the damping. We can write Newton's law as before, now including the damping force along with the restoring force. For a mass–spring system, we have

$$m\frac{d^2x}{dt^2} = -kx - b\frac{dx}{dt} \qquad (13.16)$$

We won't solve this equation, but simply state its solution:

$$x(t) = Ae^{-bt/2m}\cos(\omega t + \phi) \qquad (13.17)$$

This equation describes sinusoidal motion whose amplitude decreases exponentially with time. How fast depends on the damping constant b and mass m: When $t = 2m/b$, the amplitude has dropped to $1/e$ of its original value. When the damping is so weak that only a small fraction of the total energy is lost in each cycle, the frequency ω in Equation 13.17 is essentially equal to the undamped frequency $\sqrt{k/m}$. But with stronger damping, the damping force slows the motion, and the frequency becomes lower. As long as oscillation occurs, the motion is said to be **underdamped** (Fig. 13.23*a*). For sufficiently strong damping, though, the effect of the damping force is as great as that of the spring force. Under this condition, called **critical damping**, the system returns to its equilibrium state without undergoing any oscillations (Fig. 13.23*b*). If the damping is made still stronger, the system becomes *overdamped*. The damping force now dominates, so the system returns more slowly to equilibrium (Fig. 13.23*c*).

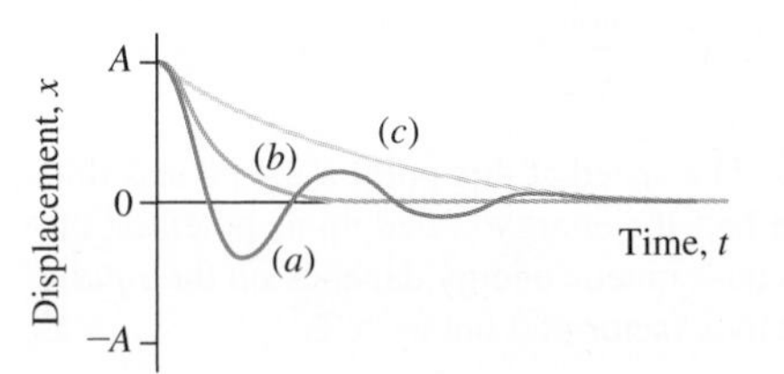

FIGURE 13.23 (a) Underdamped, (b) critically damped, and (c) overdamped oscillations.

Many physical systems, from atoms to the human leg, can be modeled as damped oscillators. Engineers often design systems with specific amounts of damping. Automobile shock absorbers, for example, coordinate with the springs to give critical damping. This results in rapid return to equilibrium while absorbing the energy imparted by road bumps.

EXAMPLE 13.6 Damped Simple Harmonic Motion: Bad Shocks

A car's suspension acts like a mass–spring system with $m = 1200$ kg and $k = 58$ kN/m. Its worn-out shock absorbers provide a damping constant $b = 230$ kg/s. After the car hits a pothole, how many oscillations will it make before the amplitude drops to half its initial value?

INTERPRET We interpret this problem as being about damped simple harmonic motion, and we identify the car as the oscillating system.

DEVELOP Our plan is to find out how long it takes the amplitude to decrease by half and then find the number of oscillation cycles in this time. Equation 13.17, $x(t) = Ae^{-bt/2m}\cos(\omega t + \phi)$, describes the motion, with the factor $e^{-bt/2m}$ giving the decrease in amplitude. At $t = 0$ this factor is 1, so we want to know when it's equal to one-half: $e^{-bt/2m} = \frac{1}{2}$.

EVALUATE Taking the natural logarithms of both sides gives $bt/2m = \ln 2$, where we used the facts that $\ln(x)$ and e^x are inverse functions and $\ln(1/x) = -\ln(x)$. Then

$$t = \frac{2m}{b}\ln 2 = \frac{(2)(1200\text{ kg})}{230\text{ kg/s}}\ln 2 = 7.23\text{ s}$$

is the time for the amplitude to drop to half its original value. For weak damping, the period is very close to the undamped period, which is

$$T = 2\pi\sqrt{\frac{m}{k}} = 2\pi\sqrt{\frac{1200\text{ kg}}{58\times 10^3\text{ N/m}}} = 0.904\text{ s}$$

Then the number of cycles during the 7.23 s it takes the amplitude to drop in half is

$$\frac{7.23\text{ s}}{0.904\text{ s}} = 8$$

ASSESS That the number of oscillations is much greater than 1 tells us that the damping is weak, justifying our use of the undamped period. It also tells us that those are really bad shocks! ■

GOT IT? 13.6 The figure shows displacement-versus-time graphs for three mass–spring systems, with different masses m, spring constants k, and damping constants b. The time on the horizontal axis is the same for all three. (1) For which system is damping the most significant? (2) For which system is damping the least significant?

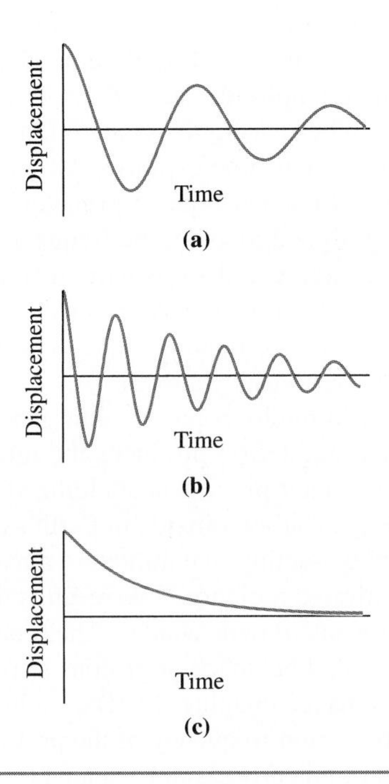

13.7 Driven Oscillations and Resonance

Pushing a child on a swing, you can build up a large amplitude by giving a relatively small push once each oscillation cycle. If your pushing were not in step with the swing's natural oscillatory motion, then the same force would have little effect.

When an external force acts on an oscillatory system, we say that the system is **driven**. Consider a mass–spring system, which you might drive as suggested in Fig. 13.24. Suppose the driving force is given by $F_d \cos \omega_d t$, where ω_d is called the **driving frequency**. Then Newton's law is

$$m\frac{d^2x}{dt^2} = -kx - b\frac{dx}{dt} + F_d \cos \omega_d t \qquad (13.18)$$

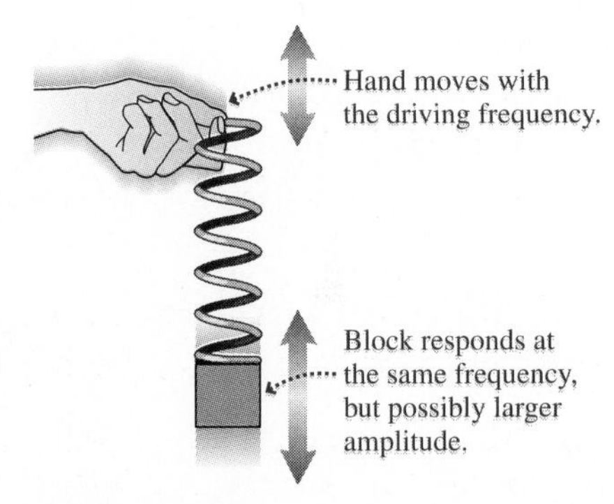

FIGURE 13.24 Driving a mass–spring system results in a large amplitude if the driving frequency is near the natural frequency $\sqrt{k/m}$.

where the first term on the right-hand side is the restoring force, the second the damping force, and the third the driving force. Since the system is being driven at the frequency ω_d, we expect it to undergo oscillatory motion at this frequency. So we guess that the solution to Equation 13.18 might have the form

$$x = A\cos(\omega_d t + \phi)$$

Substituting this expression and its derivatives into Equation 13.18 shows that the equation is satisfied if

$$A(\omega) = \frac{F_d}{m\sqrt{(\omega_d^2 - \omega_0^2)^2 + b^2\omega_d^2/m^2}} \qquad (13.19)$$

where ω_0 is the undamped **natural frequency** $\sqrt{k/m}$, as distinguished from the driving frequency ω_d.

Figure 13.25 shows **resonance curves**—plots of Equation 13.19 as a function of driving frequency—for three values of the damping constant. As long as the system is

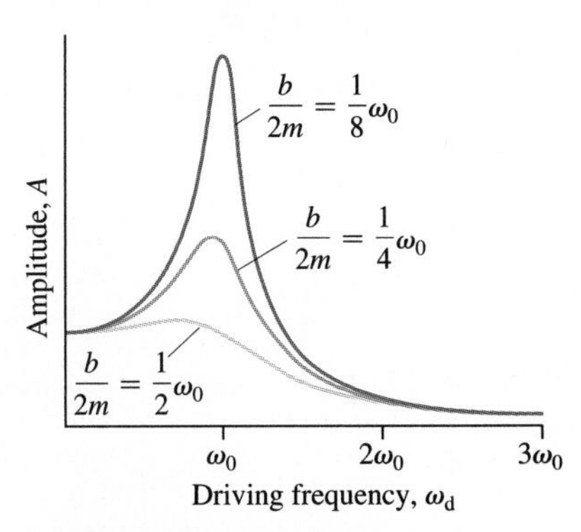

FIGURE 13.25 Resonance curves for three damping strengths; ω_0 is the undamped natural frequency $\sqrt{k/m}$.

Video Tutor Demo | **Vibrating Rods**

underdamped, the curve has a maximum at some nonzero frequency, and for weak damping, that maximum occurs at very nearly the natural frequency. The weaker the damping, the more sharply peaked is the resonance curve. Thus, in weakly damped systems, it's possible to build up large-amplitude oscillations with relatively small driving forces—a phenomenon known as **resonance**.

Most physical systems, from molecules to cars, and loudspeakers to buildings and bridges, exhibit one or more natural modes of oscillation. If these oscillations are weakly damped, then the buildup of large-amplitude oscillations through resonance can cause serious problems—sometimes even destroying the system (Fig. 13.26). Engineers designing complex structures spend a lot of their time exploring all possible oscillation modes and taking steps to avoid resonance. In an earthquake-prone area, for example, a building's natural frequencies would be designed to avoid the frequency of typical earthquake motions. A loudspeaker should be engineered so its natural frequency isn't in the range of sound it's intended to reproduce. Damping systems such as the shock absorbers of Example 13.6 or the tuned mass damper of Example 13.2 help limit resonant oscillations in cases where natural frequencies aren't easily altered.

FIGURE 13.26 Collapse of the Tacoma Narrows Bridge—only four months after its opening in 1940—followed the resonant growth of large-amplitude oscillations.

Resonance is also important in microscopic systems. The resonant behavior of electrons in a special tube called a magnetron produces the microwaves that cook food in a microwave oven; the same resonant process heats ionized gases in some experiments designed to harness fusion energy. Carbon dioxide in Earth's atmosphere absorbs infrared radiation because CO_2 molecules—acting like miniature mass–spring systems—resonate at some of the frequencies of infrared radiation. The result is the greenhouse effect, which now threatens Earth with significant climatic change. The process called nuclear magnetic resonance (NMR) uses the resonant behavior of protons to probe the structure of matter and is the basis of magnetic resonance imaging (MRI) used in medicine. In NMR, the resonance involves the natural precession frequency of the protons due to magnetic torques; we described a classical model of this process in Chapter 11.

GOT IT? 13.7 The photo shows a wineglass shattering in response to sound. What's more important here, the amplitude or the frequency of the sound?

CHAPTER 13 SUMMARY

Big Idea

The big idea here is **simple harmonic motion** (SHM), oscillatory motion that is ubiquitous and that occurs whenever a disturbance from equilibrium results in a restoring force or torque that is directly proportional to the displacement. Position in SHM is a sinusoidal function of time:

$$x(t) = A\cos\omega t$$

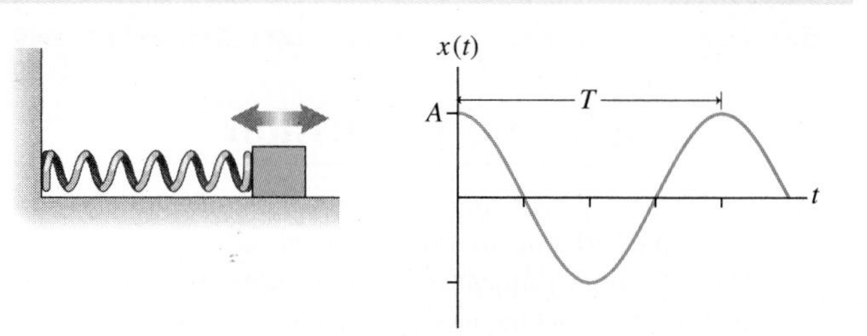

Key Concepts and Equations

Period T is the time to complete one oscillation cycle; its inverse is **frequency**, or number of oscillations per unit time:

$$f = \frac{1}{T}$$

Another measure of frequency is **angular frequency** ω, given by

$$\omega = 2\pi f = \frac{2\pi}{T}$$

Angular frequency can be understood in terms of the close relationship between circular motion and simple harmonic motion.

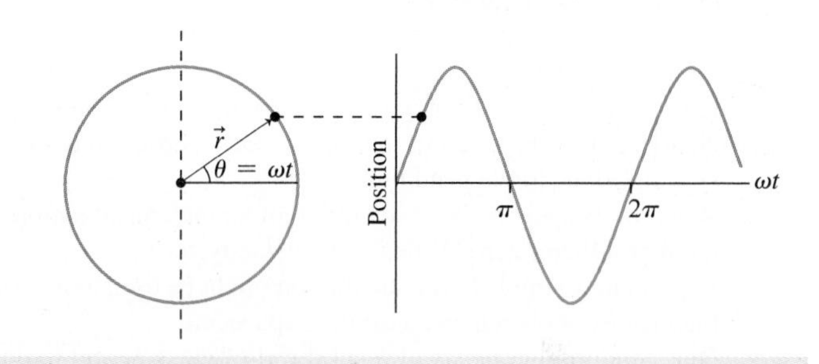

In the absence of friction and other dissipative forces, energy in SHM is conserved, although it's transformed back and forth between kinetic and potential forms:

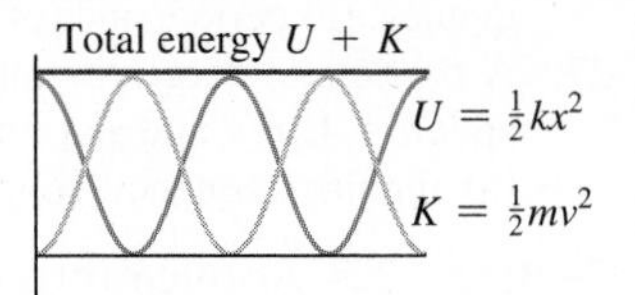

$$E = \tfrac{1}{2}mv^2 + \tfrac{1}{2}kx^2 = \text{constant}$$

When dissipative forces act, the motion is **damped**. For small dissipative forces the oscillation amplitude decreases exponentially with time:

$$x(t) = Ae^{-bt/2m}\cos(\omega t + \phi)$$

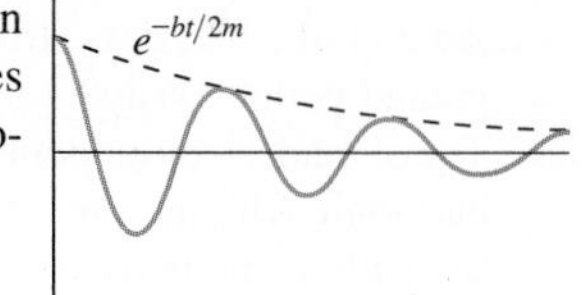

If a system is driven at a frequency near its natural oscillation frequency ω_0, then large-amplitude oscillations can build; this is **resonance**. The amplitude A depends on the driving force F_d, the driving frequency ω_d, the natural frequency $\omega_0 = \sqrt{k/m}$, and the damping constant b:

$$A(\omega) = \frac{F_d}{m\sqrt{(\omega_d^2 - \omega_0^2)^2 + b^2\omega_d^2/m^2}}$$

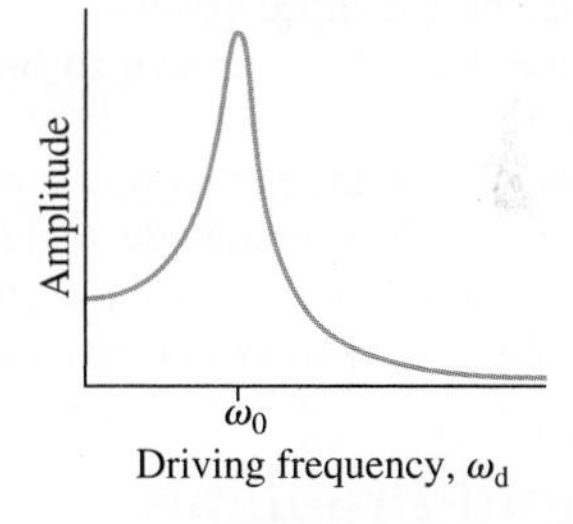

Applications

In mass–spring systems, the angular frequency is given by

$$\omega = \sqrt{\frac{k}{m}}$$

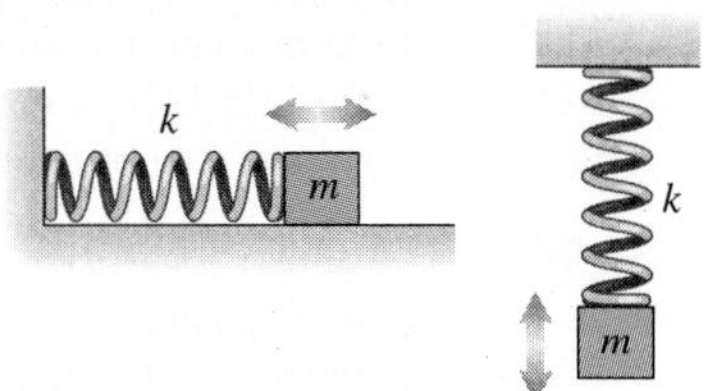

In systems involving rotational oscillations, the analogous relation involves the torsional constant and rotational inertia:

$$\omega = \sqrt{\frac{\kappa}{I}}$$

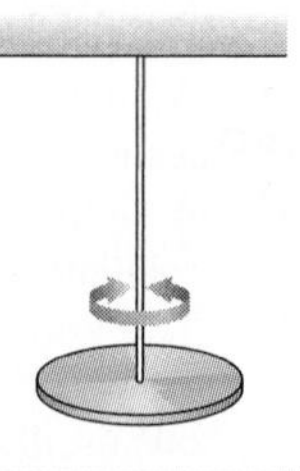

A special case is the **pendulum**, for which (with small-amplitude oscillations)

$$\omega = \sqrt{\frac{mgL}{I}}$$

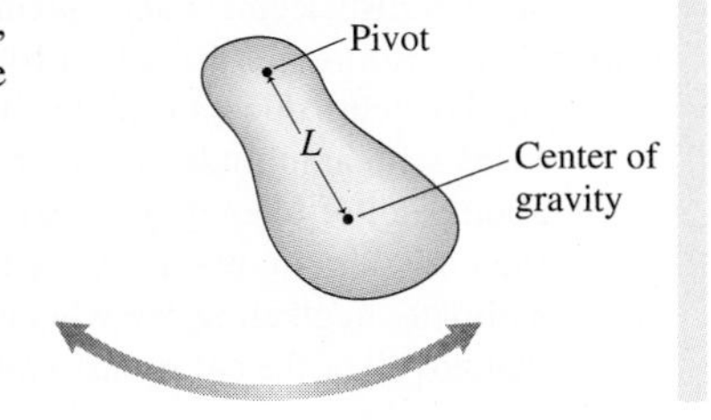

In the case of a **simple pendulum**, the angular frequency reduces to

$$\omega = \sqrt{\frac{g}{L}}$$

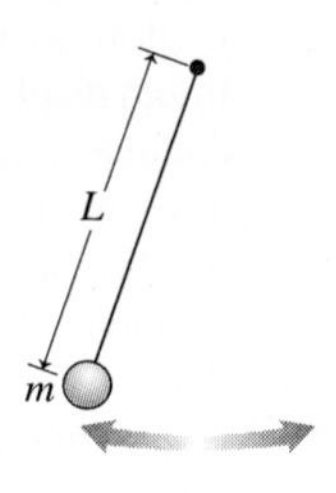

For homework assigned on MasteringPhysics, go to www.masteringphysics.com

BIO Biology and/or medicine-related problems **DATA** Data problems **ENV** Environmental problems **CH** Challenge problems **COMP** Computer problems

For Thought and Discussion

1. Is a vertically bouncing ball an example of oscillatory motion? Of simple harmonic motion? Explain.
2. The vibration frequencies of molecules are much higher than those of macroscopic mechanical systems. Why?
3. What happens to the frequency of a simple harmonic oscillator when the spring constant is doubled? When the mass is doubled?
4. If the spring of a simple harmonic oscillator is cut in half, what happens to the frequency?
5. How does the frequency of a simple harmonic oscillator depend on its amplitude?
6. How would the frequency of a horizontal mass–spring system change if it were taken to the Moon? Of a vertical mass–spring system? Of a simple pendulum?
7. When in its cycle is the acceleration of an undamped simple harmonic oscillator zero? When is the velocity zero?
8. Explain how simple harmonic motion might be used to determine the masses of objects in an orbiting spacecraft.
9. One pendulum consists of a solid rod of mass m and length L, and another consists of a compact ball of the same mass m on the end of a massless string of the same length L. Which has the greater period? Why?
10. The x- and y-components of motion of a body are both simple harmonic with the same frequency and amplitude. What shape is the path of the body if the component motions are (a) in phase, (b) $\pi/2$ out of phase, and (c) $\pi/4$ out of phase?
11. Why is critical damping desirable in a car's suspension?
12. Explain why the frequency of a damped system is lower than that of the equivalent undamped system.
13. Opera singers have been known to break glasses with their voices. How?
14. What will happen to the period of a mass–spring system if it's placed in a jetliner accelerating down a runway? What will happen to the period of a pendulum in the same situation?
15. How can a system have more than one resonant frequency?

Exercises and Problems

Exercises

Section 13.1 Describing Oscillatory Motion

16. **BIO** A doctor counts 68 heartbeats in 1.0 minute. What are the corresponding period and frequency?
17. A violin string playing the note A oscillates at 440 Hz. What's its oscillation period?
18. The vibration frequency of a hydrogen chloride molecule is 8.66×10^{13} Hz. How long does it take the molecule to complete one oscillation?
19. Write expressions for the displacement $x(t)$ in simple harmonic motion (a) with amplitude 12.5 cm, frequency 6.68 Hz, and maximum displacement when $t = 0$, and (b) with amplitude 2.15 cm, angular frequency 4.63 s^{-1}, and maximum speed when $t = 0$.
20. The top of a skyscraper sways back and forth, completing 95 full oscillation cycles in 10 minutes. Find (a) the period and (b) the frequency (in Hz) of its oscillatory motion.
21. **BIO** A hummingbird's wings vibrate at about 45 Hz. What's the corresponding period?

Section 13.2 Simple Harmonic Motion

22. A 200-g mass is attached to a spring of constant $k = 5.6$ N/m and set into oscillation with amplitude $A = 25$ cm. Determine (a) the frequency in hertz, (b) the period, (c) the maximum velocity, and (d) the maximum force in the spring.
23. An automobile suspension has an effective spring constant of 26 kN/m, and the car's suspended mass is 1900 kg. In the absence of damping, with what frequency and period will the car undergo simple harmonic motion?
24. The quartz crystal in a watch executes simple harmonic motion at 32,768 Hz. (This is 2^{15} Hz, chosen so that 15 divisions by 2 give a signal at 1.00000 Hz.) If each face of the crystal undergoes a maximum displacement of 100 nm, find the maximum velocity and acceleration of the crystal faces.
25. A 342-g mass is attached to a spring and undergoes simple harmonic motion. Its maximum acceleration is 18.6 m/s^2, and its maximum speed is 1.75 m/s. Determine (a) the angular frequency, (b) the amplitude, and (c) the spring constant.
26. A particle undergoes simple harmonic motion with amplitude 25 cm and maximum speed 4.8 m/s. Find the (a) angular frequency, (b) period, and (c) maximum acceleration.
27. A particle undergoes simple harmonic motion with maximum speed 1.4 m/s and maximum acceleration 3.1 m/s^2. Find the (a) angular frequency, (b) period, and (c) amplitude.

Section 13.3 Applications of Simple Harmonic Motion

28. How long should you make a simple pendulum so its period is (a) 200 ms, (b) 5.0 s, and (c) 2.0 min?
29. At the heart of a grandfather clock is a simple pendulum 1.45 m long; the clock ticks each time the pendulum reaches its maximum displacement in either direction. What's the time interval between ticks?
30. A 622-g basketball with 24.0-cm diameter is suspended by a wire and is undergoing torsional oscillations at 1.87 Hz. Find the torsional constant of the wire.
31. A meter stick is suspended from one end and set swinging. Find the period of the resulting small-amplitude oscillations.

Section 13.4 Circular and Harmonic Motion

32. A wheel rotates at 600 rpm. Viewed from the edge, a point on the wheel appears to undergo simple harmonic motion. What are (a) the frequency in Hz and (b) the angular frequency for this SHM?
33. The x- and y-components of an object's motion are harmonic with frequency ratio 1.75:1. How many oscillations must each component undergo before the object returns to its initial position?

Section 13.5 Energy in Simple Harmonic Motion

34. A 450-g mass on a spring is oscillating at 1.2 Hz, with total energy 0.51 J. What's the oscillation amplitude?
35. A torsional oscillator of rotational inertia 1.6 kg·m^2 and torsional constant 3.4 N·m/rad has total energy 4.7 J. Find its maximum angular displacement and maximum angular speed.
36. You're riding in a friend's 1400-kg car with bad shock absorbers, bouncing down the highway at 20 m/s and executing vertical SHM with amplitude 18 cm and frequency 0.67 Hz. Concerned about fuel efficiency, your friend wonders what percentage of the car's kinetic energy is tied up in this oscillation. Make an estimate, neglecting the wheels' rotational energy and the fact that not all of the car's mass participates in the oscillation.

Sections 13.6 and 13.7 Damped Harmonic Motion and Resonance

37. The vibration of a piano string can be described by an equation analogous to Equation 13.17. If the quantity analogous to $b/2m$ in that equation has the value $2.8\ \text{s}^{-1}$, how long will it take the amplitude to drop to half its original value?
38. A mass–spring system has $b/m = \omega_0/5$, where b is the damping constant and ω_0 the natural frequency. How does its amplitude at ω_0 compare with its amplitude when driven at frequencies 10% above and below ω_0?
39. A car's front suspension has a natural frequency of 0.45 Hz. The car's front shock absorbers are worn and no longer provide critical damping. The car is driving on a bumpy road with bumps 40 m apart. At a certain speed, the driver notices that the car begins to shake violently. What is this speed?

Problems

40. A simple model for carbon dioxide consists of three mass points (atoms) connected by two springs (electric forces), as shown in Fig. 13.27. One way this system can oscillate is if the carbon atom stays fixed and the two oxygens move symmetrically on either side of it. If the frequency of this oscillation is 4.0×10^{13} Hz, what's the effective spring constant? (*Note:* The mass of an oxygen atom is 16 u.)

FIGURE 13.27 Problem 40

41. Two identical mass–spring systems consist of 430-g masses on springs of constant $k = 2.2$ N/m. Both are displaced from equilibrium, and the first is released at time $t = 0$. How much later should the second be released so their oscillations differ in phase by $\pi/2$?
42. **BIO** The human eye and muscles that hold it can be modeled as a mass–spring system with typical values $m = 7.5$ g and $k = 2.5$ kN/m. What's the resonant frequency of this system? Shaking your head at this frequency blurs vision, as the eyeball undergoes resonant oscillations.
43. A mass m slides along a frictionless horizontal surface at speed v_0. It strikes a spring of constant k attached to a rigid wall, as shown in Fig. 13.28. After an elastic encounter with the spring, the mass heads back in the direction it came from. In terms of k, m, and v_0, determine (a) how long the mass is in contact with the spring and (b) the spring's maximum compression.

FIGURE 13.28 Problem 43

44. Show by substitution that $x(t) = A\sin\omega t$ is a solution to Equation 13.3.
45. A physics student, bored by a lecture on simple harmonic motion, idly picks up his pencil (mass 8.65 g, length 18.8 cm) by the tip with his frictionless fingers, and allows it to swing back and forth with small amplitude. If the pencil completes 5974 full cycles during the lecture, how long does the lecture last?
46. A pendulum of length L is mounted in a rocket. Find its period if the rocket is (a) at rest on its launch pad; (b) accelerating upward with acceleration $a = \frac{1}{2}g$; (c) accelerating downward with $a = \frac{1}{2}g$; and (d) in free fall.
47. **BIO** The protein dynein powers the flagella that propel some unicellular organisms. Biophysicists have found that dynein is intrinsically oscillatory, and that it exerts peak forces of about 1.0 pN when it attaches to structures called microtubules. The resulting oscillations have amplitude 15 nm. (a) If this system is modeled as a mass–spring system, what's the associated spring constant? (b) If the oscillation frequency is 70 Hz, what's the effective mass?
48. A mass is attached to a vertical spring, which then goes into oscillation. At the high point of the oscillation, the spring is in the original unstretched equilibrium position it had before the mass was attached; the low point is 5.8 cm below this. Find the oscillation period.
49. **CH** Derive the period of a simple pendulum by considering the horizontal displacement x and the force acting on the bob, rather than the angular displacement and torque.
50. A solid disk of radius R is suspended from a spring of spring constant k and torsional constant κ, as shown in Fig. 13.29. In terms of k and κ, what value of R will give the same period for the vertical and torsional oscillations of this system?

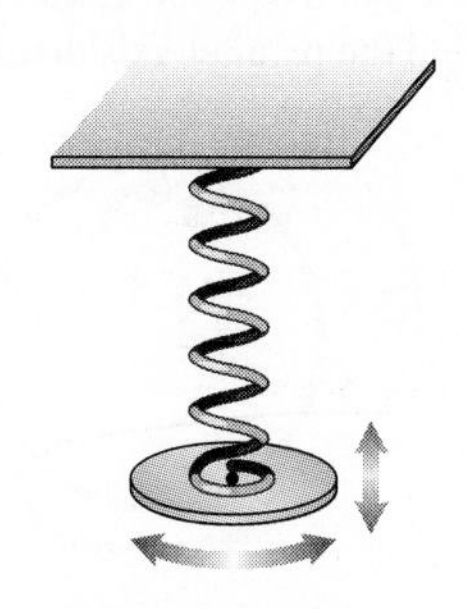

FIGURE 13.29 Problem 50

51. A thin steel beam is suspended from a crane and is undergoing torsional oscillations. Two 82.4 kg steelworkers leap onto opposite ends of the beam, as shown in Fig. 13.30. If the frequency of torsional oscillations diminishes by 21.0%, what's the beam's mass?

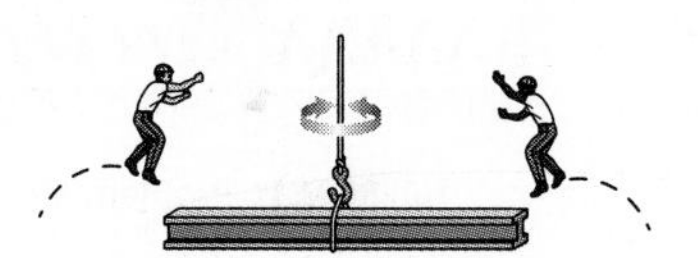

FIGURE 13.30 Problem 51

52. A cyclist turns her bicycle upside down to repair it. She then notices that the front wheel is executing a slow, small-amplitude, back-and-forth rotational motion with period 12 s. Treating the wheel as a thin ring of mass 600 g and radius 30 cm, whose only irregularity is the tire valve stem, determine the mass of the valve stem.
53. An object undergoes simple harmonic motion in two mutually perpendicular directions, its position given by $\vec{r} = A\sin\omega t\hat{\imath} + A\cos\omega t\hat{\jmath}$. (a) Show that the object remains a fixed distance from the origin (i.e., that its path is circular), and find that distance. (b) Find an expression for the object's velocity. (c) Show that the speed remains constant, and find its value. (d) Find the angular speed of the object in its circular path.
54. **BIO** The muscles that drive insect wings minimize the energy needed for flight by "choosing" to move at the natural oscillation frequency of the wings. Biologists study this phenomenon by clipping an insect's wings to reduce their mass. If the wing system is modeled as a simple harmonic oscillator, by what percent will

the frequency change if the wing mass is decreased by 25%? Will it increase or decrease?

55. CH A pendulum consists of a 320-g solid ball 15.0 cm in diameter, suspended by an essentially massless string 80.0 cm long. Calculate the period of this pendulum, treating it first as a simple pendulum and then as a physical pendulum. What's the error in the simple-pendulum approximation? (*Hint:* Remember the parallel-axis theorem.)
56. CH If Jane and Tarzan are initially 8.0 m apart in Fig. 13.12, and Jane's mass is 60 kg, what's the maximum tension in the vine, and at what point does it occur?
57. BIO A *small mass measuring device* (SMMD) used for research on the biological effects of spaceflight consists of a small spring-mounted cage. Rats or other small subjects are introduced into the cage, which is set into oscillation. Calibration of a SMMD gives a linear function for the square of the oscillation period versus the subject's mass m in kg: $T^2 = 4.0\text{ s}^2 + (5.0\text{ s}^2/\text{kg})m$. Find (a) the spring constant and (b) the mass of the cage alone.
58. A thin, uniform hoop of mass M and radius R is suspended from a horizontal rod and set oscillating with small amplitude, as shown in Fig. 13.31. Show that the period of the oscillations is $2\pi\sqrt{2R/g}$. (*Hint:* You may find the parallel-axis theorem useful.)

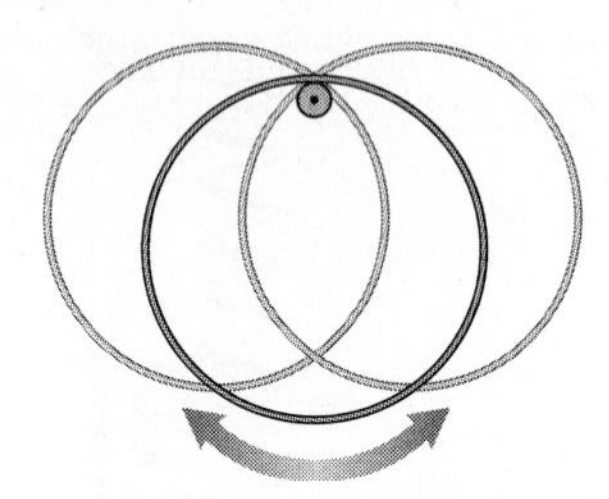

FIGURE 13.31 Problem 58

59. A mass m is mounted between two springs with constants k_1 and k_2, as shown in Fig. 13.32. Show that the angular frequency of oscillation is $\omega = \sqrt{(k_1 + k_2)/m}$.

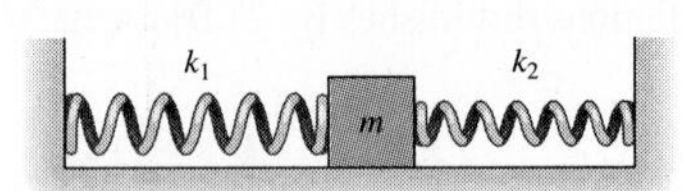

FIGURE 13.32 Problem 59

60. The equation for an ellipse is $(x^2/a^2) + (y^2/b^2) = 1$. Show that two-dimensional simple harmonic motion whose components have different amplitudes and are $\pi/2$ out of phase gives rise to elliptical motion. How are constants a and b related to the amplitudes?
61. Show that the potential energy of a simple pendulum is proportional to the square of the angular displacement in the small-amplitude limit.
62. CH The total energy of a mass–spring system is the sum of its kinetic and potential energy: $E = \frac{1}{2}mv^2 + \frac{1}{2}kx^2$. Assuming E remains constant, differentiate both sides of this expression with respect to time and show that Equation 13.3 results. (*Hint:* Remember that $v = dx/dt$.)
63. CH A solid cylinder of mass M and radius R is mounted on an axle through its center. The axle is attached to a horizontal spring of constant k, and the cylinder rolls back and forth without slipping (Fig. 13.33). Write the statement of energy conservation for this system, and differentiate it to obtain an equation analogous to Equation 13.3 (see Problem 62). Comparing your result with Equation 13.3, determine the angular frequency of the motion.

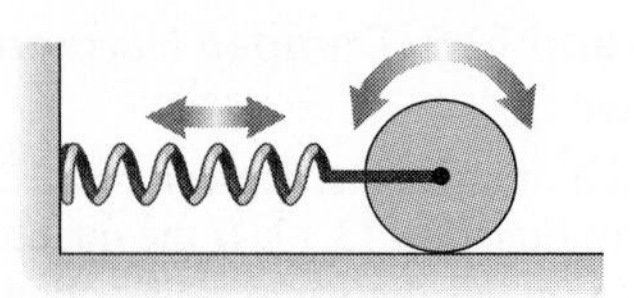

FIGURE 13.33 Problem 63

64. CH A mass m is free to slide on a frictionless track whose height y as a function of horizontal position x is $y = ax^2$, where a is a constant with units of inverse length. The mass is given an initial displacement from the bottom of the track and then released. Find an expression for the period of the resulting motion.
65. A 250-g mass is mounted on a spring of constant $k = 3.3$ N/m. The damping constant for this system is $b = 8.4\times10^{-3}$ kg/s. How many oscillations will the system undergo before the amplitude decays to $1/e$ of its original value?
66. CH A harmonic oscillator is underdamped if the damping constant b is less than $\sqrt{2}m\omega_0$, where ω_0 is the natural frequency of undamped motion. Show that for an underdamped oscillator, Equation 13.19 has a maximum at a driving frequency less than ω_0.
67. A massless spring with $k = 74$ N/m hangs from the ceiling. A 490-g mass is hooked onto the unstretched spring and allowed to drop. Find (a) the amplitude and (b) the period of the resulting motion.
68. A meter stick is suspended from a frictionless rod through a small hole at the 25-cm mark. Find the period of small-amplitude oscillations about the stick's equilibrium position.
69. A particle of mass m has potential energy given by $U = ax^2$, where a is a constant and x is the particle's position. Find an expression for the frequency of simple harmonic oscillations this particle undergoes.
70. Two balls with the same unknown mass m are mounted on opposite ends of a 1.5-m-long rod of mass 850 g. The system is suspended from a wire attached to the center of the rod and set into torsional oscillations. If the wire has torsional constant 0.63 N·m/rad and the period of the oscillations is 5.6 s, what's the unknown mass m?
71. Two mass–spring systems with the same mass are undergoing oscillatory motion with the same amplitudes. System 1 has twice the frequency of system 2. How do (a) their energies and (b) their maximum accelerations compare?
72. Two mass–spring systems have the same mass and the same total energy. The amplitude of system 1 is twice that of system 2. How do (a) their frequencies and (b) their maximum accelerations compare?
73. A 500-g mass is suspended from a thread 45 cm long that can sustain a tension of 6.0 N before breaking. Find the maximum allowable amplitude for pendulum motion of this system.
74. CH A 500-g block on a frictionless, horizontal surface is attached to a rather limp spring with $k = 8.7$ N/m. A second block rests on the first, and the whole system executes simple harmonic motion with period 1.8 s. When the amplitude of the motion is increased to 35 cm, the upper block just begins to slip. What's the coefficient of static friction between the blocks?
75. CH Repeat Problem 64 for a small solid ball of mass M and radius R that rolls without slipping on the parabolic track.
76. You're working on the script of a movie whose plot involves a hole drilled straight through Earth's center and out the other side. You're asked to determine what will happen if a person falls into the hole. You find that the gravitational acceleration *inside* Earth points toward Earth's center, with magnitude given approximately by $g(r) = g_0(r/R_E)$, where g_0 is the surface value, r is the distance from Earth's center, and R_E is Earth's radius. What

do you report for the person's motion, including equations and values for any relevant parameters?

77. **CH** A 1.2-kg block rests on a frictionless surface and is attached to a horizontal spring of constant $k = 23$ N/m (Fig. 13.34). The block oscillates with amplitude 10 cm and phase constant $\phi = -\pi/2$. A block of mass 0.80 kg moves from the right at 1.7 m/s and strikes the first block when the latter is at the rightmost point in its oscillation. The two blocks stick together. Determine the frequency, amplitude, and phase constant (relative to the *original* $t = 0$) of the resulting motion.

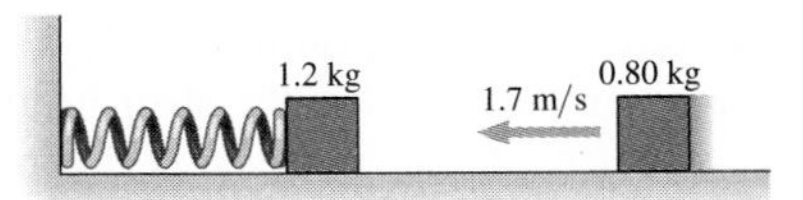

FIGURE 13.34 Problem 77

78. **CH** A disk of radius R is suspended from a pivot somewhere between its center and edge (Fig. 13.35). For what pivot point will the period of this physical pendulum be a minimum?

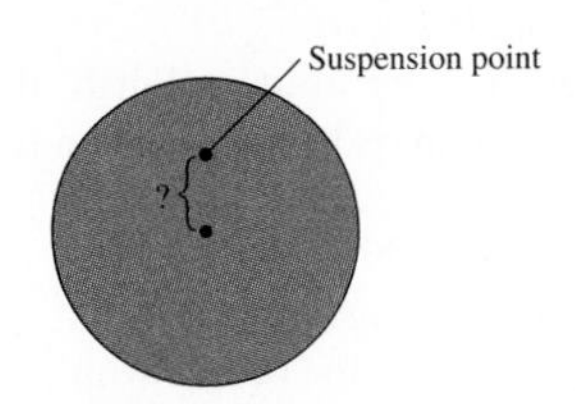

FIGURE 13.35 Problem 78

79. **CH** A simple model for a variable star considers that the outer layer of the star is subject to two forces: the inward force of gravity and the outward force due to gas pressure. As a result, Newton's law for the star's outer layer reads $m\,d^2r/dt^2 = 4\pi r^2 p - GMm/r^2$. Here m is the mass of the outer layer, M is the total mass of the star, r is the star's radius, and p is the pressure. (a) Use this equation to show that the star's equilibrium pressure and radius are related by $p_0 = GMm/4\pi r_0^4$, where the subscript 0 represents equilibrium values. (b) As you'll learn in Chapter 18, gas pressure and volume $V\,(= \frac{4}{3}\pi r^3)$ are related by $pV^{5/3} = p_0V_0^{5/3}$ (this is for an *adiabatic process*, a good approximation here, and the exponent 5/3 reflects the ionized gas that makes up the star). Let $x = r - r_0$ be the displacement of the star's surface from equilibrium. Use the binomial approximation (Appendix A) to show that, when x is small compared with r, the right-hand side of the above equation can be written $-(GMm/r_0^3)x$. (c) Since r and x differ only by a constant, the term d^2r/dt^2 in the equation above can also be written d^2x/dt^2. Make this substitution, along with substituting the result of part (b) for the right-hand side, and compare your result with Equations 13.2 and 13.7 to find an expression for the oscillation period of the star. (d) What does your simple model predict for the period of the variable star *Delta Cephei*, with radius 44.5 times that of the Sun and mass of 4.5 Sun masses? (Your answer overestimates the actual period by a factor of about 3, both because of oversimplified physics and because changes in the star's radius are too large for the assumption of a linear restoring force.)

80. **DATA** You're a structural engineer working on a design for a steel beam, and you need to know its resonant frequency. The beam's mass is 3750 kg. You test the beam by clamping one end and deflecting the other so it bends, and you determine the associated potential energy. The table below gives the results:

Beam deflection x (cm)	Potential energy U (J)
−4.54	164
−3.49	141
−2.62	71.9
−1.22	9.15
−0.448	0.162
0	0
0.730	4.13
1.29	16.3
2.13	34.0
3.39	115
4.70	225

Find a quantity which, when U is plotted against it, should give a straight line. Make your plot, determine the best-fit line, and use its slope to determine the resonant frequency of the beam.

81. Show that $x(t) = a\cos\omega t - b\sin\omega t$ represents simple harmonic motion, as in Equation 13.8, with $A = \sqrt{a^2 + b^2}$ and $\phi = \tan^{-1}(b/a)$.

82. You're working for the summer with an ornithologist who knows you've studied physics. She asks you for a noninvasive way to measure birds' masses. You propose using a bird feeder in the shape of a 50-cm-diameter disk of mass 340 g, suspended by a wire with torsional constant 5.00 N·m/rad (Fig. 13.36). Two birds land on opposite sides and the feeder goes into torsional oscillation at 2.6 Hz. Assuming the birds have the same mass, what is it?

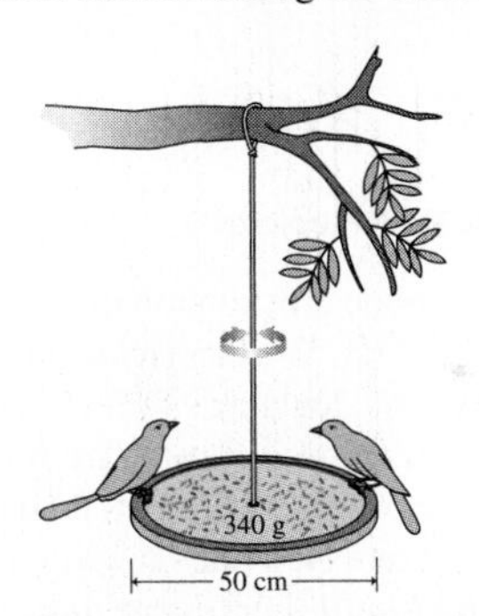

FIGURE 13.36 Problem 82

83. While waiting for your plane to take off, you suspend your keys from a thread and set the resulting pendulum oscillating. It completes exactly 90 cycles in 1 minute. You repeat the experiment as the plane accelerates down the runway, and now measure exactly 91 cycles in 1 minute. Find the plane's acceleration.

84. You're working for a playground equipment company, which wants to know the rotational inertia of its swing with a child on board; the combined mass is 20 kg. You observe the child twirling around in the swing, twisting the ropes as shown in Fig. 13.37. As a result, child and swing rise slightly, with the rise h in cm equal to the square of the number of full turns. When the child stops twisting, the swing begins torsional oscillations. You measure the period at 6.91 s. What do you report for the rotational inertia of the child–swing system?

FIGURE 13.37 Problem 84

85. You've inherited your great-grandmother's mantle clock. The clock's timekeeping is established by a pendulum consisting of a 15.0-cm-long rod and a disk 6.35 cm in diameter. The rod passes through a hole in the disk, and the disk is supported at its bottom by a decorative nut mounted on the bottom portion of the rod, which is threaded; see Fig. 13.38. As shown, the bottom of the disk is 1.92 cm above the bottom of the rod. There are 20.0 threads per cm, meaning that one full turn of the decorative nut moves the disk up or down by 1/20 cm. The clock is beautiful, but it isn't accurate; you note that it's losing 1.5 minutes per day. But you realize that the decorative nut is an adjustment mechanism, and you decide to adjust the clock's timekeeping. (a) Should you turn the nut to move the disk up or down? (b) How many times should you turn the nut? Note: The disk is massive enough that you can safely neglect the mass of the rod and nut. But you can't neglect the disk's size compared with the rod length, so you don't have a *simple* pendulum. Furthermore, note that both the effective length of the pendulum *and* its rotational inertia change as the disk moves up or down the shaft. You can either solve a quadratic or you can use calculus to get an approximate but nevertheless very accurate answer. **CH**

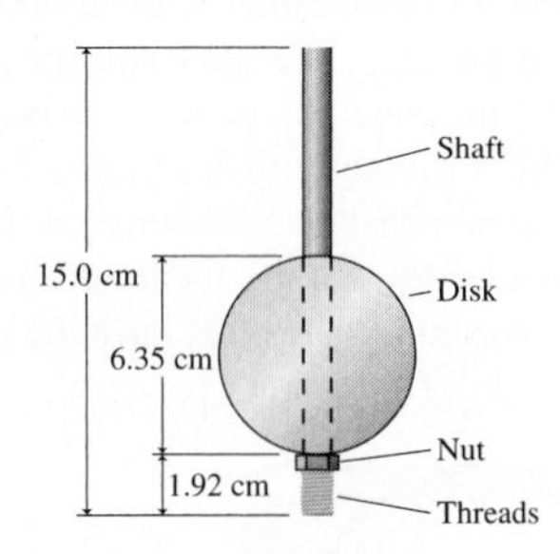

FIGURE 13.38 Problem 85

86. This problem explores the nonlinear pendulum discussed qualitatively in Conceptual Example 13.1. You can tackle this problem if you have experience with your calculator's differential-equation solving capabilities or if you've used a software program like *Mathematica* or *Maple* that can solve differential equations numerically. On page 228 we wrote Newton's law for a pendulum in the form $I\,d^2\theta/dt^2 = -mgL\sin\theta$. (a) Rewrite this equation in a form suitable for a simple pendulum, but without making the approximation $\sin\theta \cong \theta$. Although it won't affect the form of the equation, assume that your pendulum uses a massless rigid rod rather than a string, so it can turn completely upside down without collapsing. (b) Enter your equation into your calculator or software, and produce graphical solutions to the equation for the situation where you specify the initial kinetic energy K_0 when the pendulum is at its bottommost position. In particular, describe solutions for (i) $K_0 \ll U_{max}$, (ii) $K_0 \lessapprox U_{max}$, and (iii) $K_0 > U_{max}$. Here U_{max} is the maximum possible potential energy for the system, which occurs when the pendulum is completely upside down; $U_0 = 2Lmg$, where L is the pendulum's length. **COMP**

Passage Problems

Physicians and physiologists are interested in the long-term effects of apparent weightlessness on the human body. Among these effects are redistribution of body fluids to the upper body, loss of muscle tone, and overall mass loss. One method of measuring mass in the apparent weightlessness of an orbiting spacecraft is to strap the astronaut into a chairlike device mounted on springs (Fig. 13.39). This *body mass measuring device* (BMMD) is set oscillating in simple harmonic motion, and measurement of the oscillation period, along with the known spring constant and mass of the chair itself, then yields the astronaut's mass. When a 60-kg astronaut is strapped into the 20-kg chair, the time for three oscillation periods is measured to be 6.0 s.

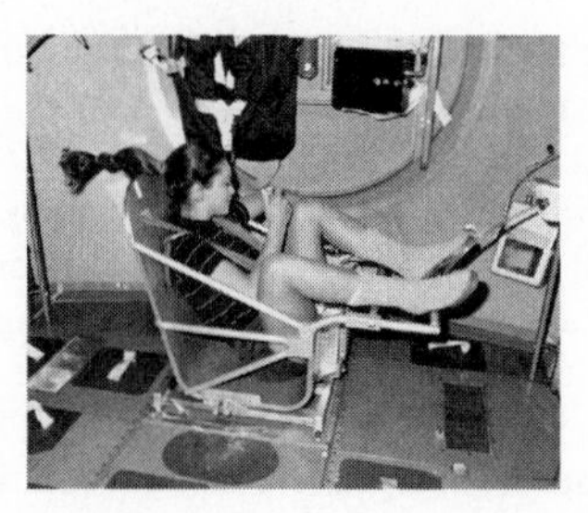

FIGURE 13.39 Astronaut Tamara Jernigan uses a body mass measuring device in the Spacelab Life Sciences Module (Passage Problems 87–90).

87. If a 90-kg astronaut is "weighed" with this BMMD, the time for three periods will be
    a. 50% longer.
    b. shorter by less than 50%.
    c. longer by less than 50%.
    d. longer by more than 50%.
88. If the same device were used on Earth, the results for a given astronaut (assuming mass hasn't yet been lost in space) would be
    a. the same.
    b. greater than in an orbiting spacecraft.
    c. less than in an orbiting spacecraft.
    d. meaningless, because the device won't work on Earth.
89. If an astronaut's mass declines linearly with time while she's in orbit, the oscillation period of the BMMD will
    a. decrease at an ever-decreasing rate.
    b. decrease linearly with time.
    c. decrease at an ever-increasing rate.
    d. increase linearly with time.
90. The spring constant for the BMMD described here is
    a. 80 N/m.
    b. 80π N/m.
    c. 2 N/m.
    d. $80\pi^2$ N/m.
    e. none of the above.

Answers to Chapter Questions

Answer to Chapter Opening Question

1 Hz is 1 cycle per second, so that's 32,768 oscillation cycles per second. This number is 2^{15}, so it takes 15 divisions by two to reduce to the one "tick" per second that drives the watch.

Answers to GOT IT? Questions

13.1 (c)
13.2 Frequencies and periods are the same; amplitudes and phase constants are different because of the different initial displacements and times of release, respectively.
13.3 (1) no change; (2) doubles; (3) doubles
13.4 (1) 1:2; (2) 3:2
13.5 The more energetic oscillator has (1) twice the mass and (2) twice the spring constant. (3) Their maximum speeds are equal.
13.6 (1) c; (2) b
13.7 The frequency, which needs to be at the glass's resonant frequency (although, even at resonance, a sound that's too weak won't break the glass).

Wave Motion

What You Know

- You understand how restoring forces lead to oscillatory motion.
- You're particularly familiar with simple harmonic motion.
- You can describe periodic motion in terms of angular frequency, frequency in hertz, or period, and you know how to relate these three measures of periodicity.

What You're Learning

- You'll learn how to describe wave motion mathematically in terms of spatial and temporal variations.
- You'll add *wavelength* and *wave number* to your vocabulary that already includes *frequency* and *period*.
- You'll see how oscillatory motion in coupled systems leads to waves.
- You'll explore the Newtonian physics behind wave motion for the special case of waves on a taut string or wire.
- You'll learn how to characterize energy carried by waves.
- You'll learn about sound waves and the decibel system for sound intensity.
- You'll learn about wave *reflection* and *interference*.
- You'll see how reflection and interference result in *standing waves*.
- You'll learn about the Doppler effect and shock waves.

How You'll Use It

- The wave concepts you learn here will help you understand electromagnetic waves, introduced in Chapter 29.
- In Part 5 you'll study optics—the science of light that's based ultimately in light's nature as an electromagnetic wave. Wave concepts will be especially important in Chapter 32.
- In Part 6 you'll see how wave concepts are at the heart of quantum physics, even in the description of matter.

Humans and other animals communicate using sound waves. Light and related waves enable us to visualize our surroundings and provide virtually all our information about the universe beyond Earth. Our cell phones keep us connected via radio waves. Physicians probe our bodies with ultrasound waves. Radio waves connect our wireless devices to the Internet and cook the food in our microwave ovens. Earthquakes trigger waves in the solid Earth and may generate dangerous tsunamis. **Wave motion** is an essential feature of our physical environment.

All these examples involve a disturbance that moves or **propagates** through space. The disturbance carries energy, but not matter. Air doesn't move from your mouth to a listener's ear, but sound energy does. Water doesn't move across the open ocean, but wave energy does. **A wave is a traveling disturbance that transports energy but not matter.**

Ocean waves travel thousands of kilometers across the open sea before breaking on shore. How much water moves with the waves?

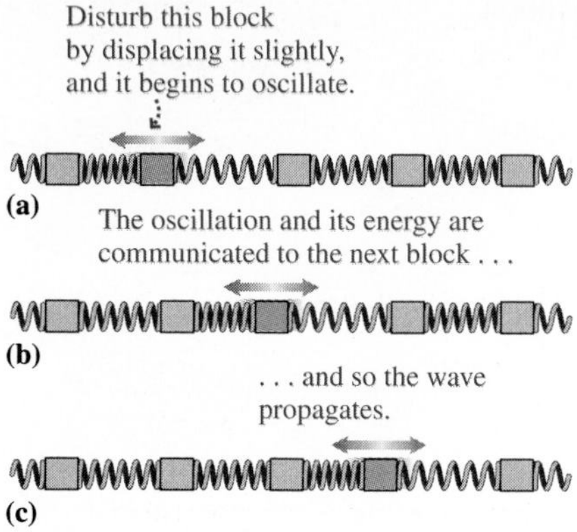

Disturb this block by displacing it slightly, and it begins to oscillate.

(a) The oscillation and its energy are communicated to the next block . . .

(b) . . . and so the wave propagates.

(c)

FIGURE 14.1 Wave propagation in a mass–spring system.

14.1 Waves and Their Properties

In this chapter we'll deal with **mechanical waves**, which are disturbances of some material **medium**, such as air, water, a violin string, or Earth's interior. Visible and infrared light waves, radio waves, ultraviolet and X rays, in contrast, are **electromagnetic waves**. They share many properties with mechanical waves, but they don't require a material medium. We'll treat electromagnetic waves in Chapters 29–32.

Mechanical waves occur when a disturbance in one part of a medium is communicated to adjacent parts. Figure 14.1 shows a multiple mass–spring system that serves as a model for many types of mechanical waves. Disturb one mass, and it goes into simple harmonic motion. But because the masses are connected, that motion is communicated to the adjacent mass. As a result, both the disturbance and its associated energy propagate along the mass–spring system, disturbing successive masses as they go.

✓TIP Wave Motions

A wave moves energy from place to place, but not matter. However, that doesn't mean that the matter making up the wave medium doesn't move. It does, undergoing localized oscillatory motion as the wave passes. But once the wave is gone, the disturbed matter returns to its equilibrium state. Don't confuse this localized motion of the medium with the motion of the wave itself. Both occur, but only the latter carries energy from one place to another.

Longitudinal and Transverse Waves

In Fig. 14.1, we disturbed the system by displacing one block so its subsequent oscillations were back and forth along the structure—in the same direction as the wave propagation. The result is a **longitudinal wave**. Sound is a longitudinal wave, as we'll see in Section 14.4. We could equally well displace a mass at right angles, as in Fig. 14.2. Then we get a **transverse wave**, whose disturbance is at right angles to the wave propagation. Some waves include both longitudinal and transverse motions, as shown for a water wave in Fig. 14.3.

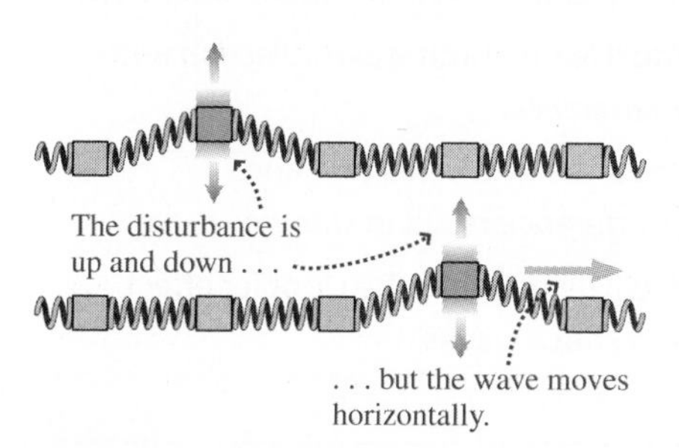

FIGURE 14.2 A transverse wave.

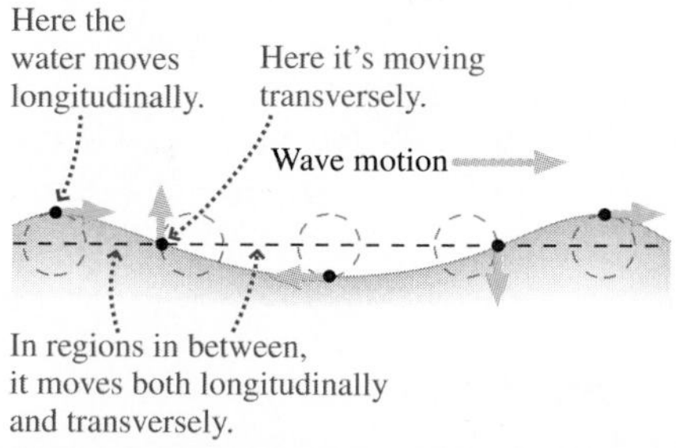

Here the water moves longitudinally.

Here it's moving transversely.

Wave motion

In regions in between, it moves both longitudinally and transversely.

FIGURE 14.3 A water wave has both longitudinal and transverse components.

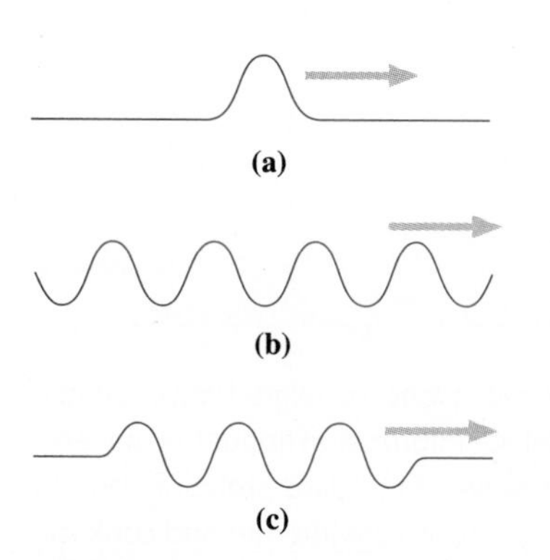

FIGURE 14.4 (a) A pulse, (b) a continuous wave, and (c) a wave train.

Amplitude and Waveform

The maximum value of a wave's disturbance is the wave **amplitude**. For a water wave, amplitude is the maximum height above the undisturbed water level; for a sound wave, it's the maximum excess air pressure; for the waves of Figs. 14.1 and 14.2, it's the maximum displacement of a mass.

Wave disturbances come in many shapes, called **waveforms** (Fig. 14.4). An isolated disturbance is a **pulse**, which occurs when the medium is disturbed only briefly. A **continuous wave** results from an ongoing periodic disturbance. Intermediate between these extremes is a **wave train**, resulting from a periodic disturbance lasting a finite time.

The wavelength can be measured between any two repeating points on the wave.

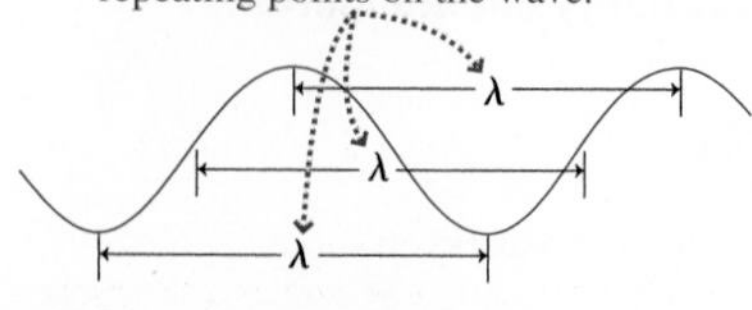

FIGURE 14.5 The wavelength λ is the distance over which the wave pattern repeats.

Wavelength, Period, and Frequency

A continuous wave repeats in both space and time. The **wavelength** λ is the *distance* over which the wave pattern repeats (Fig. 14.5). The wave **period** T is the *time* for one complete oscillation. The **frequency** f, or number of wave cycles per unit time, is the inverse of the period.

Wave Speed

A wave travels at a specific speed through its medium. The speed of sound in air is about 340 m/s. Small ripples on water move at about 20 cm/s, while earthquake waves travel at several kilometers per second. The physical properties of the medium ultimately determine the wave speed, as we'll see in Section 14.3.

Wave speed, wavelength, and period are related. In one wave period, a fixed observer sees one complete wavelength go by (Fig. 14.6). Thus, the wave moves one wavelength in one period, so its speed is

$$v = \frac{\lambda}{T} = \lambda f \qquad \text{(wave speed)} \tag{14.1}$$

where the second equality follows because period and frequency are inverses.

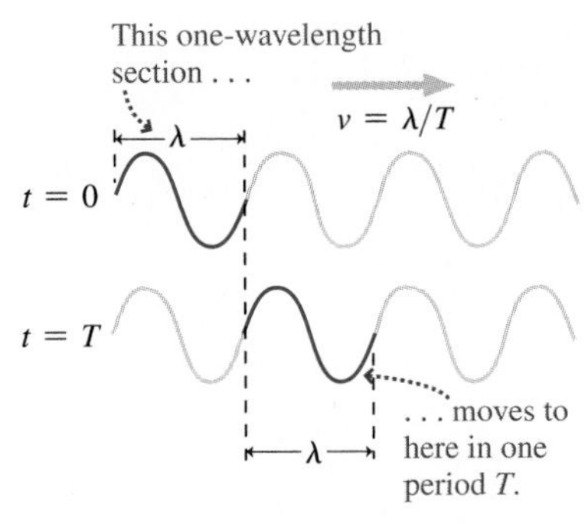

FIGURE 14.6 One full cycle passes a given point in one wave period T; the wave speed is therefore $v = \lambda/T$.

GOT IT? 14.1 A boat bobs up and down on a water wave, moving 2 m vertically in 1 s. A wave crest moves 10 m horizontally in 2 s. Is the wave speed (a) 2 m/s or (b) 5 m/s? Explain.

14.2 Wave Math

Figure 14.7 shows "snapshots" of a wave pulse at time $t = 0$ and at some later time t. Initially the wave disturbance y is some function of position: $y = f(x)$. Later the pulse has moved to the right a distance vt, but its shape, described by the function f, is the same. We can represent this displaced pulse by replacing x with $x - vt$ as the argument of the function f. Then x has to be larger—by the amount vt—to give the same value of f as it did before. For example, this particular pulse peaks when the argument of f is zero. Initially, that occurred when x was zero. Replacing x by $x - vt$ ensures that the argument becomes zero when $x = vt$, putting the peak at this new position. As time increases, so does vt and therefore the value of x corresponding to the peak. Thus $f(x - vt)$ correctly represents the moving pulse.

Although we considered a single pulse, this argument applies to *any* function $f(x)$, including continuous waves: Replace the argument x with $x - vt$, and the function $f(x - vt)$ describes a wave moving in the positive x-direction with speed v. You can convince yourself that a function of the form $f(x + vt)$ describes a wave moving in the negative x-direction.

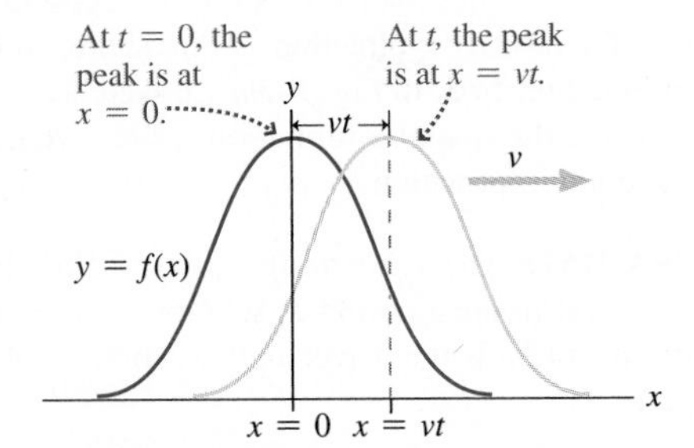

FIGURE 14.7 The wave pulse moves a distance vt in time t, but its shape stays the same.

A particularly important case is a **simple harmonic wave**, for which a "snapshot" at time $t = 0$ shows a sinusoidal function. We'll choose coordinates so that $x = 0$ is at a maximum of the wave, making the function a cosine (Fig. 14.8*a*). Then $y(x, 0) = A\cos kx$, where A is the amplitude and k is a constant, called the **wave number**. We can find k because we know that the wave repeats in one wavelength λ. Since the period of the cosine function is 2π, we therefore want kx to be 2π when x equals λ. Then $k\lambda = 2\pi$, or

$$k = \frac{2\pi}{\lambda} \qquad \text{(wave number)} \tag{14.2}$$

To describe a wave moving with speed v, we replace x in the expression $A\cos kx$ with $x - vt$, giving $y(x, t) = A\cos[k(x - vt)]$. If we now sit at the point $x = 0$, we'll see an oscillation described by $y(0, t) = A\cos(-kvt) = A\cos(kvt)$, where the last step follows because $\cos(-x) = \cos x$. But we found that $k = 2\pi/\lambda$, and Equation 14.1 shows that $v = \lambda/T$, so the argument of the cosine function becomes $kvt = (2\pi/\lambda)(\lambda/T)t = 2\pi t/T$.

In Chapter 13, we introduced the **angular frequency** $\omega = 2\pi/T$ in describing simple harmonic motion; here the same quantity arises in describing wave motion. And no wonder: At a fixed point in space, the wave medium undergoes simple harmonic motion with angular frequency $\omega = 2\pi/T$ (Fig. 14.8*b*). Putting this all together, we can write a traveling sinusoidal wave in the form

$$y(x, t) = A\cos(kx \pm \omega t) \qquad \text{(sinusoidal wave)} \tag{14.3}$$

where we've written $\pm$ so we can describe a wave going in the positive x-direction ($-$ sign) or the negative x-direction ($+$ sign). The argument of the cosine is called the

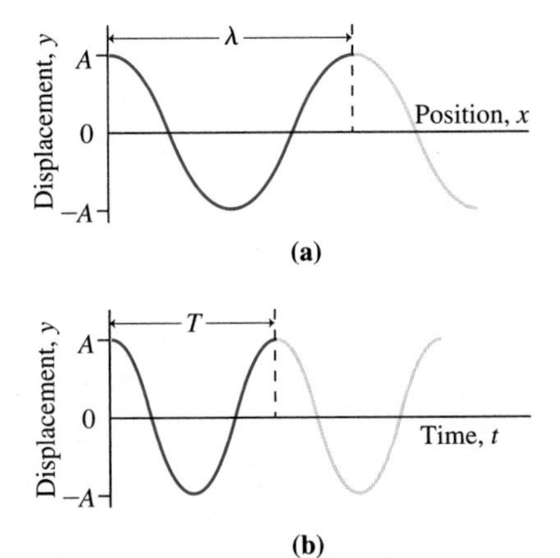

FIGURE 14.8 A sinusoidal wave (a) as a function of position at fixed time $t = 0$ and (b) as a function of time at fixed position $x = 0$.

wave's **phase**. Note that k and ω are related to the more familiar wavelength λ and period T in the same way: $k = 2\pi/\lambda$ and $\omega = 2\pi/T$. Just as ω is a measure of frequency—oscillation cycles per unit *time*, with an extra factor of 2π—so is k a measure of **spatial frequency**—oscillation cycles per unit *distance*, again with that factor of 2π to make the math simpler. The relations between k, λ and ω, T allow us to rewrite the wave speed of Equation 14.1 in terms of k and ω:

$$v = \frac{\lambda}{T} = \frac{2\pi/k}{2\pi/\omega} = \frac{\omega}{k} \tag{14.4}$$

EXAMPLE 14.1 Describing a Wave: Surfing

A surfer paddles out beyond the breaking surf to where the waves are sinusoidal in shape, with crests 14 m apart. The surfer bobs a vertical distance 3.6 m from trough to crest, a process that takes 1.5 s. Find the wave speed, and describe the wave using Equation 14.3.

INTERPRET This is a problem about a simple harmonic wave—that is, a wave with sinusoidal shape.

DEVELOP We'll take $x = 0$ at the location of a wave crest when $t = 0$, so Equation 14.3, $y(x, t) = A\cos(kx \pm \omega t)$, applies. Let's take the positive x-direction toward shore, so we'll use the minus sign in Equation 14.3. In Fig. 14.9a we sketched a "snapshot" of the wave, showing the spatial information we're given. Figure 14.9b shows the temporal information.

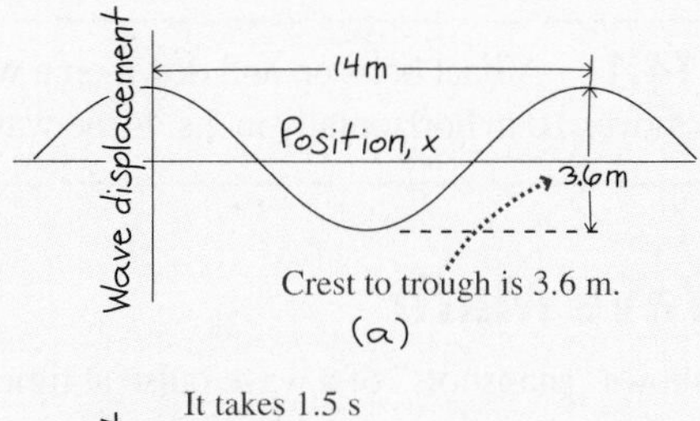

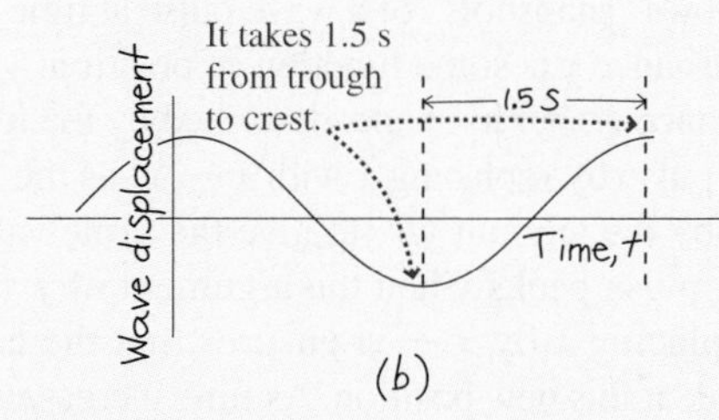

FIGURE 14.9 Our sketch of displacement versus (a) position and (b) time.

EVALUATE The 1.5-s trough-to-crest time in Fig. 14.9b is half the full crest-to-crest period T, so $T = 3.0$ s. The crest-to-crest distance in Fig. 14.9a is the wavelength λ, so $\lambda = 14$ m. Then Equation 14.1 gives

$$v = \frac{\lambda}{T} = \frac{14\text{ m}}{3.0\text{ s}} = 4.7\text{ m/s}$$

To describe the wave with Equation 14.3 we need the amplitude A, wave number k, and angular frequency ω. The amplitude is half the crest-to-trough displacement, or $A = 1.8$ m, as shown in Fig. 14.9a. The wave number k and angular frequency ω then follow from λ and T: $k = 2\pi/\lambda = 0.449\text{ m}^{-1}$ and $\omega = 2\pi/T = 2.09\text{ s}^{-1}$. Then the wave description is

$$y(x, t) = 1.8\cos(0.449x - 2.09t)$$

with y and x in meters and t in seconds.

ASSESS As a check on our answer, let's see whether our values of ω and k satisfy Equation 14.4: $v = \omega/k = 2.09\text{ s}^{-1}/0.449\text{ m}^{-1} = 4.7$ m/s. Thus the pairs λ, T and ω, k are equivalent ways to describe the same wave. ■

GOT IT? 14.2 The figure shows snapshots of two waves propagating with the same speed. Which has the greater (1) amplitude, (2) wavelength, (3) period, (4) wave number, and (5) frequency?

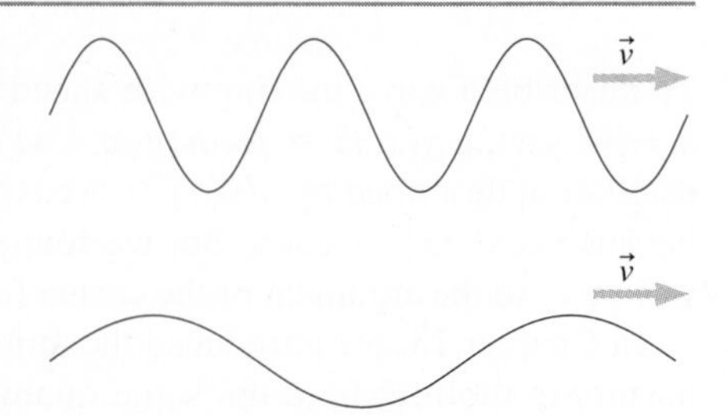

The Wave Equation

We argued our way to Equation 14.3 for a sinusoidal wave on mathematical grounds alone. Whether such waves are actually possible depends on the physical properties of the medium. Many media do, in fact, support waves as described by Equation 14.3. We'll explore one case in detail in the next section. More generally, physicists analyze the behavior of a medium in response to disturbances. Often the analysis results in an equation relating the space and time derivatives of the disturbed quantity:

$$\frac{\partial^2 y}{\partial x^2} = \frac{1}{v^2}\frac{\partial^2 y}{\partial t^2} \quad \text{(wave equation)} \tag{14.5}$$

This is the **wave equation** for waves propagating in one dimension. Here y is the wave disturbance—the height of a water wave, the pressure in a sound wave, and so on. The quantity v is the wave speed, which usually appears as a combination of quantities related to properties of the medium, and thus allows physicists to deduce the wave speed. Because the wave disturbance is a function of the two variables x (spatial position) and t (time), the derivatives here are **partial derivatives**, designated with the symbol ∂ and indicating differentiation with respect to one variable while the other is held constant. Thus the wave equation is a **partial differential equation**. Solving such equations requires more advanced math courses, but you can show directly (Problem 69) that Equation 14.3 satisfies the wave equation, with wave speed $v = \omega/k$. More generally, *any* function of the form $f(x \pm vt)$ satisfies the wave equation, as you can show in Problem 70. You'll encounter the wave equation again in Chapter 29, when you study electromagnetic waves.

14.3 Waves on a String

PhET: Wave on a String

Scientists and engineers generally explore wave possibilities in a medium by applying the laws of physics and deriving a wave equation similar to Equation 14.5. Such analysis reveals the wave speed and other wave properties. Here we'll take a simpler approach to one special case: transverse waves on a stretched string. Our results are directly applicable to musical instruments, climbing ropes, bridge cables, and other elongated structures.

Our string has mass per unit length μ in kilograms per meter, and it's stretched to a tension force F. Consider a wave pulse propagating to the right, as shown in Fig. 14.10*a*. We'll use Newton's law to analyze the string's motion and determine the speed of the pulse. It's easiest to do this in a frame of reference moving with the pulse; in that frame, the entire string moves *leftward* with the pulse speed v. At the pulse location, however, the string's motion deviates from horizontal as it rides up and down over the pulse (Fig. 14.10*b*).

Whatever the pulse shape, a small section at the top forms a circular arc of some radius R, as shown in Fig. 14.10*c*. Then the string right at the top of the pulse undergoes circular motion with speed v and radius R; if its mass is m, Newton's law requires that a force of magnitude mv^2/R act toward the center of curvature to keep the string on its circular path. This force is provided by the difference in the direction of the string tension between the two ends of the section; as Fig. 14.10*c* shows, the tension at each end contributes a downward component $F\sin\theta$. Then the net force on the segment has magnitude $2F\sin\theta$ and points toward the center of curvature.

Now we make an additional assumption: that the disturbance of the string is small, in the sense that the string remains almost horizontal even at the pulse. Then the angle θ is small, and we can apply the approximation $\sin\theta \simeq \theta$. Therefore, the net force on the string section becomes approximately $2F\theta$. Furthermore, the small-disturbance approximation means that the tension doesn't vary significantly from its undisturbed value, so F in this expression is essentially the same F we're using to characterize the tension throughout the string. Finally, our curved string section forms a circular arc whose length, from Fig. 14.10*c*, is $2\theta R$. Multiplying by the mass per unit length μ gives its mass: $m = 2\theta R\mu$. Now we can apply Newton's law, equating the net force $2F\theta$ to the mass times acceleration:

$$2F\theta = \frac{mv^2}{R} = \frac{2\theta R\mu v^2}{R} = 2\theta\mu v^2$$

Solving for the wave speed v then gives

$$v = \sqrt{\frac{F}{\mu}} \tag{14.6}$$

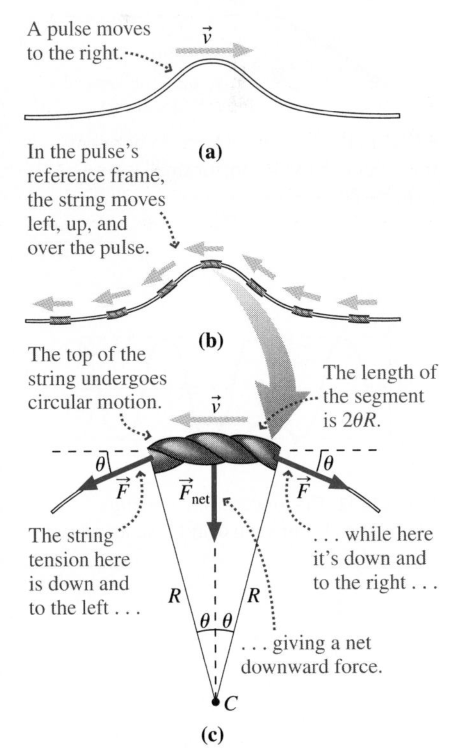

FIGURE 14.10 A wave pulse moving on a string. In (c), each of the diagonal forces shown contributes a downward component $F\sin\theta$.

Does this make sense? The greater the tension F, the greater the string's acceleration, and the more rapidly the wave should propagate. The string's inertia, on the other hand, limits the acceleration, and therefore a greater mass per unit length should slow the wave. Equation 14.6, with F in the numerator and μ in the denominator, reflects both these trends.

We've made no assumptions here other than to assume that the disturbance is small. Therefore, Equation 14.6 applies to small-amplitude pulses, continuous waves, and wave trains of any shape.

EXAMPLE 14.2 Wave Speed and Tension Force: Rock Climbing

A 43-m-long rope of mass 5.0 kg joins two climbers. One climber strikes the rope, and 1.4 s later the second climber feels the effect. What's the rope tension?

INTERPRET We're asked for the rope tension. Although wave speed isn't mentioned explicitly, we just learned to relate wave speed and rope tension. Striking the rope produces a wave, which the second climber feels. We're given the time it takes that wave to propagate along the rope.

DEVELOP Equation 14.6, $v = \sqrt{F/\mu}$, gives the relations among rope tension, mass per unit length, and wave speed. Our plan is to solve for the rope tension, but first we need to find μ and v from the given information.

EVALUATE We're given the rope's mass m and length L, so its mass per unit length is $\mu = m/L$. We're given the time t for the wave to travel the rope length L, so the wave speed is $v = L/t$. Solving Equation 14.6 for F then gives

$$F = \mu v^2 = \left(\frac{m}{L}\right)\left(\frac{L}{t}\right)^2 = \frac{mL}{t^2} = \frac{(5.0\text{ kg})(43\text{ m})}{(1.4\text{ s})^2} = 110\text{ N}$$

ASSESS Is this number reasonable? A typical adult weighs around 700 N, so the rope is supporting only a small fraction of the lower climber's weight—a reasonable situation. ■

Wave Power

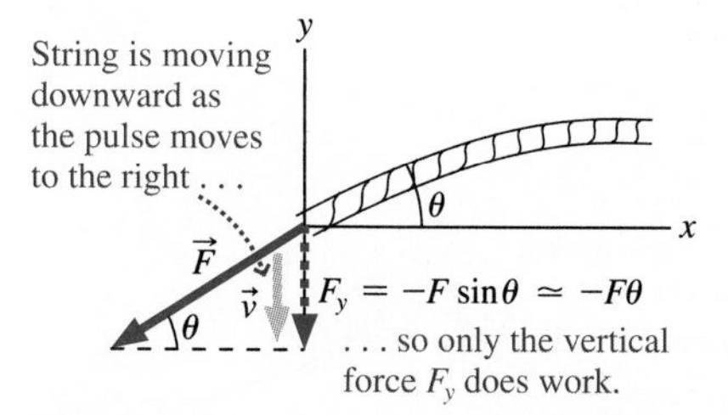

FIGURE 14.11 The vertical force component does work on the string; for small θ, $\sin\theta \simeq \theta$, so $F_y \simeq F\theta$.

Waves carry energy. For a wave on a string, the vertical component of the tension force does work that transfers energy along the string. Figure 14.11 shows that the vertical force on the string at the left side of the pulse is approximately $-F\theta$. As we showed in Chapter 6, power—the rate of doing work—is the product of force and velocity, so the power here is $P = -F\theta u$, where u is the vertical velocity of the string—*not* the wave speed. Rather, the vertical velocity is the rate of change of the string displacement y. For a simple harmonic wave, $y(x, t) = A\cos(kx - \omega t)$. We can differentiate this to get

$$u = \frac{dy}{dt} = A\omega\sin(kx - \omega t)$$

where we used the chain rule, differentiating cosine to $-$sine and then multiplying by the derivative, $-\omega$, of the cosine's argument $kx - \omega t$. As Fig. 14.11 shows, the tangent of the angle θ is the slope, dy/dx, of the string. For small angles, $\tan\theta \simeq \theta$ so $\theta \simeq dy/dx = -kA\sin(kx - \omega t)$. Putting these results for u and θ in our expression for power gives $P = -F\theta u = F\omega kA^2\sin^2(kx - \omega t)$. The sine term shows that the power fluctuates in space and time. Usually we're interested in the *average* power, $\overline{P} = \frac{1}{2}F\omega kA^2$, which follows because the average value of $\sin^2$ is $\frac{1}{2}$ (Fig. 14.12). We can give this a more physical meaning if we use Equations 14.4 and 14.6 to write $k = \omega/v$ and $F = \mu v^2$, with v the wave speed. Then we have

$$\overline{P} = \tfrac{1}{2}\mu\omega^2A^2v \qquad (14.7)$$

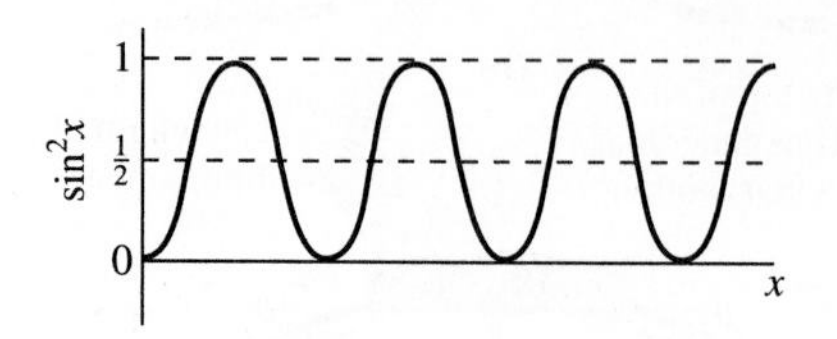

FIGURE 14.12 The function $\sin^2 x$ swings symmetrically between 0 and 1, so its average value is $\frac{1}{2}$.

This equation gives the sensible result that wave power is directly proportional to the speed v at which energy moves along the wave.

Wave Intensity

Total power is useful in describing waves confined to narrow structures like strings for mechanical waves or optical fibers for electromagnetic waves. But for waves in three-dimensional media, like sound in air, it makes more sense to talk about the **intensity**, or the rate at which the wave carries energy across a unit area perpendicular to the wave propagation. Intensity is thus power per unit area, measured in watts per square meter (W/m^2).

Wavefronts are surfaces on which the wave phase is constant—for example, wave crests. A **plane wave** is one whose wavefronts are planes. Since the wave doesn't spread out, its intensity remains constant (Fig. 14.13*a*). But as waves propagate from a localized source, they spread and their intensity drops. **Spherical waves** originate from point sources and spherical wavefronts spread in all directions. Since the area of a sphere is $4\pi r^2$, the intensity of a spherical wave decreases as the inverse square of the distance from its source:

$$I = \frac{P}{A} = \frac{P}{4\pi r^2} \quad \text{(spherical wave)} \qquad (14.8)$$

Note that energy isn't lost here; rather, the same energy is spread over ever-larger areas as the wave propagates (Fig. 14.13*b*). Table 14.1 lists some typical wave intensities.

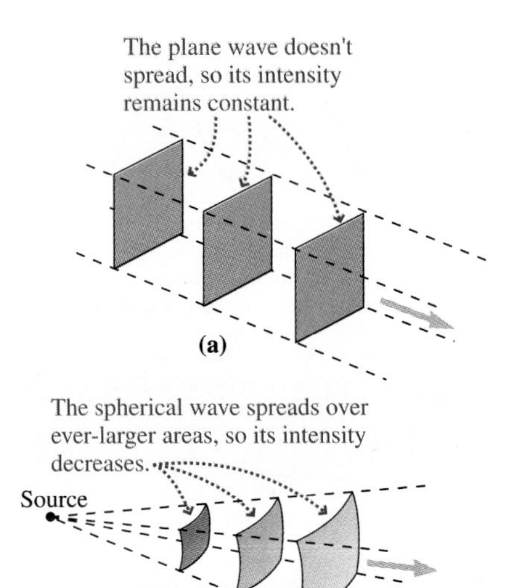

FIGURE 14.13 (a) Plane and (b) spherical waves.

Table 14.1 Wave Intensities

Wave	Intensity, W/m^2
Sound, 4 m from loud rock band	1
Sound, jet aircraft at 50 m	10
Sound, whisper at 1 m	10^{-10}
Light, sunlight at Earth's orbit	1364
Light, sunlight at Jupiter's orbit	50
Light, 1 m from typical camera flash	4000
Light, at target of laser fusion experiment	10^{18}
TV signal, 5 km from 50-kW transmitter	1.6×10^{-4}
Microwaves, inside microwave oven	6000
Earthquake wave, 5 km from Richter 7.0 quake	4×10^4

EXAMPLE 14.3 Evaluating Wave Intensity: A Reading Light

Your book is 1.9 m from a 9.2-W LED lamp, and the light is barely adequate for reading. How far from a 4.9-W LED would the book have to be to get the same intensity at the page?

INTERPRET This is a problem about wave intensity, and we identify the LEDs as sources of spherical waves.

DEVELOP Equation 14.8, $I = P/(4\pi r^2)$, gives the intensity. We want both LEDs to produce the same intensity, so we have $I = P_{9.2}/(4\pi r_{9.2}^2) = P_{4.9}/(4\pi r_{4.9}^2)$.

EVALUATE We then solve for the unknown distance r_{40}:

$$r_{4.9} = r_{9.2}\sqrt{\frac{P_{4.9}}{P_{9.2}}} = (1.9\text{ m})\sqrt{\frac{4.9\text{ W}}{9.2\text{ W}}} = 1.4\text{ m}$$

ASSESS Make sense? Although the 4.9-W LED has only about half the power output, the decrease in distance isn't as great as you might expect because the intensity depends on the inverse *square* of the distance. By the way, those energy-efficient LEDs are approximately equivalent to 75-W and 40-W incandescent bulbs, respectively. ■

GOT IT? 14.3 Two identical stars are different distances from Earth, and the intensity of the light from the more distant star as received at Earth is only 1% that of the closer star. Is the more distant star (a) twice as far away, (b) 100 times as far away, (c) 10 times as far away, or (d) $\sqrt{10}$ times as far away?

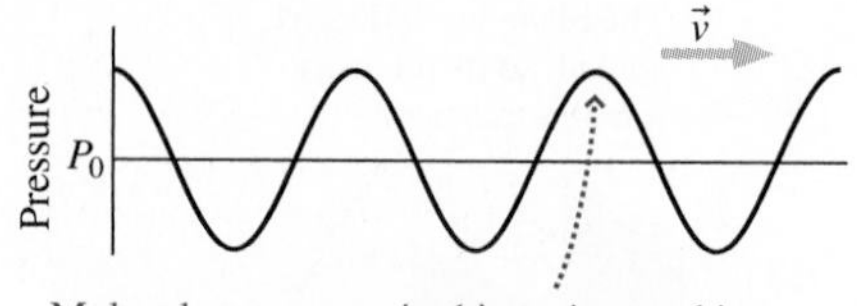

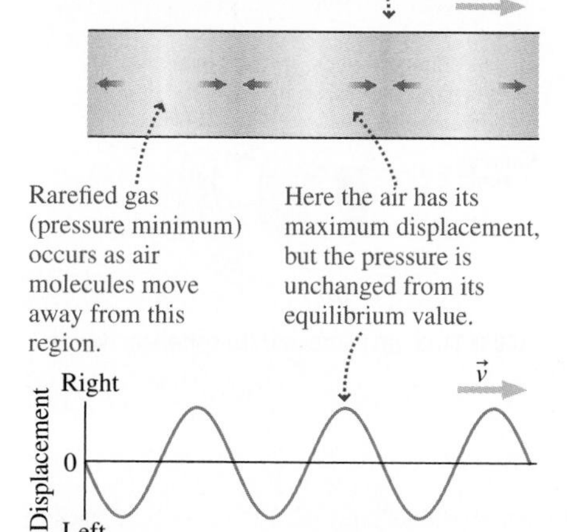

FIGURE 14.14 A sound wave consists of alternating regions of compression (higher density and pressure) and rarefaction (lower density and pressure) propagating through the air.

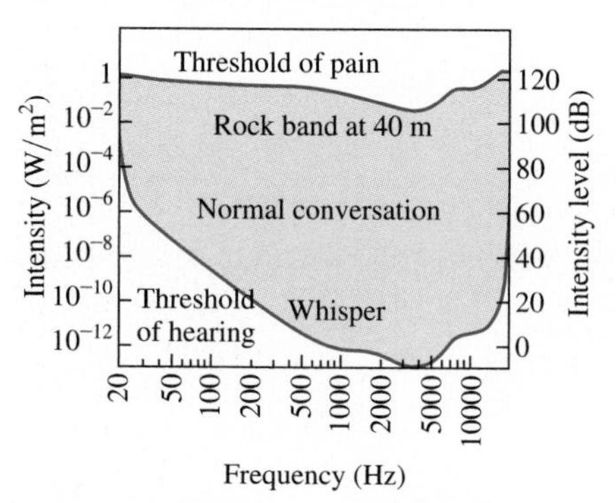

FIGURE 14.15 The human ear responds to sound whose intensity and frequency lie within the shaded region.

14.4 Sound Waves

Sound waves are longitudinal mechanical waves that propagate through gases, liquids, and solids. Most familiar is sound in air. Here the wave disturbance comprises a small change in air pressure and density accompanied by a back-and-forth motion of the air (Fig. 14.14). The speed of sound in air and other gases depends on the background pressure P (force per unit area) and density ρ (mass per unit volume):

$$v = \sqrt{\frac{\gamma P}{\rho}} \tag{14.9}$$

where γ is a constant characteristic of the gas. For air and other diatomic gases, γ is $\frac{7}{5}$; for monatomic gases like helium, it's $\frac{5}{3}$. Sound propagates faster in liquids and solids because they're less compressible.

Sound and the Human Ear

The human ear responds to a wide range of sound intensities and frequencies, as shown in Fig. 14.15. Audible frequencies range from around 20 Hz to 20 kHz, although the upper limit drops with age. Figure 14.15 shows that the minimum intensity for sound to be audible increases at high and low frequencies; that's the reason for the "loudness" switch on your stereo system, which boosts lows and highs to make the sound richer at low volumes. Dolphins, bats, and other creatures can hear much higher frequencies than we humans; bats locate their prey with sound waves at frequencies approaching 100 kHz. Medical ultrasound frequencies extend to tens of MHz.

Decibels

Figure 14.15 shows that the human ear responds to an extremely broad range of sound intensities, covering some 12 orders of magnitude; that's why Fig. 14.15 has a logarithmic scale. We therefore quantify sound levels using a logarithmic unit called the **decibel** (dB). The **sound intensity level** β in decibels is defined by

$$\beta = 10 \log\left(\frac{I}{I_0}\right) \tag{14.10}$$

where I is the intensity in W/m^2 and $I_0 = 10^{-12}$ W/m^2 is a reference level chosen as the approximate threshold of hearing at 1 kHz. Since the logarithm of 10 is 1, an increase of 10 dB corresponds to a factor-of-10 increase in the intensity I. Your ears, however, don't respond linearly, and for intensity levels above about 40 dB, you perceive a 10-dB increase as making the sound roughly twice as loud.

EXAMPLE 14.4 Decibels: Turn Down the TV!

Your sister is watching TV, the sound blasting at 75 dB. You yell to her to turn down the volume, and she lowers the intensity level to 60 dB. By what factor has the power dropped?

INTERPRET This problem is about the relation between power and sound intensity level as measured in decibels.

DEVELOP Equation 14.10, $\beta = 10\log(I/I_0)$, relates the decibel level to the intensity, or power per unit area. At a fixed distance, the sound intensity is proportional to the power from the TV speaker, so in this example we can replace I by P in Equation 14.10.

EVALUATE Call the original 75-dB level β_1; then Equation 14.10 reads $\beta_1 = 10\log(P_1/P_0) = 10\log P_1 - 10\log P_0$, where P_1 is the corresponding power and P_0 is the reference-level power. At the turned-down power P_2, the equation reads $\beta_2 = 10\log P_2 - 10\log P_0$. Subtracting our two equations gives

$$\beta_2 - \beta_1 = 10\log P_2 - 10\log P_1 = 10\log\left(\frac{P_2}{P_1}\right)$$

Therefore, $\log(P_2/P_1) = (\beta_2 - \beta_1)/10 = (60 - 75)/10 = -1.5$. The answer we want is the ratio P_2/P_1, and because logarithms and exponentials are inverses, we have $P_2/P_1 = 10^{-1.5} = 0.032$.

ASSESS Although we worked this problem using Equation 14.10, you can often do decibels in your head. Here the intensity level has dropped by 15 dB, corresponding to 1.5 orders of magnitude in actual intensity. So the intensity—and therefore the TV's power—has dropped by a factor of $10^{-1.5}$, or $1/(10\sqrt{10})$. Since $\sqrt{10}$ is about 3, that's about 1/30. Because you perceive each 10-dB change as a factor of about 2 in loudness, the reduced volume will sound somewhere between one-fourth and one-half as loud as before. ■

GOT IT? 14.4 Your band needs a new guitar amplifier, and the available models range from 25 W to 250 W of audio power. Will the sound intensity level for the most powerful amplifier compared with the least powerful be (a) 10 times greater, (b) greater by 2.25 dB, or (c) greater by 1 dB?

Video Tutor Demo | **Out-of-Phase Speakers**

MP

PhET: Sound

14.5 Interference

Figure 14.16 shows two wave trains approaching from opposite directions. Where they meet, experiment shows that the net displacement is the sum of the individual displacements. This is true for most waves, at least when the amplitude isn't too large. Waves whose displacements simply add are said to obey the **superposition principle**.

At the point shown in Fig. 14.16*b*, the wave crests coincide and so do the troughs. The resulting wave is, momentarily, twice as big. This is **constructive interference**—two waves superposing to produce a larger wave displacement. A little later, in Fig. 14.16*c*, the two waves cancel; this is **destructive interference**. Wave interference occurs throughout physics, from mechanical waves to light and even with the quantum-mechanical waves that describe matter at the atomic scale. Here we take a quick look at wave interference; we'll consider the interference of light waves in more detail in Chapter 32.

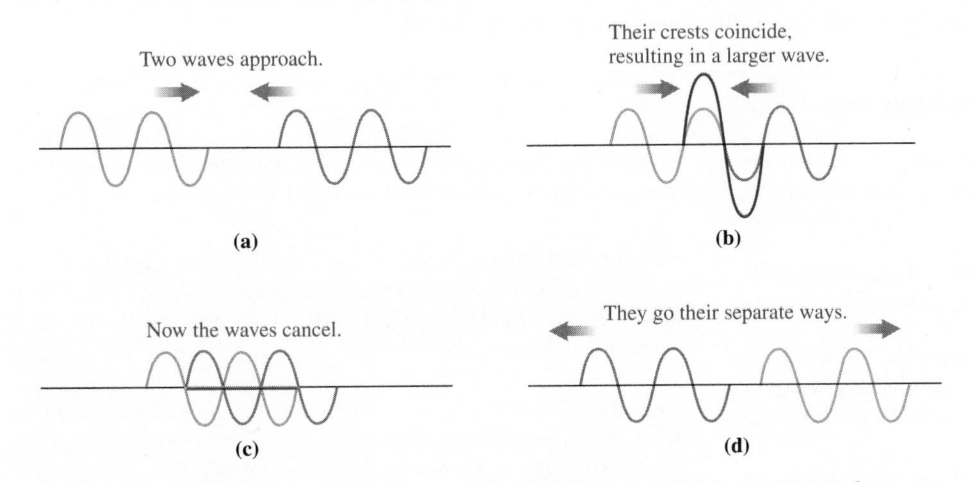

FIGURE 14.16 Wave superposition showing (b) constructive interference and (c) destructive interference.

APPLICATION Noise-Cancelling Headphones

Why does the airline passenger in the photo look so content? Because he's wearing noise-cancelling headphones. These devices exploit interference to actively cancel ambient noise, leaving the headphone signal loud and clear. Each headphone contains a tiny microphone sensing the ambient sound and an amplifier that also inverts the phase of the signal, so crests become troughs and vice versa. The phase-inverted signal is fed to the headphones along with the desired audio. Since the ambient noise delivered to the headphone is inverted—that is, out of phase—relative to the noise coming directly to the ear, the result is destructive interference that greatly reduces the listener's perception of the ambient noise. Peace and quiet!

Fourier Analysis

The superposition principle lets us build complex wave shapes by superposing simpler ones. The French mathematician Jean Baptiste Joseph Fourier (1768–1830) showed that *any* periodic wave can be written as a sum of simple harmonic waves, a process now known as **Fourier analysis**. Figure 14.17 shows a square wave—important, for example, as the "clock" signal that sets the speed of your computer—represented as a superposition of individual sine waves. Fourier analysis has applications ranging from music to structural engineering to communications because it helps us understand how a complex wave behaves if we know how its harmonic components behave. The mix of Fourier components in the waveform from a musical instrument determines the exact sound we hear and accounts for the different sounds from different instruments even when they're playing the same note (Fig. 14.18).

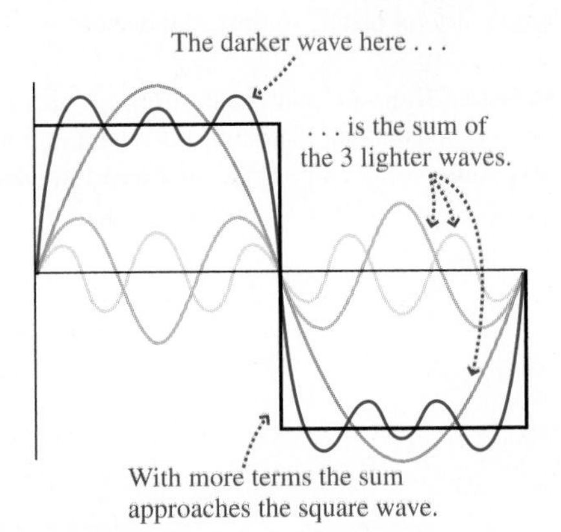

FIGURE 14.17 A square wave built up as a sum of simple harmonic waves. In this case the sum has the form $y(t) = A\sin(\omega t) + \frac{1}{3}A\sin(3\omega t) + \frac{1}{5}A\sin(5\omega t) + \cdots$. Only the first three terms are shown.

PhET: Fourier: Making Waves

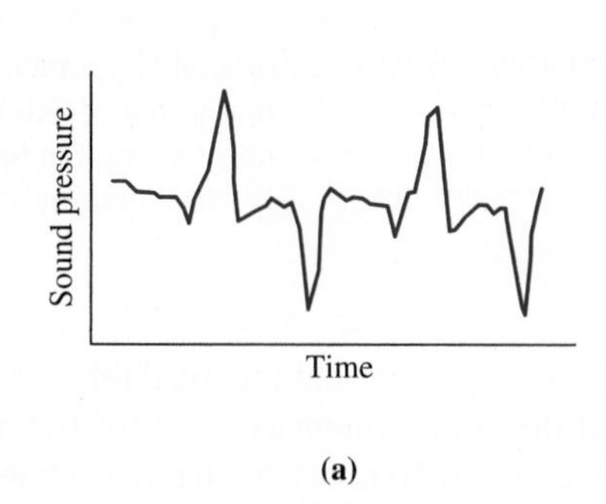

(a)

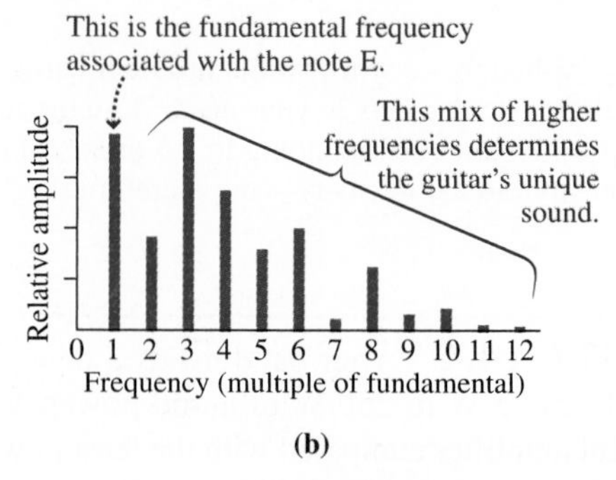

(b)

FIGURE 14.18 (a) An electric guitar plays the note E, producing a complex waveform. (b) Fourier analysis shows the relative strengths of the individual sine waves whose sum produces the waveform.

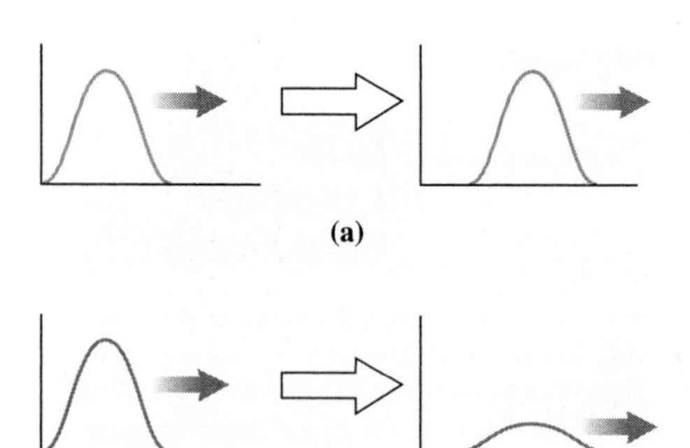

FIGURE 14.19 (a) A wave pulse in a nondispersive medium holds its shape as it propagates. (b) In a dispersive medium, the pulse shape changes.

Dispersion

When wave speed is independent of wavelength, the simple harmonic components making up a complex waveform travel at the same speed. As a result, the waveform maintains its shape. But for some media, wave speed depends on wavelength. Then, individual harmonic waves travel at different speeds, and a complex waveform changes shape as it moves. This phenomenon is called **dispersion** and is illustrated in Fig. 14.19. Waves on the surface of deep water, for example, have speed given by

$$v = \sqrt{\frac{\lambda g}{2\pi}} \qquad (14.11)$$

where λ is the wavelength and g the acceleration of gravity. Because v depends on λ, the waves are dispersive. Dispersion is also important in communications systems; for example, dispersion of the square wave pulses carrying digital data sets the maximum lengths for wires and optical fibers used in computer networks.

CONCEPTUAL EXAMPLE 14.1 Storm Brewing!

It's a lovely, sunny day at the coast, but large waves, their crests far apart, are crashing on the beach. How do these waves tell of a storm at sea that may affect you later?

EVALUATE The phrase "crests far apart" is a clue: It says we're dealing with long-wavelength waves. Equation 14.11 shows that longer-wavelength waves on the ocean surface travel faster. Most ocean waves are generated by frictional forces between wind and water, so there must be strong winds somewhere out at sea. The longest wavelengths travel faster, so they reach shore well in advance of the storm.

ASSESS High-surf warnings often go up in advance of a storm, for the very reason elucidated in this example. Incidentally, wind isn't the only source of ocean waves; so are earthquakes. But the tsunamis they produce are shallow-water waves that don't obey Equation 14.11. You can explore tsunamis further in the Passage Problems.

MAKING THE CONNECTION A storm develops 600 km offshore and starts moving toward you at 40 km/h. Large waves with crests 250 m apart are your first hint of the storm. How long after you observe these waves will the storm hit?

EVALUATE At 40 km/h, it's going to take 15 hours for the storm to reach shore. Equation 14.11 gives 71 km/h for the wave speed when $\lambda = 250$ m. So the waves took 8.4 hours to reach shore. The storm is then 6.6 hours away.

Beats

When two waves of slightly different frequencies superpose, they interfere constructively at some points and destructively at others (Fig. 14.20*a*). Quantitatively, the combined wave is the sum of the two individual waves: $y(t) = A\cos\omega_1 t + A\cos\omega_2 t$. We can express this in a more enlightening form using the identity $\cos\alpha + \cos\beta = 2\cos\left[\frac{1}{2}(\alpha - \beta)\right]\cos\left[\frac{1}{2}(\alpha + \beta)\right]$ given in Appendix A. Then we have

$$y(t) = 2A\cos\left[\tfrac{1}{2}(\omega_1 - \omega_2)t\right]\cos\left[\tfrac{1}{2}(\omega_1 + \omega_2)t\right]$$

The second cosine factor represents a sinusoidal oscillation at the average of the two individual frequencies. The first term oscillates at a lower frequency—half the difference of the individual frequencies. If we think of the entire term $2A\cos\left[\frac{1}{2}(\omega_1 - \omega_2)t\right]$ as the "amplitude" of the higher-frequency oscillation, then this amplitude itself varies with time, as Fig. 14.20*b* shows. Note that there are *two* amplitude peaks for each cycle of the slow oscillation, so the frequency with which the amplitude varies is simply $\omega_1 - \omega_2$.

For sound waves, interference of two nearly equal frequencies produces intensity variations called **beats**; the closer the two frequencies, the longer the period between beats. Pilots, for example, synchronize airplane engines by reducing the beat frequency toward zero; musicians use the same trick to tune instruments. Beating of electromagnetic waves forms the basis for some very sensitive measurements.

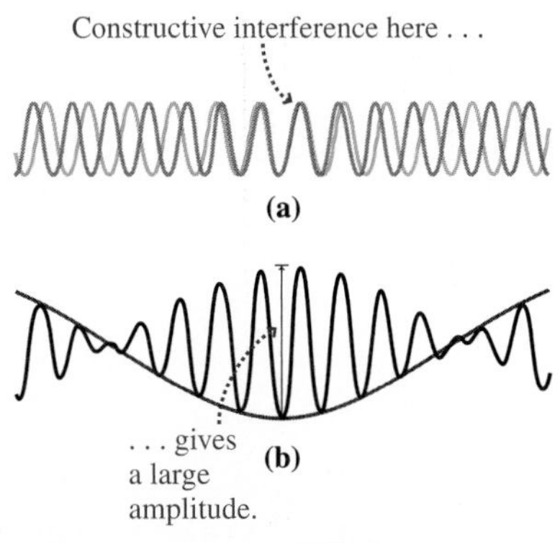

FIGURE 14.20 The origin of beats.

Interference in Two Dimensions

Waves propagating in two and three dimensions exhibit a rich variety of interference phenomena. Figure 14.21 shows one of the simplest and most important examples—the interference of waves from two point sources oscillating at the same frequency. Points on a perpendicular line midway between the sources are equidistant from both sources, and therefore waves arrive at this line in phase. Thus, they interfere constructively, producing a large amplitude. Some distance from the center line, the waves arrive exactly half a period out of phase. They therefore interfere destructively, producing a **nodal line** where the wave amplitude is very small. Since waves travel half a wavelength in half a period, the nodal line occurs where the distances to the two sources differ by half a wavelength. Additional nodal lines occur where those distances differ by $1\frac{1}{2}$ wavelengths, $2\frac{1}{2}$ wavelengths, and so forth. In practice, two-source interference is observable only when the source separation is comparable to the wavelength. If it's much larger, then the regions of constructive and destructive interference are so close that they blur together.

Two-source interference also results when plane waves pass through two closely spaced apertures that act as sources of circular or spherical wavefronts. Such two-slit interference experiments are important in optics and modern physics and are of historical interest because they were first used to demonstrate the wave nature of light.

FIGURE 14.21 Water waves from two sources interfere to produce regions of low and high amplitude.

PhET: Wave Interference

EXAMPLE 14.5 Wave Interference in Two Dimensions: Calm Water

Ocean waves pass through two small openings, 20 m apart, in a breakwater. You're in a boat 75 m from the breakwater and initially midway between the openings, but the water is pretty rough. You row 33 m parallel to the breakwater and, for the first time, find yourself in relatively calm water. What's the wavelength of the waves?

INTERPRET This is a problem about wave interference. The water is rough at your initial location because constructive interference produces large-amplitude waves. You find calm water at the first nodal line, where destructive interference reduces the wave amplitude.

DEVELOP We sketched the situation in Fig. 14.22. We've seen that the first nodal line occurs when the path lengths from two sources differ by half a wavelength. So our plan is to calculate the wavelength by applying this fact to the distances *AP* and *BP*.

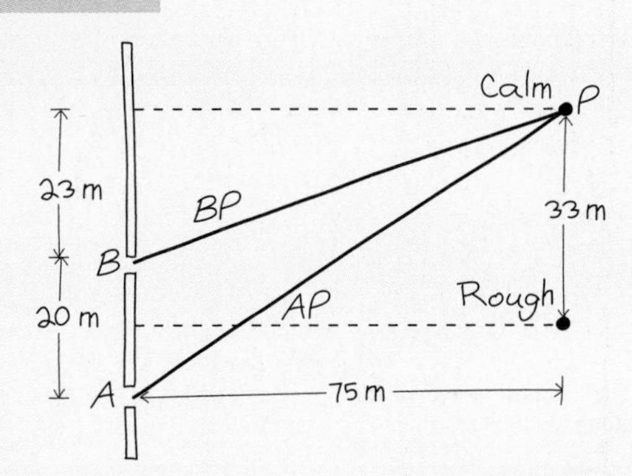

FIGURE 14.22 Calm water at *P* implies that paths *AP* and *BP* differ by half a wavelength.

EVALUATE Applying the Pythagorean theorem gives

$$AP = \sqrt{(75\text{ m})^2 + (43\text{ m})^2} = 86.5\text{ m}$$
$$BP = \sqrt{(75\text{ m})^2 + (23\text{ m})^2} = 78.4\text{ m}$$

The wavelength is twice the difference between these lengths, so

$$\lambda = 2(AP - BP) = 2(86.5\text{ m} - 78.4\text{ m}) = 16\text{ m}$$

ASSESS We expect two-source interference to be obvious when the source spacing is comparable to the wavelength. Here the 20-m spacing is indeed comparable to the 16-m wavelength, so our answer makes sense. ■

GOT IT? 14.5 Light shines through two small holes into a dark room, and a screen is mounted opposite the holes. The hole spacing is comparable to the wavelength of the light. Looking at the screen, will you see (a) two bright spots opposite the two holes or (b) a pattern of light and dark patches? Explain.

14.6 Reflection and Refraction

You shout in a mountain valley and hear echoes. You look in a mirror and see your reflection. A metal screen reflects microwaves to keep them in your oven. A physician's ultrasound probes your body, reflecting off internal structures. A bat uses reflected sound to home in on its prey. All these are examples of wave **reflection**.

You can see that wave reflection *must* occur when a wave hits a medium in which it can't propagate; otherwise, where would the wave energy go? The figures below detail the reflection process for waves on a stretched string, in the two cases where the string end is clamped at a rigid wall (Fig. 14.23) or, in contrast, free to move up and down (Fig. 14.24). In the first case, the wave amplitude must remain zero at the end, so the incident and reflected pulses interfere destructively and the reflected wave is therefore inverted. In the second case, the displacement is a maximum at the free end, and the reflected wave is not inverted.

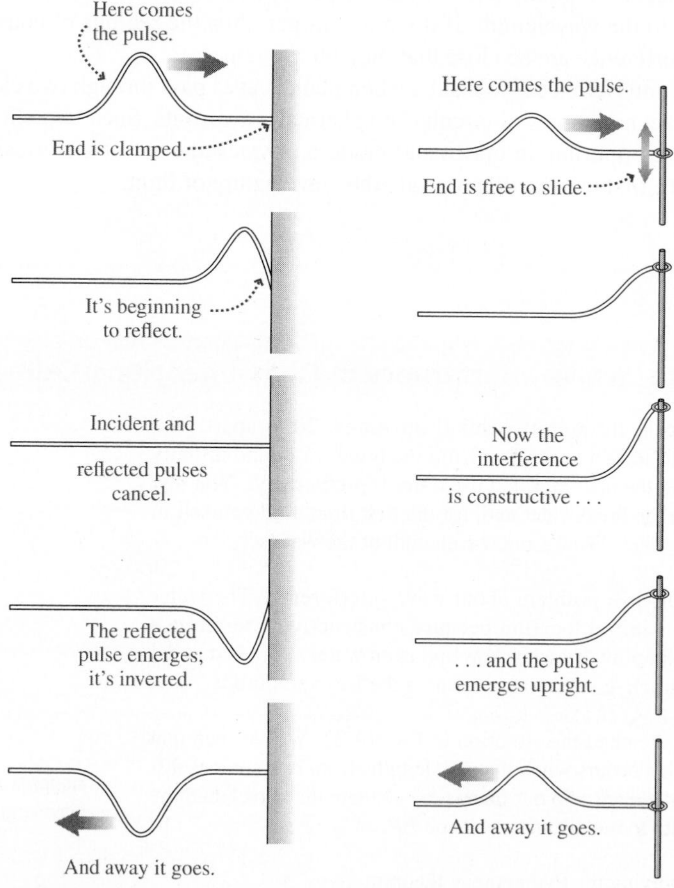

FIGURE 14.23 Reflection of a wave pulse at the rigidly clamped end of string.

FIGURE 14.24 Reflection of a wave pulse at a free end.

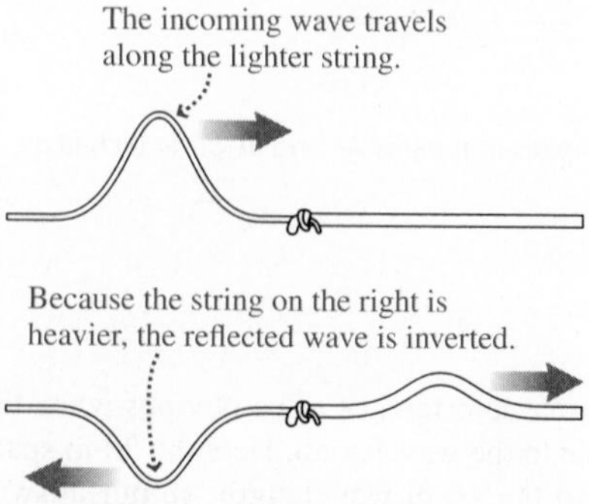

The incoming wave travels along the lighter string.

Because the string on the right is heavier, the reflected wave is inverted.

FIGURE 14.25 Partial reflection occurs at the junction between two strings.

Between the extremes of a rigid wall and a perfectly free end lies the case of one string connected to another with different mass per unit length. In this case, some wave energy is transmitted to the second string and some is reflected back along the first (Fig. 14.25).

The phenomenon of partial reflection and transmission at a junction of strings has its analog in the behavior of all sorts of waves at interfaces between different media. For example, shallow-water waves are partially reflected if the water depth changes abruptly. Light incident on even the clearest glass undergoes partial reflection because of the difference in the light-transmitting properties of air and glass (much more on this in Chapter 30). Partial reflection of ultrasound waves at the interfaces of body tissues with different densities makes ultrasound a valuable medical diagnostic.

When waves strike an interface between two media at an oblique angle and are capable of propagating in the second medium, the phenomenon of **refraction** occurs. In refraction, the direction of wave propagation changes because of a difference in wave speed between the two media (Fig. 14.26). We'll discuss the mathematics of refraction in Chapter 30.

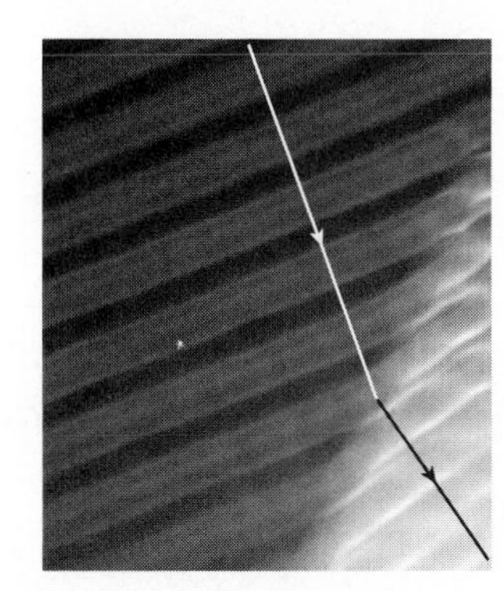

FIGURE 14.26 Waves in shallow water refract at the interface between two different water depths.

APPLICATION Probing the Earth

Waves propagating and reflecting inside the Earth help geologists deduce the planet's interior structure. That's because Earth's interior supports two types of waves. Longitudinal waves, also called P waves, propagate in both solids and liquids. Transverse, or S waves, propagate only in solids. Earthquakes generate S waves that propagate throughout the solid Earth. But as the figure suggests, they can't get through the liquid outer core, so they leave a "shadow" where seismographs don't record any S-wave activity. This effect is our clearest evidence that Earth has a liquid core.

P waves, however, do propagate through the liquid core. But they undergo partial reflections farther in—evidence for an abrupt change in core density. Careful analysis shows that wave speeds in the inner core are consistent with its being solid—giving our planet the solid–liquid–solid structure suggested in the figure.

Studies of Earth's large-scale structure generally use earthquake waves, although inner-core evidence also comes from underground nuclear explosions. At a smaller scale, explosive charges or machines that "thump" the ground produce waves whose reflections from rock layers down to a few kilometers depth help reveal oil and gas deposits.

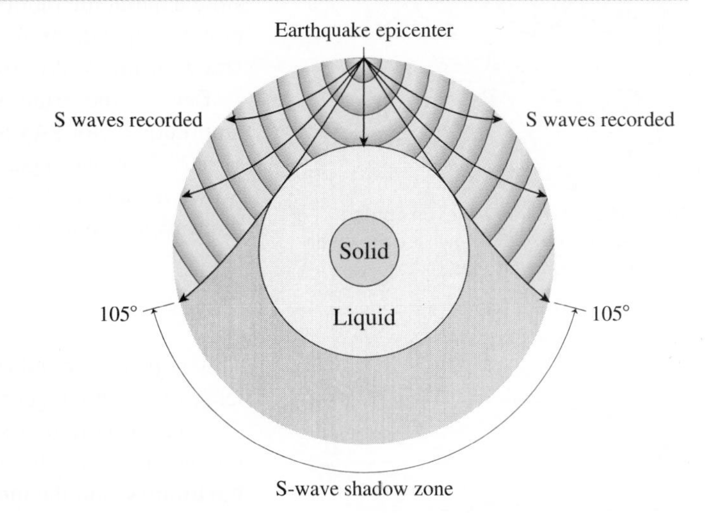

GOT IT? 14.6 You're holding one end of a taut rope, and you can't see the other end. You tweak the rope to give it an upward displacement, sending a pulse down the rope. A while later, a pulse comes back toward you. Its displacement is upward, but with considerably lower amplitude than the initial displacement you provided. Assuming there's no energy loss in the rope itself, you can conclude that the far end of the rope is (a) attached to a rigid anchor point, (b) attached in such a way that it's free to slide up and down, (c) tied to another rope with less mass per unit length, or (d) tied to another rope with more mass per unit length.

14.7 Standing Waves

PhET: Wave on a String

Imagine a string clamped tightly at both ends. Waves propagate back and forth by reflecting at the ends. But because the ends are clamped, the wave displacement at each end must always be zero. Only certain waves can satisfy this requirement; as Fig. 14.27 suggests, they're waves for which an integer number of half-wavelengths just fits the string's length L.

The waves in Fig. 14.27 are **standing waves**, so called because they essentially stand still, confined to the length of the string. At each point the string executes simple harmonic motion perpendicular to its undisturbed state. We can describe standing waves mathematically as arising from the superposition of two waves propagating in opposite directions and reflecting at the ends of the string. If we take the x-axis to coincide with the string, then we can write the string displacements in two such waves as $y_1(x, t) = A\cos(kx - \omega t)$ for the wave propagating in the $+x$-direction (recall Equation 14.3)

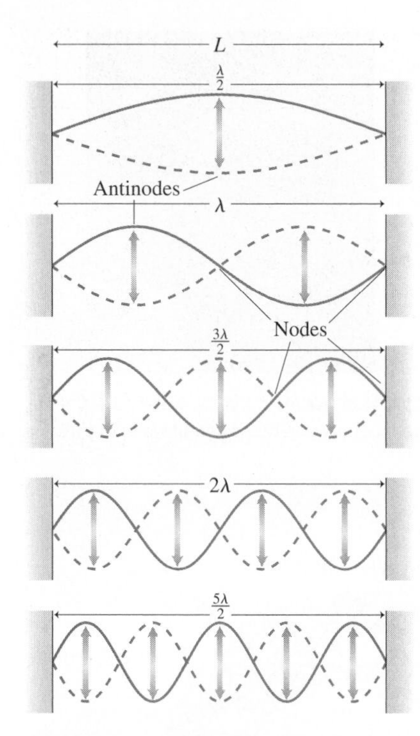

FIGURE 14.27 Standing waves on a string clamped at both ends; shown are the fundamental and four overtones.

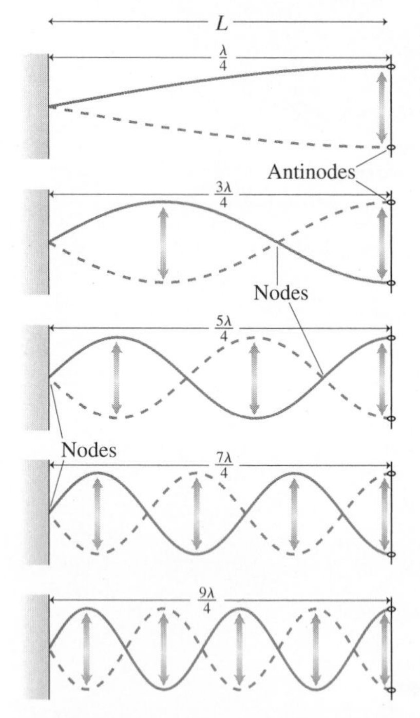

FIGURE 14.28 When one end of the string is fixed and the other free, the string can accommodate only an odd number of quarter-wavelengths.

and $y_2(x, t) = -A\cos(kx + \omega t)$ for the wave propagating in the $-x$-direction. (The minus sign in y_2 accounts for the phase change that occurs on reflection at a rigid boundary, as you saw in Fig. 14.23.) Their superposition is then

$$y(x, t) = y_1 + y_2 = A[\cos(kx - \omega t) - \cos(kx + \omega t)]$$

Appendix A lists a trig identity for the difference of two cosines:

$$\cos\alpha - \cos\beta = -2\sin\left[\tfrac{1}{2}(\alpha + \beta)\right]\sin\left[\tfrac{1}{2}(\alpha - \beta)\right]$$

Applying this identity with $\alpha = kx - \omega t$ and $\beta = kx + \omega t$ gives

$$y(x, t) = 2A\sin kx \sin\omega t \quad (14.12)$$

Equation 14.12 is the mathematical description of a standing wave, and it affirms our qualitative description that each point on the string simply oscillates up and down. Pick any point—that is, any fixed value of x—and Equation 14.12 does indeed describe simple harmonic motion in the y-direction, through the factor $\sin\omega t$. The amplitude of that motion depends on the point x you've chosen, and is given by the factor that multiplies $\sin\omega t$—namely, $2A\sin kx$.

Because the string is clamped at both ends, the amplitude at the ends must be zero. Our amplitude factor $2A\sin kx$ does give $y = 0$ in Equation 14.12 at $x = 0$, but what about at $x = L$? Here we'll get zero only if $\sin kL = 0$—and that requires kL to be a multiple of π. So we must have $kL = m\pi$, where m is any integer. But the wave number k is related to the wavelength λ by $k = 2\pi/\lambda$. Our condition $kL = m\pi$ can then be written

$$L = \frac{m\lambda}{2}, \quad m = 1, 2, 3, \ldots \quad (14.13)$$

This is just the condition we already guessed from Fig. 14.27—namely, that the string length L be an integer number of half-wavelengths.

Given a particular string length L, Equation 14.13 limits the allowed standing waves on the string to a discrete set of wavelengths. Those allowed waves are called **modes** or **harmonics**, and the integer m is the **mode number**. The $m = 1$ mode is the **fundamental** and is the longest-wavelength standing wave that can exist on the string. The higher modes are **overtones**.

Figure 14.27 shows that there are points where the string doesn't move at all. These are called **nodes**. Points where the amplitude of the wave displacement is a maximum, in contrast, are **antinodes**.

When a string is clamped rigidly at one end but is free at the other, its clamped end is a node but its free end is an antinode. Figure 14.28 shows that the string length must then be an odd multiple of a quarter-wavelength—a result that you can also get from Equation 14.12 by requiring $\sin kL = 1$ to give maximum amplitude at $x = L$.

Standing-Wave Resonance

We've discussed standing waves in terms of constraints on the wavelength λ rather than on the frequency f. But because waves on a string have a fixed speed v, and because $f\lambda = v$, Equation 14.13's discrete set of allowed wavelengths corresponds to a set of discrete frequencies. The lowest allowed frequency, the fundamental, corresponds to the longest wavelength; the overtones have higher frequencies.

Because a stretched string can oscillate in any of its allowed frequencies, the resonant behavior that we discussed in Chapter 13 can occur close to any of those frequencies. Buildings and other structures, in analogy with our simple string, support a variety of standing-wave modes. For example, a skyscraper is like the string of Fig. 14.28, with its base clamped to Earth but its top free to swing. Engineers must be sure to identify all possible modes of structures they design in order to avoid harmful resonances. The disastrous oscillations of the Tacoma Narrows Bridge shown in Fig. 13.26 are actually torsional standing waves.

Other Standing Waves

Standing waves are common phenomena. Water waves in confined spaces exhibit standing waves, and entire lakes can develop very slow oscillations corresponding to low-mode-number standing waves. Standing electromagnetic waves occur inside closed metal cavities; in microwave ovens the nodes of the standing-wave pattern would result in "cold" spots were not either the food or the source of microwaves kept in motion. Standing sound waves in the Sun help astrophysicists probe the solar interior. And even atomic structure can be understood in terms of standing waves associated with electrons.

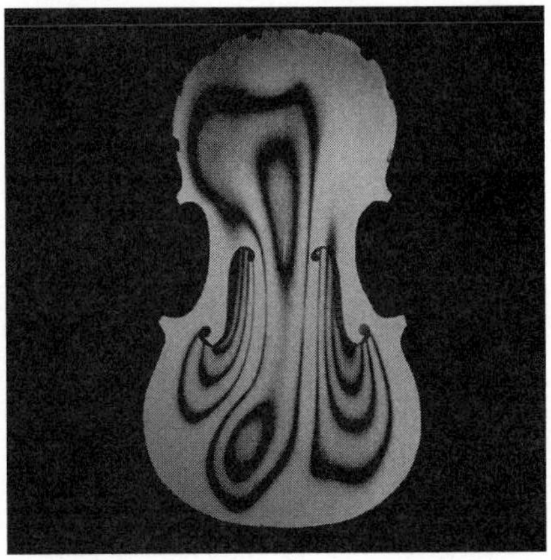

FIGURE 14.29 Standing waves on a violin, imaged using holographic interference of laser light waves.

Musical Instruments

Our analysis of standing waves on strings applies directly to stringed musical instruments such as violins, guitars, and pianos. Standing-wave vibrations in the instrument strings are communicated to the air as sound waves, usually through the intermediary of a sounding box or electronic amplifiers. For instruments in the violin family, the body of the instrument itself undergoes standing-wave vibrations, excited by the vibration of the string, that establish each individual instrument's peculiar sound quality (Fig. 14.29). Similarly, the stretched membranes of drums exhibit a variety of standing-wave patterns representing the allowed modes on these two-dimensional surfaces.

Wind instruments generate standing sound waves in air columns, as suggested in Fig. 14.30. These must be open at one end to allow sound to escape; in many instruments the column is effectively open at both ends. An open end has its pressure fixed at atmospheric pressure; it is therefore a pressure node and thus, from Fig. 14.14, a displacement antinode. As a result, an instrument open at one end supports odd-integer multiples of a quarter-wavelength (Fig. 14.30*a*), in analogy with Fig. 14.28. An instrument open at both ends, on the other hand, supports integer multiples of a half-wavelength (Fig. 14.30*b*).

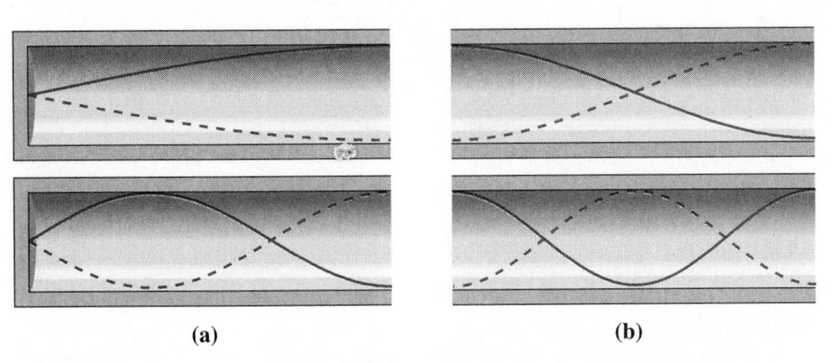

FIGURE 14.30 Standing waves in wind instruments: (a) open at one end and (b) open at both ends.

EXAMPLE 14.6 Standing-Wave Modes: The Double Bassoon

The double bassoon is the lowest-pitched instrument in a normal orchestra. The instrument is "folded" to achieve an effective air column 5.5 m long, and it acts like a pipe open at both ends. What's the frequency of the double bassoon's fundamental note? Assume the sound speed is 343 m/s.

INTERPRET This is a problem about standing-wave modes in a hollow pipe open at both ends.

DEVELOP Figure 14.30*b* applies to a pipe that's open at both ends. So our sketch of the fundamental mode in Fig. 14.31 looks like the upper of the two pictures in Fig. 14.30*b*. We can find the wavelength and then use Equation 14.1, $v = \lambda f$, to get the frequency.

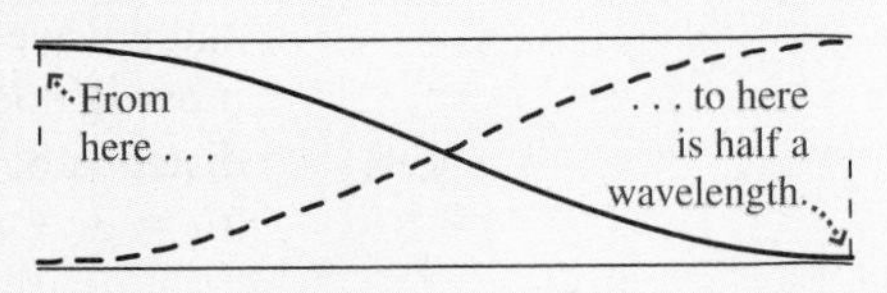

FIGURE 14.31 Sketch for Example 14.6.

EVALUATE The wavelength is twice the instrument's 5.5-m length, or 11 m. Then Equation 14.1 gives

$$f = \frac{v}{\lambda} = \frac{343 \text{ m/s}}{11 \text{ m}} = 31 \text{ Hz}$$

ASSESS This frequency is the note B_0, which lies near the low-frequency limit of the human ear. Like most wind instruments, the bassoon has a number of holes that, when uncovered, alter the positions of the antinodes and therefore change the pitch. ■

GOT IT? 14.7 A string 1 m long is clamped tightly at one end and is free to slide up and down at the other. Which of the following are possible wavelengths for standing waves on this string: $\frac{4}{5}$ m, 1 m, $\frac{4}{3}$ m, $\frac{3}{2}$ m, 2 m, 3 m, 4 m, 5 m, 6 m, 7 m, 8 m?

14.8 The Doppler Effect and Shock Waves

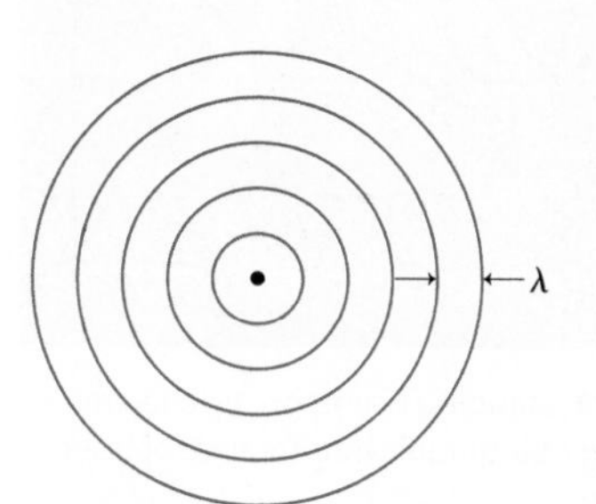

FIGURE 14.32 Circular waves from a source at rest with respect to the medium.

The speed v of a wave is its speed relative to the medium through which it propagates. A point source at rest in the medium radiates waves uniformly in all directions (Fig. 14.32). But when the source moves, wave crests bunch up in the direction toward which the source is moving, resulting in a decreased wavelength (Fig. 14.33). In the opposite direction, wave crests spread out and the wavelength increases.

The wave speed is determined by the properties of the medium, so it doesn't change with source motion. Thus the equation $v = \lambda f$ still holds. This means that an observer in front of the moving source, where λ is smaller, experiences a higher wave frequency as more wave crests pass per unit time. Similarly, an observer behind the source experiences a lower frequency. This change in wavelength and frequency from a moving source is the **Doppler effect** or **Doppler shift**, after the Austrian physicist Christian Johann Doppler (1803–1853).

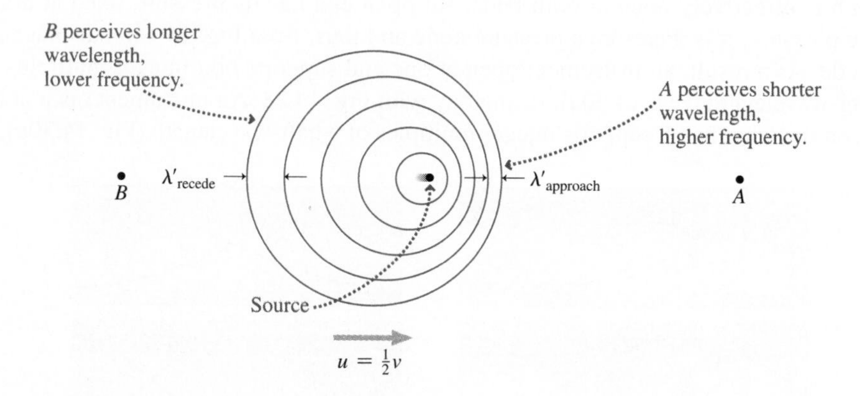

FIGURE 14.33 Origin of the Doppler effect, shown for a source moving with half the wave speed.

To analyze the Doppler effect, let λ be the wavelength measured when the source is stationary, and λ' the wavelength when the source is moving at speed u through a medium where the wave speed is v. At the source, the time between wave crests is the wave period T, and a wave crest moves one wavelength λ in this time. But during the same time T, the moving source covers a distance uT, after which it emits the next wave crest. So the distance between wave crests, as seen by an observer in front of the moving source, is $\lambda' = \lambda - uT$. Writing $T = \lambda/v$, we get

$$\lambda' = \lambda - u\frac{\lambda}{v} = \lambda\left(1 - \frac{u}{v}\right) \quad \text{(source approaching)} \qquad (14.14a)$$

The situation is similar in the direction opposite the source motion, except that now the wavelength *increases* by the amount $\lambda u/v$, giving

$$\lambda' = \lambda\left(1 + \frac{u}{v}\right) \quad \text{(source receding)} \qquad (14.14b)$$

We can recast these expressions in terms of frequency using the relations $\lambda = v/f$ and $\lambda' = v/f'$, where f' is the frequency of waves from the moving source as measured by an observer at rest in the medium. Substituting these relations in our expressions for λ' and then solving for f' gives

$$f' = \frac{f}{1 \pm u/v} \quad \text{(Doppler shift, moving source)} \qquad (14.15)$$

for the Doppler-shifted frequency, where the + and − signs correspond to receding and approaching sources, respectively.

You've probably experienced the Doppler effect for sound when standing near a highway. A loud truck approaches with a high-pitched sound "aaaaaaaaaaa." As it passes, the pitch drops abruptly: "aaaaaaaaeioooooooooo," and stays low as the truck recedes. Practical uses of the Doppler effect are numerous. The Doppler shift in reflected ultrasound measures blood flow and fetal heartbeat. Police radar uses the Doppler shift of high-frequency radio waves reflected from moving cars. The Doppler shift of starlight reveals stellar motions, and Doppler-shifted light from distant galaxies is evidence that our entire universe is expanding.

EXAMPLE 14.7 Doppler Effect: The Wrong Note

A car speeds down the highway with its stereo blasting. An observer with perfect pitch is standing by the roadside and, as the car approaches, notices that a musical note that should be G ($f = 392$ Hz) sounds like A (440 Hz). How fast is the car moving?

INTERPRET This problem is about the Doppler effect in sound from a moving source.

DEVELOP Equation 14.15, $f' = f/(1 \pm u/v)$, relates the original and shifted frequencies to the source speed u, so our plan is to solve this equation for u. We'll use the minus sign because the source is approaching. We'll also need the sound speed v, which Example 14.6 gave as 343 m/s.

EVALUATE Solving Equation 14.15 for u gives

$$u = v\left(1 - \frac{f}{f'}\right) = (343\text{ m/s})\left(1 - \frac{392\text{ Hz}}{440\text{ Hz}}\right) = 37.4\text{ m/s}$$

ASSESS Our answer—some 134 km/h or 84 mi/h—seems reasonable for a speeding car, though not a particularly safe speed! And it's a little more than 10% of the sound speed, consistent with the roughly 10% change in the sound frequency. ■

Moving Observers

A Doppler shift in frequency, but not wavelength, also occurs when a moving observer approaches a stationary source—meaning a source at rest with respect to the wave medium. An observer moving toward a stationary source passes wave crests more often than would happen if the observer were at rest, and thus measures a shorter wave period and therefore a higher frequency. The result, as you can show in Problem 78, is a shifted frequency given by

$$f' = f\left(1 \pm \frac{u}{v}\right) \quad \text{(Doppler shift, moving observer)} \qquad (14.16)$$

with the positive sign for an observer approaching the source and the negative sign for an observer receding. For observer velocities u small compared with the wave speed v, Equations 14.15 and 14.16 give essentially the same results.

Waves from a stationary source that reflect from a moving object undergo a Doppler shift *twice*. First, because the frequency as received at the reflecting object is shifted, according to Equation 14.16, due to the object's motion relative to the source. Then a stationary observer sees the reflected waves as coming from a moving source, so there's another shift, this time given by Equation 14.15. Police radar and other Doppler-based speed measurements make use of this double Doppler shift that occurs on reflection.

The Doppler Effect for Light

Although light and other electromagnetic waves do not require a material medium, they, too, are subject to the Doppler shift. Both Doppler formulas we derived here apply to electromagnetic waves, but only as approximations when the relative speed between source and observer is much lower than the speed of light.

The Doppler shift for electromagnetic waves is the same whether it's the source that moves or the observer. This reflects a profound fact at the root of Einstein's relativity: that "stationary" and "moving" are meaningful only as relative terms. Electromagnetic waves, unlike mechanical waves, do not require a medium—and therefore terms such as "stationary source" and "moving observer" are meaningless. All that matters is the relative motion between source and observer. We'll explore this point further in Chapter 33.

Shock Waves

Equation 14.14a suggests that wavelength goes to zero if a source approaches at exactly the wave speed. This happens because wave crests can't get away from the source, so they pile up just ahead of it to form a large-amplitude wave called a **shock wave** (Fig. 14.34). When the source moves faster than the wave speed, waves pile up on a cone whose half-angle is given by $\sin\theta = v/u$, as shown. The ratio u/v is called the **Mach number**, and the cone angle is the **Mach angle**.

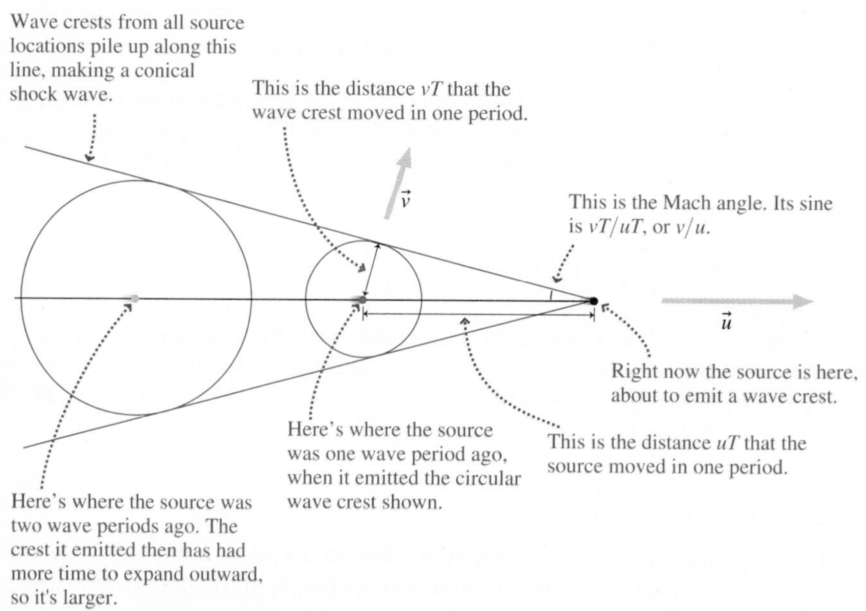

FIGURE 14.34 Shock waves form when the source speed u exceeds the wave speed v.

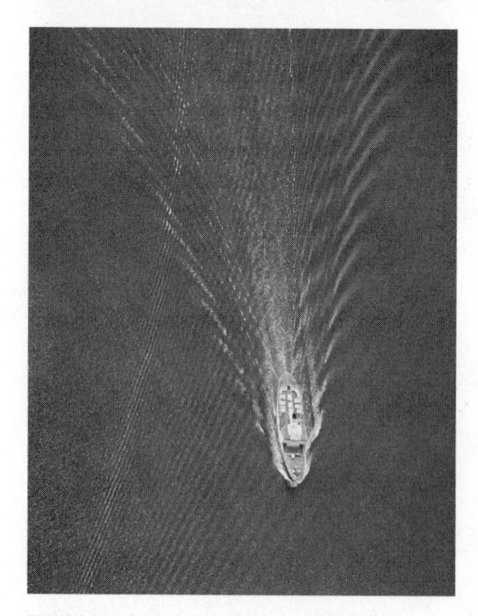

FIGURE 14.35 (a) A shock wave trails from a supersonic aircraft. The plane is flying low over the ocean, and the humid air condenses at the shock, making it visible. (b) The wake trailing from this boat is also a shock wave that arises because the boat is moving faster than the speed of water waves.

Shock waves occur in a wide variety of physical situations (Fig. 14.35). Sonic booms are shock waves from supersonic aircraft. The bow wave of a boat is a shock wave on the water surface. On a much larger scale, a huge shock wave forms in space as the solar wind—a high-speed flow of particles from the Sun—encounters Earth's magnetic field.

GOT IT? 14.8 In Fig. 14.35, which is moving faster in relation to the wave speed in the medium through which they're traveling, the airplane or the boat?

CHAPTER 14 SUMMARY

Big Idea

Waves are the big idea here. A wave is a propagating disturbance that carries energy but not matter. Waves are characterized by their amplitude, wavelength, and speed. They can be **longitudinal** or **transverse**.

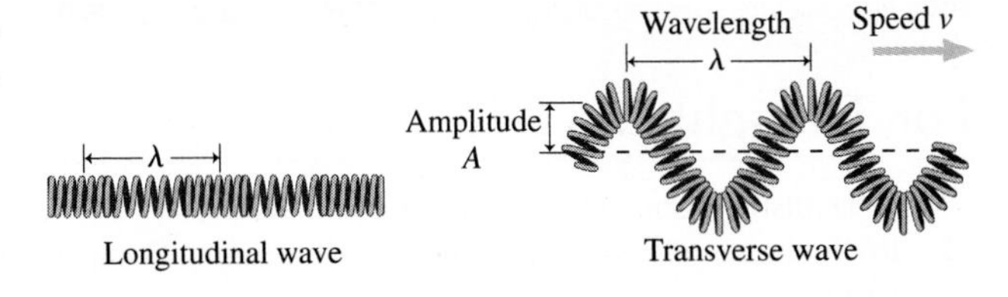

Key Concepts and Equations

Wave **period** is the time for one complete wave cycle. Period and frequency are inverses, and wavelength λ, period T or frequency f, and wave speed v are all related:

$$v = \frac{\lambda}{T} = \lambda f$$

A **simple harmonic wave** is sinusoidal in shape. The wave disturbance is a function of position and time and is most simply described in terms of its **wave number** k and **angular frequency** ω:

$$y(x, t) = A\cos(kx - \omega t)$$

They're related to wavelength and period by

$$k = \frac{2\pi}{\lambda} \quad \text{and} \quad \omega = \frac{2\pi}{T}$$

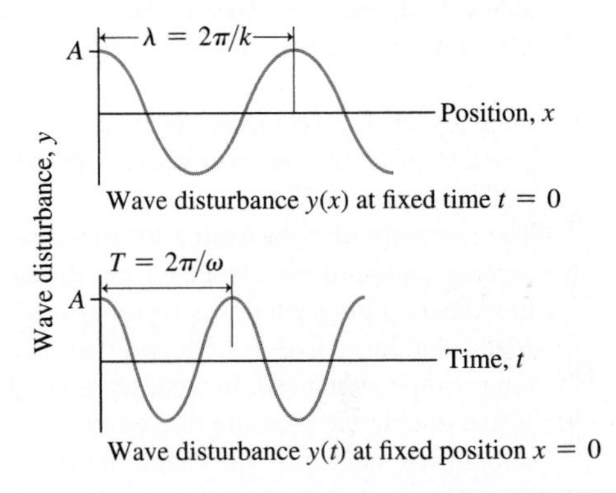

Wave **intensity** is the power per unit area carried by the wave: $I = P/A$. For a spherical wave that spreads in all directions from a localized source, intensity decreases as the inverse square of the distance from the source: $I = P/(4\pi r^2)$.

Applications

Wave speed is a characteristic of the medium.

Transverse waves on strings: $v = \sqrt{\dfrac{F}{\mu}}$

Longitudinal sound waves in a gas: $v = \sqrt{\dfrac{\gamma P}{\rho}}$, about 343 m/s in air under standard conditions

Surface waves in deep water: $v = \sqrt{\dfrac{\lambda g}{2\pi}}$

Standing waves on strings

Clamped at both ends, string length is an integer multiple of a half-wavelength: $L = m\lambda/2$

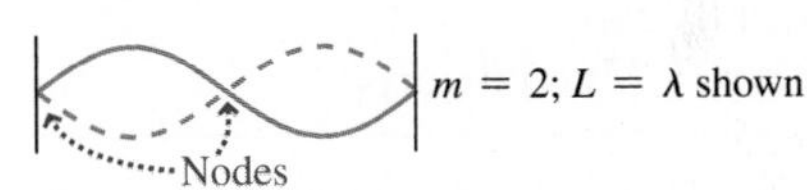

Clamped at one end, string length is an odd-integer multiple of a quarter-wavelength:

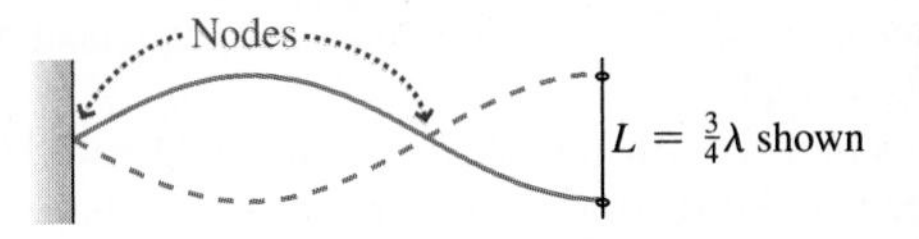

The **Doppler effect** is a frequency and/or wavelength shift due to the motion u of an observer or source relative to the medium with wave speed v.

Moving source: $f' = \dfrac{f}{(1 \pm u/v)}$, + for receding, − for approaching; λ also changes

Moving observer: $f' = f(1 \pm u/v)$, + for approaching, − for receding; no change in λ

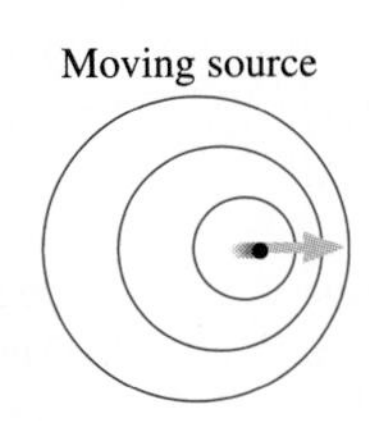

Shock waves occur when a wave source (speed u) moves through a medium at greater than the wave speed (v).

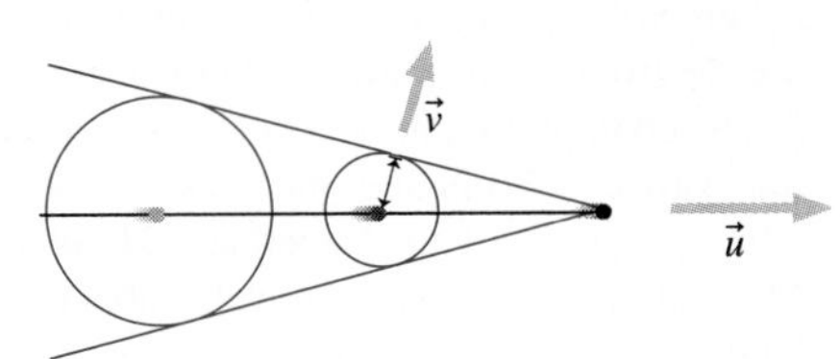

BIO *Biology and/or medicine-related problems* **DATA** *Data problems* **ENV** *Environmental problems* **CH** *Challenge problems* **COMP** *Computer problems*

For Thought and Discussion

1. What distinguishes a wave from an oscillation?
2. Red light has a longer wavelength than blue light. Compare their frequencies.
3. Consider a light wave and a sound wave with the same wavelength. Which has the higher frequency?
4. As a wave propagates on a string, the string moves back and forth sideways. Is the string speed related to the wave speed? Explain.
5. If you doubled the tension in a string, what would happen to the speed of waves on the string?
6. A heavy cable is hanging vertically, its bottom end free. How will the speed of transverse waves near the top and bottom of the cable compare? Why?
7. The intensity of light from a localized source decreases as the inverse square of the distance from the source. Does this mean that the light loses energy as it propagates?
8. **BIO** Medical ultrasound uses frequencies around 10^7 Hz, far above the range of the human ear. In what sense are these waves "sound"?
9. If you double the pressure of a gas while keeping its density the same, what happens to the sound speed?
10. Water is about a thousand times more dense than air, yet the speed of sound in water is greater than in air. Why might this be?
11. If you place a perfectly clear piece of glass in perfectly clear water, you can still see the glass. Why?
12. When a wave source moves relative to the medium, a stationary observer measures changes in both wavelength and frequency. But when the observer moves and the source is stationary, only the frequency changes. Why the difference?
13. Why can a boat easily produce a shock wave on the water surface, while only a very high-speed aircraft can produce a sonic boom?

Exercises and Problems

Exercises

Section 14.1 Waves and Their Properties

14. Ocean waves with 18-m wavelength travel at 5.3 m/s. What's the time interval between wave crests passing a boat moored at a fixed location?
15. Ripples in a shallow puddle propagate at 34 cm/s. If the wave frequency is 5.2 Hz, find (a) the period and (b) the wavelength.
16. An 89.5-MHz FM radio wave propagates at the speed of light. What's its wavelength?
17. Calculate the wavelengths of (a) a 1.0-MHz AM radio wave, (b) a channel 9 TV signal (190 MHz), (c) a police radar (10 GHz), (d) infrared radiation from a hot stove (4×10^{13} Hz), (e) green light (6.0×10^{14} Hz), and (f) 1.0×10^{18}-Hz X rays. All are electromagnetic waves that propagate at 3.0×10^8 m/s.
18. A seismograph located 1250 km from an earthquake detects seismic waves 5.12 min after the quake occurs. The seismograph oscillates in step with the waves, at 3.21 Hz. Find the wavelength.
19. **BIO** Medical ultrasound waves travel at about 1500 m/s in soft tissue. Higher frequencies provide clearer images but don't penetrate to deeper organs. Find the wavelengths of (a) 8.0-MHz ultrasound used in fetal imaging and (b) 3.5-MHz ultrasound used to image an adult's kidneys.

Section 14.2 Wave Math

20. An ocean wave has period 4.1 s and wavelength 10.8 m. Find its (a) wave number and (b) angular frequency.
21. Find the (a) amplitude, (b) wavelength, (c) period, and (d) speed of a wave whose displacement is given by $y = 1.3\cos(0.69x + 31t)$, where x and y are in centimeters and t in seconds. (e) In which direction is the wave propagating?
22. **BIO** Ultrasound used in a medical imager has frequency 4.86 MHz and wavelength 0.313 mm. Find (a) the angular frequency, (b) the wave number, and (c) the wave speed.
23. A simple harmonic wave of wavelength 18.7 cm and amplitude 2.34 cm is propagating in the negative x-direction at 38.0 cm/s. Find its (a) angular frequency and (b) wave number. (c) Write a mathematical expression describing the displacement y of this wave (in centimeters) as a function of position and time. Assume the maximum displacement occurs when $t = 0$.
24. Analysis of waves in shallow water (depth much less than wavelength) yields the following wave equation:

$$\frac{\partial^2 y}{\partial x^2} = \frac{1}{gh}\frac{\partial^2 y}{\partial t^2}$$

where h is the water depth and g the gravitational acceleration. Give an expression for the wave speed.

Section 14.3 Waves on a String

25. The main cables supporting New York's George Washington Bridge have a mass per unit length of 4100 kg/m and are under 250-MN tension. At what speed would a transverse wave propagate on these cables?
26. A transverse wave 1.2 cm in amplitude propagates on a string; its frequency is 44 Hz. The string is under 21-N tension and has mass per unit length 15 g/m. Determine its speed.
27. A transverse wave with 3.0-cm amplitude and 75-cm wavelength propagates at 6.7 m/s on a stretched spring with mass per unit length 170 g/m. Find the spring tension.
28. A rope is stretched between supports 18.3 m apart; its tension is 78.6 N. If one end of the rope is tweaked, the resulting disturbance reaches the other end 585 ms later. Find the rope's mass.
29. A rope with 280 g of mass per meter is under 550-N tension. Find the average power carried by a wave with frequency 3.3 Hz and amplitude 6.1 cm propagating on the rope.

Section 14.4 Sound Waves

30. Show that $\sqrt{P/\rho}$ from Equation 14.9 has the units of speed.
31. Find the sound speed in air under standard conditions with pressure 101 kN/m^2 and density 1.20 kg/m^3.
32. Timers in sprint races start their watches when they see smoke from the starting gun, not when they hear the sound. Why? How much error would be introduced by timing a 100-m race from the sound of the gun?
33. The factor γ for nitrogen dioxide (NO_2) is 1.29. Find the sound speed in NO_2 at 4.8×10^4-N/m^2 pressure and 0.35-kg/m^3 density.
34. A gas with density 1.0 kg/m^3 and pressure 81 kN/m^2 has sound speed 368 m/s. Are the gas molecules monatomic or diatomic?
35. **BIO** Divers in an underwater habitat breathe a special mixture of oxygen and neon to prevent the possibly fatal effects of nitrogen in ordinary air. With pressure 6.2×10^5 N/m^2 and density 4.5 kg/m^3,

the effective γ value for the mixture is 1.61. Find the frequency in this mixture for a 50-cm-wavelength sound wave, and compare with its frequency in air under normal conditions.

Section 14.5 Interference

36. You're flying in a twin-engine turboprop aircraft, with its two propellers turning at 985 and 993 rpm, respectively. How often to you hear a peak in the engine sound?
37. What's the wavelength of the ocean waves in Example 14.5 if the calm water you encounter at 33 m is the *second* calm region on your voyage from the center line?

Section 14.7 Standing Waves

38. A 2.0-m-long string is clamped at both ends. (a) Find the longest-wavelength standing wave possible on this string. (b) If the wave speed is 56 m/s, what's the lowest standing-wave frequency?
39. When a stretched string is clamped at both ends, its fundamental frequency is 140 Hz. (a) What's the next higher frequency? If the same string, with the same tension, is now clamped at one end and free at the other, what are (b) the fundamental and (c) the next higher frequency?
40. A string is clamped at both ends and tensioned until its fundamental frequency is 85 Hz. If the string is then held rigidly at its midpoint, what's the lowest frequency at which it will vibrate?
41. **BIO** A crude model of the human vocal tract treats it as a pipe closed at one end. Find the effective length of a vocal tract whose fundamental tone is 620 Hz. Take $V_{\text{sound}} = 354$ m/s at body temperature.

Section 14.8 The Doppler Effect and Shock Waves

42. A car horn emits 380-Hz sound. If the car moves at 17 m/s with its horn blasting, what frequency will a person standing in front of the car hear?
43. A fire station's siren is blaring at 85 Hz. What's the frequency perceived by a firefighter racing toward the station at 120 km/h?
44. A fire truck's siren at rest wails at 1400 Hz; standing by the roadside as the truck approaches, you hear it at 1600 Hz. How fast is the truck going?
45. Red light emitted by hydrogen atoms at rest in the laboratory has wavelength 656 nm. Light emitted in the same process on a distant galaxy is received at Earth with wavelength 708 nm. Describe the galaxy's motion relative to Earth.

Problems

46. Figure 14.36 shows a simple harmonic wave at time $t = 0$ and later at $t = 2.6$ s. Write a mathematical description of this wave.

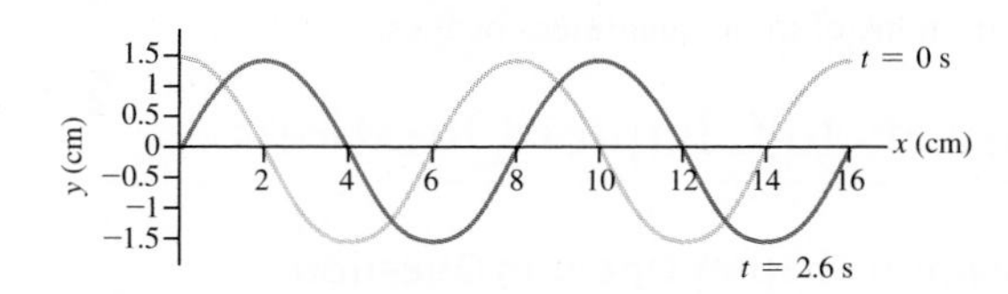

FIGURE 14.36 Problem 46

47. Transverse waves propagate at 18 m/s on a string under 14-N tension. What will be the wave speed if the tension is increased to 40 N?
48. A uniform cable hangs vertically under its own weight. Show that the speed of waves on the cable is given by $v = \sqrt{yg}$, where y is the distance from the bottom of the cable.
49. **CH** Figure 14.37 shows a wave train consisting of two sine wave cycles propagating along a string. Obtain an expression for the total energy in this wave train, in terms of the string tension F, the wave amplitude A, and the wavelength λ.

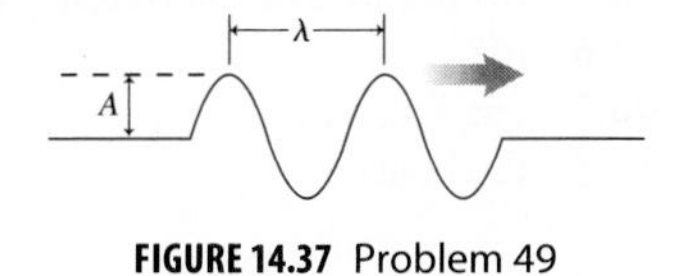

FIGURE 14.37 Problem 49

50. A loudspeaker emits energy at the rate of 50 W, spread in all directions. Find the intensity of sound 18 m from the speaker.
51. Light intensity 3.3 m from a lightbulb is 0.73 W/m^2. Find the bulb's power output, assuming it radiates equally in all directions.
52. Light emerges from a 5.0-mW laser in a beam 1.0 mm in diameter. The beam shines on a wall, producing a spot 3.6 cm in diameter. What is the beam's intensity (a) at the laser and (b) at the wall?
53. Two waves have the same angular frequency ω, wave number k, and amplitude A, but they differ in phase: $y_1 = A\cos(kx - \omega t)$ and $y_2 = A\cos(kx - \omega t + \phi)$. Show that their superposition is also a simple harmonic wave, and determine its amplitude as a function of the phase difference ϕ.
54. A wire is under 32.8-N tension, carrying a wave described by $y = 1.75\sin(0.211x - 466t)$, where x and y are in centimeters and t is in seconds. What are (a) the wave amplitude, (b) the wavelength, (c) the wave period, (d) the wave speed, and (e) the power carried by the wave?
55. A spring of mass m and spring constant k has an unstretched length L_0. Find an expression for the speed of transverse waves on this spring when it's been stretched to a length L.
56. **CH** When a 340-g spring is stretched to a total length of 40 cm, it supports transverse waves propagating at 4.5 m/s. When it's stretched to 60 cm, the waves propagate at 12 m/s. Find (a) the spring's unstretched length and (b) its spring constant.
57. At a point 15 m from a source of spherical sound waves, you measure the intensity 750 mW/m^2. How far do you need to walk, directly away from the source, until the intensity is 270 mW/m^2?
58. Figure 14.38 shows two observers 20 m apart on a line that connects them to a spherical light source. If the observer nearer the source measures a light intensity 50% greater than the other observer, how far is the nearer observer from the source?

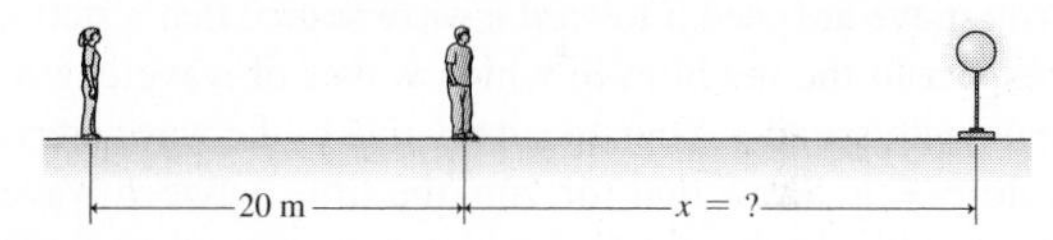

FIGURE 14.38 Problem 58

59. **CH** An ideal spring is stretched to a total length L_1. When that length is doubled, the speed of transverse waves on the spring triples. Find an expression for the unstretched length of the spring.
60. **CH** Show that the time it takes a wave to propagate up the cable in Problem 48 is $t = 2\sqrt{L/g}$, where L is the cable length.
61. You see an airplane 5.2 km straight overhead. Sound from the plane, however, seems to be coming from a point back along the plane's path at 35° to the vertical. What's the plane's speed, assuming an average sound speed of 330 ms?
62. What are the intensities in W/m^2 of sound with intensity levels of (a) 65 dB and (b) −5 dB?
63. Show that a doubling of sound intensity corresponds to approximately a 3-dB increase in the decibel level.
64. Sound intensity from a localized source decreases as the inverse square of the distance, according to Equation 14.8. If the distance from the source doubles, what happens to (a) the intensity and (b) the decibel level?
65. At 2.0 m from a localized sound source you measure the intensity level as 75 dB. How far away must you be for the perceived loudness to drop in half (i.e., to an intensity level of 65 dB)?
66. The A-string (440 Hz) on a piano is 38.9 cm long and is clamped at both ends. If the string tension is 667 N, what's its mass?
67. Show that the standing-wave condition of Equation 14.13 is equivalent to the requirement that the time it takes a wave to

make a round trip from one end of the medium to the other and back be an integer multiple of the wave period.

68. You're designing an organ for a new concert hall; the lowest note is to be 22 Hz. The architects have asked you to minimize the lengths of the organ pipes. How long will the longest pipe be if it's (a) closed at one end and (b) open at both ends?

69. CH Show by differentiation and substitution that a wave described by Equation 14.3 satisfies the wave equation (Equation 14.5), with wave speed $v = \omega/k$.

70. CH Show by differentiation and substitution that *any* function of the form $y = f(x \pm vt)$ satisfies the wave equation (Equation 14.5).

71. CH You're a marine biologist concerned with the effect of sonic booms on plankton, and you need to estimate the altitude of a supersonic aircraft flying directly over you at 2.2 times the speed of sound. You hear its sonic boom 19 s later. Assuming a constant 340 m/s sound speed, find the plane's altitude.

72. CH A 2.25-m-long pipe has one end open. Among its possible standing-wave frequencies is 345 Hz; the next higher frequency is 483 Hz. Find (a) the fundamental frequency and (b) the sound speed.

73. A wave source recedes from you at 8.2 m/s, and the wavelength you measure is 20% greater than what you would measure if the source were at rest. What's the wave speed?

74. BIO Obstetricians use ultrasound to monitor fetal heartbeat. If 5.0-MHz ultrasound reflects off the moving heart wall with a 100-Hz frequency shift, what's the speed of the heart wall? (*Hint:* You have *two* shifts to consider.)

75. You're in court, trying to argue your way out of a speeding ticket. You were stopped going 120 km/h in a 90-km/h zone. A technical expert testifies that the 70-GHz police radar signal underwent a 15.6-kHz frequency shift when it reflected off your car. You claim that corresponds to an impossible 240 km/h, so the radar must be defective. How should the judge rule?

76. You move at speed u toward a wave source that's stationary with respect to the medium in which waves of wavelength λ propagate with speed v. Your speed relative to the wave crests is therefore $v + u$. Show that for you, the time between wave crests is $T' = \lambda/(v + u)$, and from this show that you perceive a frequency given by Equation 14.16, with the + sign.

77. You're a meteorologist specifying a new Doppler radar system that determines the velocity of distant raindrops by reflecting radar signals (which travel at the speed of light) off them and measuring the Doppler shift. You need a system that will measure speeds as low as 2.5 km/h. A vendor offers a 5.0-GHz radar that can detect a frequency shift of only 50 Hz. Is that sufficient?

78. COMP Use a computer to form the sum implied in the caption of Figure 14.17, taking $\omega = 1\ \text{s}^{-1}$ and using (a) the three terms shown and (b) 10 terms (note that only odd harmonics appear in the sum). Plot your result over one cycle (t from 0 to 2π) and compare with the square wave shown in the figure.

79. Your little sister and her friend build treehouses and stretch a rope between them for sending messages. They hang a 1.4-kg mass on one end of the rope that passes over a pulley. The other end is tied to the second treehouse. When your sister plucks the rope, a wave propagates at 18 m/s. The girls deem this too slow; they want to increase the wave speed to 30 m/s. Your sister asks, "What mass should I use?" What do you reply?

80. DATA An airport neighborhood is concerned about the basing of the new F-35 jet fighter. They've got the data below for the sound intensity level measured at different distances from the plane as it takes off. They'd like to know the total sound power emitted by the plane. As a physics student, you're called to help. First, convert the sound intensity levels to actual intensity. Then find a quantity which, when you plot intensity against it, should give a straight line. Make your plot, determine the best-fit line, and use its slope to report the total sound power.

Distance (m)	1000	1200	1500	2000	3000	4000
Sound intensity level (dB)	80.7	79.4	76.9	74.2	71.6	68.8

Passage Problems

Tsunamis are ocean waves generally produced when earthquakes suddenly displace the ocean floor, and with it a huge volume of water. Unlike ordinary waves on the ocean surface, a tsunami involves the entire water column, from surface to bottom. To a tsunami, the ocean is shallow—and that makes tsunamis *shallow-water waves*, whose speed is $v = \sqrt{gd}$, where d is the water depth and g the gravitational acceleration. Tsunamis can travel thousands of kilometers across an ocean to reach the shore with their initial energy nearly intact; when they do, they can cause massive damage and loss of life (Fig. 14.39).

FIGURE 14.39 People flee as the devastating tsunami of December 2004 strikes Thailand (Passage Problems 81–84).

81. As a tsunami approaches shore, it
    a. speeds up.
    b. slows down.
    c. maintains its speed.

82. For a tsunami to behave as a shallow-water wave, its wavelength
    a. must be comparable to or longer than the ocean depth.
    b. must be shorter than the ocean depth.
    c. can have any value.

83. A tsunami is traveling at 450 km/h when the ocean depth abruptly doubles. Its new speed is roughly
    a. 225 km/h.
    b. 320 km/h.
    c. 640 km/h.
    d. 900 km/h.

84. On the open ocean, a tsunami has relatively small amplitude—typically 1 m or less. As the tsunami approaches shore, its amplitude increases and its wavelength decreases. As a result,
    a. its total energy increases.
    b. the rate at which it carries energy shoreward increases.
    c. the wave frequency increases.
    d. none of these quantities changes.

Answers to Chapter Questions

Answer to Chapter Opening Question

None. The waves transport energy, but not matter.

Answers to GOT IT? Questions

14.1 (b) 5 m/s, because that's the speed of the wave crest
14.2 (1) upper wave; (2) lower; (3) lower; (4) upper; (5) upper
14.3 (c)
14.4 (c)
14.5 (b) because of interference analogous to Fig. 14.21
14.6 (d)
14.7 $\frac{4}{5}$m, $\frac{4}{3}$m, 4 m
14.8 the boat

Fluid Motion

What You Know

- You know how to calculate the net force on an object when two or more forces are acting.
- You know how to find the gravitational force on an object.
- You're familiar with expressions for kinetic energy and gravitational potential energy.

What You're Learning

- You'll learn to apply Newtonian physics to the behavior of fluids.
- You'll learn to characterize a fluid by its *pressure* and *density*.
- You'll explore the forces on a fluid in static equilibrium.
- You'll learn about buoyancy, as described by *Archimedes' principle*.
- You'll learn to express the conservation of fluid mass and energy through the *continuity equation* and *Bernoulli's equation*.
- You'll see applications of fluid dynamics ranging from blood flow to aircraft flight to baseball.
- You'll learn about fluid friction or *viscosity*.

How You'll Use It

- You live at the bottom of an ocean of air, and you work frequently with liquids. The knowledge you gain here will help you understand the behavior of these everyday fluids, as well as fluids you'll encounter in such diverse fields as engineering, medicine, oceanography, astrophysics, climate science, and meteorology.
- You'll use concepts of pressure and density extensively in Part 3 when you study thermodynamics.

Why is only the "tip of the iceberg" above water?

A tornado whirls across a darkened sky. A plane flies, supported by air pressure on its wings. Gas from a giant star forms a cosmic whirlpool before plunging into a black hole. Fluid in your car's brake system amplifies the force of your foot on the brake pedal. Your own body is sustained by air moving into and out of your lungs, and by the flow of blood throughout your tissues. All these examples involve fluid motion.

Fluid is matter that flows under the influence of external forces. Fluids include both liquids and gases. The intermolecular forces are weaker in fluids than in solids, and as a result the molecules move around readily. In a liquid, those forces are strong enough to keep the molecules in close contact, while in a gas they're almost negligible and the molecules are usually widely spaced. Mobility of the individual molecules means that a fluid spreads out to take the shape of its container.

15.1 Density and Pressure

If we could observe a fluid on the molecular scale, we would find large numbers of molecules in continuous motion, colliding with each other and with the walls of their containers. This molecular behavior is governed by the laws of mechanics, and in principle we could study fluids by applying those laws to all the individual molecules. But even a drop of water contains about 10^{21} molecules; to calculate the motions of all those molecules would take the fastest computers many times the age of the universe!

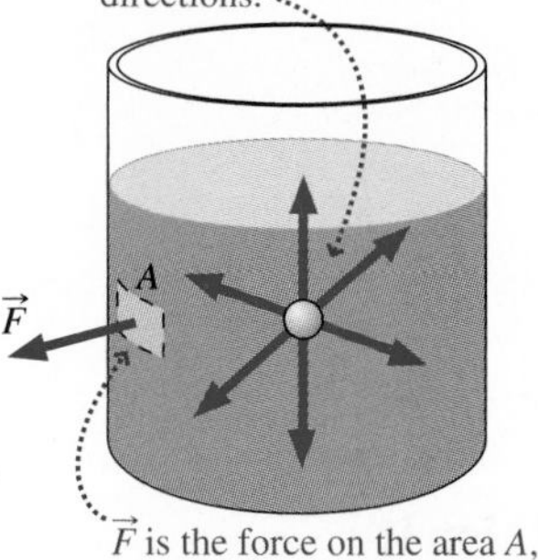

FIGURE 15.1 Pressure, the force per unit area, is exerted equally in all directions.

Because the number of molecules is so large, we approximate a fluid by treating it as continuous rather than composed of discrete particles. In this approximation, valid for fluid samples large compared with the distance between molecules, we describe the fluid by specifying macroscopic properties such as density and pressure.

Density

Density (symbol ρ, Greek rho) measures the mass per unit volume; its SI units are kg/m^3. Water's density is normally about $1000\ \text{kg/m}^3$; air's is about a factor of 1000 smaller. Because their molecules are essentially in contact, liquids are **incompressible**, meaning that their densities remain nearly constant. Gases, in contrast, are **compressible**: With relatively large intermolecular distances, their densities change readily.

Pressure

Pressure measures the normal force per unit area exerted by a fluid (Fig. 15.1):

$$p = \frac{F}{A} \quad \text{(pressure)} \tag{15.1}$$

The SI pressure unit is N/m^2, given the name **pascal** (Pa) after the French mathematician, scientist, and philosopher Blaise Pascal (1623–1662). Another commonly used pressure unit is the **atmosphere** (atm), defined as Earth's normal atmospheric pressure at sea level and equal to 101.3 kPa (in English units, that's 14.7 pounds per square inch, or psi).

Pressure is a scalar quantity; at a given point in a fluid, pressure is exerted equally in all directions (Fig. 15.1), so it makes no sense to associate a direction with it. This property explains an aspect of pressure that you may find puzzling. Although the atmosphere bears down on your body with a pressure of 14.7 pounds on every square inch, you don't feel that burden. That's because the force arising from this pressure is everywhere perpendicular to your body, and your body fluids respond by compressing until they're at the same pressure. If you've had your ears "pop" in a fast elevator or airplane, or when diving underwater, you know the pain that can develop when the pressure on your body is temporarily imbalanced.

GOT IT? 15.1 What quantity of water has the same mass as $1\ \text{m}^3$ of air under normal conditions? (a) $1\ \text{m}^3$; (b) $100\ \text{cm}^3$; (c) 1 L; (d) $0.1\ \text{m}^3$

15.2 Hydrostatic Equilibrium

For a fluid to remain at rest, the net force everywhere in the fluid must be zero; this condition is **hydrostatic equilibrium**. In the absence of any external forces, hydrostatic equilibrium requires that the pressure be constant throughout the fluid; otherwise, pressure differences would result in a net force, and the fluid would move in response. As Fig. 15.2 suggests, it's pressure *difference*, rather than pressure itself, that gives rise to forces within fluids.

Hydrostatic Equilibrium with Gravity

Hydrostatic equilibrium in the presence of gravity requires a pressure force to counteract the gravitational force. Since forces arise only from pressure differences, the fluid pressure must therefore vary with depth.

Figure 15.3 shows the forces on a fluid element of area A, thickness dh,and mass dm. A gravitational force acts downward on this fluid element; for it to be in equilibrium there must therefore be an upward pressure force—and that requires a greater pressure on the bottom. Suppose the pressures at the top and bottom are p and $p + dp$, respectively. Since pressure is force per unit area, the net pressure force is $dF_{\text{press}} = (p + dp)A - pA = A\,dp$.

$\vec{F}_{\text{net}}$
Increasing pressure
(a)

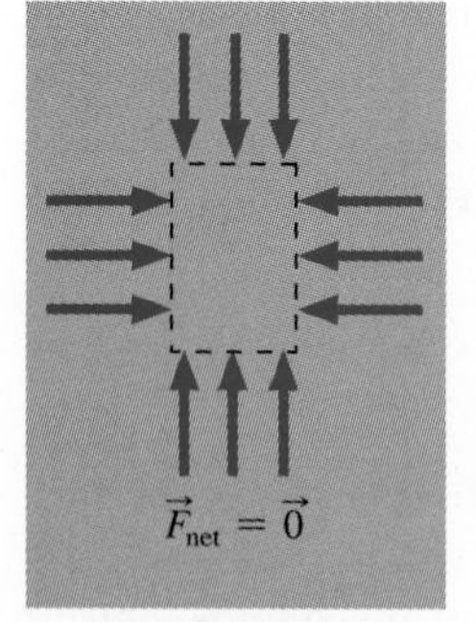

FIGURE 15.2 If pressure varies with position, then there's a net force on a volume of fluid.

The gravitational force is $dF_g = -g\,dm$, where the minus sign designates the downward direction. But the mass dm is the density times the volume, so $dF_g = -g\,dm = -g\rho A\,dh$. Hydrostatic equilibrium requires that these forces sum to zero: $A\,dp - g\rho A\,dh = 0$, or

$$\frac{dp}{dh} = \rho g \quad \text{(hydrostatic equilibrium)} \tag{15.2}$$

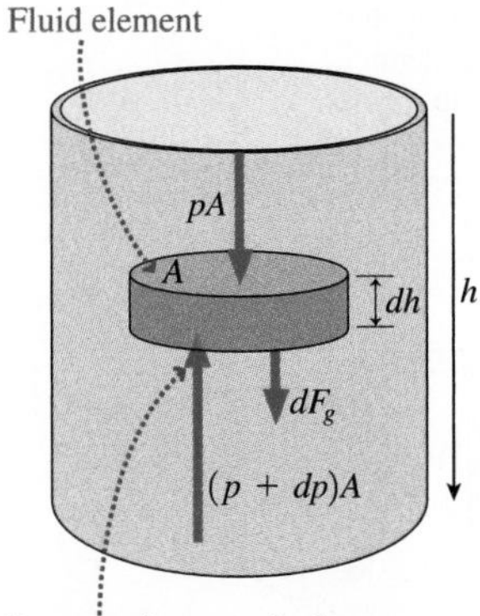

FIGURE 15.3 Forces on a fluid element in hydrostatic equilibrium.

This equation shows that dp/dh—the variation in pressure with depth h—is positive, confirming that pressure increases with depth. For a liquid, which is essentially incompressible, ρ is constant, and Equation 15.2 shows that pressure increases linearly with depth:

$$p = p_0 + \rho g h \tag{15.3}$$

where p_0 is the pressure at the liquid surface.

Equation 15.2 applies to any fluid in a uniform gravitational field; Equation 15.3 follows from Equation 15.2 for the special case of a liquid. It's also possible to integrate Equation 15.2 to find the pressure in a gas that's subject to the gravitational force. Because the gas density isn't constant, this is a little more involved mathematically. Problem 70 explores the variation of pressure with height in Earth's atmosphere.

EXAMPLE 15.1 Calculating Pressure: Ocean Depths

(a) At what water depth is the pressure twice atmospheric pressure? (b) What's the pressure at the bottom of the 11-km-deep Marianas Trench, the deepest point in the ocean? Take atmospheric pressure as 101 kPa and the density of seawater as 1030 kg/m^3.

INTERPRET This problem is about hydrostatic equilibrium, with water the fluid.

DEVELOP We determine that Equation 15.3, $p = p_0 + \rho g h$, applies, with p_0 equal to the atmospheric pressure at the water surface. Then at twice atmospheric pressure, $p = 2p_0$, and we can solve for h to answer part (a). Because pressure increases linearly with depth, we can extrapolate our result for part (a) to find the answer to part (b).

EVALUATE Solving our equation for the depth h and substituting the given numbers in, we find for part (a):

$$h = \frac{p - p_0}{\rho g} = \frac{2.02\times10^5\ \text{Pa} - 1.01\times10^5\ \text{Pa}}{(1030\ \text{kg/m}^3)(9.81\ \text{m/s}^2)} = 10.0\ \text{m}$$

This result implies that the pressure increases by 100 kPa for every 10 m of depth. In the Marianas Trench, 11×10^3 m deep, the pressure increase is then

$$(11\times10^3\ \text{m})(100\ \text{kPa}/10\ \text{m}) = 110\ \text{MPa}$$

which is our answer to (b).

ASSESS This is over a thousand times atmospheric pressure, or more than 8 tons per square inch! Creatures living at these depths are in pressure equilibrium with their surroundings. To bring them to the surface for study, scientists must maintain their natural pressure or they'll explode. A similar plight awaits scuba divers who hold their breath while ascending; air in the lungs expands, bursting the alveoli. Problem 62 involves film producer James Cameron's recent dive to the bottom of the Marianas Trench. ■

Measuring Pressure

Figure 15.4 shows a **barometer**, in which air pressure acts on the open pool of mercury, pushing the liquid into the evacuated tube. Since $p_0 = 0$ in the vacuum at the top of the tube, Equation 15.3 becomes simply $p = \rho g h$, showing that the height h of the mercury is directly proportional to atmospheric pressure p. Standard atmospheric pressure of 101.3 kPa supports a mercury column 760 mm or 29.92 in. high. Pressure varies slightly with meteorological conditions, and weather forecasters regularly report atmospheric pressure in millimeters or inches of mercury. Mercury's high density makes for a reasonable-sized barometer. Example 15.1 shows that a water-filled barometer would need to be 10 m long!

A **manometer** is a U-shaped tube filled with liquid and used to measure pressure differences. A pressure difference between the two ends results in a height difference h between the liquid surfaces (Fig. 15.5, next page). Equation 15.3 shows that h is directly proportional to the pressure difference.

Barometers and manometers are the classic pressure-measuring instruments, and understanding them will help you grasp the meaning of pressure. But pressure-measuring

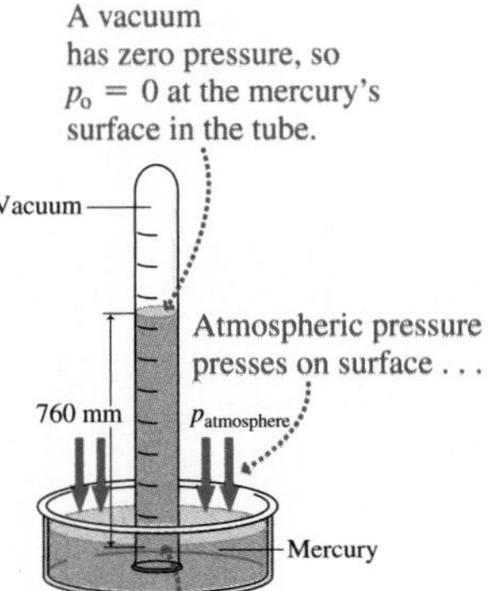

FIGURE 15.4 A mercury barometer.

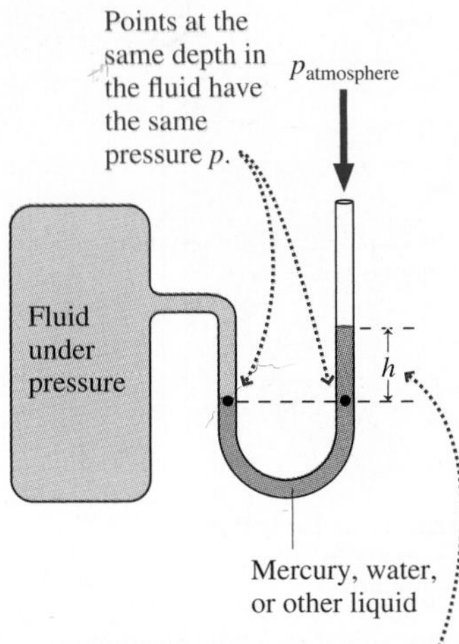

FIGURE 15.5 A manometer used to measure the pressure difference between a closed container and the atmosphere.

Video Tutor Demo | **Pressure in Water and Alcohol**

Video Tutor Demo | **Water Level in Pascal's Vases**

devices today are usually electronic, using the pressure force to alter electrical properties and produce an electrical signal proportional to pressure.

The term **gauge pressure** describes the excess pressure above atmospheric. Inflation instructions for tires and sports equipment specify gauge pressure. A tire inflated to 200 kPa (about 30 psi) has an absolute pressure of about 300 kPa because of the additional 100-kPa atmospheric pressure.

Pascal's Law

Equation 15.3 shows that an increase in surface pressure p_0 results in the same pressure increase throughout the fluid. More generally, a pressure increase anywhere is felt throughout the fluid—a fact known as **Pascal's law**. Pascal applied this principle in his invention of the hydraulic press. Today hydraulic systems, based on Pascal's law, control machinery ranging from automobile brakes to aircraft wings, bulldozers, cranes, and robots.

EXAMPLE 15.2 Applying Pascal's Law: A Hydraulic Lift

In the hydraulic lift of Fig. 15.6, a large piston supports a car; the total mass of car and piston is 3200 kg. What force must be applied to the smaller piston to support the car?

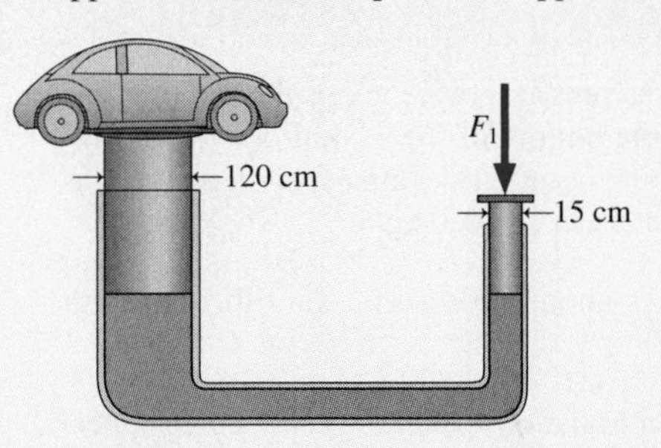

FIGURE 15.6 A hydraulic lift.

INTERPRET We interpret this as a problem involving Pascal's law. Whatever pressure results from the force on the smaller piston is transmitted through the fluid to the larger piston and thus supports the car.

DEVELOP We're given a drawing. Having determined that Pascal's law applies, and neglecting pressure variations with depth, we conclude that the pressure is the same throughout the system. Our plan, then, is to write expressions involving the pressures at both pistons and use the fact that they're equal to solve for the unknown force. We'll use the fact that the pressure on a piston is the applied force divided by the piston's area.

EVALUATE The small piston exerts a pressure $p = F_1/A_1 = F_1/\pi R_1^2$, where F_1 is the unknown force. The pressure at the large piston is the same and produces a force $F_2 = pA_2$. This force supports the weight mg of piston and car; therefore, we have

$$mg = pA_2 = p\pi R_2^2 = \frac{F_1}{\pi R_1^2}\pi R_2^2 = F_1\left(\frac{R_2}{R_1}\right)^2$$

Solving for F_1 gives our answer:

$$F_1 = mg\left(\frac{R_1}{R_2}\right)^2 = (3200\text{ kg})(9.8\text{ m/s}^2)\left(\frac{15\text{ cm}}{120\text{ cm}}\right)^2 = 490\text{ N}$$

We used the diameters from Fig. 15.3, rather than the radii, because their ratio is the same.

ASSESS How can a 490-N force—about 100 lb—support the car? Through the constant fluid pressure, this smaller force is effectively multiplied by the ratio of the piston areas. What about energy? Do we get something for nothing here? GOT IT? 15.2 explores this question. ■

GOT IT? 15.2 Neglecting friction and other nonconservative forces, does the agent applying the force $\vec{F}_1$ in Fig. 15.6 do (a) more, (b) less, or (c) the same work as is done on the car? Explain.

15.3 Archimedes' Principle and Buoyancy

Why do some objects float while others sink? Figure 15.7*a* shows the upward pressure force on an arbitrary fluid volume balancing the downward gravitational force. Now imagine replacing the fluid volume with a solid object of identical shape (Fig. 15.7*b*). The remaining fluid hasn't changed, so it continues to exert an upward force on the object—a force whose magnitude equals the weight of the *original fluid volume*. This force is the **buoyancy force**, and in giving its magnitude we've stated **Archimedes' principle**: The buoyancy force on an object is equal to the weight of the fluid displaced by the object.

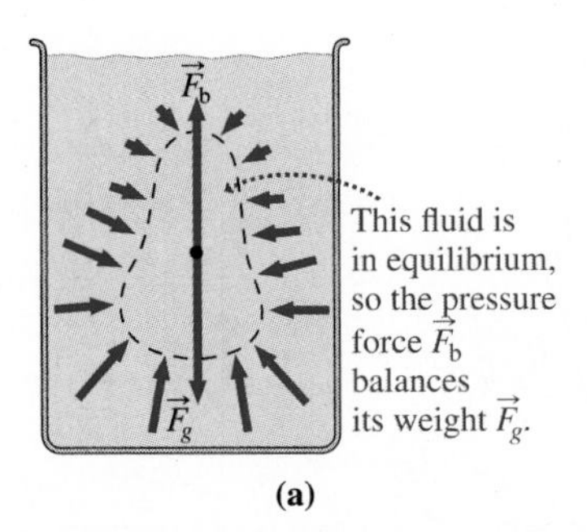

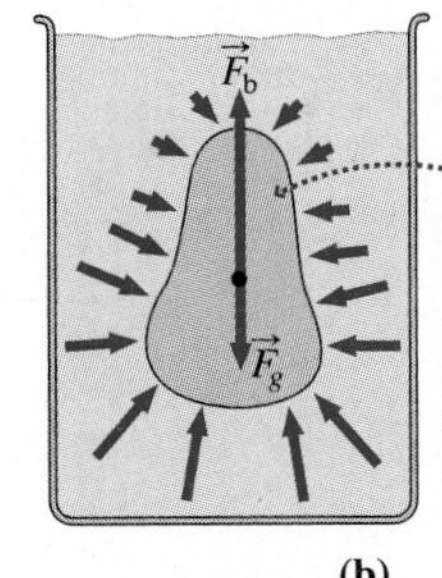

FIGURE 15.7 The buoyancy force $\vec{F}_b$ arises because pressure increases with depth.

If the submerged object weighs more than the displaced fluid, then the gravitational force exceeds the buoyancy force and the object sinks. If the object weighs less than the displaced fluid, buoyancy is greater and the object rises. Therefore, an object floats or sinks depending on whether its average density is greater than or less than that of the fluid. In between is the case of **neutral buoyancy**, when an object's average density is the same as that of the fluid. The Application on this page gives examples of neutral buoyancy.

APPLICATION Swimming Like a Fish

Fish glide through the water, maintaining depth with little effort and rising or diving at will. That's possible because they're in neutral buoyancy, with density equal to that of the surrounding water. The fish's *swim bladder*—a pair of gas-filled sacs—expands and contracts under the influence of water pressure, maintaining neutral buoyancy. Biologists believe that the lungs of terrestrial organisms may have evolved from the swim bladders of our ancestral fish. The ballast tanks of submarines serve a similar function to keep these vessels in neutral buoyancy. Analogously, the burner that's fired periodically to heat the air in a hot-air balloon serves the same function; by introducing hot, lower-density air, the balloonist can keep the craft in neutral buoyancy or induce it to rise.

PhET: Balloons and Buoyancy

EXAMPLE 15.3 Finding the Buoyancy Force: Working Underwater

You're setting up a raft in a swimming area, and you need to move a 60-kg concrete block on the lake bottom. What's the apparent weight of the block as you lift it underwater? The density of concrete is 2200 kg/m^3.

INTERPRET We interpret this as a problem about buoyancy; the concrete will seem to weigh less underwater because of the upward buoyancy force. We identify the apparent weight as the force you'll need to apply to lift the block off the lake bottom.

DEVELOP Figure 15.8 is our sketch, showing gravity and the buoyancy force on the block; you'll need to apply a force equal but opposite to their sum. Archimedes' principle applies, giving a buoyancy force equal to the weight of water that occupies the same volume as the concrete block. So our plan is to find that force and compare it with the gravitational force on the block.

EVALUATE The concrete block's mass is m_c, so its weight is the gravitational force $F_g = m_c g$. Its volume is $V_c = m_c/\rho_c$, which also equals the volume of the displaced water: $V_w = V_c = m_c/\rho_c$. Archimedes' principle says that the weight of this displaced water is the magnitude of the buoyancy force, so $F_b = m_w g = V_w \rho_w g = m_c g(\rho_w/\rho_c)$. Then

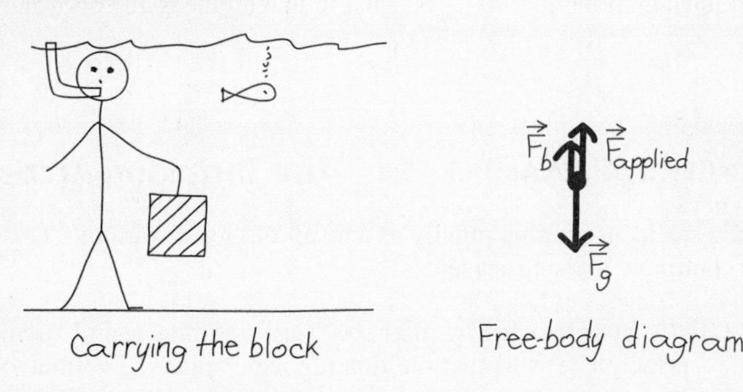

FIGURE 15.8 What's the apparent weight of the concrete block?

the upward buoyancy force and the downward gravitational force sum to give a downward force of magnitude:

$$F_g - F_b = m_c g - m_c g\left(\frac{\rho_w}{\rho_c}\right) = m_c g\left(1 - \frac{\rho_w}{\rho_c}\right)$$
$$= (60\text{ kg})(9.8\text{ m/s}^2)\left(1 - \frac{1}{2.2}\right) = 320\text{ N}$$

(continued)

You have to apply an upward force of equal magnitude to lift the block off the bottom.

ASSESS This is about 70 lb—a lot more manageable than the block's weight mg of nearly 600 N or about 130 lb in air. Knowing the apparent weight of a submerged object would let us turn this problem around to determine its density. Archimedes purportedly used his principle in this way to find the density of the king's crown, and thus show that it was not pure gold. ■

Video Tutor Demo | **Weighing Weights in Water**

Floating Objects

Archimedes' principle still holds for a floating object. But with the object in equilibrium at a liquid surface, the buoyancy force now must balance the object's weight—which will happen if the fluid displaced by the submerged part of the object weighs the same as the object. This condition determines how high in the water the object floats, as Example 15.4 illustrates.

EXAMPLE 15.4 Floating Objects: The Tip of the Iceberg

The average density of a typical arctic iceberg is 0.86 that of seawater. What fraction of an iceberg's volume is submerged?

INTERPRET We interpret this problem also as being about buoyancy, but now we have a floating object with buoyancy balancing gravity. Only the submerged portion contributes to the buoyancy force, so the condition of force balance will enable us to find how much of the iceberg is submerged.

DEVELOP Figure 15.9 is our sketch, showing gravitational and buoyancy forces of equal magnitude. Archimedes' principle applies here and states that the buoyancy force is equal to the weight of water displaced by the submerged portion of the iceberg. So our plan is to find the gravitational and buoyancy forces, and then equate their magnitudes to get the submerged volume. Since we're looking for volume, we'll write any masses as products of density and volume.

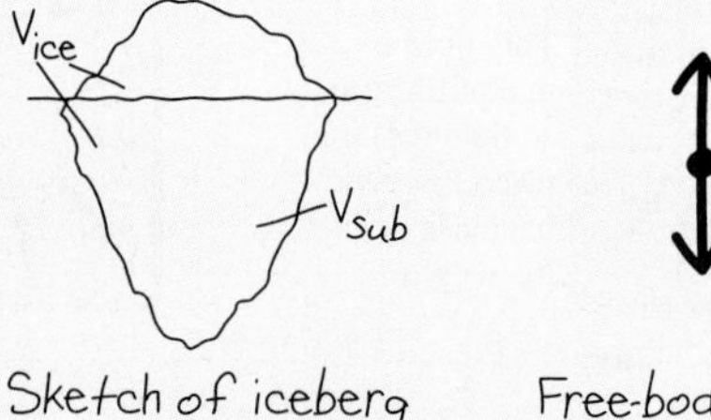

FIGURE 15.9 How much of the iceberg is submerged?

EVALUATE The iceberg's weight is $w_{ice} = m_{ice}g = \rho_{ice}V_{ice}g$, where V_{ice} is the volume of the *entire* iceberg. Only the submerged portion displaces water, so the volume of displaced water is V_{sub}, and the weight of the displaced water is therefore $w_{water} = m_{water}g = \rho_{water}V_{sub}g$. By Archimedes' principle, w_{water} is equal in magnitude to the buoyancy force, which balances gravity when the iceberg is in equilibrium. Equating the two gives $\rho_{water}V_{sub}g = \rho_{ice}V_{ice}g$, which we solve to get

$$\frac{V_{sub}}{V_{ice}} = \frac{\rho_{ice}}{\rho_{water}} = 0.86$$

ASSESS Our result means that 86% of the iceberg's volume is under water, leaving only 14% showing. Tip of the iceberg, indeed! Note that the volume ratio is just the density ratio ρ_{ice}/ρ_{water}, showing that the closer an object's density is to that of water, the lower it floats. ■

CONCEPTUAL EXAMPLE 15.1 The Shrinking Arctic

Arctic sea ice is melting rapidly as a result of global warming. Does this contribute to rising sea levels?

EVALUATE Your first answer might be "yes," but think again! Archimedes' principle tells us that the floating ice displaces a volume of water whose weight is equal to that of the *entire* ice—although only the submerged portion does the displacing. When the ice melts, it becomes water that, because it no longer sticks above the surface, displaces a volume equal to its entire weight. But since the weight hasn't changed, the amount of water displaced is the same. That means the water level is unchanged.

ASSESS Melting ice doesn't contribute to sea-level rise—as long as it's sea ice that melts. Land ice is a different story: Melting glaciers and "calving" of glaciers to form icebergs together cause about half of contemporary sea-level rise. Thermal expansion, which we'll explore in 17, causes the rest. According to the Intergovernmental Panel on Climate Change, these two processes are expected to result in sea-level rise in the range of 32 cm to almost 1 m by the year 2100.

MAKING THE CONNECTION The land-based Greenland ice cap occupies some 3 million km^3, while some 15,000 km^3 of ice are afloat in the Arctic Ocean. Compare the approximate rise in the world's oceans that would result from complete melting of these two ice volumes.

EVALUATE As this conceptual example shows, melting sea ice won't contribute to sea-level rise, but land-based ice will add water to the oceans. Its volume will be about 86% that of the ice (see Example 15.4), or about 2.6 million km^3. With oceans covering about 71% of Earth's surface area ($4\pi R_E^2$, where R_E is Earth's radius), the meltwater will spread in a layer of thickness d and therefore volume $V = (0.71)(4\pi R_E^2)d$. Setting this quantity equal to the 2.6×10^{15}-m^3 volume of meltwater and solving for d then gives $d = 7$ m—enough to inundate most of today's coastal cities.

GOT IT? 15.3 The density of a rubber ball is three-fifths that of water. When placed in water, will the ball (a) float with less than half of it out of the water, (b) float with more than half of it out of the water, or (c) sink?

Center of Buoyancy

The buoyancy force acts not at the center of mass of a floating object, but at the center of mass of the water that would be there if the object weren't. This point is called the **center of buoyancy**, and for an object to float in stable equilibrium, the center of buoyancy must lie above the center of mass. Otherwise, a net torque results that tends to tip the object. The stability of watercraft depends critically on this condition (Fig. 15.10).

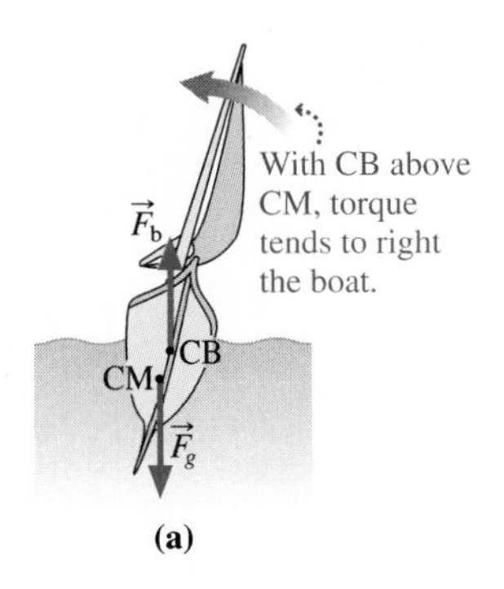

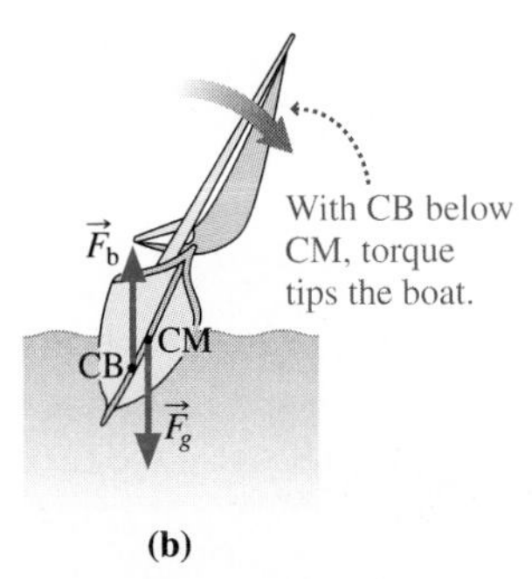

FIGURE 15.10 A boat's stability requires the center of buoyancy (CB) to be above the center of mass (CM).

15.4 Fluid Dynamics

We now turn our attention to moving fluids, described by the flow velocity at each point in the fluid and at each instant of time. We illustrate flow velocity by drawing continuous lines called **streamlines** that are everywhere tangent to the local flow direction (Fig. 15.11). Their spacing is a measure of flow speed, with closely spaced streamlines indicating higher speed. Small particles introduced into moving fluids follow streamlines and therefore give a visual indication of the flow velocity pattern.

In **steady flow**, the pattern of fluid motion remains the same at each point, even though individual fluid elements are in continuous motion. A river in steady flow always looks the same, even though you're not seeing the same water each time you look. At a given point, the water velocity is always the same. **Unsteady flow**, in contrast, involves fluid motion that changes with time. Blood flow in your arteries is unsteady; with each contraction of the heart ventricles, the pressure rises and the flow velocity increases. We'll restrict our quantitative description of fluid motion to steady flow.

Like all other motion in classical physics, fluid motion is governed by Newton's laws. It's possible to write Newton's second law in a form that involves explicitly the fluid velocity as a function of position and time. But the resulting equation is difficult to solve in any but the simplest cases. Instead of applying Newton's law directly, we'll approach fluid dynamics using energy conservation.

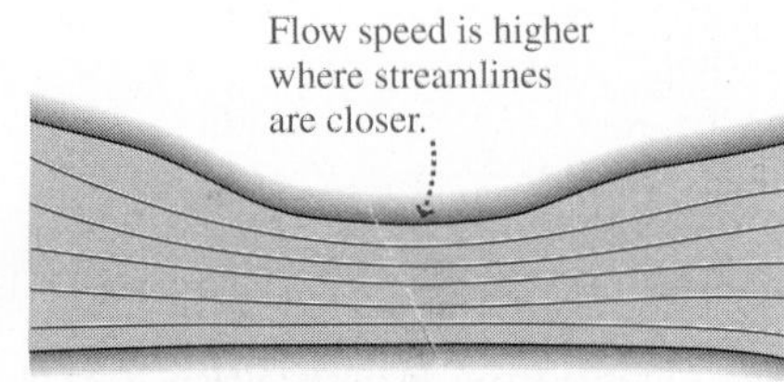

FIGURE 15.11 Streamlines represent flow velocity in a river.

GOT IT? 15.4 The photo shows smoke particles tracing streamlines in a test of a car's aerodynamic properties. Is the flow speed greater (a) over the top or (b) at the back?

Conservation of Mass: The Continuity Equation

In mechanics we had no trouble keeping track of the individual objects. But a fluid is continuous and deformable, so it's not easy to follow an individual fluid element as it moves. Yet fluid is conserved; as it moves, new fluid is neither created nor destroyed.

Consider a steady fluid flow represented by streamlines, as shown in Fig. 15.12*a*. We shaded a **flow tube**—a small tubelike region bounded on its sides by streamlines and on its ends by areas at right angles to the flow. The flow tube has a sufficiently small cross section that fluid velocity and other properties don't vary significantly over any cross section; however, fluid properties may vary along the flow tube. Although our flow tube has no physical boundaries, it nevertheless acts like a pipe because fluid flows *along*, not across, the streamlines. In steady flow, the rate at which fluid enters the tube at its left end must equal the rate at which it exits at the right.

Figure 15.12*b* shows a small fluid element just about to enter the flow tube, a process that will take some time Δt. Suppose the fluid is moving at speed v_1; since it takes time Δt

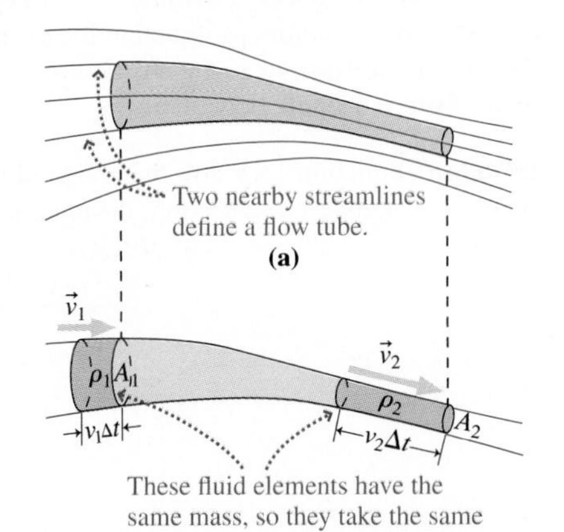

FIGURE 15.12 In steady flow, fluid enters and leaves a flow tube at the same rate.

to cross the tube end, its length is $v_1\,\Delta t$. With cross-sectional area A_1, length $v_1\,\Delta t$, and density ρ_1, the mass of the entering fluid is $m = \rho_1 A_1 v_1\,\Delta t$.

Another fluid element is shown just about to leave the tube. Suppose it has the *same* mass m as the entering fluid element. Then it must exit the tube in the *same* time Δt in order to keep the total mass in the tube constant. Its mass can be written as $m = \rho_2 A_2 v_2\,\Delta t$.

Equating our two expressions for m shows that $\rho_1 v_1 A_1 = \rho_2 v_2 A_2$. Since the endpoints of the tube are arbitrary, we conclude that the quantity ρvA must have the same value anywhere along the flow tube:

$$\rho vA = \text{constant along a flow tube} \quad \begin{pmatrix}\text{continuity equation,}\\ \text{any fluid}\end{pmatrix} \tag{15.4}$$

Equation 15.4 is the **continuity equation**, which expresses the conservation of mass in steady fluid flow. The units of ρvA here are $(\text{kg/m}^3)(\text{m/s})(\text{m}^2)$, or simply kg/s. This quantity is therefore the **mass flow rate** or mass of fluid per unit time passing through the flow tube. Equation 15.4 says that the mass flow rate is constant in steady flow.

For a liquid, the density ρ is constant, and the continuity equation becomes simply

$$vA = \text{constant along a flow tube} \quad \begin{pmatrix}\text{continuity equation,}\\ \text{liquid}\end{pmatrix} \tag{15.5}$$

Now the constant quantity is just vA, with units of $(\text{m/s})(\text{m}^2)$, or m^3/s. This is the **volume flow rate**. Equation 15.5 makes sense: Where a liquid's cross-sectional area is large, it flows slowly to transport a given volume of fluid per unit time. But in a constricted area, it must flow faster to carry the same volume. With a gas, obeying Equation 15.4 but not necessarily 15.5, the situation is slightly more ambiguous because density variations also play a role. For flow speeds below the speed of sound in a gas, it turns out that smaller area implies a higher flow speed just as for a liquid. But when the gas flow speed exceeds the sound speed, density changes become so great that flow speed actually decreases with smaller area.

EXAMPLE 15.5 Using the Continuity Equation: Ausable Chasm

The Ausable River in upstate New York is about 40 m wide. Under typical early summer conditions, it's 2.2 m deep and flows at 4.5 m/s. Just before it reaches Lake Champlain, the river enters Ausable Chasm, a deep gorge only 3.7 m wide. If the flow rate in the gorge is 6.0 m/s, how deep is the river at this point? Assume a rectangular cross section with uniform flow speed.

INTERPRET The concept behind this problem is mass conservation, embodied in the continuity equation for a liquid, Equation 15.5. Since the flow is uniform over the river's cross section, we can treat the entire river as a single flow tube.

DEVELOP Equation 15.5 says that the product vA is constant. For the river's rectangular cross section, the area A is the product of width w and depth d. Then Equation 15.5 becomes $v_1 w_1 d_1 = v_2 w_2 d_2$, where the subscripts indicate values upstream and in the gorge. Our plan is to solve for the depth d_2 in the gorge.

EVALUATE Solving gives

$$d_2 = \frac{v_1 w_1 d_1}{v_2 w_2} = \frac{(4.5\ \text{m/s})(40\ \text{m})(2.2\ \text{m})}{(6.0\ \text{m/s})(3.7\ \text{m})} = 18\ \text{m}$$

ASSESS This is about 60 feet, quite a depth for a small river! But conservation of mass requires it. In the gorge, the river is much narrower but its flow speed is only a little higher, so it's got to be a lot deeper. ■

Conservation of Energy: Bernoulli's Equation

We now turn to conservation of fluid energy. Figure 15.13 shows the same fluid element as it enters and again as it leaves a flow tube. If it enters with speed v_1 and leaves with speed v_2, the change in its kinetic energy is

$$\Delta K = \tfrac{1}{2}m(v_2^2 - v_1^2)$$

The work–kinetic energy theorem (Equation 6.14) equates this change to the net work done on the fluid element. As the element enters the tube, it's subject to a pressure force p_1A_1 from the fluid to its left. This external force acts over the length Δx_1 of the fluid element as it enters, so it does work $W_1 = p_1A_1\,\Delta x_1$. Similarly, as it leaves the tube, the fluid element experiences a force p_2A_2 from the fluid to its right. Because this force is opposite the flow direction, it does negative work $W_2 = -p_2A_2\,\Delta x_2$. External forces from adjacent flow tubes act at right angles to the flow, so they do no work. Finally, the fluid element rises a distance $y_2 - y_1$ as it traverses the tube; therefore, gravity does negative work $W_g = -mg(y_2 - y_1)$. Summing these three contributions and applying the work–kinetic energy theorem, we have $W_1 + W_2 + W_g = \Delta K$, or $p_1A_1\,\Delta x_1 - p_2A_2\,\Delta x_2 - mg(y_2 - y_1) = \frac{1}{2}m(v_2^2 - v_1^2)$. The quantities $A_1\,\Delta x_1$ and $A_2\,\Delta x_2$ are the volumes of the fluid element as it enters and leaves the flow, respectively. If we restrict ourselves to incompressible fluids, then those volumes are equal. Dividing through by this common volume $V = A\,\Delta x$ and noting that $m/V = \rho$, we get $p_1 + \frac{1}{2}\rho v_1^2 + \rho g y_1 = p_2 + \frac{1}{2}\rho v_2^2 + \rho g y_2$, or

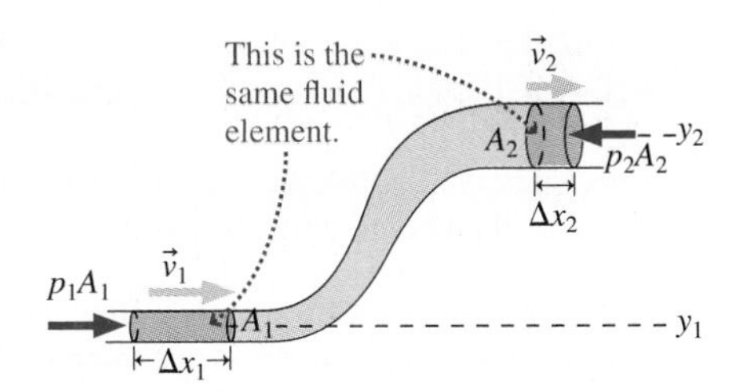

FIGURE 15.13 A flow tube showing the same fluid element entering and leaving. The work done by pressure and gravitational forces equals the change in kinetic energy of the fluid element.

$$p + \frac{1}{2}\rho v^2 + \rho g y = \text{constant along a flow tube} \quad \text{(Bernoulli's equation)} \qquad (15.6)$$

This is **Bernoulli's equation**, after the Swiss mathematician Daniel Bernoulli (1700–1782).

What do the terms in Bernoulli's equation mean? The quantity $\frac{1}{2}\rho v^2$ looks like kinetic energy $\frac{1}{2}mv^2$, except it has mass per unit volume ρ instead of mass m. It's therefore the kinetic energy per unit volume, or kinetic-energy density. Similarly, $\rho g y$ is the gravitational potential energy per unit volume. Pressure p, too, has the units of energy density and represents internal energy of the fluid. Bernoulli's equation therefore says that the total energy per unit volume of fluid is conserved as the fluid moves.

Bernoulli's equation in the form 15.6 applies to incompressible fluids. It neglects fluid friction, also called *viscosity*, that may dissipate fluid kinetic energy. It also neglects energy transfers associated with machinery such as turbines or pumps that may extract or add to the fluid's energy. Engineers often include those effects in Bernoulli's equation.

15.5 Applications of Fluid Dynamics

The laws of mass and energy conservation that we just derived for fluids allow us to analyze a wide variety of natural and technological phenomena. We'll usually need both the continuity equation and Bernoulli's equation, considering the values of the appropriate constant quantities at two points in a fluid flow. As you study the examples and applications that follow, remember that they're ultimately based in the same Newtonian principles we've been using to describe mechanical systems.

PROBLEM-SOLVING STRATEGY 15.1 Fluid Dynamics

The continuity equation and Bernoulli's equation are the keys to solving problems in fluid dynamics. Here's a strategy that will help you focus these two equations on a problem.

INTERPRET The form of Bernoulli's equation we derived applies only to incompressible fluids. So be sure you're dealing either with a liquid or with a gas flowing at speeds well below its sound speed.

DEVELOP

- Identify a flow tube. This may be a physical pipe or other structure, or a mathematical tube bounded by streamlines.
- Draw a sketch of the situation, showing the flow tube.
- Determine the point where you're interested in solving for some aspect of the flow, and another point where you know the quantities that go into the continuity equation and Bernoulli's equation. Note those quantities that you know at each point. Mark the two points on your sketch.
- Write the continuity equation and Bernoulli's equation, with the known quantities forming the terms on one side and the other side containing your unknown(s).

EVALUATE Evaluate by solving your equations for the unknown quantity or quantities. Often this will involve solving the continuity equation first and then using the result in Bernoulli's equation.

ASSESS Ask whether your result makes sense. Does flow speed increase at a constriction? Does pressure go up when flow speed drops, or vice versa? Are there any limitations that apply, or insights to be gained?

APPLICATION An Airplane Speedometer

A car speedometer works by counting rotations of its wheels as they turn on the road. But airplanes can't do that; instead, they use Bernoulli's principle to measure airspeed—the plane's speed relative to the air. The device that accomplishes this is a *Pitot tube*, which samples the pressure of air moving past the plane, as well as the pressure of air that's been stopped relative to the plane. Bernoulli's equation relates the difference of the two pressures to the relative speed of the air and plane, providing the pilots with a direct indication of their airspeed. Knowing the wind speed—often substantial at aircraft altitudes—then lets the plane's computers determine the speed relative to the ground. The photo shows a pair of external Pitot tubes on an aircraft fuselage. You can explore the physics of the Pitot tube in Problem 66, where you'll also find a simplified diagram of the device.

EXAMPLE 15.6 Bernoulli's Equation: Draining a Tank

A large, open tank is filled to height h with liquid of density ρ. Find the speed of liquid emerging from a small hole at the base of the tank.

INTERPRET We're dealing with a flow of water, an incompressible liquid. So we can apply our problem-solving strategy for fluid dynamics.

DEVELOP We take the tank to be a rather oddly shaped flow tube, and Fig. 15.14 is our sketch. We're interested in the water's velocity at the hole, so the hole is one of the points we'll use in the fluid equations. Since the hole is open to the atmosphere, the pressure at the hole is atmospheric pressure p_a. The top surface is also open to the atmosphere, so here the pressure is also p_a. Now, because the hole is very small in relation to the tank, the water level drops only slowly. Therefore, we can make the approximation $v = 0$ at the top—and thus we know both p and v at the top. Although we didn't write a formal equation here, that approximation follows from the continuity equation because the ratio of hole to top surface area is so small. We also need the potential-energy terms in Bernoulli's equation. If we take $y = 0$ at the hole, then those terms are zero at the hole and ρgh at the top. Then Bernoulli's equation, $p + \frac{1}{2}\rho v^2 + \rho gy =$ constant, becomes

$$p_a + \rho gh = p_a + \tfrac{1}{2}\rho v_{hole}^2$$

where the terms on the left are at the top surface and those on the right are at the hole. We've taken care of the continuity equation through our assumption of negligible flow speed at the top.

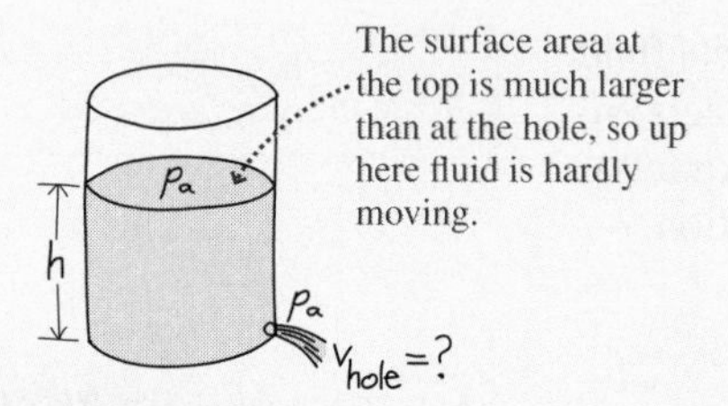

FIGURE 15.14 How fast does the liquid emerge from the tank?

EVALUATE Atmospheric pressure cancels, and we solve for the unknown flow velocity at the hole:

$$v_{hole} = \sqrt{2gh}$$

ASSESS This is the same result we would get by dropping an object from a height h—and for the same reason: conservation of energy. Draining a gram of water from the hole is energetically equivalent to removing a gram of water from the top and dropping it. Just as the speed of a falling object is independent of its mass, so the speed of the liquid is independent of its density. As the liquid drains, however, the height decreases and so does the flow rate. That's a calculus challenge you can try in Problem 69.

✓TIP Reasonable Approximations

Making reasonable approximations is often important in solving realistic problems. Look for opportunities to approximate a physical quantity, especially when other terms appear more significant. But always be sure that your approximations are reasonable. In this example, we reasoned that the fluid's speed at the top of the tank was negligible because it's proportional to the ratio of the hole to the top surface area, a very small value.

■

Video Tutor Demo | **Air Jet Blows between Bowling Balls**

Venturi Flows and the Bernoulli Effect

A constriction in a pipe carrying incompressible fluid requires that the flow speed increase in order to maintain constant mass flow. Such a constriction is a **venturi**. Because of the increased speed, Bernoulli's equation requires the pressure to be lower in the venturi. The next example shows how this effect provides a measure of fluid flow.

EXAMPLE 15.7 Measuring Flow Speed: A Venturi Flowmeter

An incompressible fluid of density ρ flows through a horizontal pipe of cross-sectional area A_1. The pipe has a venturi constriction of area A_2, and a gauge measures the pressure difference Δp between the unconstricted pipe and the venturi. Find an expression for the flow speed in the unconstricted pipe.

INTERPRET This is a problem about incompressible fluid flow, so our strategy applies.

DEVELOP For a flow tube, we choose a section of pipe that includes the venturi. Figure 15.15 is a sketch showing some streamlines through this tube. We're interested in the flow velocity in the unconstricted pipe, so any point outside the venturi will do. The other

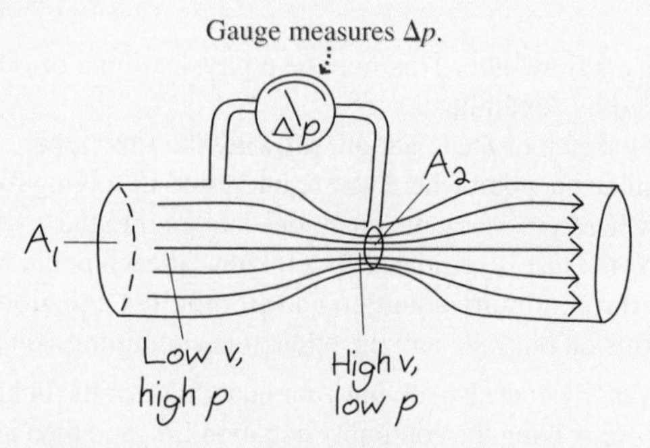

FIGURE 15.15 Our sketch of a venturi flowmeter.

point should be in the venturi. The continuity equation then reads $v_1A_1 = v_2A_2$, where the subscript 1 refers to the unconstricted pipe and 2 to the venturi. The pipe is horizontal, so the potential-energy term ρgh in Bernoulli's equation is the same on both sides, and it drops out. Bernoulli's equation then reads

$$p_1 + \tfrac{1}{2}\rho v_1^2 = p_2 + \tfrac{1}{2}\rho v_2^2$$

EVALUATE We can eliminate the velocity v_2 by solving the continuity equation: $v_2 = (A_1/A_2)v_1 = bv_1$, where we defined b as the ratio of the larger to smaller area: $b = A_1/A_2$. Using this result in Bernoulli's equation gives $p_1 + \tfrac{1}{2}\rho v_1^2 = p_2 + \tfrac{1}{2}\rho b^2 v_1^2$. In terms of the pressure difference $\Delta p = p_1 - p_2$, this becomes $\Delta p = \tfrac{1}{2}\rho b^2 v_1^2 - \tfrac{1}{2}\rho v_1^2 = \tfrac{1}{2}\rho v_1^2(b^2 - 1)$. We then solve for v_1 to get our answer:

$$v_1 = \sqrt{\frac{2\,\Delta p}{\rho(b^2 - 1)}}$$

ASSESS Make sense? The pressure difference results from the change in speed; no flow, no pressure difference. So it's reasonable that v increases with Δp. But a given pressure difference Δp is easier to get with a larger area ratio b, so flow speed depends inversely on b. Finally, the greater inertia of a denser fluid means a given pressure difference produces less acceleration, implying a lower initial speed; that's why ρ appears in the denominator. ■

FIGURE 15.16 A ping-pong ball supported by downward-flowing air. High-velocity flow is inside the narrow part of the funnel.

The occurrence of lower pressure with higher flow speeds, and vice versa—the **Bernoulli effect**—has numerous manifestations. The dirt around a prairie dog's hole is mounded up in a way that forces wind to accelerate over the hole, resulting in lower pressure above the hole. Biologists speculate that prairie dogs have evolved this design to provide natural ventilation. The Bernoulli effect can be strikingly counterintuitive. Figure 15.16 shows a ping-pong ball suspended by *downward* airflow in an inverted funnel. Rapid divergence of the flow results in lower speed and therefore higher pressure below the ball.

GOT IT? 15.5 A large tank is filled with liquid to the level h_1 shown in the figure. It drains through a small pipe whose diameter varies; emerging from each section of pipe are vertical tubes open to the atmosphere. Although the picture shows the same liquid level in each pipe, they really won't be the same. Rank levels h_1 through h_4 in order from highest to lowest.

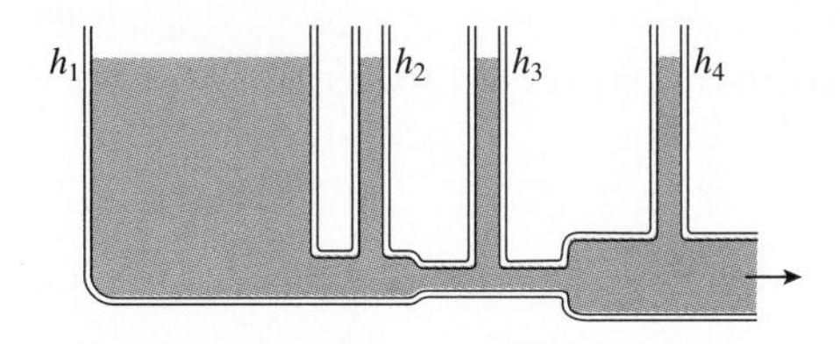

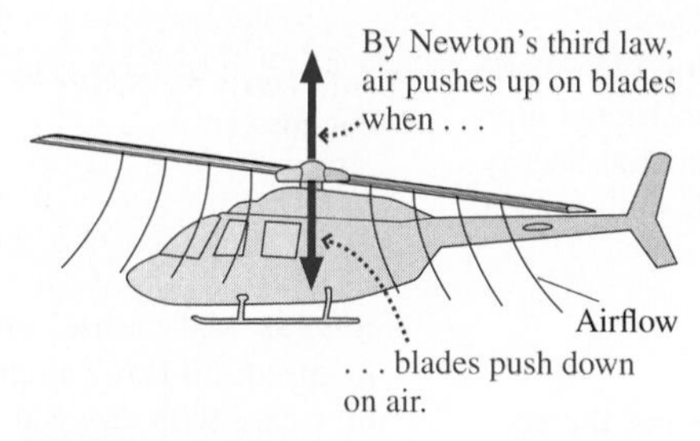

FIGURE 15.17 Newton's third law explains the helicopter's flight.

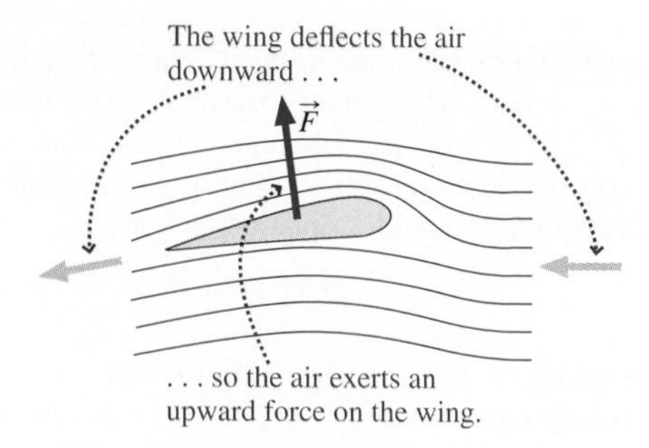

FIGURE 15.18 Flow past a wing.

Flight and Lift

Airplanes, helicopters, and birds fly using forces resulting from their dynamic interaction with the air. Hydrofoil boats, water skis, and sailboards have analogous interactions with water. Projectiles such as baseballs, though not supported by the air, have their trajectories substantially modified by aerodynamic forces.

One of the simplest examples of aerodynamic **lift** is the helicopter (Fig. 15.17). Its whirling blades are shaped and tilted so they force air downward as they move, just like a giant fan. By Newton's third law, the air exerts an upward force on the blades, ultimately supporting the helicopter. An airplane wing works in the same way, except that it moves forward in a straight line instead of describing a circle. Wings are shaped to maximize the downward deflection of the air even with the wing horizontal, but in principle even a flat board would function as a wing if it were tilted to the oncoming air. Figure 15.18 shows the airflow around a wing. Note how the flow, initially horizontal, leaves the wing moving downward—a clear indication that the wing has exerted a downward force on the air. The third law requires a corresponding upward force, and that's what supports the plane.

Baseball's "curve ball" provides another example of aerodynamic lift. Figure 15.19*a* is a top view of the airflow around a baseball that's not spinning; the flow is symmetric and the air isn't deflected. But if the ball spins as shown in Fig. 15.19*b*, air is dragged around the ball and deflected. A corresponding third-law force then acts on the ball, curving its path.

Bernoulli's equation is frequently invoked to explain lift forces. It's true, as Figs. 15.18 and 15.19*b* suggest, that flow speeds are higher, and therefore—according to Bernoulli's equation—pressures are lower on top of a wing or on one side of a spinning ball. Forces associated with that pressure difference provide the lift, so Bernoulli can help explain what's going on. But those pressure differences are manifestations of a simpler underlying phenomenon—namely, the paired forces of Newton's third law.

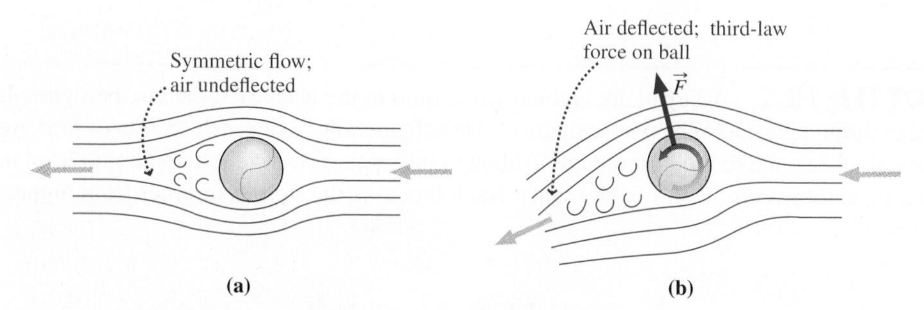

FIGURE 15.19 Top views of airflow around a baseball: (a) no spin; (b) spinning, resulting in a curve ball.

APPLICATION Wind Energy

Wind turbines extract kinetic energy from moving air. In a wind with speed v, Bernoulli's equation shows that the air has kinetic-energy density $\frac{1}{2}\rho v^2$. A chunk of air that passes through a wind turbine in time Δt has length $v\,\Delta t$ and volume $vA\,\Delta t$, where A is the area swept out by the blades. The kinetic energy in this volume is the energy density times the volume: $\Delta K = (\frac{1}{2}\rho v^2)(vA\,\Delta t) = \frac{1}{2}\rho v^3 A\,\Delta t$. Dividing by $A\,\Delta t$ gives the energy per time per unit area—that is, the power per unit area available from the wind:

$$\text{wind power per unit area} = \tfrac{1}{2}\rho v^3$$

Unfortunately, we can't extract *all* this energy because then the air would come to a complete stop behind the turbine, halting the flow. A careful analysis shows that the maximum rate for wind-energy extraction is $\frac{8}{27}\rho v^3$, about 59% of the wind's energy. Given air's density of 1.2 kg/m^3, this means a 10-m/s wind amounts to some 350 W/m^2. The factor v^3 shows that the available power increases rapidly at higher speeds. The best practical wind turbines can achieve about 80% of the theoretical maximum. Wind is the fastest-growing component of the world's energy supply, and in some European countries it provides as much as 20% of the electrical energy.

15.6 Viscosity and Turbulence

Moving fluid interacts with the surfaces it contacts, resulting in a kind of fluid friction called **viscosity**. Viscosity also results from the transfer of momentum among adjacent layers within a fluid. Viscosity is especially important right near fluid boundaries because viscous forces bring the fluid to a complete stop at the boundary (Fig. 15.20). This boundary effect produces drag forces on objects moving through fluids—but it's the same drag at the surfaces of airplane and ship propellers that exerts a force on the fluid. Without viscosity, propellers would spin uselessly and planes and ships would go nowhere.

Viscosity depends on fluid properties and dimensions. Honey is more viscous than water, but at the tiny scales of a human capillary or a bacterium wiggling its flagella for propulsion, water too can be extremely viscous. Viscosity is also important in stabilizing flows that would otherwise become **turbulent**, or chaotically unsteady. Turbulence results from the growth of waves that gain energy at the expense of the flow, turning a smooth flow into a chaotic mess (Fig. 15.21). Turbulence is still not fully understood and presents ongoing challenges to scientists and engineers.

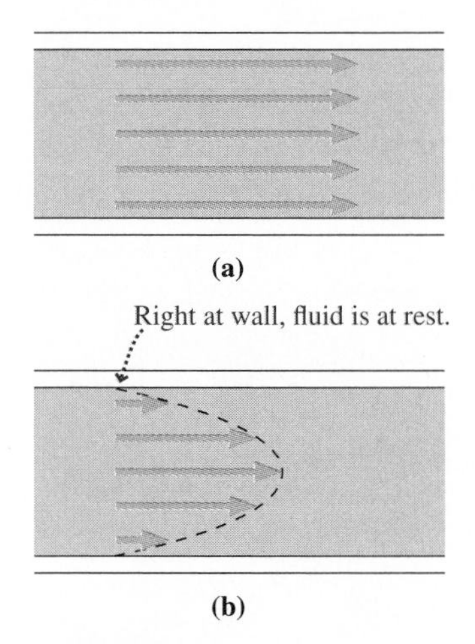

FIGURE 15.20 Velocity profiles in flows that are (a) inviscid (without viscosity) and (b) viscous.

FIGURE 15.21 Smooth flow becomes turbulent, shown here in a column of rising smoke.

CHAPTER 15 SUMMARY

Big Idea

Fluid is matter that readily deforms and flows under the influence of forces. Pressure, density, and flow velocity characterize fluids. Liquids and slowly moving gases are **incompressible**, meaning their density is essentially constant. A fluid that isn't moving is in **hydrostatic equilibrium**. In the presence of gravity, equilibrium requires that fluid pressure increase with depth.

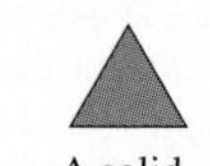

A solid maintains its shape.

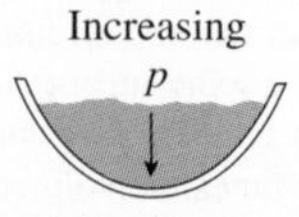

A liquid takes the shape of its container.

A gas fills a closed container.

Key Concepts and Equations

Pressure is the force per unit area: $p = F/A$. The pressure in a fluid exerts itself equally in all directions.

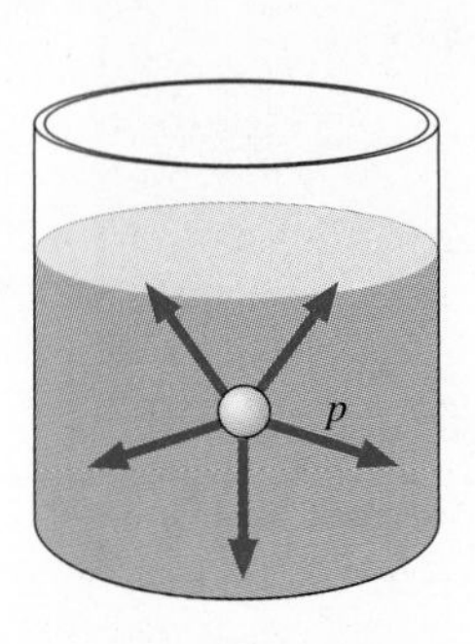

Streamlines represent a moving fluid.

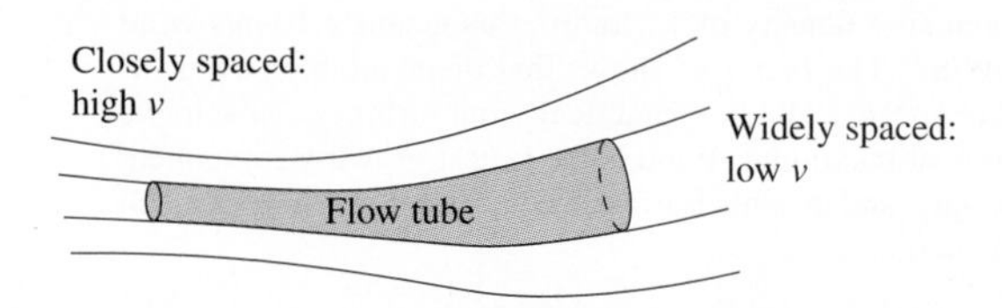

The **continuity equation** describes the conservation of mass along a flow tube:

$$\rho v A = \text{constant (any fluid)}$$

$$vA = \text{constant (incompressible fluid)}$$

Bernoulli's equation describes the conservation of energy:

$$p + \tfrac{1}{2}\rho v^2 + \rho g y = \text{constant} \quad \text{(incompressible fluid, neglecting viscosity)}$$

Viscosity, or fluid friction, is especially important when fluids interact with solid objects.

Applications

Archimedes' principle states that the **buoyancy force** $\vec{F}_b$ due to pressure on an object has the same magnitude as the weight of the displaced fluid. For an object less dense than a fluid, the buoyancy force exceeds gravity and the object floats; otherwise, it sinks or is in neutral buoyancy.

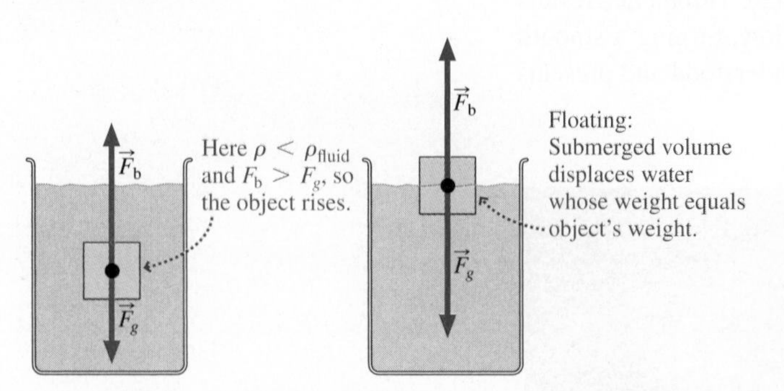

Bernoulli's principle helps explain lift forces, although ultimately these are based in Newton's third law.

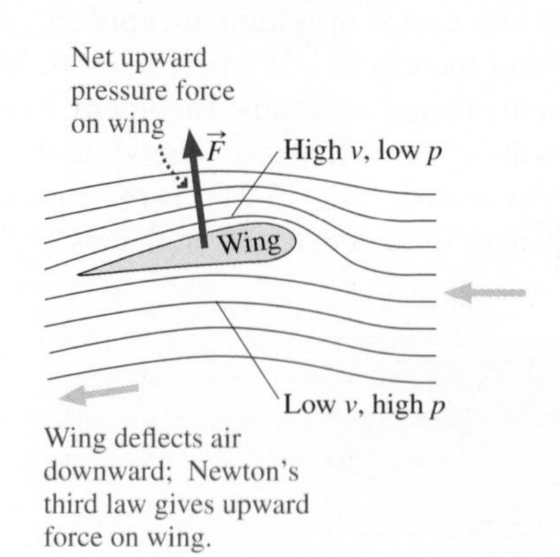

BIO *Biology and/or medicine-related problems* **DATA** *Data problems* **ENV** *Environmental problems* **CH** *Challenge problems* **COMP** *Computer problems*

For Thought and Discussion

1. Why do your ears "pop" when you drive up a mountain?
2. Commercial aircraft cabins are usually pressurized to the pressure of the atmosphere at about 2 km above sea level. Why don't you feel the lower pressure on your entire body?
3. Water pressure at the bottom of the ocean arises from the weight of the overlying water. Does this mean that the water exerts pressure only in the downward direction? Explain.
4. The three containers in Fig. 15.22 are filled to the same level and are open to the atmosphere. How do the pressures at the bottoms of the three containers compare?

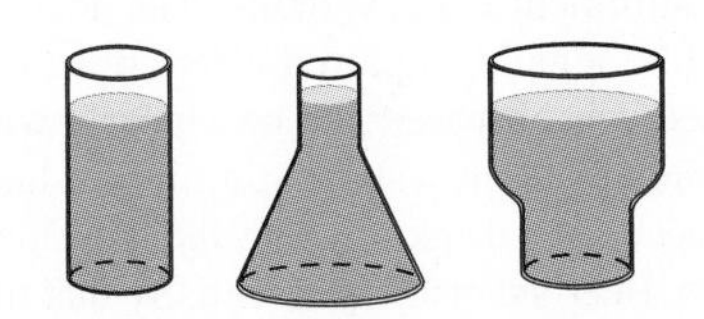

FIGURE 15.22 For Thought and Discussion 4.

5. Why is it easier to float in the ocean than in fresh water?
6. Figure 15.23 shows a cork suspended from the bottom of a sealed container of water. The container is on a turntable rotating about a vertical axis, as shown. Explain the position of the cork.

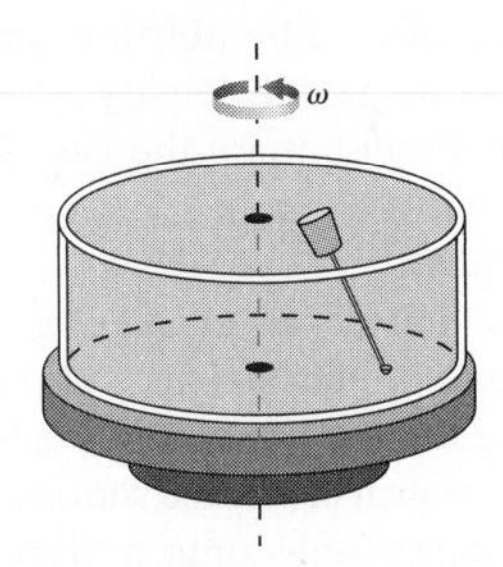

FIGURE 15.23 For Thought and Discussion 6.

7. Meteorologists in the United States usually report barometer readings in inches. What are they talking about?
8. A mountain stream, frothy with entrained air bubbles, presents a serious hazard to hikers who fall into it, for they may sink in the stream where they would float in calm water. Why?
9. Why are dams thicker at the bottom than at the top?
10. It's not possible to breathe through a snorkel from a depth greater than a meter or so. Why not?
11. A helium-filled balloon stops rising long before it reaches the "top" of the atmosphere, but a cork released from the bottom of a lake rises all the way to the surface. Why the difference?
12. A barge filled with steel beams overturns in a lake, spilling its cargo. Does the water level in the lake rise, fall, or remain the same?
13. Why do airplanes take off into the wind?
14. Is the flow speed behind a wind turbine greater or less than in front? Is the pressure behind the turbine higher or lower than in front? Is this a violation of Bernoulli's principle? Explain.

Exercises and Problems

Exercises

Section 15.1 Density and Pressure

15. The density of molasses is 1600 kg/m^3. Find the mass of the molasses in a 0.75-L jar.
16. Atomic nuclei have densities around 10^{17} kg/m^3, while water's density is 10^3 kg/m^3. Roughly what fraction of water's volume is *not* empty space?
17. Compressed air with mass 8.8 kg is stored in a 0.050-m^3 cylinder. (a) What's the density of the compressed air? (b) What volume would the same gas occupy at a typical atmospheric density of 1.2 kg/m^3?
18. The pressure unit **torr** is defined as the pressure that will support a column of mercury 1 mm high. Meteorologists often give barometric pressure in **inches of mercury**, defined analogously. Express each of these in SI units. (*Hint:* Mercury's density is 1.36×10^4 kg/m^3.)
19. Measurement of small pressure differences—for example, between the interior of a chimney and the ambient atmosphere—is often given in inches of water, where 1 in. of water is the pressure that will support a 1-in.-high water column. Express this pressure difference in SI units.
20. What's the weight of a column of air with cross-sectional area 1 m^2 extending from Earth's surface to the top of the atmosphere?
21. A 4680-kg circus elephant balances on one foot. If the foot is a circle 42.6 cm in diameter, what pressure does it exert on the ground?
22. You unbend a paper clip made from 1.5-mm-diameter wire and push the end against the wall. What force must you apply to give a pressure of 120 atm?

Section 15.2 Hydrostatic Equilibrium

23. What's the density of a fluid whose pressure increases at the rate of 100 kPa for every 6.0 m of depth?
24. A research submarine can withstand an external pressure of 62 MPa when its internal pressure is 101 kPa. How deep can it dive?
25. Scuba equipment provides a diver with air at the same pressure as the surrounding water. But at pressures higher than about 1 MPa, the nitrogen in air becomes dangerously narcotic. At what depth does nitrogen narcosis become a hazard?
26. A vertical tube open at the top contains 5.0 cm of oil with density 0.82 g/cm^3, floating on 5.0 cm of water. Find the gauge pressure at the bottom of the tube.
27. A child attempts to drink water through a 36-cm-long straw but finds that the water rises only 25 cm. By how much has the child reduced the pressure in her mouth below atmospheric pressure?
28. Barometric pressure in the eye of a hurricane is 0.91 atm (27.2 in. of mercury). How does the level of the ocean surface under the eye compare with the level under a distant fair-weather region where the pressure is 1.0 atm?

Section 15.3 Archimedes' Principle and Buoyancy

29. On land, the most massive concrete block you can carry is 25 kg. Given concrete's 2200-kg/m^3 density, how massive a block could you carry underwater?
30. A 5.4-g jewel has apparent weight 32 mN when submerged in water. Could the jewel be a diamond (density 3.51 g/cm^3)?
31. Styrofoam's density is 160 kg/m^3. What percent error is introduced by weighing a Styrofoam block in air (density 1.2 kg/m^3), which exerts an upward buoyancy force, rather than in vacuum?
32. A steel drum has volume 0.23 m^3 and mass 16 kg. Will it float in water when filled with (a) water or (b) gasoline (density 860 kg/m^3)?

Sections 15.4 and 15.5 Fluid Dynamics and Applications

33. Water flows through a 2.5-cm-diameter pipe at 1.8 m/s. If the pipe narrows to 2.0-cm diameter, what's the flow speed in the constriction?
34. Show that pressure has the units of energy density.
35. A typical mass flow rate for the Mississippi River is 1.8×10^7 kg/s. Find (a) the volume flow rate and (b) the flow speed in a region where the river is 2.0 km wide and averages 6.1 m deep.
36. A fire hose 10 cm in diameter delivers water at 15 kg/s. The hose terminates in a 2.5-cm-diameter nozzle. What are the flow speeds (a) in the hose and (b) at the nozzle?
37. **BIO** A typical human aorta, the main artery from the heart, is 1.8 cm in diameter and carries blood at 35 cm/s. Find the flow speed around a clot that reduces the flow area by 80%.

Problems

38. When a couple with total mass 120 kg lies on a water bed, pressure in the bed increases by 4700 Pa. What surface area of the two bodies is in contact with the bed?
39. A fully loaded Volvo station wagon has mass 1950 kg. If each of its four tires is inflated to a gauge pressure of 230 kPa, what's the total tire area in contact with the road?
40. You're stuck in the exit row on a long flight, and you suddenly worry that your seatmate, who's next to the window, might pull the emergency window inward while you're in flight. The window measures 40 cm by 55 cm. Cabin pressure is 0.77 atm, and atmospheric pressure at the plane's altitude is 0.22 atm. Should you worry?
41. A vertical tube 1.0 cm in diameter and open at the top contains 5.0 g of oil (density 0.82 g/cm^3) floating on 5.0 g of water. Find the gauge pressure (a) at the oil–water interface and (b) at the bottom.
42. Dam breaks present a serious risk of widespread property damage and loss of life. You're asked to assess a 1500-m-wide dam holding back a lake 95 m deep. The dam was built to withstand a force of 100 GN, which is supposed to be at least 50% over the force it actually experiences. Should the dam be reinforced? (*Hint:* You'll need your calculus skills.)
43. A U-shaped tube open at both ends contains water and a quantity of oil occupying a 2.0-cm length of the tube, as shown in Fig. 15.24. If the oil's density is 82% of water's, what's the height difference h?
44. You're a robotics engineer designing a hydraulic system to move a robotic arm. The hydraulic cylinder that drives the arm has diameter 5.0 cm and can exert a maximum force of 5.6 kN. Hydraulic tubing comes rated in multiples of 1/2 MPa, and for safety, you're to specify tubing capable of withstanding 50% greater pressure than it will ever experience in use. What pressure rating do you specify?
45. A garage lift has a 45-cm-diameter piston supporting the load. Compressed air with maximum pressure 500 kPa is applied to a small piston at the other end of the hydraulic system. What's the maximum mass the lift can support?
46. Archimedes purportedly used his principle to verify that the king's crown was pure gold by weighing the crown submerged in water. Suppose the crown's actual weight was 25.0 N. What would be its apparent weight if it were made of (a) pure gold and (b) 75% gold and 25% silver, by volume? The densities of gold, silver, and water are 19.3 g/cm^3, 10.5 g/cm^3, and 1.00 g/cm^3, respectively.
47. You're testifying in a drunk-driving case for which a blood alcohol measurement is unavailable. The accused weighs 140 lb, and would be legally impaired after consuming 36 oz of beer. The accused was observed at a beach party where a keg with interior diameter 40 cm was floating in the lake to keep it cool. After the accused's drinking stint, the keg floated 1.2 cm higher than before. Beer's density is essentially that of water. Does your testimony help or hurt the accused's case?
48. A glass beaker measures 14 cm high by 5.0 cm in diameter. Empty, it floats in water with one-third of its height submerged. How many 12-g rocks can be placed in the beaker before it sinks?
49. A typical supertanker has mass 2.0×10^6 kg and carries twice that much oil. If 9.0 m of the ship is submerged when it's empty, what's the minimum water depth needed for it to navigate when full? Assume the sides of the ship are vertical.
50. A balloon contains gas of density ρ_g and is to lift a mass M, including the balloon but not the gas. Show that the minimum mass of gas required is $m_g = M\rho_g/(\rho_a - \rho_g)$, where ρ_a is the atmospheric density.
51. (a) How much helium (density 0.18 kg/m^3) is needed to lift a balloon carrying two people, if the total mass of people, basket, and balloon (but not gas) is 280 kg? (b) Repeat for a hot-air balloon whose air density is 10% less than that of the surrounding atmosphere.
52. A 55-kg swimmer climbs onto a Styrofoam block of density 160 kg/m^3. If the water level comes right to the top of the Styrofoam, what's the block's volume?
53. **BIO** If the blood pressure in the unobstructed artery of Exercise 37 is 16 kPa gauge (about 120 mm of mercury, the unit commonly reported by doctors), what will it be at the clot? (*Note:* Blood's density is 1.06 g/cm^3.)
54. You're a consultant for maple syrup producers. They tap maple trees and collect sap with plastic tubing that connects to a common pipe delivering sap to an evaporator. There it's boiled to produce thick, tasty syrup. The system can be modeled as a pipe with one end, of cross-sectional area A, exposed to atmospheric pressure. The pipe drops through a vertical distance h_1 while its area decreases to $A/2$, as shown in Fig. 15.25. A small vertical glass tube extends from the lower portion, as shown, and is open to atmospheric pressure. You're asked to provide a formula for the volume flow rate of the sap as a function of the height h_2 of sap in the tube.

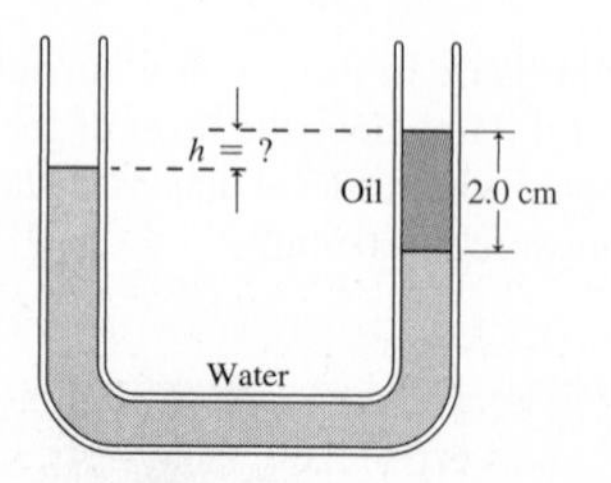

FIGURE 15.24 Problem 43

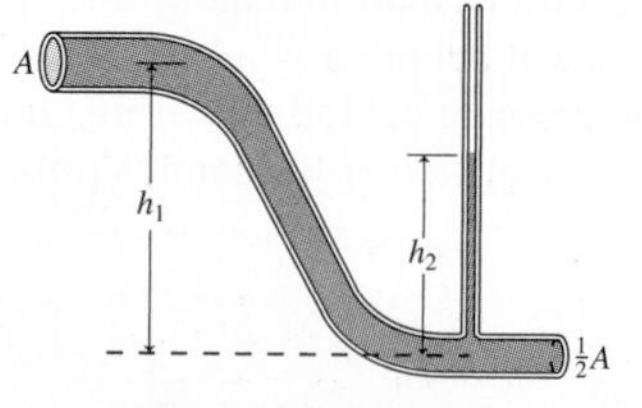

FIGURE 15.25 Problem 54

55. The water in a garden hose is at 140-kPa gauge pressure and is moving at negligible speed. The hose terminates in a sprinkler consisting of many small holes. Find the maximum height reached by the water emerging from the holes.

56. The venturi flowmeter shown in Fig. 15.26 is used to measure the flow rate of water in a solar collector system. The flowmeter is inserted in a pipe with diameter 1.9 cm; at the venturi the diameter is 0.64 cm. The manometer tube contains oil with density 0.82 times that of water. If the difference in oil levels on the two sides of the manometer tube is 1.4 cm, what's the volume flow rate?

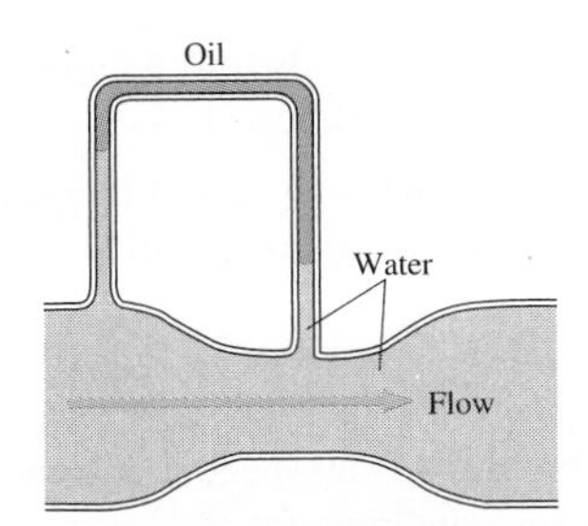

FIGURE 15.26 Problem 56

57. A 1.0-cm-diameter venturi flowmeter is inserted in a 2.0-cm-diameter pipe carrying water (density 1000 kg/m^3). Find (a) the flow speed in the pipe and (b) the volume flow rate if the pressure difference between venturi and unconstricted pipe is 17 kPa.

58. A balloon's mass is 1.6 g when it's empty. It's inflated with helium (density 0.18 kg/m^3) to form a sphere 28 cm in diameter. How many 0.63-g paper clips can you hang from the balloon before it loses buoyancy?

59. **BIO** Blood with density 1.06 g/cm^3 and 10-kPa gauge pressure flows through an artery at 30 cm/s. It encounters a plaque deposit where the pressure drops by 5%. What fraction of the artery's area is obstructed?

60. A venturi flowmeter in an oil pipeline has radius half that of the pipe. Oil flows in the unconstricted pipe at 1.9 m/s. If the pressure difference between unconstricted flow and venturi is 16 kPa, what's the oil's density?

61. A drinking straw 20 cm long and 3.0 mm in diameter stands vertically in a cup of juice 8.0 cm in diameter. A section of straw 6.5 cm long extends above the juice. A child sucks on the straw, and the juice level begins dropping at 2.0 mm/s. (a) By how much does the pressure in the child's mouth differ from atmospheric pressure? (b) What's the greatest height above the water surface from which the child could drink, assuming this same mouth pressure?

62. In 2012, film producer James Cameron (*Terminator*, *Titanic*, *Avatar*) rode his submersible *Deepsea Challenger* to the bottom of the 11-km-deep Marianas Trench, the deepest spot in Earth's oceans. Cameron could barely fit into *Deepsea Challenger*'s crew compartment, a steel sphere with inside diameter 109 cm and walls 6.4 cm thick. Find the total pressure force exerted on the sphere at the bottom of the trench. (The total force is the sum of all pressure forces without regard to direction; it's not the same as the buoyancy force, which is the *net* pressure force—a vectorial sum.)

63. **DATA** A probe descending through Mar's atmosphere records pressure as a function of altitude; the data are in the table below. Plot the natural logarithm of the pressure versus altitude and fit a line to your plotted points. Mars's atmospheric pressure is governed by the same equation that describes Earth's; see Problem 70. Use your fitted line, in connection with that equation, to determine (a) Mars's surface pressure and (b) the scale height h_0.

Altitude (km)	10	20	30	40	50	60
Pressure (Pa)	242	98.7	37.6	16.2	7.21	2.38

64. **CH** Water emerges from a faucet of diameter d_0 in steady, near-vertical flow with speed v_0. Show that the diameter of the falling water column is given by $d = d_0[v_0^2/(v_0^2 + 2gh)]^{1/4}$, where h is the distance below the faucet (Fig. 15.27).

FIGURE 15.27 Problem 64

65. Assuming normal atmospheric pressure, how massive an object can a 5.0-cm-diameter suction cup support on a vertical wall, if the coefficient of friction between cup and wall is 0.72?

66. Figure 15.28 shows a simplified diagram of a Pitot tube, used for measuring aircraft speeds. The tube is mounted on the aircraft with opening A at right angles to the flow and opening B pointing into the flow. The gauge prevents airflow through the tube. Use Bernoulli's equation to show that the plane's speed relative to the air is $v = \sqrt{2\,\Delta p/\rho}$, where Δp is the pressure difference between the tubes and ρ is the density of air. (*Hint:* The flow must be stopped at B, but continues past A with its normal speed.)

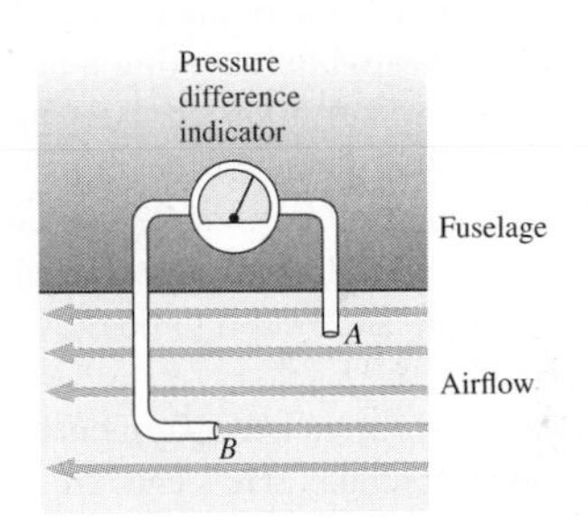

FIGURE 15.28 Problem 66

67. At a hearing on a proposed wind farm, a wind-energy advocate says an installation of 800 turbines, with blade diameter 95 m, could displace a 1-GW nuclear power plant. You're asked if that's really possible. How do you answer, given an average wind speed of 12 m/s and a turbine power output that averages 30% of the theoretical maximum?

68. **CH** A pencil is weighted so it floats vertically with length L submerged. It's pushed vertically downward without being totally submerged, then released. Show that it undergoes simple harmonic motion with period $T = 2\pi\sqrt{L/g}$.

69. **CH** A can of height h and cross-sectional area A_0 is initially full of water. A small hole of area $A_1 \ll A_0$ is cut in the bottom of the can. Find an expression for the time it takes all the water to drain from the can. (*Hint*: Call the water depth y, use the continuity equation, and integrate.)

70. **CH** Density and pressure in Earth's atmosphere are proportional: $\rho = p/h_0 g$, where $h_0 = 8.2$ km is a constant called the *scale height* and g is the gravitational acceleration. (a) Integrate Equation 15.2 for this case to show that atmospheric pressure as a function of height h above the surface is given by $p = p_0 e^{-h/h_0}$, where p_0 is the surface pressure. (b) At what height will the pressure have dropped to half its surface value?

71. **CH** (a) Use the result of Problem 70 to express Earth's atmospheric density as a function of height. (b) Use your result from (a) to

find the height below which half of Earth's atmospheric mass lies (this will require integration).

72. CH A circular pan of liquid with density ρ is centered on a horizontal turntable rotating with angular speed ω, as shown in Fig. 15.29. Atmospheric pressure is p_a. Find expressions for (a) the pressure at the bottom of the pan and (b) the height of the liquid surface, both as functions of the distance r from the axis, given that the height at the center is h_0.

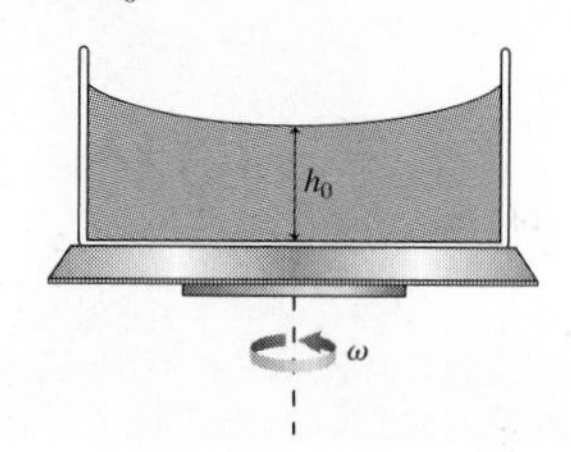

FIGURE 15.29 Problem 72

73. CH A solid sphere of radius R and mass M has density ρ that varies with distance r from the center: $\rho = \rho_0 e^{-r/R}$. Find an expression for the central density ρ_0 in terms of M and R.
74. CH The difference in air pressure between the inside and outside of a ball is a constant Δp. Show by direct integration that the net pressure force on one hemisphere is $\pi R^2 \Delta p$, with R the ball's radius.
75. Find the torque that the water exerts about the bottom edge of the dam in Problem 42.
76. CH One vertical wall of a swimming pool is a regular trapezoid, with its bottom 15 m long and its top 22 m long. The pool is 3.3 m deep, and it's full to the brim with water. Find the pressure force the water exerts on this side of the pool.
77. You're a private investigator assisting a large food manufacturer in tracking down counterfeit salad dressing. The genuine dressing is by volume one part vinegar (density 1.0 g/cm^3) to three parts olive oil (density 0.92 g/cm^3). The counterfeit dressing is diluted with water (density 1.0 g/cm^3). You measure the density of a dressing sample and find it to be 0.97 g/cm^3. Has the dressing been altered?
78. A plumber comes to your ancient apartment building where you have a part-time job as caretaker. He's checking the hot-water heating system, and notes that the water pressure in the basement is 18 psi. He asks, "How high is the building?" "Three stories, each about 11 feet," you reply. "OK, about 33 feet," he says, pausing to do some calculations in his head. "The pressure is fine," he declares. On what basis did he come to that conclusion?
79. Your class in naval architecture is working on the design for a ship with a V-shaped cross section, as shown in Fig. 15.30. The ship has total length L and keel-to-deck height h_0. When empty, the distance from waterline to keel is h_1. You're asked for the maximum load the ship can carry below deck if water is not to come over the deck. Answer in terms of h_0, h_1, L, θ, and the water density ρ.

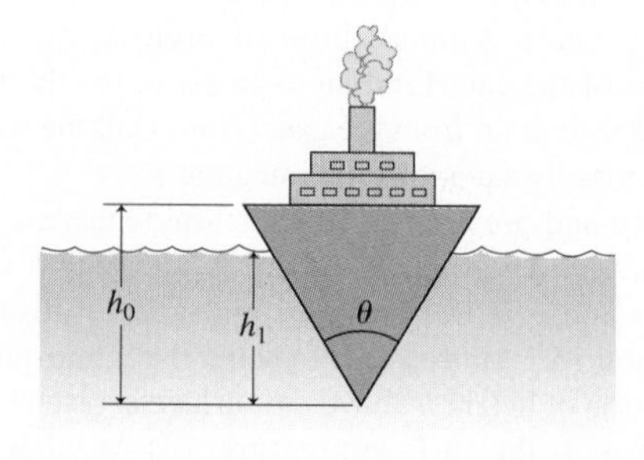

FIGURE 15.30 Problem 79

Passage Problems

Arterial stenosis is a constriction of an artery, often due to plaque buildup on the artery's inner walls. Serious medical conditions can result, depending on the affected artery. Stenosis of the carotid arteries that supply blood to the brain is a leading cause of stroke, while stenosis of the renal arteries can lead to kidney failure. Pulmonary artery stenosis results from birth defects, and can result in insufficient oxygen supply. Because the heart has to work harder to get blood through a constricted artery, stenosis can contribute to high blood pressure.

In answering the questions below, assume steady flow (which is true in arteries only on short timescales).

80. How does the volume flow rate of blood at a stenosis compare with the rate in the surrounding artery?
    a. lower
    b. the same
    c. higher
81. How does the blood flow speed at a stenosis compare with the speed in the surrounding artery?
    a. lower
    b. the same
    c. higher
82. Which of the following medical problems is more likely to occur?
    a. An artery might collapse because of lower blood pressure at the stenosis.
    b. An artery might burst because of higher blood pressure at the stenosis.
    c. Neither; pressure at the stenosis is the same as in the surrounding artery.
83. If the artery has circular cross section even at the stenosis, but the diameter at the stenosis is half that in the surrounding artery, the blood flow speed in the stenosis will be
    a. one-fourth that in the surrounding artery.
    b. one-half that in the surrounding artery.
    c. the same as in the surrounding artery.
    d. $\sqrt{2}$ times that in the surrounding artery.
    e. four times that in the surrounding artery.

Answers to Chapter Questions

Answer to Chapter Opening Question

Because the density of ice is only slightly less than that of water.

Answers to GOT IT? Questions

15.1 (c)

15.2 (c) $\vec{F}$ moves the small piston a lot farther than the upward pressure force moves the large piston; the products of force and displacement are the same for both pistons, so the work done is the same.

15.3 (a)

15.4 (a) over the top where the streamlines are closer together

15.5 $h_1 > h_4 > h_2 > h_3$ reflecting higher pressure with lower flow speed

PART TWO SUMMARY Oscillations, Waves, and Fluids

Part Two has extended Newtonian mechanics to systems that undergo oscillatory motion and wave motion or that involve the motion of fluids. Behind these more complex motions are the fundamental concepts of force, mass, and energy and their roles in characterizing motion.

Oscillatory motion describes the back-and-forth motion of a system disturbed from a stable equilibrium.

When the force or torque tending to restore equilibrium is directly proportional to the displacement, the result is simple harmonic motion.

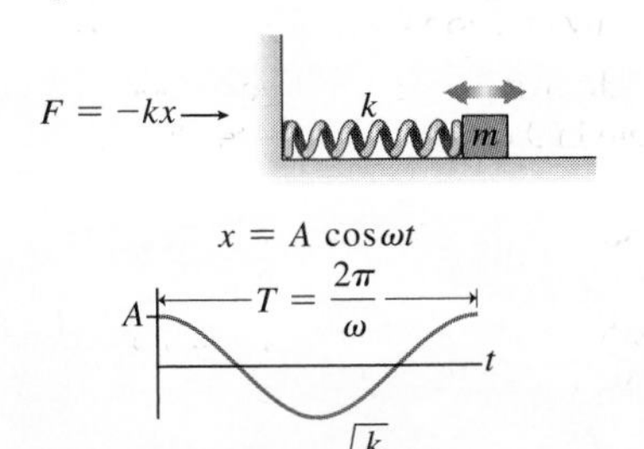

A **wave** is a propagating disturbance that carries energy but not matter.

Simple harmonic waves are sinusoidal:

$$y(x, t) = A\cos(kx - \omega t)$$

Angular frequency: $\omega = 2\pi f$

Wave number: $k = \frac{2\pi}{\lambda}$

Wave period: $T = \frac{1}{f}$

Wave speed: $v = \frac{\omega}{k} = \frac{\lambda}{T} = f\lambda$

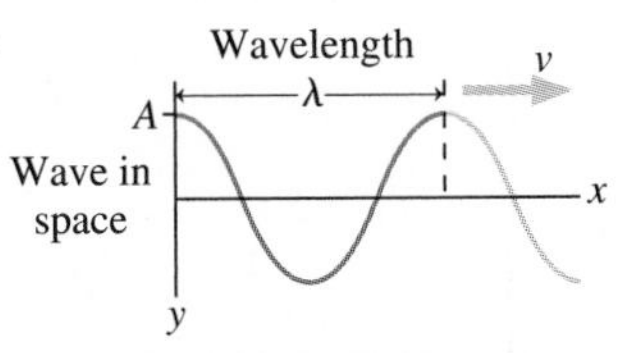

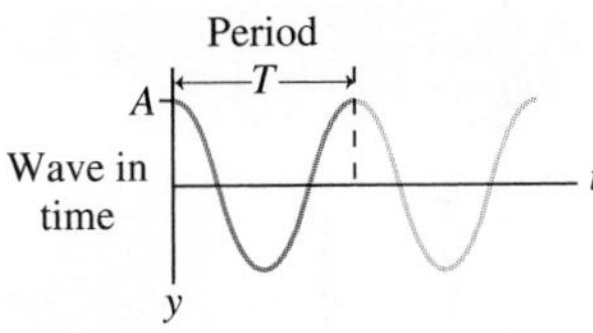

When waves overlap, the result is **interference**, which is constructive when the waves reinforce and destructive when they tend to cancel.

Nodal lines: destructive interference

Large amplitude: constructive interference

Standing waves occur when the medium has limited extent. Only certain wavelengths and frequencies are allowed, depending on the medium's length:

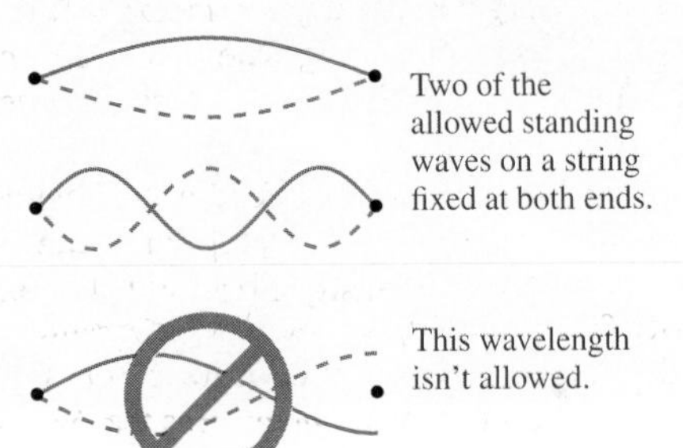

Fluids in **hydrostatic equilibrium** exhibit a depth-dependent pressure that results in an upward buoyancy force $\vec{F}_b$.

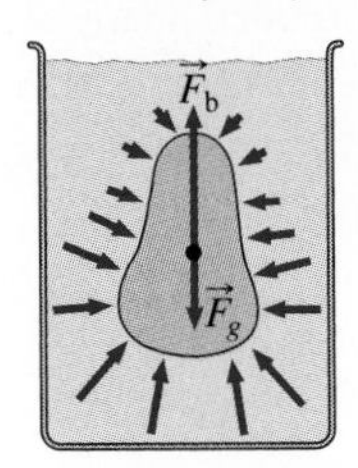

Archimedes' principle states that the buoyancy force is equal to the weight of the displaced fluid.

Moving fluids obey conservation of mass and, in the absence of fluid friction (viscosity), they also conserve energy.

In **fluid dynamics**, the continuity equation and Bernoulli's equation express these conservation laws. Both equations hold along a flow tube:

Continuity: $\rho vA = \text{constant}$

Bernoulli:

$$p + \tfrac{1}{2}\rho v^2 + \rho gy = \text{constant}$$

Part Two Challenge Problem

A cylindrical log of total mass M and uniform diameter d has an uneven mass distribution that causes it to float in a vertical position, as shown in the figure. (a) Find an expression for the length L of the submerged portion of the log when it's floating in equilibrium, in terms of M, d, and the water density ρ. (b) If the log is displaced vertically from its equilibrium position and released, it will undergo simple harmonic motion. Find an expression for the period of this motion, neglecting viscosity and other frictional effects.

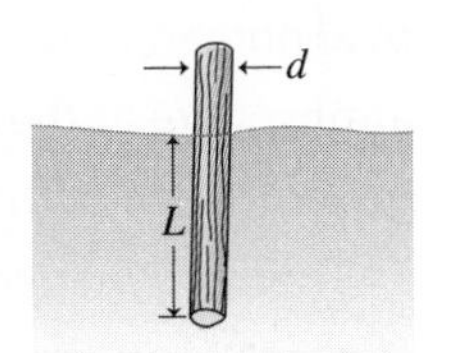

PART THREE OVERVIEW

Thermodynamics

This huge steam turbine converts the energy of high-pressure steam to mechanical energy and then, via the generator at the right end of the system, to electricity. The inset shows the turbine blades that spin when struck by high-pressure steam. Systems like this one produce nearly all the world's electrical energy, and their operation and efficiency are governed by the laws of thermodynamics.

Humanity consumes energy at the prodigious rate of some 10^{13} watts. Nearly all that energy comes from the combustion of fossil fuels—a process governed by the laws of thermodynamics. Engines that extract mechanical energy from burning fuels propel our cars, trucks, and airplanes, and produce most of our electricity. Despite the efforts of the cleverest engineers, the laws of thermodynamics set fundamental limitations on our ability to convert thermal energy to mechanical energy. Many of the energy and environmental challenges humanity faces today are grounded in thermodynamics.

Many natural systems are also thermodynamic. Without the Sun's energy, radiated across a hundred million miles of empty space, Earth would be a lifeless, frozen rock. Heat flowing throughout Earth, its oceans, and its atmosphere governs processes ranging from continental drift to ocean currents to weather and climate. Concern over human-induced climate change is rooted in thermodynamic properties of the atmosphere as they affect energy flows. On a grander scale, thermodynamic principles govern much of the energy that flows throughout the universe.

Thermodynamics—the study of heat and its connection to the all-important concept of energy—is the subject of the next four chapters.

Temperature and Heat

What You Know

- You understand the concept of *energy*, especially *kinetic* and *potential energy*.
- You've had a brief introduction to *internal energy*.
- You recognize that *power* is a rate of energy use, flow, conversion, and so on.
- You're familiar with *pressure* and *density* from your study of fluids in Chapter 15.

What You're Learning

- You'll refine your everyday notions of *temperature* and *heat*, honing them into precise physics definitions.
- You'll learn about temperature scales and how they're established.
- You'll see how a material's *specific heat* determines the energy needed to change its temperature.
- You'll learn three mechanisms of heat transfer: *conduction, convection,* and *radiation*.
- You'll see how a system's temperature is determined by *thermal-energy balance* with its surroundings, with application to systems ranging from buildings to planetary climates.

How You'll Use It

- The ideas in this chapter will serve as the groundwork for your study of thermodynamics in Chapters 17, 18, and 19.
- Thermodynamics will have many applications in subsequent work you might do in engineering, environmental science, biology, chemistry, and physics.

How does this infrared photo reveal heat loss from the house? And how can you tell that the car was recently driven?

Your own body gives you a good sense of "hot" and "cold." Questions about heat and temperature are ultimately about energy, and these concepts are crucial to understanding the energy flows that drive natural systems like Earth's climate and technologies such as engines, power plants, and refrigerators.

Properties like mass and kinetic energy apply equally to microscopic atoms and molecules and to cars and planets. But other properties, including temperature and pressure, apply only to macroscopic systems. It makes no sense to talk about the temperature or pressure of a single air molecule. **Thermodynamics** is the branch of physics that deals with these macroscopic properties. Ultimately, the thermodynamic behavior of matter follows from the motions of its constituent particles in response to the laws of mechanics. **Statistical mechanics** relates the macroscopic description of matter to the underlying microscopic processes. Historically, thermodynamics developed before the atomic theory of matter was fully established. The subsequent explanation of thermodynamics through statistical mechanics—the mechanics of atoms and molecules—was a triumph for physics.

16.1 Heat, Temperature, and Thermodynamic Equilibrium

Take a bottle of soda from the refrigerator, and eventually it reaches room temperature. At that point the soda and the room are in **thermodynamic equilibrium**, a state in which their macroscopic properties are no longer changing. To check for thermodynamic equilibrium we can consider any macroscopic property—length, volume, pressure, electrical resistance, whatever. If any macroscopic property changes

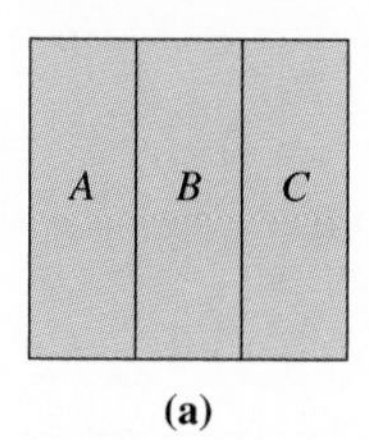

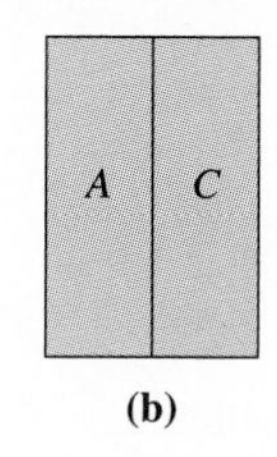

FIGURE 16.1 The zeroth law of thermodynamics.

when two systems are placed together, then they weren't originally in thermodynamic equilibrium. When changes cease, the systems have reached equilibrium.

The phrase "placed together" here has a definite meaning, stated more precisely as "placed in thermal contact." Two systems are in **thermal contact** if heating one of them results in macroscopic changes in the other. If that doesn't readily happen—for example, with a Styrofoam cup of coffee and its surroundings—then the systems are **thermally insulated**.

We can now begin to define temperature: **Two systems have the same temperature if they are in thermodynamic equilibrium**. Consider two systems A and C in thermal contact with a third system B but not with each other (Fig. 16.1a). Even though they're not in direct contact, A and C have the same temperature; that is, if you place A and C in thermal contact (Fig. 16.1b), no further changes occur. This fact—that two systems in equilibrium with a third system are therefore in equilibrium with each other—is so fundamental that it's called the **zeroth law of thermodynamics**.

A **thermometer** is a system with a conveniently observed macroscopic property that changes with temperature. It could be the length of a mercury column, gas pressure, electrical resistance, or the bending of a bimetal strip in a dial thermometer. Let the thermometer come to equilibrium with some system, and its temperature-dependent physical property provides a measure of temperature. The zeroth law assures consistency, in that two systems for which the thermometer gives the same reading must have the same temperature.

The Kelvin Scale and Gas Thermometers

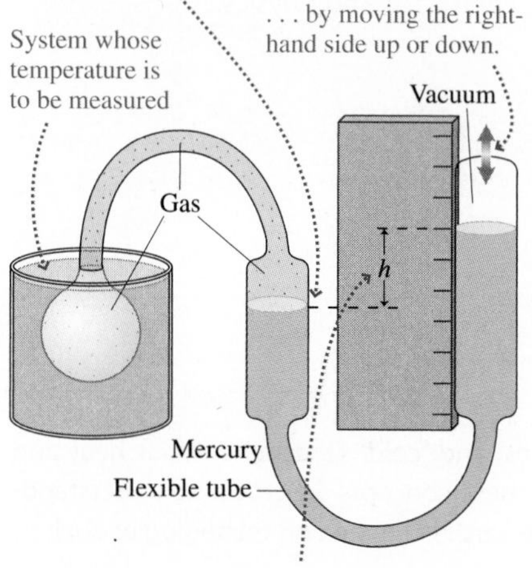

FIGURE 16.2 A constant-volume gas thermometer.

One of the most versatile thermometers is the **constant-volume gas thermometer** (Fig. 16.2), in which the pressure of a gas provides an indication of temperature. Gas thermometers function over a wide range, including very low temperatures, and they currently provide the definition of the Kelvin temperature scale used in the SI system. As Fig. 16.3 shows, the zero of the Kelvin scale is defined as the temperature at which the gas pressure would become zero. Since a gas can't have negative pressure, this point is defined as **absolute zero**—a concept whose meaning we'll explore further in Chapter 19. A second fixed temperature is provided by the so-called triple point of water, the unique temperature at which solid, liquid, and gaseous water can coexist in equilibrium (more on this in Chapter 17). In the current SI definition, water's triple point is defined to be exactly 273.16 kelvin (symbol K; *not* "degrees kelvin or °K). Other temperatures then followed by linear extrapolation, as suggested in Fig. 16.3. Although the triple-point definition of the kelvin is, in principle, a reproducible operational standard, issues with purity and the isotopic composition of water have made this standard less than ideal.

In the ongoing revision of the SI unit system, the kelvin will be given a new explicit-constant definition, by setting an exact value for the so-called Boltzmann constant. This constant establishes a direct relation between temperature and molecular energy, which we'll explore further in Chapter 17. With this new definition, the triple point of water becomes a measured quantity very close to 273.16 K but, as with all measured quantities, involving some uncertainty.

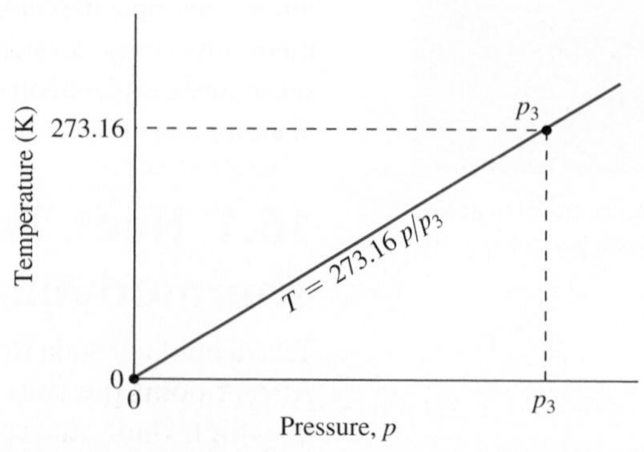

FIGURE 16.3 Two points establish a temperature scale. Until the ongoing SI revision is complete, the kelvin scale is defined by the values of absolute zero and the unique temperature of water's triple point, whose pressure is designated p_3 and whose temperature is defined as 273.16 K.

Temperature Scales

Other temperature scales include Celsius (°C), Fahrenheit (°F), and Rankine (°R) (Fig. 16.4). One Celsius degree represents the same temperature difference as one kelvin, but the zero of the Celsius scale occurs at 273.15 K, so

$$T_C = T - 273.15 \qquad (16.1)$$

where T is the temperature in kelvins. On the Celsius scale the melting point of ice at standard atmospheric pressure is exactly 0°C, while the boiling point is 100°C. The triple point of water occurs at 0.01°C, which accounts for the 273.15 difference between the kelvin and Celsius scales. Equation 16.1 shows that absolute zero occurs at −273.15°C.

The Fahrenheit and Rankine scales, from the British unit system, are used primarily in the United States. Fahrenheit has water melting at 32°F and boiling at 212°F, so the relation between Fahrenheit and Celsius temperatures is

$$T_F = \tfrac{9}{5}T_C + 32 \qquad (16.2)$$

A Rankine degree is the same size as a Fahrenheit degree, but the zero of the Rankine scale is at absolute zero (Fig. 16.4). Engineers in the United States often use Rankine.

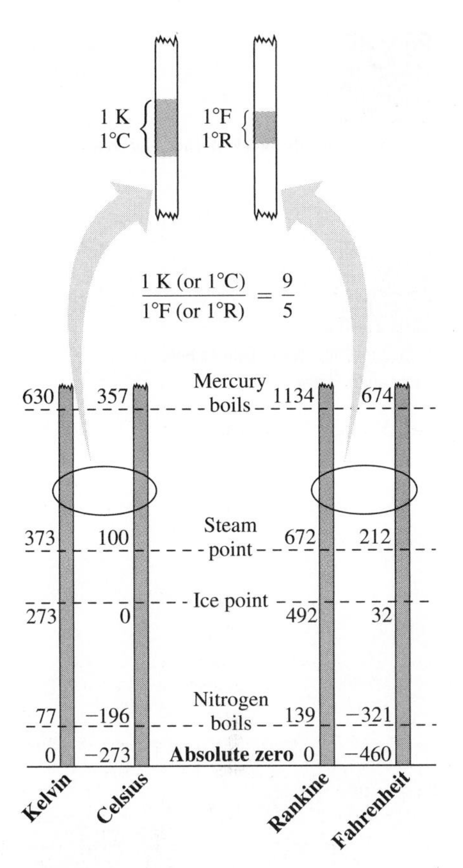

FIGURE 16.4 Relationships among four temperature scales.

Heat and Temperature

A match will burn your finger, but it doesn't provide much heat. This example shows our intuitive sense of temperature and heat: Heat measures an *amount* of "something," whereas temperature is the *intensity* of that "something."

Scientists once considered heat to be a material fluid, called **caloric**, that flowed from hot bodies to colder ones. But in the late 1700s, the American-born scientist Benjamin Thompson observed essentially limitless amounts of heat being produced in the boring of cannon, and he concluded that heat could not be a conserved fluid. Instead, Thompson suggested, heat was associated with mechanical work done by the boring tool. In the next half-century, a series of experiments confirmed the association between heat and energy. These culminated in the work of the British physicist James Joule (1818–1889), who quantified the relation between heat and energy. In so doing, Joule brought thermal phenomena under the powerful conservation-of-energy principle. In recognition of this major synthesis in physics, the SI energy unit bears Joule's name. The redefinition of the kelvin will formalize the relation between temperature and energy, since the Boltzmann constant, which will establish the kelvin's definition, has the units of J/K.

We rarely make statements about the amount of "heat" in an object; we're more concerned that the temperature be appropriate. Rather, we think of heat as something that gets transferred from one object to another, causing a temperature change. The scientific definition reflects this sense of heat as energy in transit: **Heat is energy being transferred from one object to another because of a temperature difference alone**. Strictly speaking, **heat** refers only to energy in transit. Following heat transfer, we say that the **internal energy** or **thermal energy** of the object has increased, not that it contains more heat. This distinction reflects the fact that processes other than heating—such as transfer of mechanical or electrical energy—can also change an object's temperature. We briefly explored internal energy and its relation to mechanical energy transfers when we dealt with nonconservative forces in Chapter 7.

GOT IT? 16.1 Is there (a) no temperature, (b) one temperature, or (c) more than one temperature where the Celsius and Fahrenheit scales agree?

16.2 Heat Capacity and Specific Heat

Because temperature and energy are related, it's not surprising that the heat energy Q transferred to an object and the resulting temperature change ΔT are proportional: $Q = C\,\Delta T$, where the proportionality constant C is called the **heat capacity** of the object.

Video Tutor Demo | **Heating Water and Aluminum**

Video Tutor Demo | **Water Balloon Held over Candle Flame**

Table 16.1 Specific Heats of Some Common Materials*

	Specific Heat, c	
Substance	**SI Units: J/kg·K**	**cal/g·°C, kcal/kg·°C, or Btu/lb·°F**
Aluminum	900	0.215
Concrete (varies with mix)	880	0.21
Copper	386	0.0923
Iron	447	0.107
Glass	753	0.18
Mercury	140	0.033
Steel	502	0.12
Stone (granite)	840	0.20
Water:		
Liquid	4184	1.00
Ice, −10°C	2050	0.49
Wood	1400	0.33

*Temperature range 0°C to 100°C except as noted.

Since heat is a measure of energy transfer, the units of heat capacity are J/K. The heat capacity C applies to a specific object and depends on its mass and on the substance from which it's made. We characterize different substances in terms of their **specific heat** c, or heat capacity per unit mass. The heat capacity of an object is then the product of its mass and specific heat, so we can write

$$Q = mc\,\Delta T \tag{16.3}$$

The SI units of specific heat are J/kg·K. Table 16.1 lists specific heats of common materials.

Scientists first studied thermodynamic phenomena before they knew the relation between heat and energy, and they used other units for heat. The **calorie** (cal) was defined as the heat needed to raise the temperature of 1 g of water from 14.5°C to 15.5°C; consequently, the specific heat of water is 1 cal/g·°C. Several different definitions of the calorie exist today, based on different methods for establishing the heat–energy equivalence. In this book we use the so-called thermochemical calorie, defined as exactly 4.184 J. The "calorie" used in describing the energy content of foods is actually a kilocalorie. In the British system, still widely used in engineering in the United States, the unit of heat is the **British thermal unit** (Btu). One Btu is the amount of heat needed to raise the temperature of 1 lb of water from 63°F to 64°F, and is equal to 1054 J.

EXAMPLE 16.1 Specific Heat: Waiting to Shower

Your whole family has showered before you, dropping the temperature in the water heater to 18°C. If the heater holds 150 kg of water, how much energy will it take to bring it up to 50°C? If the energy is supplied by a 5.0-kW electric heating element, how long will that take?

INTERPRET Here we're interested in the energy it takes to raise the water temperature, so we interpret this problem as involving specific heat. For the second part, we're given the heater's power output and asked for the time, so we need to recall (Chapter 6) that power is energy per time.

DEVELOP Equation 16.3, $Q = mc\,\Delta T$, relates energy and temperature change via specific heat, so our plan is to calculate the required energy from this equation. We'll then use the relation between power and energy to find the time.

EVALUATE Equation 16.3 gives the energy:

$$Q = mc\,\Delta T = (150\text{ kg})(4184\text{ J/kg·K})(50°\text{C} - 18°\text{C}) = 20\text{ MJ}$$

where we found the specific heat of water in Table 16.1. The heating element supplies energy at the rate of 5.0 kW or 5.0×10^3 J/s. At that rate the time needed to supply 20 MJ is

$$\Delta t = \frac{2.0\times10^7\text{ J}}{5.0\times10^3\text{ J/s}} = 4000\text{ s}$$

or a little over an hour.

ASSESS That's a long time to wait, but it's not an unreasonable answer!

✓TIP Is That °C or K?

It doesn't matter when we're talking about temperature *differences*. That's why we could mix units, multiplying the specific heat in J/kg·K by the difference of Celsius temperatures.

■

For common materials around room temperature, specific heat is nearly constant over a substantial temperature range. But at very low temperatures, specific heat varies significantly with temperature. When that's the case, we write Equation 16.3 in terms of infinitesimal heat flows dQ and corresponding temperature changes dT: $dQ = mc(T)\,dT$. We can then integrate to relate the overall heat flow and temperature change over a wide temperature range. Problems 73 and 74 explore this situation.

Specific heat also depends on whether an object's pressure or its volume changes when it's heated. For solids and liquids, which don't expand much, that distinction isn't very important. But it makes a big difference whether a gas is confined or allowed to expand when heated. Consequently, gases have two different specific heats, depending on whether volume or pressure is constant. We'll deal with that issue in Chapter 18, where we explore the thermodynamic behavior of gases.

The Equilibrium Temperature

When objects at different temperatures are in thermal contact, heat flows from the hotter object to the cooler one until they reach thermodynamic equilibrium. If the objects are thermally insulated from their surroundings, then all the energy leaving the hotter object ends up in the cooler one. Mathematically, this statement reads

$$m_1c_1\,\Delta T_1 + m_2c_2\,\Delta T_2 = 0 \qquad (16.4)$$

For the hotter object, ΔT is negative, so the two terms in Equation 16.4 have opposite signs. One term represents the outflow of heat from the hotter object, the other inflow into the cooler one. Example 16.2 explores the application of Equation 16.4 in finding the equilibrium temperature.

GOT IT? 16.2 A hot rock with mass 250 g is dropped into an equal mass of cool water. Which temperature changes more, that of (a) the rock or (b) the water? Explain.

EXAMPLE 16.2 Finding the Equilibrium Temperature: Cooling Down

An aluminum frying pan of mass 1.5 kg is at 180°C when it's plunged into a sink containing 8.0 kg of water at 20°C. Assuming that none of the water boils and that no heat is lost to the surroundings, find the equilibrium temperature of the water and pan.

INTERPRET Here we have two objects, initially at different temperatures, that come to thermal equilibrium. So this is a problem about the equilibrium temperature, with the system of interest comprising the pan and the water.

DEVELOP Equation 16.4, $m_1c_1\,\Delta T_1 + m_2c_2\,\Delta T_2 = 0$, applies. However, we're asked for the common equilibrium temperature T, so we write the temperature differences ΔT in terms of T and the initial temperatures T_p and T_w of pan and water. Equation 16.4 then becomes $m_pc_p(T - T_p) + m_wc_w(T - T_w) = 0$.

EVALUATE We now solve for the equilibrium temperature T:

$$T = \frac{m_pc_pT_p + m_wc_wT_w}{m_pc_p + m_wc_w}$$

Using the given values of m_p, T_p, m_w, and T_w, and taking c_p and c_w from Table 16.1, we get $T = 26°C$.

ASSESS The water has much greater mass and higher specific heat, so it makes sense that its 6°C temperature change is a lot less than the 154°C drop in the pan's temperature. ■

16.3 Heat Transfer

How is heat transferred? Engineers need to know so they can design heating and cooling systems. Scientists need to know so they can anticipate temperature changes, as in global warming. Here we'll consider three common heat-transfer mechanisms: conduction, convection, and radiation. In some situations, a single mechanism dominates; in other cases, we may need to take all three into account.

Table 16.2 Thermal Conductivities*

Material	Thermal Conductivity, k	
	SI Units: W/m·K	British Units: Btu·in/h·ft²·°F
Air	0.026	0.18
Aluminum	237	1644
Concrete (varies with mix)	1	7
Copper	401	2780
Fiberglass	0.042	0.29
Glass	0.7–0.9	5–6
Goose down	0.043	0.30
Helium	0.14	0.97
Iron	80.4	558
Steel	46	319
Styrofoam	0.029	0.20
Water	0.61	4.2
Wood (pine)	0.11	0.78

*Temperature range 0°C to 100°C.

Conduction

Conduction is heat transfer through direct physical contact. It occurs as molecules in a hotter region collide with and transfer energy to those in an adjacent cooler region. **Thermal conductivity** (symbol k; SI unit W/m·K) characterizes this process. Common materials exhibit a broad range of thermal conductivities, from about 400 W/m·K for copper—a good conductor—to 0.029 W/m·K for Styrofoam, a good thermal insulator. Table 16.2 lists some thermal conductivities; they're given in both SI and British units because the latter are widely used in heat-loss calculations for buildings. The k values in Table 16.2 reflect physical properties of the materials. Metals, for example, are good thermal conductors because they contain free electrons that move quickly. Insulators like fiberglass and Styrofoam owe their insulating properties to a physical structure that traps small volumes of air or other gas.

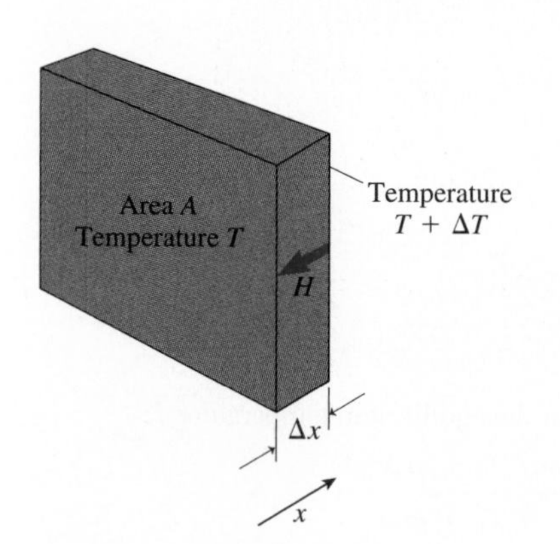

FIGURE 16.5 Heat flows from the hotter to the cooler face of the slab.

Figure 16.5 shows a slab of thickness Δx and area A. One side is at temperature T and the other at $T + \Delta T$. The temperature difference ΔT drives a conductive heat flow through the slab. That heat flow is proportional to the temperature difference, the slab area, and the thermal conductivity k. The thicker the slab, on the other hand, the more resistance to heat flow, so the flow depends inversely on thickness. Therefore,

$$H = -kA\frac{\Delta T}{\Delta x} \quad \text{(conductive heat flow)} \qquad (16.5)$$

where $H = dQ/dt$ is the rate of heat flow in watts, and where the minus sign shows that the flow is opposite the direction of increasing temperature—that is, from hotter to cooler.

EXAMPLE 16.3 Conduction: Warming a Lake

A lake with a flat bottom and steep sides has surface area 1.5 km² and is 8.0 m deep. On a summer day, the surface water is at 30°C and the bottom water at 4.0°C. What's the rate of heat conduction through the lake?

INTERPRET This is a problem about heat conduction.

DEVELOP Our sketch, Fig. 16.6, shows that we can treat the lake like the slab shown in Fig. 16.5, provided we neglect heat flow out the sides. Then Equation 16.5, $H = -kA(\Delta T/\Delta x)$, will give the heat-flow rate.

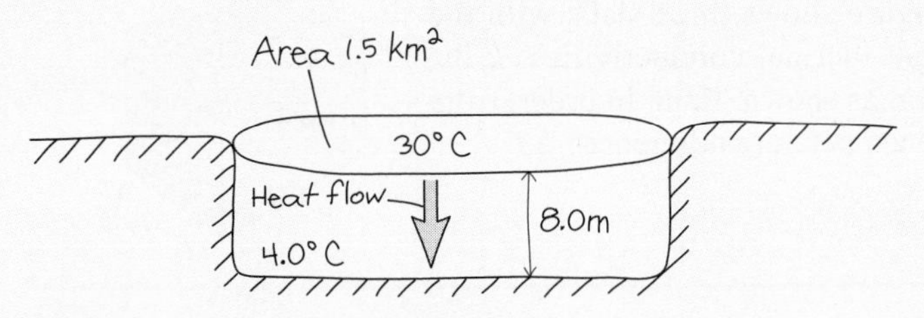

FIGURE 16.6 Our sketch for Example 16.3.

EVALUATE Substituting numerical values, including water's thermal conductivity from Table 16.2, we get

$$H = -kA\frac{\Delta T}{\Delta x}$$
$$= -(0.61\ \text{W/m}\cdot\text{K})(1.5\times 10^6\ \text{m}^2)\frac{30°\text{C} - 4.0°\text{C}}{8.0\ \text{m}} = -3.0\ \text{MW}$$

ASSESS This is a significant energy flow, but with direct sunlight averaging about 1 kW on every square meter, the lake's 1.5-km² surface area absorbs plenty of solar energy, and that's what maintains the temperature difference that drives the conductive heat flow. Figure 16.5 shows x increasing in the direction of increasing temperature, so the negative sign in our answer indicates that the flow is downward. ■

Equation 16.5 is strictly correct only when the temperature varies uniformly from one surface to the other. That's the case when two surfaces at different temperatures have the same area. With other geometries—as in the insulation surrounding a cylindrical pipe—we need to write $\Delta T/\Delta x$ as the derivative dT/dx and integrate to find the heat flow. Problems 76 and 80 explore this situation.

Often heat flows through several different materials. A building wall, for example, may contain wood, drywall, and fiberglass insulation. Figure 16.7 shows such a composite structure, with temperature T_1 on one side and T_3 on the other. The heat-flow rate H must be the same through both slabs so energy doesn't accumulate at the interface between the two. Then Equation 16.5 gives

$$H = -k_1A\frac{T_2 - T_1}{\Delta x_1} = -k_2A\frac{T_3 - T_2}{\Delta x_2}$$

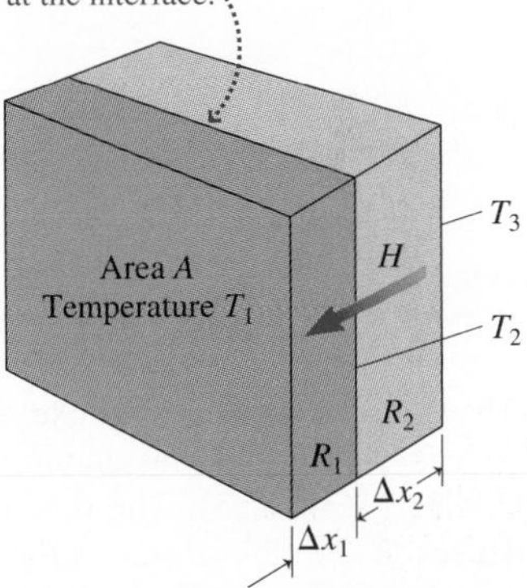

FIGURE 16.7 A composite slab.

where k_1 and k_2 are the thermal conductivities of the two materials, and T_2 is the temperature at the interface. We can express the heat-flow rate in terms of the surface temperatures T_1 and T_3 alone if we define the **thermal resistance** R of each slab:

$$R = \frac{\Delta x}{kA} \tag{16.6}$$

The SI units of R are K/W. Unlike the thermal conductivity k, which is a property of a *material*, R is a property of a *particular piece* of material, reflecting both its conductivity and its geometry. In terms of thermal resistance, our heat-flow equation becomes

$$H = -\frac{T_2 - T_1}{R_1} = -\frac{T_3 - T_2}{R_2}$$

so $R_1H = T_1 - T_2$ and $R_2H = T_2 - T_3$. Adding these two equations gives

$$(R_1 + R_2)H = T_1 - T_2 + T_2 - T_3 = T_1 - T_3$$

or

$$H = \frac{T_1 - T_3}{R_1 + R_2} \tag{16.7}$$

Equation 16.7 shows that the composite slab acts like a single slab whose thermal resistance is the sum of the resistances of the two slabs that compose it. We could easily extend this treatment to show that the thermal resistances of three or more slabs add when the slabs are arranged so the same heat flows through all of them.

FIGURE 16.8 Each square foot of this $\mathcal{R}$-19 fiberglass insulation loses $\frac{1}{19}$ Btu per hour for every °F of temperature difference ΔT.

GOT IT? 16.3 The figure shows three slabs with the same thickness but different thermal conductivities: k, $3k$, and $2k$; the left side is hotter, as shown. Rank in order, from smallest to largest, the three temperature differences ΔT.

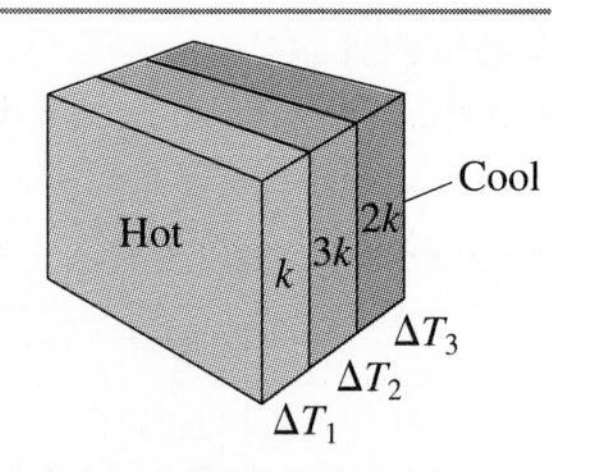

Insulating properties of building materials are described by the **$\mathcal{R}$-factor**, which is the thermal resistance for a slab of unit area:

$$\mathcal{R} = RA = \frac{\Delta x}{k} \tag{16.8}$$

The SI units of $\mathcal{R}$ are m^2·K/W, and that's how you'll find it listed if you buy insulation in Europe or other SI-based regions. In the United States, $\mathcal{R}$ is in ft^2·°F·h/Btu, although the units are almost never stated. This means that $\mathcal{R}$-19 fiberglass insulation loses $\frac{1}{19}$ Btu per hour for each square foot of insulation for each degree Fahrenheit temperature difference across the insulation (Fig. 16.8). The inverse of the $\mathcal{R}$-factor is the U value, often used in characterizing heat loss through windows.

EXAMPLE 16.4 Calculating Heat Loss: The Cost of Oil

Figure 16.9 shows a house whose walls consist of drywall ($\mathcal{R} = 0.45$), $\mathcal{R}$-11 fiberglass insulation, plywood ($\mathcal{R} = 0.65$), and cedar shingles ($\mathcal{R} = 0.55$). The roof has the same construction except it uses $\mathcal{R}$-30 fiberglass insulation. The average outdoor temperature in winter is 20°F, and the house is maintained at 70°F. The house's oil furnace produces 100,000 Btu for every gallon of oil, and oil costs \$3.48 per gallon. How much does it cost to heat the house for a month?

INTERPRET Although the problem asks for the monthly cost of oil, this isn't economics! We interpret this as a problem about heat loss and identify the walls and roof as systems for which we need to know the heat flow. This is a rare case of a problem stated in English units.

DEVELOP We're given the drawing in Fig. 16.9. We have the $\mathcal{R}$-factors; in English units, their inverses give the heat-loss rate on a square-foot basis. So our plan is to find the square footage of the walls and roof separately, calculate the total heat-loss rate, and then find the amount and cost of oil to compensate for a month's heat loss.

$A = (36\text{ ft})\left(\frac{14\text{ ft}}{\cos 30}\right)$

$h = 14\text{ ft} \times \tan 30$

30°

10 ft

36 ft

28 ft

FIGURE 16.9 House for Example 16.4.

EVALUATE The $\mathcal{R}$-factors for the wall materials sum to give $\mathcal{R}_{\text{wall}} = 12.65$; similarly, $\mathcal{R}_{\text{roof}} = 31.65$. The perimeter of the house measures $2\times 28\text{ ft} + 2\times 36\text{ ft} = 128\text{ ft}$, so the 10-ft vertical walls have area 1280 ft^2. There are also the triangular gables. Since there are two of them, each with area $\frac{1}{2}bh$, they give another bh or $(28\text{ ft})(14\text{ ft}\tan 30°) = 226\text{ ft}^2$, so $A_{\text{wall}} = 1506\text{ ft}^2$. These $\mathcal{R}$-12.65 walls lose 1/12.65 Btu/h/ft^2/°F. With 1506 ft^2 and a temperature difference of 50°F, the total heat-loss rate through the walls is

$$H_{\text{wall}} = \left(\tfrac{1}{12.65}\text{Btu/h/ft}^2/°\text{F}\right)(1506\text{ ft}^2)(50°\text{F}) = 5953\text{ Btu/h}$$

The area of the pitched roof is larger than that of a flat roof by the factor 1/cos 30°, so the heat-loss rate through the roof is

$$H_{\text{roof}} = \left(\tfrac{1}{31.65}\text{Btu/h/ft}^2/°\text{F}\right)\frac{(36\text{ ft})(28\text{ ft})}{\cos 30°}(50°\text{F}) = 1839\text{ Btu/h}$$

The total heat-loss rate is then 7792 Btu/h. In a month, this results in a heat loss of $Q = (7792\text{ Btu/h})(30\text{ days/month})(24\text{ h/day}) = 5.61\text{ MBtu}$.

Now for the oil: With 10^5 Btu (0.1 MBtu) per gallon, we'll burn 56.1 gallons per month to produce that 5.61 MBtu. At \$3.48/gal, that will cost \$195.

ASSESS If you've paid for heat in a northern climate, you know that this figure is, if anything, low. That's because we neglected heat losses through windows, doors, and the floor, as well as cold-air infiltration. On the other hand, we also left out any solar energy gained through the windows on sunny days. Problem 69 provides a more realistic look at this house. ■

Convection

Convection is heat transfer by fluid motion. It occurs as heated fluid becomes less dense and therefore rises. Figure 16.10*a* shows two plates at different temperatures, with fluid between them. Fluid heated by the lower plate rises and transfers heat to the upper plate. The cooled fluid sinks, and the process repeats. The pattern of rising and sinking fluid often acquires a striking regularity, as shown in Fig. 16.10*b*.

Convection is important in many technological and natural environments. When you heat water on a stove, convection carries heat through the water. Houses usually rely on convection from heat sources near floor level to circulate warm air throughout a room. Insulating materials trap air and thereby inhibit convection that would otherwise cause excessive heat loss. Convection associated with solar heating of Earth's surface drives the vast air movements that establish our overall climate. Violent convection, as in thunderstorms, is associated with localized temperature differences. On a much longer time scale, convection in Earth's mantle drives continental drift. Convection plays a crucial role in many astrophysical processes, including the generation of magnetic fields in stars and planets.

As with conduction, the convective heat-loss rate often is approximately proportional to the temperature difference. But the calculation of convective heat loss is complicated because of the associated fluid motion. The study of convection processes is an important research area in many fields of contemporary science and engineering.

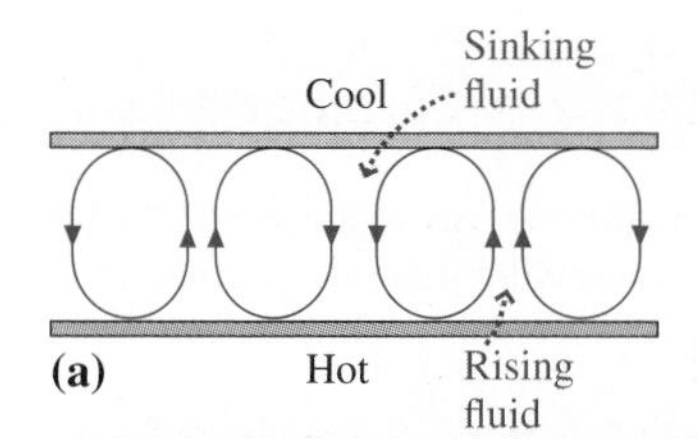

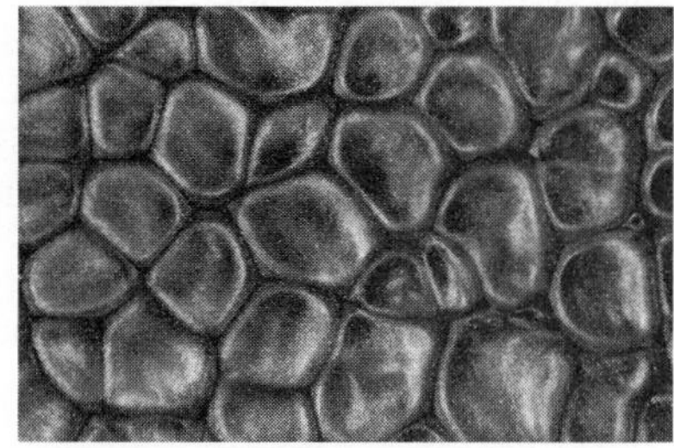

FIGURE 16.10 (a) Convection between two plates at different temperatures. (b) Top view of convection cells in a laboratory experiment. Fluid rises at the centers and sinks at the edges of the convection cells.

Radiation

Turn a stove burner to "high" and it glows brightly; turn it to "low" and you can still sense its heat although it doesn't glow visibly. Either way, the burner loses energy by emitting electromagnetic waves, or **radiation**. The radiated power P increases rapidly with temperature, as described by the **Stefan–Boltzmann law**:

$$P = e\sigma A T^4 \quad \left(\begin{array}{c}\text{Stefan–Boltzmann law;}\\ \text{radiated power}\end{array}\right) \qquad (16.9)$$

Video Tutor Demo | **Candle Chimneys**

where A is the area of the emitting surface, T the temperature in kelvins, and σ the **Stefan–Boltzmann constant**, approximately $5.67\times10^{-8}\ \text{W/m}^2\cdot\text{K}^4$. The quantity e is the **emissivity**, a number from 0 to 1 that measures the material's effectiveness in emitting radiation. For radiation of a given wavelength, a material is equally good at emitting and absorbing radiation. A perfect emitter has $e = 1$ and is also a perfect absorber. Such an object would appear black at room temperature and is therefore called a **blackbody**. A shiny object, in contrast, reflects most of the radiation that hits it and is therefore also a poor emitter. Wood stoves are often painted black to increase their emissivity; Thermos bottles, on the other hand, have a shiny coating to reduce radiation.

Because of the strong T^4 temperature dependence, radiation is generally the dominant heat-loss mechanism at high temperatures but is less important at low temperatures. Radiation also dominates for objects in vacuum, since there's no material to carry conductive or convective heat flows; that makes Equation 16.9 crucial in understanding the climates of Earth and other planets.

Objects also absorb radiant energy from their surroundings, at a rate given by Equation 16.9 using the ambient temperature T_a, so the *net* radiated power becomes $P = e\sigma A(T^4 - T_a^4)$. For an object that's much hotter than its surroundings, the second term is negligible. But for an object that's only a little warmer, like a human body, it's significant.

It's not just the amount of radiation that changes with temperature; as our stove burner example suggests, it's also the wavelength. Objects at room temperature, for example, emit mostly invisible infrared radiation, while very hot objects like the Sun emit more visible light. We'll take a quantitative look at this relation in Chapter 34.

PhET: Blackbody Spectrum

GOT IT? 16.4 Name the dominant form of heat transfer from (1) a red-hot stove burner with nothing on it, (2) a burner in direct contact with a pan of water, and (3) the bottom to the top of the water in the pan once it's begun to boil.

EXAMPLE 16.5 Calculating Radiation: The Sun's Temperature

The Sun radiates energy at the rate $P = 3.9 \times 10^{26}$ W, and its radius is 7.0×10^8 m. Treating the Sun as a blackbody ($e = 1$), find its surface temperature.

INTERPRET This is a problem about the radiation from a hot object.

DEVELOP The Stefan–Boltzmann law, Equation 16.9, gives the radiated power in terms of the temperature, emissivity, and surface area: $P = e\sigma AT^4$. Our plan is to solve this equation for T. For the Sun, radiation comes from the entire spherical surface of area $4\pi R^2$, as our sketch shows (Fig. 16.11).

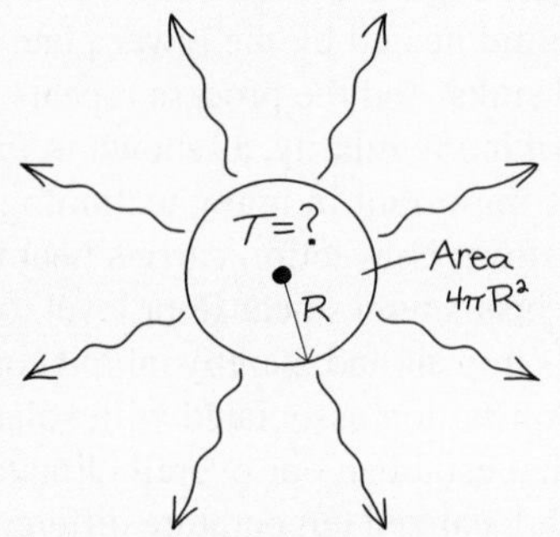

FIGURE 16.11 The Sun radiates from its spherical surface area $4\pi R^2$.

EVALUATE Using the Sun's spherical surface area and solving Equation 16.9 for T gives

$$T = \left(\frac{P}{4\pi R^2\sigma}\right)^{1/4}$$

$$= \left[\frac{3.9 \times 10^{26}\ \text{W}}{4\pi(7.0 \times 10^8\ \text{m})^2(5.7 \times 10^{-8}\ \text{W/m}^2 \cdot \text{K}^4)}\right]^{1/4} = 5.8 \times 10^3\ \text{K}$$

ASSESS Make sense? Yes: Our answer has the unit of temperature and agrees with observational measurements. Despite its bright glow, the Sun *is* essentially a blackbody, because it absorbs all radiation incident on it. But the Sun is so much hotter than its surroundings that we can neglect absorbed radiation in this calculation. ■

CONCEPTUAL EXAMPLE 16.1 Energy-Saving Windows

Why do double-pane windows reduce heat loss greatly compared with single-pane windows? Why is a window's $\mathcal{R}$-factor higher if the spacing between panes is small? And why do the best windows have "low-E" coatings?

EVALUATE Table 16.2 gives glass's thermal conductivity as around 0.8 W/m·K, while good insulators like air and Styrofoam have $k \sim 0.03$ W/m·K. That's why a layer of air between window panes greatly increases the window's $\mathcal{R}$-factor. But if the pane spacing is too great, convection currents develop between the sheets of glass, transferring heat from the warmer to the cooler surface; that's why narrower pane spacing is better. Finally, warm glass loses energy by radiation, and a thin coating of material with low emissivity ("low-E") reduces radiant heat loss.

ASSESS High-quality windows include all three features described here, so they suppress all three kinds of heat loss we've discussed. The best windows also use an inert gas—usually argon—between panes to reduce heat loss further.

MAKING THE CONNECTION Compare the $\mathcal{R}$-factor for a single-pane window made from 3.0-mm-thick glass with that of a double-pane window made from the same glass with a 5.0-mm air gap between panes.

EVALUATE Compute the $\mathcal{R}$-factors for the glass and air space, and you'll get about 0.004 m^2·K/W for the single pane and, adding two layers of glass and the air space, 0.2 m^2·K/W for the double-pane window. That's a factor of 50 improvement! In English units our answers translate into $\mathcal{R}$-factors of 0.02 and 1.1—although again they're lower than for actual windows because they neglect "dead air" layers and the other improvements discussed above. The best commercially available windows, in fact, achieve $\mathcal{R}$-factors of 5 and higher, and some multilayer windows exceed $\mathcal{R}$-10.

16.4 Thermal-Energy Balance

You keep your house at a comfortable temperature in winter by balancing heat loss with energy from your heating system (Fig. 16.12). This state of **thermal-energy balance** occurs throughout science and engineering. Understanding thermal-energy balance enables engineers to specify a building's heat sources and helps scientists predict Earth's future climate.

Engineered systems actively control the thermal-energy balance to achieve a desired temperature. But even without active control, systems with a fixed rate of energy input naturally tend toward energy balance. That's because all heat-loss mechanisms give increased loss with increasing temperature. If the rate of energy input to a system is greater than the loss rate, then the system gains energy and its temperature increases—and so, therefore, does the loss rate. Eventually the two come to balance at some fixed temperature. If the

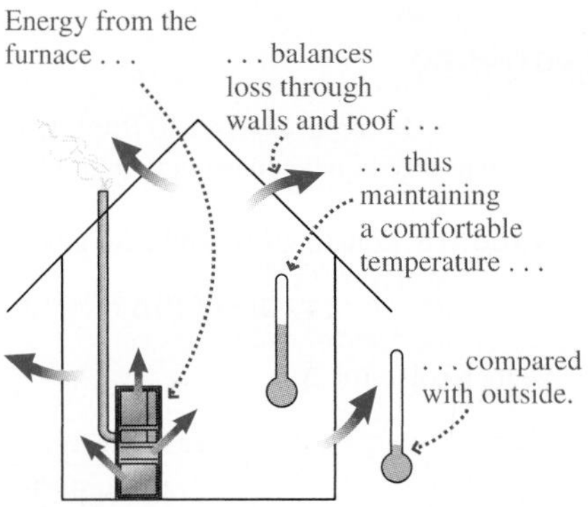

FIGURE 16.12 A house in thermal-energy balance.

loss exceeds the gain, the system cools until again it's in balance. Problems involving thermal-energy balance are similar regardless of the energy-loss mechanism or whether the application is to a technological or a natural system.

PROBLEM-SOLVING STRATEGY 16.1 Thermal-Energy Balance

INTERPRET Interpret the problem to be sure it deals with heat gains and losses. Identify the system of interest, the source(s) of energy input to the system, and the significant heat-loss mechanism(s).

DEVELOP Determine which equation(s) govern the heat loss; these will necessarily involve the system's temperature. Your plan is then to equate the rate of energy loss with the rate of energy input.

EVALUATE Write an equation that expresses equality of energy loss and input. Then evaluate by solving for the quantity the problem asks for—often the system's temperature.

ASSESS If your answer is a temperature, does it seem reasonable? Is the temperature of a heated system higher than that of its surroundings?

EXAMPLE 16.6 Thermal-Energy Balance: Hot Water

A poorly insulated electric water heater loses heat by conduction at the rate of 120 W for each Celsius degree difference between the water and its surroundings. It's heated by a 2.5-kW electric heating element and is located in a basement kept at 15°C. What's the water temperature if the heating element operates continuously?

INTERPRET The concept here is energy balance, and we identify the system of interest as the water. Its energy input comes from the heating element at the rate of 2.5 kW. The heat loss is by conduction.

DEVELOP Figure 16.13 is a sketch suggesting energy balance in the heater. We're given the conductive heat loss of 120 W/°C, meaning that the total heat-loss rate is $H = (120\ \mathrm{W/^\circ C})(\Delta T)$. We then equate the heat-loss rate to the energy-input rate: $(120\ \mathrm{W/^\circ C})(\Delta T) = 2.5\ \mathrm{kW}$.

EVALUATE Solving for ΔT gives

$$\Delta T = \frac{2.5\ \mathrm{kW}}{120\ \mathrm{W/^\circ C}} = 21^\circ\mathrm{C}$$

With the basement at 15°C, the water temperature is then 36°C.

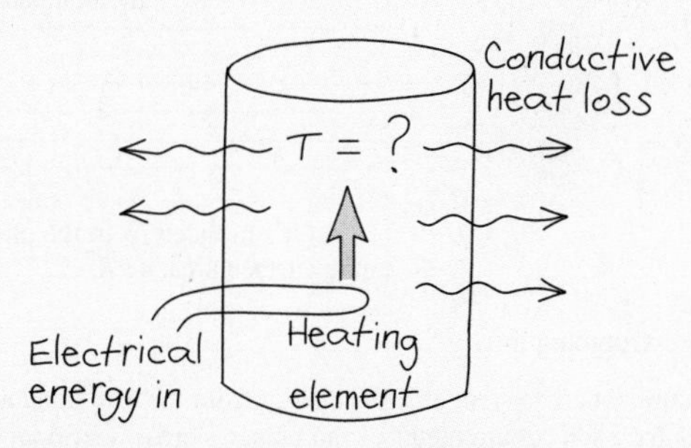

FIGURE 16.13 Balance between the heat supplied by the electric element and the conductive loss determines the water temperature.

ASSESS Is this answer reasonable? Not if you want a hot shower; our answer is 1°C below body temperature! But we're told the insulation is bad, so it's time for a new water heater! ■

EXAMPLE 16.7 Thermal-Energy Balance: A Solar Greenhouse

A solar greenhouse has 300 ft² of opaque $\mathcal{R}$-30 walls and 250 ft² of $\mathcal{R}$-1.8 double-pane glass that admits solar energy at the average rate of 40 Btu/h/ft². Find the greenhouse temperature on a day when the outdoor temperature is 15°F.

INTERPRET Again the concept is energy balance, now with the greenhouse as the system of interest. We're given $\mathcal{R}$-factors, suggesting that the energy loss is by conduction through walls and glazing. The energy input is sunlight.

DEVELOP As we saw in Example 16.4, the $\mathcal{R}$-factor determines a heat-loss rate that is related directly to area and temperature difference and inversely to the $\mathcal{R}$-factor. So we have

$$H_w = \frac{A_w\,\Delta T}{\mathcal{R}_w} = \left(\frac{300}{30}\right)\Delta T = (10\ \text{Btu/h/°F})\Delta T$$

for the heat loss through the walls and

$$H_g = \frac{A_g\,\Delta T}{\mathcal{R}_g} = \left(\frac{250}{1.8}\right)\Delta T = (139\ \text{Btu/h/°F})\Delta T$$

for the heat loss through the glass, giving a total heat loss $H = (149\ \text{Btu/h/°F})\Delta T$. Meanwhile, the energy input through the entire 250 ft² of glass is $(40\ \text{Btu/h/ft}^2)(250\ \text{ft}^2) = 1.0\times 10^4\ \text{Btu/h}$. Our plan is to equate energy input and loss and then solve for ΔT.

EVALUATE Equating loss and gain gives

$$(149\ \text{Btu/h/°F})\Delta T = 1.0\times 10^4\ \text{Btu/h}.$$

We then solve for ΔT:

$$\Delta T = \frac{1.0\times 10^4\ \text{Btu/h}}{149\ \text{Btu/h/°F}} = 67\text{°F}$$

So when it's 15°F outside, the greenhouse is at a tropical 82°F.

ASSESS This seems a reasonable greenhouse temperature. Our calculation assumes that solar input remains constant; in a real greenhouse the temperature would fluctuate as the Sun's angle changes and clouds pass over. We could minimize these fluctuations by giving the greenhouse a large heat capacity, perhaps by incorporating a massive concrete slab or concrete walls. ■

PhET: The Greenhouse Effect

GOT IT? 16.5 A house's thermostat fails, leaving the furnace running continuously. As a result, will the temperature of the house (a) increase indefinitely, (b) eventually stabilize, or (c) drop below the thermostat setting? Explain.

APPLICATION The Greenhouse Effect and Global Warming

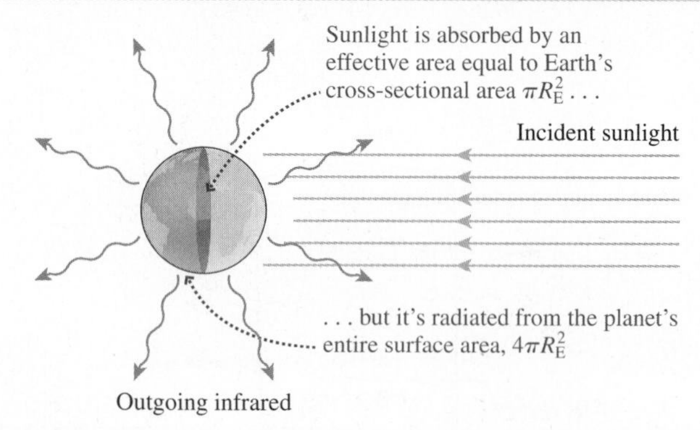

The Earth–atmosphere system absorbs energy from the Sun at an average rate of 960 watts for each square meter of the planet's cross-sectional area, πR_E^2, where R_E is Earth's radius (see the diagram). This quantity is designated S, so we write $S = 960\ \text{W/m}^2$. This value accounts for night and day; for clouds; and for the reflection of sunlight from ice, snow, deserts, and other highly reflective surfaces and especially from particulate matter in the atmosphere. Therefore, the rate at which the entire Earth–atmosphere system absorbs energy is $P_{\text{incoming}} = \pi R_E^2 S$. This incoming energy causes Earth to warm until it loses energy at the same rate. Since it's surrounded by the vacuum of space, Earth can only lose energy by radiation. Since Earth is much cooler than the Sun, that radiation is in the form of invisible infrared. Furthermore, as the diagram shows, Earth radiates from its entire surface area, $4\pi R_E^2$. Earth's emissivity for infrared radiation is essentially 1, so Earth radiates energy at a rate given by the Stefan–Boltzmann law, Equation 16.9:

$$P_{\text{outgoing}} = \sigma 4\pi R_E^2 T^4$$

where T is Earth's average temperature. Equating this outgoing power to the rate at which solar energy arrives from the Sun gives a statement of energy balance:

$$\pi R_E^2 S = \sigma 4\pi R_E^2 T^4$$

Solving for the temperature then gives $T = 255\ \text{K} = -18\text{°C}$ or 0°F. Is this reasonable? It's certainly in the right ballpark—not so hot as to boil the oceans or so cold as to freeze the atmosphere. But 0°F seems a bit cold for a global average temperature. And it is: Earth's average temperature is around 15°C or 59°F. Why the discrepancy?

The answer lies with Earth's atmosphere. The dominant atmospheric gases, nitrogen and oxygen, are largely transparent to both incoming sunlight and outgoing infrared. But others—the so-called **greenhouse gases**, especially water vapor and carbon dioxide—let sunlight pass through but impede outgoing infrared. As a result, Earth's surface temperature has to be higher to get the same total radiation to space. This is the **natural greenhouse effect**, and it explains the 33°C temperature difference between our simple calculation and Earth's actual surface temperature. Neighbor planets confirm this reasoning. Mars, with very little atmosphere, exhibits almost no greenhouse warming. Venus, whose atmosphere is 100 times denser than Earth's and largely CO_2, has a

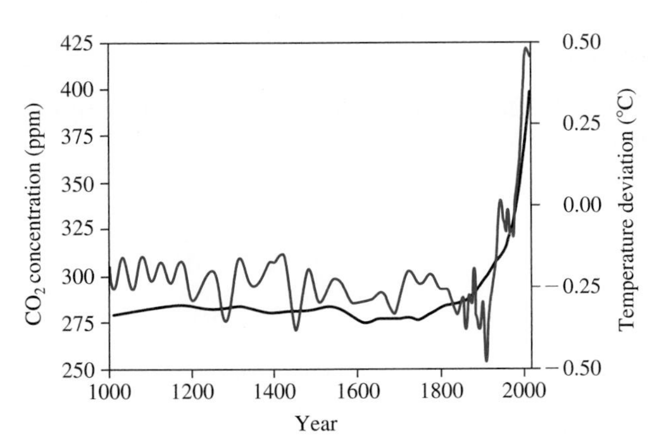

Atmospheric CO_2 concentration (black) and global temperature (color) from 1000–2015 A.D. Temperature is given as a deviation from the average for 1961–1990. Data through 1849 are reconstructed based on tree rings and other proxies; data from 1850 on are from thermometer records. The industrial era began around 1750.

"runaway" greenhouse effect that keeps its surface hotter than an oven. You can explore the climates of our neighbor planets in Problem 75.

As the graph shows, we humans have increased atmospheric carbon dioxide some 40% since the start of the industrial era, to a level—now exceeding 400 parts per million—that the planet has not seen for millions of years. Combustion of fossil fuels is the dominant source of this CO_2, although processes like deforestation also contribute, as do other greenhouse gases such as methane. Basic physics then dictates that Earth's surface temperature should rise. How much and how fast depend on complex interactions among atmosphere, surface, oceans, and life, and on future greenhouse-gas emissions. Nevertheless, a consensus among climate scientists suggests that Earth has warmed by some 0.85°C since the mid-19th century, with most of this warming attributable to human activities—especially combustion of fossil fuels and the resulting CO_2 emissions (see the graph). Further warming in the range of 1.5°C–5°C is projected by the year 2100, with the low end requiring substantial curtailing of greenhouse-gas emissions and the high end corresponding to "business as usual."

Although even a 5°C increase may seem modest, the *rate* of increase in all scenarios for the 21st century is far greater than most natural climate change. Furthermore, as the map shows, warming will not be distributed evenly over the globe but will be greatest in the arctic and over most land masses. One of many serious consequences of this rapid warming is a rise in sea level, which is already occurring substantially more rapidly than its average rate over the past 2000 years. During the last so-called interglacial warm period, some 120,000 years ago, sea level was between 5 and 10 m higher than it is today—enough to swamp Earth's coastal cities. The temperature at that time was likely only a little more than 2°C above the pre-industrial temperature of the 18th and 19th centuries. Considerations such as this have led the world's governments to adopt the goal of limiting the planet's industrial-era temperature rise to no more than 2°C. Given that we're almost halfway there already, achieving this goal will require drastic changes in the way we produce energy.

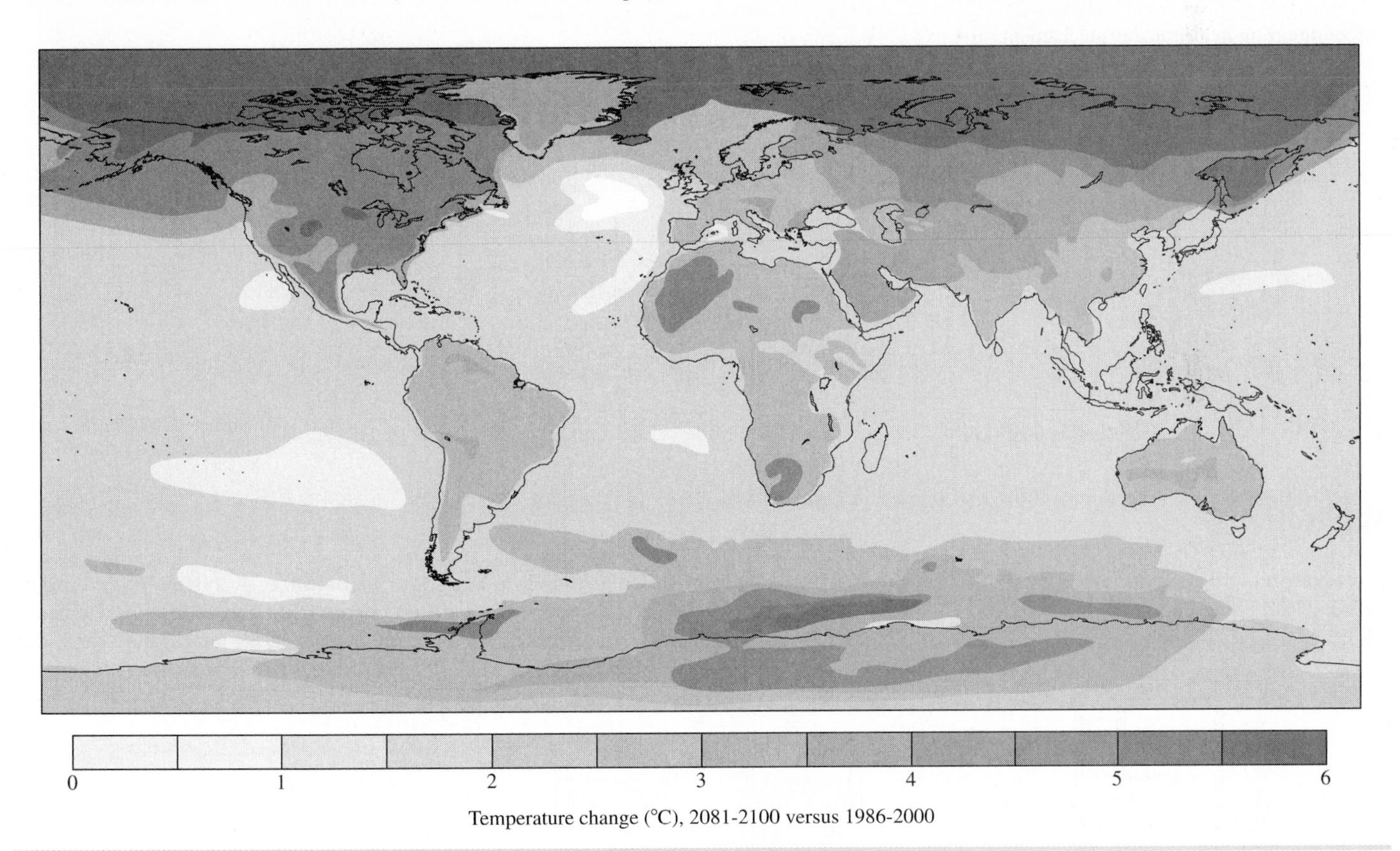

CHAPTER 16 SUMMARY

Big Ideas

The big ideas here are **temperature** and **heat**. **Temperature** is a property common to systems in **thermodynamic equilibrium**. Temperature is quantified in SI units using the **kelvin scale**, currently defined in terms of gas-based thermometers.

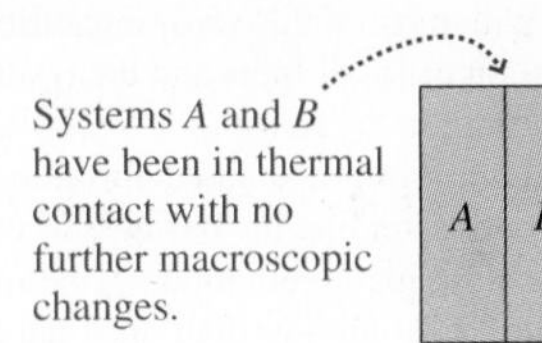

Heat is energy in transit as a result of a temperature difference.

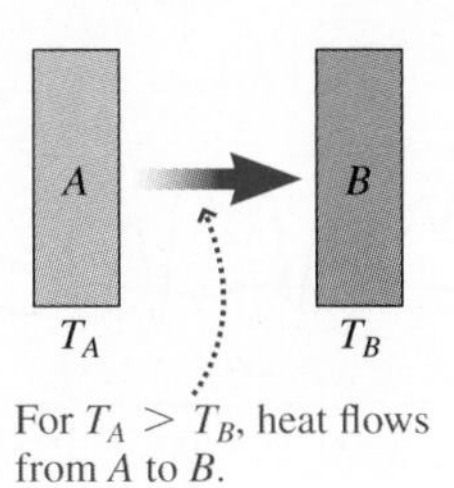

Key Concepts and Equations

Heat capacity and **specific heat** quantify the energy Q required to raise an object's temperature by ΔT:

$$Q = mc\,\Delta T$$

Mass m
Specific heat c
Temperature T
Add energy Q
Temperature increase ΔT
$Q = mc\,\Delta T$

Three important heat-transfer mechanisms are:

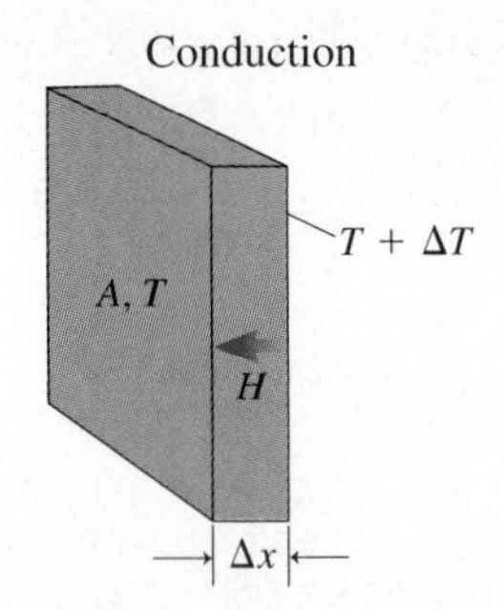

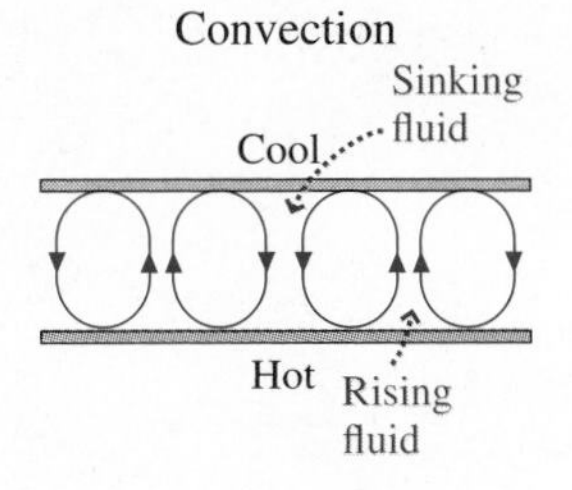

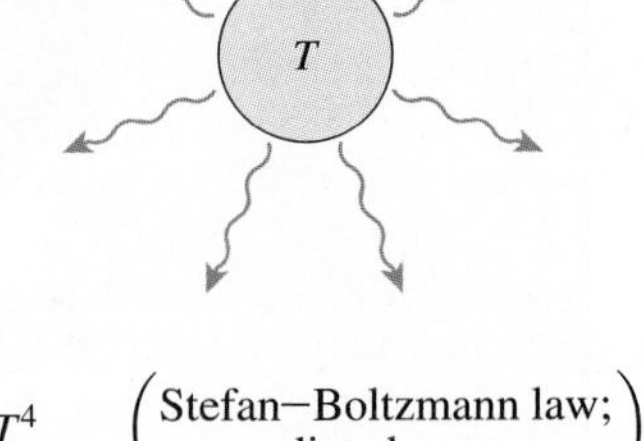

$$H = -kA\frac{\Delta T}{\Delta x} \quad \text{(conductive heat flow)}$$

$$P = e\sigma AT^4 \quad \left(\begin{array}{c}\text{Stefan–Boltzmann law;}\\ \text{radiated power}\end{array}\right)$$

Applications

Temperature scales include Kelvin (K), Celsius (°C), Fahrenheit (°F), and Rankine (°R).

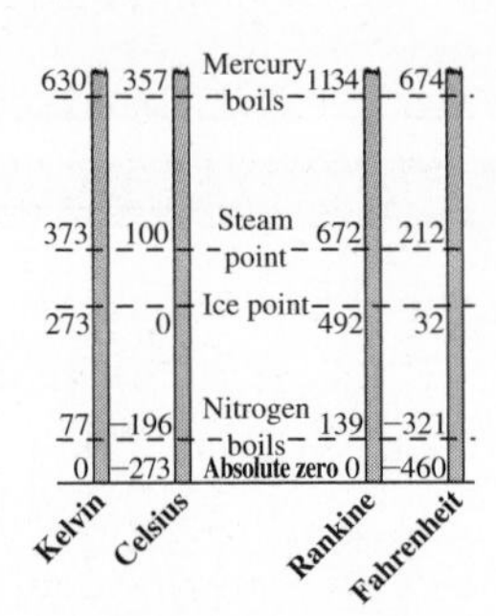

The Kelvin and Celsius scales are related by $T_C = T - 273.15$. The relation between Fahrenheit and Celsius scales is $T_F = \frac{9}{5}T_C + 32$.

Equilibrium temperature: Combining two systems at different temperatures results in a common equilibrium temperature given by $m_1c_1\,\Delta T_1 + m_2c_2\,\Delta T_2 = 0$.

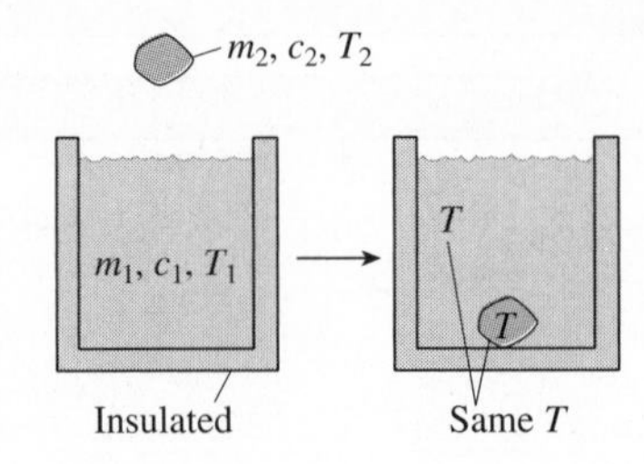

Energy balance: A system experiencing both energy input and energy loss comes to energy balance at the temperature for which the energy-loss rate equals the rate of energy input.

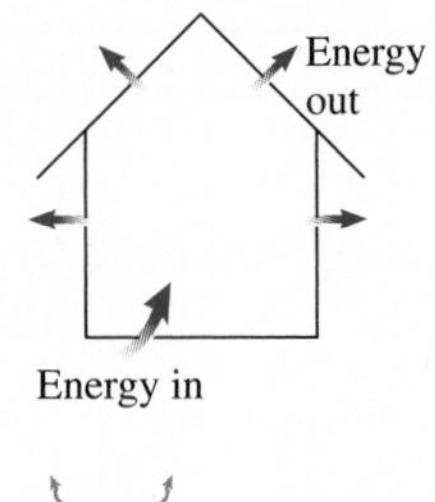

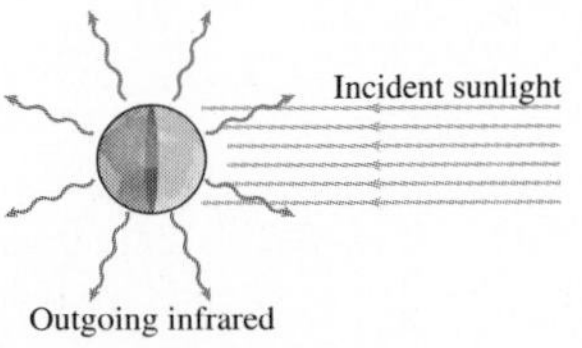

MP For homework assigned on MasteringPhysics, go to www.masteringphysics.com

BIO *Biology and/or medicine-related problems* **DATA** *Data problems* **ENV** *Environmental problems* **CH** *Challenge problems* **COMP** *Computer problems*

For Thought and Discussion

1. If system A is not in thermodynamic equilibrium with system B, and B is not in equilibrium with C, can you draw any conclusions about the temperatures of the three systems?
2. Does a thermometer measure its own temperature or the temperature of its surroundings? Explain.
3. Compare the relative sizes of the kelvin, the degree Celsius, the degree Fahrenheit, and the degree Rankine.
4. If you put a thermometer in direct sunlight, what do you measure: the air temperature, the temperature of the Sun, or some other temperature?
5. Why does the temperature in a stone building usually vary less than in a wooden building?
6. Why do large bodies of water exert a temperature-moderating effect on their surroundings?
7. A Thermos bottle consists of an evacuated, double-wall glass liner, coated with a thin layer of aluminum. How does it keep liquids hot?
8. Stainless-steel cookware often has a layer of aluminum or copper embedded in the bottom. Why?
9. What method of energy transfer dominates in baking? In broiling?
10. After a calm, cold night, the temperature a few feet above ground often drops just as the Sun comes up. Explain in terms of convection.
11. Glass and fiberglass are made from the same material, yet have dramatically different thermal conductivities. Why?
12. To keep your hands warm while skiing, you should wear mittens instead of gloves. Why?
13. Since Earth is exposed to solar radiation, why doesn't Earth have the same temperature as the Sun?
14. Global warming at Earth's surface is generally producing greater temperature rises over land than over the oceans. Why might this be?

Exercises and Problems

Exercises

Section 16.1 Heat, Temperature, and Thermodynamic Equilibrium

15. In its 2014 report, the Intergovernmental Panel on Climate Change projected a global temperature increase of 1.4°C to 3.1°C during the 21st century, for a scenario in which atmospheric carbon dioxide reaches about 700 parts per million by 2100 (it's now about 400 ppm and rising at about 2.4 ppm/year). Translate this range into Fahrenheit.
16. A Canadian meteorologist predicts an overnight low of −15°C. How would a U.S. meteorologist express that prediction?
17. Normal room temperature is 68°F. What's this in Celsius?
18. The outdoor temperature rises by 10°C. What's that rise in Fahrenheit?
19. At what temperature do the Fahrenheit and Celsius scales coincide?
20. The normal boiling point of nitrogen is 77.3 K. Express this in Celsius and Fahrenheit.
21. A sick child's temperature reads 39.1 on a Celsius thermometer. What's the temperature in Fahrenheit?

Section 16.2 Heat Capacity and Specific Heat

22. Find the heat capacity of a 55-tonne concrete slab.
23. Find the energy needed to raise the temperature of a 2.0-kg chunk of aluminum by 18°C.
24. What's the specific heat of a material if it takes 7.5 kJ to increase the temperature of a 1-kg sample by 3.0°C?
25. **BIO** The average human diet contains about 2000 kcal per day. If all this food energy is released rather than stored as fat, what's the approximate average power output of the human body?
26. **BIO** Walking at 3 km/h requires an energy expenditure rate of about 200 W. How far would you have to walk to "burn off" a 420-kcal hamburger?
27. You bring a 350-g wrench into the house from your car. The house is 15°C warmer than the car, and it takes 2.52 kJ to warm the wrench by this amount. Find (a) the heat capacity of the wrench and (b) the specific heat of the metal it's made from.
28. (a) How much heat does it take to bring a 3.4-kg iron skillet from 20°C to 130°C? (b) If the heat is supplied by a stove burner at the rate of 2.0 kW, how long will it take to heat the pan?

Section 16.3 Heat Transfer

29. Building heat loss in the United States is usually expressed in Btu/h. What's 1 Btu/h in SI units?
30. Find the heat-loss rate through a slab of (a) wood and (b) Styrofoam, each 2.0 cm thick, if one surface is at 20°C and the other at 0°C.
31. The top of a steel wood stove measures 90 cm by 40 cm and is 0.45 cm thick. The fire maintains the inside surface of the stovetop at 310°C, while the outside surface is at 295°C. Find the heat conduction rate through the stovetop.
32. **ENV** You're a builder who's advising a homeowner to have her foundation walls insulated with 2 inches of Styrofoam. To make your point, you tell her how thick the concrete walls (normally 8 inches) would have to be to have the same insulating value as 2 inches of Styrofoam. What's this thickness?
33. An 8.0 m by 12 m house is built on a concrete slab 23 cm thick. Find the heat-loss rate through the floor if the interior is at 20°C while the ground is at 10°C.
34. Find the $\mathcal{R}$-factor for a wall that loses 0.040 Btu each hour through each square foot for each °F temperature difference.
35. Compute the $\mathcal{R}$-factors for 1-inch thicknesses of air, concrete, fiberglass, glass, Styrofoam, and wood.
36. A horseshoe has surface area 50 cm^2, and a blacksmith heats it to a red-hot 810°C. At what rate does it radiate energy?

Section 16.4 Thermal-Energy Balance

37. An oven loses energy at the rate of 14 W per °C temperature difference between its interior and the 20°C temperature of the kitchen. What average power must be supplied to maintain the oven at 180°C?
38. You're having your home's heating system replaced, and the heating contractor has specified a new system that supplies energy at the maximum rate of 40 kW. You know that your house loses energy at the rate of 1.3 kW per °C temperature difference between interior and exterior, and the minimum winter temperature in your area is −15°C. You'd like to maintain 20°C (68°F) indoors. Should you go with the system your contractor recommends?

ent of a 100-W lightbulb is at 3.0 kK. What's the fila-
rface area?

human body has surface area 1.4 m^2 and skin tempera-
BIO ture 3_ C. If the body's emissivity is about 1, what's the net radiation from the body when the ambient temperature is 18°C?

Problems

41. A constant-volume gas thermometer is filled with air whose pressure is 101 kPa at the normal melting point of ice. What would its pressure be at (a) the normal boiling point of water (373 K), (b) the normal boiling point of oxygen (90.2 K), and (c) the normal boiling point of mercury (630 K)?
42. A constant-volume gas thermometer is at 55-kPa pressure at the triple point of water. By how much does its pressure change for each kelvin temperature change?
43. In Fig. 16.2's gas thermometer, the height h is 60.0 mm at the triple point of water. When the thermometer is immersed in boiling sulfur dioxide, the height drops to 57.8 mm. What is the boiling point of SO_2 in kelvins and in degrees Celsius?
44. **BIO** If your mass is 60 kg, what's the minimum number of Calories (kcal) you would "burn off" climbing a 1700-m-high mountain? (*Note:* The actual metabolic energy used would be much greater.)
45. **BIO** Typical fats contain about 9 kcal per gram. If the energy in body fat could be utilized with 100% efficiency, how much mass would a runner lose in a 26.2-mile marathon while consuming 125 kcal/mile?
46. **ENV** A circular lake 1.0 km in diameter is 10 m deep (Fig. 16.14). Solar energy is incident on the lake at an average rate of 200 W/m^2. If the lake absorbs all this energy and does not exchange heat with its surroundings, how long will it take to warm from 10°C to 20°C?

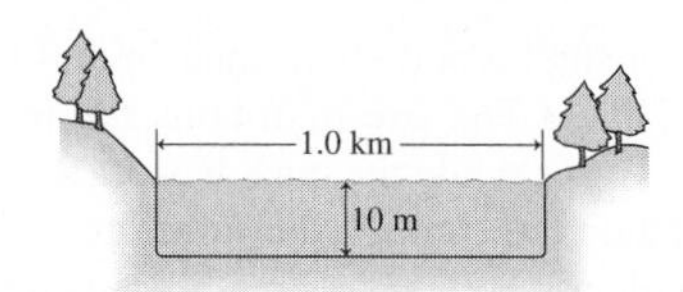

FIGURE 16.14 Problem 46

47. How much heat is required to raise an 800-g copper pan from 15°C to 90°C if (a) the pan is empty or contains (b) 1.0 kg of water and (c) 4.0 kg of mercury?
48. Initially, 100 g of water and 100 g of another substance listed in Table 16.1 are at 20°C. Heat is then transferred to each substance at the same rate for 1.0 min. At the end of that time, the water is at 32°C and the other substance at 76°C. (a) What's the other substance? (b) What's the heating rate?
49. You draw 330 mL of 10°C water from the tap and pop it into a 900-W microwave oven to heat for tea. How long should you microwave the water so it just reaches the boiling point?
50. Two neighbors return from Florida to find their houses at a frigid 35°F. Each house has a furnace that can supply 100,000 Btu/h. One house is made of stone and weighs 75 tons. The other is wood and weighs 15 tons. How long does it take each house to reach 65°F? Neglect heat loss, and assume the entire house mass reaches a uniform temperature.
51. You're arguing with your roommate about whether it's quicker to heat water on a stove burner or in a microwave. The burner supplies energy at the rate of 1.0 kW, the microwave at 625 W. You can heat water in the microwave in a paper cup of negligible heat capacity, but the stove requires a pan with heat capacity 1.4 kJ/K. How much water do you need before it becomes quicker to heat on the stovetop? Neglect energy loss to the surroundings.
52. **ENV** When a nuclear power plant's reactor is shut down, radioactive decay continues to produce heat at about 10% of the reactor's normal power level of 3.0 GW. In a major accident, a pipe breaks and all the reactor cooling water is lost. The reactor is immediately shut down, the break is sealed, and 420 m^3 of 20°C water is injected into the reactor. If the water were not actively cooled, how long would it take to reach its normal boiling point?
53. A 1.2-kg iron tea kettle sits on a 2.0-kW stove burner. If it takes 5.4 min to bring the kettle and the water in it from 20°C to the boiling point, how much water is in the kettle?
54. **BIO** The temperature of the eardrum provides a reliable measure of deep body temperature and is measured quickly with ear thermometers that sense infrared radiation. A thermometer that "views" 1 mm^2 of the eardrum requires 100 μJ of energy for a reliable reading at normal 37°C body temperature. How long does the measurement take?
55. A 1500-kg car moving at 40 km/h is brought to a sudden stop. If all the car's energy is dissipated in heating its four 5.0-kg steel brake disks, by how much do the disk temperatures increase?
56. Your young niece complains that her cocoa, at 90°C, is too hot. You pour 2 oz of milk at 3°C into the 6 oz of cocoa. Assuming milk and cocoa have the same specific heat as water, what's the cocoa's new temperature?
57. A piece of copper at 300°C is dropped into 1.0 kg of water at 20°C. If the equilibrium temperature is 25°C, what's the mass of the copper?
58. While camping, you boil water to make spaghetti. Your pot contains 2.5 kg of water initially at 10°C. You stoke up the campfire, and as a result the water gains energy at an increasing rate: $P = a + bt$, where $a = 1.1$ kW, $b = 2.3$ W/s, and t is the time in s. To the nearest minute, how long will it take to bring the water to a boil?
59. A biology lab's walk-in cooler measures 3.0 m by 2.0 m by 2.3 m and is insulated with 8.0-cm-thick Styrofoam. If the surrounding building is at 20°C, at what average rate must the cooler's refrigeration unit remove heat in order to maintain 4.0°C in the cooler?
60. One end of an iron rod 40 cm long and 3.0 cm in diameter is in ice water, the other in boiling water (Fig. 16.15). The rod is well insulated so no heat is lost out the sides. Find the heat-flow rate along the rod.

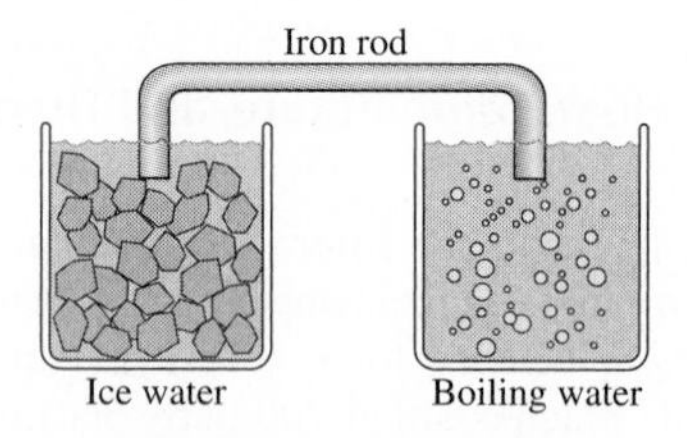

FIGURE 16.15 Problem 60

61. You arrive for a party on a night when it's 8°C outside. Your hosts meet you at the door and say the party may need to be cancelled, because the heating system has failed and they don't want to discomfort their guests. You say, "Not so fast!" A total of 36 people are expected, the average power output of a human body is 100 W, and the house loses energy at the rate 320 W/°C. Will the house remain comfortable?
62. An electric stove burner has surface area 325 cm^2 and emissivity $e = 1$. The burner consumes 1500 W and is at 900 K. If room temperature is 300 K, what fraction of the burner's heat loss is from radiation?

63. An electric current passes through a metal strip 0.50 cm by 5.0 cm by 0.10 mm, heating it at a rate of 50 W. The strip has emissivity $e = 1$ and its surroundings are at 300 K. What will be the strip's temperature if it's enclosed in (a) a vacuum bottle transparent to all radiation and (b) an insulating box with thermal resistance $R = 8.0$ K/W that blocks all radiation?

64. **BIO** You're considering purchasing a new sleeping bag whose manufacturer claims will keep you warm to -10°F. The bag has down insulation with 4.0-cm loft (thickness). Your body produces heat at the rate of 100 W and has area 1.5 m^2. Considering only conductive heat loss, will you be able to maintain normal body temperature in the bag at -10°F?

65. A blacksmith heats a 1.1-kg iron horseshoe to 550°C, then plunges it into a bucket containing 15 kg of water at 20°C. What's the equilibrium temperature?

66. What's the power output of a microwave oven that can heat 430 g of water from 20°C to the boiling point in 2.5 min? Neglect the container's heat capacity.

67. A cylindrical log 15 cm in diameter and 65 cm long is glowing red hot in a fireplace. The log's emissivity is essentially 1. If it's emitting radiation at the rate of 34 kW, what's its temperature?

68. A blue giant star whose surface temperature is 23 kK radiates energy at the rate of 3.4×10^{30} W. Find the star's radius, assuming it behaves like a blackbody.

69. **ENV** Rework Example 16.4, now assuming the house has 10 single-glazed windows, each measuring 2.5 ft by 5.0 ft. Four of the windows are on the south, and each admits solar energy at the average rate of 30 Btu/h·ft^2. *All* the windows lose heat; their $\mathcal{R}$-factor is 0.90. (a) Find the total heating cost for the month. (b) How much is the solar gain worth?

70. **CH** A black wood stove with surface area 4.6 m^2 is made from cast iron 4.0 mm thick. Its interior wall is at 650°C, while the exterior is at 647°C. (a) What's the rate of heat conduction through the stove wall? (b) What's the rate of heat loss by radiation from the stove? (c) Use the results of (a) and (b) to find how much heat the stove loses by a combination of conduction and convection in the surrounding air.

71. Estimate the average temperature on Pluto, treating the dwarf planet as a blackbody whose great distance from the Sun means that it receives energy from the Sun at the rate of only 0.876 W/m^2.

72. **DATA** The table below shows temperature versus time for 500 g of water heated in a microwave oven. In a microwave, essentially all the microwave energy goes into the water-containing food in the oven. Plot the data, determine a best-fit line, and use the slope of your line to determine the microwave power of this particular oven. Assume that water's specific heat is independent of temperature (which is only approximately true; see Problem 73).

Time (s)	0	25	60	95	125	160	190
Temperature (°C)	12	20	39	53	64	83	93

73. **CH** Water's specific heat in the range from 0°C to 100°C is given very nearly by $c(T) = c_0 + aT + bT^2$, where $c_0 = 4207.9$ J/kg·K, $a = -1.292$ J/kg·K^2, and $b = 0.01330$ J/kg·K^3. Use this expression to find the heat required to raise the temperature of 1.000 kg of water from 0°C to 100°C. By what percentage does this differ from the result you would get using the value of c in Table 16.1 over the entire temperature range?

74. **CH** At low temperatures the specific heats of solids are approximately proportional to the cube of the temperature: $c(T) = a(T/T_0)^3$. For copper, $a = 31$ J/g·K and $T_0 = 343$ K. Find the heat required to bring 40 g of copper from 10.0 K to 25.0 K.

75. **ENV** The Application on global warming (page 296) gives 960 W/m^2 as the average rate at which solar energy reaches Earth. You can approximate the solar energy rate reaching other planets by scaling this quantity by the inverse square of the planet's distance from the Sun (see Appendix E)—although what you'll get is only an approximation because that 960 W/m^2 includes effects of clouds and reflection that are unique to Earth and, more importantly, it neglects the greenhouse effect. Follow the procedure used in the Application to find approximations to the temperatures of Mars and Venus, and compare with their mean measured surface temperatures (you'll have to research those). Your results suggest that Mars has very little greenhouse effect, while Venus exhibits a "runaway" greenhouse effect resulting in a very high surface temperature.

76. **CH** In a cylindrical pipe where area isn't constant, Equation 16.5 takes the form $H = -kA(dT/dr)$, where r is the radial coordinate measured from the pipe axis. Use this equation to show that the heat-loss rate from a cylindrical pipe of radius R_1 and length L is

$$H = \frac{2\pi kL(T_1 - T_2)}{\ln(R_2/R_1)}$$

where the pipe is surrounded by insulation of outer radius R_2 and thermal conductivity k and where T_1 and T_2 are the temperatures at the pipe surface and the outer surface of the insulation, respectively. (*Hint:* Consider the heat flow through a thin section of pipe, with thickness dr, as shown in Fig. 16.16. Then integrate.)

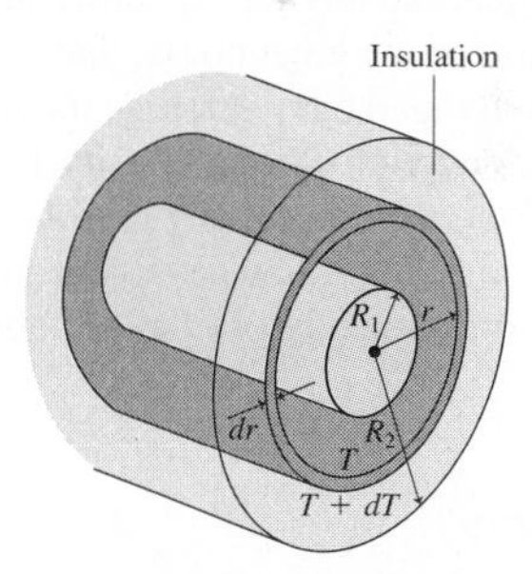

FIGURE 16.16 Problem 76

77. **ENV** A friend who's skeptical about climate change argues that the roughly 0.85°C increase in Earth's temperature during the industrial era could be caused by an increase in the Sun's power output. The Sun's average power has, in fact, increased by about 0.05% during this time. Could your friend be right?

78. Your family is winterizing its lakefront camp, and you want at least $\mathcal{R}$-19 insulation in the walls. You've got some European-made insulation with $\mathcal{R}$-factor 3.5 m^2·K/W. Will it do?

79. Your niece from Problem 56 keeps her pet rabbit in a backyard hutch with thermal resistance 0.25 K/W. On a day when the outside temperature is -15°C, she's worried that the rabbit's water will freeze, so you put a 50-W heat lamp in the hutch. Will the bunny be able to drink its water? Neglect the heat due to the animal's metabolism.

80. CH Use the method outlined in Problem 76 to show that the steady heat-flow rate in the direction of the axis of a truncated cone with conductivity k, faces of radii R_1 and R_2, and length L is $H = \pi k R_1 R_2 (T_1 - T_2)/L$. Here, T_1 and T_2 are the temperatures on the two faces, and insulation prevents any heat flow out the sides (Fig. 16.17).

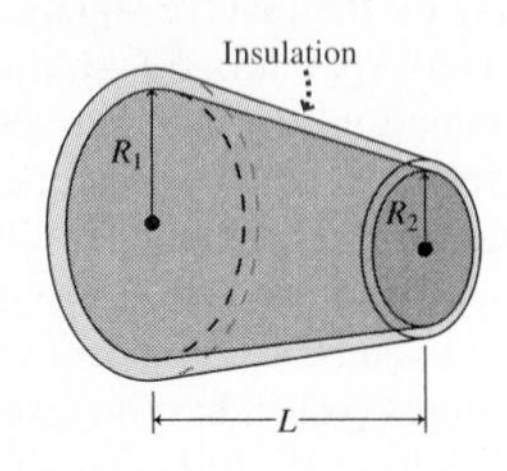

FIGURE 16.17 Problem 80

81. CH A house is at 20°C on a winter night when the outside temperature is a steady −15°C. The house's heat capacity is 6.5 MJ/K and its thermal resistance is 6.67 mK/W. If the furnace suddenly fails, how long will it take the house temperature to reach the freezing point? (*Hint:* Combine the differential forms of Equations 16.3 and 16.5 to show that the rate of temperature change is proportional to the temperature difference between the house and its surroundings. This relation is known as *Newton's law of cooling*.)

82. COMP ENV A more realistic approach to the solar greenhouse of Example 16.7 considers the time dependence of the solar input. A function that approximates the solar input is $(40\ \text{Btu/h/ft}^2)\sin^2(\pi t/24)$, where t is the time in hours, with $t = 0$ at midnight. Then the greenhouse is no longer in energy balance, but is described instead by the differential form of Equation 16.3 with Q the time-varying energy input. Use computer software or a calculator with differential-equation-solving capability to find the time-dependent temperature of the greenhouse, and determine the maximum and minimum temperatures. Assume the same numbers as in Example 16.7, along with a heat capacity $C = 1500$ Btu/°F for the greenhouse. You can assume any reasonable value for the initial temperature, and after a few days your greenhouse temperature should settle into a steady oscillation independent of the initial value.

Passage Problems

Fiberglass is a popular, economical, and fairly effective building insulation. It consists of fine glass fibers—often including recycled glass—formed loosely into rectangular slabs or rolled into blankets (Fig. 16.18). One side is often faced with heavy paper or aluminum foil. Fiberglass insulation comes in thicknesses compatible with common building materials—for example, 3.5 inch and 6 inch for wood-framed walls. Standard 6-inch fiberglass has an $\mathcal{R}$-factor of 19.

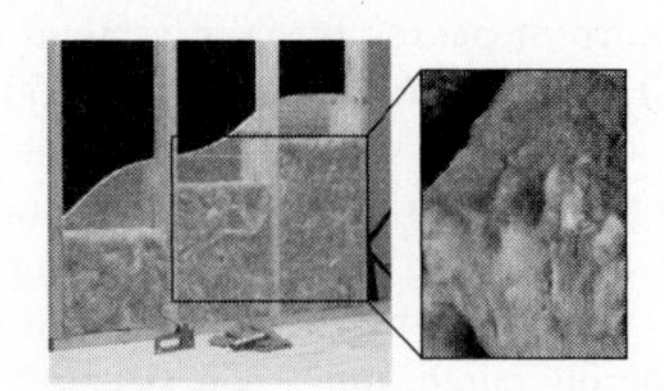

FIGURE 16.18 End view of a slab of fiberglass insulation (Passage Problems 83–86).

83. Fiberglass insulation owes its insulating quality primarily to
    a. the low thermal conductivity of glass.
    b. its ability to block cold air infiltration.
    c. the low thermal conductivity of air trapped between the glass fibers.

84. One purpose of foil facing on fiberglass insulation is to reduce heat loss by
    a. conduction.
    b. convection.
    c. radiation.

85. Fiberglass insulation for attics is available in 12-inch thickness. Its $\mathcal{R}$-factor is
    a. 38.
    b. 76.
    c. 29.

86. Since fiberglass insulation is readily compressible, you could squash two slabs initially 6 inches wide into a 6-inch wall space. This would
    a. double the overall $\mathcal{R}$-factor.
    b. increase the overall $\mathcal{R}$-factor but not double it.
    c. decrease the overall $\mathcal{R}$-factor.
    d. not change the overall $\mathcal{R}$-factor.

Answers to Chapter Questions

Answer to Chapter Opening Question

The photo is taken in infrared light, and the amount of infrared radiation increases rapidly with increasing temperature. The car's wheels are glowing with infrared, a result of frictional heating when the brakes were recently applied.

Answers to GOT IT? Questions

16.1 (b)
16.2 (a) The rock's temperature changes more because its specific heat is lower.
16.3 $\Delta T_2 < \Delta T_3 < \Delta T_1$; Since H and Δx are the same for each slab, the product $k\,\Delta T$ must be constant, so a higher conductivity means a lower ΔT.
16.4 (1) Radiation; (2) conduction; (3) convection
16.5 (b) Because as the temperature rises so does the heat-loss rate—eventually bringing the house into energy balance.

The Thermal Behavior of Matter

What You Know

- You understand the concepts of temperature, heat, and internal energy.
- You can use specific heats to determine the heat involved in temperature changes.
- You recognize three heat-transfer mechanisms: conduction, convection, and radiation.
- You can analyze systems in thermal-energy balance.

What You're Learning

- You'll gain an understanding of ideal gases based on both experiments and Newtonian physics applied to gas molecules.
- You'll see how temperature and molecular energy are intimately related.
- You'll learn about the heat involved in *phase changes* among solid, liquid, and gaseous forms of a substance and how to interpret *phase diagrams*.
- You'll explore the phenomenon of thermal expansion.

How You'll Use It

- In Chapter 18 you'll see how heat and work combine into a new statement of energy conservation.
- You'll also explore the thermodynamics of ideal gases.
- You'll see how details of molecular structure affect the thermodynamic behavior of gases.

What unusual property of water is evident in this photo?

Matter responds to heating in several ways. It may get hotter or may melt. It may change size, shape, or pressure. This chapter explores the thermal behavior of matter. We start with a simple gaseous state, whose behavior follows from Newtonian mechanics at the molecular level. We then move to liquids and solids, whose behavior is still grounded in the molecular properties of matter, but whose description is more empirical.

17.1 Gases

Gases are simple because their molecules are far apart and only rarely interact. That makes gas behavior and its physical explanation particularly straightforward. Developing that explanation will clarify the relation between macroscopic properties—such as temperature and pressure—and the underlying microscopic properties of gas molecules.

The Ideal-Gas Law

The macroscopic state of a gas in thermodynamic equilibrium is determined by its temperature, pressure, and volume. Moreover, it turns out that all gases exhibit, to a very good approximation, the same relation among these three quantities.

A simple system for studying gas behavior consists of a gas-filled cylinder sealed by a movable piston (Fig. 17.1). This is not just a pedagogical abstraction: Practical devices including engines, pumps, and air compressors contain piston–cylinder systems, while lungs, balloons, gas bubbles, and many other natural systems are analogous to our piston–cylinder system.

If we maintain the system of Fig. 17.1 at constant temperature and move the piston to vary the gas volume, we find that the pressure varies inversely with the volume. If we increase the temperature while holding the volume fixed, the pressure rises in direct

PhET: Gas Properties

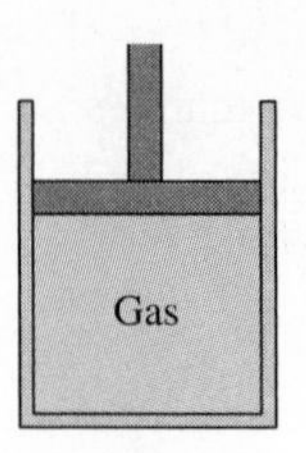

FIGURE 17.1 A piston–cylinder system.

proportion to the temperature. If we double the amount of gas while holding temperature and volume constant, the pressure doubles. Putting all these results together, we can write

$$pV = NkT \quad \text{(ideal-gas law)} \tag{17.1}$$

with p, V, and T the pressure, volume, and temperature, respectively, and N the number of molecules in the gas. Equation 17.1 is the **ideal-gas law**. Most real gases obey this law to a very good approximation. The constant k that appears in the ideal-gas law is **Boltzmann's constant**, named for the Austrian physicist Ludwig Boltzmann (1844–1906), who was instrumental in developing the microscopic description of thermal phenomena. In the upcoming revision of the SI unit system, k will be given an the exact value, close to its current value of approximately 1.38×10^{-23} J/K. Since the joule (J) can be expressed in terms of the fundamental units kilograms, meters, and seconds, the value of k will then provide an explicit-constant definition of the kelvin. That definition—relating joules and kelvins—reflects a fundamental relationship between temperature and energy, which we'll develop very soon.

Because the number of molecules N in a typical gas sample is astronomically large, we often express the ideal-gas law in terms of the number of **moles** (mol) of gas molecules. One mole is an SI unit equal to Avogadro's number, $N_A = 6.022 \times 10^{23}$, of atoms or molecules.

If we have n moles of a gas, then $N = nN_A$ is the number of molecules, so the ideal-gas law becomes

$$pV = nN_AkT = nRT \tag{17.2}$$

where $R = N_Ak = 8.314$ J/K·mol is called the **universal gas constant**.

EXAMPLE 17.1 The Ideal-Gas Law: STP

What volume is occupied by 1.00 mol of an ideal gas at standard temperature and pressure (STP), where $T = 0°\text{C}$ and $p = 101.3$ kPa?

INTERPRET We're dealing with an ideal gas, and we're given the amount of gas, the temperature, and the pressure.

DEVELOP Because we're given the number of moles n, we'll use the ideal-gas law in the form of Equation 17.2, $pV = nRT$, to find the volume.

EVALUATE Solving for V gives

$$V = \frac{nRT}{p} = \frac{(1.00\text{ mol})(8.314\text{ J/K}\cdot\text{mol})(273.15\text{ K})}{1.013 \times 10^5\text{ Pa}}$$
$$= 22.4 \times 10^{-3}\text{ m}^3 = 22.4\text{ L}$$

where we expressed $T = 0°\text{C}$ as 273.15 K.

ASSESS This result may be familiar from earlier chemistry or physics courses: 1 mole of any ideal gas—no matter what its chemical composition—occupies 22.4 L at standard temperature and pressure. ■

The ideal-gas law is remarkably simple. Neither its form nor the constants k and R depend on the substance making up the gas or on the mass of the gas molecules. Yet most real gases follow the ideal-gas law very closely over a wide range of pressures. This nearly ideal behavior is what gives gas thermometers their high precision over a wide temperature range.

Kinetic Theory of the Ideal Gas

Why do gases obey such a simple relation among temperature, pressure, and volume? Here we answer that question with an analysis based ultimately on Newtonian mechanics.

We start with some simplifying assumptions:

1. The gas consists of many identical molecules, each with mass m but negligible size and no internal structure. This assumption is approximately true for real gases when the distance between molecules is large compared with their size. This allows us to neglect intermolecular collisions, an assumption that simplifies our analysis but isn't crucial to the ideal gas.

2. The molecules don't exert action-at-a-distance forces on each other. Thus there's no intermolecular potential energy, and therefore the molecules have only kinetic energy. This assumption is fundamental to an ideal gas.
3. The molecules move in random directions with a distribution of speeds that's independent of direction.
4. Collisions with the container walls are elastic, conserving the molecules' energy and momentum. Here's where we tie our gas model to Newtonian mechanics.

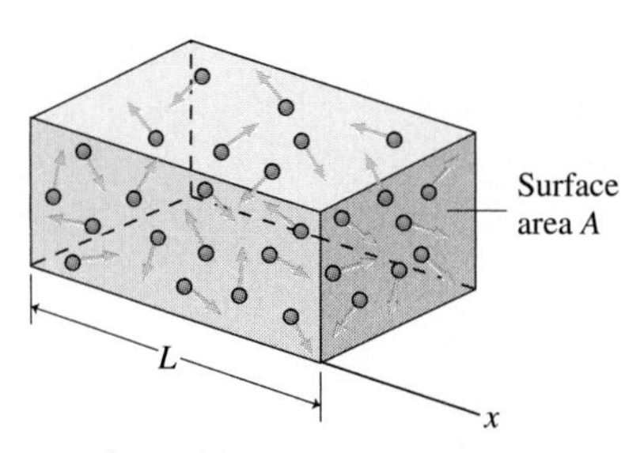

FIGURE 17.2 Gas molecules confined to a rectangular box.

Consider N molecules confined to a rectangular box with length L (Fig. 17.2). Each molecule that collides with a wall exerts a force. There are so many molecules that individual collisions aren't evident; instead the wall experiences an essentially constant average force. The gas pressure p is a measure of this force on a unit area. We're going to find an expression for p and use it to gain deep insights into the ideal-gas law and the meaning of temperature.

Figure 17.3 shows one molecule colliding with the right-hand wall. Since the collision is elastic, the y-component of the molecule's velocity is unchanged, while the x-component reverses sign. Thus the molecule undergoes a momentum change of magnitude $2mv_{xi}$, where i labels this particular molecule. After the molecule collides with the right-hand wall, nothing will change its x velocity until it hits the left-hand wall and its x velocity again reverses. So it will be back at the right-hand wall in the time $\Delta t_i = 2L/v_{xi}$ that it takes to go back and forth along the container.

Now each time our molecule collides with the right-hand wall, it delivers momentum $2mv_{xi}$ to the wall. Newton's second law says that force is the rate of change of momentum. So we can calculate the average force $\overline{F_i}$ due to one molecule by dividing the momentum delivered, $2mv_{xi}$, by the time, $2L/v_{xi}$, between collisions:

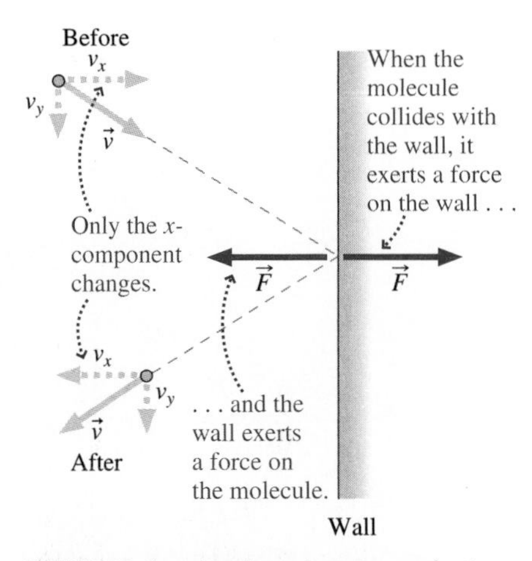

FIGURE 17.3 A molecule undergoes an elastic collision, reversing its x-component and transferring momentum $2mv_x$ to the wall.

$$\overline{F_i} = \frac{2mv_{xi}}{2L/v_{xi}} = \frac{mv_{xi}^2}{L}$$

To get the total force on the wall, we sum over all N molecules with their different x velocities. Dividing by the wall area A then gives the pressure:

$$p = \frac{\overline{F}}{A} = \frac{\sum \overline{F_i}}{A} = \frac{\sum mv_{xi}^2/L}{A} = \frac{m\sum v_{xi}^2}{AL}$$

The last step follows because the box length L and molecular mass m are the same for all molecules, so they factor out of the sum. We can simplify by noting that the denominator AL is just the volume V. Let's also multiply by 1 in the form N/N, with N the number of molecules. Then we have

$$p = \frac{m\sum v_{xi}^2}{AL} = \frac{mN}{V}\frac{\sum v_{xi}^2}{N}$$

In the final expression here, the term $\sum v_{xi}^2/N$ is the average of the squares of all the x velocity components of all the molecules; we designate this quantity $\overline{v_x^2}$. So the pressure becomes

$$p = \frac{mN}{V}\overline{v_x^2}$$

We still haven't used assumption 3—that the molecules move in random directions with speeds independent of direction. If we grab a molecule at random, that means we're just as likely to find it moving in the x-direction, the y-direction, the z-direction, or any direction in between—and its speed, on average, won't depend on its direction of motion. So the average quantities $\overline{v_x^2}$, $\overline{v_y^2}$, and $\overline{v_z^2}$ must be equal. Since the three directions x, y, and z are perpendicular, the average of the molecular speeds squared is $\overline{v^2} = \overline{v_x^2} + \overline{v_y^2} + \overline{v_z^2}$. We've just argued that all three terms on the right are equal, so we can write $\overline{v^2} = 3\overline{v_x^2}$, or $\overline{v_x^2} = \frac{1}{3}\overline{v^2}$. Then our expression for pressure becomes

$$p = \frac{mN}{3V}\overline{v^2}$$

Multiplying through by V and by 1 in the form 2/2, we have

$$pV = \tfrac{2}{3}N\left(\tfrac{1}{2}m\overline{v^2}\right)$$

This looks a lot like the ideal-gas law (Equation 17.1), except that instead of kT we have $\frac{2}{3}\left(\frac{1}{2}m\overline{v^2}\right)$. Take a good look at the quantity in parentheses: You'll see that it's just the average kinetic energy of a gas molecule.

Think about what we've done here. We applied the fundamental laws of mechanics to an ideal gas and came up with an equation that looks like the experimentally verified ideal-gas law, except that it's expressed in terms of a microscopic quantity—molecular kinetic energy—rather than the macroscopic quantity temperature. Since our equation describes the behavior of an ideal gas, it *must be* the ideal-gas law. Comparing with the ideal-gas law in the form 17.1, we must therefore have

$$\tfrac{1}{2}m\overline{v^2} = \tfrac{3}{2}kT \quad \text{(temperature and molecular energy)} \qquad (17.3)$$

Our derivation shows why, in terms of Newtonian mechanics, a gas obeying our four assumptions should obey the ideal-gas law. In Equation 17.3 we get an added bonus—a microscopic understanding of the meaning of temperature: **Temperature measures the average kinetic energy associated with random translational motion of the molecules.**

This fundamental connection between temperature and energy is what lies behind the upcoming redefinition of the kelvin in terms of Boltzmann's constant. In Chapter 18 you'll see how, with more complex molecules, we need to broaden energy here to include other forms of molecular energy in addition to translational kinetic energy.

EXAMPLE 17.2 Molecular Energy and Speed: An Air Molecule

Find the average kinetic energy of a molecule in air at room temperature (20°C or 293 K), and determine the speed of a nitrogen molecule (N_2) with this energy.

INTERPRET This problem asks about the linkage between thermodynamic quantities and molecular energy. We just found that linkage: The temperature of a gas is a measure of the average kinetic energy of its molecules.

DEVELOP Equation 17.3, $\frac{1}{2}m\overline{v^2} = \frac{3}{2}kT$, quantifies the relation between temperature and molecular kinetic energy. Once we find the molecular kinetic energy, we'll need the molecular mass to determine the speed. We can get that using the atomic weight of nitrogen and the fact that an N_2 molecule contains two atoms.

EVALUATE We first evaluate the average molecular kinetic energy:

$$\overline{K} = \tfrac{1}{2}m\overline{v^2} = \tfrac{3}{2}kT = \tfrac{3}{2}(1.38\times10^{-23}\ \text{J/K})(293\ \text{K}) = 6.07\times10^{-21}\ \text{J}$$

We can solve for the corresponding speed if we know the molecular mass m. A nitrogen molecule consists of two atoms each with mass 14 u (see Appendix D), so its mass is

$$m = 2(14\ \text{u})(1.66\times10^{-27}\ \text{kg/u}) = 4.65\times10^{-26}\ \text{kg}$$

Since $\overline{K} = \frac{1}{2}m\overline{v^2}$, the speed corresponding to this kinetic energy is

$$v = \sqrt{\frac{2K}{m}} = \sqrt{\frac{2(6.07\times10^{-21}\ \text{J})}{4.65\times10^{-26}\ \text{kg}}} = 511\ \text{m/s}$$

ASSESS Make sense? Not surprisingly, the answer is the same order of magnitude as the speed of sound ($\sim$340 m/s) in air at room temperature. At the microscopic level, the speed of the individual molecules limits the rate at which information can be transmitted by disturbances—sound waves—propagating through the gas. ■

We call the speed calculated in Example 17.2 the **thermal speed**. In terms of temperature, Equation 17.3 shows

$$v_{\text{th}} = \sqrt{\frac{3kT}{m}} \qquad (17.4)$$

GOT IT? 17.1 If you double the kelvin temperature of a gas, what happens to the thermal speed of the gas molecules? (a) it doubles; (b) it quadruples; (c) it increases by $\sqrt{2}$

The Distribution of Molecular Speeds

The thermal speed v_{th} is a typical molecular speed, but it doesn't tell us much about the distribution of speeds. Are molecular speeds limited to a narrow band about v_{th}? Or are lots of molecules moving much faster or much slower?

In the 1860s, the Scottish physicist James Clerk Maxwell showed that elastic collisions among molecules result in a speed distribution that peaks near the thermal speed but may extend considerably higher. Figure 17.4 plots this **Maxwell–Boltzmann distribution** for two different temperatures. Note that increasing temperature results in a higher thermal speed, as expected, but that it also broadens the distribution so there are more molecules at lower and higher speeds. The high-speed "tail" of the distribution is especially important to chemists because high-energy molecules participate most readily in chemical reactions. The rapid extension of the high-energy tail with increasing temperature shows why reaction rates are strongly temperature sensitive, and therefore explains why foods keep much longer with even modest refrigeration. High-energy molecules are also the first to evaporate from a liquid, leaving slower, cooler molecules behind. This explains the phenomenon of evaporative cooling, which your own body uses as you sweat. Without evaporative cooling, Earth's atmosphere would be much drier and it would rain far less frequently. You can explore the Maxwell–Boltzmann distribution quantitatively in Problem 76.

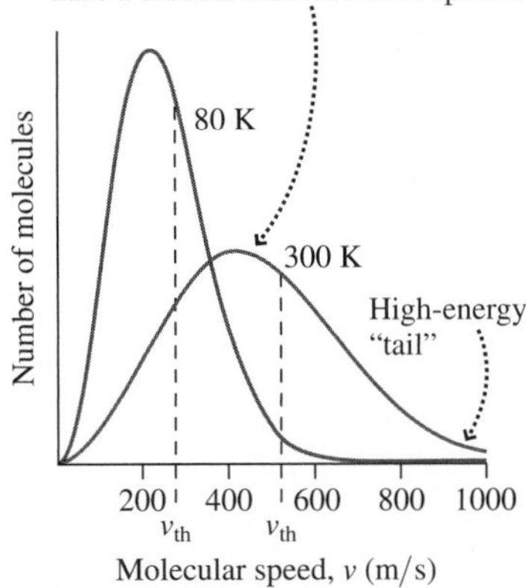

FIGURE 17.4 Maxwell–Boltzmann distribution of molecular speeds for nitrogen (N_2) at temperatures of 80 K and 300 K.

Real Gases

The ideal-gas law is a good approximation to the behavior of most real gases, but it's not perfect because our assumptions aren't entirely realistic. Two factors are especially important. First, real molecules take up space. This reduces the available volume, altering the ideal-gas law. Second, electrical effects that we'll explore in Chapter 20 result in a weak attractive force between nearby molecules. As they move apart, molecules do work against this **van der Waals force**, and their kinetic energy drops. This, too, results in a deviation from ideal-gas behavior. You can learn more about these effects by working Problem 77.

17.2 Phase Changes

PhET: States of Matter

Step out of a steamy shower, and you'll find the mirror fogged with water condensed on the cool glass. Climb a mountain in winter, and you'll be treated to the lovely spectacle of every branch and pine needle covered with a delicate coating of frost that's formed right from the air. Burn a rewritable CD or DVD, and you've stored information with a laser that melts tiny spots on the spinning disc. These examples involve **phase changes** between gas and liquid, gas and solid, and solid and liquid.

Heat and Phase Changes

Drop ice cubes into a drink and stir. What's the temperature of the drink? It's 0°C, and it stays at 0°C as long as any ice remains. The melting of a pure solid occurs at a fixed temperature. During the process, energy goes into breaking the molecular bonds that hold the material in its solid form. This increases the molecules' potential energy but not their kinetic energy. Since temperature is a measure of molecular kinetic energy, that means the temperature doesn't change either.

The energy per unit mass required to change phase is called a **heat of transformation** and is given the symbol L; for the solid–liquid change it's the **heat of fusion** L_f, and for liquid–gas it's the **heat of vaporization** L_v, Less familiar is the **heat of sublimation** for the transition from solid directly to gas. These quantities have units of J/kg, so the energy required to change the phase of a mass m is

$$Q = Lm \quad \text{(heat of transformation)} \tag{17.5}$$

To reverse the change requires removing the same energy. Table 17.1 lists heats of transformation for some common materials. These quantities are typically quite large; water's heat of fusion, for example, is 334 kJ/kg or 80 cal/g—meaning it takes as much energy to melt 1 gram of ice as to heat the resulting water to 80°C.

Table 17.1 Heats of Transformation (at Atmospheric Pressure)

Substance	Melting Point (K)	L_f (kJ/kg)	Boiling Point (K)	L_v (kJ/kg)
Alcohol, ethyl	159	109	351	879
Copper	1357	205	2840	4726
Lead	601	24.7	2013	858
Mercury	234	11.3	630	296
Oxygen	54.8	13.8	90.2	213
Sulfur	388	53.6	718	306
Water	273.15	334	373.15	2257
Uranium dioxide	3120	259	3815	1533

CONCEPTUAL EXAMPLE 17.1 Water Phases

You put a block of ice initially at $-20°C$ in a pan on a hot stove with a constant power output, and heat it until it has melted, boiled, and evaporated. Make a sketch of temperature versus time for this experiment.

EVALUATE As the ice starts heating, its temperature goes up, so our graph (Fig. 17.5) begins with an upward slope. At 0°C the ice starts melting, and while that's happening its temperature doesn't change, so the graph stays horizontal for a while. When the ice is all melted, the water starts to warm. Table 16.1 shows that liquid water's specific heat is about twice that of ice; given the same power input, that means the water heats more slowly than the ice. So our graph has a lower slope as the water goes from 0°C to the boiling point at 100°C. Then it starts turning to vapor, and stays at 100°C until it's all evaporated. Table 17.1 shows that water's heat of vaporization is much greater than its heat of fusion, so it takes much more time to boil the water away than it did to melt the ice. Our graph reflects that time difference.

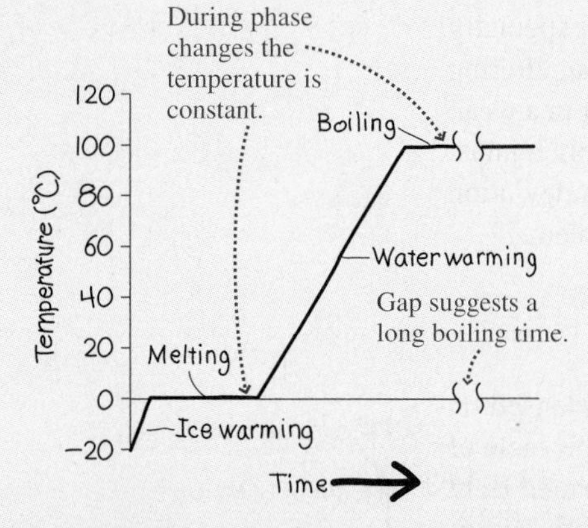

FIGURE 17.5 Temperature versus time for what's initially a block of ice at $-20°C$, supplied with energy at a constant rate. The process takes place at atmospheric pressure.

ASSESS Makes sense: It takes a lot longer to boil a pan dry than to bring it to a boil.

MAKING THE CONNECTION If you start with 0.95 kg of ice at $-20°C$ and supply heat at the rate of 1.6 kW, how much time will it take until you're left with only water vapor?

EVALUATE Use Equation 16.3 for heating, with specific heats from Table 16.1. Use Equation 17.4 for phase changes, with heats of transformation from Table 17.1. The result is 2.9 MJ of heat required for the whole process; at 1.6 kW or 1.6 kJ/s, that takes 1.8 ks, or half an hour.

GOT IT? 17.2 You bring a pot of water to boil and then forget about it. Ten minutes later you come back to the kitchen to find the water still boiling. Is its temperature (a) less than, (b) greater than, or (c) equal to 100°C?

EXAMPLE 17.3 The Heat of Fusion: Meltdown!

A nuclear power plant's reactor vessel cracks, and all the cooling water drains out. Although nuclear fission stops, radioactive decay continues to heat the reactor's 2.5×10^5 kg of uranium-dioxide fuel at the rate of 120 MW. Once the melting point is reached, how much energy will it take to melt the fuel? How long will this take?

INTERPRET Since this problem is about melting, it must involve the heat of fusion. We identify the material as uranium dioxide (UO_2).

DEVELOP Our plan is to find UO_2's heat of fusion in Table 17.1 and then use Equation 17.5, $Q = Lm$, to calculate the energy required for melting. We're given the rate of energy generation by radioactive decay, and from that we'll be able to get the time.

EVALUATE Using UO_2's L_f value from Table 17.1 in Equation 17.5, we have

$$Q = L_f m = (259 \text{ kJ/kg})(2.5 \times 10^5 \text{ kg}) = 65 \text{ GJ}$$

With a heating rate of 120 MW or 0.12 GJ/s, the time to melt the fuel is $(65 \text{ GJ})/(0.12 \text{ GJ/s}) = 542$ s.

ASSESS The time to meltdown is just under 10 minutes! Failsafe emergency cooling systems are essential to prevent nuclear meltdowns. ■

Often we're interested in the total energy needed to bring a material to its transition point and then to make the phase transition. Then we need to combine specific-heat considerations of Chapter 16 with the heats of transformation introduced here.

EXAMPLE 17.4 Heating and Phase Change: Enough Ice?

When 200 g of ice at $-10°C$ are added to 1.0 kg of water at 15°C, is there enough ice to cool the water to 0°C? If so, how much ice is left in the mixture?

INTERPRET This problem involves both a temperature rise and a phase change. We identify water as the substance involved.

DEVELOP Equation 16.3, $Q = mc\,\Delta T$, determines the energy for the temperature rise, and Equation 17.5, $Q = Lm$, determines the phase-change energy. But we don't know whether all the ice melts. So our plan is to find the energy that it *would* take to heat the ice to 0°C and then melt all of it; if *more* than that much is available in cooling the water to 0°C, we'll know that we end up with all water at $T > 0°C$. But if there isn't sufficient energy, then we'll have a mixture with both ice and water at 0°C, and we can use the energy extracted in cooling the water to find out how much ice melts.

EVALUATE We begin by evaluating the energy Q_1 to heat the ice and then melt it all, adding the energies from Equations 16.3 and 17.5 and then getting the specific heat and heat of fusion from Tables 16.1 and 17.1, respectively:

$$\begin{aligned} Q_1 &= m_{\text{ice}}c_{\text{ice}}\,\Delta T_{\text{ice}} + m_{\text{ice}}L_{\text{f}} \\ &= (0.20\text{ kg})(2.05\text{ kJ/kg}\cdot\text{K})(10\text{ K}) + (0.20\text{ kg})(334\text{ kJ/kg}) \\ &= 4.1\text{ kJ} + 66.8\text{ kJ} = 70.9\text{ kJ} \end{aligned}$$

Cooling the water to 0°C would extract energy Q_2 given by Equation 16.3:

$$Q_2 = m_{\text{water}}c_{\text{water}}\Delta T_{\text{water}} = (1.0\text{ kg})(4.184\text{ kJ/kg}\cdot\text{K})(15\text{ K}) = 62.8\text{ kJ}$$

This is far more than the 4.1 kJ needed to bring the ice to 0°C, but not quite the 70.9 kJ needed to leave it all melted. So there's enough ice to cool the water to 0°C, with some left over. How much? Our calculation of Q_1 shows that 4.1 kJ go into raising the ice temperature. Of the 62.8 kJ extracted from the water, the remaining 58.7 kJ go to melting ice. From Equation 17.5, the amount of ice melted is then

$$m_{\text{melted}} = \frac{Q}{L_{\text{f}}} = \frac{58.7\text{ kJ}}{334\text{ kJ/kg}} = 0.176\text{ kg} = 176\text{ g}$$

So we're left with 24 g of ice in 1176 g of water, all at 0°C.

ASSESS Make sense? Our 62.8 kJ was nearly enough to bring all the ice to the liquid phase, so it makes sense that only a small fraction of the ice remains. ■

Phase Diagrams

Why can't mountaineers enjoy piping hot coffee? Because water's boiling point drops with the decreasing pressure at high altitudes. In general, the temperatures at which phase changes occur depend on pressure. A **phase diagram** shows the different phases on a plot of pressure versus temperature. Figure 17.6 is a phase diagram for a typical substance. Most phase diagrams are similar, although water's is slightly unusual for reasons we'll discuss in the next section.

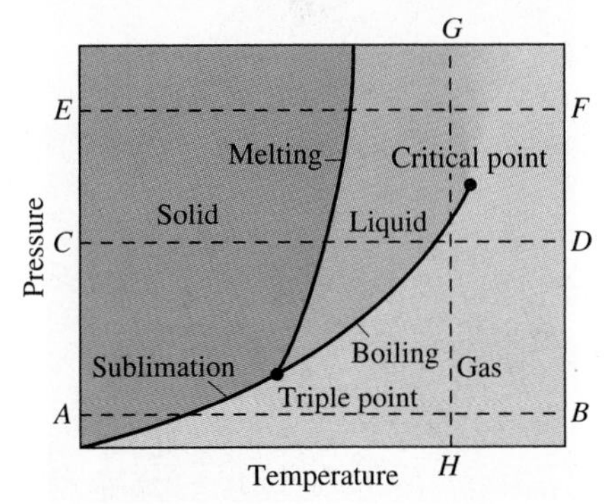

FIGURE 17.6 A phase diagram showing solid, liquid, and gas phases on a plot of pressure versus temperature.

The phase diagram divides pressure–temperature space into regions corresponding to solid, liquid, and gas phases. Lines separating these regions mark the phase transitions. Everyday experience suggests that heating takes a substance from solid, to liquid, to gas—as with water in Fig. 17.5. But Fig. 17.6 shows that this sequence doesn't always occur. At low pressure (line *AB* in Fig. 17.6) the substance goes directly from solid to gas. This is **sublimation**. We don't see this with water because normal atmospheric pressure is too high. For carbon dioxide, though, atmospheric pressure is low in the phase diagram, which is why "dry ice" turns directly into gaseous CO_2 without becoming liquid. At higher pressures (line *CD*) we get the familiar solid–liquid–gas sequence. Higher still (line *EF*), we're above the **critical point**, where the abrupt distinction between liquid and gas disappears. Instead, the substance starts out as a thick fluid whose properties change gradually from liquidlike to gaslike as it's heated.

We think of changing phase by applying heat, but Fig. 17.6 shows we can also change phase by changing pressure. Lowering pressure along line *GH*, for example, takes the substance from liquid to gas while the temperature remains constant. Since heat requires a temperature difference, there's no heat involved in this constant-temperature phase transition. You may have seen a demonstration of water boiling vigorously at room temperature in a closed container pumped down to low pressure.

Don't let Fig. 17.6 fool you into thinking that phase transitions occur instantaneously. Those heats of transformation are large, and a substance moving, say, along line CD in response to heating will linger at each phase transition until all of it has changed phase; that's what the level portions of Fig. 17.5 showed.

The dividing curves in Fig. 17.6 show where two phases can coexist simultaneously, like ice floating in water at 0°C and atmospheric pressure. It's because phase changes occur along curves that terms like "melting point" and "boiling point" are meaningless unless pressure is specified. But there's one unique **triple point** where solid, liquid, and gas all coexist in equilibrium. Here temperature and pressure have unique, unambiguous values—which is why the 273.16-K triple point of water helps provide an operational definition of the kelvin.

17.3 Thermal Expansion

We've seen how heating causes changes in temperature and phase. But heating also results in pressure or volume changes. For a gas at constant pressure, for example, the ideal-gas law shows that volume increases in direct proportion to temperature. The volume and pressure relations for liquids and solids aren't so simple. Because their molecules are closely spaced, liquids and solids aren't very compressible, so thermal expansion is less pronounced.

We characterize the change in the volume with temperature using the **coefficient of volume expansion** β, defined as the fractional change in volume when a substance undergoes a small temperature change ΔT:

$$\beta = \frac{\Delta V/V}{\Delta T} \tag{17.6}$$

This equation assumes that β is independent of temperature; if it varies significantly, then we would need to define β in terms of the derivative dV/dT (Problem 68). Our definition of β also assumes constant pressure; we could entirely inhibit thermal expansion with appropriate pressure increases.

Often we want to know how one linear dimension of a solid changes with temperature. This is especially true with long structures, where the absolute change is greatest along the long dimension (Fig. 17.7). We then speak of the **coefficient of linear expansion** α, defined by

FIGURE 17.7 Thermal expansion distorted these tracks, causing a derailment. Expansion of long structures like this is best described using the coefficient of linear expansion.

$$\alpha = \frac{\Delta L/L}{\Delta T} \tag{17.7}$$

The volume- and linear-expansion coefficients are related in a simple way: $\beta = 3\alpha$, as you can show in Problem 71. However, the linear-expansion coefficient α is really meaningful only with solids, because liquids and gases deform and don't expand proportionately in all directions. Table 17.2 lists the expansion coefficients for some common substances.

Table 17.2 Expansion Coefficients*

Solids	α (K^{-1})	Liquids and Gases	β (K^{-1})
Aluminum	24×10^{-6}	Air	3.7×10^{-3}
Brass	19×10^{-6}	Alcohol, ethyl	75×10^{-5}
Copper	17×10^{-6}	Gasoline	95×10^{-5}
Glass (Pyrex)	3.2×10^{-6}	Mercury	18×10^{-5}
Ice	51×10^{-6}	Water, 1°C	-4.8×10^{-5}
Invar†	0.9×10^{-6}	Water, 20°C	20×10^{-5}
Steel	12×10^{-6}	Water, 50°C	50×10^{-5}

*At approximately room temperature unless noted.
†Invar, consisting of 64% iron and 36% nickel, is an alloy designed to minimize thermal expansion.

GOT IT? 17.3 The figure shows a donut-shaped object. If it's heated, will the hole get (a) larger or (b) smaller?

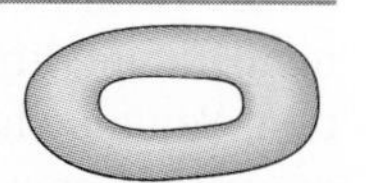

EXAMPLE 17.5 Thermal Expansion: Spilled Gasoline

A steel gas can holds 20 L at 10°C. It's filled to the brim with gas at 10°C. If the temperature now increases to 25°C, by how much does the can's volume increase? How much gas spills out?

INTERPRET This is a problem about thermal expansion. Since it involves volume, we identify the relevant quantity as the coefficient of volume expansion β.

DEVELOP Equation 17.6, $\beta = (\Delta V/V)/\Delta T$, determines the volume change. Our plan is to calculate the expanded volume of the tank and then of the gasoline. The difference will be the amount that spills out. Table 17.2 lists β for gasoline but α for steel; therefore, we'll use the equation $\beta = 3\alpha$ for the steel.

EVALUATE First we use Equation 17.6 to evaluate the volume change ΔV of the steel can. Using $\beta = 3\alpha$, we have

$$\Delta V_{\text{can}} = \beta V \Delta T = (3)(12\times10^{-6}\,\text{K}^{-1})(20\,\text{L})(15\,\text{K}) = 0.0108\,\text{L}$$

Similarly, for the gasoline,

$$\Delta V_{\text{gas}} = \beta V \Delta T = (95\times10^{-5}\,\text{K}^{-1})(20\,\text{L})(15\,\text{K}) = 0.285\,\text{L}$$

We therefore lose 0.275 L.

ASSESS Make sense? The thermal-expansion coefficient for gasoline is so much greater than for steel that the can's expansion is negligible and the gas has nowhere to go. By the way, that spill wastes nearly 10 MJ of energy! ■

Thermal Expansion of Water

The entry for water at 1°C in Table 17.2 is remarkable, the negative expansion coefficient showing that water at this temperature actually *contracts* on heating. This unusual behavior occurs because ice has a relatively open crystal structure (Fig. 17.8) and therefore is less dense than liquid water. That's why ice floats. Immediately above the melting point, the intermolecular forces that bond H_2O molecules in ice still exert an influence, giving cold liquid water a lower density than at slightly higher temperatures. At 4°C water reaches its maximum density, and above this temperature the effect of molecular kinetic energy in keeping molecules apart wins out over intermolecular forces. From there on, water exhibits the more normal behavior of expansion with increasing temperature.

This unusual property of water near its melting point is reflected in its phase diagram, shown in Fig. 17.9. Note that the solid–liquid boundary extends leftward from the triple point, in contrast to the more typical behavior in Fig. 17.6. That means that ice at a fixed temperature will melt if the pressure is *increased*—an unusual property known as pressure melting.

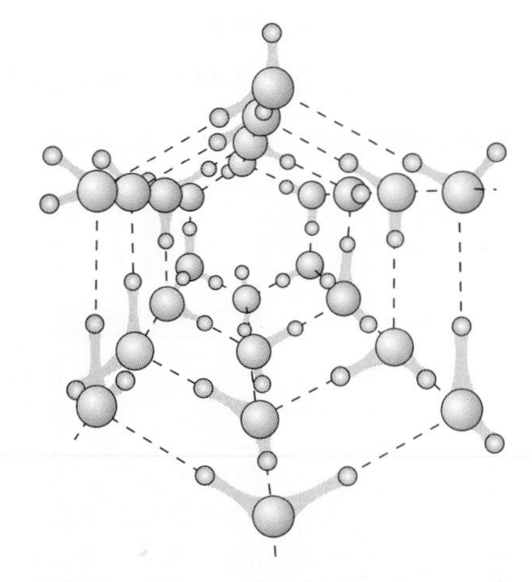

FIGURE 17.8 Water molecules in an ice crystal form an open structure, giving solid water a lower density than the liquid.

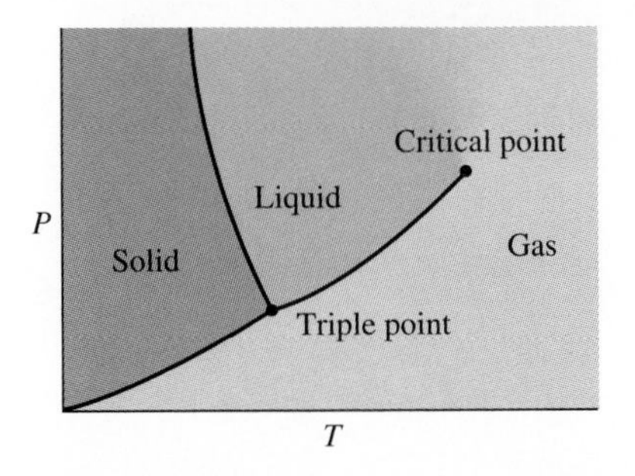

FIGURE 17.9 Phase diagram for water. Compare the solid–liquid boundary with that of Fig. 17.6.

APPLICATION Aquatic Life and Lake Turnover

The anomalous behavior of water has important consequences for life. If ice didn't float, then ponds, lakes, and even oceans would freeze solid from the bottom up, making aquatic life impossible. What actually happens, instead, is that a thin layer of ice forms on the surface, insulating the water below and keeping it liquid; as a result, ice cover in temperate climates rarely exceeds a meter or so. Because water's density is greatest at 4°C, water at this temperature sinks to the bottom. At lake depths greater than a few meters, sunlight is inadequate to raise the temperature, which therefore remains year-round at 4°C.

Water's unusual density behavior also causes the twice-yearly turnover of lakes in temperate climates. In the summer, a lake's surface water is warm, but deep water remains at 4°C. In the winter, water just beneath the ice is at 0°C, while the bottom water is still at 4°C. Both situations are stable, with less dense and therefore more buoyant water at the surface. But in the spring, ice melts and the surface water warms. When that water reaches 4°C, there's no density variation and the lake water mixes freely. This is the spring overturning. A similar overturning occurs in the fall, as the surface water cools through 4°C. Turnover is important to aquatic life because it brings up nutrients that would otherwise be trapped in the deep water.

CHAPTER 17 SUMMARY

Big Idea

The big idea here is that matter responds to heating in a variety of ways in addition to changing temperature. Other responses include changes of phase and of volume and/or pressure. The ideal gas provides a particularly simple system for understanding volume and pressure changes. Analyzing ideal-gas behavior provides a link between the Newtonian mechanics of molecules and macroscopic thermodynamics, showing that temperature is a measure of the average molecular kinetic energy.

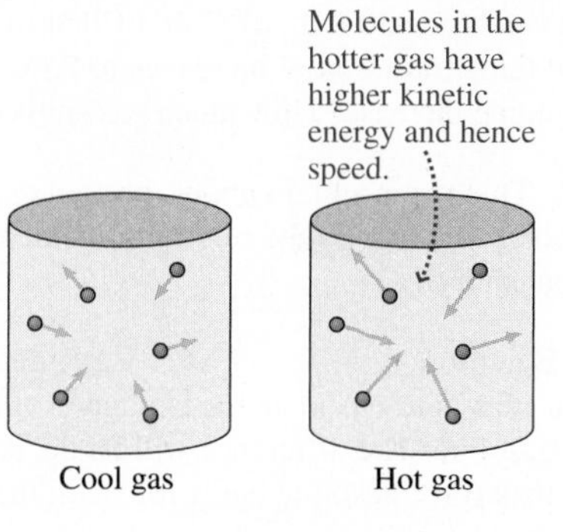

Key Concepts and Equations

The **ideal-gas law** relates pressure, volume, temperature, and the number of molecules in a gas:

$$pV = NkT \quad \text{(ideal-gas law)}$$

where **Boltzmann's constant** k is approximately 1.381×10^{-23} J/K.

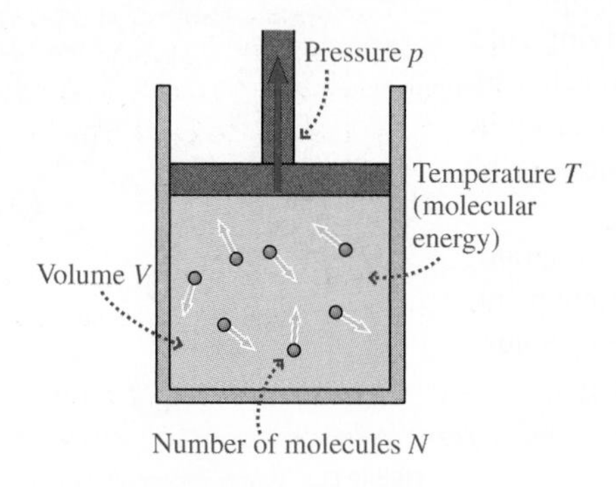

In terms of the number of moles n, the ideal-gas law is

$$pV = nN_AkT = nRT$$

where the **universal gas constant** $R = N_Ak = 8.314$ J/K·mol.

The temperature of an ideal gas is a measure of the gas molecules' average kinetic energy:

$$\tfrac{1}{2}m\overline{v^2} = \tfrac{3}{2}kT \quad \text{(temperature and molecular energy)}$$

Heats of transformation L describe the energy per unit mass needed to effect phase changes. The total energy required to change the phase of a mass m is given by

$$Q = Lm \quad \text{(heat of transformation)}$$

Phase diagrams plot solid, liquid, and gas phases against temperature and pressure, and reveal the **triple point**, where all three phases can coexist, and the **critical point**, where the liquid–gas distinction disappears.

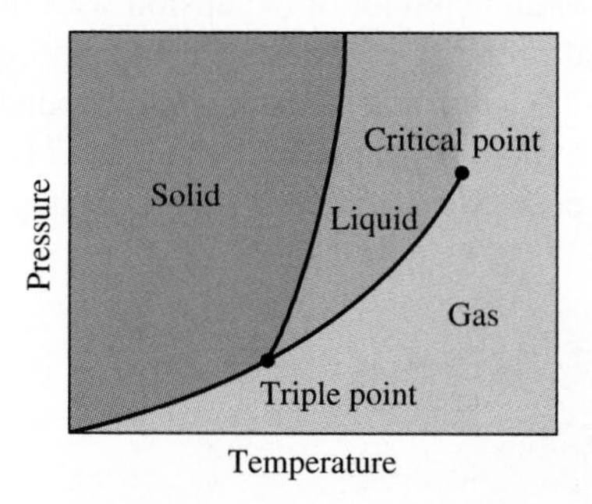

Applications

Thermal expansion is characterized by the **coefficient of volume expansion** and its linear counterpart. The volume-expansion coefficient relates the fractional volume change $\Delta V/V$ to the temperature change ΔT:

$$\beta = \frac{\Delta V/V}{\Delta T} \quad \text{(volume-expansion coefficient)}$$

while the **coefficient of linear expansion** relates the fractional length $\Delta L/L$ change to ΔT:

$$\alpha = \frac{\Delta L/L}{\Delta T} \quad \text{(linear-expansion coefficient)}$$

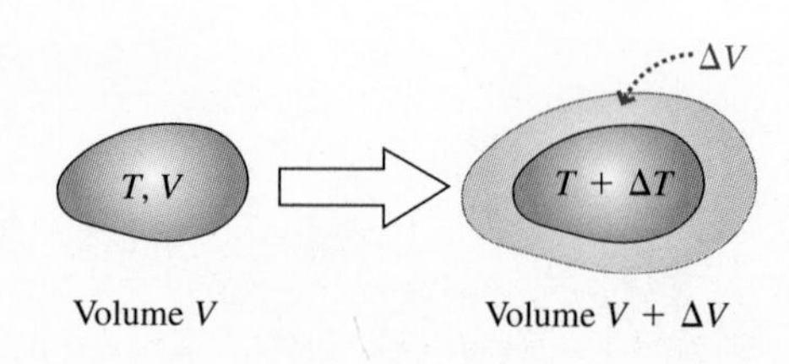

BIO *Biology and/or medicine-related problems* **DATA** *Data problems* **ENV** *Environmental problems* **CH** *Challenge problems* **COMP** *Computer problems*

For Thought and Discussion

1. If the volume of an ideal gas is increased, must the pressure drop proportionately? Explain.
2. According to the ideal-gas law, what should be the volume of a gas at absolute zero? Why is this result absurd?
3. Why are you supposed to check tire pressure when your tires are cold?
4. The average *speed* of the molecules in a gas increases with increasing temperature. What about the average *velocity*?
5. Suppose you start running while holding a closed jar of air. Do you change the average speed of the air molecules? The average velocity? The temperature?
6. Two different gases are at the same temperature, and both have low enough densities that they behave like ideal gases. Do their molecules have the same thermal speeds? Explain.
7. Your roommate claims that ice and snow must be at 0°C. Is that true?
8. What's the temperature of water just under the ice layer of a frozen lake? At the bottom of a deep lake?
9. Ice and water have been together in a glass for a long time. Is the water hotter than the ice?
10. Which takes more heat: melting a gram of ice already at 0°C, or bringing the melted water to the boiling point?
11. The atmospheres of relatively low-mass planets like Earth don't contain much hydrogen (H_2), while more massive planets like Jupiter have considerable atmospheric hydrogen. What factors might account for the difference?
12. The triple point of water defines a precise temperature, but the freezing point doesn't. Why the difference?
13. How is it possible to have boiling water at a temperature other than 100°C?
14. How does a pressure cooker work?
15. Suppose mercury and glass had the same coefficient of volume expansion. Could you build a mercury thermometer?
16. A bimetallic strip consists of thin pieces of brass and steel bonded together (Fig. 17.10). What happens when the strip is heated? (*Hint:* Consult Table 17.2.)

FIGURE 17.10 For Thought and Discussion 16

Exercises and Problems

Exercises

Section 17.1 Gases

17. Mars's atmospheric pressure is about 1% that of Earth, and its average temperature is around 215 K. Find the volume of 1 mol of the Martian atmosphere.
18. How many molecules are in an ideal-gas sample at 350 K that occupies 8.5 L when the pressure is 180 kPa?
19. What's the pressure of an ideal gas if 3.5 mol occupy 2.0 L at −150°C?
20. Your professor asks you to order a tank of argon gas for a lab experiment. You obtain a "type C" gas cylinder with interior volume 6.88 L. The supplier claims it contains 45 mol of argon. You measure its pressure to be 14 MPa at room temperature (20°C). Did you get what you paid for?
21. (a) If 2.0 mol of an ideal gas are initially at temperature 250 K and pressure 1.5 atm, what's the gas volume? (b) The pressure is now increased to 4.0 atm, and the gas volume drops to half its initial value. What's the new temperature?
22. A pressure of 10^{-10} Pa is readily achievable with laboratory vacuum apparatus. If the residual air in this "vacuum" is at 0°C, how many air molecules are in 1 L?
23. What's the thermal speed of hydrogen molecules at 800 K?
24. In which gas are the molecules moving faster: hydrogen at 75 K or sulfur dioxide at 350 K?

Section 17.2 Phase Changes

25. How much energy does it take to melt a 65-g ice cube?
26. It takes 200 J to melt an 8.0-g sample of one of the substances in Table 17.1. What's the substance?
27. If it takes 840 kJ to vaporize a sample of liquid oxygen, how large is the sample?
28. Carbon dioxide *sublimes* (changes from solid to gas) at 195 K. The heat of sublimation is 573 kJ/kg. How much heat must be extracted from 250 g of CO_2 gas at 195 K in order to solidify it?
29. Find the energy needed to convert 28 kg of liquid oxygen at its boiling point into gas.

Section 17.3 Thermal Expansion

30. A copper wire is 20 m long on a winter day when the temperature is −12°C. By how much does its length increase on a 26°C summer day?
31. You have exactly 1 L of ethyl alcohol at room temperature (20°C). You put it in a refrigerator at 2°C. What's its new volume?
32. A Pyrex glass marble is 1.00000 cm in diameter at 20°C. What will be its diameter at 85°C?
33. At 0°C, the hole in a steel washer is 9.52 mm in diameter. To what temperature must it be heated in order to fit over a 9.55-mm-diameter bolt?
34. Suppose a single piece of welded steel railroad track stretched 5000 km across the continental United States. If the track were free to expand, by how much would its length change if the entire track went from a cold winter temperature of −25°C to a hot summer day at 40°C?

Problems

35. The solar corona is a hot (2 MK) extended atmosphere surrounding the Sun's cooler visible surface. The coronal gas pressure is about 0.03 Pa. What's the coronal density in particles per cubic meter? Compare with Earth's atmosphere.
36. A helium balloon occupies 8.0 L at 20°C and 1.0-atm pressure. The balloon rises to an altitude where the air pressure is 0.65 atm and the temperature is −10°C. What's its volume when it reaches equilibrium at the new altitude?
37. A compressed air cylinder stands 100 cm tall and has internal diameter 20.0 cm. At room temperature, the pressure is 180 atm. (a) How many moles of air are in the cylinder? (b) What volume would this air occupy at 1.0 atm and room temperature?
38. You're a lawyer with an unusual case. A whipped-cream can burst at a wedding, damaging the groom's expensive tuxedo.

The can warned against temperatures in excess of 50°C, and the manufacturer has evidence that it reached 60°C. You don't contest this, but you point out that the can was only half full of cream when it burst, meaning that the gas propellant had available more than twice the volume it would in a full can, and that some of the propellant had already been used. You argue that the real safety criterion is pressure, and that the can's maximum pressure wasn't exceeded. Who's right?

39. A 3000-mL flask is initially open in a room containing air at 1.00 atm and 20°C. The flask is then closed and immersed in boiling water. When the air in the flask has reached thermodynamic equilibrium, the flask is opened and air is allowed to escape. The flask is then closed and cooled back to 20°C. Find (a) the maximum pressure reached in the flask, (b) the number of moles that escape when air is released, and (c) the final pressure in the flask.

40. **BIO** The recommended treatment for frostbite is rapid heating in a water bath. Suppose a frostbitten hand with mass 120 g is immersed in water that conducts energy into the hand at the rate of 800 W. Treating the hand as essentially water, initially frozen solid, how long will it take for it to thaw and return to body temperature (37°C)?

41. A stove burner supplies heat to a pan at the rate of 1500 W. How long will it take to boil away 1.1 kg of water, once the water reaches its boiling point?

42. If a 1-megaton nuclear bomb were exploded deep in the Greenland ice cap, how much ice would it melt? Assume the ice is initially at about its freezing point, and consult Appendix C for the appropriate energy conversion.

43. You're winter camping and are melting snow for drinking water. The snow temperature is right around 0°C. You set a pot containing 5.0 kg of snow on your campfire, and you keep stoking up the fire. As a result, the snow gains energy at an increasing rate: $P = a + bt$, where $a = 1.1$ kW, $b = 2.3$ W/s, and t is the time in s. To the nearest minute, how long will it take to melt the snow?

44. **ENV** At winter's end, Lake Superior's surface is frozen to a depth of 1.3 m; the ice density is 917 kg/m^3. (a) How much energy does it take to melt the ice? (b) If the ice disappears in 3 weeks, what's the average power supplied to melt it?

45. A refrigerator extracts energy from its contents at the rate of 95 W. How long will it take to freeze 750 g of water already at 0°C?

46. **ENV** Climatologists have recently recognized that black carbon (soot) from burning fossil fuels and biomass contributes significantly to arctic warming. You're asked to determine whether this effect might cause ice to melt that would normally stay frozen year-round. Consider an ice layer 2.5 m thick that normally reflects 90% of the incident solar energy and absorbs the rest. Suppose black carbon darkens the ice so it now reflects only 50% of the incident solar energy. The arctic summertime solar input averages 300 W/m^2. You can assume 0°C for the initial ice temperature, and an ice density of 917 kg/m^3. What do you conclude?

47. Repeat Example 17.4 with an initial ice mass of 50 g.

48. How much energy does it take to melt 10 kg of ice initially at −10°C?

49. Water is brought to its boiling point and then allowed to boil away completely. If the energy needed to raise the water to the boiling point is one-tenth of that needed to boil it away, what was the initial temperature?

50. **ENV** During a nuclear accident, 420 m^3 of emergency cooling water at 20°C are injected into a reactor vessel where the reactor core is producing heat at the rate of 200 MW. If the water is allowed to boil at normal atmospheric pressure, how long will it take to boil the reactor dry?

51. What's the minimum amount of ice in Example 17.4 that will ensure a final temperature of 0°C?

52. A bowl contains 16 kg of punch (essentially water) at a warm 25°C. What's the minimum amount of ice at 0°C needed to cool the punch to 0°C?

53. A 50-g ice cube at −10°C is placed in an equal mass of water. What must the initial water temperature be if the final mixture still contains equal amounts of ice and water?

54. **BIO** Evaporation of sweat is the human body's cooling mechanism. At body temperature, it takes 2.4 MJ/kg to evaporate water. Marathon runners typically lose about 3 L of sweat each hour. How much energy gets lost to sweating during a 3-hour marathon?

55. What power is needed to melt 20 kg of ice in 6.0 min?

56. You put 300 g of water at 20°C into a 500-W microwave oven and accidentally set the time for 20 min instead of 2.0 min. How much water is left at the end of 20 min?

57. **ENV** If 4.5×10^5 kg of emergency cooling water at 10°C are dumped into a malfunctioning nuclear reactor whose core is producing energy at the rate of 200 MW, and if no circulation or cooling occurs, how long will it take for half the water to boil away?

58. Describe the composition and temperature of the equilibrium mixture after 1.0 kg of ice at −40°C is added to 1.0 kg of water at 5.0°C.

59. A glass marble 1.000 cm in diameter is to be dropped through a hole in a steel plate. At room temperature the hole diameter is 0.997 cm. By how much must the plate's temperature be raised so the marble will fit through the hole?

60. A 2000-mL graduated cylinder is filled with liquid at 350 K. When the liquid is cooled to 300 K, the cylinder is full to only the 1925-mL mark. Use Table 17.2 to identify the liquid.

61. A steel ball bearing is encased in a Pyrex glass cube 1.0 cm on a side. At 330 K, the ball bearing fits tightly inside the cube. At what temperature will it have a clearance of 1.0 μm all around?

62. **ENV** Fuel systems of modern cars are designed so thermal expansion of gasoline doesn't result in wasteful and polluting fuel spills. As an engineer, you're asked to specify the size of an expansion tank that will handle this overflow. You know that gasoline comes from its underground storage at 10°C, and your expansion tank must handle the expansion of a full 75-L gas tank when the gas reaches a hot summer day's temperature of 35°C. How large an expansion tank do you specify?

63. A rod of length L_0 is clamped rigidly at both ends. Its temperature increases by ΔT and in the ensuing expansion, it cracks to form two straight pieces, as shown in Fig. 17.11. Find an expression for the distance d shown in the figure, in terms of L_0, ΔT, and the linear expansion coefficient α.

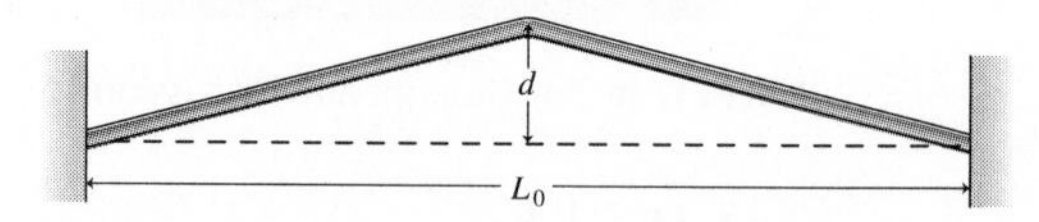

FIGURE 17.11 Problem 63

64. You're home from college on vacation, and there's a power failure. The power company says it will be 15 hours before it's repaired. Your parents send you out to buy ice to keep the 'fridge cold. You look up the thermal resistance of the refrigerator's walls; it's 0.12 K/W. If room temperature is 20°C, how much ice should you buy?

65. **ENV** A solar-heated house stores energy in 5.0 tons of Glauber salt ($Na_2SO_4 \cdot 10H_2O$), which melts at 90°F. The heat of fusion of Glauber salt is 104 Btu/lb and the specific heats of the solid and liquid are, respectively, 0.46 Btu/lb·°F and 0.68 Btu/lb·°F. After a week of sunny weather, the storage medium is all liquid at 95°F. Then comes a cloudy period during which the house loses heat at

an average of 20,000 Btu/h. (a) How long is it before the temperature of the storage medium drops below 60°F? (b) How much of this time is spent at 90°F?

66. Show that the coefficient of volume expansion of an ideal gas at constant pressure is the reciprocal of its kelvin temperature.

67. **CH** Water's coefficient of volume expansion in the temperature range from 0°C to about 20°C is given approximately by $\beta = a + bT + cT^2$, where T is in Celsius and $a = -6.43 \times 10^{-5}\,°\text{C}^{-1}$, $b = 1.70 \times 10^{-5}\,°\text{C}^{-2}$, and $c = -2.02 \times 10^{-7}\,°\text{C}^{-3}$. Show that water has its greatest density at approximately 4.0°C.

68. **CH** When the expansion coefficient varies with temperature, Equation 17.6 is written $\beta = (1/V)(dV/dT)$. If a sample of water occupies 1.00000 L at 0°C, find its volume at 12°C. (*Hint*: Use the information from Problem 67, and integrate the equation above.)

69. Ignoring air resistance, find the height from which to drop an ice cube at 0°C so it melts completely on impact. Assume no heat exchange with the environment.

70. **CH** The timekeeping of a grandfather clock is regulated by a brass pendulum 1.35 m long. If the clock is accurate at 20°C but is in a room at 17°C, how soon will the clock be off by 1 minute? Will it be fast or slow?

71. Prove the equation $\beta = 3\alpha$ (Section 17.3) by considering a cube of side s and therefore volume $V = s^3$ that undergoes a small temperature change dT and corresponding length and volume changes ds and dV.

72. You're on a team planning a mission to Venus to collect atmospheric samples for analysis. The design specs call for a 1-L sample container, while the scientists want at least 1 mol of gas. Venus's atmospheric pressure is 90 times that of Earth's, and its average temperature is 730 K. Will the design work?

73. **CH DATA** Figure 17.12 shows an apparatus used to determine the linear expansion coefficient of a metal wire. The wire is attached to two points a distance d apart (you don't know d). A mass hangs from the middle of the wire. The wire's total length is 100.00 cm at 0°C. The distance y from the suspension points to the top of the mass is measured, and the results are given in the table below. (a) Find an expression for y as a function of temperature, and manipulate your expression to get a linear relation between some function of y and some function of temperature T. You'll encounter the expression L^2, where L is the length of the wire, and, because the change in length is small, you can drop terms involving α^2 when you expand L^2. (b) Calculate the quantities in your relation from the given data, and plot. Determine a best-fit line and use it to determine the coefficient of linear expansion α and the separation d. (c) Consult Table 17.2 to identify the metal the wire is made of. Ignore any stretching of the wire due to its "springiness"; that is, consider only thermal expansion.

Temperature, T (°C)	0	20	40	60	80	100	120
y (cm)	30.00	30.05	30.07	30.11	30.16	30.19	30.24

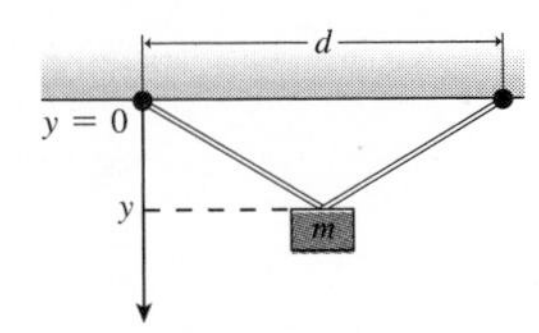

FIGURE 17.12 Problem 73

74. **ENV** The Intergovernmental Panel on Climate Change estimates that Greenland is losing ice, as a result of global warming, at approximately 250 Pg/year. (a) Find the energy needed to melt 250 Pg of ice. (b) Greenland's ice melt results most immediately from an imbalance between incoming and outgoing energy—an imbalance created largely by the absorption of infrared radiation by human-produced greenhouse gases. Use your answer to part (a) to express Greenland's energy imbalance in watts per square meter of the Greenland ice sheet's surface area. That your result is larger than the global imbalance of somewhat less than 1 W/m² shows that the impact of global warming is greater in the Arctic.

75. (a) Show that, for an ideal gas, the speed of sound given by Equation 14.9 can be written $v_{\text{sound}} = \sqrt{\gamma kT/m}$. (b) For diatomic gases like N_2 and O_2 that are the dominant constituents of air, $\gamma = 7/5$. Use your result to show that, for diatomic gases, the speed of sound is about 68% of the thermal speed given by Equation 17.4.

76. **CH** The Maxwell–Boltzmann distribution, plotted in Fig. 17.4, is given by

$$N(v)\,\Delta v = 4\pi N\left(\frac{m}{2\pi kT}\right)^{3/2} v^2 e^{-mv^2/2kT}\,\Delta v$$

where $N(v)\Delta v$ is the number of molecules in a small speed range Δv around speed v, N is the total number of molecules in the gas, m is the molecular mass, k is Boltzmann's constant, and T is the temperature. Use this equation to show that the most probable speed for a gas molecule—the speed at the peak of the curves in Fig. 17.4—is $\sqrt{2kT/m}$. Note that the thermal speed (Equation 17.4), which is the *average* molecular speed, is a factor of $\sqrt{3/2}$ or about 20% greater than the most probable speed—a fact that reflects the long, high-energy "tail" of the Maxwell–Boltzmann distribution.

77. At high gas densities, the *van der Waals equation* modifies the ideal-gas law to account for nonzero molecular volume and for the van der Waals force that we discussed in Section 17.1. The van der Waals equation is

$$\left(p + \frac{n^2 a}{V^2}\right)(V - nb) = nRT$$

where a and b are constants that depend on the particular gas. For nitrogen (N_2), $a = 0.14\ \text{Pa}\cdot\text{m}^6/\text{mol}^2$ and $b = 3.91 \times 10^{-5}\ \text{m}^3/\text{mol}$. For 1.000 mol of N_2 at 10.00 atm pressure, confined to a volume of 2.000 L, find the temperatures predicted (a) by the ideal-gas law and (b) by the van der Waals equation.

Passage Problems

A *pressure cooker* is a sealed pot that cooks food much faster than most other methods because increased pressure allows water to reach higher temperatures than the normal boiling point (Fig. 17.13). Pressure cookers afford many advantages: faster cooking, lower energy consumption, and less vitamin loss. The pressure-cooker principle is also used in autoclaves for sterilizing surgical instruments in hospitals and equipment in biology labs.

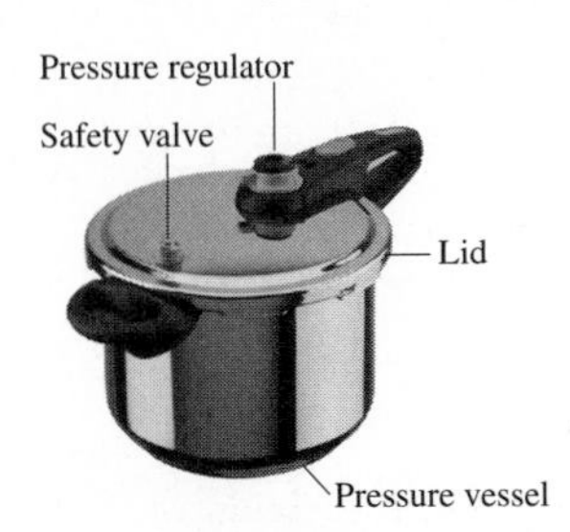

FIGURE 17.13 A pressure cooker (Passage Problems 78–81)

78. In water's phase diagram (Fig. 17.9), normal boiling occurs at a point on the line between the triple point and the critical point. In a pressure cooker, boiling occurs
    a. at a point in the diagram directly above where it normally occurs.
    b. higher up on the line between the triple and critical points.
    c. at a point directly to the right of where it normally occurs.
    d. beyond the critical point.
79. A typical pressure cooker operates at twice normal atmospheric pressure, raising water's boiling point to about 120°C. Compared with steam at 1 atm and the normal 100°C boiling point, the density of steam in a pressure cooker is
    a. double.
    b. somewhat more than double.
    c. somewhat less than double.
    d. quadruple.
80. **BIO** Because some pathogens can survive 120°C temperatures, medical autoclaves typically operate at 3 atm pressure, where water boils at 134°C. Based on this information and that given in the preceding problem, you can conclude that
    a. Fig. 17.9's depiction of the liquid–gas interface for water is correct in being concave upward.
    b. Fig. 17.9's liquid–gas interface should actually be concave downward.
    c. autoclaves operate above the critical point.
    d. at its operating temperature, there can't be any liquid water in the autoclave.
81. A pressure cooker has a regulating mechanism that releases steam so as to maintain constant pressure. If that mechanism became clogged
    a. the pressure would nevertheless level off once water in the cooker began to boil.
    b. the pressure would continue to rise although the temperature would remain constant.
    c. both temperature and pressure would continue to rise.
    d. the density of the steam would decrease.

Answers to Chapter Questions

Answer to Chapter Opening Question

Water's solid phase is less dense than the liquid, which causes ice to float. Our world would be a very different place if ice were denser than water.

Answers to GOT IT? Questions

17.1 (a)
17.2 (c)
17.3 (a) The hole gets larger because all of the object's linear dimensions expand equally.

Heat, Work, and the First Law of Thermodynamics

What You Know

- You understand the ideal-gas law.
- You know the relation between temperature and molecular energy.
- You understand the concepts of work and heat as ways of transferring energy.

What You're Learning

- You'll see a new statement of energy conservation: the first law of thermodynamics.
- You'll learn to calculate the work done as an ideal gas changes volume.
- You'll come to understand the work and heat involved in basic thermodynamic processes: isothermal, isobaric, constant-volume, and adiabatic.
- You'll see how molecular structure determines the specific heats of ideal gases.

How You'll Use It

- In Chapter 19 you'll apply the first law of thermodynamics to simple heat engines.
- Heat engines will give you insights into a deeper fact of thermodynamics: the second law, which limits our ability to extract mechanical energy from fuels.
- You'll come to understand entropy as a measure of disorder, and how entropy changes in thermodynamic processes.

A jet engine converts the energy of burning fuel into mechanical energy. How does energy conservation apply in this process?

In Chapter 7, we introduced the powerful idea that energy is conserved, and we developed the principle of energy conservation as a quantitative statement for mechanical energy in the presence of conservative forces. We also introduced nonconservative forces and briefly described their role in converting mechanical energy into the random molecular energy that we call **internal energy**. In Chapters 16 and 17 you've learned that thermal processes involve energy—a realization that sets the stage for us to extend the conservation-of-energy principle to encompass thermodynamic systems. In this chapter, we'll explore this broader principle of energy conservation and see how it describes energy interchanges in systems ranging from engines to atmospheres.

18.1 The First Law of Thermodynamics

Figure 18.1 shows two ways to raise the temperature in a beaker of water: by heating with a flame and by stirring vigorously with a spoon. Using the flame involves heat—energy in transit because of the temperature difference between flame and water. But there's no temperature difference between spoon and water; here the energy transfer occurs because the spoon does mechanical work on the water. We already know that doing work can increase the kinetic or potential energy of a macroscopic object; here we see it, instead, changing the internal energy associated with the motions of individual molecules. The point is that both processes—heating and mechanical work—result in exactly the same final state—namely, water with a higher temperature and therefore greater internal energy. It's this common result that made possible Joule's quantitative identification of heat as a form of energy (Fig. 18.2).

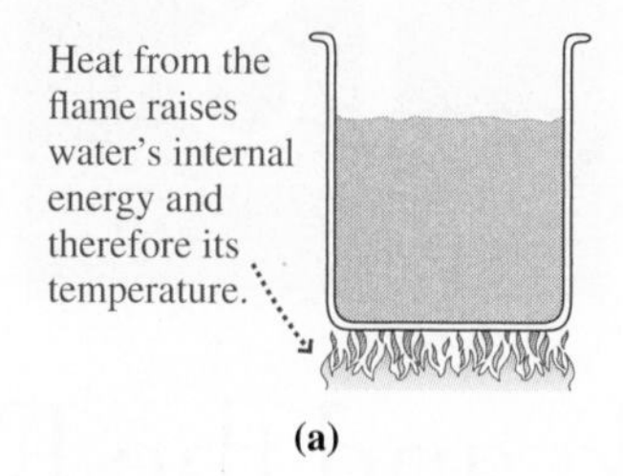

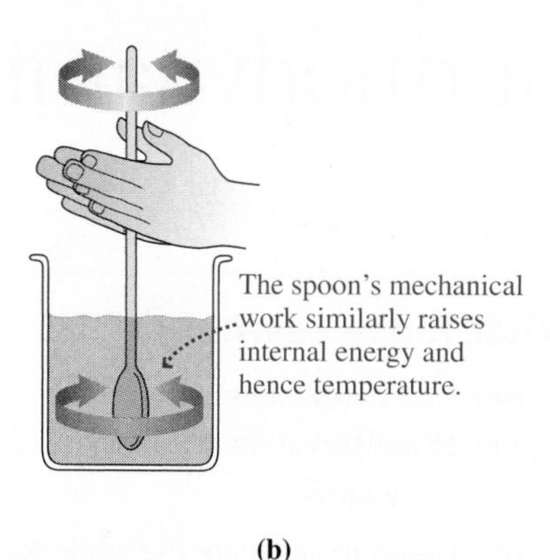

FIGURE 18.1 Two ways to raise temperature: (a) by heat transfer and (b) by mechanical work.

FIGURE 18.2 Joule's apparatus for determining what he called "the mechanical equivalent of heat."

Keep track of all the energy entering and leaving a system—both heat and work—and you'll find that the change in the system's internal energy depends only on the net energy transferred. In one sense this is hardly surprising; it just extends the idea of energy conservation to include heat. But in another way it's remarkable; it doesn't matter at all *how* the energy gets into the system—heat, work, or some combination of the two. This statement constitutes the **first law of thermodynamics**:

First law of thermodynamics The change in the internal energy of a system depends only on the net heat transferred to the system and the net work done on the system, independent of the particular processes involved.

Mathematically, the first law is

$$\Delta E_{\text{int}} = Q + W \quad \text{(first law of thermodynamics)} \tag{18.1}$$

where ΔE_{int} is the change in a system's internal energy, Q the heat transferred to the system, and W the work done on the system.* The first law says that the change in a system's internal energy doesn't depend on how the energy gets transferred, but only on the net energy. Internal energy is therefore a **thermodynamic state variable**, meaning a quantity whose value doesn't depend on how a system got into its particular state. Temperature and pressure are also thermodynamic state variables; heat and work are not.

We're frequently concerned with *rates* of energy flow. Differentiating the first law with respect to time gives a statement about rates:

$$\frac{dE_{\text{int}}}{dt} = \frac{dQ}{dt} + \frac{dW}{dt} \tag{18.2}$$

where dE_{int}/dt is the rate of change of a system's internal energy, dQ/dt the rate of heat transfer to the system, and dW/dt the rate at which work is done on the system.

*Some older books define W as the work done *by* the system, in which case there's a minus sign in the first law. This is because the law was first introduced in connection with engines, which take in heat and put out mechanical work.

EXAMPLE 18.1 The First Law of Thermodynamics: Thermal Pollution

The reactor in a nuclear power plant supplies energy at the rate of 3.0 GW, boiling water to produce steam that turns a turbine-generator. The spent steam is then condensed through thermal contact with water taken from a river. If the power plant produces electrical energy at the rate of 1.0 GW, at what rate is heat transferred to the river?

INTERPRET This problem is about heat and mechanical energy, which are related by the first law of thermodynamics. We identify the system as the entire power plant, comprising the nuclear reactor, including its fuel, and the turbine-generator. We identify E_{int} as the energy in the fuel, W as the mechanical work that ends up as electrical energy, and Q as the heat transferred to the river.

DEVELOP Since we're dealing here with *rates*, Equation 18.2, $dE_{\text{int}}/dt = dQ/dt + dW/dt$, applies. The reactor extracts energy from its fuel, so the rate dE_{int}/dt is negative. The power plant delivers electrical energy to the outside world, so it's *doing* work; since W in the first law is the work done *on* the system, dW/dt is therefore *negative*. Our plan is then to solve for dQ/dt, the rate of energy transfer to the river.

EVALUATE Solving, we have

$$\frac{dQ}{dt} = \frac{dE_{\text{int}}}{dt} - \frac{dW}{dt} = -3.0\text{ GW} - (-1.0\text{ GW}) = -2.0\text{ GW}$$

ASSESS Make sense? Since positive Q represents heat transferred *to* the system, the minus sign shows that heat is transferred *from* the power plant to the river at the rate of 2 GW. The numbers here are typical for large nuclear and coal-burning power plants, and show that about two-thirds of the energy extracted from the fuel is wasted in heating the environment. We'll see in the next chapter just why this waste occurs.

✓TIP Identify the System

The first law of thermodynamics deals with energy flows into and out of a system. We first introduced the system concept in the context of energy in our discussion surrounding Fig. 6.1. Here, as there, it's up to you to define the system. How you do so affects the meanings of the terms in the first law. In this Example we included the nuclear reactor, with the internal energy of its fuel, as part of the system. If we had considered only the turbine-generator, then we would have had 3 GW of heat coming in from the reactor and no change in internal energy. But the result would be the same: 1 GW going out as electricity and 2 GW of heat dumped into the river.

18.2 Thermodynamic Processes

Although the first law applies to *any* system, it's easiest to understand when applied to an ideal gas. The ideal-gas law relates the temperature, pressure, and volume of a given gas sample: $pV = nRT$. The thermodynamic state is completely determined by any two of the quantities p, V, or T. We'll find it convenient to represent different states as points on a ***pV* diagram**—a graph whose vertical and horizontal axes represent pressure and volume, respectively.

Reversible and Irreversible Processes

Imagine a gas sample immersed in a large reservoir of water and allowed to come to equilibrium (Fig. 18.3). If we then raise the reservoir temperature very slowly, both water and gas temperatures will rise essentially in unison, and the gas will remain in equilibrium. Such a slow change is called a **quasi-static process**. Because a system undergoing a quasi-static process is always in thermodynamic equilibrium, its evolution from one state to another is described by a continuous sequence of points—a curve—in its pV diagram (Fig. 18.4).

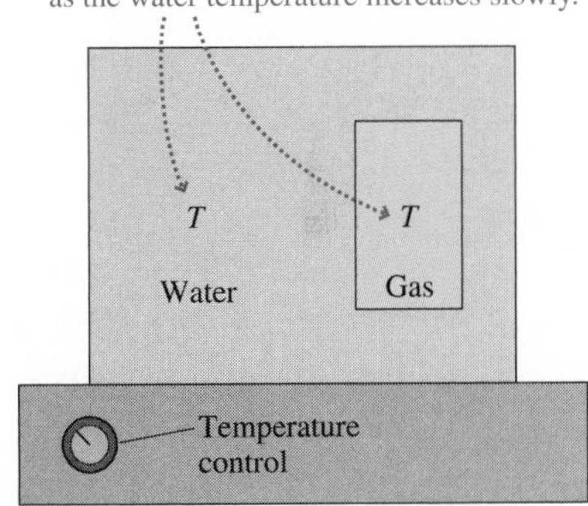

FIGURE 18.3 A quasi-static, or reversible, process keeps water and gas always in equilibrium.

We could reverse this heating process by slowly lowering the reservoir temperature; the gas would cool, reversing its path in the pV diagram. For that reason, a quasi-static process is also a **reversible process**. A process like suddenly plunging a cool gas sample into hot water is, in contrast, **irreversible**. During an irreversible process the system isn't in equilibrium, and thermodynamic variables like temperature and pressure don't have well-defined values. It therefore makes no sense to think of a path in the pV diagram. A process may be irreversible even though it returns a system to its original state. The distinction lies not in the end states but in the *process* that takes the system between states.

There are many ways to change a system's thermodynamic state. Here we consider important special cases involving an ideal gas. These illustrate the physical principles behind a myriad of technological devices and natural phenomena, from the operation of a gasoline engine to the propagation of a sound wave to the oscillations of a star.

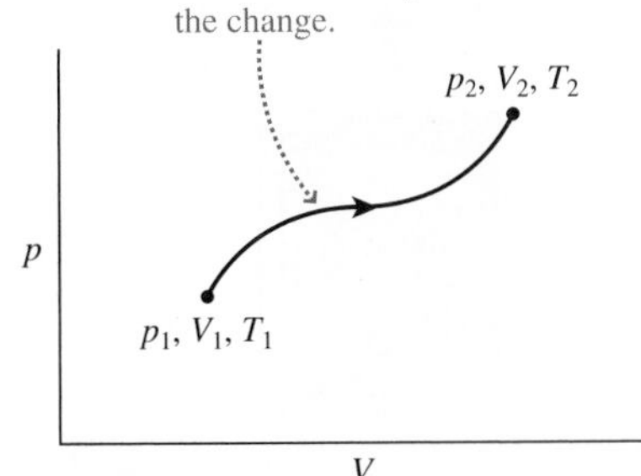

FIGURE 18.4 The pV diagram of a system undergoing quasi-static change.

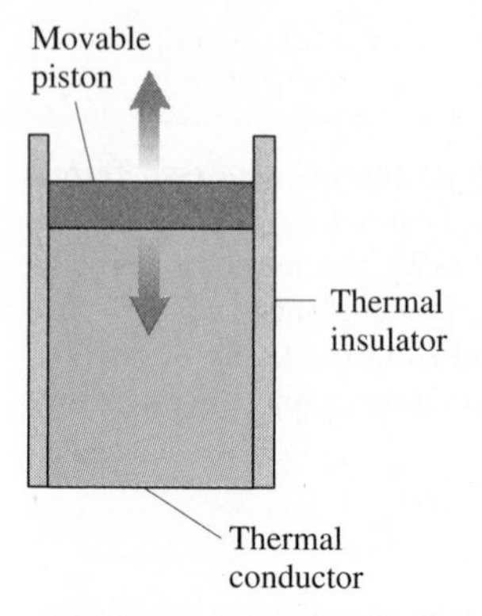

FIGURE 18.5 A gas–cylinder system with insulating walls and a conducting bottom.

Our system consists of an ideal gas confined to a cylinder sealed with a movable piston (Fig. 18.5). The piston and cylinder walls are perfectly insulating—they block all heat transfer—and the bottom is a perfect conductor of heat. We can change the thermodynamic state of the gas mechanically by moving the piston, or thermally by transferring heat through the bottom. We'll consider only reversible processes, which we can describe by paths in the pV diagram for the gas.

Work and Volume Changes

We begin by developing an expression for the work done on a gas that holds for all processes. If our piston–cylinder system has cross-sectional area A and gas pressure p, then $F_{gas} = pA$ is the force the gas exerts on the piston. If the piston moves a small distance Δx, the gas does work $\Delta W_{gas} = F_{gas}\,\Delta x = pA\,\Delta x = p\,\Delta V$, where $\Delta V = A\,\Delta x$ is the change in gas volume (Fig. 18.6*a*). Our expression for the first law of thermodynamics involves the work done *on* the gas; by Newton's third law, the piston exerts a force on the gas that's equal but opposite to F_{gas}, so the work done *on* the gas is $\Delta W = -F_{gas}\,\Delta x = -p\,\Delta V$. Pressure may vary with volume, so we find the total work done as the gas goes from volume V_1 to volume V_2 by replacing ΔV with the infinitesimal quantity dV and integrating:

$$W = \int dW = -\int_{V_1}^{V_2} p\,dV \quad \text{(work done on gas during volume change)} \tag{18.3}$$

Figure 18.6*b* shows that the work done on the gas is the negative of the area under the pV curve. That work is positive if the gas is compressed ($V_2 < V_1$) and negative if it expands ($V_2 > V_1$).

We'll now explore several basic thermodynamic processes, in each case holding one thermodynamic variable constant.

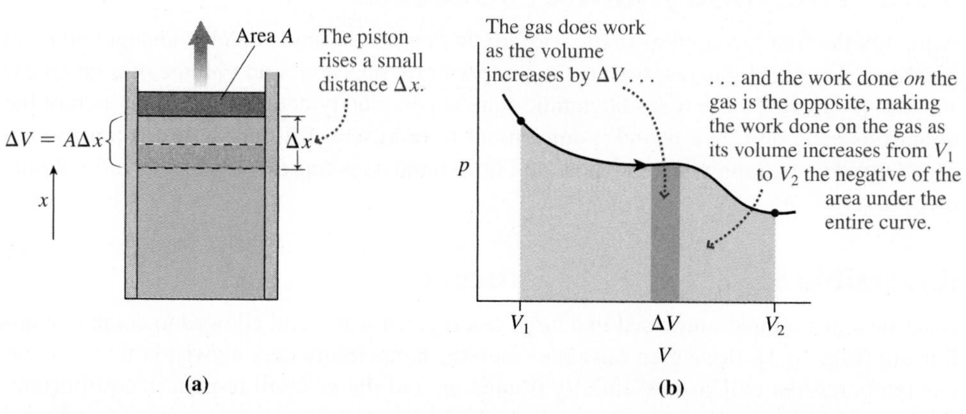

FIGURE 18.6 Work done *on* the gas as the piston rises (a) is the negative of the area under the pV curve (b).

GOT IT? 18.1 Two identical gas–cylinder systems are taken from the same initial state to the same final state, but by different processes. Which of the following is or are the same in both cases? (a) the work done on or by the gas; (b) the heat added or removed; or (c) the change in internal energy

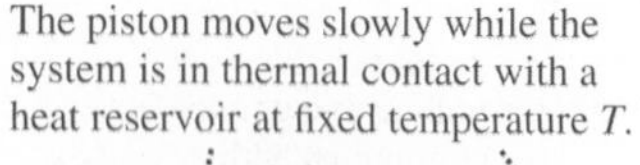

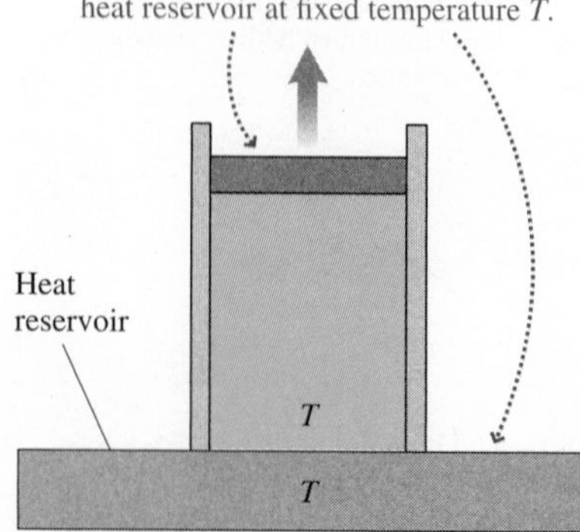

FIGURE 18.7 An isothermal process.

Isothermal Processes

An **isothermal process** occurs at constant temperature. Figure 18.7 shows one way to effect an isothermal process: Place a gas cylinder in thermal contact with a heat reservoir whose temperature is constant. Then move the piston to change the gas volume, slowly enough that the gas remains in equilibrium with the heat reservoir. The system moves from its initial state to its final state along a curve of constant temperature—an **isotherm**—in the pV diagram (Fig. 18.8). The work done on the gas is given by Equation 18.3 and is the negative of the area under the isotherm.

To find that work, we relate pressure and volume through the ideal-gas law: $p = (nRT)/V$. Then Equation 18.3 becomes

$$W = -\int_{V_1}^{V_2} \frac{nRT}{V} dV$$

For an isothermal process, the temperature T is constant, giving

$$W = -nRT\int_{V_1}^{V_2} \frac{dV}{V} = -nRT \ln V \bigg|_{V_1}^{V_2} = -nRT \ln\left(\frac{V_2}{V_1}\right)$$

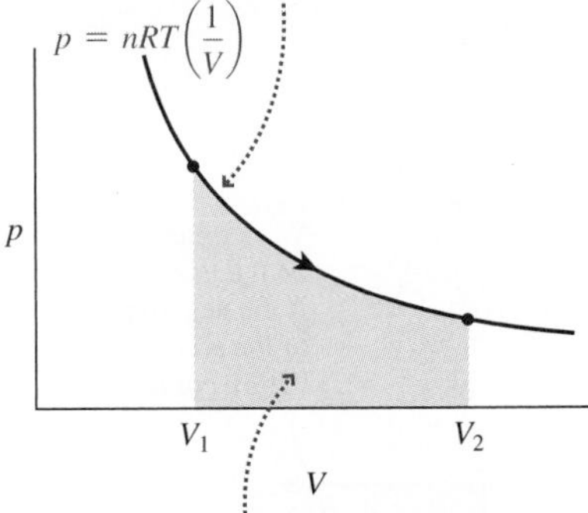

FIGURE 18.8 A pV diagram for an isothermal process.

The internal energy of an ideal gas consists only of the kinetic energy of its molecules, which, in turn, depends only on temperature. That dependence of internal energy on temperature alone is a defining feature of the ideal gas. Thus, there's no change in the internal energy of an ideal gas during an isothermal process. The first law of thermodynamics then gives $\Delta E_{\text{int}} = 0 = Q + W$, so

$$Q = -W = nRT \ln\left(\frac{V_2}{V_1}\right) \quad \text{(isothermal process)} \qquad (18.4)$$

Does this result $Q = -W$ make sense? Recall that Q is the heat transferred to the gas and W is the work done on it. So $-W$ is the work done *by* the gas, and our result shows that for a gas to do work without its temperature changing, it must absorb an equal amount of heat. Similarly, if work is done on the gas, it must transfer an equal amount of heat to its surroundings if it's to maintain a constant temperature.

EXAMPLE 18.2 An Isothermal Process: Bubbles!

A scuba diver is 25 m down, where the pressure is 3.5 atm or about 350 kPa. The air she exhales forms bubbles 8.0 mm in radius. How much work does each bubble do as it rises to the surface, assuming the bubbles remain at the uniform 300 K temperature of the water?

INTERPRET The constant 300 K temperature tells us we're dealing with an isothermal process.

DEVELOP Equation 18.4 determines the work: $-W = nRT\ln(V_2/V_1)$. Here $-W$ is just what we're after: the work done *by* the gas in the bubble. To use this equation, we need the quantity nRT and the volume ratio V_2/V_1. We know p and V (actually the radius, from which we can get V) at the 25-m depth, so we can use the ideal-gas law $pV = nRT$ to get nRT and also the bubble volume just before it reaches the surface. Then we'll have everything we need to apply Equation 18.4.

EVALUATE The ideal-gas law gives $nRT = pV = \frac{4}{3}\pi r^3 p$. The number of moles n doesn't change and R is a constant, so pV is itself constant in the isothermal process. That means $p_1V_1 = p_2V_2$, showing that the volume expands by a factor of 3.5 as the pressure drops from 3.5 atm to 1 atm at the surface—so $V_2/V_1 = 3.5$. Then Equation 18.4 gives

$$-W = nRT \ln\left(\frac{V_2}{V_1}\right) = \tfrac{4}{3}\pi r^3 p \ln 3.5$$

Using the 8-mm bubble radius and the 350-kPa pressure gives 0.94 J for the work. Note that we needed to use pressure in SI units here; to find the volume ratio, any units would do because V_2/V_1 followed from the pressure *ratio* p_1/p_2.

ASSESS Make sense? The work $-W$ done *by* the gas is positive because an expanding bubble pushes water outward and ultimately upward. It therefore raises the ocean's gravitational potential energy. When the bubble breaks, this excess potential energy becomes kinetic energy, appearing as small waves on the water surface. The bubble, in turn, gets its energy from heat that flows in to keep it at constant temperature. Energy is conserved! ■

Constant-Volume Processes and Specific Heat

A **constant-volume process** (also called isometric, isochoric, or isovolumic) occurs in a rigid closed container whose volume can't change. We could tightly clamp the piston in Fig. 18.5 for a constant-volume process. Because the piston doesn't move, the gas does no work, and the first law becomes simply $\Delta E_{\text{int}} = Q$. To express this result in terms of a temperature change ΔT, we introduce the **molar specific heat at constant volume** C_V, defined by

$$Q = nC_V\,\Delta T \quad \text{(constant-volume process)} \qquad (18.5)$$

where n is the number of moles. This molar specific heat is like the specific heat defined in Chapter 16, except it's per mole rather than per unit mass. Using Equation 18.5 for Q in our first-law statement $\Delta E_{int} = Q$ gives

$$\Delta E_{int} = nC_V \Delta T \quad \text{(any process)} \tag{18.6}$$

For an ideal gas, the internal energy is a function of temperature alone, so $\Delta E_{int}/\Delta T$ has the same value no matter what process the gas undergoes. Therefore, Equation 18.6, relating the temperature change ΔT and internal-energy change ΔE_{int}, applies not only to a constant-volume process but to *any* ideal-gas process. Why, then, have we been so careful to label C_V the specific heat *at constant volume*? Although Equation 18.6, $\Delta E_{int} = nC_V \Delta T$, holds for any process, it's only when there's no work that the first law lets us write $Q = \Delta E_{int}$, and therefore only for a constant-volume process that Equation 18.5 holds.

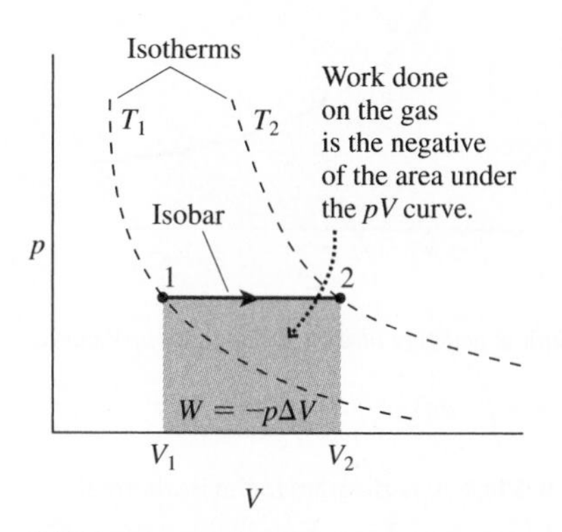

FIGURE 18.9 A pV diagram for an isobaric process; also shown are isotherms for the initial and final temperatures.

Isobaric Processes and Specific Heat

Isobaric means constant pressure. Processes occurring in systems exposed to the atmosphere are essentially isobaric. In a reversible isobaric process, a system moves along an isobar, or curve of constant pressure, in its pV diagram (Fig. 18.9). The work done on the gas as the volume changes from V_1 to V_2 is the negative of the rectangular area under the isobar, or

$$W = -p(V_2 - V_1) = -p\,\Delta V \tag{18.7}$$

a result we could obtain formally by integrating Equation 18.3.

Solving the first law (Equation 18.1) for Q and using our expression for work gives $Q = \Delta E_{int} - W = \Delta E_{int} + p\,\Delta V$. For an ideal gas, we've just found that the change in internal energy is $\Delta E_{int} = nC_V \Delta T$ for *any* process. Therefore, $Q = nC_V \Delta T + p\,\Delta V$ for an ideal gas undergoing an isobaric process. We define the **molar specific heat at constant pressure** C_p as the heat required to raise 1 mol of gas by 1 K at constant pressure, or $Q = nC_p \Delta T$. Equating our two expressions for Q gives

$$nC_p \Delta T = nC_V \Delta T + p\,\Delta V \quad \text{(isobaric process)} \tag{18.8}$$

This is a useful form for calculating temperature changes in an isobaric process if we know both specific heats C_p and C_V. However, we really need only one of these specific heats because a simple relation holds between the two. The ideal-gas law, $pV = nRT$, allows us to write $p\,\Delta V = nR\,\Delta T$ for an isobaric process. Using this expression in Equation 18.8 gives $nC_p \Delta T = nC_V \Delta T + nR\,\Delta T$, so

$$C_p = C_V + R \quad \text{(molar specific heats)} \tag{18.9}$$

Does this make sense? Specific heat measures the heat needed to cause a given temperature change. In a constant-volume process, no work is done and all the heat goes into raising the internal energy and thus the temperature of an ideal gas. In a constant-pressure process, work *is* done and some of the added heat ends up as mechanical energy, leaving less available for raising the temperature. Therefore, a constant-pressure process requires *more* heat for a given temperature change. Thus the specific heat at constant pressure is greater than at constant volume, as reflected in Equation 18.9.

Why didn't we distinguish specific heats at constant volume and constant pressure earlier? Because we were concerned mostly with solids and liquids, whose coefficients of expansion are far lower than those of gases. As a result, much less work is done by a solid or liquid than by a gas. Since work is what gives rise to the difference between C_V and C_p, the distinction is less significant for solids and liquids. As a practical matter, measured specific heats are usually at constant pressure.

APPLICATION Boiling Water

You slip a mug of water into the microwave to boil for tea, or you put a pot of water on the stove to cook pasta. Boiling water is an example of an isobaric process, because the water is exposed to atmospheric pressure as it boils. At its normal 100°C boiling point, water's volume increases some 2000-fold as it goes from liquid to vapor. According to Equation 18.7, the work done *by* the gas as it expands is $p\,\Delta V$. That 2000-fold expansion means that ΔV is very nearly the same as the final volume V, so the work done by the gas is essentially pV. Then the ideal-gas law in the form of Equation 17.2, $pV = nRT$, implies that the work done per mole of gas is RT. With $T = 100°C$ or 373 K, that amounts to some 3.1 kJ/mol. Converting moles to kg of H_2O gives an equivalent of 170 kJ/kg. The energy needed to do that work must be included in the heat of vaporization, which we introduced in Chapter 17. There, Table 17.1 gave 2257 kJ/kg as water's heat of vaporization at its boiling point. Our 170 kJ/kg shows that only about 8% of the energy supplied to boil water goes into expanding the vaporized water against atmospheric pressure. The rest is due largely to the breaking of the hydrogen bonds that keep H_2O molecules close together in the liquid state.

Adiabatic Processes

In an **adiabatic process**, no heat flows between a system and its environment. The way to achieve this is to surround the system with perfect thermal insulation. Even without insulation, processes that occur quickly are often approximately adiabatic because they're over before significant heat transfer has had time to occur. In a gasoline engine, for example, compression of the gasoline–air mixture and expansion of the combustion products are nearly adiabatic because they occur so rapidly that little heat flows through the cylinder walls.

Since the heat Q is zero in an adiabatic process, the first law becomes simply

$$\Delta E_{\text{int}} = W \quad \text{(adiabatic process)} \qquad (18.10)$$

This says that if we do work on a system and there's no heat transfer, then the system must gain an equal amount of internal energy. Conversely, if the system does work on its environment, then it loses internal energy (Fig. 18.10).

As a gas expands adiabatically, its volume increases while its internal energy and temperature decrease. The ideal-gas law, $pV = nRT$, then requires that the pressure decrease as well—and by more than it would in an isothermal process where T remains constant. In a pV diagram, the path of an adiabatic process—called an **adiabat**—is therefore steeper than the isotherms (Fig. 18.11).

Tactics 18.1 details the math involved in finding the adiabatic path; the result is

$$pV^{\gamma} = \text{constant} \quad \text{(adiabatic process)} \qquad (18.11a)$$

where $\gamma = C_p/C_V$ is the ratio of the specific heats. Because $C_p = C_V + R$, the ratio $\gamma = C_p/C_V$ is always greater than 1. As expected, an adiabatic process therefore results in a greater pressure change than would a comparable isothermal process, as reflected in the steeper adiabatic path in Fig. 18.11. Physically, the adiabatic path is steeper because the gas loses internal energy as it does work, so its temperature drops. Problem 71 shows how to rewrite Equation 18.11a in terms of temperature:

$$TV^{\gamma-1} = \text{constant} \quad \text{(adiabatic process)} \qquad (18.11b)$$

Molecules rebound with the same speed, and the gas's internal energy doesn't change.

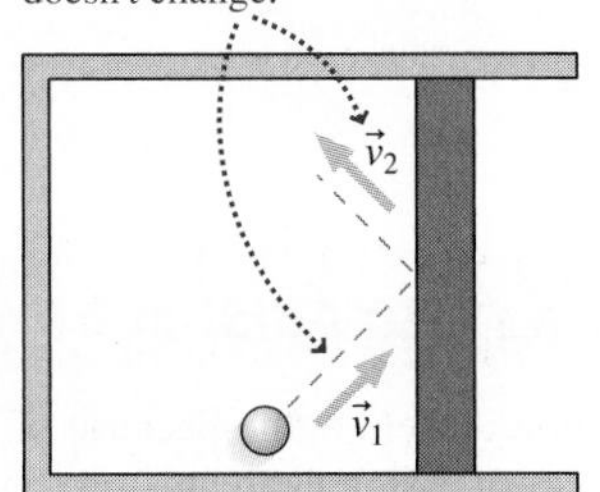

(a) Stationary piston

Rebounding molecules have lower speed as energy is transferred to the outward-moving piston. With the decrease in internal energy comes a drop in temperature.

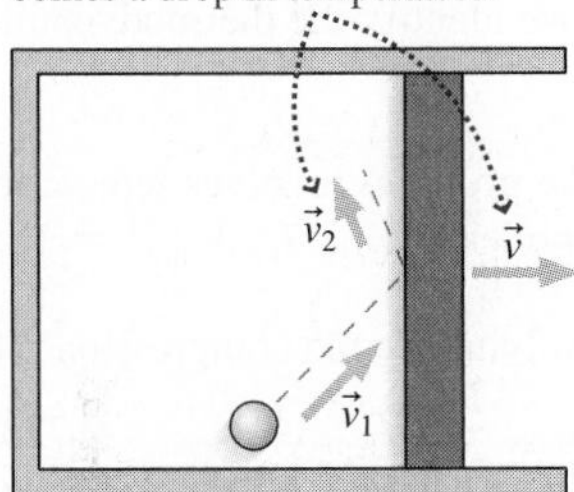

(b) Moving piston

FIGURE 18.10 In an adiabatic expansion, a gas does work on the piston and its internal energy decreases. Part (b) shows microscopically how this occurs.

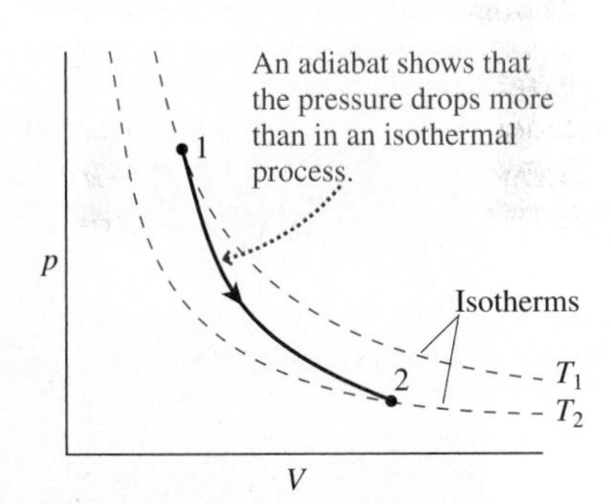

FIGURE 18.11 A pV curve for an adiabatic expansion (*dark curve*).

TACTICS 18.1 Deriving the Adiabatic Equation

Equation 18.6 gives the infinitesimal change in internal energy for *any* process: $dE_{\text{int}} = nC_V\,dT$. The corresponding work is $dW = p\,dV$ so, with $Q = 0$ in an adiabatic process, the first law becomes $nC_V\,dT = -p\,dV$. We can eliminate dT by differentiating the ideal-gas law, now letting *both* p and V change: $nR\,dT = d(pV) = p\,dV + V\,dp$. Solving for dT, substituting in our first-law statement, and multiplying through by R leads to $C_V V\,dp + (C_V + R)p\,dV = 0$. But $C_V + R = C_p$; substituting this and dividing through by $C_V pV$ gives

$$\frac{dp}{p} + \frac{C_p}{C_V}\frac{dV}{V} = 0$$

Defining $\gamma \equiv C_p/C_V$ and integrating gives

$$\ln p + \gamma \ln V = \ln(\text{constant})$$

where we've chosen to call the constant of integration ln(constant). Since $\gamma \ln V = \ln V^{\gamma}$, it follows by exponentiation that

$$pV^{\gamma} = \text{constant}$$

CONCEPTUAL EXAMPLE 18.1 Ideal-Gas Law versus the Adiabatic Equation

The ideal-gas law says $pV = nRT$, but, seemingly in contrast, Equation 18.11a says pV^{γ} = constant for an ideal gas undergoing an adiabatic process. Which is right?

EVALUATE The ideal-gas law is fundamental, so we know it's right. And we derived Equation 18.11a based on the behavior of an ideal gas. So *both* must be right. But how can that be, when one equation talks about pV and the other about pV^{γ}? The answer lies in the right-hand side of the ideal-gas law: nRT. For an adiabatic process, T isn't constant and therefore pV isn't constant—but pV^{γ} is.

ASSESS Compare the adiabatic process with an isothermal process. In the isothermal case, T is constant and we would write pV = constant. Both processes obey the ideal-gas law, but the relation of p and V differs, so there's no contradiction.

MAKING THE CONNECTION Suppose you halve the volume of an ideal gas with $\gamma = 1.4$. What happens to the pressure if the process is (a) isothermal and (b) adiabatic?

EVALUATE For the isothermal process pV = constant, so halving the volume doubles the pressure. For the adiabatic process it's pV^{γ} that's constant. Setting $p_1V_1^{\gamma} = p_2V_2^{\gamma}$ with $V_2 = V_1/2$ gives $p_2 = 2^{\gamma}p_1$. With $\gamma = 1.4$, that means the pressure increases by a factor of 2.64. The pressure increase is greater than in the isothermal case because the temperature goes up.

It's another exercise in calculus to integrate Equation 18.3 for the work done on the gas in an adiabatic process. You can do this in Problem 69; the result is

$$W = \frac{p_2V_2 - p_1V_1}{\gamma - 1} \tag{18.12}$$

EXAMPLE 18.3 An Adiabatic Process: Diesel Power

Fuel ignites in a diesel engine because of the temperature rise that results from compression as the piston moves toward the top of the cylinder; there's no spark plug as in a gasoline engine. Compression is fast enough that the process is essentially adiabatic. If the ignition temperature is 500°C, what compression ratio V_{max}/V_{min} is needed (Fig. 18.12)? Air's specific-heat ratio is $\gamma = 1.4$, and before compression the air is at 20°C.

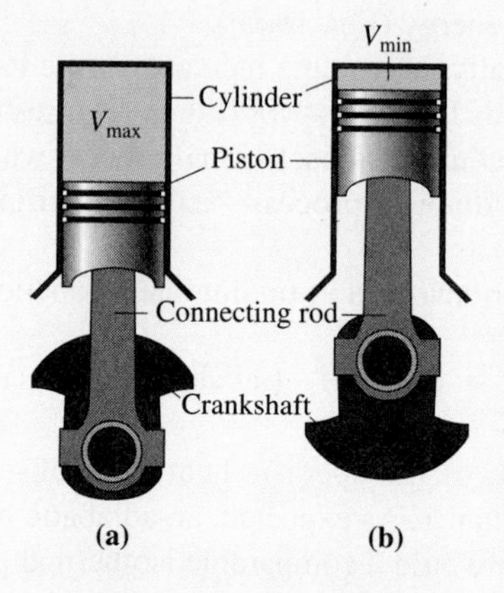

FIGURE 18.12 One cylinder of a diesel engine, shown with the piston (a) at the bottom of its stroke and (b) at the top. The compression ratio is V_{max}/V_{min}.

INTERPRET We identify the thermodynamic process here as adiabatic compression.

DEVELOP The problem involves temperature and volume, so Equation 18.11b applies, giving $T_{min}V_{min}^{\gamma-1} = T_{max}V_{max}^{\gamma-1}$.

EVALUATE Solving for the compression ratio V_{max}/V_{min} gives

$$\frac{V_{max}}{V_{min}} = \left(\frac{T_{min}}{T_{max}}\right)^{1/(\gamma-1)} = \left(\frac{773\text{ K}}{293\text{ K}}\right)^{1/0.4} = 11$$

ASSESS Practical diesel engines have higher ratios to ensure reliable ignition. Their high compression makes diesels heavier than their gasoline counterparts, but also more fuel efficient. You can explore the diesel engine further in Chapter 19 ■.

APPLICATION Smog Alert!

The smog that blankets urban areas is an unfortunate manifestation of our prolific fossil-fueled energy consumption. Adiabatic processes in the atmosphere determine whether or not smog lingers over a city. Consider a volume of air that's heated, perhaps because it's over hot pavement that absorbs solar energy. The air becomes less dense, and its buoyancy makes it rise. As it ascends into regions of lower pressure, it expands, doing work against the surrounding atmosphere. Air is a poor heat conductor, so the process is essentially adiabatic. Therefore, the gas cools as it does work.

Now, temperature in the atmosphere normally decreases with altitude. So here's the crucial question: Does the rising air cool faster or slower than the surrounding atmosphere? If it cools more slowly, then it continues to be warmer, and it continues to rise. Any pollution is carried high into the atmosphere where it's dispersed. But if the decrease in air temperature with altitude isn't so great, or in an **inversion** where it's actually warmer aloft, the rising air will soon reach equilibrium with its surroundings and won't rise any higher. The effect is to trap air and its entrained pollutants near the surface, as shown in this photo of Los Angeles. Smog alert!

GOT IT? 18.2 Name the basic thermodynamic process involved when each of the following is done to a piston–cylinder system containing ideal gas, and tell also whether temperature, pressure, volume, and internal energy increase or decrease: (1) The piston is locked in place and a flame is applied to the bottom of the cylinder; (2) the cylinder is completely insulated and the piston is pushed downward; (3) the piston is exposed to atmospheric pressure and is free to move, while the cylinder is cooled by placing it on a block of ice.

Table 18.1 Ideal-Gas Processes

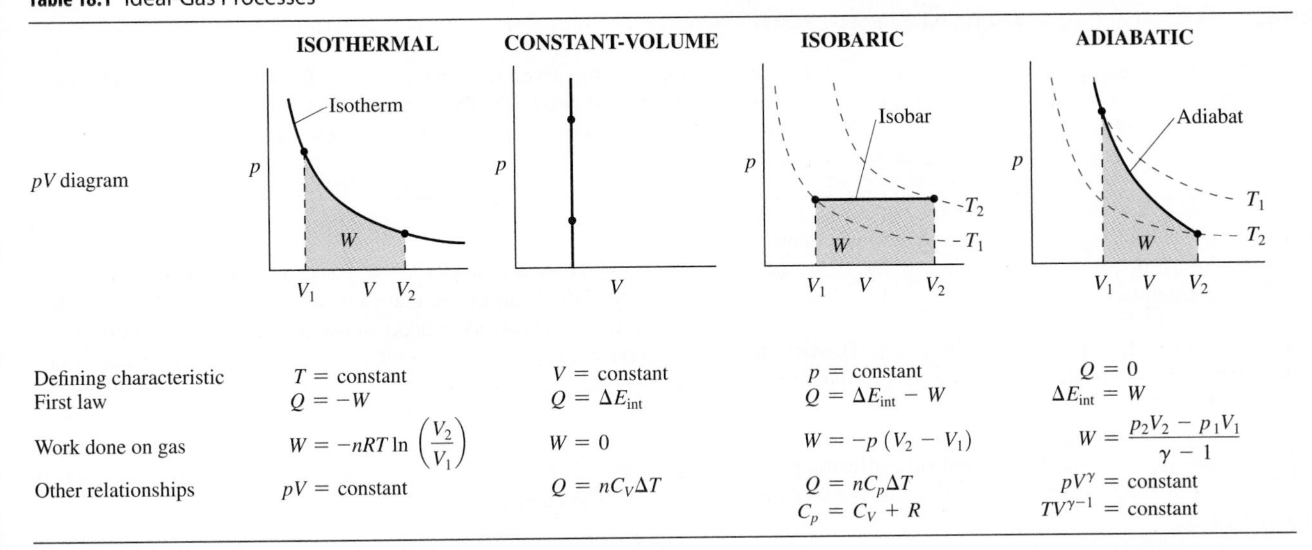

	ISOTHERMAL	CONSTANT-VOLUME	ISOBARIC	ADIABATIC
pV diagram				
Defining characteristic	$T = \text{constant}$	$V = \text{constant}$	$p = \text{constant}$	$Q = 0$
First law	$Q = -W$	$Q = \Delta E_{\text{int}}$	$Q = \Delta E_{\text{int}} - W$	$\Delta E_{\text{int}} = W$
Work done on gas	$W = -nRT\ln\left(\frac{V_2}{V_1}\right)$	$W = 0$	$W = -p\,(V_2 - V_1)$	$W = \frac{p_2V_2 - p_1V_1}{\gamma - 1}$
Other relationships	$pV = \text{constant}$	$Q = nC_V\Delta T$	$Q = nC_p\Delta T$ $C_p = C_V + R$	$pV^{\gamma} = \text{constant}$ $TV^{\gamma-1} = \text{constant}$

Cyclic Processes

Many natural and technological systems undergo **cyclic processes**, in which the system returns periodically to the same thermodynamic state. Engineering examples include engines and refrigerators whose mechanical construction ensures cyclic behavior. Many natural oscillations, like those of a sound wave or a pulsating star, are essentially cyclic.

Cyclic processes often involve the four basic processes we've just explored, as summarized in Table 18.1. We've seen that the work done in any reversible process is just the area under the pV curve. A cyclic process returns to the same point in the pV diagram, so it involves both expansion and compression (Fig. 18.13). During compression, work is done on the gas; during expansion, the gas does work on its surroundings. The net work done on the gas is the difference between the two, shown in Fig. 18.13 as the area enclosed by the cyclic path in the pV diagram.

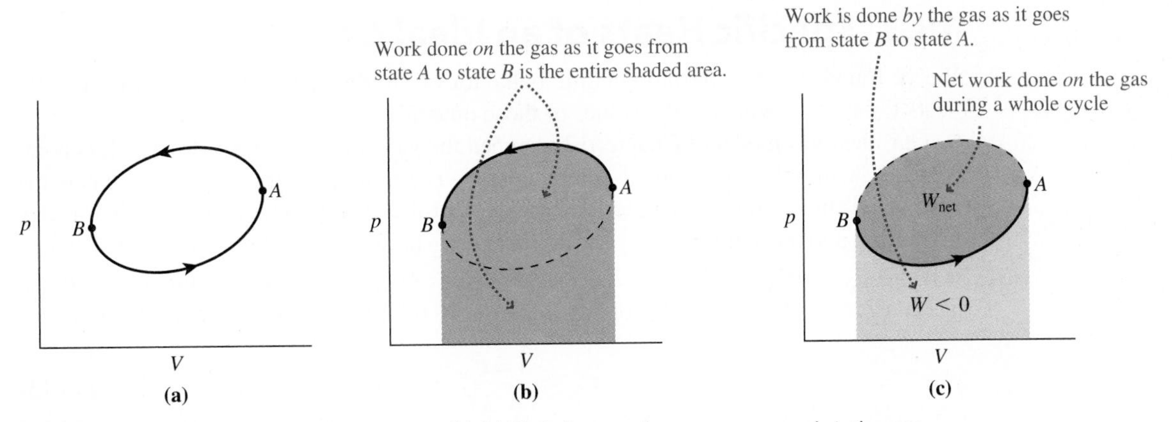

FIGURE 18.13 (a) A pV diagram for a cyclic process. (b), (c) Work done on the gas over one cycle is the area inside the closed path.

EXAMPLE 18.4 A Cyclic Process: Finding the Work

An ideal gas with $\gamma = 1.4$ occupies 4.0 L at 300 K and 100 kPa pressure. It's compressed adiabatically to one-fourth of its original volume, then cooled at constant volume back to 300 K, and finally allowed to expand isothermally to its original volume. How much work is done on the gas?

INTERPRET This problem involves a cyclic process, and we identify three separate thermodynamic processes that make up the cycle: adiabatic, constant-volume, and isothermal.

DEVELOP Here it helps to draw a pV diagram, shown in Fig. 18.14. Our plan is to use equations in Table 18.1 to determine the work for each of the basic processes and then combine them to get the net work. For the adiabatic process AB, Table 18.1 gives $W_{AB} = (p_B V_B - p_A V_A)/(\gamma - 1)$; for the constant-volume process BC, $W_{BC} = 0$; and for the isothermal process CA, the work is $W_{CA} = -nRT\ln(V_A/V_C)$.

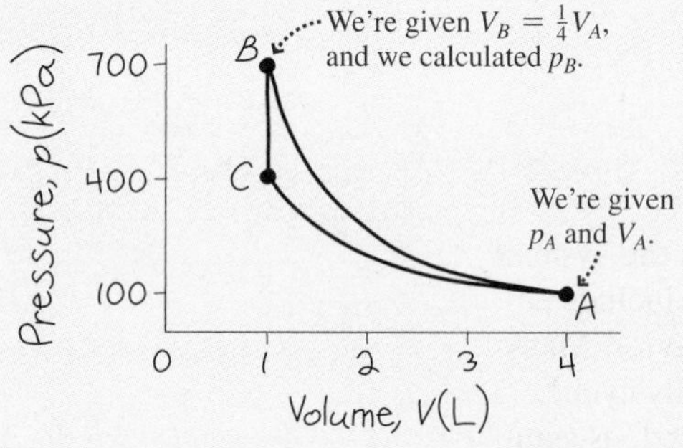

FIGURE 18.14 The cyclic process *ABCA* of Example 18.4 includes adiabatic (*AB*), constant-volume (*BC*), and isothermal (*CA*) sections.

EVALUATE For the adiabatic process AB we're given all quantities except p_B. This we get from the adiabatic equation $pV^\gamma = \text{constant}$, or $p_B V_B^\gamma = p_A V_A^\gamma$. Solving gives $p_B = p_A(V_A/V_B)^\gamma = 696.4$ kPa, where we used the given information $p_A = 100$ kPa, $\gamma = 1.4$, and a compression to one-fourth the original volume ($V_A/V_B = 4$). We now have enough information to find the work done over the adiabatic path:

$$W_{AB} = \frac{p_B V_B - p_A V_A}{\gamma - 1} = 741 \text{ J}$$

where, with pressures in kPa ($=10^3$ Pa) and volumes in L ($=10^{-3}$ m^3), the factors $10^{\pm 3}$ cancel and there's no need to convert. The work W_{AB} is positive because work is done *on* the gas when it's compressed.

In the expression $W_{CA} = -nRT\ln(V_A/V_C)$ for the isothermal work, we can evaluate the quantity nRT at *any* point on the isothermal curve because T is constant. The ideal-gas law says that $nRT = pV$, and we know both p and V at point A. So $nRT = p_A V_A = 400$ J, where again we could multiply $p_A = 100$ kPa by $V_A = 4.0$ L to get an answer in SI units. The isothermal work is then

$$W_{CA} = -nRT\ln\left(\frac{V_A}{V_C}\right) = -(400\text{ J})(\ln 4) = -555 \text{ J}$$

This is negative because the gas does work in expanding from C to A.

Combining our results for all three segments gives the net work:

$$W_{ABCA} = W_{AB} + W_{BC} + W_{CA} = 741\text{ J} + 0\text{ J} - 555\text{ J} = 186\text{ J}$$

ASSESS Make sense? The final answer is positive because we've done net work *on* the gas; that's always the case in going counterclockwise around a cyclic path in a pV diagram. Since the system returns to its original state, its internal energy undergoes no net change. That means all the work that's done on it must be transferred to its surroundings as heat. Since no heat flows during the adiabatic process AB, and since the gas *absorbs* heat during the isothermal expansion CA, the only time it transfers heat *to* its surroundings is during the constant-volume cooling process BC. ■

18.3 Specific Heats of an Ideal Gas

We've found that the thermodynamic behavior of an ideal gas depends on the specific heats C_V and C_p. What are the values of those quantities?

Our ideal-gas model of Chapter 17 assumed the gas molecules were structureless point particles with only translational kinetic energy. The internal energy E_{int} of the gas is the sum of all those molecular kinetic energies. But the average kinetic energy is directly proportional to the temperature: $\frac{1}{2}m\overline{v^2} = \frac{3}{2}kT$. If we have n moles of gas, the internal energy is then $E_{\text{int}} = nN_A(\frac{1}{2}m\overline{v^2}) = \frac{3}{2}nN_A kT$, where N_A is Avogadro's number. But $N_A k = R$, the gas constant, so $E_{\text{int}} = \frac{3}{2}nRT$. Solving Equation 18.6 for the molar specific heat then gives

$$C_V = \frac{1}{n}\frac{\Delta E_{\text{int}}}{\Delta T} = \frac{3}{2}R \qquad (18.13)$$

For this gas of structureless particles, the adiabatic exponent γ is therefore

$$\gamma = \frac{C_p}{C_V} = \frac{C_V + R}{C_V} = \frac{\frac{5}{2}R}{\frac{3}{2}R} = \frac{5}{3} = 1.67$$

Some gases, notably the inert gases helium (He), neon (Ne), argon (Ar), and others in the last column of the periodic table, have adiabatic exponents and specific heats given by these equations. But others do not. At room temperature, for example, hydrogen (H_2), oxygen (O_2), and nitrogen (N_2) obey adiabatic laws with γ very nearly $\frac{7}{5}$ $(=1.4)$ and, correspondingly, specific heat $C_V = \frac{5}{2}R$. On the other hand, sulfur dioxide (SO_2) and nitrogen dioxide (NO_2) have specific-heat ratios close to 1.3 and therefore C_V of about $3.4R$.

What's going on here? A clue lies in the structure of individual gas molecules, reflected in their chemical formulas. The inert-gas molecules are **monatomic**, consisting of single atoms. To the extent that these atoms behave like structureless mass points, the only energy they can have is kinetic energy of translational motion. We can think of that kinetic energy as being a sum of *three* terms, each associated with motion in one of the three mutually perpendicular directions. We call each separate term in the energy of a system a **degree of freedom**, meaning a way that system can take on energy. So a monatomic molecule has three degrees of freedom.

In contrast, hydrogen, oxygen, and nitrogen molecules are **diatomic**, as shown in Fig. 18.15. Although a gas of such molecules should still obey the ideal-gas law $PV = nRT$, these molecules can have rotational as well as translational kinetic energy. Then the kinetic energy of a diatomic molecule consists of *five* terms, three for the three directions of translational motion and two for rotational motions about the two mutually perpendicular axes shown in Fig. 18.15. So a diatomic molecule has five degrees of freedom. You'll now see how this difference between *three* degrees of freedom for monatomic molecules and *five* for diatomic molecules accounts for the difference between their specific heats.

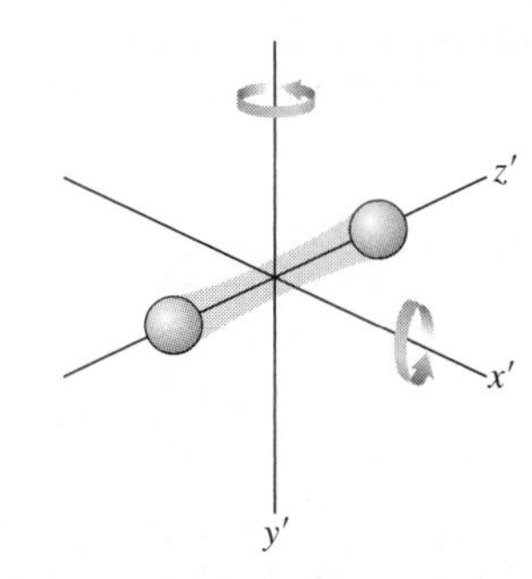

FIGURE 18.15 A diatomic molecule can have significant rotation about two perpendicular axes.

The Equipartition Theorem

We showed in Chapter 17 that the average kinetic energy associated with a gas molecule's motion in one direction is $\frac{1}{2}kT$. We then argued that all three directions are equally probable, making the molecular kinetic energy, on average, $\frac{3}{2}kT$. The argument from one direction to three is based on the assumption that random collisions will share energy equally among the possible motions. When a molecule can rotate as well as translate, energy should be shared also among possible rotational motions. The 19th-century Scottish physicist James Clerk Maxwell first proved this fact, which is known as the **equipartition theorem**:

Equipartition theorem When a system is in thermodynamic equilibrium, the average energy per molecule is $\frac{1}{2}kT$ for each degree of freedom.

We've just seen that a diatomic molecule has five degrees of freedom: three translational and two rotational. The average energy of such a molecule is then $5\left(\frac{1}{2}kT\right) = \frac{5}{2}kT$, so the total internal energy in n moles of a diatomic gas is $E_{\text{int}} = nN_A\left(\frac{5}{2}kT\right) = \frac{5}{2}nRT$. Equation 18.6 then gives the molar specific heat at constant volume:

$$C_V = \frac{1}{n}\frac{\Delta E_{\text{int}}}{\Delta T} = \frac{5}{2}R \quad \text{(diatomic molecule)}$$

Our result $C_p = C_V + R$ still holds, since it was derived from the first law of thermodynamics without regard to molecular structure, so $C_p = \frac{7}{2}R$ and $\gamma = C_p/C_V = \frac{7}{5} = 1.4$. These results describe the observed behavior of diatomic gases like hydrogen, oxygen, and nitrogen at room temperature.

A polyatomic molecule like NO_2 can rotate about any of three perpendicular axes (Fig. 18.16). It then has a total of six degrees of freedom, giving $E_{\text{int}} = 3nRT$ and corresponding specific heats $C_V = 3R$ and $C_p = C_V + R = 4R$. The adiabatic exponent is then $\gamma = \frac{4}{3} \simeq 1.33$, reasonably close to the experimental value $\gamma = 1.29$ for NO_2.

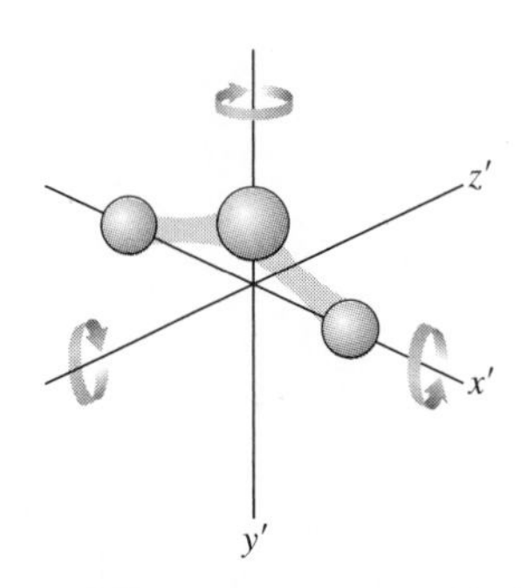

FIGURE 18.16 A triatomic molecule like NO_2 has three rotational degrees of freedom.

EXAMPLE 18.5 Specific Heat: A Gas Mixture

A gas mixture consists of 2.0 mol of oxygen (O_2) and 1.0 mol of argon (Ar). Find the volume specific heat of the mixture.

INTERPRET This problem is about specific heat and molecular structure. We identify the molecules involved as diatomic O_2 and monatomic Ar.

DEVELOP Equation 18.6, $\Delta E_{int} = nC_V\,\Delta T$, determines the volume specific heat, so we need to find how the internal energy E_{int} depends on temperature. Our plan is to use the equipartition theorem to get the energy per molecule for each gas, then find the total energy as a function of temperature, and from that the specific heat.

EVALUATE Being diatomic, O_2 has five degrees of freedom, so the equipartition theorem gives the average energy per molecule as $\frac{5}{2}kT$. Then the total energy in $n = 2$ moles of oxygen is $E_{int\,O_2} = nN_A(\frac{5}{2}kT) = \frac{5}{2}nRT = 5.0RT$, where we used $N_Ak = R$. Monatomic Ar has three degrees of freedom, so the internal energy in our 1 mole of argon is, similarly, $E_{int\,Ar} = \frac{3}{2}nRT = 1.5RT$. The total internal energy is then $E_{int} = 6.5RT$, so Equation 18.6 gives

$$C_V = \frac{1}{n}\frac{\Delta E_{int}}{\Delta T} = \frac{6.5R}{3.0\text{ mol}} = 2.2R$$

ASSESS Make sense? Our answer lies between the values 1.5*R* and 2.5*R* that we found for monatomic and diatomic gases, respectively. It's closer to 2.5*R* because there's more oxygen in the mixture. ■

GOT IT? 18.3 The same amount of heat flows into equal volumes of nitrogen (N_2) and nitrogen dioxide (NO_2), while both are held at constant pressure. Is the resulting temperature rise (a) greater for N_2, (b) the same for both, or (c) greater for NO_2?

Quantum Effects

Relating molecular structure and gas behavior is a remarkable triumph for Newtonian physics. But hidden in our analysis is an assumption that Newtonian physics can't justify. Real atoms have size, so even monatomic molecules should rotate. Why not more degrees of freedom? The answer lies in quantum physics, which requires a certain minimum energy for a periodic motion such as rotation. At normal temperatures, the average thermal energy is too low to excite rotation of monatomic molecules, or of diatomic molecules about their long axis. So these molecules exhibit three and five degrees of freedom, respectively. That results in the volume specific heats $\frac{3}{2}R$ and $\frac{5}{2}R$ that we've seen. For diatomic molecules at higher temperatures, still another motion comes into play—the simple harmonic oscillation of the two atoms due to the springlike bond between them. That adds two more degrees of freedom, corresponding to the kinetic and potential energies of this oscillation, and the specific heat increases correspondingly. At very low temperatures, in contrast, there isn't enough thermal energy to excite any rotation in a diatomic gas, and it then exhibits the specific heat $C_V = \frac{3}{2}R$ that we normally associate with a monatomic gas. Figure 18.17 shows these effects for diatomic hydrogen (H_2).

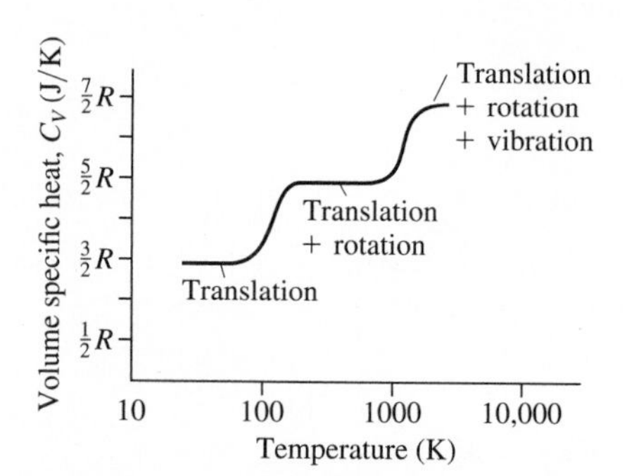

FIGURE 18.17 Volume specific heat of H_2 gas as a function of temperature. Below 20 K hydrogen is liquid, and above 3200 K it dissociates into individual atoms.

Are you bothered by the strange restrictions quantum mechanics imposes on molecular rotation and vibration? You should be! Nothing in your experience suggests that a rotating object can't have any amount of energy you care to give it. But quantum mechanics deals with a realm much smaller than that of our daily experience. The quantization of energy is only one of many unusual things that occur in the quantum realm. We'll explore more quantum phenomena in Part 6.

CHAPTER 18 SUMMARY

Big Idea

The big idea here is conservation of energy, now expanded to include heat. The expanded statement of energy conservation is the **first law of thermodynamics**, which relates the change in a system's internal energy to the heat flowing into the system and the work done on the system. The first law can be used with the ideal-gas law to give a quantitative description of basic thermodynamic processes applied to ideal gases; these are described graphically using ***pV* diagrams**. The **equipartition theorem** states that in thermodynamic equilibrium, internal energy is shared equally among the possible energy modes of a system.

Key Concepts and Equations

Quantitatively, the first law of thermodynamics states

$$\Delta E_{\text{int}} = Q + W$$

Meaning of terms in the first law:

- ΔE_{int} is the change in a system's internal energy.
- Q is the heat transferred *to* the system.
  - Positive Q means a net heat input to the system.
  - Negative Q means heat leaves the system.
- W is the work done *on* the system.
  - Positive W means work is done on the system.
  - Negative W means the system does work on its surroundings.

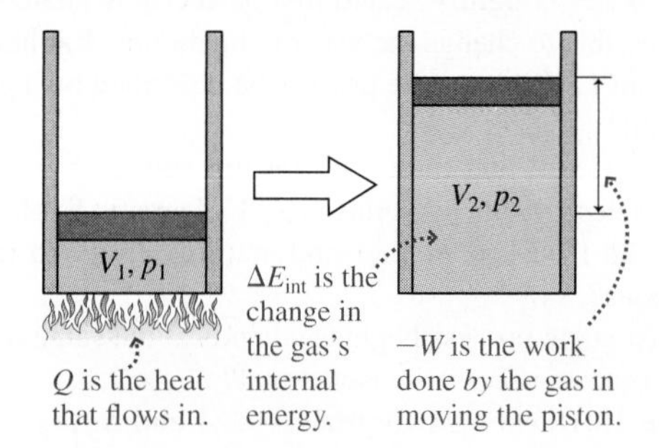

In general, the work done by a system is related to the changes in pressure and volume:

$$W = -\int_{V_1}^{V_2} p\, dV$$

Applications

Ideal-gas processes:

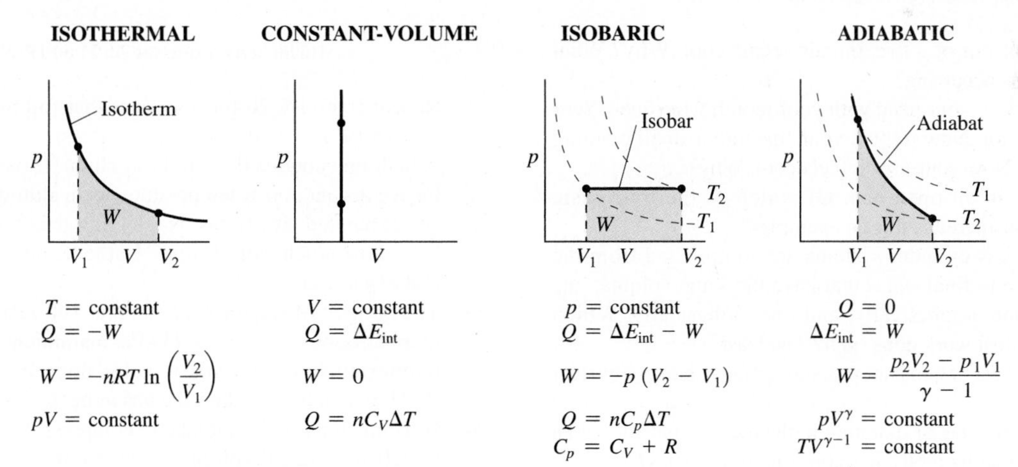

The specific heats of an ideal gas follow from the **degrees of freedom** of each molecule:

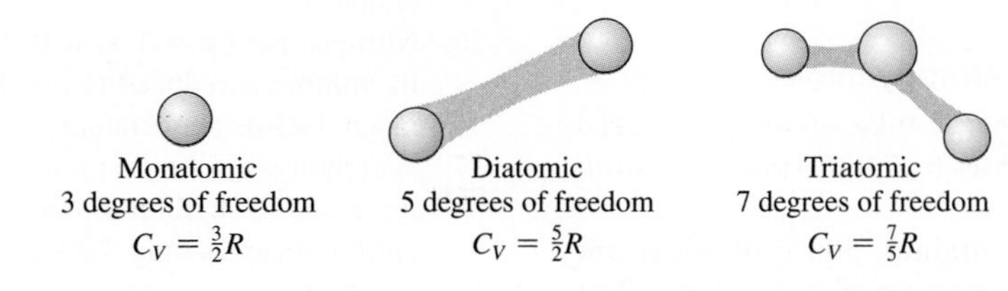

MP For homework assigned on MasteringPhysics, go to www.masteringphysics.com

BIO *Biology and/or medicine-related problems* **DATA** *Data problems* **ENV** *Environmental problems* **CH** *Challenge problems* **COMP** *Computer problems*

For Thought and Discussion

1. The temperature of the water in a jar is raised by violently shaking the jar. Which of the terms Q and W in the first law of thermodynamics is involved in this case?
2. What's the difference between heat and internal energy?
3. Some water is tightly sealed in a perfectly insulated container. Is it possible to change the water temperature? Explain.
4. Why can't an irreversible process be described by a path in a pV diagram?
5. Are the initial and final equilibrium states of an irreversible process describable by points in a pV diagram? Explain.
6. Does the first law of thermodynamics apply to irreversible processes?
7. A quasi-static process begins and ends at the same temperature. Is the process necessarily isothermal?
8. Figure 18.18 shows two processes, A and B, that connect the same initial and final states, 1 and 2. For which process is more heat added to the system?

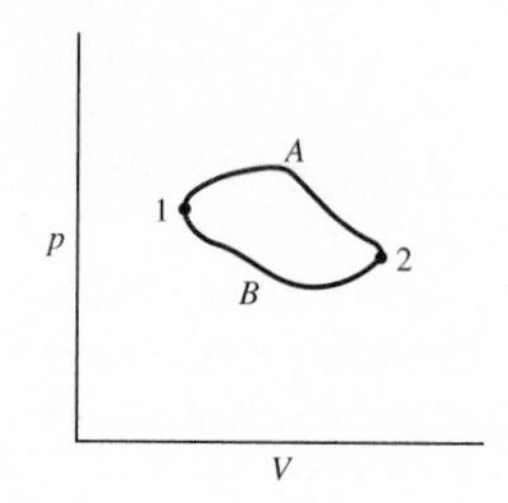

FIGURE 18.18 For Thought and Discussion 8

9. When you let air out of a tire, the air seems cool. Why? What kind of process is occurring?
10. Blow on the back of your hand with your mouth wide open. Your breath will feel hot. Now tighten your lips into a small opening and blow again. Now your breath feels cool. Why?
11. You boil water in an open pan. Of which of the four basic processes we considered is this an example?
12. Three identical gas-cylinder systems are compressed from the same initial state to final states that have the same volume, one isothermally, one adiabatically, and one isobarically. Which system has the most work done on it? The least?
13. Why is specific heat at constant pressure greater than at constant volume?
14. In what sense can a gas of diatomic molecules be considered an ideal gas, given that its molecules aren't point particles?

Exercises and Problems

Exercises

Section 18.1 The First Law of Thermodynamics

15. In a perfectly insulated container, 1.0 kg of water is stirred vigorously until its temperature rises by 7.0°C. How much work is done on the water?
16. In a closed but uninsulated container, 500 g of water are shaken violently until the temperature rises by 3.0°C. The mechanical work done in the process is 9.0 kJ. (a) How much heat is transferred to the surroundings during the shaking? (b) How much mechanical energy would have been required if the container had been perfectly insulated?
17. A 40-W heat source is applied to a gas sample for 25 s, during which time the gas expands and does 750 J of work on its surroundings. By how much does the internal energy of the gas change?
18. Find the rate of heat flow into a system whose internal energy is increasing at the rate of 45 W, given that the system is doing work at the rate of 165 W.
19. In a certain automobile engine, 17% of the total energy released in burning gasoline ends up as mechanical work. What's the engine's mechanical power output if its heat output is 68 kW?

Section 18.2 Thermodynamic Processes

20. An ideal gas expands from the state (p_1, V_1) to the state (p_2, V_2), where $p_2 = 2p_1$ and $V_2 = 2V_1$. The expansion proceeds along the diagonal path AB in Fig. 18.19. Find an expression for the work done *by* the gas during this process.

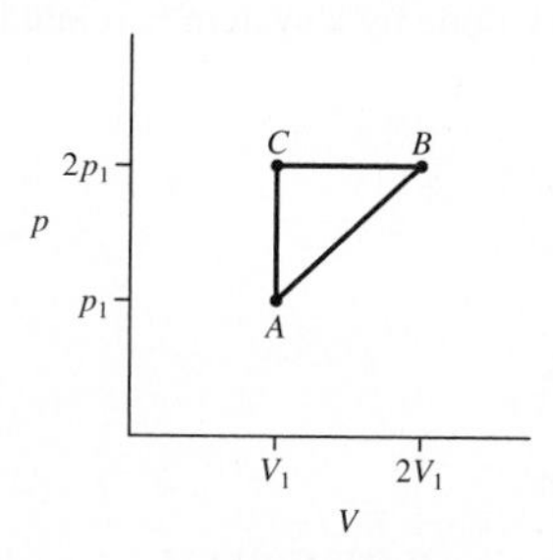

FIGURE 18.19 Exercises 20, 21 and Problem 75

21. Repeat Exercise 20 for a process that follows the path ACB in Fig. 18.19.
22. A balloon contains 0.30 mol of helium. It rises, while maintaining a constant 300-K temperature, to an altitude where its volume has expanded five times. Neglecting tension forces in the balloon, how much work is done *by* the helium during this isothermal expansion?
23. The balloon of Exercise 22 starts at 100 kPa pressure and rises to an altitude where $p = 75$ kPa, maintaining a constant 300 K temperature. (a) By what factor does its volume increase? (b) How much work does the gas in the balloon do?
24. How much work does it take to compress 2.5 mol of an ideal gas to half its original volume while maintaining a constant 300 K temperature?
25. By what factor must the volume of a gas with $\gamma = 1.4$ be changed in an adiabatic process if the kelvin temperature is to double?
26. Nitrogen gas ($\gamma = 1.4$) at 18°C is compressed adiabatically until its volume is reduced to one-fourth of its initial value. By how much does its temperature increase?
27. **ENV** A carbon-sequestration scheme calls for isothermally compressing 6.8 m^3 of carbon dioxide, initially at atmospheric pressure, until it occupies only 5.0% of its original volume. Find the work required.

Section 18.3 Specific Heats of an Ideal Gas

28. A gas mixture contains 2.5 mol of O_2 and 3.0 mol of Ar. What are this mixture's molar specific heats C_V and C_p at constant volume and constant pressure?
29. A mixture of monatomic and diatomic gases has specific-heat ratio $\gamma = 1.52$. What fraction of its molecules are monatomic?
30. What should be the approximate specific-heat ratio of a gas consisting of 50% NO_2 molecules ($\gamma = 1.29$), 30% O_2 ($\gamma = 1.40$), and 20% Ar ($\gamma = 1.67$)?
31. By how much does the temperature of (a) an ideal monatomic gas and (b) an ideal diatomic gas (with molecular rotation but no vibration) change in an adiabatic process in which 2.5 kJ of work are done on each mole of gas?

Problems

32. An ideal gas expands to 10 times its original volume, maintaining a constant 440 K temperature. If the gas does 3.3 kJ of work on its surroundings, (a) how much heat does it absorb, and (b) how many moles of gas are there?
33. **BIO** During cycling, the human body typically releases stored energy from food at the rate of 500 W, and produces about 120 W of mechanical power. At what rate does the body produce heat during cycling?
34. A 0.25-mol sample of ideal gas initially occupies 3.5 L. If it takes 61 J of work to compress the gas isothermally to 3.0 L, what's the temperature?
35. **BIO** As the heart beats, blood pressure in an artery varies from a high of 125 mm of mercury to a low of 80 mm. These values are gauge pressures—that is, excesses over atmospheric pressure. An air bubble trapped in an artery has diameter 1.52 mm when blood pressure is at its minimum. (a) What will its diameter be at maximum pressure? (b) How much work does the blood (and ultimately the heart) do in compressing this bubble, assuming the air remains at the same 37.0°C temperature as the blood?
36. It takes 1.5 kJ to compress a gas isothermally to half its original volume. How much work would it take to compress it by a factor of 22 starting from its original volume?
37. A gas undergoes an adiabatic compression during which its volume drops to half its original value. If the gas pressure increases by a factor of 2.55, what's its specific-heat ratio γ?
38. A gas with $\gamma = 1.40$ occupies 6.25 L when it's at 98.5 kPa pressure. (a) What's the pressure after the gas is compressed adiabatically to 4.18 L? (b) How much work does that compression require?
39. A gas sample undergoes the cyclic process *ABCA* shown in Fig. 18.20, where *AB* is an isotherm. The pressure at *A* is 60 kPa. Find (a) the pressure at *B* and (b) the net work done on the gas.

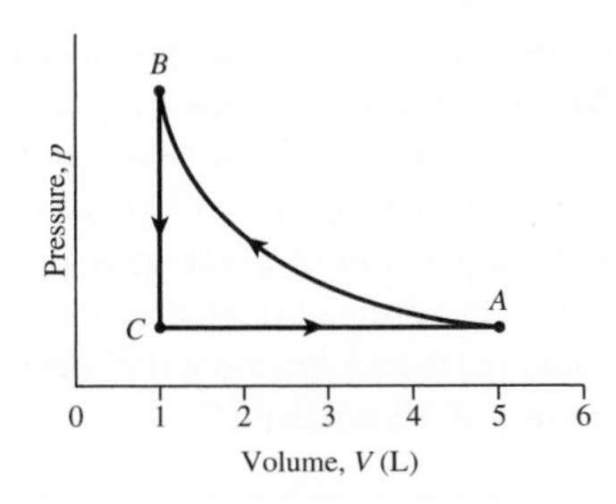

FIGURE 18.20 Problems 39 and 40

40. Repeat Problem 39 taking *AB* as an adiabat and using specific-heat ratio $\gamma = 1.4$.
41. A gasoline engine has compression ratio 8.5 (see Example 18.3 for the meaning of this term), and the fuel–air mixture compresses adiabatically with $\gamma = 1.4$. If the mixture enters the engine at 30°C, what will its temperature be at maximum compression?
42. By what factor must the volume of a gas with $\gamma = 1.4$ be changed in an adiabatic process if the pressure is to double?
43. Volvo's B5340 engine, used in the V70 series cars, has compression ratio 10.2, and the fuel–air mixture undergoes adiabatic compression with $\gamma = 1.4$. If air at 320 K and atmospheric pressure fills an engine cylinder at its maximum volume, what will be (a) the temperature and (b) the pressure at the point of maximum compression?
44. A research balloon is prepared for launch by pumping into it 1.75×10^3 m^3 of helium gas at 12°C and 1.00 atm pressure. It rises high into the atmosphere to where the pressure is only 0.340 atm. Assuming the balloon doesn't exchange significant heat with its surroundings, find (a) its volume and (b) its temperature at the higher altitude.
45. Monatomic argon gas is initially at a chilly 28 K. By what factor would you have to increase its pressure, adiabatically, to bring it to room temperature (293 K)?
46. By what factor does the internal energy of an ideal diatomic gas change when it's compressed to half its original volume (a) isothermally, (b) isobarically, or (c) adiabatically?
47. An ideal monatomic gas is compressed to half its original volume. (a) By what factor is the work greater when the compression is adiabatic as compared with isothermal? (b) Where does the extra work go?
48. A gas expands isothermally from state *A* to state *B*, in the process absorbing 35 J of heat. It's then compressed isobarically to state *C*, where its volume equals that of state *A*. During the compression, 22 J of work are done on the gas. The gas is then heated at constant volume until it returns to state *A*. (a) Draw a *pV* diagram for this process. (b) How much work is done on or by the gas during the complete cycle? (c) How much heat is transferred to or from the gas as it goes from *B* to *C* to *A*?
49. A 3.50-mol sample of ideal gas with molar specific heat $C_V = \frac{5}{2}R$ is initially at 255 K and 101 kPa pressure. Determine the final temperature and the work done *by* the gas when 1.75 kJ of heat are added to the gas (a) isothermally, (b) at constant volume, and (c) isobarically.
50. Prove that the slope of an adiabat at a given point in a *pV* diagram is γ times the slope of the isotherm passing through the same point.
51. An ideal gas with $\gamma = 1.67$ starts at point *A* in Fig. 18.21, where its volume and pressure are 1.00 m^3 and 250 kPa, respectively. It undergoes an adiabatic expansion that triples its volume, ending at *B*. It's then heated at constant volume to *C*, and compressed isothermally back to *A*. Find (a) the pressure at *B*, (b) the pressure at *C*, and (c) the net work done on the gas.

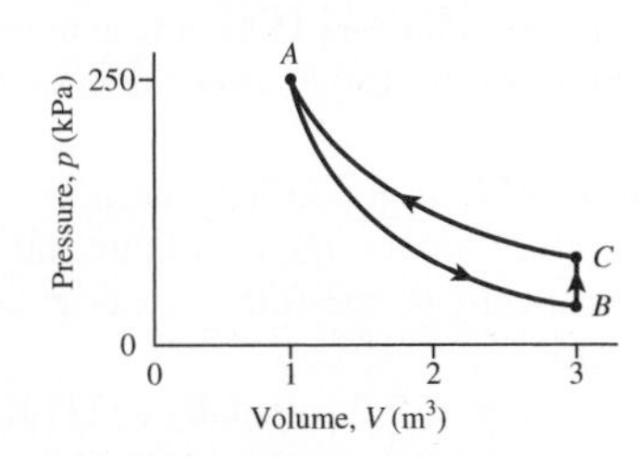

FIGURE 18.21 Problem 51

52. The gas of Example 18.4 starts at state A in Fig. 18.14 and is compressed adiabatically until its volume is 2.0 L. It's then cooled at constant pressure until it reaches 300 K, then allowed to expand isothermally back to state A. Find (a) the net work done on the gas and (b) the minimum volume of the gas. **CH**

53. The gas of Example 18.4 starts at state A in Fig. 18.14 and is heated at constant volume until its pressure has doubled. It's then compressed adiabatically until its volume is one-fourth its original value, then cooled at constant volume to 300 K, and finally allowed to expand isothermally to its original state. Find the net work done on the gas.

54. A 25-L sample of ideal gas with $\gamma = 1.67$ is at 250 K and 50 kPa. The gas is compressed isothermally to one-third of its original volume, then heated at constant volume until its state lies on the adiabatic curve that passes through its original state, and then allowed to expand adiabatically to that original state. Find the net work involved. Is net work done *on* or *by* the gas?

55. Show that the relation between pressure and temperature in an adiabatic process is $p^{1-\gamma}T^{\gamma} = \text{constant}$.

56. A 25-L sample of ideal gas with $\gamma = 1.67$ is at 250 K and 50 kPa. The gas is compressed adiabatically until its pressure triples, then cooled at constant volume back to 250 K, and finally allowed to expand isothermally to its original state. (a) How much work is done on the gas? (b) What is the gas's minimum volume? (c) Sketch this cyclic process in a pV diagram.

57. You're the product safety officer for a company that makes cycling accessories. You're given a new design for a bicycle pump that includes a cylinder 32 cm long when the pump handle is all the way out. To keep the pump from getting too hot, you specify that the temperature rise should not exceed 75°C when the handle is pushed rapidly, with the outlet blocked, until the internal length of the cylinder is 16 cm. Assuming air initially at 18°C, does the pump meet your temperature-rise criterion?

58. Figure 18.22 shows data and a fit curve from an experimental measurement of the pressure–volume curve for a human lung. Estimate the work involved in fully inflating the lung. **BIO**

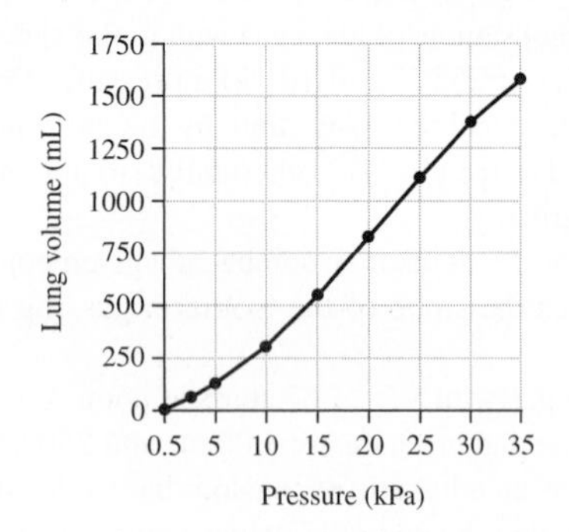

FIGURE 18.22 Problem 58

59. External forces compress 21 mol of ideal monatomic gas. During the process, the gas transfers 15 kJ of heat to its surroundings, yet its temperature rises by 160 K. How much work was done on the gas?

60. A gas with $\gamma = 7/5$ is at 273 K when it's compressed isothermally to one-third of its original volume and then further compressed adiabatically to one-fifth of its original volume. Find its final temperature.

61. An ideal gas with $\gamma = 1.3$ is initially at 273 K and 100 kPa. The gas is compressed adiabatically to 240-kPa pressure. Find its final temperature.

62. The curved path in Fig. 18.23 lies on the 350-K isotherm for an ideal gas with $\gamma = 1.4$. (a) Calculate the net work done on the gas as it goes around the cyclic path $ABCA$. (b) How much heat flows into or out of the gas on the segment AB? **CH**

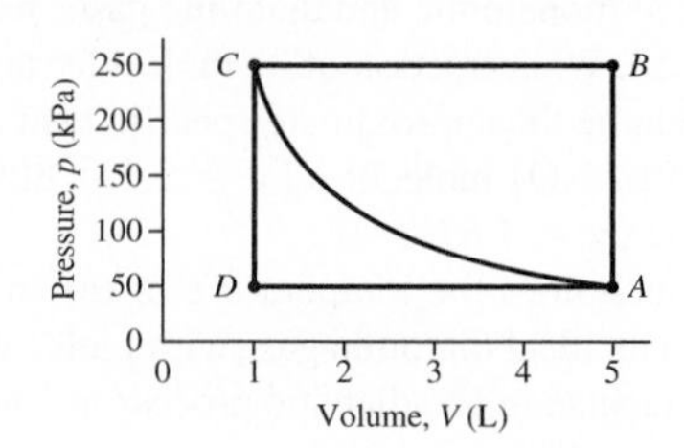

FIGURE 18.23 Problems 62 and 63

63. Repeat part (a) of Problem 62 for the path $ACDA$ in Fig. 18.23. (b) How much heat flows into or out of the gas on the segment CD?

64. A gas mixture contains monatomic argon and diatomic oxygen. An adiabatic expansion that doubles its volume results in the pressure dropping to one-third of its original value. What fraction of the molecules are argon?

65. How much of a triatomic gas with $C_V = 3R$ would you have to add to 10 mol of monatomic gas to get a mixture whose thermodynamic behavior was like that of a diatomic gas?

66. An 8.5-kg rock at 0°C is dropped into a well-insulated vat containing a mixture of ice and water at 0°C. When equilibrium is reached, there are 6.3 g less ice. From what height was the rock dropped?

67. A piston–cylinder arrangement containing 0.30 mol of nitrogen at high pressure is in thermal equilibrium with an ice–water bath containing 200 g of ice. The pressure of the ambient air is 1.0 atm. The gas is allowed to expand isothermally until it's in pressure balance with its surroundings. After the process is complete, the bath contains 210 g of ice. What was the original gas pressure? **CH**

68. Experimental studies show that the pV curve for a frog's lung can be approximated by $p = 10v^3 - 67v^2 + 220v$, with v in mL and p in Pa. Find the work done when such a lung inflates from zero to 4.5 mL volume. **BIO**

69. Show that the application of Equation 18.3 to an adiabatic process results in Equation 18.12.

70. A horizontal piston–cylinder system containing n mol of ideal gas is surrounded by air at temperature T_0 and pressure p_0. If the piston is displaced slightly from equilibrium, show that it executes simple harmonic motion with angular frequency $\omega = Ap_0/\sqrt{MnRT_0}$, where A and M are the piston area and mass, respectively. Assume the gas temperature remains constant. **CH**

71. Use the ideal-gas law to eliminate pressure in Equation 18.11a, and show that the result can be written as Equation 18.11b.

72. The table below shows measured values of pressure versus volume for an ideal gas undergoing a thermodynamic process. Make a log–log plot (logarithm of p versus logarithm of V) of these data and use it to determine (a) whether the process is isothermal or adiabatic and (b) the temperature if it's isothermal and the adiabatic exponent γ if it's adiabatic. **DATA**

Volume, V (L)	1.1	1.27	1.34	1.56	1.82	2.14	2.37
Pressure, p (atm)	0.998	0.823	0.746	0.602	0.493	0.372	0.344

73. In a reversible process, a volume of air $V_0 = 17\,\text{m}^3$ at pressure $p_0 = 1.0$ atm is compressed such that the pressure and volume are related by $(p/p_0)^{-2} = V/V_0$. How much work is done by the gas in reaching a final pressure of 1.4 atm? **CH**
74. A real gas is more accurately described using the van der Waals equation: $[p + a(n/V)^2](V - nb) = nRT$, where a and b are constants. Find an expression, corresponding to Equation 18.4, for the work done by a van der Waals gas undergoing an isothermal expansion from V_1 to V_2. **CH**
75. Repeat Exercise 20 for an expansion along the path $p = p_1[1 + (V - V_1)^2/V_1^2]$.
76. The *adiabatic lapse rate* is the rate at which air cools as it rises and expands adiabatically in the atmosphere (see Application: Smog Alert, on page 324). Express dT in terms of dp for an adiabatic process, and use the hydrostatic equation (Equation 15.2) to express dp in terms of dy. Then, calculate the lapse rate dT/dy. Take air's average molecular weight to be 29 u and $\gamma = 1.4$, and remember that the altitude y is the negative of the depth h in Equation 15.2. **ENV**
77. The nuclear power plant at which you're the public affairs manager has a backup gas-turbine system. The backup system produces electrical energy at the rate of 360 MW, while extracting energy from natural gas at the rate of 670 MW. The local town council has raised concern over waste thermal energy dumped into the environment. Their standards state the thermal waste power must not exceed 400 MW and that all power generation must be at least 50% efficient. Does the backup turbine meet this standard? **ENV**
78. Your class on alternative habitats is designing an underwater habitat. A small diving bell will be lowered to the habitat. A hatch at the bottom of the bell is open, so water can enter to compress the air and thus keep the air pressure inside equal to the pressure of the surrounding water. The bell is lowered slowly enough that the inside air remains at the same temperature as the water. But the water temperature increases with depth in such a way that the air pressure and volume are related by $p = p_0\sqrt{V_0/V}$, where $V_0 = 17\,\text{m}^3$ and $p_0 = 1.0$ atm are the surface values. Suppose the diving bell's air volume cannot be less than $8.7\,\text{m}^3$ and the pressure must not exceed 1.5 atm when submerged. Are these criteria met? **CH**
79. One scheme for reducing greenhouse-gas emissions from coal-fired power plants calls for capturing carbon dioxide and pumping it into the deep ocean, where the pressure is at least 350 atm. You're called to assess the energy cost of such a scheme for a power plant that produces electrical energy at the rate of 1.0 GW while at the same time emitting CO_2 at the rate of 1100 tonnes/hour. If CO_2 is extracted from the plant's smokestack at 320 K and 1 atm pressure and then compressed adiabatically to 350 atm, what fraction of the plant's power output would be needed for the compression? Take $\gamma = 1.3$ for CO_2. (Your answer is a rough estimate because CO_2 doesn't behave like an ideal gas at very high pressures; also, it doesn't include the energy cost of separating the CO_2 from other stack gases or of transporting it to the compression site.) **ENV**

Passage Problems

ENV Warm winds called Chinooks (a Native-American term meaning "snow eaters") sometimes sweep across the plains just east of the Rocky Mountains. These winds carry air from high in the mountains down to the plains rapidly enough that the air has no time to exchange heat with its surroundings (Fig. 18.24). On a particular Chinook day, temperature and pressure high in the Colorado Rockies are 60 kPa and 260 K (−13°C), respectively; the plain below is at 90 kPa.

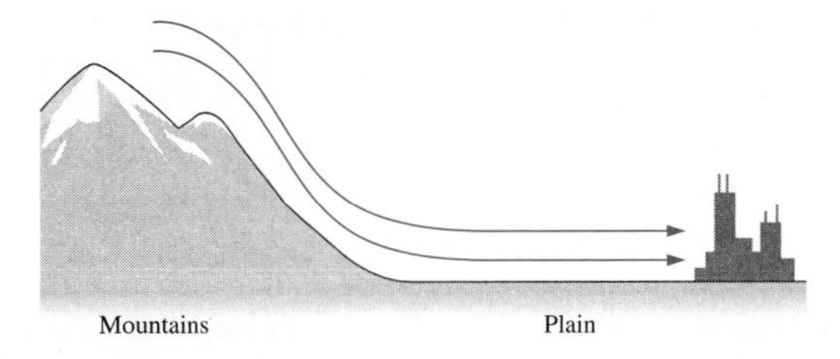

FIGURE 18.24 Chinooks (Passage Problems 80–83)

80. The process the air undergoes as it descends the mountains is
    a. isothermal.
    b. isovolumic.
    c. isobaric.
    d. adiabatic.
81. As the air descends, its internal energy
    a. increases.
    b. decreases.
    c. is unchanged.
82. As the air descends, its volume
    a. increases by 50%.
    b. increases by less than 50%.
    c. decreases by 50%.
    d. decreases by less than 50%.
    e. is unchanged.
83. When the air reaches the plain, its temperature is approximately
    a. 240 K.
    b. 260 K.
    c. 290 K.
    d. 390 K.

Answers to Chapter Questions

Answer to Chapter Opening Question

Energy is conserved, provided thermal energy is included. The engine produces both mechanical energy and thermal energy of its exhaust gases; together, they sum to the energy released in combustion.

Answers to GOT IT? Questions

18.1 (c) Only the internal energy is the same, since it's a thermodynamic state variable unique to a point in the pV diagram.

18.2 (1) Constant-volume, T and p increase, V doesn't change, E_{int} increases as heat flows into the gas; (2) Adiabatic, T and p increase, V decreases, E_{int} increases as work is done on the gas; (3) Isobaric, T decreases, p doesn't change, V decreases, E_{int} decreases as heat flows out of the gas

18.3 (a) because the energy is spread over fewer degrees of freedom

The Second Law of Thermodynamics

What You Know

- You're familiar with the first law of thermodynamics, and how it relates heat, work, and internal energy.
- You can describe basic thermodynamic processes, especially isothermal and adiabatic processes.

What You're Learning

- The big idea here is the second law of thermodynamics, which limits our ability to extract useful work from thermal energy.
- You'll learn about heat engines, both as practical devices and as conceptual tools for exploring the second law of thermodynamics.
- You'll see how refrigerators and heat pumps operate as engines in reverse, subject to similar thermodynamic limitations.
- You'll learn the concept of *entropy* and how it provides a measure of disorder in systems.
- You'll come to understand entropy in statistical terms, and you'll see how to express the second law of thermodynamics as a statement about entropy.

How You'll Use It

- If you go into engineering or science, the second law of thermodynamics will always limit the use of thermal energy in any systems you design or study.
- Even if you don't do engineering or science professionally, the second law will continue to limit the efficiency of your car, of the power plants that produce your electricity, and even of natural processes like the hydrologic cycle that turns sunlight energy into the mechanical energy of water.

Most of the energy extracted from fuel in power plants is discarded as waste heat. The large cooling tower shown here dumps this waste heat into the environment. Why is so much energy wasted?

The first law of thermodynamics relates heat and other forms of energy. Much of our world depends on this relationship. Cars extract energy from the heat of burning gasoline. Most of our electricity originates in heat released by burning fuels or fissioning uranium. Our own bodies run on energy that begins as heat in the Sun's core. But the first law doesn't tell the whole story. Heat and mechanical energy aren't the same, and the difference makes the conversion of heat to work a more subtle task than the first law would imply.

19.1 Reversibility and Irreversibility

Figure 19.1 shows a movie of a bouncing ball. Play it backward and it still makes sense. Figure 19.2 shows a block sliding along a table, slowing because of friction—and warming in the process. Play this film backward and it makes no sense. You'll never see a block at rest suddenly start to move, cooling as it goes. Yet energy would be conserved if it did, so the first law of thermodynamics would be satisfied. Beat an egg, blending yolk and white. Reverse the beater, and you'll never see them separate again. Put cups of cold and hot water in contact; the hot water cools and the cold water warms. The opposite never occurs—although energy would still be conserved.

Why are these events **irreversible**? In each case we start with matter in an organized state. The molecules of the sliding block share a common motion. The yolk molecules are all in one place. The hot water has more energetic molecules. Of all possible states, these *organized* ones are rare. There are many more *disorganized* states—for example, all the possible arrangements of molecules in a scrambled egg. As a system evolves, chances are it will end up less organized, simply because there are far more such states available to it. It's very unlikely to assume spontaneously a more organized state.

A key word here is "spontaneous." We could restore organization—for example, by putting one cup of water in the refrigerator and the other in the microwave—but that requires a rather deliberate and energy-consuming process.

Irreversibility is a probabilistic notion. Events that *could* occur without violating the principles of Newtonian physics nevertheless *don't* occur because they're too improbable. As a practical consequence, harnessing the internal energy associated with random molecular motions is difficult because those motions won't spontaneously become organized. That makes much of the world's energy unavailable for doing useful work.

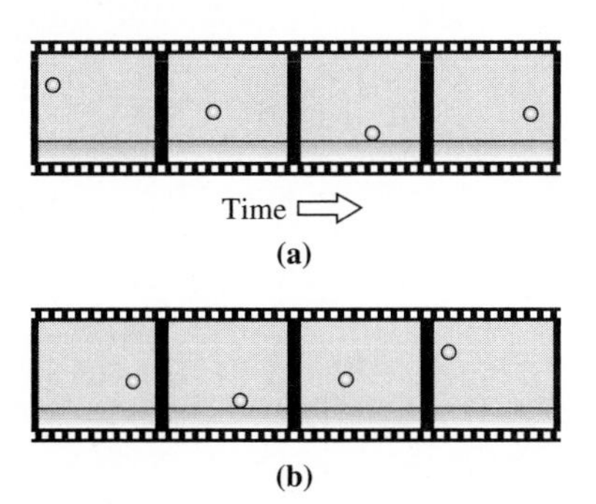

FIGURE 19.1 A movie of a bouncing ball makes sense whether it's shown (a) forward or (b) backward.

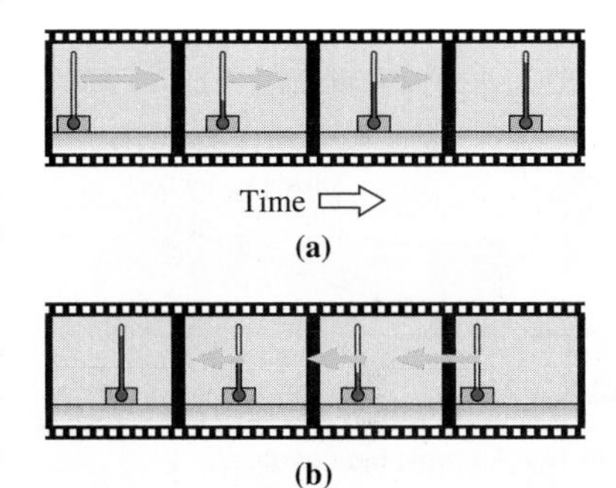

FIGURE 19.2 (a) A block warming (note thermometer) as friction dissipates its kinetic energy and it slows to a stop. (b) The reverse sequence would never happen, even though it doesn't violate energy conservation.

GOT IT? 19.1 Which of these processes is irreversible? (a) stirring sugar into coffee; (b) building a house; (c) demolishing a house with a wrecking ball; (d) demolishing a house by taking it apart piece by piece; (e) harnessing the energy of falling water to drive machinery; (f) harnessing the energy of falling water to heat a house

19.2 The Second Law of Thermodynamics

Heat Engines

It's impossible to convert *all* the internal energy of a system to useful work. But **heat engines** extract *some* of that internal energy. Examples include gasoline and diesel engines, fossil-fueled and nuclear power plants, and jet aircraft engines.

Figure 19.3*a* is an energy-flow diagram for a "perfect" heat engine—one that extracts heat from a heat reservoir and converts it all to work. Such an engine would do exactly what we've just argued against: It would convert the random energy of thermal motion entirely to the ordered motion associated with mechanical work. In fact a perfect heat engine is impossible, for the same reason that we can't unscramble an egg or make a block accelerate spontaneously using its internal energy. This fact leads to one statement of the **second law of thermodynamics**:

> **Second law of thermodynamics (Kelvin–Planck statement)** It is impossible to construct a heat engine operating in a cycle that extracts heat from a reservoir and delivers an equal amount of work.

The phrase "in a cycle" means that a practical engine goes through a repeated sequence of steps, as in the back-and-forth motions of the pistons in a gasoline engine.

A simple heat engine consists of a gas–cylinder system and a heat reservoir, the latter kept hot, perhaps, by burning a fuel. With the gas initially at high pressure, we place the cylinder in contact with the heat reservoir. The gas expands and does work W on the piston. In this isothermal process, the gas extracts heat $Q = W$ from the reservoir. Eventually the gas reaches pressure equilibrium and stops expanding. The piston must then be returned to its original position if it's to do more work.

If we just push the piston back, we'll have to do as much work as we got during the expansion, and our engine won't produce any net work. Instead we can cool the gas to reduce its volume, through contact with a cool reservoir. But then some energy leaves the

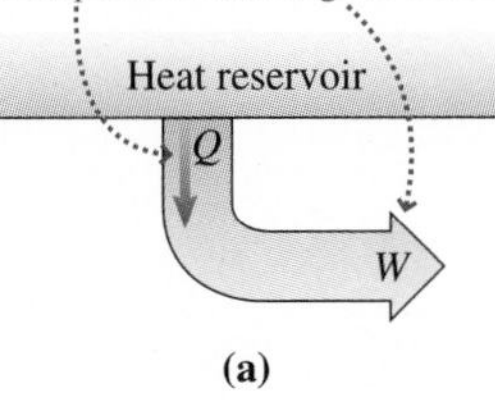

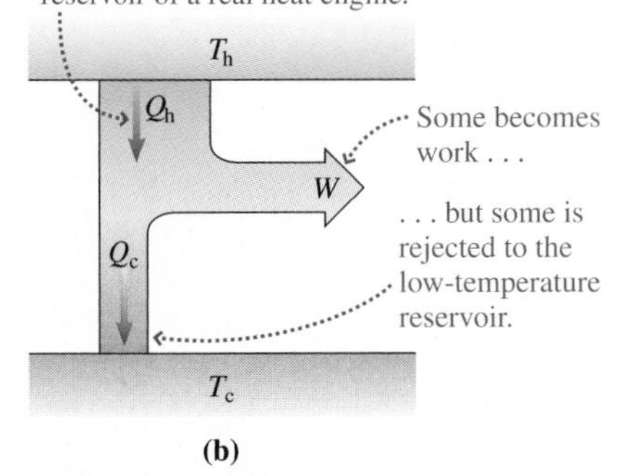

FIGURE 19.3 (a) Energy-flow diagram for a perfect heat engine. (b) A real engine delivers as work only a fraction of the energy extracted from the high-temperature reservoir.

system as heat rather than work, as shown conceptually in Fig. 19.3*b*. Our engine extracts heat from a source and delivers mechanical work, but over a full cycle the work delivered is less than the heat extracted. The remaining energy is rejected to the lower-temperature reservoir, usually the environment. That's why much of the energy released from fuels in car engines and power plants ends up as waste heat.

The second law of thermodynamics says we can't build a perfect heat engine. But how close can we come? We define the **efficiency** e of an engine as the ratio of the work W we get from it to what we have to supply—namely, the heat Q_h: $e = W/Q_h$. Since the process is cyclic, there's no net change in internal energy over one cycle. The first law of thermodynamics then shows that the work W done *by* the engine is the difference between the heat Q_h extracted from the high-temperature reservoir and the heat Q_c rejected to the cool reservoir:

$$e = \frac{W}{Q_h} = \frac{Q_h - Q_c}{Q_h} = 1 - \frac{Q_c}{Q_h} \qquad (19.1)$$

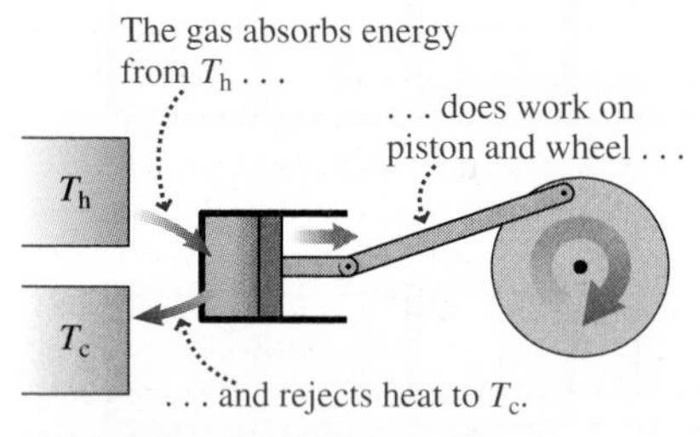

FIGURE 19.4 A simple heat engine.

In this chapter we'll often use W for the work done *by* an engine; in the first law it's the work done *on* a system. That's why W here is equal to the net heat $Q_h - Q_c$.

Figure 19.4 shows a heat engine whose efficiency we can calculate. The engine consists of a cylinder containing an ideal gas, sealed by a movable piston. The piston is connected to a rod that turns a wheel. The engine gets its energy from a heat reservoir at a high temperature T_h, and it rejects heat to a cooler reservoir at temperature T_c. Figure 19.5 shows how the engine works in a cycle of four steps, starting with the piston in its leftmost position (state A in Fig. 19.5), where the gas volume is a minimum:

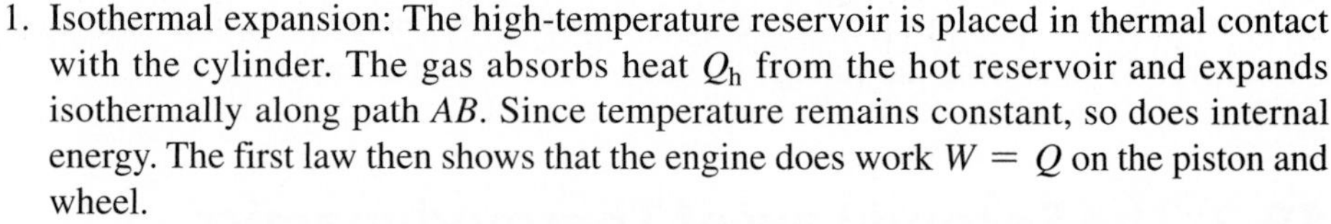

1. Isothermal expansion: The high-temperature reservoir is placed in thermal contact with the cylinder. The gas absorbs heat Q_h from the hot reservoir and expands isothermally along path AB. Since temperature remains constant, so does internal energy. The first law then shows that the engine does work $W = Q$ on the piston and wheel.

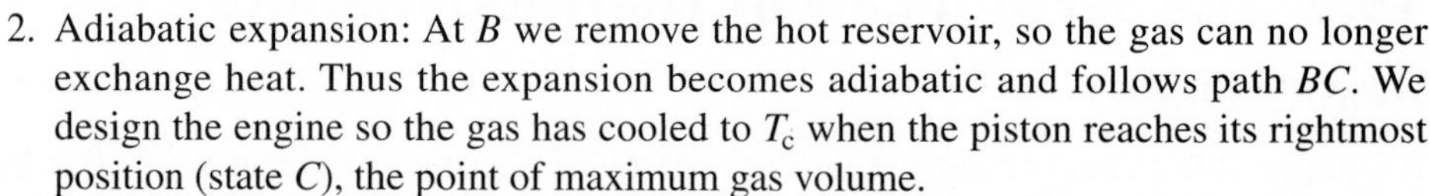

2. Adiabatic expansion: At B we remove the hot reservoir, so the gas can no longer exchange heat. Thus the expansion becomes adiabatic and follows path BC. We design the engine so the gas has cooled to T_c when the piston reaches its rightmost position (state C), the point of maximum gas volume.
3. Isothermal compression: At C we bring the cool reservoir into thermal contact with the cylinder. The wheel's inertia keeps it turning, so the piston does work on the gas, compressing it isothermally from state C to D. This work ends up as heat rejected to the cool reservoir.
4. Adiabatic compression: At D we remove the cool reservoir and the compression continues adiabatically until the gas temperature is once again at T_h and the engine is back at state A.

This cyclic process of two isothermal and two adiabatic steps is the **Carnot cycle** and the engine a **Carnot engine**, after the French engineer Sadi Carnot (1796–1832). The

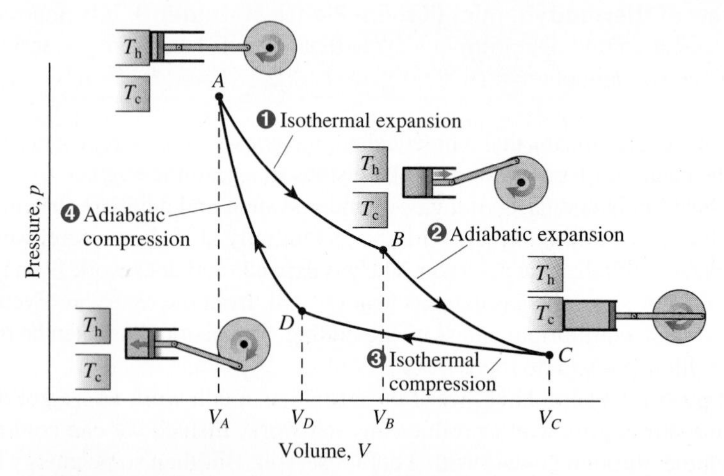

FIGURE 19.5 A pV diagram for the Carnot engine.

particular configuration of the engine isn't important, nor is the choice of an ideal gas as the engine's **working fluid**. What distinguishes the Carnot cycle from others is the sequence of thermodynamic processes and the fact that these processes are reversible. The Carnot engine is an example of a **reversible engine**—one in which thermodynamic equilibrium is maintained so that all steps could, in principle, be reversed.

What's the efficiency of a Carnot engine? To find out, we need the heats Q_h and Q_c absorbed and rejected during the isothermal parts of the cycle shown in Fig. 19.5. Equation 18.4 gives the heat Q_h absorbed during the isothermal expansion *AB*:

$$Q_h = nRT_h \ln\left(\frac{V_B}{V_A}\right)$$

and the heat Q_c rejected during the isothermal compression *CD*:

$$Q_c = -nRT_c \ln\left(\frac{V_D}{V_C}\right) = nRT_c \ln\left(\frac{V_C}{V_D}\right)$$

We put the minus sign here because the first law takes Q to be the heat *absorbed*, while Equation 19.1 for the engine efficiency requires that Q_c be the heat *rejected*. To calculate engine efficiency according to Equation 19.1, we need the ratio Q_c/Q_h:

$$\frac{Q_c}{Q_h} = \frac{T_c \ln(V_C/V_D)}{T_h \ln(V_B/V_A)} \qquad (19.2)$$

This expression can be simplified by applying Equation 18.11b to the adiabatic processes *BC* and *DA* in the Carnot cycle: $T_h V_B^{\gamma-1} = T_c V_C^{\gamma-1}$ and $T_h V_A^{\gamma-1} = T_c V_D^{\gamma-1}$. Dividing the first of these two equations by the second gives

$$\left(\frac{V_B}{V_A}\right)^{\gamma-1} = \left(\frac{V_C}{V_D}\right)^{\gamma-1} \quad \text{or} \quad \frac{V_B}{V_A} = \frac{V_C}{V_D}$$

so Equation 19.2 becomes simply $Q_c/Q_h = T_c/T_h$. Using this result in Equation 19.1 then gives the efficiency of the Carnot engine:

$$e_{\text{Carnot}} = 1 - \frac{T_c}{T_h} \quad \text{(Carnot engine efficiency)} \qquad (19.3)$$

where the temperatures are measured on an absolute scale (Kelvin or Rankine). Equation 19.3 shows that the Carnot engine's efficiency depends only on the highest and lowest temperatures of its working fluid. In practice, the low temperature is usually that of the environment; then maximizing efficiency requires making the high temperature as high as possible. Real engines trade off efficiency with the ability of materials to withstand high temperature and pressure.

APPLICATION **Internal Combustion Engines**

Internal combustion engines (ICEs) power most of the world's cars and trucks and will continue to do so for decades despite inroads by electric propulsion systems. Their name refers to the fact that combustion in an ICE takes place within the engine itself, as opposed to external combustion in systems like power plants (look ahead to Fig. 19.10), industrial boilers, and old-fashioned steam locomotives. Today's ICEs build on more than a century of engineering development, and coupled with modern electronic sensors and control systems, they represent a pinnacle of engineering design.

ICEs include the common gasoline and diesel engines. Both these engines undergo cycles involving back-and-forth motion of pistons that's converted to rotary motion that usually drives a vehicle's wheels. ICEs are heat engines, but standard gasoline and diesel engines aren't Carnot engines. Gasoline engines, for example, operate on a cycle that consists approximately of two adiabatic and two constant-volume segments. Because heat transfer doesn't occur at fixed high and low temperatures, the efficiency is less than the Carnot limit of Equation 19.1. The diesel cycle, consisting approximately of adiabatic, isobaric, and constant-volume segments, is, for the same reason, also less efficient than the Carnot limit. You learned about the adiabatic compression phase of a diesel engine in Example 18.3, and you can explore and compare gasoline and diesel engines further in Problems 54–58. The image shows a cutaway view of a modern gasoline engine.

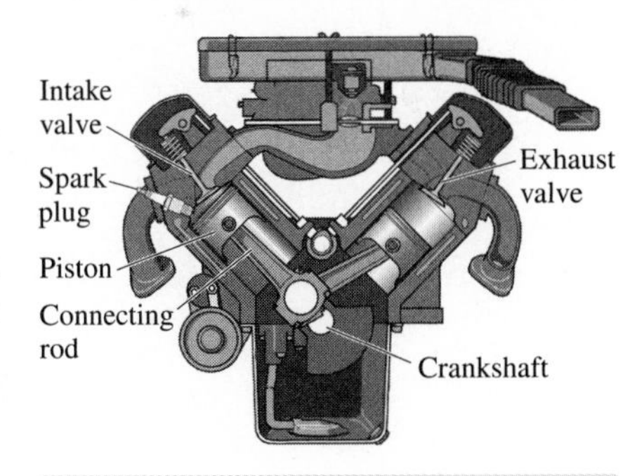

EXAMPLE 19.1 Calculating Efficiency: A Carnot Engine

A Carnot engine extracts 240 J from its high-temperature reservoir during each cycle, and rejects 100 J to the environment at 15°C. How much work does the engine do in one cycle? What's its efficiency? What's the temperature of the hot reservoir?

INTERPRET This problem is about a Carnot engine, which operates via the Carnot cycle.

DEVELOP Equation 19.3, $e_{\text{Carnot}} = 1 - (T_c/T_h)$, relates the two temperatures and the efficiency. Here $Q_h = 240$ J, $Q_c = 100$ J, and $T_c = 15$°C or 288 K. The first law of thermodynamics relates work and heat flows. So our plan is to use the first law to find the work, then find the efficiency, and then use Equation 19.3 to find T_h.

EVALUATE Since there's no change in internal energy over one cycle, the first law requires that the work W done *by* the engine be equal to the net heat absorbed—namely, 240 J − 100 J. So $W = 140$ J. The efficiency is the ratio of work delivered to heat extracted, so $e = W/Q_h = 140 \text{ J}/240 \text{ J} = 58.3\%$. Knowing the efficiency, we solve Equation 19.3 for T_h:

$$T_h = \frac{T_c}{1 - e} = \frac{288 \text{ K}}{1 - 0.583} = 691 \text{ K} = 418°\text{C}$$

ASSESS Make sense? The engine rejects somewhat less than half the 240 J as waste heat, so we should expect efficiency somewhat over 50%. And T_h must be greater than T_c, as our calculation confirms. ■

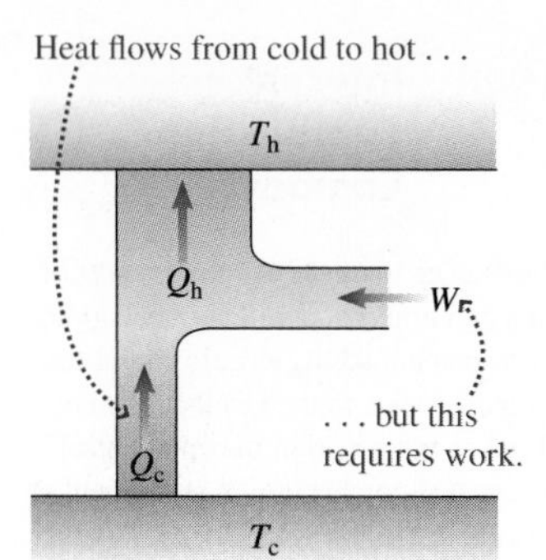

FIGURE 19.6 Energy-flow diagram for a real refrigerator.

Engines, Refrigerators, and the Second Law

Why this emphasis on the Carnot engine? Because understanding this device will help answer the broader question of how much work we can hope to extract from thermal energy. That, in turn, will help you understand practical limitations on humankind's attempts to harness ever more energy and will lead to a deeper understanding of the second law of thermodynamics.

Why is Carnot's engine special? Couldn't you build a better engine with greater efficiency? The answer is no. The special role of the Carnot cycle is embodied in **Carnot's theorem**:

Carnot's theorem All Carnot engines operating between temperatures T_h and T_c have the same efficiency, $e_{Carnot} = 1 - (T_c/T_h)$, and no other engine operating between the same two temperatures can have a greater efficiency.

To prove Carnot's theorem, we introduce the **refrigerator**. A refrigerator is the opposite of an engine: It extracts heat from a cool reservoir and rejects it to a hotter one, using work in the process (Fig. 19.6). A refrigerator forces heat to flow from cold to hot, but to do so it requires work. A household refrigerator cools its contents and warms the house (you can feel the heat coming out the back), but it uses electricity. That heat doesn't flow spontaneously from cold to hot leads to another statement of the second law of thermodynamics:

Second law of thermodynamics (Clausius statement) It is impossible to construct a refrigerator operating in a cycle whose sole effect is to transfer heat from a cooler object to a hotter one.

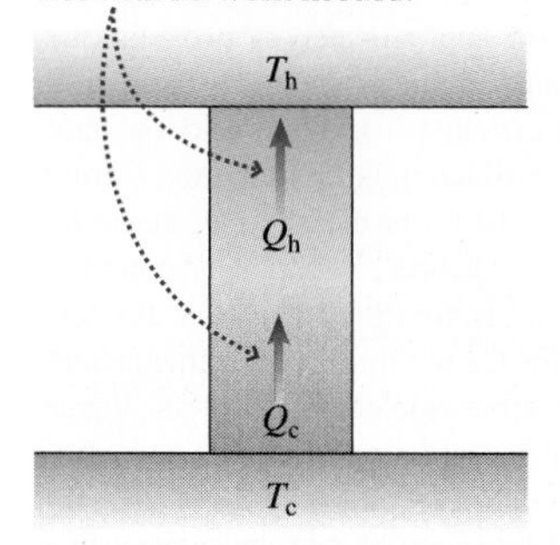

FIGURE 19.7 A perfect refrigerator is impossible.

The Clausius statement rules out a perfect refrigerator (Fig. 19.7).

Suppose the Clausius statement were false. Then we could build the device of Fig. 19.8*a*, consisting of a reversible Carnot engine and a perfect refrigerator. In each cycle the engine would extract, say, 100 J from the hot reservoir, put out 60 J of useful work, and reject 40 J to the cool reservoir. The perfect refrigerator could transfer the 40 J back to the hot reservoir. The net effect would be to extract 60 J from the hot reservoir and convert it entirely to work (Fig. 19.8*b*)—and we would have a perfect heat engine, in violation of the Kelvin–Planck statement of the second law. A similar argument (Problem 38) shows that if a perfect heat engine is possible, then so is a perfect refrigerator. So the Clausius and Kelvin–Planck statements of the second law are equivalent, in that if one is false, then so is the other.

Because the Carnot engine is reversible, we could run it backward and reverse its path in Fig. 19.5. The engine would extract heat from the cool reservoir, take in work, and reject heat to the hot reservoir. It would be a refrigerator. Although real refrigerators aren't designed exactly like engines, the two are, in principle, interchangeable.

We're now ready to prove Carnot's assertion that Equation 19.3 gives the maximum engine efficiency. Consider again the Carnot engine in Fig. 19.8*a*. It extracts 100 J of heat

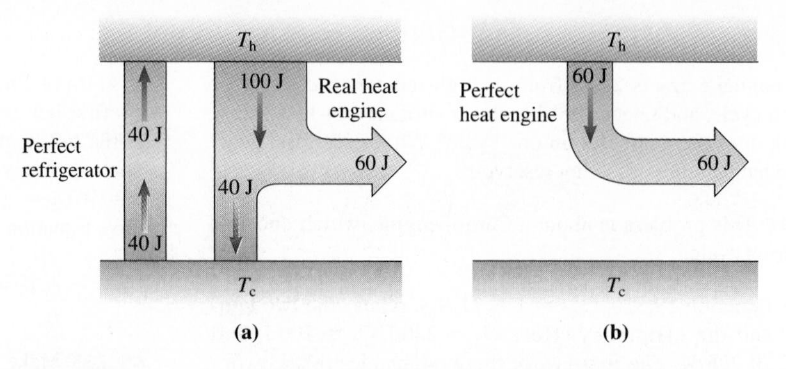

FIGURE 19.8 (a) A real heat engine combined with a perfect refrigerator is equivalent to (b) a perfect heat engine.

and delivers 60 J of work, so it's 60% efficient. Suppose we had another engine operating between the same two reservoirs, but with 70% efficiency. Since the Carnot engine is reversible, we can run it as a refrigerator. If we put the two together, we get the device of Fig. 19.9*a*. Its net effect is to extract 10 J from the cool reservoir and deliver 10 J of work—so it's a perfect heat engine, in violation of the second law (Fig. 19.9*b*). It's therefore impossible to make an engine that's more efficient than a Carnot engine, and thus Equation 19.3 gives the maximum possible efficiency for *any* heat engine operating between the same two fixed temperatures. For that reason the Carnot efficiency of Equation 19.3 is also called the **thermodynamic efficiency**.

Irreversible engines, because they involve processes that dissipate organized motion, are necessarily *less* efficient. So are reversible engines, if their heat exchange doesn't take place solely at the highest and lowest temperatures. The ordinary gasoline engine is a case in point; even if it could be made perfectly reversible, its efficiency would be less than that of a comparable Carnot engine (see Problem 54 and the Application on page 337).

GOT IT? 19.2 The low temperature for a practical heat engine is generally set by the ambient environment, at about 300 K. With that value for T_c, what will happen to the efficiency of a Carnot engine if you re-engineer it so its high temperature T_h doubles? (a) efficiency will double; (b) efficiency will quadruple; (c) efficiency will increase by an amount that depends on the original value of T_h; (d) efficiency will decrease

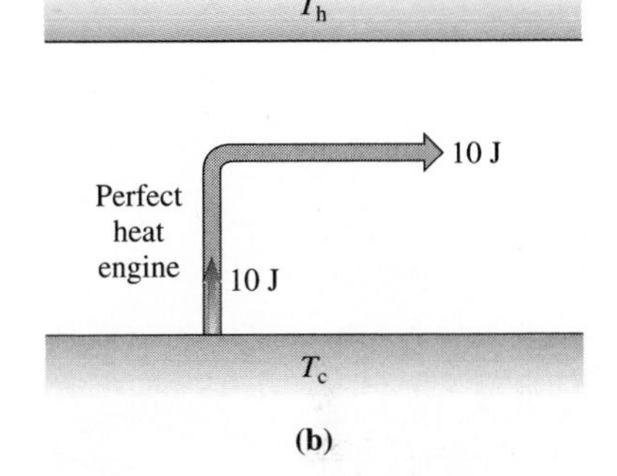

FIGURE 19.9 (a) A 60% efficient reversible engine run as a refrigerator, along with a hypothetical engine with 70% efficiency. (b) The combination is equivalent to a perfect heat engine.

19.3 Applications of the Second Law

The world abounds with thermal energy, but the second law of thermodynamics limits our ability to use that energy. Any device we construct that involves the interchange of heat and work is a heat engine or refrigerator, subject to the second law.

Limitations on Heat Engines

Most of our electricity is produced in large power plants that are heat engines powered by the fossil fuels coal, oil, or natural gas, or by nuclear fission. Figure 19.10 diagrams such a power plant. The working fluid is water, heated in a boiler and converted to steam at high pressure. The steam expands adiabatically to spin a fanlike turbine. The turbine turns a generator that converts mechanical work to electrical energy.

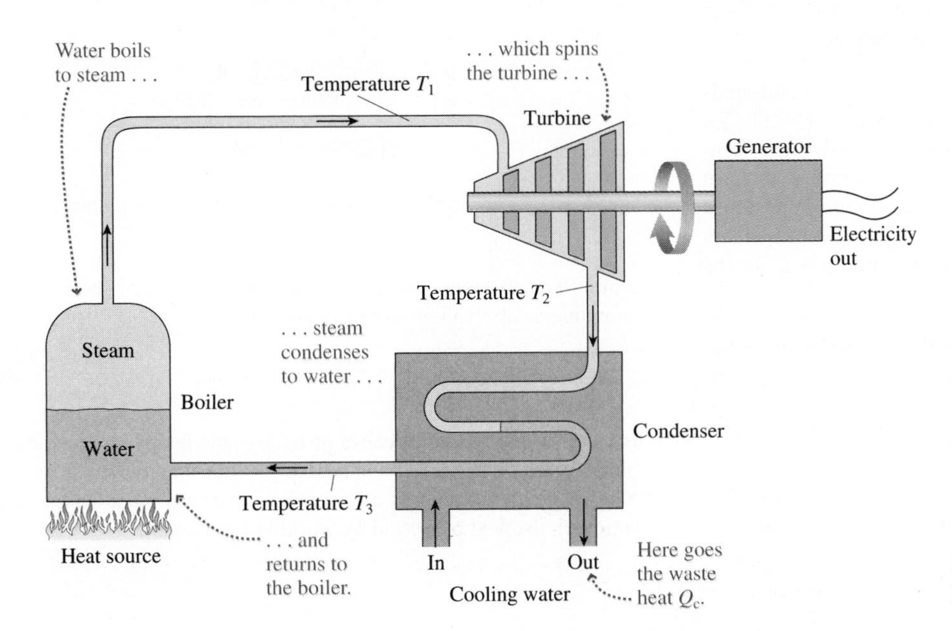

FIGURE 19.10 Schematic diagram of an electric power plant.

Steam leaving the turbine is still gaseous and is hotter than the water supplied to the boiler. Here's where the second law applies! Had the water returned from the turbine in its original state, we would have extracted as work all the energy acquired in the boiler, in violation of the second law. Therefore, we must run the steam through a **condenser**, where it contacts pipes carrying cool water, typically from a river, lake, or ocean. The condensed steam, now cool water, is fed back into the boiler to repeat the cycle.

The maximum steam temperature in a power plant is limited by the materials used in its construction. For a conventional fossil-fuel plant, current technology permits high temperatures of around 650 K. Potential damage to nuclear fuel rods limits the temperature in a nuclear plant to around 570 K. The average temperature of the cooling water is about 40°C (310 K), so the maximum possible efficiencies for these power plants, given by Equation 19.3, are

$$e_{\text{fossil}} = 1 - \frac{310\text{ K}}{650\text{ K}} = 52\% \quad \text{and} \quad e_{\text{nuclear}} = 1 - \frac{310\text{ K}}{570\text{ K}} = 46\%$$

Temperature differences between steam and cooling water, mechanical friction, and energy needed for pumps and pollution-control devices all reduce efficiency further, to about 33% for both nuclear and coal-fired plants—the latter being the world's dominant source of electricity. So roughly two-thirds of the fuel energy we use to make electricity ends up as waste heat.

A typical large power plant produces 1 GW of electricity, so another 2 GW of waste heat goes into the cooling water. The resulting temperature rise can cause serious ecological problems. The huge cooling towers you see at power plants reduce such "thermal pollution" by transferring much of the waste heat to the atmosphere (see this chapter's opening photo). Even so, a substantial fraction of all rainwater falling on the United States eventually finds its way through the condensers of power plants (see Problem 31).

EXAMPLE 19.2 Improving Efficiency: A Combined-Cycle Power Plant

The gas turbine in a combined-cycle power plant (see the Application on the next page) operates at 1450°C. Its waste heat at 500°C is the input for a conventional steam cycle, with its average condenser temperature at 40°C. Find the thermodynamic efficiency of the combined cycle, and compare with the efficiencies of the individual components if they were operated independently.

INTERPRET This problem is about the thermodynamic efficiency of a combined-cycle power plant. As described in the Application, that means a plant using a high-temperature gas turbine whose waste heat becomes the energy input to a conventional steam turbine.

DEVELOP Figure 19.11 is a conceptual diagram of the combined-cycle plant, based on the Application. Equation 19.3, $e = 1 - (T_c/T_h)$, gives the thermodynamic efficiencies of each cycle and of the combination. We identify the 1450°C = 1723 K temperature as T_h in Equation 19.3 for the gas turbine. The intermediate temperature 500°C = 773 K serves as T_c for the gas turbine but as T_h for the steam cycle. Finally, the 40°C or 313-K condenser temperature is T_c for the steam cycle.

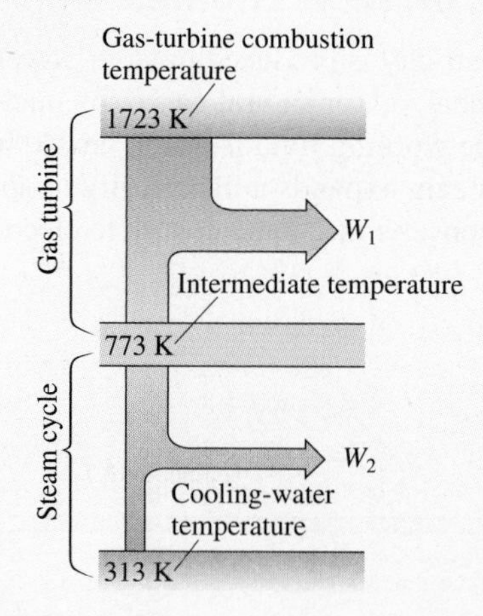

FIGURE 19.11 Conceptual diagram of a combined-cycle power plant.

EVALUATE To treat the entire plant as a single heat engine in Equation 19.3, we use the highest and lowest temperatures:

$$e_{\text{combined}} = 1 - \frac{T_c}{T_h} = 1 - \frac{313\text{ K}}{1723\text{ K}} = 0.82 = 82\%$$

Friction and other losses would reduce this figure substantially, but a combined-cycle plant operating at these temperatures could have a practical efficiency near 60%. The efficiencies of the individual components also follow from Equation 19.3:

$$e_{\text{gas turbine}} = 1 - \frac{773\text{ K}}{1723\text{ K}} = 55\% \quad \text{and} \quad e_{\text{steam}} = 1 - \frac{313\text{ K}}{773\text{ K}} = 60\%$$

ASSESS Make sense? Because of its extreme temperatures, the combined cycle gives an efficiency that's better than either of its parts! You can learn more about combined-cycle power plants in the Application on the next page, and by working Problem 32. ■

APPLICATION Combined-Cycle Power Plants

Improving power-plant efficiency helps reduce air pollution and greenhouse-gas emissions, not to mention the cost of electricity. Modern *combined-cycle* power plants achieve efficiencies approaching 60% by combining a conventional steam system like that of Fig. 19.10 with a *gas turbine* similar to a jet aircraft engine. Gas turbines operate at high temperatures—between 1000 K and 2000 K—but they aren't very efficient because their exhaust temperature (T_c in Equation 19.3) is also high. In a combined-cycle plant, exhaust from a gas turbine drives a conventional steam cycle. The overall effect is the same as that of a single heat engine operating between the gas turbine's high combustion temperature and the low temperature of the environment (see Problem 32). The second law still limits the efficiency, but the high T_h and low T_c make for greater efficiency than in a conventional plant. The photo shows a gas-fired combined-cycle plant.

Gasoline and diesel engines provide another pervasive example of heat engines, discussed in the Application on page 337. A typical automobile engine has a theoretical maximum efficiency of around 50%, but irreversible thermodynamic processes make the actual efficiency much lower. Mechanical friction dissipates additional energy, with the end result that less than 20% of the fuel energy reaches the driving wheels. Problems 54 and 55 explore the gasoline engine.

We wouldn't be so concerned with efficiency if we didn't have to pay for fuel or worry about the environment. Engines with "free" fuel include solar–thermal power plants that concentrate sunlight to boil a fluid that drives a turbine, and ocean thermal-energy conversion (OTEC) schemes that extract useful work from the modest temperature difference between tropical surface waters and the deep ocean. Neither provides significant energy today, but that could change as the world moves away from fossil fuels.

Refrigerators and Heat Pumps

A refrigerator works like an engine in reverse; it takes in mechanical work and transfers heat from its cooler interior to its warmer surroundings. An air conditioner is a refrigerator whose "interior" is the building being cooled. A close cousin is the **heat pump**, which transfers heat either way, cooling a building in the summer and warming it in the winter (Fig. 19.12). In warmer climates, heat pumps exchange energy between a building and the outside air; in cooler climates they use groundwater, typically at about 10°C year-round. Heat pumps require electricity, but they transfer more heat energy than they consume in electricity. That makes heat pumps potentially energy-saving devices for winter heating. However, some of that gain is offset by the inefficiency of the power plant producing the electricity.

An efficient refrigerator (or any other device, for that matter) should maximize what we want from the device compared with what we have to put in. The **coefficient of performance** (COP) quantifies this ratio:

$$\text{COP} = \frac{\text{What we want}}{\text{What we put in}}$$

For a refrigerator or summertime heat pump, "what we want" is cooling, so the numerator is Q_c. For a wintertime heat pump, "what we want" is heating, so the numerator is Q_h. For either, "what we put in" is mechanical work, W, or its equivalent in electricity. Thus we have

$$\text{COP}_{\text{refrigerator}} = \frac{Q_c}{W} = \frac{Q_c}{Q_h - Q_c} \qquad \text{COP}_{\text{heat pump}} = \frac{Q_h}{W} = \frac{Q_h}{Q_h - Q_c}$$

In summer the heat pump cools the house by extracting energy and rejecting it to the outdoor environment.

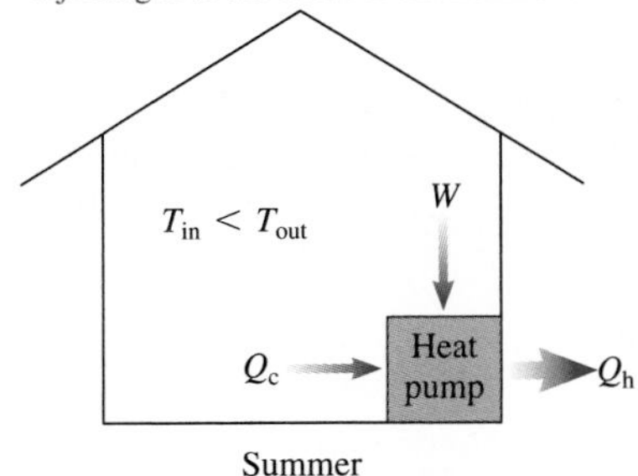

In winter the pump extracts energy from outside and transfers it to the inside.

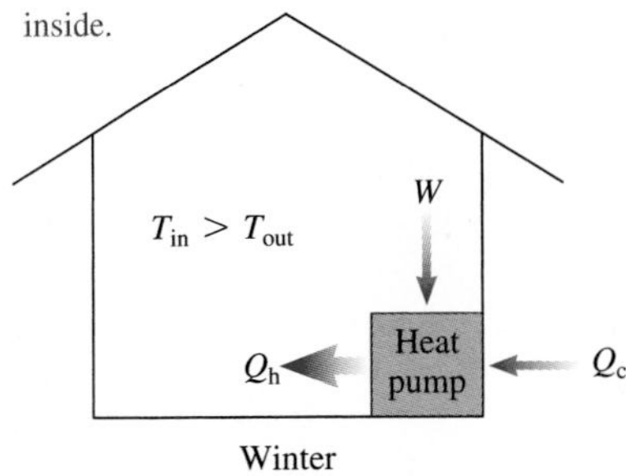

FIGURE 19.12 A heat pump.

In both cases the second equality follows from the first law of thermodynamics. In deriving the maximum efficiency of a heat engine, we found that $Q_c/Q_h = T_c/T_h$. Therefore the maximum possible COPs are

$$\mathrm{COP}_{\text{refrigerator}} = \frac{T_c}{T_h - T_c} \quad (19.4a)$$

$$\mathrm{COP}_{\text{heat pump}} = \frac{T_h}{T_h - T_c} \quad (19.4b)$$

When the temperatures T_h and T_c are close, Equations 19.4 give high COPs—meaning the refrigerator or heat pump takes relatively little work to do its job. But as the difference increases, the COP drops and we have to supply more work. Incidentally, our COP expression works for engines as well, if we take "what we want" to be mechanical work W and "what we put in" to be the heat Q_h.

EXAMPLE 19.3 The COP: A Home Freezer

A typical home freezer operates between a low of 0°F (−18°C or 255 K) and a high of 86°F (30°C or 303 K). What's its maximum possible COP? With this COP, how much electrical energy would it take to freeze 500 g of water initially at 0°C?

INTERPRET This problem is about a refrigerator—in this case a freezer. We identify T_h and T_c with the values 303 K and 255 K, respectively.

DEVELOP Equation 19.4a, $\mathrm{COP} = T_c/(T_h - T_c)$, will determine the COP. Then we'll use Equation 17.5, $Q = Lm$, to find the heat Q_c that the freezer must extract to freeze the water. From there we'll be able to use $\mathrm{COP} = Q_c/W$ to find the work—equivalently, the electrical energy—required.

EVALUATE Equation 19.4a gives

$$\mathrm{COP} = \frac{T_c}{T_h - T_c} = \frac{255\text{ K}}{303\text{ K} - 255\text{ K}} = 5.31$$

From Equation 17.5 and Table 17.1, we find the heat that needs to be removed in freezing 500 g of ice: $Q_c = Lm = (334\text{ kJ/kg})(0.50\text{ kg}) = 167\text{ kJ}$. The COP is the ratio of the heat removed to the work or electrical energy required, so we have $W = Q_c/\mathrm{COP} = 167\text{ kJ}/5.31 = 31\text{ kJ}$.

ASSESS Make sense? A COP of 5.3 means that each unit of work transfers 5.3 units of heat from inside the freezer to its surroundings—so the electrical-energy requirement is modest. A practical freezer operating between these temperatures would have a lower COP and require more electrical energy. ■

GOT IT? 19.3 A clever engineer decides to increase the efficiency of a Carnot engine by cooling the low-temperature reservoir using a refrigerator with the maximum possible COP. Will the overall efficiency of this system (a) exceed, (b) be less than, or (c) equal that of the original engine alone?

19.4 Entropy and Energy Quality

If offered a joule of energy, would you rather have it in the form of mechanical work, heat at 1000 K, or heat at 300 K? Your answer might depend on what you want to do. To lift or accelerate a mass, you'd be smart to take your energy as work. But if you want to keep warm, heat at 300 K would be perfectly acceptable.

But which should you choose if you want to keep all your options open, making the energy available for the most possible uses? The second law of thermodynamics answers clearly: You should take the work. Why? Because you could use it directly as mechanical energy, or you could, through friction or other irreversible processes, use it to raise the temperature of something.

If you chose 300 K heat for your joule of energy, then you could supply a full joule only to objects cooler than 300 K. You couldn't do mechanical work unless you ran a heat engine. With its T_h only a little above the ambient temperature, your engine would be inefficient, and you could extract only a small fraction of a joule of mechanical energy. You'd be better off with 1000-K heat since you could transfer it to anything cooler than 1000 K, or you could run a heat engine to produce up to 0.7 joule of mechanical energy (because $1 - T_c/T_h = 1 - 300/1000 = 0.7$).

CONCEPTUAL EXAMPLE 19.1 Energy Quality and Cogeneration

You need a new water heater, and you're trying to decide between gas and electric. The gas heater is 85% efficient, meaning 85% of the fuel energy goes into heating water. The electric heater is essentially 100% efficient. Thermodynamically, which heater makes the most sense?

EVALUATE Your electricity is energy of the highest quality. It probably comes from a thermal power plant, which typically discards as waste heat twice as much energy as it produces in electricity. The electric heater may be 100% efficient in your home, but when you consider the big picture, only about one-third of the fuel energy consumed at the power plant ends up heating your water. With 85% efficiency, the gas heater is the wiser choice.

ASSESS It makes sense to match energy sources to their end uses. Electricity is high-quality energy, so it's best for running motors, light sources, electronics, and other devices requiring high-quality energy. Turning it into low-grade heat is a thermodynamic folly! A really smart strategy is **cogeneration**, in which the waste heat from electric power generation is used to heat buildings. In Europe, whole communities are heated that way, and institutions in the United States are increasingly turning to cogeneration to reduce energy costs and carbon emissions.

MAKING THE CONNECTION If the electricity comes from a more efficient gas-fired power plant with $e = 48\%$, compare the gas consumption of your two heater choices.

EVALUATE The gas heater turns 1 unit of fuel energy into 0.85 unit of thermal energy in the water. The power plant turns 1 unit of fuel energy into 0.48 unit of electrical energy, which the electric heater converts to 0.48 unit of thermal energy. The electric heater is therefore responsible for $0.85/0.48 = 1.8$ times as much gas consumption.

Taking your energy in the form of work gives you the most options. Anything you can do with a joule of energy, you can do with the work. Heat is less versatile, with 300 K heat the least useful of the three. We're not talking here about the quantity of energy—we have exactly 1 joule in each case—but about **energy quality** (Fig. 19.13). We can readily convert an entire amount of energy from higher to lower quality, but the second law precludes going in the opposite direction with 100% efficiency.

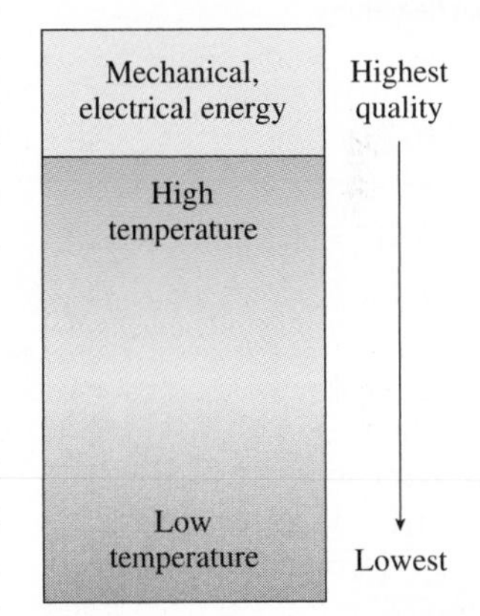

FIGURE 19.13 Energy quality measures the versatility of different energy forms.

Entropy

Mix hot and cold water, and you get lukewarm water. There's no energy loss, but you *have* lost something—namely, the ability to do useful work. In the initial state, you could have run a heat engine using the temperature difference ΔT between the hot and cold water. In the final state, there's no temperature difference, so you couldn't run a heat engine. The *quantity* of energy hasn't changed, but its *quality* has decreased. **Entropy**, symbol S, quantifies the loss of quality associated with energy transformations. In his Ninth Memoir, Clausius coined the term *entropy* for its similarity to the word "energy" and its Greek root "troph," meaning *transformation*.

To motivate the definition of entropy, consider an ideal gas undergoing a Carnot cycle. Recall that a Carnot cycle consists of two isothermal and two adiabatic processes (Fig. 19.5). In deriving Equation 19.3 for the Carnot efficiency, we found that $Q_c/Q_h = T_c/T_h$, where Q_c was the heat *rejected* from the system to the low-temperature reservoir at T_c, and Q_h the heat *added* from the reservoir at T_h.

Let's focus on the ideal gas itself and define all heats as the heat *added* to the gas, so Q_c changes sign. The relationship $Q_c/Q_h = T_c/T_h$ between heats and temperatures can now be expressed as

$$\frac{Q_c}{T_c} + \frac{Q_h}{T_h} = 0 \quad \text{(Carnot cycle)}$$

We can generalize this result to *any* reversible cycle by approximating the cycle as a sequence of Carnot cycles, as shown in Fig. 19.14. For each segment, we have $\sum Q/T = 0$. As we increase the number of cycles, the volume change associated with each isothermal segment shrinks and the edges get less jagged. We can approximate the closed cycle ever closer by using more and more Carnot cycles. In the limit, the approximation becomes exact and the sum becomes an integral:

$$\oint \frac{dQ}{T} = 0 \quad \text{(any reversible cycle)} \tag{19.5}$$

where the circle indicates integration over a *closed* path.

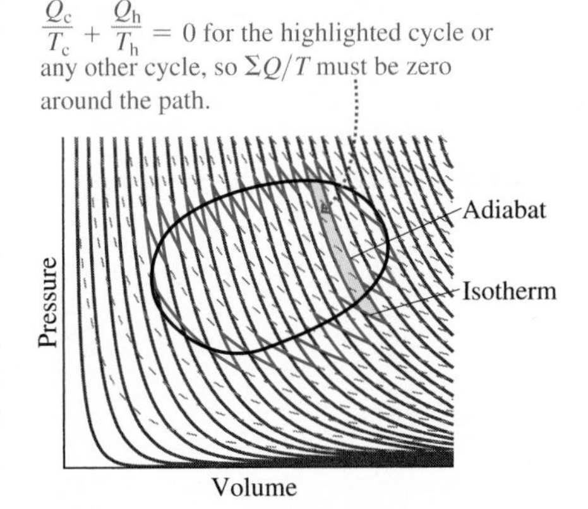

FIGURE 19.14 An arbitrary cycle approximated by isothermal (dashed curves) and adiabatic (solid curves) steps. Heat transfer occurs only during the isothermal steps.

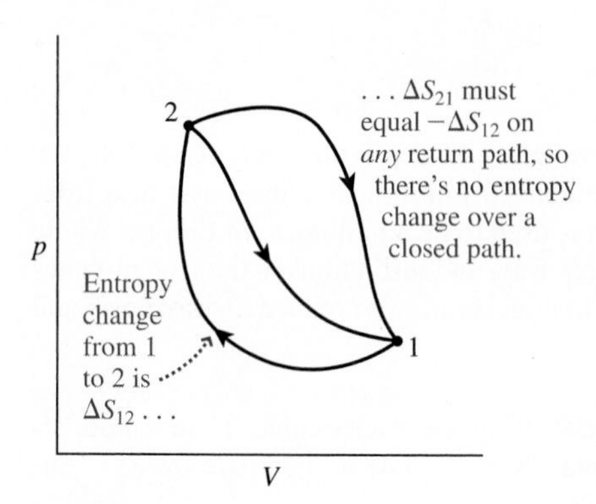

FIGURE 19.15 Entropy change is path-independent.

Equation 19.5 holds for *any* closed path in the pV diagram—that is, for any *reversible cycle*. That means we can define the *entropy change*, ΔS, between an initial state 1 and a final state 2 as

$$\Delta S_{12} = \int_1^2 \frac{dQ}{T} \quad \text{(entropy change)} \tag{19.6}$$

Note that entropy has the units J/K, the same units as Boltzmann's constant k_B.

Take a system along a path from state 1 to state 2 in its pV diagram; Equation 19.6 gives the corresponding entropy change ΔS_{12}. Go back to state 1 by any other reversible path, and the resulting entropy change ΔS_{21} must be $-\Delta S_{12}$ so that there's no entropy change around the closed path (Fig. 19.15). Thus the entropy change of Equation 19.6 is independent of path; it depends only on the initial and final states. The only restriction is that we integrate over a reversible path. Like pressure and temperature, entropy is therefore a thermodynamic *state variable*—a quantity that characterizes a given state independently of how the system got into that state.

PhET: Reversible Reactions

We restricted ourselves to reversible paths in Equation 19.6 since irreversible processes take a system out of thermodynamic equilibrium and therefore aren't described by paths in the pV diagram. But because entropy depends only on the initial and final states, we can calculate the entropy change in an *irreversible* process by using Equation 19.6 for a *reversible* process that goes between the same two states.

Note also that Equation 19.6 gives the entropy change of *just* the working fluid. The fluid—perhaps in an engine—is thermally connected to its surroundings, and if we're interested in the *total* entropy change resulting from the engine's operation, we'll need to add the entropy changes for its environment—in this case the hot and cold reservoirs.

Adiabatic Free Expansion

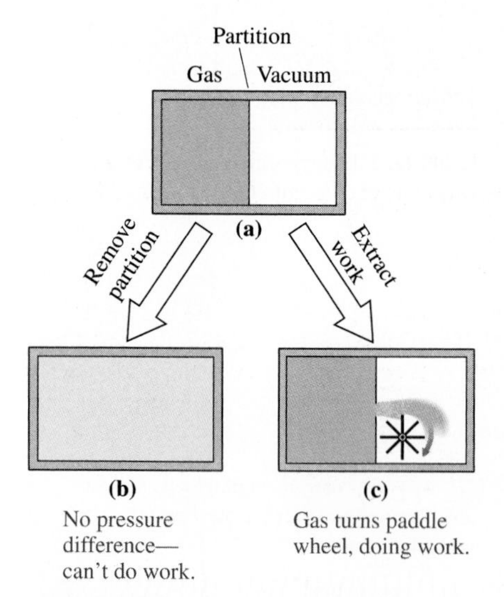

FIGURE 19.16 Two ways for a gas to expand into a vacuum.

In Fig. 19.16*a*, a partition confines an ideal gas to one side of a box; the other side is vacuum. Remove the partition, and the gas undergoes a **free expansion**, filling the box. Consider the box to be insulated, so there's no heat flow and the expansion is therefore adiabatic. But this expansion is *irreversible*, so it's significantly different from the adiabatic expansions we considered in Chapter 18. In our free expansion, the vacuum doesn't exert pressure to oppose the gas, so the gas does no work and therefore its internal energy doesn't change. Figure 19.16*c* shows how we could have used the expanding gas to turn a paddle wheel, extracting useful work. We can't do that with the uniform-pressure gas of Fig. 19.16*b*, so the free expansion results in the system's losing its ability to do work.

Let's determine the entropy change for this irreversible process. We do that by finding a reversible process that takes the gas between the same two states. Since the gas's internal energy doesn't change, neither does its temperature. So the corresponding reversible process is an isothermal expansion, for which Equation 18.4 gives the heat added: $Q = nRT\ln(V_2/V_1)$. With the temperature constant, the entropy change of Equation 19.6 becomes

$$\Delta S = \int \frac{dQ}{T} = \frac{1}{T}\int dQ = \frac{Q}{T} = nR\ln\left(\frac{V_2}{V_1}\right)$$

The final volume V_2 is larger than V_1, so entropy has *increased*. Although we computed this result for the reversible process, it holds for *any* process that takes the system between the same initial and final states—including our irreversible free expansion.

Entropy and the Availability of Work

Entropy increases during irreversible expansion—and energy quality decreases, in that the system loses its ability to do work. Had we let the gas in Fig. 19.16 undergo a reversible isothermal expansion instead of free expansion, it would have done work equal to the heat gained:

$$W = Q = nRT\ln\left(\frac{V_2}{V_1}\right)$$

After the irreversible free expansion, the gas can no longer do this work, even though its energy is unchanged. Comparing W with the entropy change ΔS we calculated above, we see that the energy that becomes unavailable to do work is $E_{\text{unavailable}} = T\,\Delta S$. This illustrates a more general relation between entropy and energy quality:

> During an irreversible process in which the entropy of a system increases by ΔS, energy $E = T_{\text{min}}\Delta S$ becomes unavailable to do work, where T_{min} is the lowest temperature available to the system.

This statement shows that entropy provides our measure of energy quality. Given two systems with identical energy content, the one with the lower entropy contains the higher-quality energy. An entropy increase corresponds to a degradation in energy quality, as energy becomes unavailable to do work.

EXAMPLE 19.4 Increasing Entropy: The Loss of Energy Quality

A 2.0-L cylinder contains 5.0 mol of compressed gas at 290 K. If the cylinder is discharged into a 150-L vacuum chamber and its temperature remains 290 K, how much energy has become unavailable to do work?

INTERPRET This problem asks about the loss of energy quality during an irreversible and therefore entropy-increasing process—namely, an adiabatic free expansion.

DEVELOP Figure 19.17 is a sketch of the situation, similar to Fig. 19.16 except that here the gas is initially confined to a small cylinder, so its volume changes more dramatically as it expands into the large, empty chamber. In analyzing the free expansion of Fig. 19.16, we found $\Delta S = nR\ln(V_2/V_1)$. Our statement relating entropy and energy quality says that the energy made unavailable to do work is $T_{\text{min}}\Delta S$. So our plan is to calculate ΔS and multiply by T_{min} to find that unavailable energy.

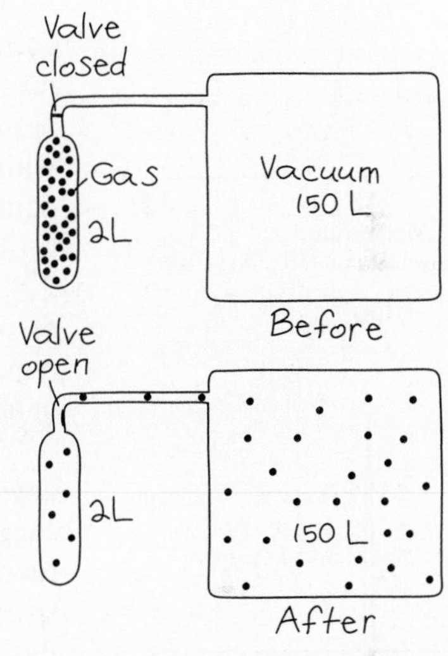

FIGURE 19.17 Our sketch for Example 19.4. Note that the final volume is 152 L.

EVALUATE Because the temperature doesn't change, T_{min} is the 290-K temperature we're given, and we have

$$E_{\text{unavailable}} = T\,\Delta S = nRT\ln\left(\frac{V_2}{V_1}\right)$$

$$= (5.0\text{ mol})(8.314\text{ J/K}\cdot\text{mol})(290\text{ K})\ln\left(\frac{152\text{ L}}{2.0\text{ L}}\right) = 52\text{ kJ}$$

ASSESS Make sense? Yes: This is the work we could have extracted from a reversible isothermal expansion. By letting the gas undergo an irreversible process, we gave up the possibility of extracting this work. ■

A Statistical Interpretation of Entropy

We began this chapter arguing that systems naturally evolve from ordered to disordered states. Entropy increase measures that loss of order, which is what makes energy unavailable to do work. Here we'll explore the meaning of entropy further, based on the partitioned box we used for adiabatic free expansion.

Suppose we have a gas with just two identical molecules. The left side of Fig. 19.18 shows that, with the partition removed, there are four possible **microstates**—specific arrangements of the individual molecules in the box. But say we only care about the number of molecules in each side of the box. Then two of these arrangements are indistinguishable, because they both have one molecule in each half of the box. Those two correspond to a single **macrostate**, specified by giving the number of molecules in each half of the box, without regard to which molecules they are. This is shown on the right in Fig. 19.18.

With four available microstates, the probability of being in any one microstate is $\frac{1}{4}$. There's only one microstate with both molecules on the left, so the chances of being in the macrostate with two molecules on the left is also $\frac{1}{4}$; the same is true for the macrostate with

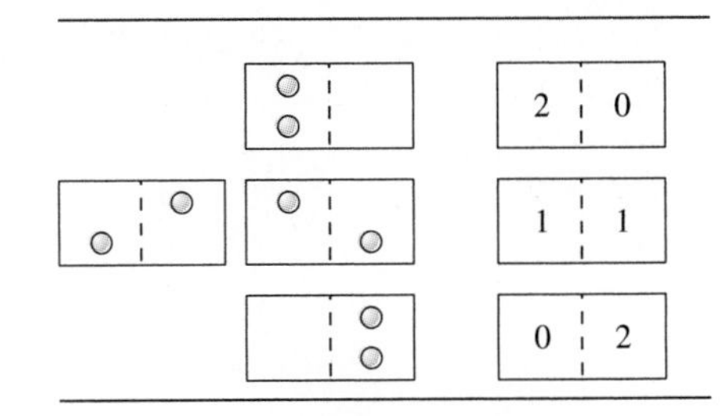

FIGURE 19.18 A gas of two molecules has four possible microstates and three macrostates.

Microstates (16 total)	Macrostates	Probability of macrostate
	4 ¦ 0	$\frac{1}{16} = 0.06$
	3 ¦ 1	$\frac{4}{16} = 0.25$
	2 ¦ 2	$\frac{6}{16} = 0.38$
	1 ¦ 3	$\frac{4}{16} = 0.25$
	0 ¦ 4	$\frac{1}{16} = 0.06$

FIGURE 19.19 Microstates, macrostates, and probabilities for a gas of four molecules.

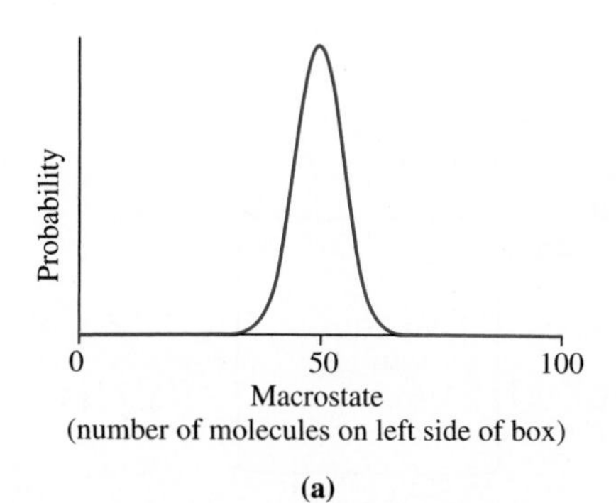

(a)

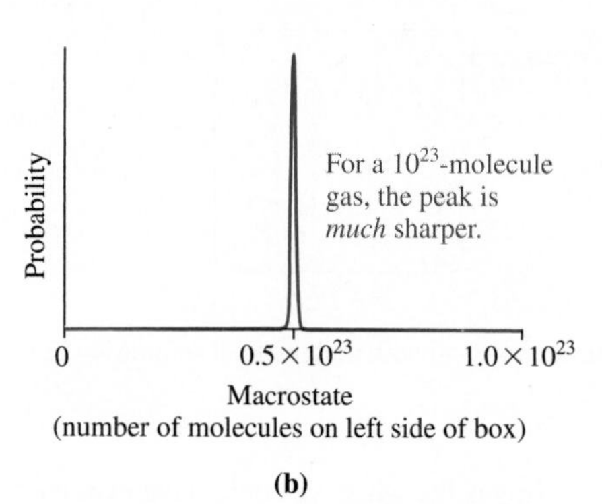

(b)

FIGURE 19.20 Probability distributions for a gas of (a) 100 molecules and (b) 10^{23} molecules.

two molecules on the right. But two of the possible microstates have one molecule on each side, so the probability for this macrostate is $\frac{1}{2}$.

Now consider a gas of four molecules. Figure 19.19 shows 16 possible microstates, corresponding to five macrostates. Again, the probability of finding the system in a given macrostate depends on the associated number of microstates; Fig. 19.19 enumerates these probabilities. The figure shows that we're most likely to find the system in the macrostate with the molecules evenly divided; the states with all the molecules on one side are now quite improbable.

Raise the number of molecules to 100, and the number of microstates becomes huge—2^{100}, or more than 10^{30}. That makes the macrostates with all or nearly all the molecules on one side extremely improbable. The macrostate with half the molecules on each side remains the most likely, although states with nearly equal divisions of molecules are also quite probable. Rather than enumerate these probabilities, we graph them (Fig. 19.20a).

Typical gas samples have roughly 10^{23} molecules, and that makes macrostates with anything other than a nearly equal distribution of molecules extremely unlikely—as suggested by the spike-like probability distribution in Fig. 19.20b. You could sit in your room for many times the age of the universe, and you'd never see all the air molecules spontaneously end up on one side of the room!

Entropy and the Second Law of Thermodynamics

The statistical improbability of more ordered states—in our example, those with significantly more molecules on one side of the box—is at the root of the second law of thermodynamics. Although we defined entropy in terms of heat flow and temperature (Equation 19.6), a more fundamental definition involves the probabilities of individual microstates. In that sense, entropy is indeed a measure of disorder.

Systems naturally evolve toward disordered or higher-entropy states simply because there are far more of these states available. So a general statement of the second law is:

> **Second law of thermodynamics** The entropy of a closed system can never decrease.

At best, the entropy of a closed system remains constant—and that's only in an ideal, reversible process. If anything irreversible occurs—friction, or any deviation from thermodynamic equilibrium—then entropy increases. As it does, energy becomes unavailable to do work, and nothing within the closed system can restore that energy to its original quality. This new statement of the second law subsumes our previous statements about the impossibility of perfect heat engines and refrigerators, for their operation would require an entropy decrease.

We *can* decrease the entropy of a system that isn't closed—but only by supplying high-quality energy from outside. Running a refrigerator decreases the entropy of its contents, but this requires electrical energy to make heat flow from cold to hot. That high-quality electrical energy deteriorates into additional heat that's rejected to the refrigerator's environment. If we consider the entire system, not just the refrigerator's contents, the overall entropy has increased.

Any system whose entropy seems to decrease—that gets more rather than less organized—can't be closed. If we enlarge a system's boundaries to encompass the entire universe, then we have the ultimate statement of the second law:

Second law of thermodynamics The entropy of the universe can never decrease.

Examples include the growth of a living thing from the random mix of molecules in its environment, the construction of a skyscraper from materials that were originally dispersed about Earth, and the appearance of ordered symbols on a printed page from a bottle of ink. All these are entropy-decreasing processes in which matter goes from near chaos to a highly organized state—akin to separating yolk and white from a scrambled egg. But Earth isn't a closed system. It gets high-quality energy from the Sun, energy that's ultimately responsible for life. If we consider the Earth–Sun system, the entropy decrease associated with life and civilization is more than balanced by the entropy increase associated with the degradation of high-quality solar energy. We living things represent a remarkable phenomenon—the organization of matter in a universe governed by a tendency toward disorder. But we can't escape the second law of thermodynamics. Our highly organized selves and society, and the entropy decreases they represent, come into being only at the expense of greater entropy increases elsewhere.

GOT IT? 19.4 In each of the following processes, does the entropy of the named system alone increase, decrease, or stay the same? (1) a balloon deflates; (2) cells differentiate in a growing embryo, forming different physiological structures; (3) an animal dies, and its remains gradually decay; (4) an earthquake demolishes a building; (5) a plant utilizes sunlight, carbon dioxide, and water to manufacture sugar; (6) a power plant burns coal and produces electrical energy; (7) a car's friction-based brakes stop the car

---

CHAPTER 19 SUMMARY

Big Idea

The big idea behind this chapter is the **second law of thermodynamics**—ultimately, the statement that systems tend naturally toward disorder, or states of higher **entropy**. The second law is manifest in the real world by forbidding the construction of perfect heat engines and perfect refrigerators—therefore preventing us from extracting as useful work all the energy that's contained in random thermal motions. Ultimately, the second law says that the entropy of any closed system, including the entire universe, cannot decrease.

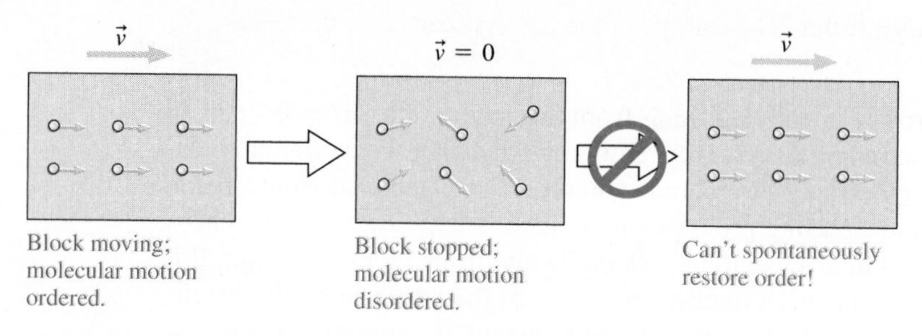

Key Concepts and Equations

Entropy is a quantitative measure of energy quality and of disorder; the higher the entropy, the lower the energy quality and the greater the disorder. The highest-quality energy is mechanical or electrical energy, followed by the internal energy of systems at high temperature, and finally low-temperature internal energy. Whenever entropy increases, energy becomes unavailable to do work.

- $\Delta S = \int_1^2 \frac{dQ}{T}$ gives the entropy change as a system goes from state 1 to state 2.
- $E_{\text{unavailable}} = T_{\text{min}}\Delta S$ is the energy that becomes unavailable as a result of entropy increase ΔS.

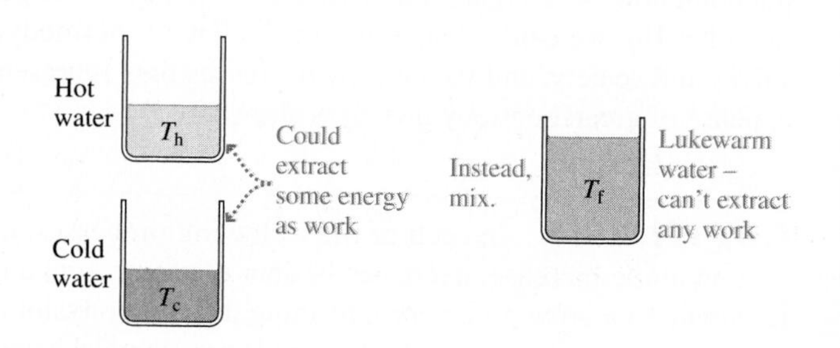

Applications

The second law sets the maximum possible efficiency of any heat engine as that of the **Carnot engine**, an engine that combines adiabatic and isothermal processes.

$$\underbrace{e = \frac{W}{Q_h}}_{\text{This defines an engine's efficiency.}} \le \underbrace{e_{\text{max}} = 1 - \frac{T_c}{T_h}}_{\text{This is the maximum possible efficiency.}}$$

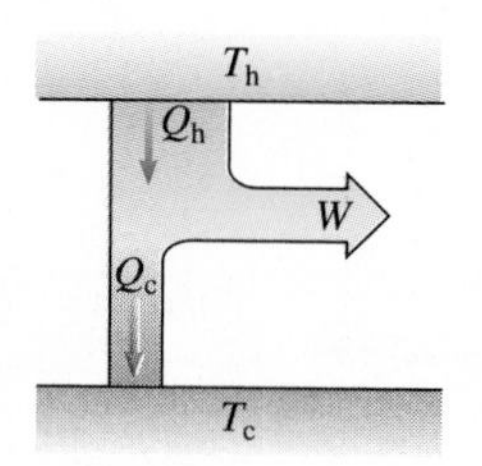

Energy-flow diagram for an engine

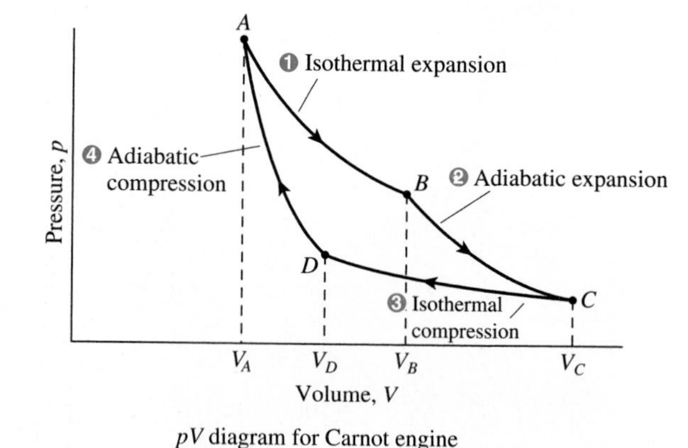

pV diagram for Carnot engine

Similarly, the second law limits the **coefficient of performance (COP)** of refrigerators and heat pumps:

$$\text{COP}_{\text{refrigerator}} = \frac{T_c}{T_h - T_c} \qquad \text{COP}_{\text{heat pump}} = \frac{T_h}{T_h - T_c}$$

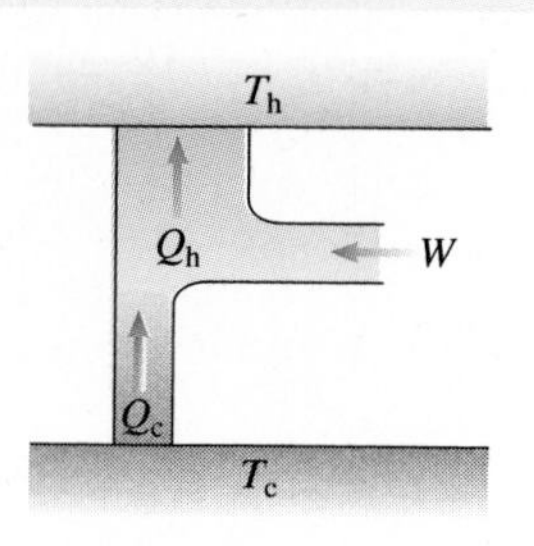

MP For homework assigned on MasteringPhysics, go to www.masteringphysics.com

BIO Biology and/or medicine-related problems DATA Data problems ENV Environmental problems CH Challenge problems COMP Computer problems

For Thought and Discussion

1. Could you cool the kitchen by leaving the refrigerator open? Explain.
2. Could you heat the kitchen by leaving the oven open? Explain.
3. Should a car get better mileage in the summer or the winter? Explain.
4. Is there a limit to the maximum temperature that can be achieved by focusing sunlight with a lens? If so, what is it?
5. Name some irreversible processes that occur in a real engine.
6. Your power company claims that electric heat is 100% efficient. Discuss.
7. A hydroelectric power plant, using the energy of falling water, can operate with efficiency arbitrarily close to 100%. Why?
8. A heat-pump manufacturer claims the device will heat your home using only energy already available in the ground. Is this true?
9. Why do refrigerators and heat pumps have different definitions of COP?
10. The heat Q added during adiabatic free expansion is zero. Why can't we then argue from Equation 19.6 that the entropy change is zero?
11. Energy is conserved, so why can't we recycle it as we do materials?
12. Why doesn't the evolution of human civilization violate the second law of thermodynamics?

Exercises and Problems

Exercises

Sections 19.2 and 19.3 The Second Law of Thermodynamics and Its Applications

13. What are the efficiencies of reversible heat engines operating between (a) the normal freezing and boiling points of water, (b) the 25°C temperature at the surface of a tropical ocean and deep water at 4°C, and (c) a 1000°C flame and room temperature?
14. A cosmic heat engine might operate between the Sun's 5800 K surface and the 2.7 K temperature of intergalactic space. What would be its maximum efficiency?
15. A reversible Carnot engine operating between helium's melting point and its 4.25 K boiling point has an efficiency of 77.7%. What's the melting point?
16. A Carnot engine absorbs 900 J of heat each cycle and provides 350 J of work. (a) What's its efficiency? (b) How much heat is rejected each cycle? (c) If the engine rejects heat at 10°C, what's its maximum temperature?
17. Find the COP of a reversible refrigerator operating between 0°C and 30°C.
18. How much work does a refrigerator with COP = 4.2 require to freeze 670 g of water already at its freezing point?
19. BIO The human body can be 25% efficient at converting chemical energy of fuel to mechanical work. Can the body be considered a heat engine, operating on the temperature difference between body temperature and the environment?

Section 19.4 Entropy and Energy Quality

20. Calculate the entropy change associated with melting 1.0 kg of ice at 0°C.
21. BIO You metabolize a 650-kcal burger at your 37°C body temperature. What's the associated entropy increase?
22. You heat 250 g of water from 10°C to 95°C. By how much does the entropy of the water increase?
23. Melting a block of lead already at its melting point results in an entropy increase of 900 J/K. What's the mass of the lead? (*Hint:* Consult Table 17.1.)
24. How much energy becomes unavailable for work in an isothermal process at 440 K, if the entropy increase is 25 J/K?
25. For a gas of six molecules confined to a box, find the probability that (a) all the molecules will be found on one side of the box and (b) half the molecules will be found on each side.

Problems

26. A Carnot engine extracts 745 J from a 592-K reservoir during each cycle and rejects 458 J to a cooler reservoir. It operates at 18.6 cycles per second. Find (a) the work done during each cycle, (b) its efficiency, (c) the temperature of the cool reservoir, and (d) its mechanical power output.
27. ENV The maximum steam temperature in a nuclear power plant is 570 K. The plant rejects heat to a river whose temperature is 0°C in the winter and 25°C in the summer. What are the maximum possible efficiencies for the plant during these seasons?
28. ENV You're engineering an energy-efficient house that will require an average of 6.85 kW to heat on cold winter days. You've designed a photovoltaic system for electric power, which will supply on average 2.32 kW. You propose to heat the house with an electrically operated groundwater-based heat pump. What should you specify as the minimum acceptable COP for the pump if the photovoltaic system supplies its energy?
29. A power plant's electrical output is 750 MW. Cooling water at 15°C flows through the plant at 2.8×10^4 kg/s, and its temperature rises by 8.5°C. Assuming that the plant's only energy loss is to the cooling water, which serves as its low-temperature reservoir, find (a) the rate of energy extraction from the fuel, (b) the plant's efficiency, and (c) its highest temperature.
30. ENV A power plant extracts energy from steam at 280°C and delivers 880 MW of electric power. It discharges waste heat to a river at 30°C. The plant's overall efficiency is 29%. (a) How does this efficiency compare with the maximum possible at these temperatures? (b) Find the rate of waste-heat discharge to the river. (c) How many houses, each requiring 23 kW of heating power, could be heated with the waste heat from this plant?
31. ENV The electric power output of all the thermal electric power plants in the United States is about 2×10^{11} W, and these plants operate at an average efficiency of around 33%. Find the rate at which all these plants use cooling water, assuming an average 5°C rise in cooling-water temperature. Compare with the 1.8×10^7 kg/s average flow at the mouth of the Mississippi River.
32. Consider a Carnot engine operating between temperatures T_h and T_i, where T_i is intermediate between T_h and the ambient

temperature T_c (Fig. 19.21). It should be possible to operate a second engine between T_i and T_c. Show that the maximum overall efficiency of such a two-stage engine is the same as that of a single engine operating between T_h and T_c (which is why combined-cycle power plants achieve high efficiencies).

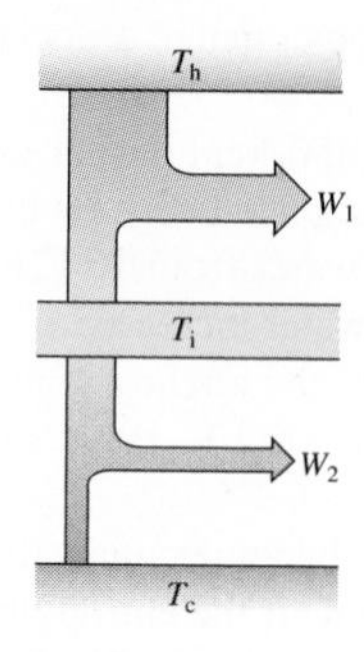

FIGURE 19.21 Problem 32

33. An industrial freezer operates between 0°C and 32°C, consuming electrical energy at the rate of 12 kW. Assuming the freezer is perfectly reversible, (a) what's its COP? (b) How much water at 0°C can it freeze in 1 hour?
34. Use appropriate energy-flow diagrams to analyze the situation in GOT IT? 19.3; that is, show that using a refrigerator to cool the low-temperature reservoir can't increase the overall efficiency of a Carnot engine when the work input to the refrigerator is included.
35. ENV It costs \$230 to heat a house with electricity in a winter month. (Electric heat converts all the incoming electrical energy to heat.) What would the monthly heating bill be after converting to an electrically powered heat pump with COP = 3.4?
36. A refrigerator maintains an interior temperature of 4°C while its exhaust temperature is 30°C. The refrigerator's insulation is imperfect, and heat leaks in at the rate of 340 W. Assuming the refrigerator is reversible, at what rate must it consume electrical energy to maintain a constant 4°C interior?
37. ENV You operate a store that's heated by an oil furnace supplying 30 kWh of heat from each gallon of oil. You're considering switching to a heat-pump system. Oil costs \$1.75/gallon, and electricity costs 16.5¢/kWh. What's the minimum heat-pump COP that will reduce your heating costs?
38. Use energy-flow diagrams to show that the existence of a perfect heat engine would permit the construction of a perfect refrigerator, thus violating the Clausius statement of the second law.
39. ENV A heat pump extracts energy from groundwater at 10°C and transfers it to water at 70°C to heat a building. Find (a) its COP and (b) its electric power consumption if it supplies heat at the rate of 20 kW. (c) Compare the pump's hourly operating cost with that of an oil furnace if electricity costs 15.5¢/kWh and oil costs \$3.60/gallon and releases about 30 kWh/gal when burned.
40. A reversible engine contains 0.350 mol of ideal monatomic gas, initially at 586 K and confined to a volume of 2.42 L. The gas undergoes the following cycle:
    - Isothermal expansion to 4.84 L
    - Constant-volume cooling to 292 K
    - Isothermal compression to 2.42 L
    - Constant-volume heating back to 586 K

    Determine the engine's efficiency, defined as the ratio of the work done to the heat *absorbed* during the cycle.
41. (a) Determine the efficiency for the cycle shown in Fig. 19.22, using the definition given in the preceding problem. (b) Compare with the efficiency of a Carnot engine operating between the same temperature extremes. Why are the two efficiencies different?

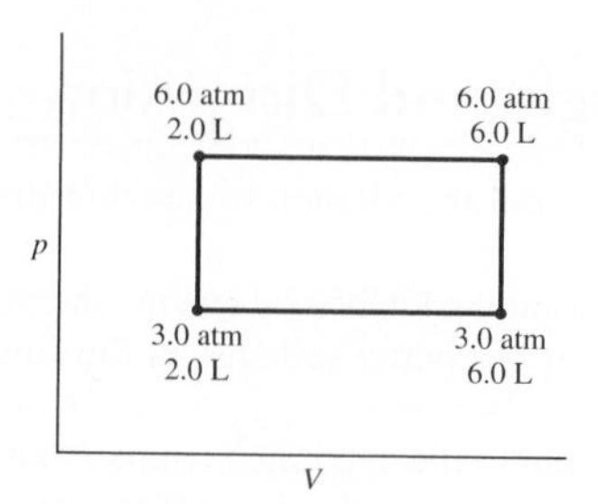

FIGURE 19.22 Problem 41

42. A 0.20-mol sample of an ideal gas goes through the Carnot cycle of Fig. 19.23. Calculate (a) the heat Q_h absorbed, (b) the heat Q_c rejected, and (c) the work done. (d) Use these quantities to determine the efficiency. (e) Find the maximum and minimum temperatures, and show explicitly that the efficiency as defined in Equation 19.1 is equal to the Carnot efficiency of Equation 19.3.

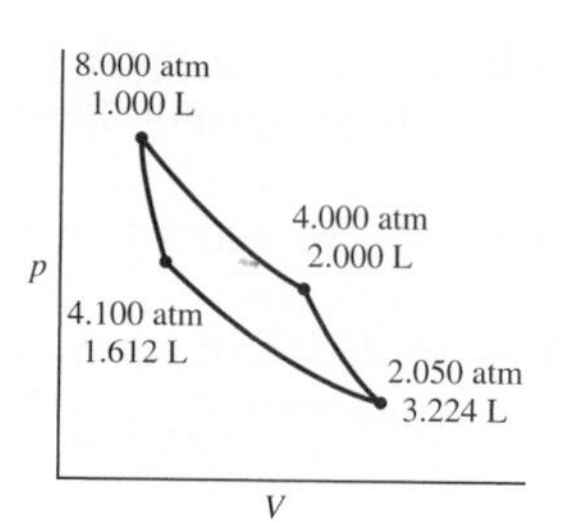

FIGURE 19.23 Problem 42

43. A shallow pond contains 94 Mg of water. In winter, it's entirely frozen. By how much does the entropy of the pond increase when the ice, already at 0°C, melts and then heats to its summer temperature of 15°C?
44. BIO Estimate the rate of entropy increase associated with your body's normal metabolism.
45. The temperature of n moles of ideal gas is changed from T_1 to T_2 at constant volume. Show that the corresponding entropy change is $\Delta S = nC_V \ln(T_2/T_1)$.
46. The temperature of n moles of ideal gas is changed from T_1 to T_2 with pressure held constant. Show that the corresponding entropy change is $\Delta S = nC_p \ln(T_2/T_1)$.
47. A 6.36-mol sample of ideal diatomic gas is at 1.00 atm pressure and 288 K. Find the entropy change as the gas is heated reversibly to 552 K (a) at constant volume, (b) at constant pressure, and (c) adiabatically.
48. A 250-g sample of water at 80°C is mixed with 250 g of water at 10°C. Find the entropy changes for (a) the hot water, (b) the cool water, and (c) the system.
49. An ideal gas undergoes a process that takes it from pressure p_1 and volume V_1 to p_2 and V_2, such that $p_1V_1^\gamma = p_2V_2^\gamma$, where γ is the specific heat ratio. Find the entropy change if the process consists of constant-pressure and constant-volume segments. Why does your result make sense?
50. In an adiabatic free expansion, 6.36 mol of ideal gas at 305 K expands 15-fold in volume. How much energy becomes unavailable to do work?
51. Find the entropy change when a 2.4-kg aluminum pan at 155°C is plunged into 3.5 kg of water at 15°C.

52. An engine with mechanical power output 8.5 kW extracts heat from a source at 420 K and rejects it to a 1000-kg block of ice at its melting point. (a) What's its efficiency? (b) How long can it maintain this efficiency if the ice isn't replenished?
53. Find the change in entropy as 2.00 kg of H_2O at 100°C turns to vapor at the same temperature.
54. Gasoline engines operate approximately on the Otto cycle, consisting of two adiabatic and two constant-volume segments. Figure 19.24 shows the Otto cycle for a particular engine. (a) If the gas in the engine has specific heat ratio γ, find the engine's efficiency, assuming all processes are reversible. (b) Find the maximum temperature in terms of the minimum temperature $T_{\min}$. (c) How does the efficiency compare with that of a Carnot engine operating between the same temperature extremes?

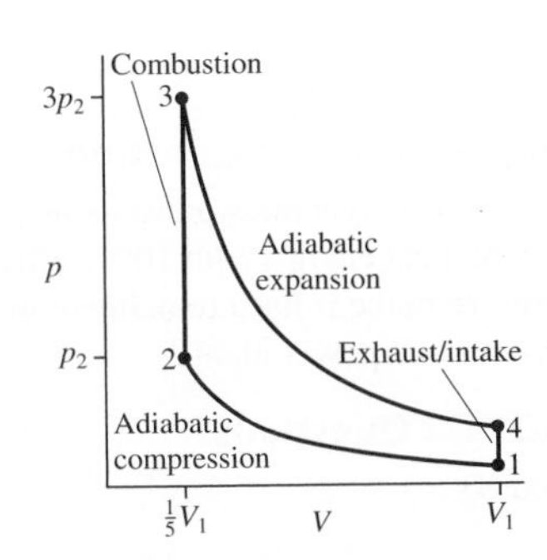

FIGURE 19.24 Problem 54

55. The compression ratio r of an engine is the ratio of maximum to minimum gas volume. (For the gasoline engine of the preceding problem, Fig. 19.24 shows that the compression ratio is 5.) Find a general expression for the engine efficiency of an Otto-cycle engine as a function of compression ratio.
56. **CH** In a diesel cycle, gas at volume V_1 and pressure p_1 undergoes adiabatic compression to a smaller volume V_2. It is then heated at constant pressure while it expands to volume V_3. The gas then expands adiabatically until it's again at volume V_1, whereupon it cools, at constant volume, until it's back to its initial state of p_1 and V_1. Show that the work done by the engine over one cycle can be written as $W = \dfrac{p_1V_1[r^{\gamma-1}(\alpha-1)\gamma-\alpha^{\gamma}+1]}{\gamma-1}$. Here $r = V_1/V_2$ is the *compression ratio* and $\alpha = V_3/V_2$ is the so-called *cutoff ratio*.
57. **CH** (a) Show that the heat flowing *into* the diesel engine of Problem 56 during the cycle (*not* the *net* heat flow) is given by $Q_{\text{in}} = \dfrac{p_1V_1r^{\gamma-1}(\alpha-1)\gamma}{\gamma-1}$. (b) Use this quantity, along with the result of Problem 56, to show that the diesel engine's efficiency can be written as $e_{\text{diesel}} = 1 - \dfrac{r^{1-\gamma}(\alpha^{\gamma}-1)}{(\alpha-1)\gamma}$.
58. You're considering buying a car that comes in either gasoline or diesel versions. The gas engine has compression ratio $r = 8.3$, while the diesel has compression ratio $r = 19$ and cutoff ratio $\alpha = 2.4$. Use the results of Problems 55 and 57 to determine which engine is more efficient. Both engines "breathe" air, which is diatomic.
59. **ENV** The 54-MW wood-fired McNeil Generating Station in Burlington, Vermont, produces steam at 950°F to drive its turbines, and condensed steam returns to the boiler as 90°F water. (Note the temperatures in °F, used in U.S. engineering situations.) Find McNeil's maximum thermodynamic efficiency, and compare with its actual efficiency of 25%.
60. A 500-g copper block at 80°C is dropped into 1.0 kg of water at 10°C. Find (a) the final temperature and (b) the entropy change of the system.
61. An object's heat capacity is inversely proportional to its absolute temperature: $C = C_0(T_0/T)$, where C_0 and T_0 are constants. Find the entropy change when the object is heated from T_0 to T_1.
62. **CH** A Carnot engine extracts heat from a block of mass m and specific heat c initially at temperature T_{h0} but without a heat source to maintain that temperature. The engine rejects heat to a reservoir at constant temperature T_{c}. The engine is operated so its mechanical power output is proportional to the temperature difference $T_{\text{h}} - T_{\text{c}}$:

$$P = P_0\frac{T_{\text{h}} - T_{\text{c}}}{T_{\text{h0}} - T_{\text{c}}}$$

where T_{h} is the instantaneous temperature of the hot block and P_0 is the initial power. (a) Find an expression for T_{h} as a function of time, and (b) determine how long it takes for the engine's power output to reach zero.
63. **CH** In an alternative universe, you've got the impossible: an infinite heat reservoir, containing infinite energy at temperature T_{h}. But you've only got a finite cool reservoir, with initial temperature T_{c0} and heat capacity C. Find an expression for the maximum work you can extract if you operate an engine between these two reservoirs.
64. **ENV** You're the environmental protection officer for a 35% efficient nuclear power plant that produces 750 MW of electric power, situated on a river whose minimum flow rate is 110 m³/s. State environmental regulations limit the rise in river temperature from your plant's cooling system to 5°C. Can you achieve this standard if you use river water for all your cooling, or will you need to install cooling towers that transfer some of your waste heat to the atmosphere?
65. Find an expression for the entropy gain when hot and cold water are irreversibly mixed. A corresponding reversible process you can use to calculate this change is to bring each water sample slowly to their common final temperature T_{f} and then mix them. Express your answer in terms of the initial temperatures T_{h} and T_{c}. Assume equal masses of hot and cold water, with constant specific heat c. What's the sign of your answer?
66. **CH** Problem 74 of Chapter 16 provided an approximate expression for the specific heat of copper at low absolute temperatures: $c = 31(T/343\text{ K})^3$ J/kg·K. Use this to find the entropy change when 40 g of copper are cooled from 25 K to 10 K. Why is the change negative?
67. **CH** The molar specific heat at constant pressure for a certain gas is given by $C_p = a + bT + cT^2$, where $a = 33.6$ J/mol·K, $b = 2.93\times10^{-3}$ J/mol·K², and $c = 2.13\times10^{-5}$ J/mol·K³. Find the entropy change when 2.00 moles of this gas are heated from 20.0°C to 200°C.
68. **CH** Consider a gas containing an even number N of molecules, distributed among the two halves of a closed box. Find expressions for (a) the total number of microstates and (b) the number of microstates with half the molecules on each side of the box. (You can either work out a formula, or explore the term "combinations" in a math reference source.) (c) Use these results to find the ratio of the probability that all the molecules will be found on one side of the box to the probability that there will be equal numbers on both sides. (d) Evaluate for $N = 4$ and $N = 100$.
69. **DATA** Energy-efficiency specialists measure the heat Q_{h} delivered by a heat pump and the corresponding electrical energy W needed to run the pump, and they compute the pump's COP as the ratio

Q_h/W. They also measure the outdoor temperature, and they know that the pump produces hot water at $T_h = 52°C$. The table below shows their results for Q and T. (a) Determine a quantity that, when you plot the COP against it, should give a straight line. (b) Make your plot, fit a straight line, and from it determine how the heat pump's COP compares with the theoretical maximum COP.

T_c (°C)	−18	−10	−5	0	10
COP Q_h/W	2.7	3.2	3.6	3.7	4.7

Passage Problems

Refrigerators remain among the greatest consumers of electrical energy in most homes, although mandated efficiency standards have decreased their energy consumption by some 80% in the past four decades. In the course of a day, one kitchen refrigerator removes 30 MJ of energy from its contents, in the process consuming 10 MJ of electrical energy. The electricity comes from a 40% efficient coal-fired power plant.

70. The electrical energy
    a. is used to run the light bulb inside the refrigerator.
    b. wouldn't be necessary if the refrigerator had enough insulation.
    c. retains its high-quality status after the refrigerator has used it.
    d. ends up as waste heat rejected to the kitchen environment.
71. The refrigerator's COP is
    a. $\frac{1}{3}$.
    b. 2.
    c. 3.
    d. 4.
72. The fuel energy consumed at the power plant to run this refrigerator for the day is
    a. 12 MJ.
    b. 25 MJ.
    c. 40 MJ.
    d. 75 MJ.
73. The total energy rejected to the surrounding kitchen during the course of the day is
    a. 10 MJ.
    b. 30 MJ.
    c. 40 MJ.
    d. 75 MJ.

Answers to Chapter Questions

Answer to Chapter Opening Question

The second law of thermodynamics prevents us from converting thermal energy to mechanical energy with 100% efficiency, and practical limits on temperature make it hard to achieve more than about 50% efficiency in conventional power plants.

Answers to GOT IT? Questions

19.1 (a), (c), and (f)
19.2 (c)
19.3 (c) see Problem 34 for a proof
19.4 (1) increase; (2) decrease; (3) increase; (4) increase; (5) decrease; (6) increase; (7) increase

PART THREE SUMMARY Thermodynamics

Thermodynamics is the study of heat, temperature, and related phenomena—and their relation to the all-important concept of energy. Thermodynamics provides a macroscopic description in terms of parameters like temperature and pressure.

This contrasts with **statistical mechanics**, which provides a microscopic description in terms of the properties and behavior of molecules.

Thermodynamic equilibrium occurs when two systems are brought into thermal contact and no further changes occur in any macroscopic properties. The **zeroth law of thermodynamics** says that two systems each in thermodynamic equilibrium with a third are also in thermodynamic equilibrium with each other. This law allows us to establish temperature scales and construct thermometers.

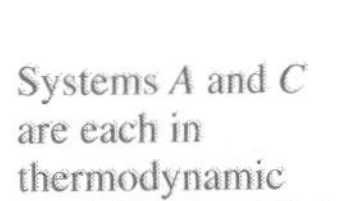

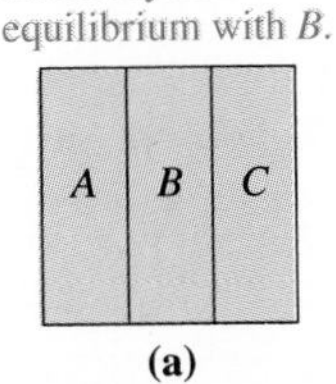

(a)

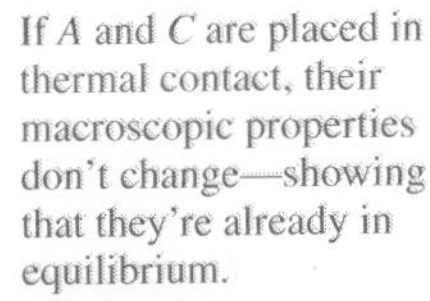

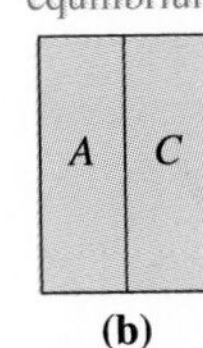

(b)

Heat is energy that's flowing because of a temperature difference. Important heat-transfer mechanisms include **conduction**, **convection**, and **radiation**. A system is in **thermal-energy balance** at a fixed temperature when its energy input balances heat transfer to its surroundings.

Earth's energy balance

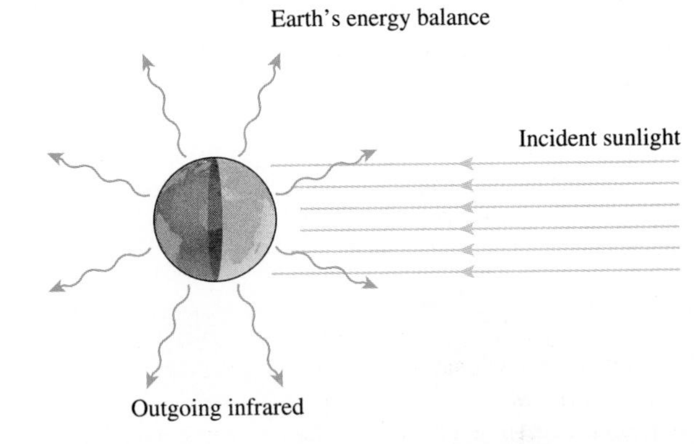

Ideal gases exhibit a simple relation among temperature, pressure, and volume:

$$pV = NkT = nRT$$

This is the **ideal gas law**, with $k = 1.381 \times 10^{-23}$ J/K and $R = 8.314$ J/K·mol.

Real substances undergo **phase changes** among liquid, solid, and gaseous phases. Substantial **heats of transformation** describe the energies involved in phase changes.

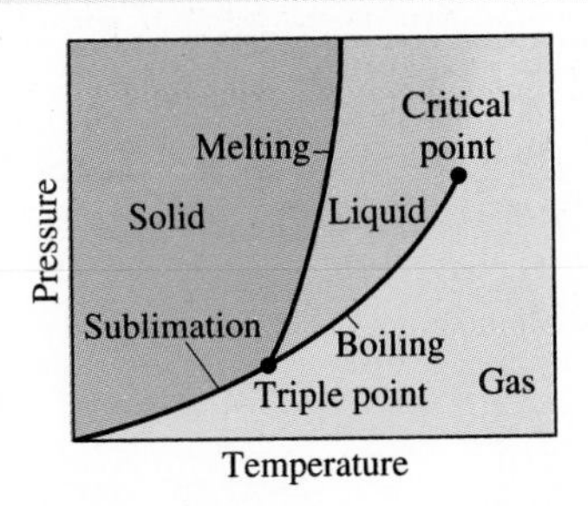

The **first law of thermodynamics** relates the change ΔE_{int} in a system's internal energy to the heat Q added *to* the system and the work W done *by* the system:

$$\Delta E_{int} = Q - W$$

For an ideal gas, **reversible thermodynamic processes** are described by curves in the pressure–volume diagram. Common processes include **isothermal** (constant temperature), **constant volume**, **constant pressure**, and **adiabatic** (no heat flow).

Entropy is a measure of disorder. The **second law of thermodynamics** states that the entropy of a closed system can never decrease. Applied to the heat engines that provide most of humankind's electrical and transportation energy, the second law shows that it's impossible to extract as useful work all the random internal energy of hot objects.

Maximum efficiency (Carnot):

$$e = \frac{W}{Q_h} = 1 - \frac{Q_c}{Q_h} = 1 - \frac{T_c}{T_h}$$

Part Three Challenge Problem

The ideal Carnot engine shown in the figure at right operates between a heat reservoir and a block of ice with mass M. An external energy source maintains the reservoir at a constant temperature T_h. At time $t = 0$, the ice is at its melting point T_0, but it's insulated from everything except the engine, so it's free to change state and temperature. The engine is operated in such a way that it extracts heat from the reservoir at a constant rate P_h. (a) Find an expression for the time t_1 at which the ice is all melted, in terms of the quantities given and any other appropriate thermodynamic parameters. (b) Find an expression for the mechanical power output of the engine as a function of time for times $t > t_1$. (c) Your expression in part (b) holds up only to some maximum time t_2. Why? Find an expression for t_2.

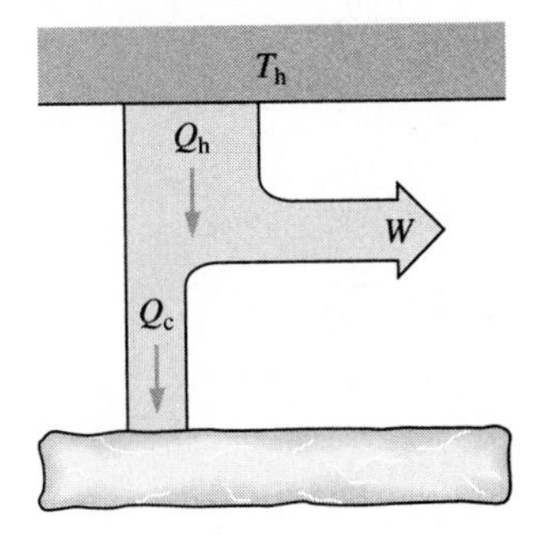

Mathematics

APPENDIX A

A-1 Algebra and Trigonometry

Quadratic Formula

If $ax^2 + bx + c = 0$, then $x = \dfrac{-b \pm \sqrt{b^2 - 4ac}}{2a}$.

Circumference, Area, Volume

Where $\pi \simeq 3.14159\ldots$:

circumference of circle	$2\pi r$
area of circle	πr^2
surface area of sphere	$4\pi r^2$
volume of sphere	$\frac{4}{3}\pi r^3$
area of triangle	$\frac{1}{2}bh$
volume of cylinder	$\pi r^2 l$

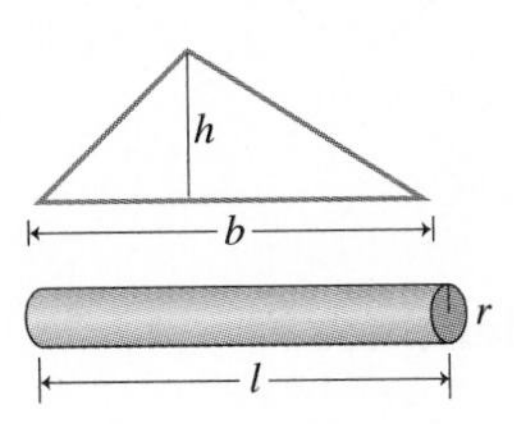

Trigonometry

definition of angle (in radians): $\theta = \dfrac{s}{r}$

2π radians in complete circle

1 radian $\simeq 57.3°$

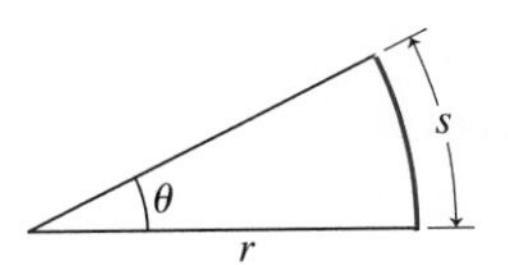

Trigonometric Functions

$\sin\theta = \dfrac{y}{r}$

$\cos\theta = \dfrac{x}{r}$

$\tan\theta = \dfrac{\sin\theta}{\cos\theta} = \dfrac{y}{x}$

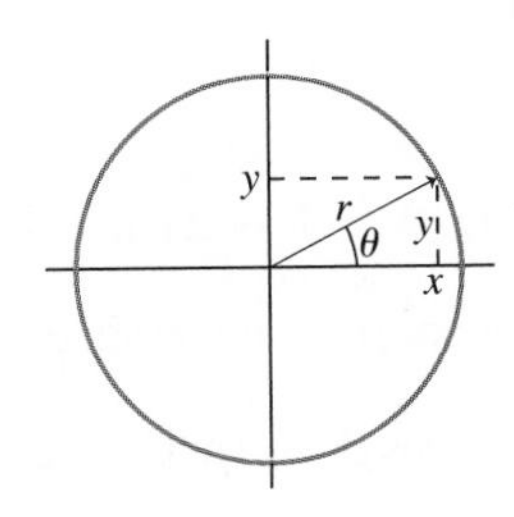

Values at Selected Angles

$\theta \rightarrow$	**0**	$\frac{\pi}{6}$ **(30°)**	$\frac{\pi}{4}$ **(45°)**	$\frac{\pi}{3}$ **(60°)**	$\frac{\pi}{2}$ **(90°)**
$\sin\theta$	0	$\frac{1}{2}$	$\frac{\sqrt{2}}{2}$	$\frac{\sqrt{3}}{2}$	1
$\cos\theta$	1	$\frac{\sqrt{3}}{2}$	$\frac{\sqrt{2}}{2}$	$\frac{1}{2}$	0
$\tan\theta$	0	$\frac{\sqrt{3}}{3}$	1	$\sqrt{3}$	∞

Graphs of Trigonometric Functions

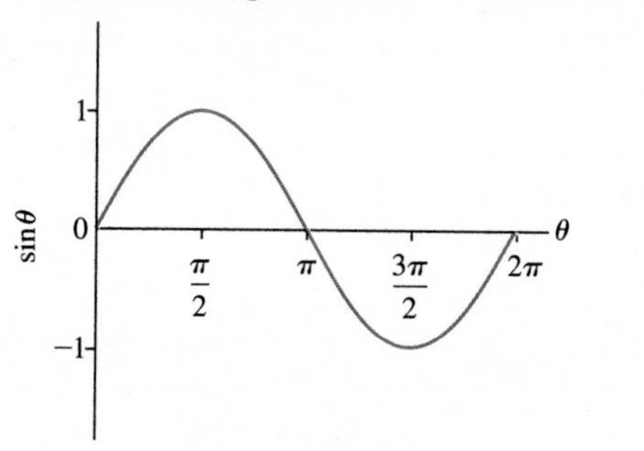

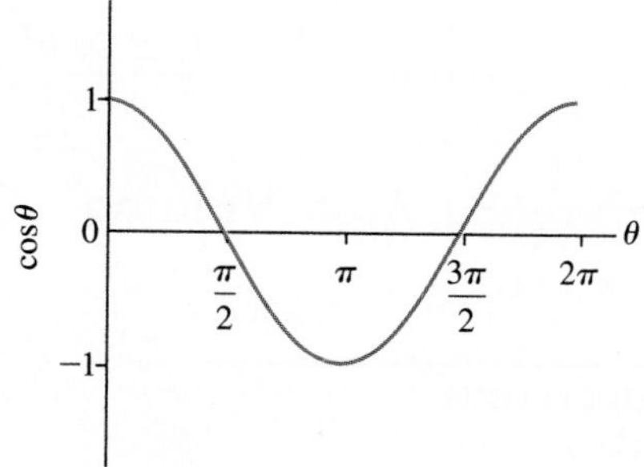

Trigonometric Identities

$$\sin(-\theta) = -\sin\theta$$

$$\cos(-\theta) = \cos\theta$$

$$\sin\left(\theta \pm \frac{\pi}{2}\right) = \pm\cos\theta$$

$$\cos\left(\theta \pm \frac{\pi}{2}\right) = \mp\sin\theta$$

$$\sin^2\theta + \cos^2\theta = 1$$

$$\sin 2\theta = 2\sin\theta\cos\theta$$

$$\cos 2\theta = \cos^2\theta - \sin^2\theta = 1 - 2\sin^2\theta = 2\cos^2\theta - 1$$

$$\sin(\alpha \pm \beta) = \sin\alpha\cos\beta \pm \cos\alpha\sin\beta$$

$$\cos(\alpha \pm \beta) = \cos\alpha\cos\beta \mp \sin\alpha\sin\beta$$

$$\sin\alpha \pm \sin\beta = 2\sin\left[\tfrac{1}{2}(\alpha \pm \beta)\right]\cos\left[\tfrac{1}{2}(\alpha \mp \beta)\right]$$

$$\cos\alpha + \cos\beta = 2\cos\left[\tfrac{1}{2}(\alpha + \beta)\right]\cos\left[\tfrac{1}{2}(\alpha - \beta)\right]$$

$$\cos\alpha - \cos\beta = -2\sin\left[\tfrac{1}{2}(\alpha + \beta)\right]\sin\left[\tfrac{1}{2}(\alpha - \beta)\right]$$

Laws of Cosines and Sines

Where A, B, C are the sides of an arbitrary triangle and α, β, γ the angles opposite those sides:

Law of cosines

$$C^2 = A^2 + B^2 - 2AB\cos\gamma$$

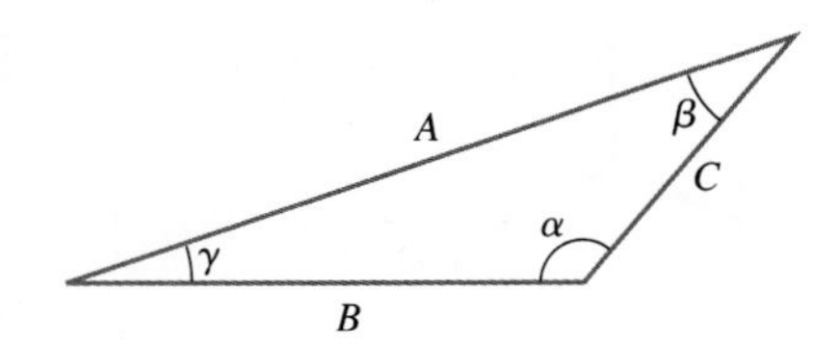

Law of sines

$$\frac{\sin\alpha}{A} = \frac{\sin\beta}{B} = \frac{\sin\gamma}{C}$$

Exponentials and Logarithms

$$e^{\ln x} = x, \quad \ln e^x = x \quad e = 2.71828\ldots$$

$$a^x = e^{x\ln a} \qquad \ln(xy) = \ln x + \ln y$$

$$a^x a^y = a^{x+y} \qquad \ln\left(\frac{x}{y}\right) = \ln x - \ln y$$

$$(a^x)^y = a^{xy} \qquad \ln\left(\frac{1}{x}\right) = -\ln x$$

$$\log x \equiv \log_{10} x = \ln(10)\ln x \simeq 2.3\ln x$$

Approximations

For $|x| \ll 1$, the following expressions provide good approximations to common functions:

$$e^x \simeq 1 + x$$

$$\sin x \simeq x$$

$$\cos x \simeq 1 - \tfrac{1}{2}x^2$$

$$\ln(1 + x) \simeq x$$

$$(1 + x)^p \simeq 1 + px \quad \text{(binomial approximation)}$$

Expressions that don't have the forms shown may often be put in the appropriate form. For example:

$$\frac{1}{\sqrt{a^2 + y^2}} = \frac{1}{a\sqrt{1 + \dfrac{y^2}{a^2}}} = \frac{1}{a}\left(1 + \frac{y^2}{a^2}\right)^{-1/2} \simeq \frac{1}{a}\left(1 - \frac{y^2}{2a}\right) \quad \text{for} \quad y^2/a^2 \ll 1, \text{ or } y^2 \ll a^2$$

Vector Algebra

Vector Products

$$\vec{A}\cdot\vec{B} = AB\cos\theta$$

$|\vec{A}\times\vec{B}| = AB\sin\theta$, with direction of $\vec{A}\times\vec{B}$ given by the right-hand rule:

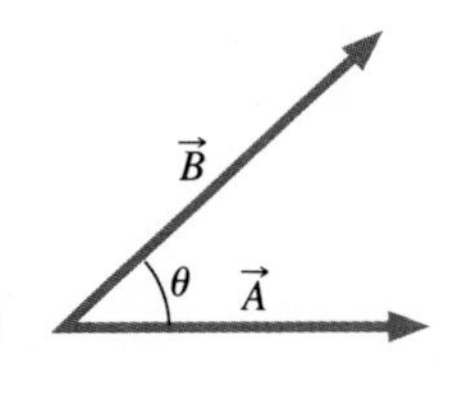

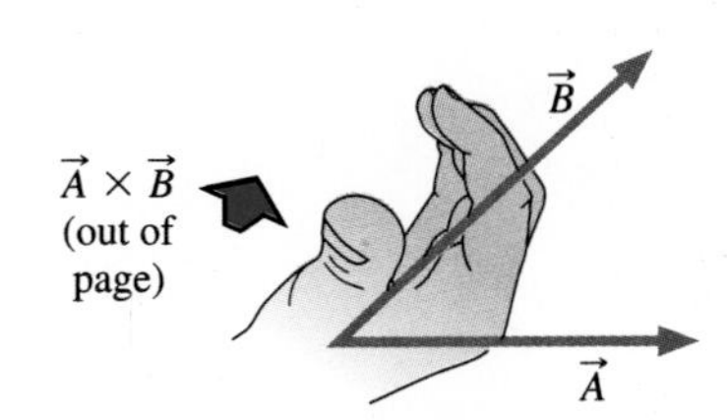

Unit Vector Notation

An arbitrary vector $\vec{A}$ may be written in terms of its components A_x, A_y, A_z and the unit vectors $\hat{\imath}$, $\hat{\jmath}$, $\hat{k}$ that have magnitude 1 and lie along the x-, y-, z-axes:

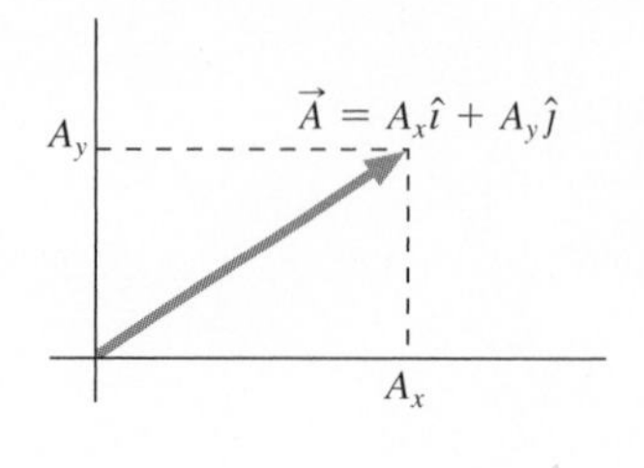

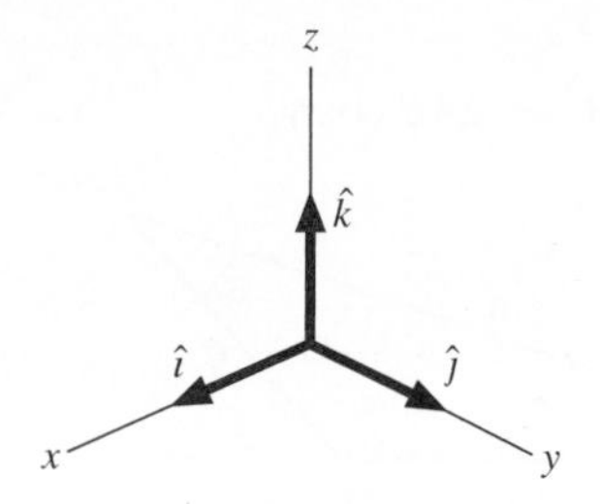

In unit vector notation, vector products become

$$\vec{A}\cdot\vec{B} = A_xB_x + A_yB_y + A_zB_z$$

$$\vec{A}\times\vec{B} = (A_yB_z - A_zB_y)\hat{\imath} + (A_zB_x - A_xB_z)\hat{\jmath} + (A_xB_y - A_yB_x)\hat{k}$$

Vector Identities

$$\vec{A}\cdot\vec{B} = \vec{B}\cdot\vec{A}$$

$$\vec{A}\times\vec{B} = -\vec{B}\times\vec{A}$$

$$\vec{A}\cdot(\vec{B}\times\vec{C}) = \vec{B}\cdot(\vec{C}\times\vec{A}) = \vec{C}\cdot(\vec{A}\times\vec{B})$$

$$\vec{A}\times(\vec{B}\times\vec{C}) = (\vec{A}\cdot\vec{C})\vec{B} - (\vec{A}\cdot\vec{B})\vec{C}$$

A-2 Calculus

Derivatives

Definition of the Derivative

If y is a function of x, then the **derivative of y with respect to x** is the ratio of the change Δy in y to the corresponding change Δx in x, in the limit of arbitrarily small Δx:

$$\frac{dy}{dx} = \lim_{\Delta x\to 0}\frac{\Delta y}{\Delta x}$$

Algebraically, the derivative is the rate of change of y with respect to x; geometrically, it is the slope of the y versus x graph—that is, of the tangent line to the graph at a given point:

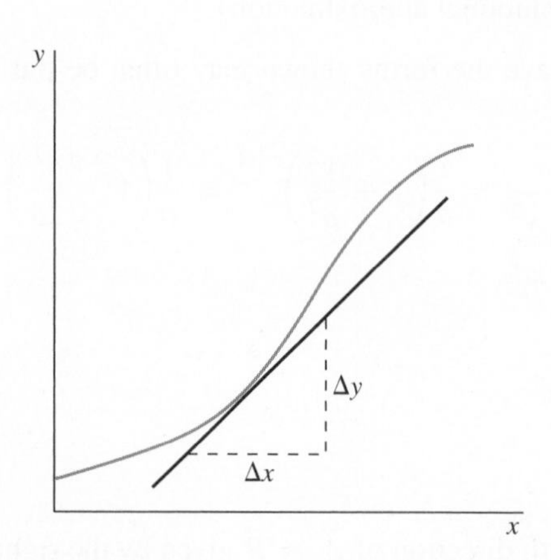

Derivatives of Common Functions

$$\frac{da}{dx} = 0 \quad (a \text{ is a constant})$$

$$\frac{dx^n}{dx} = nx^{n-1} \quad (n \text{ need not be an integer})$$

$$\frac{d}{dx}\sin x = \cos x$$

$$\frac{d}{dx}\cos x = -\sin x$$

$$\frac{d}{dx}\tan x = \frac{1}{\cos^2 x}$$

$$\frac{de^x}{dx} = e^x$$

$$\frac{d}{dx}\ln x = \frac{1}{x}$$

Derivatives of Sums, Products, and Functions of Functions

1. Derivative of a constant times a function

$$\frac{d}{dx}[af(x)] = a\frac{df}{dx} \qquad (a \text{ is a constant})$$

2. Derivative of a sum

$$\frac{d}{dx}[f(x) + g(x)] = \frac{df}{dx} + \frac{dg}{dx}$$

3. Derivative of a product

$$\frac{d}{dx}[f(x)g(x)] = g\frac{df}{dx} + f\frac{dg}{dx}$$

Examples

$$\frac{d}{dx}(x^2\cos x) = \cos x\frac{dx^2}{dx} + x^2\frac{d}{dx}\cos x = 2x\cos x - x^2\sin x$$

$$\frac{d}{dx}(x\ln x) = \ln x\frac{dx}{dx} + x\frac{d}{dx}\ln x = (\ln x)(1) + x\left(\frac{1}{x}\right) = \ln x + 1$$

4. Derivative of a quotient

$$\frac{d}{dx}\left[\frac{f(x)}{g(x)}\right] = \frac{1}{g^2}\left(g\frac{df}{dx} - f\frac{dg}{dx}\right)$$

Example

$$\frac{d}{dx}\left(\frac{\sin x}{x^2}\right) = \frac{1}{x^4}\left(x^2\frac{d}{dx}\sin x - \sin x\frac{dx^2}{dx}\right) = \frac{\cos x}{x^2} - \frac{2\sin x}{x^3}$$

5. Chain rule for derivatives

If f is a function of u and u is a function of x, then

$$\frac{df}{dx} = \frac{df}{du}\frac{du}{dx}$$

Examples

a. Evaluate $\frac{d}{dx}\sin(x^2)$. Here $u = x^2$ and $f(u) = \sin u$, so

$$\frac{d}{dx}\sin(x^2) = \frac{d}{du}\sin u\frac{du}{dx} = (\cos u)\frac{dx^2}{dx} = 2x\cos(x^2)$$

b. $\frac{d}{dt}\sin\omega t = \frac{d}{d\omega t}\sin\omega t\frac{d}{dt}\omega t = \omega\cos\omega t \qquad (\omega \text{ is a constant})$

c. Evaluate $\frac{d}{dx}\sin^2 5x$. Here $u = \sin 5x$ and $f(u) = u^2$, so

$$\frac{d}{dx}\sin^2 5x = \frac{d}{du}u^2\frac{du}{dx} = 2u\frac{du}{dx} = 2\sin 5x\frac{d}{dx}\sin 5x$$
$$= (2)(\sin 5x)(5)(\cos 5x) = 10\sin 5x\cos 5x = 5\sin 2x$$

Second Derivative

The second derivative of y with respect to x is defined as the derivative of the derivative:

$$\frac{d^2y}{dx^2} = \frac{d}{dx}\left(\frac{dy}{dx}\right)$$

Example

If $y = ax^3$, then $dy/dx = 3ax^2$, so

$$\frac{d^2y}{dx^2} = \frac{d}{dx}3ax^2 = 6ax$$

Partial Derivatives

When a function depends on more than one variable, then the partial derivatives of that function are the derivatives with respect to each variable, taken with all other variables held constant. If f is a function of x and y, then the partial derivatives are written

$$\frac{\partial f}{\partial x} \quad \text{and} \quad \frac{\partial f}{\partial y}$$

Example

If $f(x, y) = x^3 \sin y$, then

$$\frac{\partial f}{\partial x} = 3x^2 \sin y \quad \text{and} \quad \frac{\partial f}{\partial y} = x^3 \cos y$$

Integrals

Indefinite Integrals

Integration is the inverse of differentiation. The **indefinite integral**, $\int f(x)\,dx$, is defined as a function whose derivative is $f(x)$:

$$\frac{d}{dx}\left[\int f(x)\,dx\right] = f(x)$$

If $A(x)$ is an indefinite integral of $f(x)$, then because the derivative of a constant is zero, the function $A(x) + C$ is also an indefinite integral of $f(x)$, where C is any constant. Inverting the derivatives of common functions listed in the preceding section gives the integrals that follow (a more extensive table appears at the end of this appendix).

$$\int a\,dx = ax + C \qquad \int \cos x\,dx = \sin x + C$$

$$\int x^n\,dx = \frac{x^{n+1}}{n+1} + C, \quad n \neq -1 \qquad \int e^x\,dx = e^x + C$$

$$\int \sin x\,dx = -\cos x + C \qquad \int x^{-1}\,dx = \ln x + C$$

Definite Integrals

In physics we're most often interested in the **definite integral**, defined as the sum of a large number of very small quantities, in the limit as the number of quantities grows arbitrarily large and the size of each arbitrarily small:

$$\int_{x_1}^{x_2} f(x)\,dx \equiv \lim_{\substack{\Delta x \to 0 \\ N \to \infty}} \sum_{i=1}^{N} f(x_i)\,\Delta x$$

where the terms in the sum are evaluated at values x_i between the limits of integration x_1 and x_2; in the limit $\Delta x \to 0$, the sum is over all values of x in the interval.

The key to evaluating the definite integral is provided by the **fundamental theorem of calculus**. The theorem states that, if $A(x)$ is an *indefinite* integral of $f(x)$, then the *definite integral* is given by

$$\int_{x_1}^{x_2} f(x)\,dx = A(x_2) - A(x_1) \equiv A(x)\Big|_{x_1}^{x_2}$$

Geometrically, the definite integral is the area under the graph of $f(x)$ between the limits x_1 and x_2:

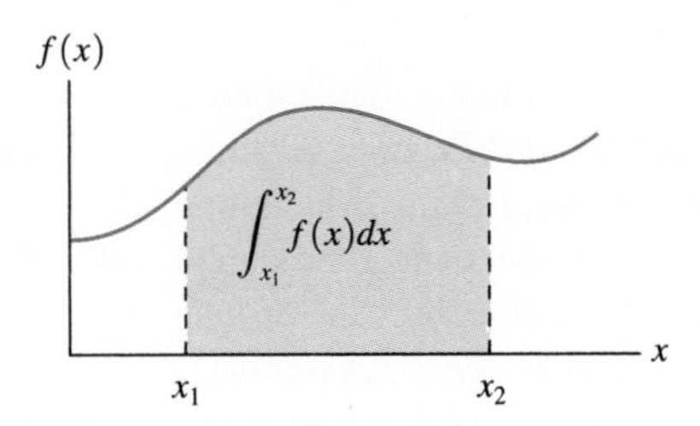

Evaluating Integrals

The first step in evaluating an integral is to express all varying quantities within the integral in terms of a single variable; Tactics 9.1 in Chapter 9 outlines a general strategy for setting up an integral. Once you've set up an integral, you can evaluate it yourself or look it up in tables. Two common techniques can help you evaluate integrals or convert them to forms listed in tables:

1. **Change of variables**
   An unfamiliar integral can often be put into familiar form by defining a new variable. For example, it is not obvious how to integrate the expression

$$\int \frac{x\,dx}{\sqrt{a^2 + x^2}}$$

where a is a constant. But let $z = a^2 + x^2$. Then

$$\frac{dz}{dx} = \frac{da^2}{dx} + \frac{dx^2}{dx} = 0 + 2x = 2x$$

so $dz = 2x\,dx$. Then the quantity $x\,dx$ in our unfamiliar integral is just $\frac{1}{2}dz$, while the quantity $\sqrt{a^2 + x^2}$ is just $z^{1/2}$. So the integral becomes

$$\int \frac{1}{2} z^{-1/2}\,dz = \frac{\frac{1}{2} z^{1/2}}{\frac{1}{2}} = \sqrt{z}$$

where we have used the standard form for the integral of a power of the independent variable. Substituting back $z = a^2 + x^2$ gives

$$\int \frac{x\,dx}{\sqrt{a^2 + x^2}} = \sqrt{a^2 + x^2}$$

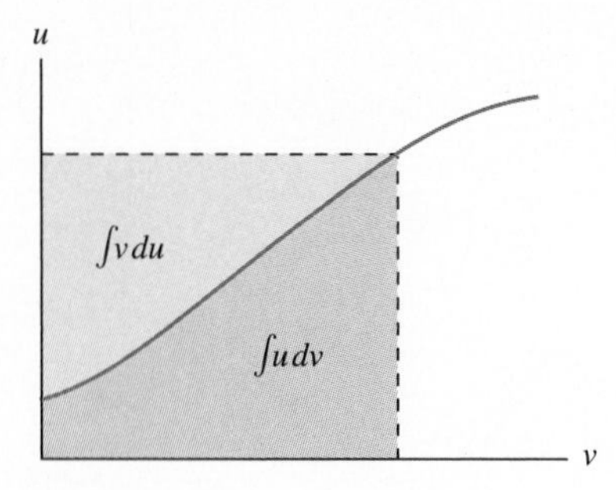

2. Integration by parts

The quantity $\int u\,dv$ is the area under the curve of u as a function of v between specified limits. In the figure, that area can also be expressed as the area of the rectangle shown minus the area under the curve of v as a function of u. Mathematically, this relation among areas may be expressed as a relation among integrals:

$$\int u\,dv = uv - \int v\,du \qquad \text{(integration by parts)}$$

This expression may often be used to transform complicated integrals into simpler ones.

Example

Evaluate $\int x \cos x\,dx$. Here let $u = x$, so $du = dx$. Then $dv = \cos x\,dx$, so we have $v = \int dv = \int \cos x\,dx = \sin x$. Integrating by parts then gives

$$\int x \cos x\,dx = (x)(\sin x) - \int \sin x\,dx = x \sin x + \cos x$$

where the + sign arises because $\int \sin x\,dx = -\cos x$.

Table of Integrals

More extensive tables are available in many mathematical and scientific handbooks; see, for example, ***Handbook of Chemistry and Physics*** (Chemical Rubber Co.) or Dwight, ***Tables of Integrals and Other Mathematical Data*** (Macmillan). Some math software, including *Mathematica* and *Maple*, can also evaluate integrals symbolically. Wolfram Research provides *Mathematica*-based integration both at *integrals.wolfram.com* and through WolframAlpha at *www.wolframalpha.com/calculators/integral-calculator.*

In the expressions below, a and b are constants. An arbitrary constant of integration may be added to the right-hand side.

$$\int e^{ax}\,dx = \frac{e^{ax}}{a}$$

$$\int \sin ax\,dx = -\frac{\cos ax}{a}$$

$$\int \cos ax\,dx = \frac{\sin ax}{a}$$

$$\int \tan ax\,dx = -\frac{1}{a}\ln(\cos ax)$$

$$\int \sin^2 ax\,dx = \frac{x}{2} - \frac{\sin 2ax}{4a}$$

$$\int \cos^2 ax\,dx = \frac{x}{2} + \frac{\sin 2ax}{4a}$$

$$\int x \sin ax\,dx = \frac{1}{a^2}\sin ax - \frac{1}{a}x\cos ax$$

$$\int x \cos ax\,dx = \frac{1}{a^2}\cos ax + \frac{1}{a}x\sin ax$$

$$\int \frac{dx}{\sqrt{a^2 - x^2}} = \sin^{-1}\left(\frac{x}{a}\right)$$

$$\int \frac{dx}{\sqrt{x^2 \pm a^2}} = \ln\left(x + \sqrt{x^2 \pm a^2}\right)$$

$$\int \frac{dx}{x^2 + a^2} = \frac{1}{a}\tan^{-1}\left(\frac{x}{a}\right)$$

$$\int \frac{x\,dx}{\sqrt{a^2 - x^2}} = -\sqrt{a^2 - x^2}$$

$$\int \frac{x\,dx}{\sqrt{x^2 \pm a^2}} = \sqrt{x^2 \pm a^2}$$

$$\int \frac{dx}{(x^2 \pm a^2)^{3/2}} = \frac{\pm x}{a^2\sqrt{x^2 \pm a^2}}$$

$$\int x e^{ax}\,dx = \frac{e^{ax}}{a^2}(ax - 1)$$

$$\int x^2 e^{ax}\,dx = \frac{x^2 e^{ax}}{a} - \frac{2}{a}\left[\frac{e^{ax}}{a^2}(ax - 1)\right]$$

$$\int \frac{dx}{a + bx} = \frac{1}{b}\ln(a + bx)$$

$$\int \frac{dx}{(a + bx)^2} = -\frac{1}{b(a + bx)}$$

$$\int \ln ax\,dx = x \ln ax - x$$

The International System of Units (SI)

APPENDIX B

The International System of Units (SI) system is undergoing a major overhaul, which should be completed by 2017. The new SI will give explicit-constant definitions to six of the seven base units, defining those units by setting exact values for appropriate physical constants. Here we list SI unit definitions informally to reflect this ongoing transition.

length (meter): The meter is defined so that the speed of light in vacuum is exactly 299,792,458 m/s. In effect since 1983, this definition is reworded but not otherwise changed.

mass (kilogram): In the new SI, the definition of the kilogram changes from one based on a physical standard, the international prototype kilogram, to an explicit-constant definition based on a defined value for Planck's constant h.

time (second): The second is defined as the duration of 9,192,631,770 periods of the radiation corresponding to the transition between the two hyperfine levels of the ground state of the cesium-133 atom. In effect since 1967, this definition is reworded but not otherwise changed.

electric current (ampere): Since 1948 the ampere has been defined in terms of the force between two current-carrying wires, but that changes to a definition based on an exact value for the elementary charge.

temperature (kelvin): Since 1967 the kelvin has been defined such that the triple point of water is at 273.16 K. That changes to a definition based on an exact value for Boltzmann's constant k.

amount of substance (mole): The 1971 definition of the mole in terms of carbon-12 atoms changes to one based on an exact value for Avogadro's number.

luminous intensity (candela): The 1979 definition defines the candela as the luminous intensity of a 540-THz source emitting (1/683) watt per steradian. This is reworded but not otherwise changed.

SI Base and Supplementary Units

	SI Unit	
Quantity	**Name**	**Symbol**
Base Unit		
Length	meter	m
Mass	kilogram	kg
Time	second	s
Electric current	ampere	A
Thermodynamic temperature	kelvin	K
Amount of substance	mole	mol
Luminous intensity	candela	cd
Supplementary Units		
Plane angle	radian	rad
Solid angle	steradian	sr

SI Prefixes

Factor	Prefix	Symbol
10^{24}	yotta	Y
10^{21}	zetta	Z
10^{18}	exa	E
10^{15}	peta	P
10^{12}	tera	T
10^{9}	giga	G
10^{6}	mega	M
10^{3}	kilo	k
10^{2}	hecto	h
10^{1}	deka	da
10^{0}	—	—
10^{-1}	deci	d
10^{-2}	centi	c
10^{-3}	milli	m
10^{-6}	micro	μ
10^{-9}	nano	n
10^{-12}	pico	p
10^{-15}	femto	f
10^{-18}	atto	a
10^{-21}	zepto	z
10^{-24}	yocto	y

Some SI Derived Units with Special Names

	SI Unit			
Quantity	**Name**	**Symbol**	**Expression in Terms of Other Units**	**Expression in Terms of SI Base Units**
Frequency	hertz	Hz		s^{-1}
Force	newton	N		$m \cdot kg \cdot s^{-2}$
Pressure, stress	pascal	Pa	N/m^2	$m^{-1} \cdot kg \cdot s^{-2}$
Energy, work, heat	joule	J	$N \cdot m$	$m^2 \cdot kg \cdot s^{-2}$
Power	watt	W	J/s	$m^2 \cdot kg \cdot s^{-3}$
Electric charge	coulomb	C		$s \cdot A$
Electric potential, potential difference, electromotive force	volt	V	J/C	$m^2 \cdot kg \cdot s^{-3} \cdot A^{-1}$
Capacitance	farad	F	C/V	$m^{-2} \cdot kg^{-1} \cdot s^4 \cdot A^2$
Electric resistance	ohm	Ω	V/A	$m^2 \cdot kg \cdot s^{-3} \cdot A^{-2}$
Magnetic flux	weber	Wb	$T \cdot m^2$, $V \cdot s$	$m^2 \cdot kg \cdot s^{-2} \cdot A^{-1}$
Magnetic field	tesla	T	Wb/m^2	$kg \cdot s^{-2} \cdot A^{-1}$
Inductance	henry	H	Wb/A	$m^2 \cdot kg \cdot s^{-2} \cdot A^{-2}$
Radioactivity	becquerel	Bq	1 decay/s	s^{-1}
Absorbed radiation dose	gray	Gy	J/kg, 100 rad	$m^2 \cdot s^{-2}$
Radiation dose equivalent	sievert	Sv	J/kg, 100 rem	$m^2 \cdot s^{-2}$

Conversion Factors

APPENDIX C

The listings below give the SI equivalents of non-SI units. To convert from the units shown to SI, multiply by the factor given; to convert the other way, divide. For conversions within the SI system, see the table of SI prefixes in Appendix B, Chapter 1, or the inside front cover. Conversions that are not exact by definition are given to, at most, four significant figures.

Length

1 inch (in) $= 0.0254$ m

1 foot (ft) $= 0.3048$ m

1 yard (yd) $= 0.9144$ m

1 mile (mi) $= 1609$ m

1 nautical mile $= 1852$ m

1 angstrom (Å) $= 10^{-10}$ m

1 light year (ly) $= 9.46 \times 10^{15}$ m

1 astronomical unit (AU) $= 1.496 \times 10^{11}$ m

1 parsec $= 3.09 \times 10^{16}$ m

1 fermi $= 10^{-15}$ m $= 1$ fm

Mass

1 slug $= 14.59$ kg

1 metric ton (tonne; t) $= 1000$ kg

1 unified mass unit (u) $= 1.661 \times 10^{-27}$ kg

Force units in the English system are sometimes used (incorrectly) for mass. The units given below are actually equal to the number of kilograms multiplied by g, the acceleration of gravity.

1 pound (lb) = weight of 0.454 kg

1 ton = 2000 lb = weight of 908 kg

1 ounce (oz) = weight of 0.02835 kg

Time

1 minute (min) $= 60$ s

1 hour (h) $= 60$ min $= 3600$ s

1 day (d) $= 24$ h $= 86{,}400$ s

1 year (y) $= 365.2422$ d* $= 3.156 \times 10^{7}$ s

Area

1 hectare (ha) $= 10^{4}$ m^2

1 square inch (in^2) $= 6.452 \times 10^{-4}$ m^2

1 square foot (ft^2) $= 9.290 \times 10^{-2}$ m^2

1 acre $= 4047$ m^2

1 barn $= 10^{-28}$ m^2

1 shed $= 10^{-30}$ m^2

Volume

1 liter (L) $= 1000$ cm^3 $= 10^{-3}$ m^3

1 cubic foot (ft^3) $= 2.832 \times 10^{-2}$ m^3

1 cubic inch (in^3) $= 1.639 \times 10^{-5}$ m^3

1 fluid ounce $= 1/128$ gal $= 2.957 \times 10^{-5}$ m^3

1 barrel (bbl) $= 42$ gal $= 0.1590$ m^3

1 gallon (U.S.; gal) $= 3.785 \times 10^{-3}$ m^3

1 gallon (British) $= 4.546 \times 10^{-3}$ m^3

Angle, Phase

1 degree (°) $= \pi/180$ rad $= 1.745 \times 10^{-2}$ rad

1 revolution (rev) $= 360° = 2\pi$ rad

1 cycle $= 360° = 2\pi$ rad

*The length of the year changes very slowly with changes in Earth's orbital period.

Speed, Velocity

1 km/h = (1/3.6) m/s = 0.2778 m/s

1 mi/h (mph) = 0.4470 m/s

1 ft/s = 0.3048 m/s

1 ly/y = 3.00×10^8 m/s

Angular Speed, Angular Velocity, Frequency, and Angular Frequency

1 rev/s = 2π rad/s = 6.283 rad/s (s^{-1})

1 Hz = 1 cycle/s = $2\pi\ s^{-1}$

1 rev/min (rpm) = 0.1047 rad/s (s^{-1})

Force

1 dyne = 10^{-5} N

1 pound (lb) = 4.448 N

Pressure

1 dyne/cm^2 = 10^{-1} Pa

1 atmosphere (atm) = 1.013×10^5 Pa

1 torr = 1 mm Hg at 0°C = 133.3 Pa

1 bar = 10^5 Pa = 0.987 atm

1 lb/in^2 (psi) = 6.895×10^3 Pa

1 in H_2O (60°F) = 248.8 Pa

1 in Hg (60°F) = 3.377×10^3 Pa

Energy, Work, Heat

1 erg = 10^{-7} J

1 calorie* (cal) = 4.184 J

1 electronvolt (eV) = 1.602×10^{-19} J

1 foot-pound (ft · lb) = 1.356 J

1 Btu* = 1.054×10^3 J

1 kWh = 3.6×10^6 J

1 megaton (explosive yield; Mt) = 4.18×10^{15} J

Power

1 erg/s = 10^{-7} W

1 horsepower (hp) = 746 W

1 Btu/h (Btuh) = 0.293 W

1 ft · lb/s = 1.356 W

Magnetic Field

1 gauss (G) = 10^{-4} T

1 gamma (γ) = 10^{-9} T

Radiation

1 curie (ci) = 3.7×10^{10} Bq

1 rad = 10^{-2} Gy

1 rem = 10^{-2} Sv

Energy Content of Fuels

Energy Source	Energy Content
Coal	29 MJ/kg = 7300 kWh/ton = 25×10^6 Btu/ton
Oil	43 MJ/kg = 39 kWh/gal = 1.3×10^5 Btu/gal
Gasoline	44 MJ/kg = 36 kWh/gal = 1.2×10^5 Btu/gal
Natural gas	55 MJ/kg = 30 kWh/100 ft^3 = 1000 Btu/ft^3
Uranium (fission)	
Normal abundance	5.8×10^{11} J/kg = 1.6×10^5 kWh/kg
Pure U-235	8.2×10^{13} J/kg = 2.3×10^7 kWh/kg
Hydrogen (fusion)	
Normal abundance	7×10^{11} J/kg = 3.0×10^4 kWh/kg
Pure deuterium	3.3×10^{14} J/kg = 9.2×10^7 kWh/kg
Water	1.2×10^{10} J/kg = 1.3×10^4 kWh/gal = 340 gal gasoline/gal water
100% conversion, matter to energy	9.0×10^{16} J/kg = 931 MeV/u = 2.5×10^{10} kWh/kg

*Values based on the thermochemical calorie; other definitions vary slightly.

The Elements

The atomic weights of stable elements reflect the abundances of different isotopes; values given here apply to elements as they exist naturally on Earth. For stable elements, parentheses express uncertainties in the last decimal place given. For elements with no stable isotopes (indicated in **boldface**), at most three isotopes are given; for elements 99 and beyond, only the longest-lived isotope is given. (Exceptions are the unstable elements thorium, protactinium, and uranium, for which atomic weights reflect natural abundances of long-lived isotopes.) See also the periodic table inside the back cover.

Atomic Number	Names	Symbol	Atomic Weight
1	Hydrogen	H	1.00794 (7)
2	Helium	He	4.002602 (2)
3	Lithium	Li	6.941 (2)
4	Beryllium	Be	9.012182 (3)
5	Boron	B	10.811 (5)
6	Carbon	C	12.011 (1)
7	Nitrogen	N	14.00674 (7)
8	Oxygen	O	15.9994 (3)
9	Fluorine	F	18.9984032 (9)
10	Neon	Ne	20.1797 (6)
11	Sodium (Natrium)	Na	22.989768 (6)
12	Magnesium	Mg	24.3050 (6)
13	Aluminum	Al	26.981539 (5)
14	Silicon	Si	28.0855 (3)
15	Phosphorus	P	30.973762 (4)
16	Sulfur	S	32.066 (6)
17	Chlorine	Cl	35.4527 (9)
18	Argon	Ar	39.948 (1)
19	Potassium (Kalium)	K	39.0983 (1)
20	Calcium	Ca	40.078 (4)
21	Scandium	Sc	44.955910 (9)
22	Titanium	Ti	47.88 (3)
23	Vanadium	V	50.9415 (1)
24	Chromium	Cr	51.9961 (6)
25	Manganese	Mn	54.93805 (1)
26	Iron	Fe	55.847 (3)
27	Cobalt	Co	58.93320 (1)
28	Nickel	Ni	58.69 (1)
29	Copper	Cu	63.546 (3)
30	Zinc	Zn	65.39 (2)
31	Gallium	Ga	69.723 (1)
32	Germanium	Ge	72.61 (2)
33	Arsenic	As	74.92159 (2)

(continued)

Atomic Number	Names	Symbol	Atomic Weight
34	Selenium	Se	78.96 (3)
35	Bromine	Br	79.904 (1)
36	Krypton	Kr	83.80 (1)
37	Rubidium	Rb	85.4678 (3)
38	Strontium	Sr	87.62 (1)
39	Yttrium	Y	88.90585 (2)
40	Zirconium	Zr	91.224 (2)
41	Niobium	Nb	92.90638 (2)
42	Molybdenum	Mo	95.94 (1)
43	**Technetium**	**Tc**	**97, 98, 99**
44	Ruthenium	Ru	101.07 (2)
45	Rhodium	Rh	102.90550 (3)
46	Palladium	Pd	106.42 (1)
47	Silver	Ag	107.8682 (2)
48	Cadmium	Cd	112.411 (8)
49	Indium	In	114.82 (1)
50	Tin	Sn	118.710 (7)
51	Antimony (Stibium)	Sb	121.75 (3)
52	Tellurium	Te	127.60 (3)
53	Iodine	I	126.90447 (3)
54	Xenon	Xe	131.29 (2)
55	Cesium	Cs	132.90543 (5)
56	Barium	Ba	137.327 (7)
57	Lanthanum	La	138.9055 (2)
58	Cerium	Ce	140.115 (4)
59	Praseodymium	Pr	140.90765 (3)
60	Neodymium	Nd	144.24 (3)
61	**Promethium**	**Pm**	**145, 147**
62	Samarium	Sm	150.36 (3)
63	Europium	Eu	151.965 (9)
64	Gadolinium	Gd	157.25 (3)
65	Terbium	Tb	158.92534 (3)
66	Dysprosium	Dy	162.50 (3)
67	Holmium	Ho	164.93032 (3)
68	Erbium	Er	167.26 (3)
69	Thulium	Tm	168.93421 (3)
70	Ytterbium	Yb	173.04 (3)
71	Lutetium	Lu	174.967 (1)
72	Hafnium	Hf	178.49 (2)
73	Tantalum	Ta	180.9479 (1)
74	Tungsten (Wolfram)	W	183.85 (3)
75	Rhenium	Re	186.207 (1)
76	Osmium	Os	190.2 (1)
77	Iridium	Ir	192.22 (3)
78	Platinum	Pt	195.08 (3)
79	Gold	Au	196.96654 (3)
80	Mercury	Hg	200.59 (3)
81	Thallium	Tl	204.3833 (2)
82	Lead	Pb	207.2 (1)
83	Bismuth	Bi	208.98037 (3)

Atomic Number	Names	Symbol	Atomic Weight
84	Polonium	Po	209, 210
85	Astatine	At	210, 211
86	Radon	Rn	211, 220, 222
87	Francium	Fr	223
88	Radium	Ra	223, 224, 226
89	Actinium	Ac	227
90	Thorium	Th	232.0381 (1)
91	Protactinium	Pa	231.03588 (2)
92	Uranium	U	238.0289 (1)
93	Neptunium	Np	237, 239
94	Plutonium	Pu	239, 242, 244
95	Americium	Am	241, 243
96	Curium	Cm	245, 247, 248
97	Berkelium	Bk	247, 249
98	Californium	Cf	249, 250, 251
99	Einsteinium	Es	252
100	Fermium	Fm	257
101	Mendelevium	Md	258
102	Nobelium	No	259
103	Lawrencium	Lr	262
104	Rutherfordium	Rf	263
105	Dubnium	Db	268
106	Seaborgium	Sg	266
107	Bohrium	Bh	272
108	Hassium	Hs	277
109	Meitnerium	Mt	276
110	Darmstadtium	Ds	281
111	Roentgenium	Rg	280
112	Copernicium	Cn	285
113	—	—	284
114	Flerovium	Fl	289
115	—	—	288
116	Livermorium	Lv	292
117	—	—	294
118	—	—	294

APPENDIX E

Astrophysical Data

Sun, Planets, Principal Satellites

Body	Mass (10^{24} kg)	Mean Radius (10^6 m except as noted)	Surface Gravity (m/s^2)	Escape Speed (km/s)	Sidereal Rotation Period* (days)	Mean Distance from Central Body† (10^6 km)	Orbital Period	Mean Orbital Speed (km/s)
Sun	1.99×10^6	696	274	618	36 at poles 27 at equator	2.6×10^{11}	200 My	250
Planets								
Mercury	0.330	2.44	3.70	4.25	58.6	57.9	88.0 d	47.4
Venus	4.87	6.05	8.87	10.4	−243	108	225 d	35.0
Earth	5.97	6.37	9.81	11.2	0.997	149.6	365.2 d	29.8
Moon	0.0735	1.74	1.62	2.38	27.3	0.3844	27.3 d	1.02
Mars	0.642	339	3.71	5.03	1.03	228	1.88 y	24.1
Phobos	1.07×10^{-8}	9–13 km	0.0057	0.0114	0.319	9.4×10^{-3}	0.319 d	2.14
Deimos	1.48×10^{-9}	5–8 km	0.003	0.00556	1.26	23×10^{-3}	1.26 d	1.35
Jupiter	1.90×10^3	69.9	24.8	60.2	0.414	779	11.9 y	13.1
Io	0.0893	1.82	1.80	2.38	1.77	0.422	1.77 d	17.3
Europa	0.480	1.56	1.32	2.03	3.55	0.671	3.55 d	13.7
Ganymede	0.148	2.63	1.43	2.74	7.15	1.07	7.15 d	10.9
Callisto	0.108	2.41	1.24	2.44	16.7	1.88	16.7 d	8.20
and 13 smaller satellites								
Saturn	568	58.2	10.4	36.1	0.444	1.43×10^3	29.5 y	9.69
Tethys	0.0007	0.53	0.2	0.4	1.89	0.294	1.89 d	11.3
Dione	0.00015	0.56	0.3	0.6	2.74	0.377	2.74 d	10.0
Rhea	0.0025	0.77	0.3	0.5	4.52	0.527	4.52 d	8.5
Titan	0.135	2.58	1.35	2.64	15.9	1.22	15.9 d	5.6
and 12 smaller satellites								
Uranus	86.8	25.4	8.87	21.4	−0.720	2.87×10^3	84.0 y	6.80
Ariel	0.0013	0.58	0.3	0.4	2.52	0.19	2.52 d	5.5
Umbriel	0.0013	0.59	0.3	0.4	4.14	0.27	4.14 d	4.7
Titania	0.0018	0.81	0.2	0.5	8.70	0.44	8.70 d	3.7
Oberon	0.0017	0.78	0.2	0.5	13.5	0.58	13.5 d	3.1
and 11 smaller satellites								
Neptune	102	24.6	11.2	23.5	0.673	4.50×10^3	165 y	5.43
Triton	0.134	1.9	2.5	3.1	5.88	0.354	5.88 d	4.4
and 7 smaller satellites								
Dwarf Planets								
Ceres	0.000945	0.476	0.27	0.51	0.38	414	4.60 y	17.9
Pluto	0.0131	1.20	0.58	1.2	−6.39	5.91×10^3	248 y	4.67
Charon	0.00162	0.604	0.278	0.580	−6.39	0.00196	6.39 d	0.23
and 4 smaller satellites								
Eris	0.0167	1.16	0.827	1.38	1.1	1.02×10^4	560 y	3.43
and 1 small satellite, Dysnomia								

*Negative rotation period indicates retrograde motion, in opposite sense from orbital motion. Periods are sidereal, meaning the time for the body to return to the same orientation relative to the distant stars rather than the Sun.

†Central body is galactic center for Sun, Sun for planets, and planet for satellites.

Answers to Odd-Numbered Problems

Chapter 1

13. 10^5
15. 108.783 ps
17. 10^8
19. 0.62 rad = 35°
21. 30 g
23. 10^6
25. 8.6 m^2/L
27. 3.6 km/h
29. 57.3°
31. 24 Zm
33. 7.4×10^6 m/s^2
35. 4×10^6
37. 41 m
39. (a) 5.18 (b) 5.20
41. 3×10^6
43. About 0.08%
45. (a) $\sim 3 \times 10^3$ m^3 (b) ~100 days
47. 10^5
49. ~ 250 μm
51. (a) 40 nm (b) 5×10^5 calculations per second
53. $\Delta = 100(\pm 0.05/N)\%$
55. in the U.S.
57. about 2000
59. about 1–2 m^2
61. (a) 1.0 m (b) 0.001 m^2 (c) 0.0 m (d) 1.0
63. $10.10
65. slope = 4.09 g/cm^3
67. b
69. c

Chapter 2

13. (a) 375 yd/min (b) 5.72 min
15. 21 h
17. (a) 3.0×10^4 m/s (b) 19 mi/s
21. (a) $v = b - 2ct$ (b) 8.4 s
23. 0.35 m/s^2
25. falling: 9.82 m/s^2, stopping: 84.0 m/s^2
27. 17 m/s^2
29. $v = dx/dt = d/dt(x_0 + v_0t + at^2/2) = v_0 + at$
31. (a) 46 m/s^2 (b) 61 s
33. 27 ft/s^2
35. 15 s
37. 95 m
39. (a) 123 m (b) 39 m/s, 40 m (c) 9.8 m/s, 100 m; (d) −20 m/s, 100 m
41. 11 m/s
43. 48 mi/h
45. 2.2 s
47. (a) 9.82 m/s (b) 9.34 m/s (c) 9.18 m/s (d) 9.18 m/s
49. (a) 39.95 m/s (b) 0.13%
51. 4.3 m/s^2
53. 2.75 s
55. 55%
57. (a) 0.014 s (b) 51 cm
59. 0.89 km
61. (a) 25 m/s (b) 180 m
63. 0.0051 m/s^2
65. 11 m/s
67. 270 m
69. $-\frac{1}{2}\sqrt{hg}$
71. (a) 7.88 m/s, 7.67 m/s (b) 0.162 s
73. 3.9 s, 6.2 m/s
75. 4.8 m/s (17 km/h)
77. (a) $\bar{v} = (v_1 + v_2)/2$ (b) $\bar{v} = (2v_1v_2)/(v_1 + v_2)$ (c) arithmetic mean
79. 70.7 %
81. −0.3 m/s
83. $\frac{h}{4}\left(\frac{2h}{g\Delta t^2}\right)\left(\frac{g\Delta t^2}{2h} - 1\right)^2$
85. 15 s^{-1}
89. (a) $-(a_0/\omega)\sin\omega t$ (b) $(a_0/\omega^2)\cos\omega t$ (c) $v_{max} = a_0/\omega$, $x_{max} = a_0/\omega^2$
91. (a) $v_0 > \sqrt{gh_0/2}$ (b) $h_0 - gh_0^2/2v_0$
93. c
95. c

Chapter 3

13. 270 m, 150°
15. 700 km, 110°
17. $105\hat{\imath} + 58\hat{\jmath}$ km
19. 1.414, $\theta = 45°$
21. (−14 m/s, −12 m/s)
23. $3ct^2\hat{\imath}$
25. (a) $\vec{v} = -2.2 \times 10^{-6}\hat{\jmath}$ m/s (b) $\vec{a} = -3.2 \times 10^{-10}\,\hat{\imath}$ m/s^2
27. $\vec{v}_2 = 1.3\hat{\imath} + 2.3\hat{\jmath}$ m/s
29. (a) 26° upstream (b) 53.9 s
31. 42.8° west of south
33. 49 m, 6.4° to your original direction
35. (a) 1.3 s (b) 15 m
37. 34 nm
39. 1090 m
41. 2.28×10^{-7}m/s^2
43. 2.8 mm/s^2
45. (a) $A\sqrt{5}$ (b) $A\sqrt{10}$
47. $\vec{C} = -15\hat{\imath} + 9\hat{\jmath} - 18\hat{k}$
49. (a) $4c/3d$ (b) $c/3d$
51. 96 m
53. (a) 0.249 m/s (b) 7.00×10^{-4} m/s^2 (c) 7.21×10^{-4} m/s^2, about 3% difference
55. $A = B$
57. 0.50 m/s^2
59. 5.7 m/s
61. (a) $x_1 = x_2$ implies $y_1 = h\left(1 - \frac{gh}{v_0^2}\right) = y_2$ (b) $v_0 \geq \sqrt{gh}$
63. 8.3 m/s, 61°
67. semi-circle of radius 2.5 cm; 6.54 m/s, 17.1 m/s^2
69. Yes
71. 66°
73. (a) $v\sqrt{2/\sqrt{3}} \approx 1.07v$ (b) $\sqrt[4]{3}t \approx 1.3t$
75. 2.3 km
77. $2h$
79. 19 m
81. $dx/d\theta_0 = 2v_0^2/g\cos(2\theta_0) = 0 \Rightarrow \theta_0 = 45°$.
83. $\frac{1}{2}\cos^{-1}(1/(1 + v_0^2/gh))$
87. $2va_t/r$
89. c
91. c

Chapter 4

15. (a) 2.0 m/s^2 (b) 0.082 m/s^2
17. −13 kN
19. 2.0×10^6 m/s^2
21. 22 cm
25. 210 kg
27. 9000 kg
29. 490 N
31. 380 N
35. 55 kN
37. 130 N
39. 19 cm
41. 2.94 m/s^2, downward
43. 4.9 m/s^2
45. 0.53 s
47. 6.0 N to the right
49. 1.62×10^{-7}N/m
51. (a) 5.3 kN (b) 1.1 kN (c) 0.49 kN (d) 0.59 kN
53. 680 m
55. 0.96 m

57. 950 N
59. (a) $-0.40mg$ (b) $2.40mg$ (c) $1.40mg$
61. F-35A: yes, 0.81 m/s^2; A-380: no
65. 1.96 m/s^2
67. (a) $(m_f - m_s)g/m_s$ (b) $\dfrac{m_f a_s h_0}{(m_f - m_s)(g + a_s)}$
69. (a) 60.0 m/s (b) 0.672 m
71. 11.8 m/s^2
73. 0.92 kg, 1.4 kg
75. $\omega F_0/M$
77. a
79. b

Chapter 5

13. $5.40\hat{\imath} + 11.0\hat{\jmath}$ N
15. 22.4°
17. 880 N
19. 34°
21. $m_R/m_L = 2.5$
23. (a) 3.9 m/s^2 (b) 530 N
27. 58 km/h
29. 490 km/h
31. 0.18
33. 0.53
35. 0.43 m
37. about 2.62 times
39. $T = m_2 g$, $\tau = 2\pi\sqrt{(m_1 R)/(m_2 g)}$
41. 310 N downward (b) $-m_{SB}v^2/R$ (c) nothing
43. 8.5 km
45. 0.15
47. 25 s
49. Yes
51. $0.23 \le \mu_s \le 0.30$
53. 4.2 m/s^2
55. 0.62
57. (a) 10 cm (b) no
59. 100 km/h
63. 17 min^{-1}
65. Brake, don't swerve
67. 28 cm
69. $T' = u_k/\sqrt{1 + \mu_k^2}$
73. Yes
75. 7.6 km
77. a
79. b

Chapter 6

13. 900 J
15. 150 kJ
17. 190 MN
19. $\vec{A}\cdot(\vec{B} + \vec{C}) = AB\cos(\theta_{AB}) + AC\cos(\theta_{AC}) = \vec{A}\cdot\vec{B} + \vec{A}\cdot\vec{C}$
21. 1.9 m
23. (a) 1 J (b) 3 J
25. 30 cm
27. 7.5 GJ
29. ± 120 km/h
31. 110 m/s
33. 97 W
35. (a) 60 kW (b) 1 kW (c) 41.7 W
37. 9.4×10^6 J
39. 0 W
41. 22 s
43. (a) 400 J (b) 31 kg
45. (a) 76,000 (b) 14 kW
47. 25°
49. (a) 0 (b) 90°
51. 622 J
53. $k_B = 8k_A$
55. $W = F_0\left(x - \dfrac{x^2}{2L_0} + \dfrac{L_0^2}{L_0 + x} - L_0\right)$
57. $v_2 = \pm 2v_1$
59. (a) 1.3×10^{-17}W (b) 1.4×10^{-14}J
61. 9.6 kW
63. $F_0 x_0/3$
65. 70.5°
67. 370×10^6 gal/day
69. 26 m/s
71. 0.60
73. (a) 0.45 kW (b) 7.99 kJ
75. 42 kJ
77. 6.0 years
81. $W_{x_1\to x_2} = 2b(\sqrt{x_2} - \sqrt{x_1})$, $W(x_1 = 0) = 2b\sqrt{x_2}$
83. (a) $\frac{1}{2}kL_0^2 + \frac{1}{3}bL_0^3 + \frac{1}{4}cL_0^4 + \frac{1}{5}dL_0^5$ (b) 12 kJ
85. 135 J
87. 30 people
89. Stopping force is 35 times weight of leg
91. c
93. c

Chapter 7

11. $W_a = W_b = -mgL$
13. (a) 1.3 MJ (b) -59 kJ
15. 840 m
17. 55 cm
19. ± 22 m/s, ± 35 m/s
21. 92 m
23. 2.3 kN/m
25. 0.75
27. ± 2.0 m
29. (a) 4.4×10^{13}J (b) 11 h
31. (a) 1.07 J (b) 1.12 J
33. 778 J, 4.90%
35. 2.5 J
37. $U(x) = -\frac{1}{3}ax^3 - bx$
39. $r = \dfrac{kx^2}{2mg\sin\theta}$
43. (a) -11 cm (b) ± 4 m/s
45. $h \ge 5R/2$
49. (a) $U(x) = -\dfrac{a}{2}x^2 + \dfrac{b}{4}x^4$ (b) 0.7 m and 2 m
51. 20 m/s, 30 m/s
53. 1.4 m
55. 62.5 cm
57. 2.9 m
59. 14 m
61. $v = 2x^{3/4}\sqrt{\dfrac{a}{3m}}$
63. 5.8 s
65. $\dfrac{mgh}{2d}\sqrt{2g(h - d)}$
67. 185 N/m
69. d
71. b

Chapter 8

11. $R_P = R_E/\sqrt{2}$
13. 57.5%
15. 8.6 kg
17. 542 m
19. 3070 m/s
21. 1.77 d
23. 0.28×10^6 m
25. 3.17 GJ
27. 4.29 km/s
29. -2.64×10^{33} J
31. (a) 2.44 km/s (b) 2.10×10^8 m/s
33. 10 m/s^2
35. $g(h)/g(0) = 0.414$
37. 2.73×10^{-3} m/s^2, $a_c/g = 2.78\times10^{-4}$
39. 60.5 min
41. 2.6×10^{41} kg
43. $T^2 = \dfrac{4\pi^2L^3}{3GM}$
45. 2.79 AU
47. $E > 0$, hyperbolic path
51. The comet is going faster than the escape velocity from the Sun, so it will not return to Earth's vicinity.
55. (a) 2.06×10^6 m (b) 0.805×10^6 m
57. (a) 4.59 km/s (b) 14.2 km/s
59. 4.17 km/s
61. 4.60×10^{10} m
63. 1.42×10^3 km
65. 1.58×10^{16} kg
67. 3.8 m/century
69. No danger, since the puck needs at least 6100 km/h to go into orbit.
71. 1.5×10^6 km
73. d
75. d

Chapter 9

15. $2m$
17. $(0, 0.289L)$
19. $\vec{v}_2 = -67\hat{\imath}$ cm/s
21. 0.268 Mm/s
23. 1.21 J
25. The impulse imparted by gravity is 0.08% of the collision impulse.
27. 41.8 s
31. The second truck's load was about 7600 kg—close to, but slightly below, the legal limit.
33. 46 m/s

35. $v_{1f} = -11$ Mm/s, $v_{2f} = +6.9$ Mm/s, velocities are exchanged
37. (0, 0.115 (*a*)
39. $\vec{r}_{cm} = (2t^2 + 4t + \frac{8}{3})\hat{\imath} + (\frac{5}{3}t + \frac{4}{3})\hat{\jmath}$; $\vec{v}_{cm} = (4t + 4)\hat{\imath} + (\frac{5}{3})\hat{\jmath}$; $\vec{a}_{cm} = 4\hat{\imath}$
41. $m_b = 4m_m$
43. $(0, 0, h/4)$
45. (a) 0.99 m (b) 3.9 m/s
47. (a) $\vec{a}_c = \frac{v_0}{M}\left(\frac{dm}{dt}\right)\hat{\imath}$ (b) v_0
51. (a) (0, 0, 13 m) (b) (0, 0, 11 m)
53. $\vec{v}_3 = 4.4\hat{\imath} + 3.0\hat{\jmath}$ m/s
55. 9.4 m/s
57. $\frac{2}{5}v$; $\frac{7}{5}v$
59. (a) 37.7° (b) −65.8 cm/s
61. 5.8 s
63. 0.92 m/s
65. If $v_{Buick} = 55$ km/h, then $v_{Toyota} = 90$ km/h; if $v_{Toyata} = 55$ km/h, then $v_{Buick} = 65$ km/h
67. 120°
69. 5.83
73. 18.6%
75. $J = 2F_0/a$
77. (a) 12.0 m/s (b) 15.4 m/s
79. $v_1 = v/6$, $v_2 = 5v/6$
81. 8.3 kg
83. The peak force of 327 kN occurs at 165 ms.
87. The center of mass lies along line through the middle of the slice, at a distance of $(4R/3\theta)\sin(\frac{1}{2}\theta)$ from the tip.
89. 3.75 min
91. (a) $\frac{M}{1+a}$; (b) $\frac{1+a}{2+a}L$ (c) M and $\frac{1}{2}L$
93. 3 collisions, final speeds $0.26v_0$ and $0.31v_0$
95. b
97. a

Chapter 10

13. (a) $7.27\times10^{-5}\,\text{s}^{-1}$ (b) $1.75\times10^{-3}\,\text{s}^{-1}$ (c) $1.45\times10^{-4}\,\text{s}^{-1}$ (d) $31.4\,\text{s}^{-1}$
15. (a) 75 rad/s (b) 2.4×10^{-4} rad/s (c) 6×10^3 rad/s (d) 2×10^{-7} rad/s
17. (a) 0.068 rpm/s (b) $7.1\times10^{-3}\,\text{s}^{-2}$
19. (a) 0.16 rev (b) 0.07 rad/s
21. 1.2 m
23. 7.9×10^{-2} N·m
25. (a) $2mL^2$ (b) mL^2
27. (a) $4.4\times10^{-4}\,\text{kg}\cdot\text{m}^2$ (b) 3.7×10^{-3} N·m
29. (a) mL^2 (b) $\frac{1}{2}mL^2$
31. (a) $3.2\times10^{38}\,\text{kg}\cdot\text{m}^2$ (b) 1.8×10^{34} N·m
33. 20 min
35. 12,000 y
37. (a) 1.6×10^8 J (b) 16 MW
39. 1/3
41. (a) 6.9 rad/s (b) 3.7 s
43. (a) 1.1 rad/s (b) 1.1 m/s
45. (a) $170\,\text{s}^{-2}$ (b) $2.9\,\text{m/s}^2$ (c) 150 revolutions
47. 570 rev
49. (a) $2ML^2/3$ (b) $2ML^2/3$ (c) $4ML^2/3$
51. $Ma^2/3$
53. 33 pN
55. (a) 7.2 h (b) 1900 rev
57. 0.36
59. ±2.1 rad/s
61. $v = \sqrt{\frac{6}{5}gd\sin\theta}$
63. 17%
65. $0.494\,MR^2$
67. 33 m
69. (a) $M = \frac{2\pi\rho_0 wR^2}{3}$ (b) $I = 3MR^2/5$
71. yes for spin-up time (53 s), but no for efficiency (94%)
73. $3MR^2/10$
75. $\tau = \frac{1}{2}MGL\sin\theta$
77. The specs are incorrect. The storage capacity is 3 MJ below what's claimed.
79. $5.2\times10^{-5}\,\text{kg}\cdot\text{m}^2$
81. a
83. b

Chapter 11

15. $\vec{\omega} = 63\,\text{s}^{-1}$ west
17. (a) $1.1\times10^6\,\text{s}^{-2}$ (b) −37°
19. (a) $-12\hat{k}$ N·m (b) $36\hat{k}$ N·m (c) $12\hat{\imath} + 36\hat{\jmath}$ N·m
21. 3.1 N·m, out of the page
23. $414\,\text{kg}\cdot\text{m}^2\cdot\text{s}^{-1}$
25. 2.3 J·s along axis
27. 17.4 rpm
29. 2.5 days.
31. $-9.0\hat{k}$ N·m
33. 1600 N·m
35. 37 J·s
37. 2.66×10^5 J·s, out of plane of figure
39. 3.1×10^{-16} J·s
41. $0.21\,\text{kg}\cdot\text{m}^2$
43. 63%
45. 5.5 m/s
47. 3.1 rpm
51. (a) 0.25 rad/s (b) 6.4 kJ
53. (a) $\omega d(\frac{1}{2} - I/2md^2)$ (b) ωd (c) $\omega d(2 + I/md^2)$
55. 2.8%, orbital angular momentum of Jupiter
57. (a) 140 rpm (b) 27%
59. (a) $2\omega_0/7$ (b) $t = \frac{2R\omega_0}{\mu_k g}$
63. 9.2×10^{26}N·m
65. d
67. d

Chapter 12

15. (a) $\tau = mgL/2$ (b) $\tau = 0$ (c) $\tau = -mgL/2$
17. 16 m relative to the wall
19. (a) 0.61 m from left end (b) 1.42 m from left end
21. 480 N
23. −0.797 m, unstable; 1.46 m, stable
25. (a) 40 N·m (b) 1.3 kN
27. 500 N
29. 79 kg
31. 1.4 W
35. 87 kg
37. $\tan^{-1}(L/w)$
39. (a) $\frac{mg}{2}[L\sin\theta - w(1 - \cos\theta)]$ (b) $\tan^{-1}(L/w)$ (c) concave down, unstable
41. 74 kg
43. 0.366 *mgs*
47. $F_{app} = Mg\tan(\theta/2)$
49. $\mu_s < \tan\alpha = 1/2$
53. $\mu \geq \frac{\tan\theta}{2 + \tan^2\theta}$
55. 840 N
57. 170 N
59. (a) $F = G\frac{M_E m}{R_E^2}(1.229)$, 21.3° (b) $\tau = G\frac{M_E m}{R_E}(-0.0356)$
61. The tie beam will not hold under the 10 kN of tension.
63. stable equilibrium at ~6 nm and ~14 nm, unstable equilibrium at ~11 nm
65. a
67. b

Chapter 13

17. 2.27×10^{-3} s
19. (a) $x(t) = (12.5\text{ cm})\cos[(42.0\,\text{s}^{-1})t]$ (b) $x(t) = (2.15\text{ cm})\sin[(4.63\,\text{s}^{-1})t]$
21. 22 ms
23. 0.59 Hz; 1.7 s
25. (a) $10.6\,\text{s}^{-1}$ (b) 16.5 cm (c) 38.6 N/m
27. (a) 2.2 rad/s (b) 2.8 s (c) 0.63 m
29. 1.21 s
31. 1.6 s
33. 7 oscillations in *x* direction for 4 oscillations in the *y* direction
35. ±1.7 rad, ±15 rad/s
37. 0.25 s
39. 65 km/h
41. 0.70 s
43. (a) $t = \pi\sqrt{m/k}$ (b) $A = v_0\sqrt{m/k}$
45. 50 min
47. (a) 67 μN/m (b) 3.4×10^{-10} kg
51. 821 kg
53. (a) $|\vec{r}| = A$ (b) $\vec{v} = (\omega A\cos\omega t)\hat{\imath} - (\omega A\sin\omega t)\hat{\jmath}$ (c) $|\vec{v}| = \omega A$ (d) ω
55. 0.147%
57. (a) 1.3 N/m (b) 0.80 kg
59. $\omega = \sqrt{(k_1 + k_2)/m}$
63. $\omega = \sqrt{2k/3M}$
65. 34
67. (a) 6.5 cm (b) 0.51 s
69. $f = \frac{1}{2\pi}\sqrt{2a/m}$
71. (a) $E_1 = 4E_2$ (b) $a_{max,1} = 4a_{max,2}$
73. 27°
75. $T = 2\pi\sqrt{7l/(10ga)}$

77. 0.54 Hz; 22 cm; −0.11 rad
79. c) $T = 2\pi\sqrt{\frac{r_0^3}{GM}}$ (d) 16 days
83. 2.1 m/s^2
85. (a) up (b) 0.46 turn
87. c
89. c

Chapter 14

15. (a) 0.19 s (b) 6.5 cm
17. (a) 300 m (b) 1.58 m (c) 3.0 cm (d) 8 μm (e) 500 nm (f) 3.0 Å
19. (a) 0.19 mm (b) 0.43 mm
21. (a) 1.3 cm (b) 9.1 cm (c) 0.20 s (d) 45 cm/s (e) $-x$ direction
23. (a) 12.8 rad/s (b) 0.336 cm^{-1} (c) $(2.34\text{ cm})\cos[(0.336\text{ cm}^{-1})x + (12.8\text{ s}^{-1})t]$
25. 250 m/s
27. 7.6 N
29. 9.9 W
31. 343 m/s
33. 420 m/s
35. 940 Hz
37. 5.4 m
39. (a) 280 Hz (b) 70 Hz (c) 210 Hz
41. 14 cm
43. 93 Hz
45. Galaxy receding
47. 30 m/s
49. $\overline{E} = \frac{4\pi^2 F A^2}{\lambda}$
51. 1.0×10^2 W
55. $v = \sqrt{\frac{kL(L - L_0)}{m}}$
57. 10 m
59. $L_0 = 5L_1/7$
61. 440 mph
65. 6.3 m
71. 7.3 km
73. 41 m/s
75. radar worked properly
77. Not sufficient: The minimum measurable speed is 5.4 km/h.
79. 3.9 kg
81. b
83. c

Chapter 15

15. 1.2 kg
17. (a) 180 kg/m^3 (b) 7.3 m^3
19. 249 kPa
21. 322 kPa
23. 1.7×10^3 kg/m^3
25. 92 m
27. 2.4%
29. 46 kg
31. 0.75%
33. 2.8 m/s
35. (a) 1.8×10^4 m^3/s (b) 1.5 m/s
37. 1.8 cm/s
39. 830 cm^2
41. (a) 620 Pa (b) 1.2 kPa
43. 3.6 mm
45. 8100 kg
47. The accused apparently drank 51 oz.
49. 27 m
51. (a) 49 kg (b) 2500 kg
53. 14 kPa
55. 14 m
57. (a) 1.5 m/s (b) 0.47 L/s
59. 70%
61. (a) 98% less (b) 17 cm
63. (a) 603 Pa (b) 11.0 km
65. 15 kg
67. Yes, the wind farm could produce 1 GW of power.
69. $t = \frac{A_0}{A_1}\sqrt{\frac{2h}{g}}$
71. (b) 5.8 km
73. $\frac{M}{4\pi R^3(1 - 2e^{-1})}$
75. 2.1×10^{12} N·m
77. Yes
79. $\rho_{H_2O} L \tan\frac{\theta}{2}(h_0^2 - h_1^2)$
81. c
83. e

Chapter 16

15. 2.5°F to 5.6°F
17. 20°C
19. −40°C = −40°F
21. 102°F
23. 32 kJ
25. 100 W
27. (a) 170 J/K (b) 480 J/(kg·K)
29. 0.293 W
31. 55 kW
33. 4 W
35. $\mathcal{R}_{air} = 0.98\ m^2\cdot K/W$, $\mathcal{R}_{concrete} = 0.03\ m^2\cdot K/W$, $\mathcal{R}_{fiberglass} = 0.60\ m^2\cdot K/W$, $\mathcal{R}_{glass} = 0.03\ m^2\cdot K/W$, $\mathcal{R}_{Styrofoam} = 0.88\ m^2\cdot K/W$, $\mathcal{R}_{pine} = 0.23\ m^2\cdot K/W$
37. 2.2 kW
39. 2×10^{-5} m^2
41. (a) 138 kPa (b) 33.4 kPa (c) 233 kPa
43. 263 K = −10°C
45. 364 g
47. (a) 23.2 kJ (b) 337 kJ (c) 65.2 kJ
49. 138 s
51. 0.56 kg
53. 1.8 kg
55. 9.2 K
57. 0.20 kg
59. 2.0×10^2 W
61. The house will remain at a comfortable 19°C
63. (a) 1200 K (b) 700 K
65. 24°C
67. 1200 K
69. (a) $319/month (b) $37.58/month
71. 44 K
73. 418.76 kJ, 0.09% higher
75. Mars: 207K vs. ~210 K measured; Venus: 301 K vs. ~740 K measured
77. The solar increase accounts for only 4% of recent warming.
79. The hutch temperature will be −2.5°C, so the water will freeze.
81. 10 h
83. c
85. a

Chapter 17

17. 1.8 m^3
19. 1.8×10^6 Pa
21. (a) 27 L (b) 330 K
23. 3.16 km/s
25. 22 kJ
27. 3.9 kg
29. 6.0 MJ
31. 0.987 L
33. 263°C
35. 1×10^{15} m^{-3}, which is over 10 billion times less dense than Earth's atmosphere
37. (a) 235 mol (b) 5.65 m^3
39. (a) 1.27 atm (b) 0.980 mol (c) 0.786 atm
41. 27.6 min
43. 14 min
45. 43.9 min
47. 10°C
49. 46.1°C
51. 177 g
53. 4.9°C
55. 19 kW
57. 56 min
59. 251 K
61. 307 K
63. $d = \frac{L_0}{2}\sqrt{2\alpha\Delta T + \alpha^2\Delta T^2}$
65. (a) 61 h (b) 52 h
67. 3.97°C
69. 34.1 km
73. (a) $y^2 = \frac{1}{4}(L_0^2 - d^2) + \frac{1}{2}L_0^2\alpha\Delta T$ (b) $\alpha = 2.35\times10^{-5}$/C°, d = 80.00 cm (c) aluminium
77. (a) 244 K (b) 247 K
79. c
81. c

Chapter 18

15. 29.3 kJ
17. 250 J
19. −14 kW
21. $2p_1V_1$
23. (a) 4/3 (b) 220 J
25. 0.177
27. 2.1 MJ
29. 57.7%
31. (a) 200 K (b) 120 K
33. 380 W
35. (a) 1.49 mm (b) 10.7 μJ
37. 1.35
39. (a) 300 kPa (b) 240 J
41. 440°C

43. (a) 810 K (b) 25.8 atm
45. 354
47. (a) 1.27 (b) internal energy to raise gas temperature
49. (a) 255 K, 1.75 kJ (b) 279 K (c) 272 K, 500 J
51. (a) 40 kPa (b) 83 kPa (c) 80 kJ
53. 930 J
57. The temperature rises 75°C, missing the criteria.
59. 57 kJ
61. 330 K
63. (a) 202 J (b) 500 J transferred out of the gas
65. 20 mol
67. 140 atm
73. 2.0 mJ
75. $4p_1V_1/3$
77. Yes
79. 18%
81. a
83. c

Chapter 19

13. (a) 26.8% (b) 7.05% (c) 77.0%
15. 0.948 K
17. 9.10
19. No
21. 8.8 kJ/K
23. 21.9 kg
25. (a) 1/64 (b) 5/16
27. 52.1% (winter), 47.7% (summer)
29. (a) 1.75 GW (b) 43.0% (c) 232°C
31. 2×10^{11} kg/s
33. (a) 8.53 (b) 1.10×10^3 kg
35. $68
37. 2.83
39. (a) 5.7 (b) 3.5 kW (c) pump: 54¢/h; oil furnace; $2.40/h
41. (a) 17.4% (b) 83.3%
43. 140 MJ/K
47. (a) 86.0 J/K (b) 120 J/K (c) 0
49. 0
51. 160 J/K
53. 12.1 kJ/K
55. $1-r^{1-\gamma}$
59. 61%
61. $C_0(1 - T_0/T_1)$
63. $W = CT_h(\ln x - 1 + 1/x)$
65. $\Delta S_{tot} = mc\ln\left(1 + \frac{(T_h - T_c)^2}{4T_hT_c}\right)$
67. 36.2 J/K
69. 62%
71. c
73. c

Chapter 20

13. 3 C, or about 0.05 C/kg
15. (a) *uud* (b) *udd*
17. 1.1×10^9
19. 5.1 m
21. (a) $\hat{\jmath}$ (b) $-\hat{\imath}$ (c) $0.316\hat{\imath} + 0.949\hat{\jmath}$
23. 3.8×10^9 N/C
25. (a) 2.2×10^6 N/C (b) 77 N
27. $-1.6\hat{\imath}$ pN
29. (a) −26 MN/C (b) 5.2 MN/C (c) −58 MN/C
31. 1.1 kN/C
33. $E = kQ/(\sqrt{8}a^2)$
35. 5.1×10^4 N/C
37. 980 N/C
39. (a) 22.3 μC (b) no
41. $16\hat{\imath} - 9.1\hat{\jmath}$ N
43. $4q$
45. (a) 20 μC (b) $-1.6\hat{\imath}$ N
47. (a) $2.3\hat{\imath}$ MN/C (b) $0.82\hat{\imath} + 0.82\hat{\jmath}$ MN/C (c) $\vec{E} = 0.30\hat{\imath} - 0.89\hat{\jmath}$ MN/C
49. $-4e$
51. (a) $8.0\hat{\jmath}$ GN/C (b) $190\hat{\jmath}$ MN/C (c) $220\hat{\jmath}$ kN/C
53. 0
55. $q_1 = \pm40$ μC, $q_2 = \mp6.9$ μC
59. −14 μC/m
61. The device doesn't work because its two halves depend on charge-to-mass ratio in the same way.
63. 1.3×10^{-30} C·m
65. (a) $2kQqa/x^2$ (b) $2kQqa/x^3$ (c) upward
67. $0.4e$, $0.03e$
69. (a) $\vec{E}(x) = 2kqa^2\frac{(3x^2 - a^2)}{x^2(x^2 - a)^2}\hat{\imath}$ (b) $\vec{E}(x) \approx \frac{6kqa^2}{x^4}\hat{\imath}$
71. (a) 2.5 μC/m (b) 300 kN/C (c) 1.8 N/C
73. (b) $dq = 2\pi\sigma r dr$ (c) $dE_x = \frac{2\pi k\sigma xr}{(x^2 + r^2)^{3/2}}dr$
77. $y = a/\sqrt{2}$
79. $E = -\frac{k\lambda_0\hat{\imath}}{L}\left[\frac{1}{2} + 2\ln(2)\right]$
81. mdv^2/qL^2
83. a
85. a

Chapter 21

17. 3 μC
19. $Q_C = 2Q = -Q_B$
21. 650 kN/C
23. ±1.5 kN·m²/C
25. 69 N·m²/C
27. (a) $-q/\epsilon_0$ (b) $-2q/\epsilon_0$ (c) 0 (d) 0
29. 49 kN·m²/C
31. (a) 1.2 MN/C (b) 2.0 MN/C (c) 50×10^4 N/C
33. Line symmetry
35. 49×10^3 N/C
37. (a) 5.1×10^6 N/C (b) 34 N/C
39. (a) 2.0×10^6 N/C (b) 7.2×10^3 N/C
41. (a) 0 (b) 4.0×10^{-3} C/m²
43. 1.8 MN/C
45. $\pm E_0a^2/2$
47. 7.0 MN/C; 17 MN/C
49. (a) 2.8 cm (b) 3.5 nC
51. ±154 nC
53. (a) $3.6\hat{r}$ MN/C (b) $3.8\hat{r}$ MN/C (c) $7.8\hat{r}$ MN/C
55. (a) $20\hat{r}$ kNC (b) $1.7\hat{r}$ kN/C
57. 6.3 μC/m³
59. (a) $\rho x/\epsilon_0$ (b) $\rho d/2\epsilon_0$; away from the center plane of slab if $\rho > 0$, toward center plane if $\rho < 0$
61. 18 N/C
63. (b) $-Q$
67. (a) $Q = \pi\rho_0a^3$ (b) $E(r) = \rho_0r^2/(4\epsilon_0a)$
69. $a = 5\rho_0/(3R^2)$
71. $R^3\frac{\rho_0}{\epsilon_0}(e - 2)$
73. $\frac{\rho_0r^2}{3\epsilon_0R}$
75. $E_{in} = \frac{\rho_0x^2}{2\epsilon_0d}, E_{out} = \frac{\rho_0d}{8\epsilon_0}$
77. c
79. d

Chapter 22

15. 600 μJ
17. 3.0 kV
19. 910 V
21. Proton, ionized He atom: 1.6×10^{-17} J, proton: 3.2×10^{-17} J
23. $-E_0y$
25. 53 nC
27. (a) 440 kV, 9.2×10^6 m/s
31. (a) 4 V (b) $E_x = 1$ V/m, $E_y = -12$ V/m, $E_z = 3$ V/m
33. 3 kV
35. 5.6 kV/m
37. 4.5 V
41. 6.1 μC
43. $\sqrt{2keQ/(mR)}$
45. kQ/R
47. $-ax^2/2$
49. −52 nC/m
51. $-a/2$, $a/4$
53. (a) 2.6 kV (b) 1.8 kV (c) 0
55. $V = 2kQ/R$
57. $2\pi k\sigma(\sqrt{x^2 + b^2} - \sqrt{x^2 + a^2})$
61. $(V/R)\hat{r}$
63. (a) 43 kV (b) 1.7 MN/C (c) 540 V (d) 0
65. $-E_0R/3$
67. (a) 7.2 kV (b) 14 kV
69. 14 cm, 1.7 nC
71. 0.12 J
73. $\omega = 232$ nC/m², $q = 3.75$ nC, $r = 7.18$ cm
77. (a) $\pi k\sigma_0a[\sqrt{1 + (x/a)^2} - (x/a)^2\ln(a/x + \sqrt{1 + (a/x)^2})]$
79. $-\frac{k\lambda_0}{L^2}\left[Lx + x^2\ln\left(\frac{2x - L}{2x + L}\right)\right]$
81. 8.0 mm
83. d
85. b

Chapter 23

13. 4.4 kJ
15. 0
17. −48.5 eV
19. (a) 1.4 J (b) 4.2 J
21. 22 nF
23. 740 pF
25. 1.5 J
27. 3.0 μF, 0.67 μF
29. (a) 1.20 μF (b) $Q_1 = 14.4\ \mu C$, $Q_2 = 4.80\ \mu C$, $Q_3 = 9.60\ \mu C$ (c) $V_1 = 7.2$ V, $V_2 = V_3 = 4.8$ V
31. 8.2×10^5 V/m
33. No
35. $Q_y = 4Q_0/(\sqrt{2} + 1) \approx 1.66Q_0$
37. 2.8 μC
41. (a) 4.4 kV (b) 120 kW
43. 129 F
45. 0.86 μF
49. (a) 4.1 nF (b) 1.3 kV
51. 2.7 nm
53. 24 μJ
55. $U = kQ^2/(2R)$
57. 6.0×10^{-4} J
59. 13 min
61. $C = \dfrac{4\pi\epsilon_0 ab}{b - a}$
63. $\dfrac{1}{6}$
65. (b) $\dfrac{C_0V_0^2}{2}\left(\dfrac{\kappa x + L - x}{L}\right)$ (c) $\dfrac{C_0V_0^2(\kappa - 1)}{2L}$
67. $\dfrac{\pi\rho^2R^4}{8\epsilon_0}$
69. (b) 4.3 μF
71. a
73. c

Chapter 24

13. 9.4×10^{18}
15. 1.9×10^{11}
17. 3.2×10^6 A/m^2
19. 6.8 cm
21. (a) $5.95 \times 10^7 (\Omega \cdot \text{m})^{-1}$ (b) $4.55\ (\Omega \cdot \text{m})^{-1}$
23. 360 V
25. 32 mΩ
27. $4R$
29. (a) 6.0 V (b) 8.0 Ω
31. 230 V
33. 300 Ω
35. (a) 0.12 mA (b) no
37. (a) 420 A/mm^2 (b) 0.24 A/mm^2
39. greater in Cu by factors of (a) 7.6; (b) 4
41. 9.7 μC
43. (a) 5.8 MA/m^2 (b) 97 mV/m
45. Ge
47. 50 ft
49. $R_1 = 388\ \mu\Omega$, $R_2 = 0.971\ \mu\Omega$, and $R_3 = 0.243\ \mu\Omega$
51. (a) 81 miles (b) 7.3 h at 3.3 kW, 3.6 h at 6.6 kW, 33 min at 44 kW (c) 203A
53. ~ 2 TW
55. 2.8 min
57. $d_1 = \sqrt{2}d_2$.
59. 0.63 A
61. Aluminum at \$3.30/m is more economical than copper at \$14/m.
63. 2.5 A
65. 250°C
69. $2\pi J_0 a^2/3$
71. 19°
73. a
75. c

Chapter 25

17. 1.4 h
19. 43 kΩ
21. 10 V
23. 50 Ω
25. $I_1 = 2$ A, $I_2 = 0.2$ A, $I_3 = 2$ A
27. 0 A
29. −0.66%
35. $\mathcal{E}R_2/(R_1 + R_2)$
37. 1.5 mA
39. 30 A
41. 14 W
43. 120 mA, so yes, possibility fatal
45. (a) $\mathcal{E}R_1/(R + 2R_1)$
47. 2.4 W
49. $7R/5$
51. (a) 48 V (b) 57 V (c) 60 V
53. (a) 0.992 A (b) 0.83%
55. 360 μF; 1200 V
57. 3.4 μJ
59. a. $V_C = 0, I_1 = 25$ mA, $I_2 = 0$
 b. $V_C = 60$ V, $I_1 = I_2 = 10$ mA
 c. $V_C = 60$ V, $I_1 = 0, I_2 = 10$ mA
 d. $V_C = 0, I_1 = I_2 = 0$
61. (a) 5.015 V (b) 66.53 Ω
63. 1.07 A, left to right
65. 2.15 μF
67. 80 μs
69. 8 Ω; 89 W
71. (a) R_1 (b) R_1 (c) R_1
75. (a) 9 V (b) 1.5 ms (c) 0.3 μF
77. 220 mV
79. $\tau = \dfrac{R_1R_2C}{R_1 + R_2}$
81. Yes
83. b
85. c

Chapter 26

15. (a) 16 G (b) 23 G
17. (a) 2.0×10^{-14} N (b) 1.0×10^{-14} N (c) 0
19. 400 km/s
21. 360 ns
23. (a) 87.6 mT (b) 1.25 keV
25. 0.373 N
27. 12,500 lb, so clamping down the bar is a good idea.
29. (a) 9.85 cm (b) 14.8 μT
31. 1.2 mT
33. 5 mN/m
35. 4.05×10^{-2} A·m^2 (b) 7.78×10^{-2} N·m
37. 7.0 A
39. (a) 0.569 mT (b) 3.90 mT (c) 2.85 mT
41. 17 T
43. 2.3×10^{27} A·m^2
45. 3.8 GA
49. (a) 71 μm (b) 440 μm
51. 0.53 A
53. 8.5×10^{22} cm^{-3}
55. (a) 4.6 A·m^2 (b) 0.43 N·m
57. 0.021 N, 45° above horizontal
59. $(1 + \pi)\dfrac{\mu_0 I}{2\pi a}$, out of page
61. $\dfrac{\mu_0 I}{4a}$, into page
63. 16 μN, toward long wire
65. (a) 0 (b) $B = \mu_0 I/(2\pi r)$
67. (a) 2300 (b) 3.3 kW
71. (a) 8.0 μT (b) 4.0 μT (c) 0
73. $\dfrac{\mu_0 J_s x}{d}$
75. (a) $B \approx \dfrac{\mu_0 I}{2w}$ (b) $B \approx \dfrac{\mu_0 I}{2\pi r}$
77. (a) $\pi R^2 J_0/3$ (b) $B = \dfrac{\mu_0 J_0 R^2}{6r}$ (c) $B = \dfrac{\mu_0 J_0 r}{2}\left(1 - \dfrac{2r}{3R}\right)$
79. Since $\tau \propto 1/N$, more torque from a 1-turn loop.
81. $\dfrac{\mu_0 I^2}{2\pi w}\ln\left(\dfrac{a + w}{a}\right)$
83. $\mu_0 nIl/\sqrt{l^2 + 4a^2}$
85. No; the force between each meter of the two conductors is 150 N.
87. The hall potential is 10,000 times smaller than bioelectric potentials.
89. d
91. d

Chapter 27

15. 1.2×10^{-4} Wb
17. 160 T/s
19. 6.5 mH
21. 42 kV
23. 330 mH
25. 3.1 kJ
27. 66 mJ
31. 4.4 T
33. $-rb/2$
35. (a) −0.30 A (b) −0.20 A
37. 15 mT
39. (a) 3 s (b) clockwise
41. (a) 2.0 mA (b) 4.4 mA
43. −42 mA, clockwise
45. 130
47. (a) Upper bar (b) 0

49. (a) 25 mA (b) 1.3 mN (c) 2.5 mW (d) 2.5 mW
51. 58 T/ms
55. 0.76 s
57. 20 s
59. (a) 5 Ω (b) 500 J
61. (a) 1.0 A (b) 0.43 A (c) −1.7 A
63. 190 mΩ
65. 3.4×10^{21} J/m^3
67. $\dfrac{\mu_0 I^2}{16\pi}$
69. 3×10^8 m/s (speed of light)
71. (a) $-br/(2\rho)$ (b) $\dfrac{\pi b^2 h a^4}{8\rho}$
73. 3.69 H
75. $v(t) = \dfrac{FR}{B^2l^2}\left[1 - \exp\left(-\dfrac{B^2l^2}{Rm}t\right)\right]$
77. (a) $\dfrac{\mu_0 I^2}{4\pi}\ln(b/a)$
81. c
83. a

Chapter 28

15. (a) 294 V (b) 2.51×10^3 s^{-1}
17. (a) $V(0) \approx V_p/\sqrt{2}$, 45°
(b) $V(0) = 0, \phi_b = 0$
(c) $V(0) = V_p, \phi_c = 90°$
(d) $V(0) = 0, \phi_d = \pm\pi$
(e) $V(0) = -V_p, \phi_e = -90°$
19. $I_{R,\text{rms}} = 13$ mA, $I_{C,\text{rms}} = 24$ mA, $I_{L,\text{rms}} = 22$ mA
21. (a) 250 V (b) 15 V
23. 16 kHz
25. 78.1 H
27. (a) 32 mH (b) 1.0 V
29. 3.5 kΩ
31. 5.0 mA
33. 390 mA
35. 1
37. (a) 150 mA (b) 330 mA
41. 4.3 kHz
43. (a) 52 nF (b) 350 Hz
45. 0.199 μH
47. (a) $1/\sqrt{2}$ (b) 1/2 (c) $-1/\sqrt{2}$ (d) 1/2
49. 50
51. 6.2 Ω
53. (a) Above resonance; (b) ~50°
55. (a) 0.369 (b) 6.43 W
57. (a) 5.5% (b) 9.1%
59. 3.7 mF
61. 2.7 V
63. 1620 Hz
67. $R = 400\ \Omega, L = 68$ mH, $C = 94$ nF
71. 910 Hz, 36 V
73. a
75. d

Chapter 29

13. 1.3 nA
15. $-\hat{k}$
19. 11.2 km
21. 2.57 s
23. 5.00×10^6 m
25. x-direction
27. 12%
29. 1×10^{10} W/m^2
31. The radio has a minimum intensity of 0.27 nW/m^2, so it will work at the cabin.
33. 20 kW
35. 3.1 cm
37. 0.94 PHz -1.0 PHz
39. 1.07 pT
41. 91%
43. 19%
45. 0.00004%
47. Quasar power is greater by factor of 4×10^{10}
49. (a) 4.6 kW (b) 53 mV/m
51. (a) $1/r$ (b) $1/r^2$
53. (a) 8.9×10^6 W/m^2 (b) 58×10^3 V/m
55. 6.2×10^3 y
57. 2.52 kPa
59. (a) 1.12 MV/m (b) 4.14 mm (c) 91.0 mJ (d) 3.03×10^{-10} kg·m/s (e) 86.0 W
61. 6
65. 2.75 m
67. 2.2 km
69. (a) 51 MV/m (b) 0.17 T (c) 96 TW
71. b
73. d

Chapter 30

11. 15°
13. 0.5°
15. Ice
17. 77.7°
19. 14.2°
21. 1.9
23. 79.1°
25. 1.66
27. 6.41°
29. (a) 18° (b) 390 nm
31. Ethyl alcohol
33. 1.83
35. 5.1 m
37. 139 nm
39. 35°
41. Diagonal face, 23°
43. 1.07
47. 63.8°
49. red: 72.3°, violet: total internal reflection
51. 2.7 m
53. 1.9 m
57. c) 50.9°
61. $\dfrac{d}{c}\left(\dfrac{2}{3}n_1 + \dfrac{1}{3}n_2\right)$
63. c
65. b

Chapter 31

15. 35°
17. (a) −1/4 (b) real, inverted
19. (a) $3f$ (b) $3f/2$ (c) real
21. −2
23. 21 cm
25. 27 cm
27. 40 cm
29. 0.86 mm
31. 2.2 diopters
33. −1.3 diopters
35. −200
37. (a) −24 cm (b) 29 mm (c) virtual, upright, enlarged
39. 18 cm
41. 16 cm
43. 7.59 cm
45. 12 cm
47. (a) −7.7 cm, inverted, real (b) +7.7 cm, upright, virtual
49. 29 cm or 41 cm
51. 11 cm
53. $s' = 1.1$ m, inverted real image
55. −67.9 cm
57. 2.0
59. 2
61. Choose plastic, because it meets requirements and is cheaper.
63. (a) Real, inverted image (b) −2.82
65. 3.3 diopters
67. 0.3°
69. 72 cm
79. (a) $dn = -\dfrac{2c}{\lambda^3}d\lambda$ (b) 0.858 mm
81. c
83. d

Chapter 32

11. 1.7 cm
13. 420 nm
15. 4
17. (a) 4.8°, 9.7° (b) 2.9°, 6.8°
19. (a) 2 (b) 1
21. 103 nm
23. 594 nm, 424 nm
25. The top 1.5-cm of the film
27. 29.3°
29. 1.62%
31. 37 cm
33. 3×10^{-4} rad
35. 96°
37. (a) 38 (b) 3
39. 44 μm
43. 2
45. Not feasible because a 2-km-diameter telescope is needed
47. 3.3 Å
49. 5
51. 236
53. 128.8 μm
55. $1 + 2.93\times10^{-4}$
57. 34 m
59. 2.0 μm
61. 6.9 km
63. Rep is correct, but microscope won't resolve rhinovirus.
65. $n_{\text{gas}} = 1 + \dfrac{m\lambda}{2L}$

67. $\Delta y = D\lambda/2d$
69. 92
71. c
73. a

Chapter 33

13. (a) 4.50 h (b) 4.56 h (c) 4.62 h
15. 33 ly
17. 40 m
19. $0.14c$
21. (a) 2.0 (b) 2.5
23. $0.14c$
25. (a) 2.1 MeV (b) 1.6 MeV
29. (a) $0.86c$ (b) 9.7 min
31. $c/\sqrt{2}$
33. Twin A = 83.2 years old, twin B = 39.7 years old
35. $0.96c$
39. Civilization B, 3.8×10^5 y
41. yes, from A toward B at $0.45c$
43. earlier by 5.2 min
47. $0.94c$
49. (a) 10 ly, 13 y (b) 0 ly, 7.5 y
53. (a) 4.2 ly (b) −2.4 ly
55. (a) $0.758c$ (b) 1.09 GeV/c
57. 25 h
59. (a) 0.26 eV (b) 1.3 keV (c) 3.1 MeV
65. $0.866c$
67. $0.95c$
71. $\gamma ma\left[1+\left(\dfrac{\gamma u}{c}\right)^2\right]$
73. 0.31 c; 27 kV
75. (a) 2.976×10^8 m/s
(b) 9.46×10^{-31} kg, 4% higher than known electron mass
77. a
79. a

Chapter 34

15. 16
17. $\lambda_{peak} = 10.1\ \mu m$, $\lambda_{median} = 14.3\ \mu m$
19. (a) 500.0 nm (b) 708.6 nm
21. 2.8×10^{-19} J to 5.0×10^{-19} J
23. 1.44
25. 122 nm, 103 nm, 97.2 nm
27. 91.2 nm
29. (a) 3.7×10^{-63} m (b) 73 nm
31. The electron moves 1836 times faster than the proton.
33. 6×10^7 m/s
35. 130 nm
37. 23 keV
39. UV is smaller by factor 5.4×10^{-2}
41. (a) 5.19×10^3 K (b) 0.748
43. (a) $1.7\times10^{28}\ s^{-1}$ (b) $3.2\times10^{15}\ s^{-1}$
(c) $1.3\times10^{18}\ s^{-1}$
45. (a) 1.12×10^{15} Hz (b) 2.79 eV
47. (a) 2.9 eV and 1.9 eV (b) Plants absorb blue and red, reflect green.
49. 440 nm
51. (a) 154 pm (b) 222 eV
53. No
55. (a) 313 m/s (b) 96 km/s
57. 0.22 meV
59. (a) 26.4 cm (b) 4.70 μeV
61. 229
63. 3.40 eV
65. (a) 0.0265 nm (b) 40.8 eV
67. 1.62 km/s
69. 2.5 km/s
71. 1 ps
75. $E_0 = \frac{1}{2}m_ec^2[(\gamma-1)+\sqrt{(\gamma-1)(\gamma+3)}]$
83. (a) 6.65×10^{-34} J·s (b) 2.3 eV
(c) potassium
85. b
87. c

Chapter 35

13. (a) 0 (b) $\pm a\sqrt{\ln 2/2}$
15. 5
17. 3.8 meV
19. (a) 1.6 eV (b) 6.5 eV
21. Electron
23. 0.2 MeV
25. 33 eV
27. 8.0 eV
29. $E \rightarrow E/4$
31. 930 pm
35. (a) 2.2 eV (b) 570 nm
37. 21 μm
39. (a) 6 (b) $\lambda_{4\to1} = 153$ nm, $\lambda_{4\to2} = 191$ nm, $\lambda_{4\to3} = 328$ nm, $\lambda_{3\to1} = 287$ nm, $\lambda_{3\to2} = 459$ nm, $\lambda_{2\to1} = 765$ nm (c) UV, visible, and IR
41. (a) $\psi_{n\text{-odd}}(x) = \sqrt{\dfrac{2}{L}}\cos\left(\dfrac{n\pi x}{L}\right)$,
$\psi_{n\text{-even}}(x) = \sqrt{\dfrac{2}{L}}\sin\left(\dfrac{n\pi x}{L}\right)$
(b) $E_n = n^2h^2/(8mL^2)$
43. 0.759 nm
45. 2.5×10^{-17} eV; quantization is insignificant
47. (a) 0.30 (b) 0.15
53. 4
55. (c) $A_0 = (\alpha^2/\pi)^{1/4}$
57. 2.23 nm
61. b
63. a

Chapter 36

15. 3
17. d
19. 5
21. 3d
23. 2.58×10^{-34} J·s
25. 3/2, 5/2
27. $11.5\hbar\omega$
29. $1s^22s^22p^63s^23p^64s^23d^1$
33. 0.6934 meV
35. $n = 4, l = 3$
37. 2.67×10^{68}
39. 90°, 65.9°, 114°, 35.3°, 145°
41. 0, ±1, ±2, ±3
45. (a) $E_1/16$ (b) $\sqrt{12}\hbar$ (c) $\frac{1}{2}\sqrt{35}\hbar$
47. (a) $16\hbar\omega$ (b) $4\hbar\omega$
49. $1s^22s^22p^63s^23p^64s^13d^{10}$
51. 3.0×10^{17}
53. 0.1
55. 2.50 meV
57. even N: $\hbar\omega(N-1)/2$; odd N: $\hbar\omega N/2$
59. (a) 5 (b) $9E_1$
61. $3\hbar$, $6h$
63. (a) 0.966 (b) 0.0595
65. $P(r)dr = 4\pi r^2\psi_{2s}^2dr$, $3+\sqrt{5}$
67. (b) 54.4 eV, 870 eV, 91.4 keV, 115 keV
69. (b) 141 eV, 65.8 eV, 47.0 eV, 28.2 eV
71. $3a_0/2$
77. a
79. c

Chapter 37

17. 3.48 mm
19. 9.41×10^{-46} kg·m²
21. 7.08×10^{13} Hz
23. 181 kcal/mol
25. 549 nm
27. 3.54 μm
29. 1.34 meV
31. $l\hbar^2/I$
33. 0.121 nm
35. (a) 0.179 eV (b) 0.358 eV
37. 14.95 μm
39. (a) 15.09 meV (b) 82.22 μm
(c) far infrared
41. 35.8 μm
43. 10.2
45. −8.40 eV
49. 4.68 eV
51. 6.36×10^4 K, ~200 times room temperature
53. 709 nm, no
55. 1.8 kA
57. 508 nm
59. $I = m_1m_2R^2/(m_1+m_2)$; 0.128 nm
63. (a) $(2^{9/2}\pi m^{3/2}L^3/3h^3)E^{3/2}$
65. 64 kA
67. 2.75×10^{-47} kg·m²
69. a
71. a

Chapter 38

17. $^{211}_{86}Ra$, $^{220}_{86}Ra$, and $^{222}_{86}Ra$.
19. (a) $A = 35$ for both (b) $Z_K = Z_{Cl} + 2$
21. 5.9 fm
23. $^{64}_{29}Cu \rightarrow {}^{64}_{30}Zn + e^- + \bar{\nu}$
$^{64}_{29}Cu \rightarrow {}^{64}_{28}Ni + e^+ + \nu$
$^{64}_{29}Cu + e^- \rightarrow {}^{64}_{28}Ni + \nu$
25. 17 Bq/L
27. (a) 190 y (b) 290 y
29. 4×10^{-30} kg

31. 59.930 u
33. 5.612 MeV
35. 2
37. $1.0 \times 10^{20}\ s^{-1}$
39. $2 \times 10^{20}\ m^{-3}$
41. 10^3 s
43. 5.3×10^{-12} eV
45. 8.80 MeV
47. 0 atoms; 5×10^5 atoms; 8×10^4 atoms; U-238 and K-40 are suitable
49. 9.6 d
51. (a) $^{228}_{90}Th$
53. 8.9×10^3 y
55. Poland: 8.04 d; Austria: 16.2 d, Germany: 10.0 d
57. 3.0×10^9 y
59. 3.31%
61. 3×10^{-13}
63. 1.3×10^3 kg
65. 88.9%
67. 580 kg
69. 0.461 s
71. (a) $4 \times 10^{38}\ s^{-1}$ (b) 7×10^9 y
73. 8×10^{17} s, which is about 20 billion years longer than the Sun will shine
75. Bohrium-262 ($^{262}_{107}Bh$)
77. (a) $^{65}_{29}Cu$ (b) 4 h
79. (a) 210 MJ (b) $14\ s^{-1}$ (c) 450 kg
81. Yes
85. (b) 1.4 μs
87. b
89. d

Chapter 39

21. 0.336 fs
23. $\pi^+ \rightarrow \mu^+ + \nu_\mu$
25. $\eta \rightarrow \pi^+ + \pi^- + \pi^0$
27. No, violates conservation of baryon number and angular momentum
29. *sss*
31. 4.54×10^7 L
33. 10^{28} K
35. 1.2 Gly
37. 1.32×10^{10} yr
39. Reaction (a) is not possible because it violates conservation of baryon number and angular momentum.
41. (a) No (b) yes
43. $c\bar{c}$
45. (a) 0.16 μJ (b) 0.02 mm
47. (a) essentially no change (b) essentially *c* (c) 90 μs
49. 313 ly
51. (a) 256 fm (b) −2.81 keV
53. (a) 5.740×10^3 km/s (b) 253 Mly
55. 2.6×10^{-25} s
57. older value gives 22% larger age, 17.6 Gy vs. 14.4 Gy
59. 5.0 km/s/Mly
61. b
63. c

Credits

Front Matter

Page i: (volume 1): 68/Ocean/Corbis. Page i: (volume 2): Argus/Shutterstock. Page vi: Robert Keren.

Chapter 1

Page 1: David Parker/Science Source. Page 3: NASA. Page 5: The Photos/Fotolia. Page 6: NASA. Page 6: Andrew Syred/Science Source. Page 9: cesc_assawin/Shutterstock. Page 11: NASA Images. Page 11: cesc_assawin/Shutterstock. Page 14: F1online digitale Bildagentur GmbH/Alamy. Page 14: Paul Fleet/Alamy.

Chapter 2

Page 15: Bob Thomas/Getty Images. Page 21: DARPA. Page 24: Richard Megna/Fundamental Photographs. Page 26: National Institute of Standards and Technology.

Chapter 3

Page 32: Martha Holmes/Nature Picture Library. Page 38: Richard Megna/Fundamental Photographs. Page 39: Paul Whitfield/Rough Guides/Dorling Kindersley ltd. Page 41: Jamie Squire/Getty Images.

Chapter 4

Page 51: NASA/JPL-Caltech. Page 56: Dudarev Mikhail/Shutterstock. Page 57: Marvin Dembinsky Photo Associates/Alamy. Page 61: NASA. Page 64: David Scharf/Science Source. Page 69: Source: From WP 3e Fig05-40.

Chapter 5

Page 71: Kevin Maskell/Alamy. Page 76: NIC BOTHMA/EPA /Landov Media. Page 86: Source: From EUP 2e ISM CH5 Exercise 17 Solution. Page 89: AP Photo/Lionel Cironneau.

Chapter 6

Page 90: PAOLO COCCO/AFP/Getty Images. Page 92: Source: K. Trenberth et al., Bulletin of American Meteorological Society. March 2009. Page 314. Page 107: Jeffrey Liao/Shutterstock.

Chapter 7

Page 109: Kennan Ward/Encyclopedia/Corbis. Page 113: United States Geological Survey. Page 118: Pearson Education.

Chapter 8

Page 129: NASA/JPL-Caltech. Page 130: Lowell Observatory. Page 134: David Hardy/Science Source. Page 137: Yekaterina Pustynnikova/Chelyabinsk.ru/AP Images.

Chapter 9

Page 144: Peter Muller/Cultura RM/Alamy. Page 148: Wally McNamee/Corbis Sports/Corbis. Page 153: NASA/Bill Ingalls. Page 166: Fundamental Photographs.

Chapter 10

Page 168: IM_photo/Shutterstock. Page 179: Mark Flynn/University of Texas-Austin.

Chapter 11

Page 189: Goddard Space Flight Center/NASA. Page 196: Courtesy of Moog, Inc. Page 196: Courtesy of Chipworks. Page 198: Stefano Rellandini/Reuters.

Chapter 12

Page 204: Michael Jenner/Robert Harding Picture Library Ltd/Alamy.

Chapter 13

Page 221: Tischenko Irina/Shutterstock. Page 222: Tommy/Fotolia. Page 222: Sheila Terry/Science Source. Page 227: Peter Tsai Photography/Alamy. Page 227: Shutterstock. Page 236: AP Images. Page 236: Marc Wuchner/Corbis. Page 242: NASA Johnson Space Center (NASA-JSC).

Chapter 14

Page 243: AGE Fotostock. Page 251: Gene Chutka/E+/Getty Images. Page 253: Science Source. Page 255: Richard Megna/Fundamental Photographs. Page 257: Michael Freeman Photography. Page 260: SVSimagery/Shutterstock. Page 260: Drazen Vukelic/E+/Getty Images. Page 264: David Rydevik.

Chapter 15

Page 265: Ralph A. Clevenger/Crush/Corbis. Page 269: Bojan Pavlukovic/Fotolia. Page 271: culture-images/MPI/Frank Herzog/Alamy. Page 273: Pascal Rossignol/Reuters. Page 275: Pearson Science/Pearson Education. Page 277: Maciej Noskowski/Getty Images. Page 277: Alekss/Fotolia. Page 281: Richard Wolfson. Page 283: Science Source.

Chapter 16

Page 284: Mark Antman/The Image Works. Page 284: Arogant/Shutterstock. Page 285: Ted Kinsman/Science Source. Page 292: Branko Miokovic/E+/Getty Images. Page 293: Scott Camazine/Alamy. Page 297: Source: This is Figure 15.12 from the W.W. Norton book Energy, Environment, and Climate 2e by Rich Wolfson. It was adapted from Meehl et al, "Climate System Response to External Forcings and Climate Change Projections in CCSM4", J. Climate, 25, 3661-3683, doi: 10.1175/JCLI-D-11-00240.1. Page 302: Branko Miokovic/E+/Getty Images.

Chapter 17

Page 303: MIMOTITO/Getty Images. Page 310: AP Photo. Page 311: Brian J. Skerry/National Geographic/Getty Images. Page 315: samoshkin/Shutterstock.

Chapter 18

Page 317: Don Farrall/Photodisc/Getty Images. Page 322: Steven Coling/Shutterstock.com. Page 324: Walter Bibikow/Agency Jon Arnold Images/AGE Fotostock.

Chapter 19

Page 334: James Hardy/PhotoAlto/AGE Fotostock. Page 337: Encyclopaedia Britannica/Universal Images Group Limited/Alamy. Page 341: Peter Bowater/Photo Researchers, Inc.

Chapter 20

Page 354: NASA. Page 355: Anna Omelchenko/Shutterstock. Page 356: Fundamental Photographs, NYC. Page 360: Steven Puetzer/Getty Images. Page 369: Helen Sessions/Alamy. Page 369: Courtesy of Bohdan Senyuk. Page 369: CB2/ZOB/WENN.com/Newscom.

Chapter 21

Page 375: Peter Menzel/Science Source. Page 393: Fox Photos/Stringer/Hulton Archive/. Page 398: Eric Schrader/Pearson Education.

Chapter 22

Page 399: Mark Graham/AP Images. Page 412: Kevin Cruff/Getty Images.

Chapter 23

Page 418: Rob Kim/Stringer/Getty Images Entertainment/Getty Images. Page 422: Pearson Education. Page 425: Cindy Charles/PhotoEdit. Page 431: Lawrence Livermore National Laboratory.

Chapter 24

Page 432: Daniel Sambraus/Science Source. Page 443: AP Images. Page 444: Pearson Education.

Chapter 25

Page 449: villorejo/Shutterstock. Page 452: Pearson Education.

Chapter 26

Page 469: NASA. Page 470: Cordelia Molloy/Science Source. Page 473: Photo by Ivan Massar. Page 485: Jerry Lodriguss/Science Source. Page 490: 1990 Richard Megna/Fundamental Photographs.

Chapter 27

Page 497: NASA Images. Page 499: Source: Albert Einstein. Page 516: NASA Images. Page 519: Copyright: © 1988 Richard Megna-Fundamental Photographs. Page 521: Source: A. Einstein, ON THE ELECTRODYNAMICS OF MOVING BODIES, June 30, 1905.

Chapter 28

Page 525: Li Ding/Alamy. Page 533: Pearson Education. Page 537: B.S.P.I./Documentary/Corbis. Page 537: Larry Lawhead/Getty Images. Page 537: ROBERT ROBINSON/Getty Images. Page 537: jlsohio/Getty Images. Page 537: EricVega/E+/Getty Images. Page 537: Arogant/Shutterstock. Page 537: TebNad/Shutterstock.

Chapter 29

Page 543: Arthur S. Aubry/Photodisc/Getty Images. Page 543: Brent Bossom/Getty Images. Page 543: Neustockimages/Getty Images. Page 553: Fundamental Photographs, NYC. Page 555: Babak Tafreshi/Science Source. Page 555: NASA E/PO, Sonoma State University, Aurore Simonnet. Page 563: NASA.

Chapter 30

Page 565: Craig Tuttle/Documentary/Corbis. Page 566: Charles Hood/Alamy. Page 567: D. Scott/NASA. Page 568: Paul Carstairs/Alamy. Page 571: Universal Images Group Limited/Alamy. Page 571: Richard Megna /Fundamental Photographs. Page 572: Spencer Grant/Science Source. Page 574: Richard Megna/Fundamental Photographs. Page 574: Library of Congress Prints and Photographs Division Washington, D.C. 20540 USA http://hdl.loc.gov/loc.pnp/pp.print. Page 575: Universal Images Group Limited/Alamy.

Chapter 31

Page 579: Hunckstock inc./Alamy. Page 580: Fundamental Photographs, NYC. Page 581: Space Telescope Science Institute/NASA. Page 582: Pearson Education. Page 582: Scott Leigh/Getty Images. Page 584: Science Source. Page 587: Pearson Education. Page 592: Rolf Vennenbernd/AP Images. Page 594: TMT Observatory Corporation. Page 598: ajt /Shutterstock.

Chapter 32

Page 599: PRNewsFoto/GeoEye, Inc/AP Images. Page 601: Andrew Syred/Science Source. Page 602: Courtesy of Dr. Christopher Jones. Page 604: Courtesy of Dr. Christopher Jones. Page 608: Jay M. Pasachoff. Page 609: Fundamental Photographs, NYC. Page 609: Pearson Education. Page 610: Ames Research Center/Ligo Project/NASA. Page 613: M. Cagnet et al. Atlas of Otical Phenomena, Springer-Verlag, 1962. Page 613: M. Cagnet et al. Atlas of Otical Phenomena, Springer-Verlag, 1962. Page 613: Courtesy of Dr. Christopher Jones. Page 614: Courtesy of Dr. Christopher Jones. Page 615: Courtesy of the Blu-ray Disc Association. Page 615: Courtesy of the Blu-ray Disc Association. Page 615: Courtesy of the Blu-ray Disc Association. Page 616: Courtesy of Dr. Christopher Jones. Page 616: M. Cagnet et al. Atlas of Otical Phenomena, Springer-Verlag, 1962. Page 616: Courtesy of Dr. Christopher Jones. Page 616: Courtesy of Dr. Christopher Jones. Page 616: Courtesy of Dr. Christopher Jones. Page 616: Courtesy of Dr. Christopher Jones. Page 620: M. Cagnet et al. Atlas of Otical Phenomena, Springer-Verlag, 1962. Page 620: Courtesy of Dr. Christopher Jones. Page 620: Courtesy of Dr. Christopher Jones. Page 620: Courtesy of Dr. Christopher Jones. Page 620: Courtesy of Dr. Christopher Jones.

Chapter 33

Page 621: Delft University of Technology/Science Source. Page 622: ASSOCIATED PRESS/AP Photos. Page 623: Source: Albert Einstein; Leopold Infeld, The evolution of physics : the growth of ideas from the early concepts to relativity and quanta, Cambridge : Cambridge University Press, 1938. Page 625: The Hebrew University of Jerusalem. Page 631: SLAC National Accelerator Laboratory/AP Images. Page 638: WDCN/Univ. College London/Science Source. Page 642: NASA.

Chapter 34

Page 647: Dr. Wolfgang Ketterle. Page 658: Andrew Syred/ Science Source. Page 659: Millie LeBlanc/Courtesy of The Educational Development Center. Page 661: Kim Steele/Stockbyte/Getty Images. Page 661: Source: Werner Heisenberg, Physics and Philosophy: The Revolution in Modern Science (New York: Harper & Brothers, 1962).

Chapter 35

Page 667: IBM Corporation. Page 669: RIchard Wolfson. Page 671: Source: Albert Einstein, Max Born, The Born-Einstein Letters 1916-55, Palgrave Macmillan, 1971. Page 677: IBM Corporation. Page 683: Courtesy of Dr. Stephan Diez.

Chapter 36

Page 684: Fundamental Photographs. Page 692: Wolfgang Ketterle. Page 698: Wolfgang Ketterle. Page 701: Medical Body Scans/Science Source.

Chapter 37

Page 702: Alfred Pasieka/Science Source. Page 712: Hemis/ Alamy. Page 714: Koichi Kamoshida/Getty Images.

Chapter 38

Page 720: TEPCO/AFLO/Newscom. Page 724: David Job/The Image Bank/Getty Images. Page 740: LLNL/Science Source.

Chapter 39

Page 747: CMS Experiment at the LHC, CERN. Page 749: Brookhaven National Laboratory. Page 755: European Organization for Nuclear Research. Page 755: Kyodo News/Newscom. Page 757: CERN/European Organization for Nuclear Research. Page 758: NASA. Page 759: European Space Agency (ESA). Page 759: Source: http://arxiv.org/pdf/1303.5062v1.pdf. Page 765: Richard Megna/Fundamental Photographs. Page 766: European Space Agency (ESA).

Index

F

Q

R

T